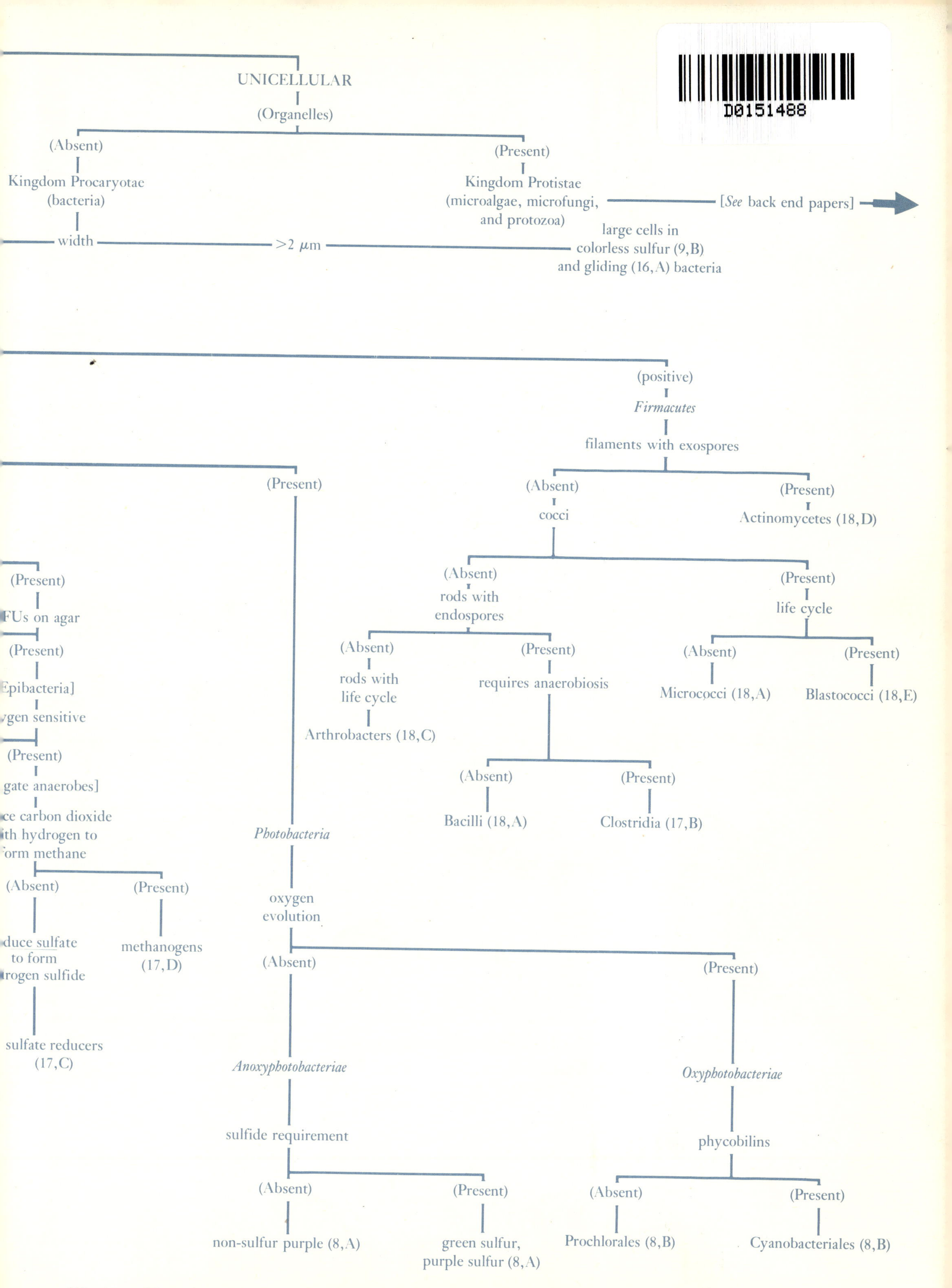

*Numbers and letters in parentheses refer to the chapter and section for that group.

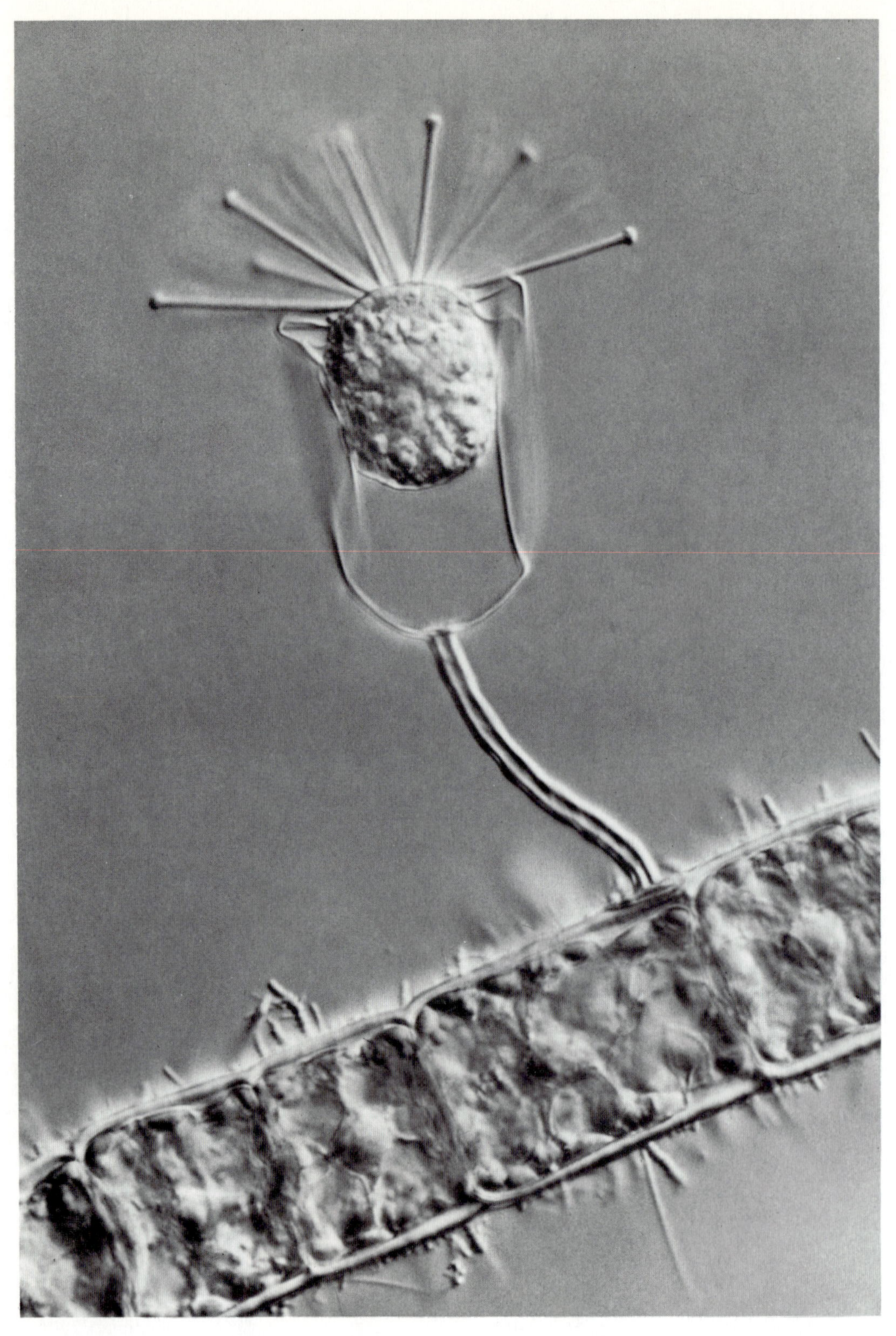

A multitrophic microcosm of the intertidal zone showing a multicellular phototroph, the filamentous green alga *Spongomorpha lanosa* (containing chloroplasts and nuclei), which is the substrate for osmotrophic and oxyphototrophic bacteria (attached endwise to the algal filament), and a stalked phagotrophic suctorian of the loricate genus *Paracineta*. Nomarski differential interference-contrast, 1260× (jacket micrograph, 1900×) Courtesy of Paul W. Johnson.

SEA MICROBES

John McNeill Sieburth
UNIVERSITY OF RHODE ISLAND

New York   Oxford University Press   1979

Library of Congress Cataloging in Publication Data

Sieburth, John McNeill.
Sea microbes.

Includes bibliographical references and index.
1. Marine microbiology. I. Title.
QR106.S56 576'.19'2 78-3618
ISBN 0-19-502419-2

Printed in the United States of America

To the creative researchers who down through the years have produced brilliant and sustaining works that have provided both the excitement of discovery and the fabric upon which this book is based.

Preface

Sea microbes were the ancestors of all life on earth, but many of their myriad forms persist today in the diverse microenvironments of the sea. These microorganisms form the food base for the pyramid of animal life in the sea that developed from them. They also recycle the excrement and debris of both unicellular and multicellular life, thereby keeping life and other processes in the sea in a long-term steady state. Microscopic life in the sea constitutes a seemingly endless array of beautiful and fascinating microorganisms whose processes hold the key to understanding the function of the community or ecosystem. But before ecological problems and principles can be defined and understood, the sea microbes must first be observed and identified.

My ignorance of the many forms that kept introducing themselves, both in class and in the course of research, kept me busy running down colleagues or authors who might have the answers. What I needed was a single source, which would include all microbial groups known to live in the sea, logically organized according to appearance and function. This book is for me and any student, teacher, or practitioner in the diverse fields of microbiology and the marine sciences who has a similar need.

The introductory material introduces the marine scientist to microorganisms and the microbiologist to the ecosystems of the sea. The methodology necessary to obtain, observe, quantify, and culture microorganisms is presented before the groups of microorganisms are considered. The microalgae, bacteria, fungi, and protozoa evolved together, live together, and depend upon one another; they are seen together by microscopy. It is only natural to study this assemblage as it occurs. Emphasis is placed on the free-living forms. Convenient vernacular or taxonomic groups are placed into appropriate morphologic–trophic divisions to eliminate disciplinary boundaries which can cause territorial disputes and undue redundancy. This volume, which assumes that the reader has a basic knowledge of microbiology and chemistry, is intended to bridge the gaps between contemporary texts on microbiology, the taxonomic treatises, and the ecological monographs, but not to replace them.

The flaws in a single-author book dealing with so wide an area will be obvious to the specialists who would never have attempted such a book. I hope the advantages outweigh the flaws and that flagrant errors, misinterpretations, and inadvertent omissions will be gently brought to my attention for revision. The illustrations were selected to show how the microorganisms look in different modes of microscopy, and the plate format was used to tell a story whenever possible. The descriptive legends are intended to make them independent of the text. The copious references reflect this author's conviction that the reader must have access to the primary literature. This includes many fundamental works, which are ageless and inspirational. Virtually all of them are available through the superb biological journal library of the Marine Biological Laboratory at Woods Hole, Massachusetts (02543) where the working draft of this book was prepared.

I would like to record a few of the milestones in the creation of this book and to acknowledge those who gave me inspiration and help. The compulsion to write the book arose from the frustration of teaching and researching without a text. The book was conceived during a visit to Soviet colleagues in 1969 and was inspired by the revolutionary scientist Ivan Morozov of Jaroslavl who wrote a number of scientific treatises in several fields while imprisoned by the Tsar before the revolution. The opportunity to prepare a working draft was afforded by a sabbatical leave during 1973-74 and an understanding wife who permitted this bittersweet period of bachelorhood. An abortive start at Dalhousie University in Halifax, Nova Scotia established the concept and style of the book through the interest and counsel of Jim Craigie. The phycological knowledge of Dag Klaveness and Bob Guillard was invaluable during the early bewildering months at Woods Hole. Chats with Paul Hargraves, Lynn Margulis, Holger Jannasch, and Stjepko Golubic, among others, have left their imprint. The friendship and enthusiasm of Molly Bang and Dick Campbell urged on a tiring spirit toward the end of the first draft. The use of the working draft and its revisions for four successive years in the lecture room—and in the laboratory examination of natural and enriched microcosms—has shown its value and its deficiencies, and led to the final format.

I would like to gratefully acknowledge the peer reviewers who pointed out glaring errors, eliminated trivia, and added useful information. These were L. Ralph Berger, Marvin Bryant, John Corliss, Francis Drouet, Jack Fell, Paul Hargraves, Terry Johnson, Galen Jones, Jan Kohlmeyer, John J. Lee, Lynn Margulis, Ted Moore, Scott Nixon, Norbert Pfennig, F.J.R. Taylor, Chase Van Baalen, and Stanley Watson. I would also like to gratefully acknowledge the many individuals who dug through their files and generously provided illustrative material. Above all, I must recognize with sincere appreciation the individuals who helped me see the book through to completion. They are Sara Finnegan of Williams & Wilkins who had the courage to acquire this title and then bully me to meet deadlines until Oxford University Press obtained this book; my secretary, Jean Knapp, who deciphered and typed and hunted references through five years of manuscript preparation and then helped remove the errors in the proofs; my wife, Janice, a science librarian at the University of Rhode Island, who read the penultimate draft for clarity; my research associate, Paul W. Johnson, who created many of the micrographs and photocopied much of the illustrative material; and at Oxford University Press, Carol Miller who syntaxed, undangled, clarified, questioned, and mothered the manuscript.

Kingston Village
1 May 1978

J. McN. S.

Contents

Sea Microbes

PART I

Introduction

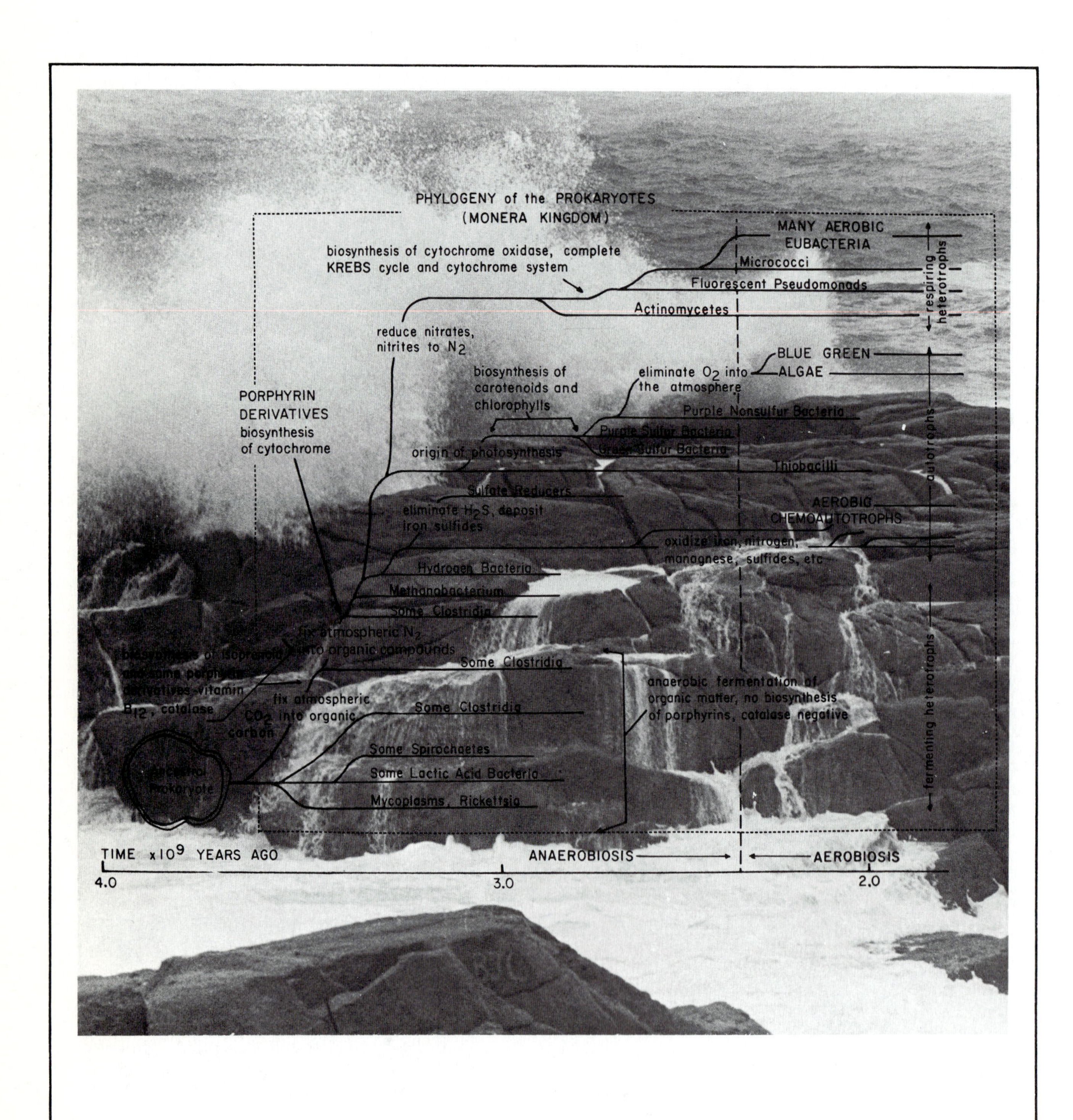

PHYLOGENY of the PROKARYOTES
(MONERA KINGDOM)
MANY AEROBIC EUBACTERIA
biosynthesis of cytochrome oxidase, complete KREBS cycle and cytochrome system
Micrococci
Fluorescent Pseudomonads
Actinomycetes
respiring heterotrophs
reduce nitrates, nitrites to N_2
BLUE GREEN ALGAE
biosynthesis of carotenoids and chlorophylls
eliminate O_2 into the atmosphere
PORPHYRIN DERIVATIVES
biosynthesis of cytochrome
Purple Nonsulfur Bacteria
Purple Sulfur Bacteria
Green Sulfur Bacteria
origin of photosynthesis
Thiobacilli
autotrophs
Sulfate Reducers
eliminate H_2S, deposit iron sulfides
AEROBIC CHEMOAUTOTROPHS
oxidize iron, nitrogen, manganese, sulfides, etc
Hydrogen Bacteria
Methanobacterium
Some Clostridia
fix atmospheric N_2 into organic compounds
Some Clostridia
vitamin B_{12}, catalase
fix atmospheric CO_2 into organic carbon
Some Clostridia
anaerobic fermentation of organic matter, no biosynthesis of porphyrins, catalase negative
fermenting heterotrophs
Some Spirochaetes
Some Lactic Acid Bacteria
Mycoplasms, Rickettsia
TIME $\times 10^9$ YEARS AGO
4.0
3.0
2.0
ANAEROBIOSIS
AEROBIOSIS

Hypothetical evolutionary pathway of the procaryotic microorganisms, superimposed upon the edge of the sea near Flat Rocks, Narragansett, Rhode Island. Photograph courtesy of H. L. "Wes" Pratt.

Microbial life in the sea is ubiquitous. It thrives in the thin sunlit (photic) zone where it lives a free-floating (planktonic) existence, providing food for the larger members of the plankton and the swimming animals (nekton), such as squid, fish, and whales, which are plankton feeders. It carpets all types of surfaces, providing food for grazing and nibbling animals. It is in constant association with most plants and animals as fouling organisms, symbionts, pathogens, and members of the enteric microbiota. It also lives in the deep lightless (aphotic) zone on fragmenting excrement and debris; these provide surfaces and organic matter for an ever-present population of heterotrophic microorganisms, which utilize preformed organic matter. The excrement and debris that descend to become sediment on the sea floor support a heterotropic microbiota, which in turn sustains other forms of life on the bottom (benthos). Throughout the depth and breadth of the seas, this microbial biomass is the foundation of the pyramid of animal life in the sea. This microbial biomass recycles the "waste" materials of both unicellular and multicellular forms metabolically, thereby ensuring a long-term steady state in the sea. The first three chapters give an overview of these microorganisms and their environments.

The first chapter reviews the evolution of unicellular microbial life and shows the differences between the two basic types of cells, the procaryotes and the eucaryotes, and how the latter could have originated from the former. This review is followed by a discussion of the microbial disciplines and their classification systems to show why a pragmatic approach has been taken in this text. The second chapter shows how the major marine ecosystems can be divided and discusses the exposed habitats of neustonic, planktonic, and epibiotic microorganisms dwelling in the sea-air interface, floating free in the water, and attaching to surfaces, respectively. The third chapter discusses the protected habitats where microorganisms live within the food and feces of animals, within the tissues of plants, animals, and other microorganisms, and in the sediment of the sea floor.

CHAPTER 1

Microorganisms

The most plausible and aesthetically pleasing myths and hypotheses will be briefly reviewed to show how life could have developed in the sea. The purpose is to show that the great diversity of form and function of the microorganisms found in the sea today resulted from prototypes that had to change their life-style as conditions in the developing biosphere changed. The wide range of habitats and microhabitats in the sea today provides niches for the staggering diversity of life forms that resulted from these endless experiments in adaptation and selection.

A. PREBIOLOGICAL ORIGIN

The earth, which is assumed to have been formed around five billion years ago, holds the remains of the more primitive procaryotic cells that date back about three billion years. During the first two billion years, changes in the atmosphere and lithosphere (the rocky crust of the earth) produced the ancient seas (hydrosphere) and an environment suitable for the creation of microbial life. This was due to the out-gassing of water and chlorine, which extracted soluble ions from the eroding crust to form the seas, as well as such other gaseous substances as hydrogen, nitrogen, methane, ammonia, and carbon dioxide. These substances were the starting materials for organic synthesis. The absence of oxygen and therefore its allotrope ozone in the early reducing atmosphere would have permitted ultraviolet light from the sun to penetrate the upper layers of the ancient seas. One mechanism for the abiotic synthesis of organic matter would be the utilization of the ultraviolet energy from the sun as well as the energy derived from x-irradiation, heat, and electrical discharge. Experiments during the last two decades (sparked by the work of Garrison in Calvin's laboratory and Miller in Urey's laboratory) have shown that these energy sources in a number of simulated primordial atmospheres can produce a variety of biochemical building blocks, including sugars, lipids, amino acids, purines, peptides, and polypeptides (Fox et al., 1970; Fox and Dose, 1972). An alternative mechanism for the abiotic synthesis of organic matter is one that could have occurred at low temperatures (0 to −21 °C) in a cool, not a warm, ancient ocean (Miller and Orgel, 1974). A cool sea would have slowed the decomposition of organic compounds and polymers, thereby greatly extending their half-life. It would also have provided a method of eutectic concentration whereby compounds, including volatile compounds, would be concentrated by freezing out at temperatures below that of ice. The ice crystals could have also served as templates to direct the reactions required for polynucleotide synthesis.

Regardless of the mechanism for the abiotic synthesis of organic matter, at some point in time, as the ancient seas became a "rich soup" due to the accumulation of organic matter in the absence of life, life originated. The site must have been one where organic compounds accumulated. Dissolved organic compounds adsorb onto surfaces, such as those of clay minerals; they become concentrated at the sea-air interface and at the sea's edge as a result of water circulation and winds and condense and polymerize to form colloids, which readily precipitate out of solution. Such precipitated organic compounds would accumulate at the thermocline (an isolating layer of water below the upper mixed layer where temperature changes rapidly with depth) and on the sea floor (Weyl, 1968). All these sites in the sea today are characterized by accumulations of organic matter and more intense activity of heterotrophic microorganisms. Any one of these could have been the site where all the parts of the puzzle finally fell into place over a billion or more years to yield a metabolizing and replicating ancestor of the procaryotes. But once the process was started, these primitive cells began changing the environment that had produced them, utilizing the organic matter from which they were formed, and were therefore diminishing the chances of this event ever occurring again. As this ancestor procaryote changed its environment, it had to adapt or die. The great experiment had begun.

B. THE PROCARYOTIC CELL

The ancestral procaryote had to have certain minimum cellular components for a self-replicating system (Margulis, 1970). It had to have a simple membrane to enclose the compound cytoplasm that contained enzymes for catalysis, RNA for protein synthesis, and DNA for replication. It could have been a mycoplasma-like microorganism with just a simple membrane. Such forms are still present in the marine environment. There have been two major changes since life first formed in the Precambrian seas of our young planet. The microbe-free seas were very rich in inorganically derived organic matter, and they were anoxic (free of dissolved oxygen). The creation of life capable of growing in the absence of oxygen (anaerobic) had several advantages. The absence of oxygen and its toxic derivatives would have prevented autoxidation of the accumulating organic matter as well as damage to the oxygen-sensitive, anaerobic microorganisms. As the environment slowly changed to one that produced oxygen, the microorganisms had to become aerobes and facultative anaerobes by developing protective mechanisms against oxygen and its toxic products such as peroxide and super oxide, remain in an anaerobic niche, or perish. An advantage for life starting in a rich organic soup was that it would develop an osmotrophic mode of nutrition, with simple nutrients obtained by diffusion through the cell envelope and only dissimilatory enzymes would be required.

Before we consider how this anaerobic, osmotrophic ancestral procaryote diversified prior to the development of an ancestral eucaryote, the basic differences in cellular structure that separate the procaryotes from all other forms of life should be made clear. Procaryotes, which are all bacteria, are mostly small microorganisms that do not possess a nuclear membrane or form a mitotic figure during cell division; they also lack most of the organelles present in the eucaryotes. In contrast, the eucaryotes, which can be large organisms as well as microorganisms, have a well-defined nucleus and a variety of organelles and divide by mitosis. A more detailed list of differences is shown in Table 1-1. The differences between procaryotic and eucaryotic microorganisms are not superficial and distinguish two groups that vary markedly in size, form, and function.

For the ancestral procaryotes to have led to the wide spectrum of procaryotes and eucaryotes we know today, there had to be a succession of mutations beginning with the ancestral fermenting heterotrophs (which required preformed organic matter) to the autotrophic phototrophs (which produce organic matter and evolve oxygen by photosynthesis) and the aerobic heterotrophs (which

TABLE 1-1. Major differences between procaryotic and eucaryotic cells (adapted from Margulis, 1970).

CELL FEATURE	PROCARYOTE	EUCARYOTE
Size	0.1 to 2 μm wide; all are bacteria	2 to 100 μm wide; some are protists, others are cells of multicellular organisms
Nuclear material	Nucleoid, no membrane	Membrane-bounded nucleus
Organelles	No mitochondria, chloroplasts, or endoplasmic reticulum	Mitochondria, chloroplasts, and endoplasmic reticulum present.
Cell division	Direct, mostly binary fission. Chromatin body contains DNA and polyamines. No centrioles, mitotic spindles, or microtubules	Classical mitosis; many chromosomes containing DNA, RNA, and proteins. Centrioles, mitotic spindles, and 250 Å microtubules present
Sex	Absent in most forms; when present, is unidirectional transfer of genetic material from donor to host	Present in most forms; both male and female partners participate in meiotic production of gametes
Multicellular	No diploid zygote, no tissue differentiation	Develop from diploid zygotes, extensive tissue differentiation
Oxygen requirement	In addition to aerobic forms, have microaerophiles as well as facultative and obligate anaerobes	All aerobic, few exceptions are clearly secondary modifications
Metabolic pattern	Enormous variation, mitochondria absent, oxidative enzymes bound to cell membrane	Similar metabolic pattern, mitochondria present, oxidative enzyme "packeted" within membrane-bound sac
Flagella	If present, simple bacterial flagella	If present, complex (9+2) flagella or cilia
Photosynthetic enzymes	In chromatophores (bound to cell membrane)	In chloroplasts or other plastids (membrane-bound packets)
Type of photosynthesis	Anaerobic in anoxyphotobacteria, aerobic in oxyphotobacteria	Aerobic, oxygen elimination in all forms
DNA synthesis	Throughout life cycle of cell	Only during specific part of life cycle
Ribosomes	Generally small and sensitive to chloramphenicol and streptomycin	Generally larger and insensitive to chloramphenicol but sensitive to cycloheximide

use preformed organic matter and consume oxygen by chemoorganotrophy) that dominate the oxygenated seas today. Hypothetical evolutionary pathways for some of the main groups of procaryotic microorganisms are shown in Plate 1-1. As evolution proceeded, the biosynthetic pathways lengthened. When, as a result, a new compound became necessary, the microorganism developed a mechanism to convert an available compound to the required one, adding this ability to its replicating system to ensure survival. The first group, the fermenting anaerobes, developed the anaerobic glycolytic pathway, the Embden-Meyerhoff-Parnas scheme for the dissimilation of glucose. Some clostridia, spirochetes, and mycoplasmas that can utilize just this system still require the original conditions that were necessary for the creation of life, an oxygen-free environment rich in organic matter. Such an environment exists in marshes, inshore sediments, and anoxic (oxygen-free) basins isolated from oxygenated water by high sills. A second line of microorganisms fixes both carbon dioxide and nitrogen and synthesizes porphyrins and ferrodoxins as well. These synthetic activities permitted the development of anaerobic autotrophs, which included an ancestral photobacterium. This organism led not only to the anoxyphotobacteria but also to the aerobic oxygen-evolving cyanobacteria that started the process of oxygenating the atmosphere. A third line of microorganisms developed the ability to use nitrate as an electron acceptor in denitrification, then the cytochrome system, and finally the tricarboxylic acid (Krebs) cycle, which permitted the aerobic osmotrophs to develop in an oxygen-containing atmosphere. The scheme in Plate 1-1 indicates the possible phylogeny of many groups of bacteria. Until organisms with such a diversity of form and function, including cell wall-defective bacteria (mycoplasmas), motile spirochetes, oxygen-evolving cyanobacteria, and aerobic osmotrophs had developed, the evolution of the eucaryotic cell as we know it could not occur.

C. EUCARYOTIC EVOLUTION

The marked differences between the procaryotic and eucaryotic cells shown in Table 1-1 are very difficult to explain if we follow the classical and, until recently, prevalent concept of the primitive photosynthetic bacteria evolving into algae and plants, some of which then lost their photosynthetic abilities to form the fungi and the animals. Margulis (1970) calls this concept the "botanical myth." Hereditary endosymbiosis is a more plausible, alternative hypothesis. The concept behind this hypothesis is that the fusion of procaryotic endosymbionts and ectosymbionts with a procaryotic host cell resulted in a hereditary endosymbiosis to yield eucaryotic cells.

Symbiotic hypotheses in various contexts have been offered every ten or twenty years since 1882. But not until Margulis (1968, 1970, 1971) consolidated current information from molecular biology, biochemistry, paleontology, and other sources and presented the hypothesis forcefully in a "plausible, intriguing and esthetically pleasing" (McMahon, 1971) way did it receive the attention it deserves.

As the biosphere became oxygenated and the ozone in the atmosphere absorbed much of the ultraviolet radiation at wavelengths that had earlier produced the first organic matter, the osmotrophic microorganisms became more dependent upon the organic matter pro-

PLATE 1-1. Hypothetical evolutionary pathways and phylogenetic relationships of the procaryotes. From L. Margulis, 1970, *Origin of Eukaryotic Cells*, Yale University Press, New Haven, Conn.

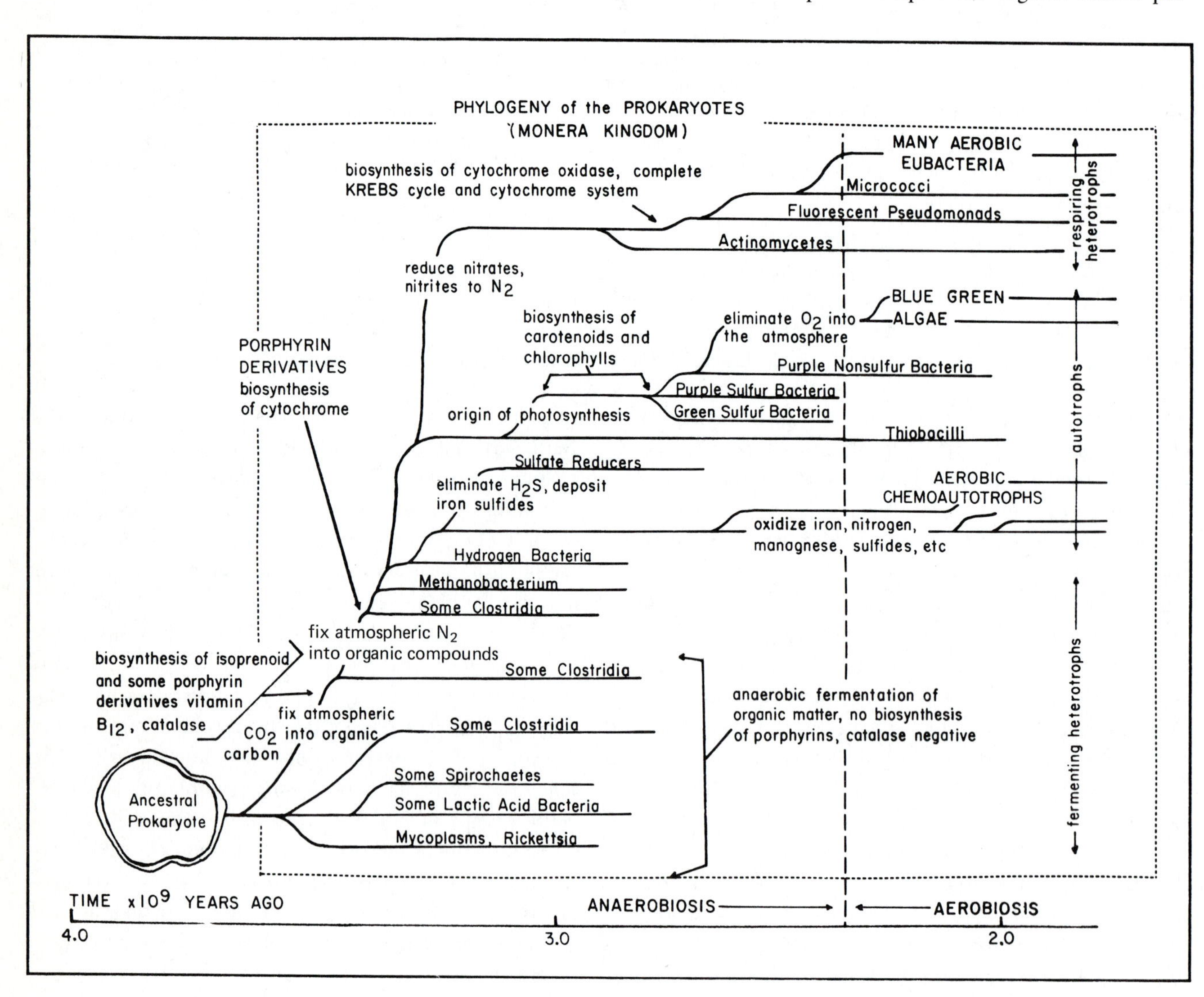

duced by the phototrophic microorganisms. All groups of microorganisms became more interdependent for nutrition and favorable microenvironments. The development of various mutualistic associations, such as symbiosis, commensalism, parasitism, and predation, led to the three basic trophic modes and the competition and cooperation seen in the microbial world today. The hypothesis of hereditary endosymbiosis, as advanced by Margulis, is that a very specific series of symbioses led to the formation of an ancestral eucaryote, a spectrum of protists, and eventually to the Plant, Animal, and Fungal kingdoms. This evolution of the eucaryotic cell by serial symbiosis is illustrated in Plate 1-2.

The procaryotic host that provided the cytoplasm was presumably a large, pleomorphic, mycoplasma-like cell capable of fermenting glucose only as far as pyruvate. When this host cell fused with a smaller aerobic procaryote, the endosymbiont conferred upon its host the biosynthetic ability to form cytochromes, flavins, and ubiquinone and to oxidize all energy sources to carbon dioxide; it also provided a ground nucleus. The endosymbiont was, by definition, the promitochondrion and the symbiotic pair became the proto-eucaryote which led to the primitive amoebae, many of which exist today. Since the aerobic symbiont required more surface membranes for oxidation-reduction reactions, it evolved into the chrystae-containing mitochondria of today. The evolution of extensive cytoplasmic membrane systems and food vacuoles then led to the predatory amoebae.

The next stage was the evolution of motility leading to the successors of the primitive amoebae, the amoeboflagellates. This motility is hypothesized to have been conferred by a spirochete-like microorganism, which became an ectosymbiont or a proto-flagellum. In the weakest part of this model, it is assumed that the (9 + 2) fibrillar arrangement of protein, as seen in cross section of this ectosymbiont, later became the (9 + 2) flagella and the cilia of eucaryotic cells. Although the amoeboid cell became motile by the incorporation of the motile spirochete, a long series of differentiations and de-differentiations could have led to other structures, such as the mitotic apparatus, the oral apparatus, and the contractile vesicle in the protozoa.

The basic characteristic of living organisms is their ability to reproduce. When a microorganism has grown to a certain size, it reproduces. In the procaryotes, this is usually by binary fission into two equal-size daughter cells; more rarely it occurs by budding. The eucaryotes also reproduce by fission and budding and, like the procaryotes, some may reproduce indefinitely by vegetative or asexual processes, including multiple fission or sporogamy. During the asexual reproduction of eucaryotic cells, the diploid nucleus is reduplicated and distributed into the nuclei of each daughter cell through the process known as mitosis. Sexual reproduction is not a regular occurrence in some eucaryotic microorganisms and is triggered by certain environmental conditions. This facultative alternation of generations (between haploid and diploid states) can occur in all trophic modes in the eucaryotes. In other eucaryotes, there is an obligate alternation of generations when, at maturity, one cell and nuclear state must by necessity form the other cell and nuclear state. In many forms, the daughter cells are gametes with the ability to fuse to form a zygote. A necessary part of the fertilization of this zygote is the fusion of two gametic nuclei (karyogamy). The resulting doubling in chromosome number is later compensated for in the life cycle by a reduction of chromosome number through nuclear division (meiosis). Sexual reproduction, in addition to nuclear fusion and reduction, usually involves morphological changes. These morphological and nuclear changes produce some fascinating but very intricate life cycles. This is further complicated in foraminifera and in ciliates that have different nuclear types within the same cell (nuclear dimorphism). These protozoa have generative nuclei (micronuclei) involved in reproduction and somatic nuclei (macronuclei) that have other functions, and if they are involved in reproduction, it is only briefly. This nuclear dimorphism is better understood in context with the life cycles of the foraminifera (Chapter 23,C) and the ciliates (Chapter 24).

The plants then arose after mitosis and meiosis became perfected in the protozoa. The ingestion and maintenance of photosynthesizing cyanobacteria by the amoeboflagellates gave rise to protoplastids. As these oxygen-evolving bodies were modified and incorporated into their hosts, the endosymbionts became more obligate; this led to the development of a number of algal lines and eventually to the green plants.

The phylogeny of the lower eucaryotes in the Kingdom Protista is shown in Plate 1-3. The hereditary endosymbiosis hypothesis claims to explain, unambiguously, the relationship of all the lower eucaryotes. The algae, fungi, and such miscellaneous groups as the labyrinthulids and slime molds are taken to be examples of

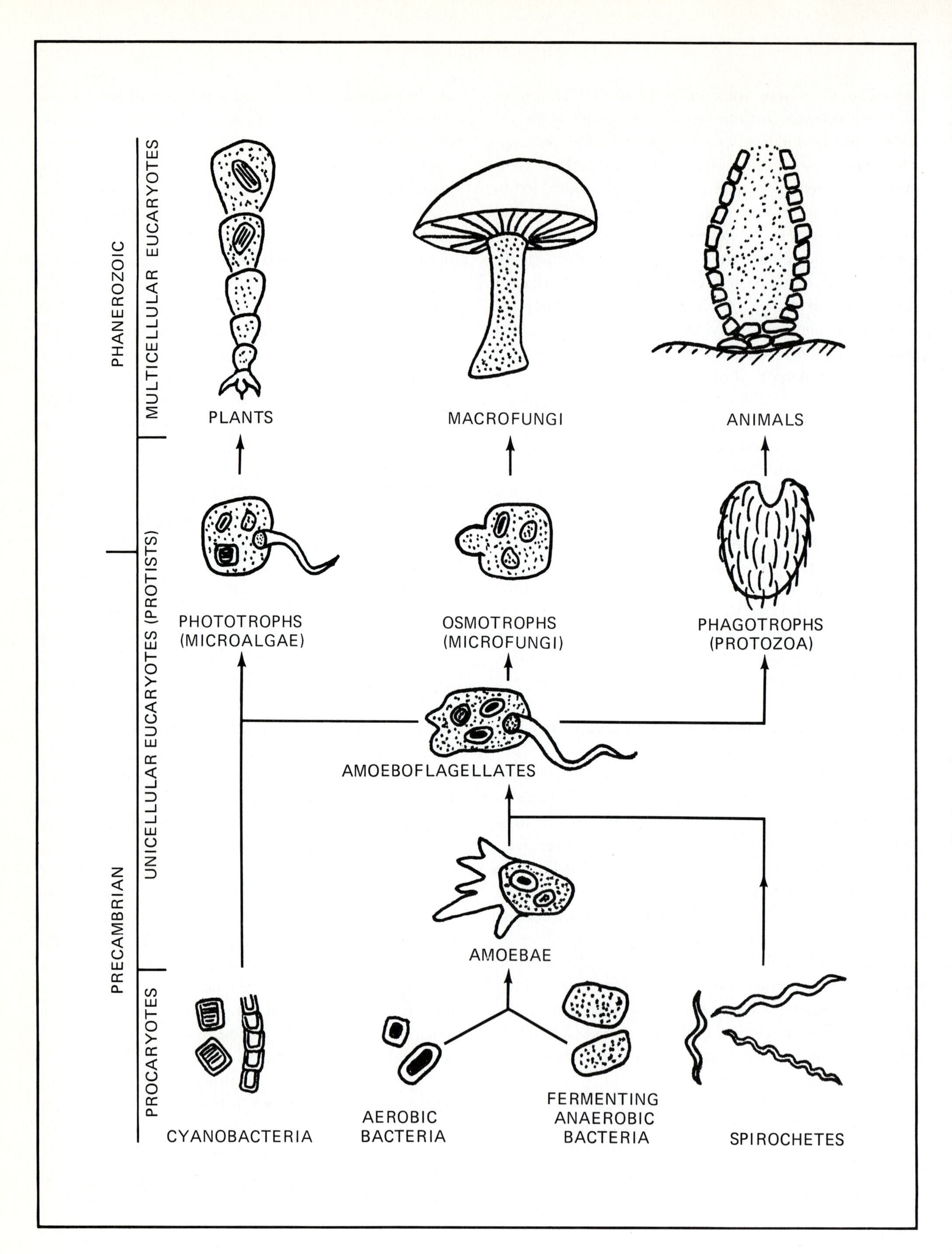
PHANEROZOIC
PRECAMBRIAN
MULTICELLULAR EUCARYOTES
UNICELLULAR EUCARYOTES (PROTISTS)
PROCARYOTES
PLANTS
MACROFUNGI
ANIMALS
PHOTOTROPHS
(MICROALGAE)
OSMOTROPHS
(MICROFUNGI)
PHAGOTROPHS
(PROTOZOA)
AMOEBOFLAGELLATES
AMOEBAE
CYANOBACTERIA
AEROBIC
BACTERIA
FERMENTING
ANAEROBIC
BACTERIA
SPIROCHETES

PLATE 1-2. A hypothetical series of events that explains how four different types of procaryotic cells could have undergone serial endosymbioses to form the eucaryotic microalgae, microfungi, and protozoa, as well as the cells of the plant, animal, and fungal kingdoms. Adapted from Lynn Margulis, Department of Biology, Boston University.

different approaches to the evolution of a mitotic system. The vast majority of the eucaryotes are believed to have evolved directly from symbiotic complexes of heterotrophic procaryotic forms. Only a few fungal and protozoan genera appear to have evolved secondarily by the loss of plastids to form "apochlorotic" (colorless) algae. The endosymbiotic hypothesis gives a workable explanation of how things could have been, and yields a natural and pleasing explanation of the relationships between the present-day members of the lower eucaryotes (protists).

A modification of the above, two-kingdom system of cell organization is the inclusion of a third kingdom, the Mesocaryota, proposed by Dodge (1965, 1966). It consists of the dinoflagellates, which have a nuclear organization intermediate between that of the procaryotes and the eucaryotes. Dodge also suggested that the heli-

PLATE 1-3. Hypothetical evolutionary pathways and phylogenetic relationships of the lower eucaryotes (protists). From L. Margulis, 1970, *Origin of Eukaryotic Cells*, Yale University Press, New Haven, Conn.

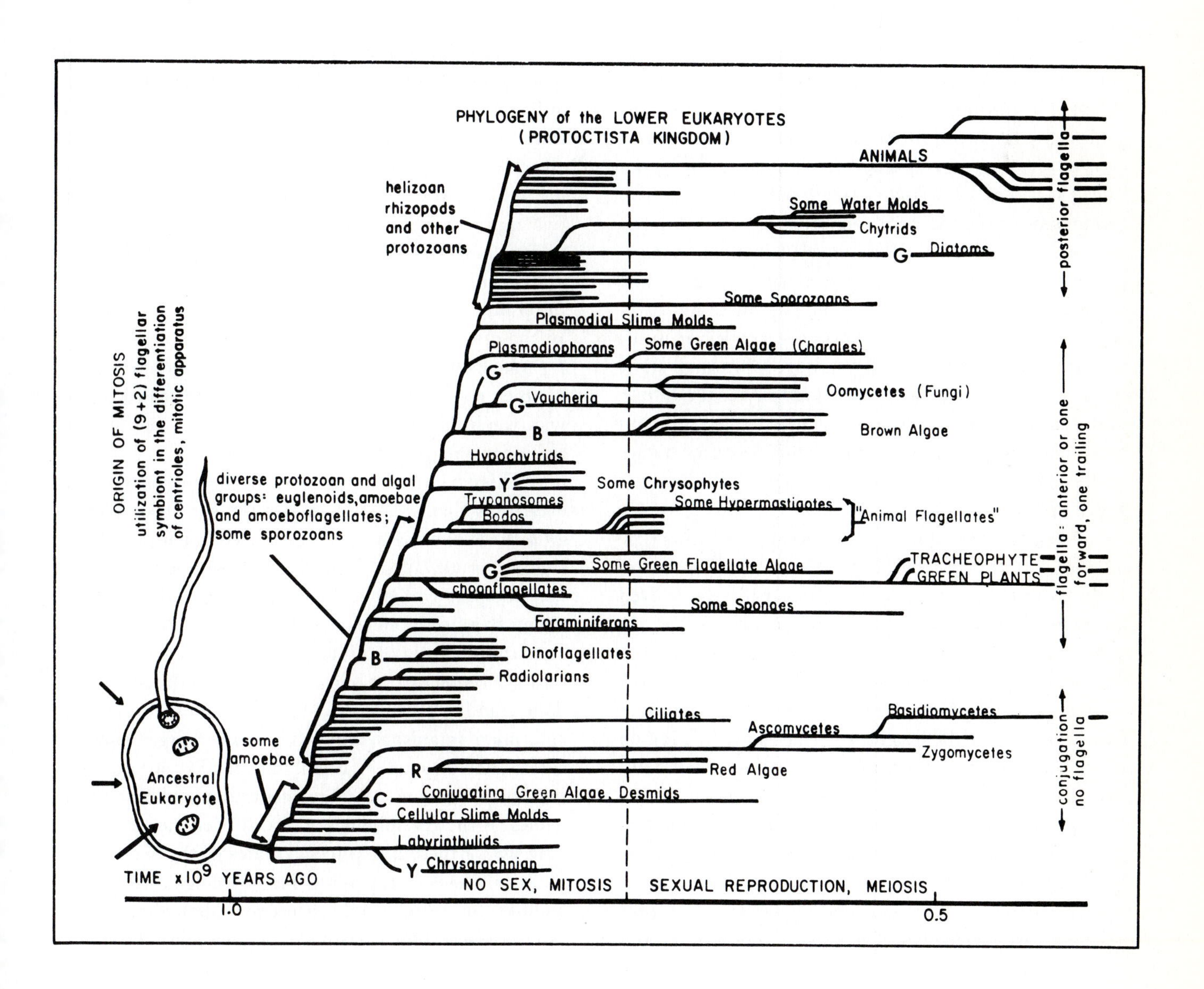

ozoans, radiolarians, and other microorganisms with unusual nuclear characteristics could be included. Allsopp (1969) saw no real justification for creating a new kingdom only for microorganisms with unusual nuclei. Zingmark (1970) has rejected the Heliozoa and the Radiolaria from the Mesocaryota because they have a typical eucaryotic nuclei and defines the "typical" nuclear organization of the interphase nucleus of the Mesocaryota.

D. CLASSIFICATION

The existence of microorganisms with such widely diverse forms and functions has always posed a problem in classification. The inclusion of bacteria and fungi with the algae in the division Thallophyta of the Plant Kingdom never made any sense. The recognition by Haeckel (1866) of a third kingdom, the Protista, for the microorganisms, was an improvement. Although Haeckel recognized the monerans (bacteria including the blue-green "algae"), it was Copeland (1938, 1956) who gave them "Kingdom" status and developed a rather successful system of four kingdoms. These are (1) Kingdom Monera: procaryotic microorganisms, unicellular or with simple colonial organization (bacteria including the cyanobacteria); (2) Kingdom Protoctista: lower eucaryotic organisms without tissue differentiation (algae, protozoa, slime molds, and fungi); (3) Kingdom Metaphyta: the higher multicellular eucaryotic organisms with walled cells and green plastids (plants); and (4) Kingdom Metazoa: higher multicellular eucaryotic organisms with wall-less cells and no plastids (animals).

Although this system of Copeland, as refined by others, is quite workable, it does have two drawbacks. Only two of the three trophic modes that have evolved are recognized (photosynthesis and ingestion); the absorptive or osmotrophic mode characteristic of the bacteria and fungi is not recognized. The second drawback is that the Kingdom Protoctista (Protista) is not as clearly defined as the other three kingdoms. In addition to the unicellular microorganisms, it includes such multicellular organisms as some algae and sponges.

These objections to the Copeland four-kingdom system are largely overcome in the five-kingdom system of Whittaker (1969), shown in broad outline in Plate 1-4. It is based on three levels of organization and three principal trophic modes. The three levels of organization are (1) the unicellular procaryotic microorganisms in the Kingdom Procaryotae (Monera), which includes both photosynthetic and absorptive trophic modes; (2) the unicellular eucaryotic microorganisms in the Kingdom Protista, which includes all three trophic modes (phototrophic, osmotrophic, and phagotrophic); and (3) the multicellular eucaryotic and multinucleated organisms, which fall in the remaining three kingdoms: the Kingdom of Plants with phototrophic organisms, the Kingdom of Fungi with osmotrophic organisms, and the Kingdom of Animals with ingestive organisms. The classification of organisms is greatly simplified by putting the fungi on a par with the plants and animals and elevating the seaweeds and such metazoa as the sponges into the higher kingdoms. The microorganisms and their trophic modes are recognized in this system.

A remaining problem with the Whittaker five-kingdom system is how to further divide the protists. Margulis (1974, 1976) has refined and modified the scheme from an evolutionary viewpoint but has not yet undertaken the division of the protists. Leedale (1974) in asking "How many are the kingdoms of organisms?" offers two approaches to this problem. One is the "pteropod," or four-kingdom system, which sidesteps the issue by not considering the Kingdom Protista and by including the protists in the Plant, Fungal, and Animal kingdoms. In the other approach, each phylum, or even class, is a separate member of a fan-shaped, multi-kingdom scheme. Neither approach helps the student or textbook writer who would like to organize the groups in a logical, nonredundant manner.

This text attempts to cover all the groups of microorganisms that are free-living in the sea. Microbiology *sensu stricto* is not a discipline but a consortium of bacteriology, protozoology, and the microscopic branches of the disciplines of phycology (algology) and mycology. When these disciplines or subdisciplines are dealt with separately, an overlap into another discipline is not too serious. But in a text such as this, it would be chaotic. This is indicated in Plate 1-5,*A* in which the organisms and microorganisms, and the submicroorganisms (viruses) that develop in them, are divided according to their trophic mode, cellular organization and host. If the names of the disciplines are placed over the theoretical coverage of these disciplines, an idea of the overlap and potential chaos is apparent. This coverage of these disciplines, however, is only theoretical. Biologists work in discrete areas. One can divide the theoretical disciplines shown in Plate 1-5,*A* into the practical subdisciplines shown in Plate 1-5,*B*. This is a much truer approximation of how microorganisms and submicroorganisms are studied. Even then, most workers only become familiar with a small group within one of these subdisciplines.

If the organisms and the microorganisms and the submicroorganisms living in association with them are divided into subdisciplines in practice, then we can also put them into pragmatic subdivisions in taxonomic works if we want to. The recent trend in taxonomy is the pragmatic approach, as exemplified by the 8th Edi-

PLATE 1-4. Whittaker's five-kingdom concept based on three levels of organization: the unicellular procaryotes (Monera), the unicellular eucaryotes (Protista), and the multicellular and multinucleate eucaryotes (plants, fungi, and animals). At each level there is a divergence into three trophic modes (photosynthetic, absorptive, and ingestive) except for the Monera, which lack the ingestive trophic mode. From R.H. Whittaker, 1969, *Science* 163:150–60, copyright 1969 by the American Association for the Advancement of Science.

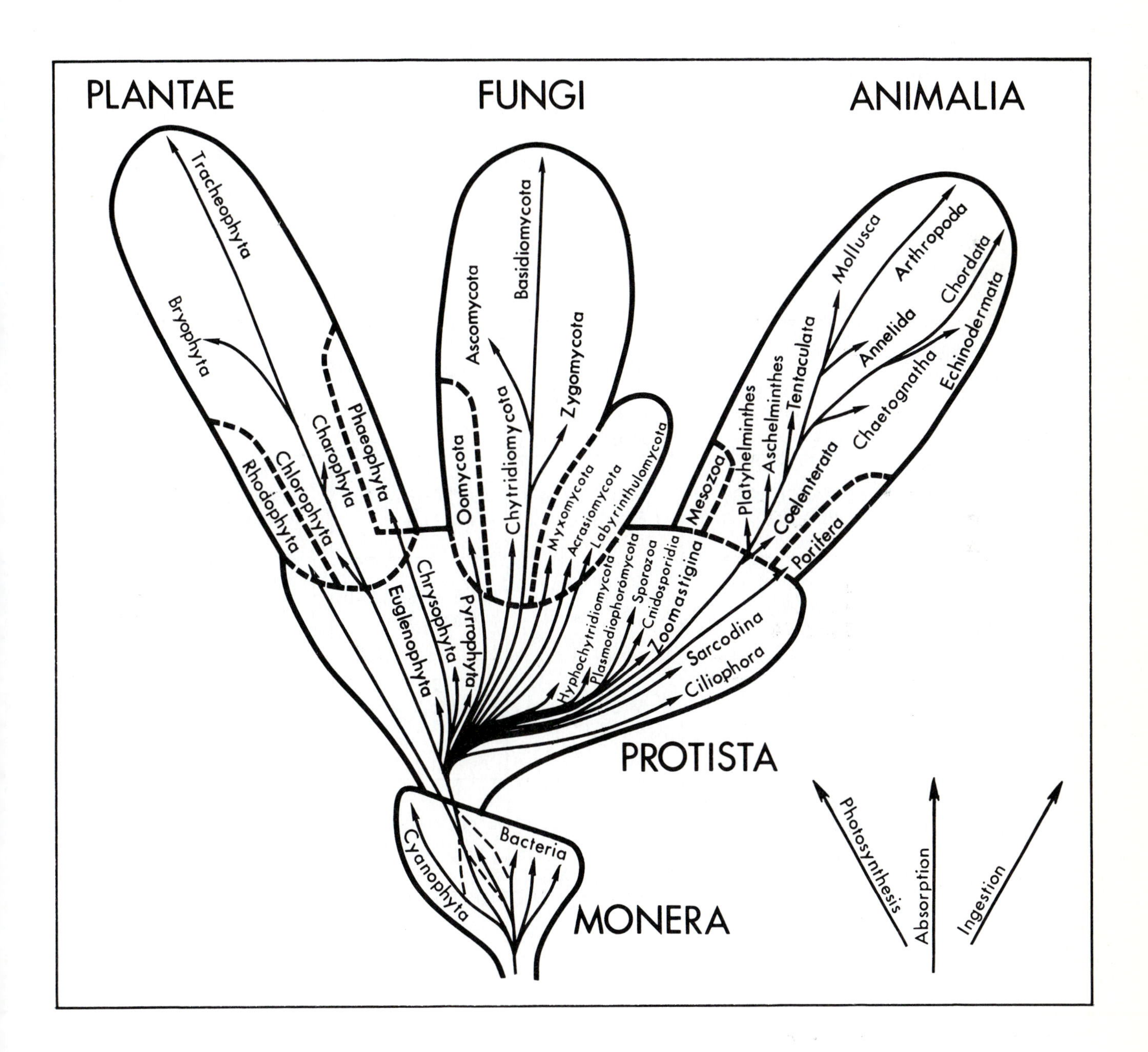

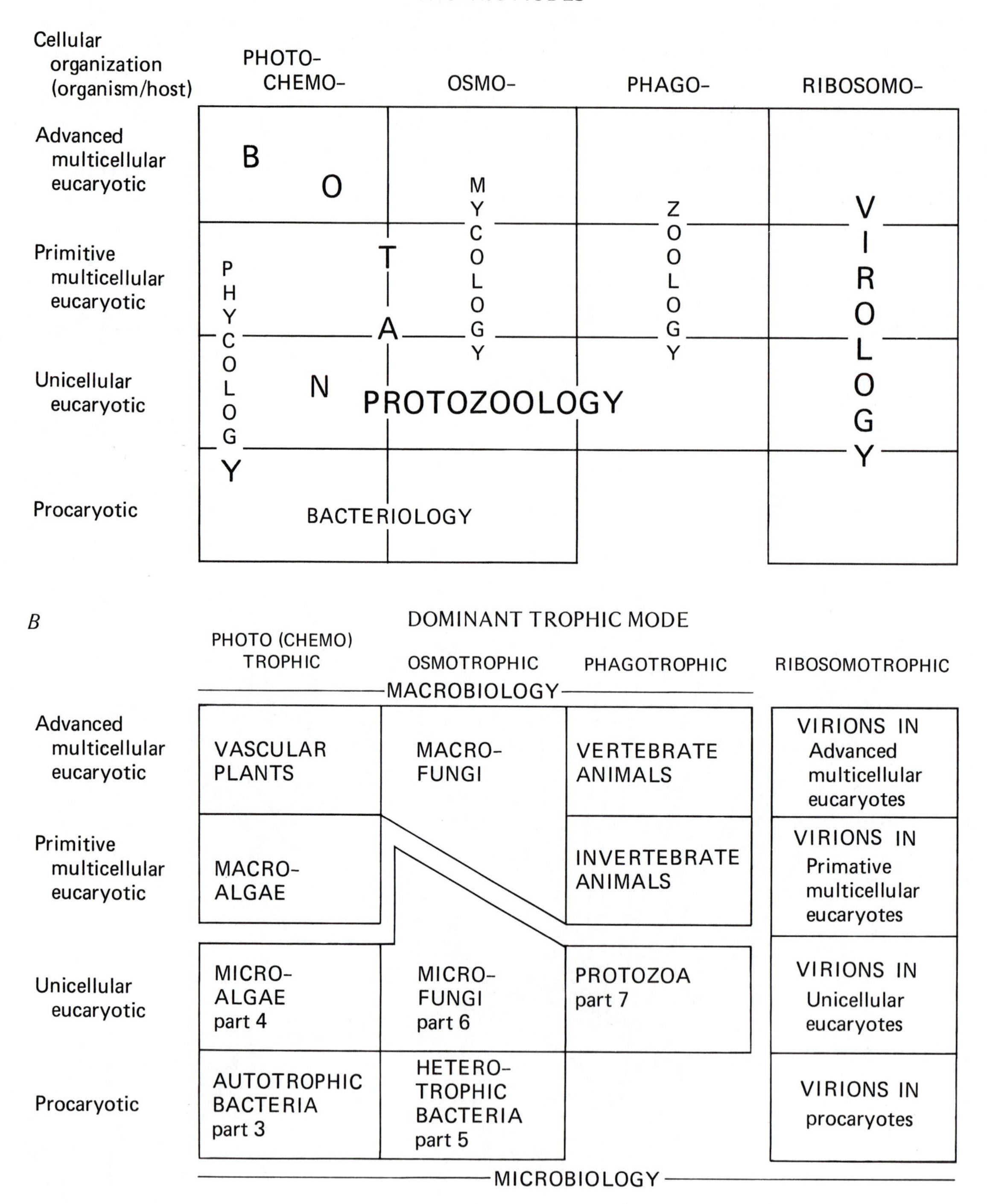
A
TROPHIC MODES
Cellular organization (organism/host)
PHOTO- CHEMO-
OSMO-
PHAGO-
RIBOSOMO-
Advanced multicellular eucaryotic
Primitive multicellular eucaryotic
Unicellular eucaryotic
Procaryotic
BOTANY
MYCOLOGY
ZOOLOGY
VIROLOGY
PHYCOLOGY
PROTOZOOLOGY
BACTERIOLOGY
B
DOMINANT TROPHIC MODE
PHOTO (CHEMO) TROPHIC
OSMOTROPHIC
PHAGOTROPHIC
RIBOSOMOTROPHIC
MACROBIOLOGY
Advanced multicellular eucaryotic
Primitive multicellular eucaryotic
Unicellular eucaryotic
Procaryotic
VASCULAR PLANTS
MACRO- FUNGI
VERTEBRATE ANIMALS
VIRIONS IN Advanced multicellular eucaryotes
MACRO- ALGAE
INVERTEBRATE ANIMALS
VIRIONS IN Primative multicellular eucaryotes
MICRO- ALGAE part 4
MICRO- FUNGI part 6
PROTOZOA part 7
VIRIONS IN Unicellular eucaryotes
AUTOTROPHIC BACTERIA part 3
HETERO- TROPHIC BACTERIA part 5
VIRIONS IN procaryotes
MICROBIOLOGY

PLATE 1-5. The division of replicating forms. (*A*) The theoretical coverage of replicating forms, by discipline, showing overlap. (*B*) An interdisciplinary subdivision of replicating forms into practical groupings based on ultrastructural and dominant trophic modes as followed in the text.

tion of *Bergey's Manual of Determinative Bacteriology* (Buchanan and Gibbons, 1974), which has dispensed with assumed phylogenies and their associated hierarchies and has placed the microorganisms into arbitrary ecological groupings. The subdisciplines in Plate 1-5,*B* are also working groups. The types of organisms, microorganisms, and submicroorganisms occurring in each of these biomodules are shown within the framework of four trophic modes and four states of cellular organization. The five biomodules of free-living microorganisms shown are each dealt with in a separate part of this text, as noted. To do this, a number of arbitrary decisions have been made, although such decisions have not yet been made by the "big picture" taxonomists.

The protists (unicellular eucaryotes) covered in Parts IV, VI, and VII were assigned to a dominant trophic mode. Such an assignment is complicated by the flagellates, which occur in all three modes. The chloroplast-containing forms are considered to be microalgae, but for convenience, the osmotrophic forms are included with the phagotrophic forms in Part VII. The microfungi are dealt with in Part VI. The one remaining redundancy is the appearance of certain classes of flagellates in both the phototrophic (Part IV) forms and the phagotrophic forms (Part VII). The result of these arbitrary decisions is that the phototrophic flagellates, which are claimed by the protozoologists (Honigberg et al., 1964) but are not studied by them, are included in the microalgae, which are studied by microalgologists. Having made these pragmatic divisions of the protists, we can now discuss the microorganisms without prolonging the nonproductive arguments over territorial claims or entering into discussions on possible phylogenetic relationships. For the student of marine microorganisms interested in the environmental problems of today, the important relationships are those between microorganisms and those between the larger organisms and their associated microorganisms. The restraints of disciplines and kingdoms should probably give way to a biomodular approach for classifying the life and processes in our ecosystems.

REFERENCES

Allsopp, A. 1969. Phylogenetic relationships of the Procaryota and the origin of the eucaryotic cell. *New Phytol. 68*:591–12.

Buchanan, R.E., and N.E. Gibbons (eds.). 1974. *Bergey's Manual of Determinative Bacteriology*. 8th Edition. Williams & Wilkins, Baltimore, Md., 1246 p.

Copeland, H.F. 1938. The kingdoms of organisms. *Quart. Rev. Biol.* 13:383–420.

Copeland, H.F. 1956. *The Classification of Lower Organisms*. Pacific Books, Palo Alto, Ca., 302 p.

Dodge, J.D. 1965. Chromosome structure in the dinoflagellates and the problem of the mesocaryotic cell. *Excerp. Med. Int. Cong. Ser.* 91:339 (Abstr.).

Dodge, J.D. 1966. The Dinophyceae. In: *The Chromosomes of the Algae* (M. Godward, ed.). St. Martin's Press, New York and Edward Arnold, London. pp. 96–115.

Fox, S.W., and K. Dose. 1972. *Molecular Evolution and the Origin of Life*. W.H. Freeman, San Francisco, Ca., 359 p.

Fox, S.W., K. Harada, G. Krampitz, and G. Mueller. 1970. Chemical origins of cells. *C&E News*, June 1970:80–94.

Haeckel, E. 1886. *Generelle Morphologie der Organismen*. Reimer, Berlin.

Honigberg, B.M., W. Balamuth, E.C. Bovee, J.O. Corliss, M. Gojdisc, R.P. Hall, R.R. Kudo, N.D. Levine, A.R. Loeblich, Jr., J. Weiser, and D. H. Wenrich. 1964. A revised classification of the Phylum Protozoa. *J. Protozool.* 11:7–20.

Leedale, G.F. 1974. Origin and evolution of the eukaryotic cell. IV. How many are the kingdoms of organisms? *Taxon* 23:267–70.

Margulis, L. 1968. Evolutionary criteria in the Thallophytes: A radical alternative. *Science* 161:1020–21.

Margulis, L. 1970. *Origin of Eukaryotic Cells*. Yale University Press, New Haven, Conn., 349 p.

Margulis, L. 1971. Symbiosis and evolution. *Sci. American* 225(2):48–57.

Margulis, L. 1974. The classification and evolution of prokaryotes and eukaryotes. In: *Handbook of Genetics*, Vol. 1 (R.C. King, ed.), Plenum Press, New York, pp. 1–41.

Margulis, L. 1976. The theme (mitotic cell division) and the variations (protists): Implications for higher taxa. *Taxon* 25:391–403.

McMahon, D. 1971. The colonization hypothesis (review of *Origin of Eukaryotic Cells* by L. Margulis). *Science* 172:675–76.

Miller, S.L., and L.E. Orgel. 1974. *The Origins of Life on Earth*. Prentice-Hall, Englewood Cliffs, N.J., 229 p.

Weyl, P.K. 1968. Precambrian marine environment and the development of life. *Science* 161:158–60.

Whittaker, R.H. 1969. New concepts of Kingdoms of organisms. *Science* 163:150–60.

Zingmark, R.G. 1970. Ultrastructural studies on two kinds of mesocaryotic dinoflagellate nuclei. *Amer. J. Bot.* 57:586–92.

CHAPTER 2

Exposed Microbial Habitats

A. SUBDIVIDING THE SEA

Each microorganism requires specific conditions for its growth. The broad spectrum of microorganisms ensures that some types will be present, in detectable numbers, throughout the various habitats in the sea, except for a very few highly specialized niches with a reduced microbiota, such as the hot brines of the Red Sea (Watson and Waterbury, 1969) or the gut of penguins who ingest inhibitory substances (Sieburth, 1960). How can the sea be divided into manageable units to describe the niche of microorganisms?

The usual classification of marine environments is the one given by Hedgepeth (1957); it is shown in Plate 2-1,*A*. Since the intertidal area is exposed at low tide and the shallow neritic waters along the shoreline can be easily worked with small craft and SCUBA, these are the areas studied most intensively by most marine biologists and microbiologists. Coastal waters are strongly influenced by the weather and the presence of the adjacent landmass, as well as by the currents of the sea. One such area with great seasonal variation in temperature is the intensively studied waters of southern New England, extending from Cape Cod past Narragansett Bay and Block Island Sound to Long Island Sound. The cold winter air from the central and northern parts of the continent chills the water to as low as −2 °C, the temperature of polar seawater. By contrast, the warm summer air from the land and the warm Gulf Stream waters elevate the water temperatures to 25 to 30 °C during the late summer and early fall. For the macroorganisms such as seaweeds, boreal (arctic) species grow during the winter, whereas tropical–subtropical

PLATE 2-1. Useful terminology for classifying marine environments and habitats. (*A*) The traditional environments of the sea: (*B*) The types of habitats of marine organisms that span the marine environments. (*A*) after J. Hedgepeth, 1957, *Treatise on Marine Ecology and Paleoecology*, Geological Society of America Mem. 67, Ch. 2.

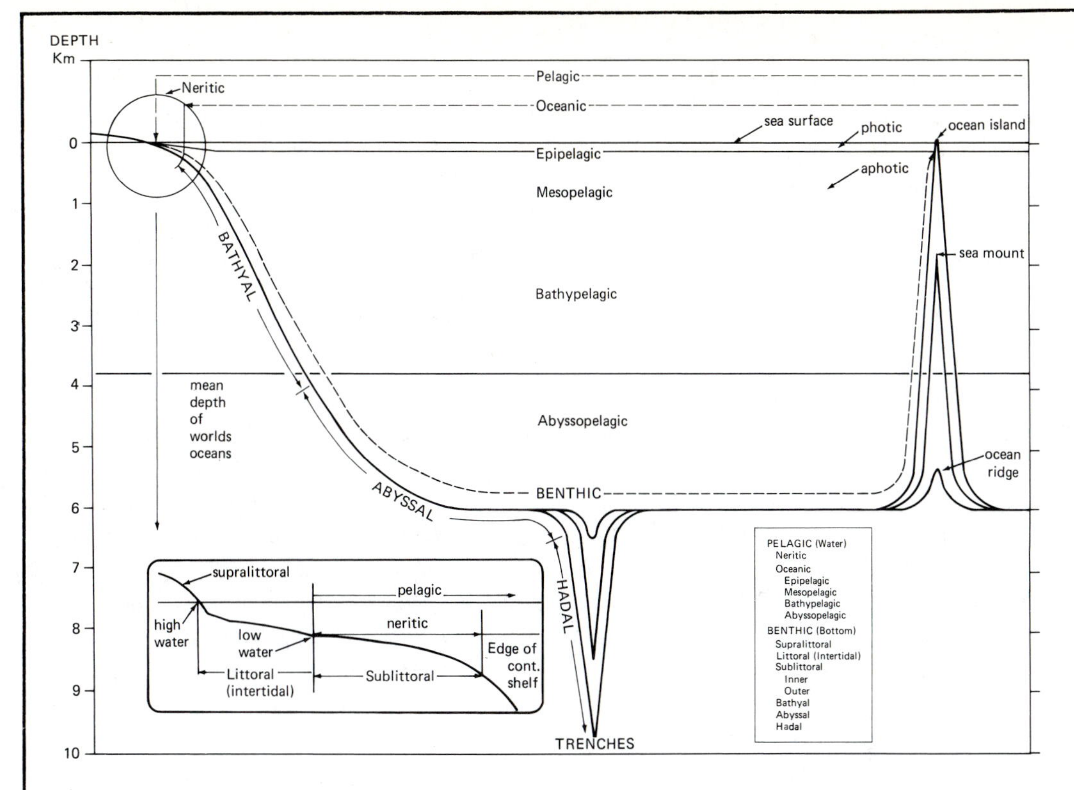
DEPTH
Km
Pelagic
Neritic
Oceanic
sea surface
photic
ocean island
Epipelagic
aphotic
Mesopelagic
BATHYAL
sea mount
Bathypelagic
mean depth of worlds oceans
Abyssopelagic
ocean ridge
BENTHIC
ABYSSAL
HADAL
TRENCHES
supralittoral
pelagic
high water
low water
neritic
Edge of cont. shelf
Littoral (intertidal)
Sublittoral
PELAGIC (Water)
Neritic
Oceanic
Epipelagic
Mesopelagic
Bathypelagic
Abyssopelagic
BENTHIC (Bottom)
Supralittoral
Littoral (Intertidal)
Sublittoral
Inner
Outer
Bathyal
Abyssal
Hadal

A

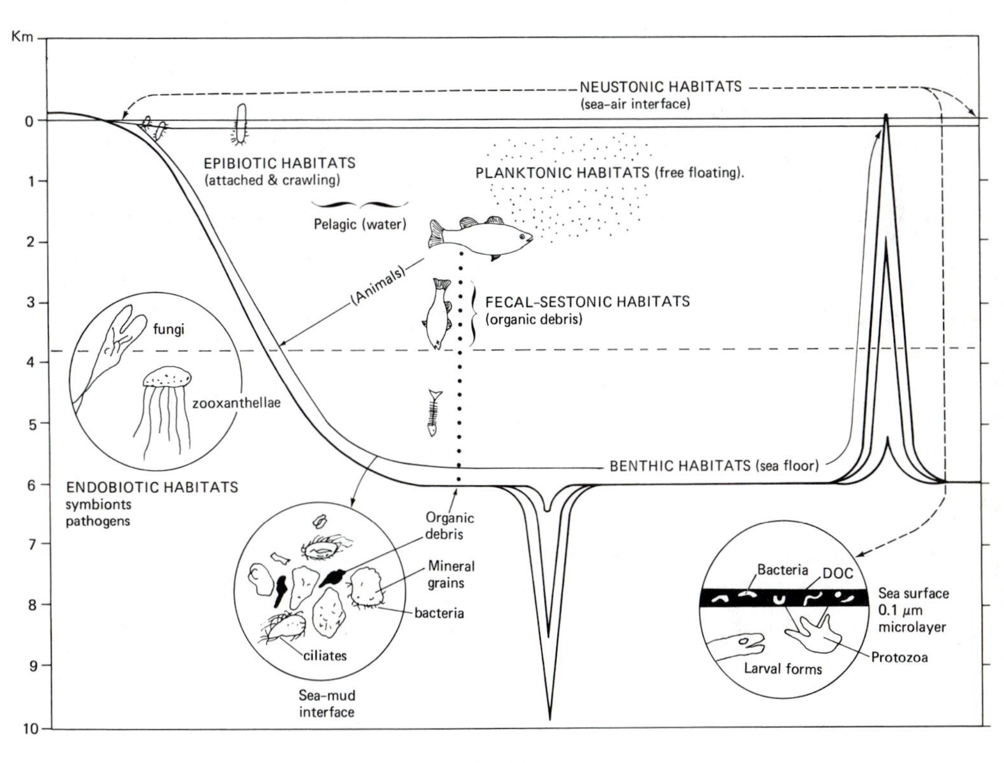
Km
NEUSTONIC HABITATS
(sea-air interface)
EPIBIOTIC HABITATS
(attached & crawling)
PLANKTONIC HABITATS (free floating).
Pelagic (water)
(Animals)
FECAL-SESTONIC HABITATS
(organic debris)
fungi
zooxanthellae
ENDOBIOTIC HABITATS
symbionts
pathogens
BENTHIC HABITATS (sea floor)
Organic debris
Mineral grains
bacteria
ciliates
Sea-mud interface
Bacteria
DOC
Sea surface 0.1 μm microlayer
Protozoa
Larval forms

B

species flourish in the late summer. Such wide seasonal excursions in water temperature also have a profound effect on the microorganisms and select a sequence of thermal types of bacteria during the yearly temperature cycle (Sieburth, 1967).

The influence of land is ever present in nearshore waters. In the intertidal area, the biota can be exposed to unseasonal and rapid changes in air temperature during emersion at low tide, which can interrupt the seasonal changes occurring in the water column (Seshadri and Sieburth, 1975). The estuaries can also undergo marked seasonal changes in salinity due to the input of freshwater from their river systems. The effect of such physical-chemical stresses as temperature and salinity on marine organisms is fully discussed in the comprehensive monograph series edited by Kinne (1972, 1975). The same rivers that reduce the salinity also carry terrestrial organic and mineral debris into the sea. The margins of the sea also have a large input from the seagrasses of marsh communities and from the seaweeds of rocky coasts, which die, fragment, and add to the organic debris. The nearbottom turbulence of the water caused by each tidal flow resuspends the sedimented organic debris, fecal debris, and mineral fragments (Rhoads, 1973) thereby superimposing the benthos upon the plankton. Such suspended particles are shown in Plate 2-2. The combination of all these factors makes it very difficult to study the microbiology of shallow water ecosystems. This is true for many nearshore environments, including ocean islands, coral reefs, marshes, sandy beaches, mud flats, and estuaries.

As one leaves the margins of the sea and the influence of the land, the conditions for marine life become more stable and the habitats more distinct. These offshore areas with water columns exceeding 2000 m in depth include some 85% of the world's oceans. The habitats for life in offshore waters are similar to those in nearshore waters except that the biota concentrated in the euphotic zone and on the sea floor are greatly separated by the very deep aphotic water mass with its more constant salinity, low temperature, and high hydrostatic pressure. In fact, this repetition of a series of habitats throughout the major environments of the sea is probably sufficient reason to look at the sea from a habitat viewpoint, rather than from an environmental one (Sieburth, 1976).

All waters can be thought of as possessing six basic habitats, which are shown in Plate 2-1,*B*. The planktonic habitat, consisting of suspended but poorly motile

PLATE 2-2. A comparison of the larger particles suspended in nearshore water with those retained by a plankton net. (*A*) Net (#20) plankton concentrated on a HA Millipore membrane. (*B*) Water sample concentrated directly on a HA Millipore membrane (Sc, *Skeletonema costatum;* Tn, *Thalassiosira nordenskjoldii;* Th, *Thalassionema nitzschioides;* Dc, *Detonula confervacea;* mg, mineral grains; ff, fecal fragments; bf, bacteria in film; bars = 10 μm). From J.McN. Sieburth, 1975, *Microbial Seascapes*, University Park Press, Baltimore, Md.

forms, concentrates in the thin sunlit photic zone where radiant energy from the sun is utilized by photosynthetic microorganisms to produce organic matter. This soluble and particulate organic matter produced by the phytoplankton, which is moved up the food chain by bacterioplankton, protozooplankton, and zooplankton, is the basis for virtually all life in the sea. Part of the organic matter released by the activities of the plankton is concentrated at the sea–air interface to form a thin skin or surface microlayer. Organisms adapted to this specialized microhabitat of the sea surface are termed neuston or pleuston. The larger swimming animals that live upon the plankton, such as squid, fish, and baleen whales, are known as nekton. The debris resulting from the excrement of zooplankton and nekton, as well as that from the die-off of organisms of all sizes, produces substantial amounts of suspended and sedimenting particulate organic matter called seston. In terms of the quantity of organic carbon present in the sea, seston is second only to dissolved organic carbon. This suspended and sedimenting particulate organic matter is part of the enteric–fecal–sestonic habitat. The smaller particles decompose in the water column, sustaining threshold levels of life in abyssal waters, while the larger particles settle to the sea floor. The larger particles, such as fecal pellets are concentrated at the sea–mud interface and develop a microbiota that nourishes animal life. These are the benthic habitats and they extend from the sea margins to the deepest trenches at a depth of 10,000 m. Epibiotic habitats occur on inanimate, microbial, plant, and animal surfaces where there is a surface suitable for microorganisms of all trophic modes and of many sizes to attach to or to crawl around on. Certain microorganisms have found conditions conducive for life within the tissues of macroorganisms or other microorganisms, and form endobiotic habitats. Those that debilitate the host are pathogens; those that benefit the host are mutualistic symbionts.

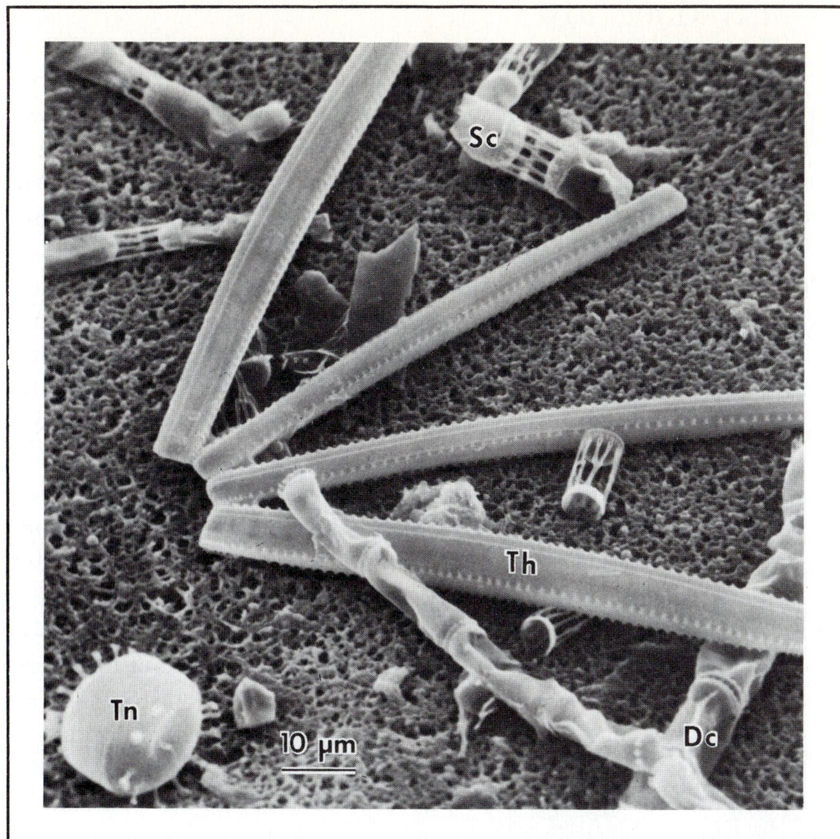

A

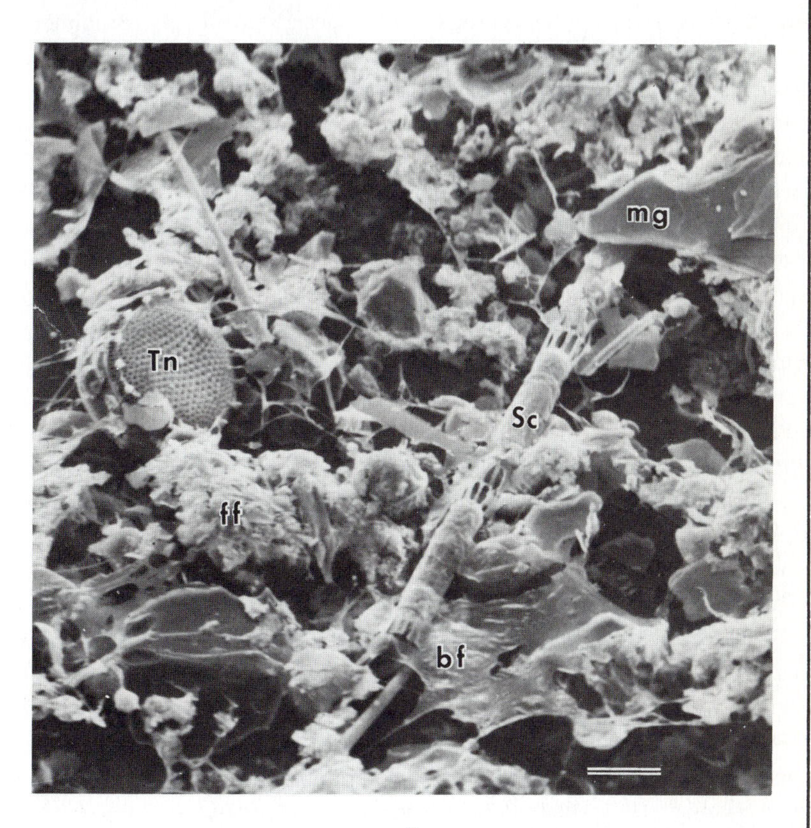

B

Together these six habitat groupings encompass the wide range of microenvironments in which sea microbes are seen. Each of the habitats and some of the specific microenvironments they encompass will be discussed and illustrated so that, as the types of microorganisms are discussed, their neighborhood and niche will be recognized. The types of macroorganisms that occur in each habitat are also important, since they help to shape the environment. The three exposed habitats are discussed in this chapter and the three protected habitats in Chapter 3.

B. NEUSTONIC HABITATS

The sea surface with its accumulations of flotsam and jetsam (Beebe, 1926), pelagic gulfweed (*Sargassum*), and such surface-adapted animals (David, 1965b; Cheng, 1975) as the nudibranch mollusc *Glaucus*, the purple storm snail *Janthina*, the by-the-wind sailor *Velella*, the Portuguese man-of-war *Physalia*, the ocean sunfish *Mola*, and the sea snakes (Kropach, 1975) have attracted the attention of seafarers for centuries. The specialized biota at the air–water boundary, with a spectrum of organisms from bacteria to vertebrate larvae, has been called neuston (Naumann, 1917). Although neuston plays an important role in the developmental stages of pelagic fish and in the maintenance of seabirds, it is only recently that the sea–air surface has received attention as a habitat (Zaitsev, 1970). The development of special nets to sample the upper 10 cm of the macroneuston (David, 1965a) has shown the dominance of blue-pigmented copepods, which were once considered rare, as well as the larger blue forms that make the neustonic habitat a fascinating microcosm (David, 1965b; see Plate 2-3,*B*).

The physical–chemical basis for the existence of this extraordinary layer and its effect on marine, atmospheric, and terrestrial processes is being actively studied (Duce et al., 1972; Chesselet et al., 1974; MacIntyre, 1974a; see Plate 2-3,*C*). The sea–air surface is often marked by parallel rows of floating debris and slicks formed by the coalescing of surface material where the counterrotating surface currents generated by the wind meet. These phenomena, known as Langmuir circulations (Faller, 1971; Plate 2-3,*A*), can have a profound effect upon sea surface chemistry (Sutcliffe, Baylor, and Menzel, 1963). Concentrated surface films are observed as slicks or smooth water in the presence of wind-generated internal waves (Dietz and LaFond, 1950; Ewing, 1950) known to sailors as "catspaws." In the presence of light airs (winds having a speed of 1 to 3 miles per hour), these films or surface skins are pushed over the underlying water, dampening wave action, and the leeward or moving edge is clearly apparent as a ridge (McCutchen, 1970), a phenomenon which has been rediscovered sporadically since its description in 1854 by Henry David Thoreau (McDowell and McCutchen, 1971).

The surfactant nature of such films and an *a priori* presumption that they were composed of hydrocarbons and fatty acids or their esters have led to the selective extraction of these materials (Garrett, 1967; Jarvis et al., 1967; Duce et al., 1972). The marked lipolytic activity of bacterial isolates from the surface film of upwelling areas (Sieburth, 1971a) would indicate that fatty acids may be important in some waters. A non-destructive infrared technique for characterizing the chemical nature of monolayers of films (Baier, 1972b) has shown that, in addition to "dry surfactants" composed of hydrophobic molecules such as fatty acids, which extend into the air from the water, there are "wet surfactants" composed of large molecular weight complexes with a hydrophobic group that form a 1000 Å (0.1 μm) thick film below the sea–air boundary (MacIntyre, 1974b). Thousands of observations in freshwaters and marine waters indicate that, except for hydrocarbon pollution in heavily used recreational waters (Baier, 1972b), the wet surfactants composed of polysaccharide–protein complexes are the major components of sea surface films, which vary quantitatively, but not qualitatively (Baier et al., 1974). Carbohydrates alone account for some 8 to 25% of the dissolved organic carbon in subsurface waters (Burney, Johnson, and Sieburth, in prep), with 13 to 46% in the surface microlayer (Sieburth et al., 1976).

The concentration of dissolved organic carbon in the 0.1 μm thick surface monolayer or skin (Baier, 1972b; Baier et al., 1974) where bacteria might be expected to concentrate can be estimated from data obtained from water collected with screen samplers. Although other methods of film collection have been described (Goering and Menzel, 1965; Tsyban, 1971; MacIntyre, 1974a), the screen method of film collection (Garrett, 1965; Sieburth, 1965) has proved to be a useful technique for

PLATE 2-3. The neustonic habitat at the sea–air interface. (*A*) Wind reacting with waves sets up Langmuir circulations that mix the water and concentrate materials at the surface. (*B*) Neuston net collection near the Azores, showing the blue fauna consisting of the siphonophore *Porpita* and the copepod *Pontella*, as well as "tarballs" arising from petroleum discharged into the sea. (*C*) A diagram of the ocean with an exaggerated vertical scale to show the microstructure of the top layers; the enriched concentrations of dissolved organic carbon in the microlayer supports a correspondingly large secondary production of bacterioneuston that nurture the microfauna of the sea-air interface. (*A*) from A.J. Faller, 1971, *Ann. Rev. Ecol. Syst.* 2:201–36; (*B*) photograph courtesy of Kenneth R. Hinga; (*C*) from F. MacIntyre, 1974a, The Top Millimeter of the Ocean. Copyright by Scientific American, Inc. All rights reserved. (Sielab data superimposed.)

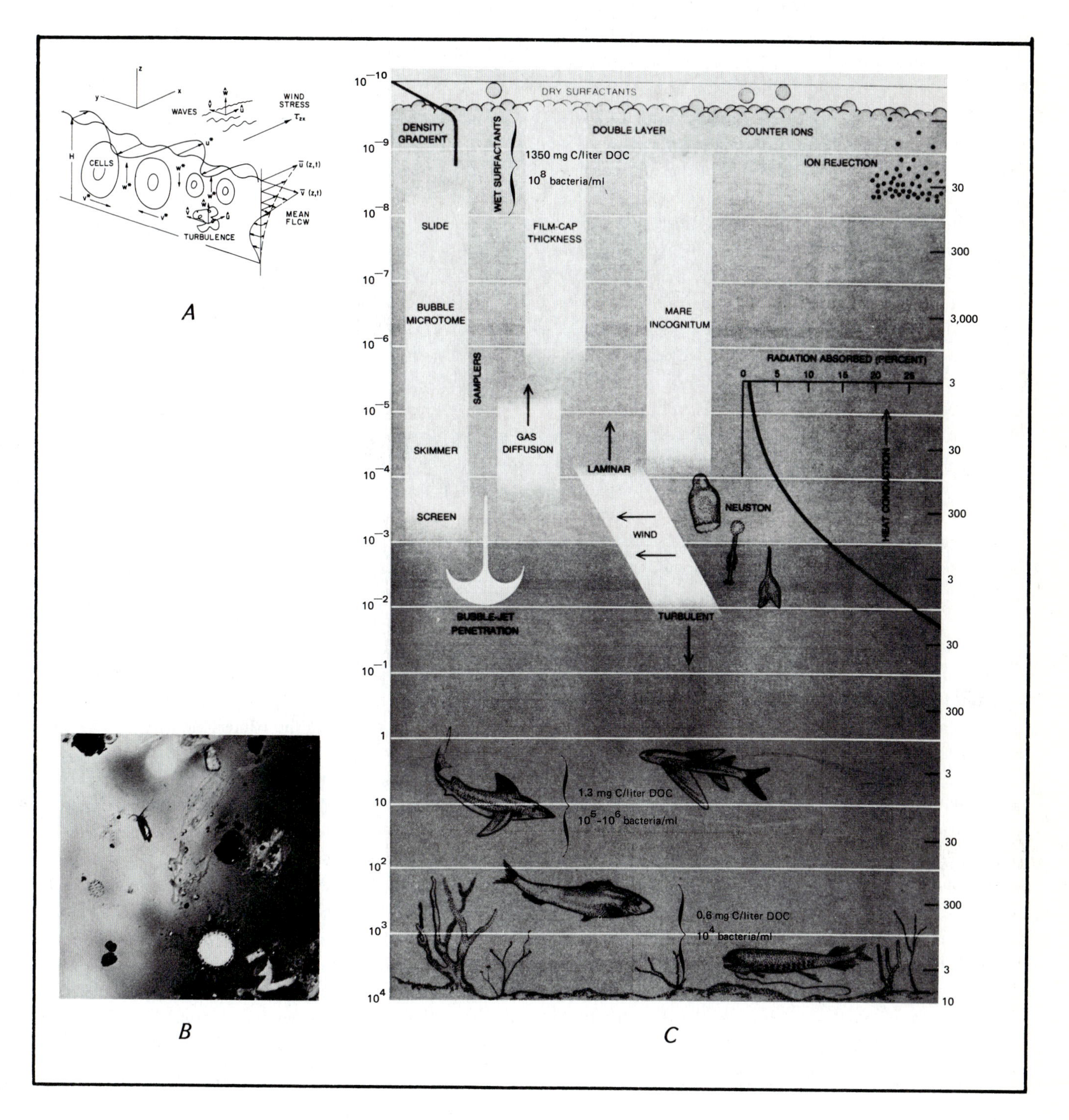

obtaining quantities of the upper μm of film (Duce et al., 1972; MacIntyre, 1974a). The actual concentration of organic carbon in the surface skin is affected by a 70% collection efficiency of the screen and a dilution of at least 1:2,500 by the subsurface waters. Therefore, the average difference between the oceanic values for the screen sample and subsurface samples of 0.63 mg of carbon/liter (Sieburth et al., 1976) would indicate an actual concentration of dissolved organic carbon of 1,427 mg of carbon/liter in the surface skin. These concentrations of 2.9 g of organic matter equal those used in bacterial media. The enrichment of the sea surface with dissolved organic nitrogen, ammonia, nitrate, and phosphate (Williams, 1967) indicates that nutrients in the surface films do not limit bacterial growth. Cultivable bacterial populations for sea surface samples are at least an order of magnitude higher than those for the subsurface maxima (Sieburth, 1965, 1971a; Tsyban, 1971; Sieburth et al., 1976) with populations up to 10^4 bacteria/ml. Corrections to account for dilution with subsurface water would give values of 10^5 to 10^7/ml. Adsorption of bacteria on Nuclepore membranes with a minimal adherence of water and subsequent desorption and culture has indicated values of 10^5 to 10^8 bacteria/ml or 10^1 to 10^4 bacteria/cm^2 in nearshore waters (Crow, Ahearn, and Cook, 1975). Such populations approach those of laboratory cultures. In addition to these cultivable epibacteria (Part V), there is a much larger population of non-cultivable forms that can be detected by epifluorescence microscopy (Chapter 5,C) and by the ATP assay of selectively filtered preparations (Chapter 6,C).

What is the specific mechanism for the formation of the surface skin of the sea? Langmuir circulations account for the circulation of water, and its dissolved organic carbon, in the productive upper 100 m. A popular hypothesis for the concentration of dissolved organic carbon in the upper 0.1 μm as a film has arisen from a series of observations on the sea-to-air transfer of organic matter, salts, and trace elements (Parker and Barsom, 1970; MacIntyre, 1974c). Bubbles forced into the water by turbulent seas are coated by this carbon, which is brought to the surface and released as the bubble bursts (MacIntyre, 1970). These bursting bubbles eject dissolved organic carbon, microorganisms, and salts into the atmosphere at a high rate of efficiency (Blanchard and Syzdek, 1970, 1972, 1974a,b, 1975; Bezdek and Carlucci, 1972, 1974). The efficiency of bubble scavenging, however, appears to be too low to

PLATE 2-4. The relationship between the biomass of phytoplankton and bacterioplankton and their rates of production for offshore and nearshore tropical waters. (*A*) Vertical distribution of photosynthesis (Ph, daily rate of photosynthesis in mg of carbon/m^3; K_r, relative distribution of living phytoplankton; K_T, relative dependence of photosynthetic rate upon submarine illumination). (*B*) Vertical relationship of living phytoplankton (K_r) to bacterial biomass (*B*, mg/m^3), bacterial numbers (*N*, 10^3/ml), and bacterial production (*P*, mg/m^3 daily) ($t°$, temperature). (*C*) Horizontal relationship between daily phytoplankton photosynthesis (*Ph*, mg of carbon/m^2) and bacterial biomass (*B*, mg/m^3), daily bacterial production (*P*, mg/m^3), coefficients of *P*/*B* for both phytoplankton and bacteria, and depth (*H*, m). From Y. Sorokin, 1971, *Int. revue ges. Hydrobiol.* 56:1–48.

account for the concentration of bacteria from the water column (Carlucci and Williams, 1965; Blanchard and Syzdek, 1974b) and could not account for the high concentrations of bacteria in natural waters (Bezdek and Carlucci, 1972). These populations must be largely due to the *in-situ* growth of bacteria developing on the nutrient-rich organic film. Nor does bubble scavenging explain the presence of concentrated films during calm sea states. Is there another mechanism for the concentration of dissolved organic carbon in the surface film that could also operate in calm sea states? The diurnal migration of phagotrophic protists through phytoplankton aggregations (see Plate 2-4) where they feed and release their wastes could be one carbon source in the upper subsurface waters (Sieburth et al., 1976). Hydrophobic groups on wet surfactants such as polysaccharides (MacIntyre, 1974b) might be a sufficiently buoyant force for small-scale migration to, and maintenance in, the surface skin.

The importance of the bacterioneuston in the open sea must be that in the usual absence of a significant biomass of phytoneuston (Taguchi and Nakajima, 1971; Sieburth et al., 1976) it provides a food source for the significant populations of such protozoans as tintinnids, radiolaria, foraminifera (Zaitsev, 1970), and amoebae (Sieburth et al., 1976) concentrated in the neuston. The absence of bacterial enrichment in the surface layers of turbid nearshore waters (Dietz, Albright, and Tuominen, 1976) may be due to the consumption of dissolved organic matter above threshold levels by the bacteria associated with the suspended debris. The anomalous observation that cells in the bacterioneuston were dead (Marumo, Taga and Nakai, 1971) may be due to the

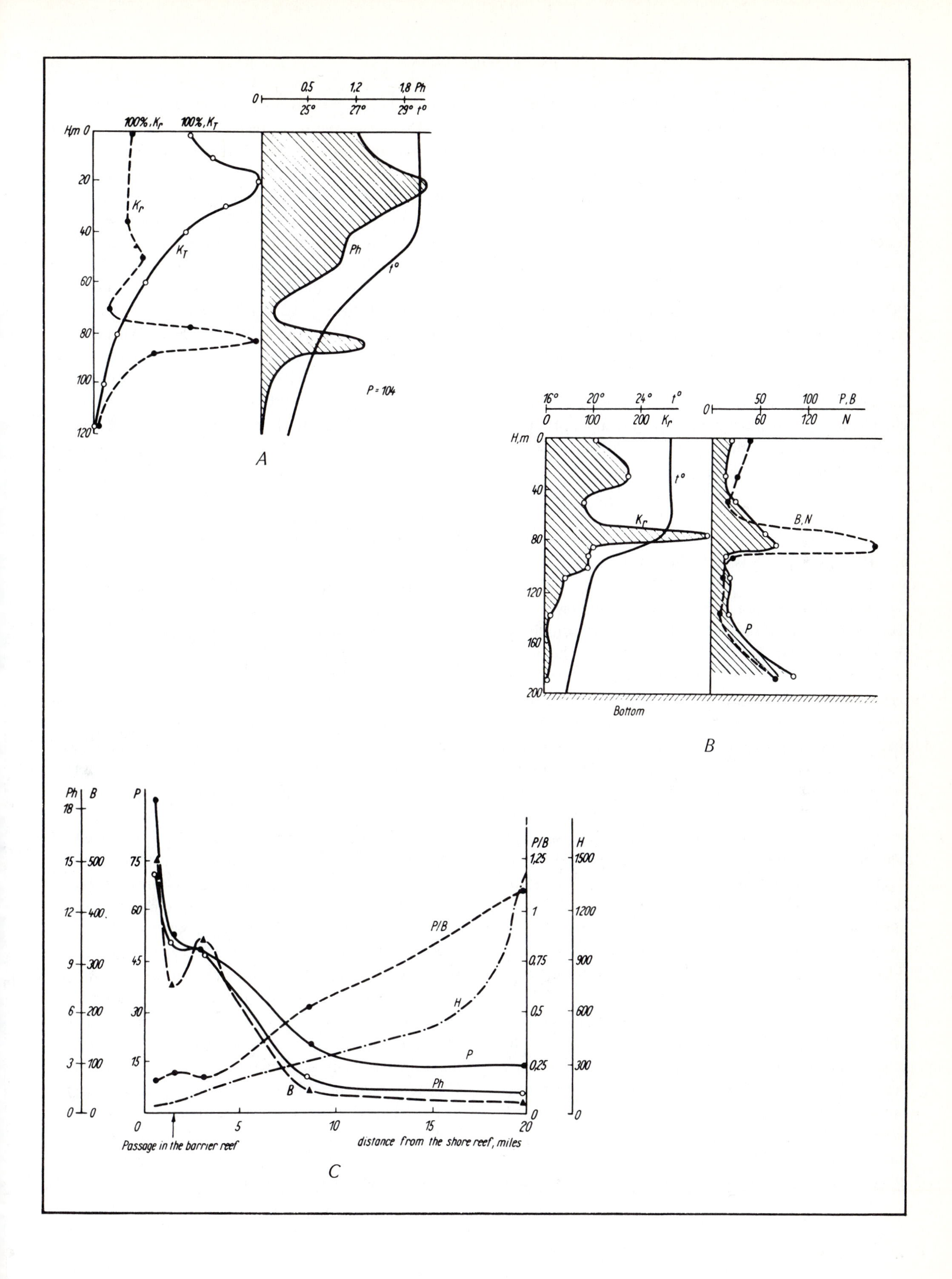

0.5
1.2
1.8 Ph
25°
27°
29° t°
100%, K_r
100%, K_T
H,m
K_r
K_T
Ph
t°
P = 104
A
16°
20°
24°
t°
K_r
50
100
P, B
60
120
N
H,m
t°
K_r
B, N
P
Bottom
B
Ph
B
P
P/B
H
Passage in the barrier reef
distance from the shore reef, miles
C

sporadic occurrence of inhibitory substances (Sieburth, 1971b; Sieburth et al., 1976) rather than to a phototoxicity as assumed by these authors. Phototoxicity does explain the usual lack of phytoneuston. Exceptions are *Trichodesmium* and *Sargassum* in the open sea and some light-resistant pennate diatoms in protected nearshore waters (Hardy, 1973; Gallagher, 1975); these appear to originate on the surface of exposed mud and to be lifted into the surface film by the incoming tide (Gallagher, 1975). Choanoflagellates attach to the surface film and are part of the neuston in protected tidepools (Norris, 1965). The substantial populations of bacterioneuston (Sieburth, 1971a; Tsyban, 1971; Crow, Ahearn, and Cook, 1975; Sieburth et al., 1976) and the amoebae (Davis, Caron, and Sieburth, 1978), heliozoans, tintinnids, and other protozoans that graze upon the bacterioneuston have been largely overlooked.

C. PLANKTONIC HABITAT

The plankton are unattached floating organisms that are buoyant, through morphological adaptation, or are weakly motile for small-scale migration. They extend throughout the water column from the sea–air interface to the sea–mud interface. Planktonic life, however, is centered on the photosynthetic microorganisms that are both dependent upon and limited by solar radiation. Excessive light can be inhibitory or phototoxic for photosynthetic microorganisms. Chlorophyll maxima are often near the compensation point or 1% light level, which varies from 2 to 100 m depending upon whether the location is in the gray–green and turbid waters of a temperate-zone estuary (see Plate 2-2) or the clear blue waters of a tropical open ocean. Maximal photosynthesis occurs above the compensation point (Plate 2-4,*A*) and is indicated by an oxygen maximum (Plate 6-3). As in all photic habitats, microbial life is divided into three trophic modes. The photosynthetic phytoplankton is a mixture of species of diatoms and phytoflagellates whose composition varies with the water mass, season, and grazing pressure, among other factors. Bacterioplankton are suspended bacteria nurtured by the soluble organic matter released by metabolizing and decaying organisms throughout the water column. The protozooplankton are ingesting forms that graze upon the phytoplankton and bacterioplankton as well as upon other protozoa. These three trophic components of the microbial plankton are the food of the copepods and larval forms that comprise much of the metazooplankton. The small size, large numbers, and chitinous bodies of the copepods give them the distinction of being the "insects of the sea," just as the phytoplankton has been called the "grass of the sea." The overall processes of planktonic life are summarized by Raymont (1963), Parsons, Takahashi, and Hargrave (1977), and Cushing and Walsh (1976).

Phytoplankton productivity (Koblents-Mishke and Vedernikov, 1973), which has been used to estimate the fisheries they support (Ryther, 1969), apparently grossly underestimates the stocks of fish that can actually be counted (Steele, 1974; Mills, 1975); it is in itself underestimated. The first problem with primary productivity measurements may be that growth rates determined in the confines of an unstirred bottle, as in the standard ^{14}C productivity methods (Vollenweider, 1969; Strickland and Parsons, 1972), are inhibited by nutrient exhaustion and by-product accumulation. The second problem is that light bottle uptake values are corrected for a very large dark bottle uptake due primarily to bacteria; this would underestimate primary production if a suspected light-inhibition of bacterial activities occurs (Lavoie, 1975). The third problem is that the sample also contains predators, and the extent of primary production may be masked by consumption. The fourth problem is that biomass estimates needed for productivity calculations are based on chlorophyll *a* (Holm-Hansen et al., 1965), which varies greatly from species to species (Strickland, 1968a), and that nanoplankton, which are very fragile and easily missed or destroyed, are much more productive than diatoms (Williams and Yentsch, 1976). And the fifth problem, the distribution of phytoplankton as indicated by plant pigment fluorescence is not only quite variable vertically (Holm-Hansen et al., 1965; Lorenzen, 1967; Strickland, 1968b) but also quite patchy horizontally. And sixth, the ubiquitous but unrecognized presence of chroococcoid cyanobacteria, which are even smaller than the nanoplankton, can occur as a significant biomass in the bacterioplankton of oceanic waters (see Chapters 8, B and Plate 8-5). If phototrophic biomass and productivity are really more abundant in the open sea than current estimates indicate, this should be seen in both a high production of dissolved organic matter and a corresponding abundance of bacterioplankton.

Dissolved organic carbon in the open sea is much greater than one would anticipate from primary productivity measurements being about one-quarter to one-half that in the estuaries ranging from 1.5 mg of carbon/liter at the surface to 0.5 mg of carbon/liter in deep water. It is apparently largely cycled in the upper zone of the sea, whereas the dissolved organic carbon in the deep water masses is believed to be several centuries old (Menzel and Ryther, 1970; Wangersky, 1976). In general, it has been calculated that known substances, such as amino acids, sugars, fatty acids, and other organic compounds, account for only 10% (Degens, 1970) to 30 to 40% (Ogura, 1974) of the dissolved organic carbon, leaving some 60 to 90% as unidentified or very old humic fractions (Wagner, 1969). What is the source, nature, and concentration of the organic carbon that is being cycled in the euphotic zone?

The major source of dissolved organic carbon must be, directly or indirectly, the phytoplankton. Although most axenic cultures of algae release some 3 to 6% of their photosynthate under artificial illumination, with only a few releasing 10 to 15%, the range in natural sunlight increased to 9 to 52%, with a mean of 23% (Hellebust, 1965). Values for natural populations of phytoplankton indicate a similar release of some 24% of the photosynthate (Antia et al., 1963; Horne, Fogg, and Eagle, 1969; Anderson and Zeutschel, 1970; Sorokin, 1971). The per cent release of photosynthesized organic matter from natural populations of *Skeletonema costatum* in the surface waters of Scottish Sea lochs was found to vary seasonally, increasing from 3.5% in March to 16 to 19% in July to 46% in October (Ignatiades, 1973). There is also a marked variation in release with depth (Anderson and Zeutschel, 1970). Although the release of dissolved organic carbon from phytoplankton has been customarily reported as per cent of photosynthate produced, it has been observed that the release rate is constant in photosynthesizing microalgae and that the amount of photosynthate is the factor that varies (Smith and Wiebe, 1976).

Neutral compounds make up a sizable fraction of released materials, including monomeric reducing sugars and the sugar alcohol mannitol (Hellebust, 1965). Polysaccharides are also part of the extracellular products released (Myklestad, Haug, and Larsen, 1972; Smestad, Haug, and Myklestad, 1974, 1975), as well as the reserve and structural components of the cells (Myklestad and Haug, 1972; Myklestad, 1974). These carbohydrates, to some degree, must be released as a result of phytoplankton metabolism and phytoplankton consumption by the phagotrophic protists and filter-feeding zooplankton. One might expect a diurnal input of carbohydrates into the euphotic zone, with a highly variable vertical distribution that would correlate with biological activity. Tests for the determination of dissolved monosaccharides (Johnson and Sieburth, 1977), total dissolved carbohydrates, and the estimation of polysaccharides by difference (Burney and Sieburth, 1977) have permitted an investigation of the distribution of carbohydrates in the North Atlantic (Burney et al., in press; Plate 6-3,*B*), as well as the estimation of their rates of uptake (Sieburth et al., 1977). Minimum threshold values of approximately 50 μg of carbon/liter of monosaccharides were found in predawn subsurface waters all the way down to the sea floor. Such a result had been predicted by chemostat experiments (Jannasch, 1970) and can be explained by positive feedback effects in the bacterial population below a critical level of substrate. The critical levels observed in the sea are much lower than those observed in the chemostat with pure bacterial cultures. During the day, pronounced maxima up to 140 μg with a mean of 65 μg of carbon/liter developed at the depths where the phagotrophic protists accumulated. Polysaccharide maxima occurred during both the daytime and predawn sampling and were associated with phagotrophic protists or nitrogen-impoverished phytoplankton as indicated by the presence of ciliates and low chlorophyll *a* to ATP ratios. The carbohydrate fraction of the dissolved organic carbon appears to be in a state of flux above a minimum threshold value, accounting for some 8 to 24% of the dissolved organic carbon, some 32% of the labile carbohydrate, and 40% of the labile dissolved organic carbon. About 15% of the total dissolved carbon is labile.

What are the *in-situ* rates of bacterioplankton production on these labile substrates? The problem has been how to maintain natural populations of bacterioplankton so that their rate of growth could be measured while they grow on dissolved substrates at natural levels and of natural compositions, as they are released *in situ*. The studies of Monod (1942) show how an ever-changing growth rate is obtained in batch cultures due to a dwindling supply of nutrients and the accumulation of inhibitory by-products. The inherent problems in using isolated bottled samples of natural multicultures to estimate the productivity of phytoplankton (Vollenweider,

1969) and bacterioplankton (Sorokin, 1971) are largely ignored. The chemostat (Chapter 7,C,2) is a fundamental tool for studying the kinetics of pure cultures (Herbert, Elsworth, and Telling, 1956) and the competition between strains (Jannasch, 1968b; Bell and Mitchell, 1972), but it does not simulate the cultural conditions in the sea. Chemostats select monocultures, whereas natural ecosystems maintain multicultures. Doubling times in excess of 100 hours (Jannasch, 1968a) and threshold levels of 0.5 to over 100 mg of substrate/liter are far in excess of those that apparently occur in the sea. Holding bacterial populations in dialysis bags or behind membranes has been an often-tried but little understood technique. In aquatic environments, it has been used mainly for studies on the survival of enteric bacteria (Slanetz and Bartley, 1965; McFeters and Stuart, 1972; Graham and Sieburth, 1973; Fliermans and Gordon, 1977). The principles of dialysis culture for captive populations of industrially useful microorganisms have only recently been formulated (Schultz and Gerhardt, 1969). The first application of this technique to ecological studies was on marine phytoplankton, conducted by Jensen, Rystad, and Skoglund (1972) and Prakash et al. (1973). But the dialysis membranes used as culture vessels are quickly attacked and perforated by cellulolytic bacteria (Lavoie, 1975), and their surfaces are rapidly fouled (Vargo, Hargraves, and Johnson, 1975). The impossibility of adequate agitation on either side of the membrane and the limiting porosity of the membranes may also have imposed conditions intermediate between batch and idealized caged culture upon the unialgal cultures (see Chapter 7,C,3). Even worse, this procedure may not be used to approximate the growth of natural populations of phytoplankton, since it is impossible to remove the phagotrophic protists that are similar in size. The sieve-like qualities of Nuclepore filters (Sheldon, 1972), however, make them ideal not only for the selective filtration of bacteria (Salonen, 1974), but as inert barriers for caging indigenous bacterioplankton (Chapter 14). A diffusion culture apparatus has been developed (see Chapter 7,C,3c) that holds captive populations behind a 0.1 μm porosity membrane where they are nurtured by nutrients that actively diffuse through the membrane due to active stirring on both sides to reduce the Nerntz effect of boundary layers (Lavoie, 1975). Such an apparatus has been used to observe the growth patterns and estimate growth rates of the bacterioplankton (picoplankton size fraction) exposed to *in-situ* water in Narragansett Bay and also *in situ* in the blue waters of the North Atlantic (Sieburth et al., 1977). Bacterioplankton from water in the phytoplankton maxima at 40 and 80 m showed doubling times of 4 hours associated, but out of phase with, diel periodicity, to give an overall three doublings per day. A standing crop of 6 mg of carbon/m^3 gave an estimated productivity of a daily 42 mg. These doubling times are similar to those obtained in unsupplemented seawater with bacteria isolated from unenriched seawater agar (Carlucci and Shimp, 1974). These preliminary results are somewhat higher than bacterioplankton productivity estimates by ^{14}C uptake measurements in bottle experiments (Sorokin, 1971), which have been criticized as being an order of magnitude too high (Banse, 1974). But if our observed rates of hourly dissolved organic carbon uptake of 9.8 mg of carbon/liter and hourly bacterial growth rates of 5.1 mg of cellular carbon/liter are real, then something is basically wrong with the accepted values for phytoplankton biomass and productivity (Sieburth, 1977). Further studies are required to determine the true nature and extent of phytoplankton and bacterioplankton productivity.

A diel cycle has been observed for the bacterioplankton in which growth occurs during the late afternoon and evening hours, with a resting or death phase during the morning hours (Sieburth et al., 1977). Growth appears to be dependent upon, but out of phase with, the phototrophs. Bell, Lang, and Mitchell (1974) used the term "phycosphere" to describe the zone in which the phytoplankton both nourish and inhibit the heterotrophic bacteria with primary and secondary substances, respectively (Sieburth, 1968a,b), to form blooms of specific bacterial floras. The bacterial flora of the phycosphere synthesizes factors utilized by the phytoplankton, such as vitamin B_{12} (Sorokin, 1971) and the siderchromes, which are required by them to chelate and utilize iron (Gonye and Carpenter, 1974). The supply of B_{12} to diatoms by an associated bacterioplankton may also be inferred from the observations of Swift (1973). The B_{12} levels in the cells of natural populations of diatoms in the Gulf of Maine approximated the levels in cells grown on B_{12}-supplemented media rather than those grown on B_{12}-deficient media (Carlucci and Silbernagel, 1966). The concept of a phycosphere should be enlarged to include the phototrophic microalgae, phagotrophic protozoa, and osmotrophic bacteria, fungi, and protozoa responsible for the three-part trophic sequence in the primary production and transfer of organic matter up the food chain. Microorganisms in each trophic mode are dependent upon one another for their well-being.

The study of some of the protozooplankton is almost a virgin field. Occasionally, such larger forms as *Noctiluca* have been included with the zooplankton. The planktonic foraminifera, radiolaria, and acantharia (Chapter 23) are well known by their tests (shells) of calcium carbonate, silica, and strontium sulfate, respectively. But they are more often studied in the sediments by marine geologists than in the plankton by marine microbiologists. Other protozoa are also present as either stable or sporadic members of the plankton. Fragile and small microorganisms, such as bicoeca, heliozoa, amoebae, and choanoflagellates, are all but ignored. Amoebae have been studied in nearshore waters in recent years, but only very recently have observations been made on their occurrence in offshore waters (Davis, Caron, and Sieburth, 1978). The occurrence of cultivable amoebae at concentrations of 1 organism/liter in 75% of the samples throughout the water column in the North Atlantic indicates that the threshold levels of readily utilizable carbohydrates and the threshold levels of 10^3 bacteria/ml in the water column are apparently maintaining threshold concentrations of bacteria-ingesting forms throughout the water column. Ciliates are generally regarded as members of the benthos, despite reports by Fauré-Fremiet (1967), Beers and Stewart (1969, 1971) and Eriksson, Sellei, and Wallström (1977) on their occurrence in the plankton. In the open sea, ciliates are observed in reverse-flow concentrates, in culture, and within dying and dead copepods. Planktonologists who have only studied phytoplankton and zooplankton apparently will not have a complete picture of their biology until the protozooplankton and bacterioplankton and their roles in releasing and recycling organic matter are also studied.

D. EPIBIOTIC HABITAT

A great variety of organisms have adapted to a sessile mode of life for at least part of their life cycles. These epibiotic organisms turn naked surfaces of all types into jungles of attached and crawling forms, to foul most submerged marine surfaces (Sieburth, 1975). Aside from the unsightly appearance and economic loss caused by such fouling, the biology of these fouled surfaces is as fascinating and important as the biology of planktonic habitats. It is this community on piers, reefs (both artificial and natural) (Unger, 1966; Steimle and Stone, 1973; Colunga and Stone, 1974), and rocks that attracts baitfish and, in turn, migrating schools of larger fish (Nixon, Oviatt, and Northby, 1973). Epibiotic ecosystems as such are largely ignored, however, except by specialists studying the biology of a specific sedentary species (Crisp and Ryland, 1960; Meadows and Williams, 1963) or the economic aspects of fouling, biodeterioration, and corrosion (Acker et al., 1972).

1. Inanimate Surfaces

Most submerged surfaces of the sea offer sites for colonization by a variety of organisms. The fouling process is a fairly definite sequence of events, which produces epibiotic ecosystems, with all the strata of the food pyramid represented (Acker et al., 1972). Bacteria are the first detectable microorganisms to colonize new surfaces and they apparently do this by two processes, as first suggested by ZoBell (1943) and further elaborated by Marshall (1976) in his monograph on the subject. The first process is an instantaneous but reversible adsorption of the bacteria onto a surface. The cells are apparently held loosely by two opposing forces, attraction by van der Waals forces and repulsion by electrical charge (Marshall, Stout, and Mitchell, 1971b). Such a phenomenon would explain the immediate adsorption of bacteria onto glass (DiSalvo, 1973) and polycarbonate surfaces (Crow, Ahearn, and Cook, 1975) from which the bacteria can be easily removed and counted.

The second process is irreversible adsorption, in which a population develops with time. The bacterial types are those that adhere with polymeric fibrils (fimbrae) (Marshall, Stout, and Mitchell, 1971a,b; Costerton, Geesey, and Cheng, 1978). This time-dependent formation of a bacterial film apparently involves a sequence of events. Different surfaces are colonized at different rates, which are apparently determined by the wettability or critical surface tensions and modification of the surfaces by the adhesion of dissolved organic substances, such as polysaccharide–protein complexes (Baier, 1970, 1972a; Dexter et al., 1975). The modified surface actively attracts bacteria through positive chemotaxis

(Young and Mitchell, 1972), and the bacteria attach by forming fibrils of polyanionic polysaccharides (Corpe, 1970, 1972; Marshall, Stout, and Mitchell, 1971a,b). The accumulation and growth of adsorbed bacteria is rapid, and a film is detectable after a day's immersion (O'Neill and Wilcox, 1971). This rapid bacterial growth plateaus after several days (Skerman, 1956; O'Neill and Wilcox, 1971; Sechler and Gunderson, 1972). There is a marked reduction in diversity with film development (Baier, 1972a), to yield a bacterial flora dominated by short rods (Marshall, Stout, and Mitchell, 1971a), prosthecate forms (Baier, 1972a), and filaments of *Leucothrix mucor* (Persoone, 1968). The rotational movement of bacteria during settlement (Marshall, Stout, and Mitchell, 1971b) and the development of lawns of irreversibly adsorbed populations in either an edgewise or an endwise manner (Sieburth, 1975) require further study. Diatoms, which settle with the bacteria but have a longer generation time, soon overtake the bacteria by their sheer bulk, to become the dominant biomass (Skerman, 1956; O'Neill and Wilcox, 1971; Sechler and Gunderson, 1972; Plate 2–5). The appearance of protozoa at about the time bacterial growth plateaus indicates that the protozoa are enriched by the bacteria and are presumably keeping the bacteria in a healthy state by limiting them to numbers that can balance the diatom productivity. The microbial film appears to periodically slough off and be renewed. After these microbial films have formed, spores and larvae of such macroscopic forms as seaweeds and invertebrates attach to cause the gross fouling observed with extended periods of immersion. The microbial films can have a negligible, positive, or negative effect on larval settlement (Crisp, 1972).

Quantitative and qualitative studies on the rapid fouling of the epoxy gel coat of reinforced fiber glass, which is identical to that used for the tanks of a marine ecosystem and the hulls of pleasure craft, have been made by Caron and Sieburth (in prep.). These studies have not only verified the fouling sequence described above but have also shown two other important findings. The marked fouling process of the warm water periods (above 10 °C) slows markedly during the cold water periods (below 10 °C). Regular brushing of the surface once or twice a week, depending upon the season, is enough to interrupt the fouling process and restrict it to a bacterial film equivalent to a 3- to 4-day exposure and to prevent the accumulation of diatom, protozoan, and metazoan populations. This finding could alter our thinking on fouling prevention in an age of environmental concern and encourage mechanical rather than chemical methods of prevention.

The environmental conditions in the fouling habitat will dictate the composition of the fouling microorganisms. Phototrophic forms disappear from aphotic surfaces. In the dark, ammonia-enriched environment of high-density, closed-system aquaculture (Meade, 1974), where there are no phototrophic diatoms, another autotrophic component, the chemolithotrophic bacteria that oxidize ammonia and nitrite (Chapter 9,A), is present. Ammonia, which is excreted rapidly by invertebrates and fish after feeding, would soon reach toxic levels if it were not removed. Under most conditions, ammonia-oxidizing (nitrifying) bacteria are soon enriched for in such systems. Although this population is small (some 5% of the bacterial forms), it is important in that it maintains the animal population in a healthy state (Johnson and Sieburth, 1976). A microbial film that developed in a freshwater salmon culture system is shown in Plate 2-6. The scanning electron micrograph (Plate 2-6,*A*) gives an indication of the horizontal microbial seascape but distorts the vertical distribution, since the microbial film, which is largely water (held by a delicate latticework of polysaccharide molecules), collapses when it is dried during the preparation process. The thin section of the bacterial film observed by transmission electron microscopy (Plate 2-6,*B*) retains the vertical dimension and shows a seascape of bacteria and a protozoan living in the loose, permeable "jelly" of the film. The ultrastructure of the individual cells gives hard clues to their taxonomic identity, their trophic modes, and their ecological relationships. The nitrifying bacteria of the genus *Nitrosomonas* usually occur in microcolonies (Plate 2-6,*C*) as both cyst and zoogleal forms. The lamellar structure characteristic of these bacteria becomes highly invaginated when the bacteria are embedded in the film, which might indicate an attempt to increase their surface area when the natural levels of ammonia are low. The consumers in these films are ciliates and amoebae, including those with a lorica (house) as shown in Plate 2-6,*D*.

2. Living Surfaces

Macroorganisms, both attached and free living, also provide surfaces that can support microbial films and

PLATE 2-5. The colonization of nonbiodegradable inanimate surfaces (plastic objects) by bacteria and pennate diatoms. (*A*) Bacteria developing on a braided nylon fishing line after 4 days' immersion, Pigeon Key, Fla. (*B*) The pennate diatom *Cocconeis scutellum* developing on monofilament nylon after 30 days in a turbulent intertidal area of Narragansett Bay, R.I. The surface is badly scoured as can be seen by the prevalence of bottom valves and the damage to upper valves. (*C*) The selective colonization by an assortment of pennate diatoms in the depression of a polypropylene ribbon held 30 days in a turbulent intertidal area in Narragansett Bay. (*D*) The more normal, mixed colonization of diatoms supporting a film dominated by filamentous bacteria on a subtidal polyethylene surface, Narragansett Bay. (*A*) from Sieburth, 1975, *Microbial Seascapes*, University Park Press, Baltimore, Md.; (*B*, *C*,) scanning electron micrographs courtesy of J. Lawton Tootle; (*D*) micrograph courtesy of Paul W. Johnson.

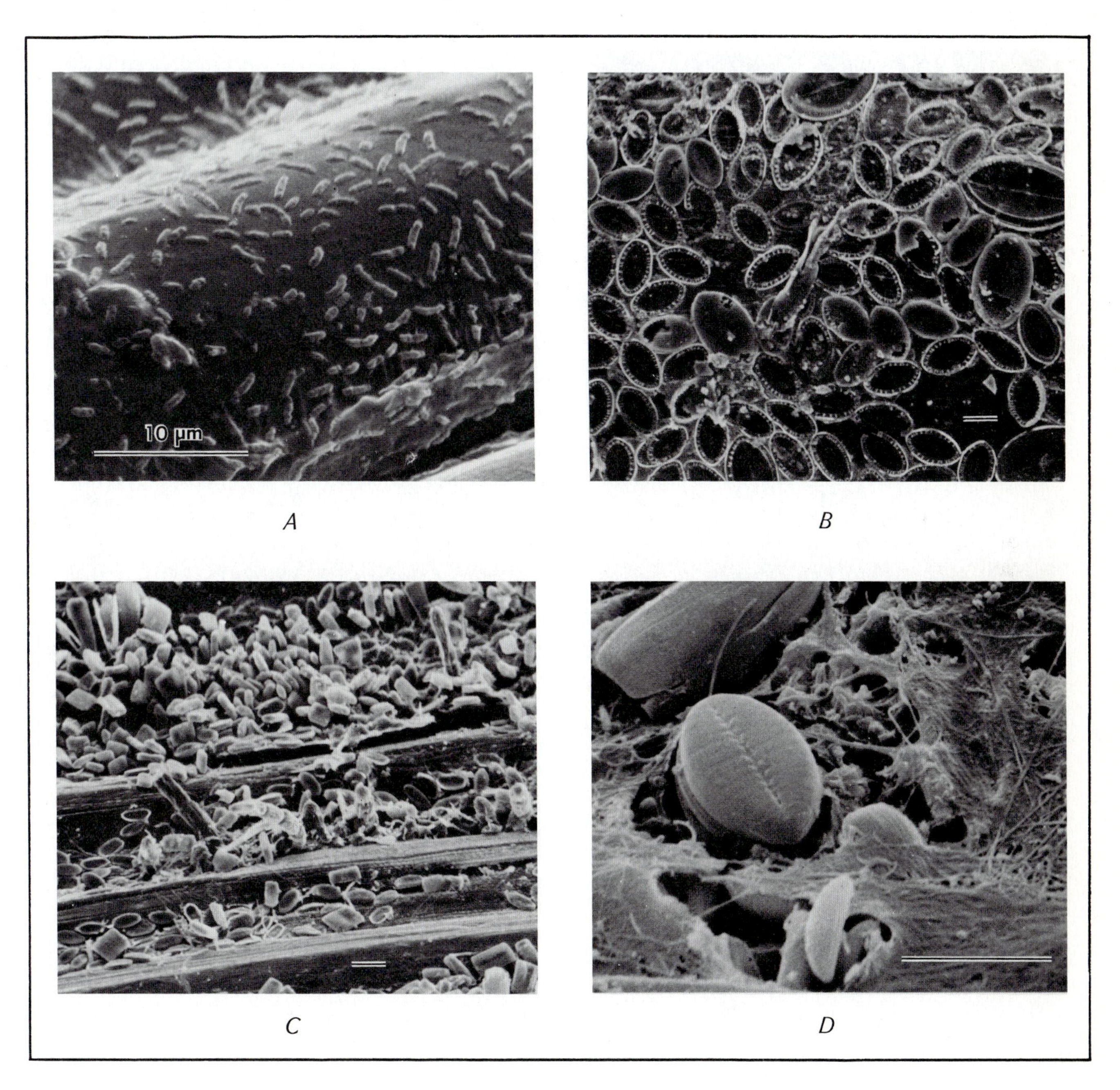

PLATE 2-6. A microbial-film microcosm based on ammonia and other metabolic by-products of salmon culture that enrich for nitrifying bacteria, heterotrophic bacteria, and protozoans; the latter consume excess bacterial biomass and keep the bacteria healthy. (*A*) Scanning electron micrograph of the dehydrated film shows the horizontal arrangement of the individual cells and microcolonies of bacteria as well as loricate amoebae (arrows) with extended pseudopodia. (*B*) Transmission electron micrograph of a thin section of the microbial film developing on a Nuclepore membrane (bottom) shows the vertical distribution of microorganisms (an empty amoeban lorica is in the lower left corner). (*C*) Cyst-type microcolony of *Nitrosomonas* deep within the film shows the invaginated lamellae, with an increased surface area apparently induced by ecological levels of ammonia diffusing through the film. (*D*) Loricate amoeba, with an opening in the lorica through which its pseudopods can grasp bacteria. (*A,B,D*) courtesy of Paul W. Johnson; (*C*) from P.W. Johnson and J.McN. Sieburth, 1976, *Appl. Environ. Microbiol.* 31:423–32.

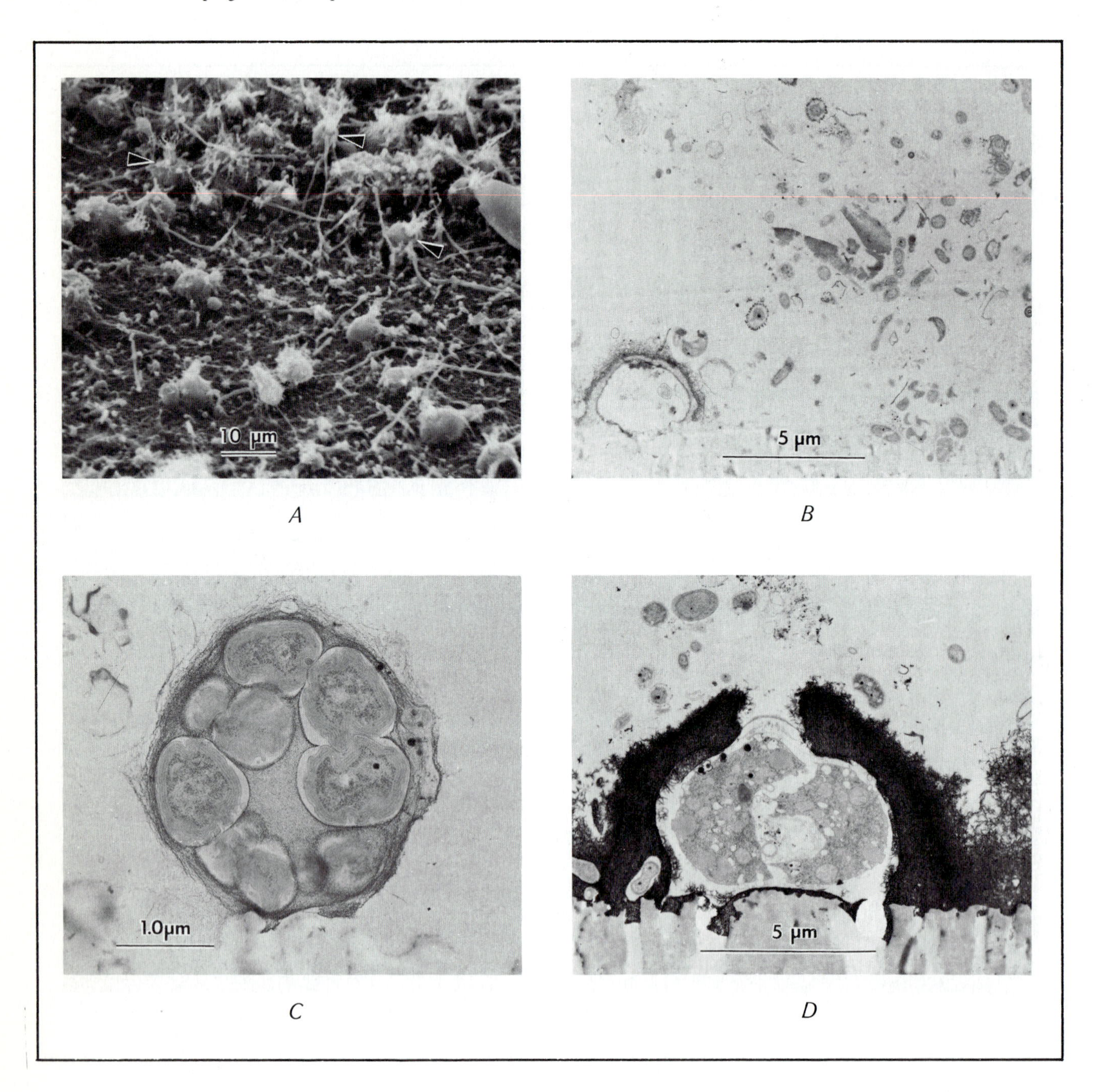

visible fouling. Just as the bacteria closely associated with phytoplankton are nurtured by and, in turn, nurture the phytoplankton, bacterial films may play a vital role in the growth and development of macroscopic marine plants. A number of seaweeds, including *Ulva* and *Monostroma*, have distinctive epibacterial floras (see Plate 2-7,*A*). When *Ulva* and the similar *Enteromorpha* are brought into axenic culture they have aberrant and overlapping growth patterns (Bonneau, 1977), which can be restored to typical growth by the addition of bacteria (Provasoli and Pintner, 1964, and in press; Kapraun, 1970; Fries, 1975). Possible explanations for the role of bacteria in this phenomenon are the supply of growth factors and the bacterial destruction of algal autoinhibitory substances. Other algae, such as the kelp *Laminaria*, actively grow at the base of the blade (flat foliose part) near the stipe (basal stalk), and several standing crops of blade are produced and released at the decaying distal end (Mann, 1973). The seasonal variation in bacterial numbers and types associated with the different parts of *Laminaria* has been described by Laycock (1974). The usual rod and coccoid forms of bacteria that grow on algal surfaces are masked by dense filaments of the bacterium *Leucothrix mucor*, which can cover up to 80% of the algal surface (Tootle, 1974; Plate 3-7,*D*). The extent of microbial colonization of seaweed surfaces is apparently regulated by antifouling substances, such as inhibitory polyphenols (Conover and Sieburth, 1964; Sieburth and Cónover, 1965; Sieburth, 1968a), which inhibit colonization seasonally (Sieburth et al., 1974; Tootle, 1974), as well as by a cuticle in some species (Hanic and Craigie, 1969), which sloughs periodically to expose a clean algal surface and prevents an excessive buildup of fouling film (Tootle, 1974; Sieburth, 1975).

The seagrasses can also be heavily colonized by epibiotic microorganisms. The eelgrass *Zostera marina*, when growing in protected waters of marshes and salt ponds, can become ensheathed in a unialgal film of the pennate diatom *Cocconeis scutellum*. This film is then nonselectively colonized to form a thick fouling crust (Sieburth and Thomas, 1973) whose biomass can rival or exceed the biomass of the host and the phytoplankton in the surrounding water. These microbially enriched grasses provide food for diverse species, such as the brant (Ranwell and Downing, 1959), and for shellfish and shrimp, which feed upon their fragments. In southern waters, the turtlegrass *Thalassia testudinum* undergoes similar microbial colonization and utilization in the food chain, both as living plants and as microbially enriched fragments (Fenchel, 1970). The cordgrass *Spartina alterniflora* undergoes an initial fungal colonization (Gessner, Goos, and Sieburth, 1972) before falling into the marsh where it fragments and is further colonized. The microbiota of the deteriorating grass serves as the principal diet of the grass shrimp *Palaemonetes pugio* (Welsh, 1975). The importance of seagrasses and their microbiota to benthic animals (Harrison and Mann, 1975) is further discussed in Chapter 3,B.

Animal surfaces, like plant surfaces, also appear to selectively enhance or inhibit microbial colonization (Sieburth, 1975). The surfaces of some nereid worms and fish, among others, are characterized by a regular pattern of proteinaceous material and pores and are virtually free of microorganisms (Plate 2-8,*A*). The marked absence of bacteria on such surfaces may be due to cleansing by ablative slime layers or to inhibition by mucoproteins or peptides. Many animal surfaces appear to be biologically inactive and undergo fouling similar to that on inanimate surfaces. Motile and stalked pennate diatoms are conspicuous on these surfaces, especially in protected crevices (Plate 2-8,*B*). In eutrophic waters, such particle-ingesting protozoans as choanoflagellates attach to diatoms and other suspended fragments. The vorticellids settle on the textured surfaces of bryozoan colonies and along the protected areas of hydroid rhizoids (Sieburth, 1975) as well as on the exoskeleton of such estuarine copepods as *Acartia* (Herman and Mihursky, 1964; Hirche, 1974). In the neuston, with its rich microbial film, suctorians can infest the blue pontellid copepods (Sieburth et al., 1976). Other animal surfaces, such as egg masses, larval forms, and especially the chitinous skeleton of crustaceans, appear to be nutrient surfaces that encourage bacterial attachment and growth and undergo extensive colonization (Plate 2-8,*C*,*D*; Sieburth, 1975). Such populations can affect the results of physiological observations (Anderson and Stephens, 1969) and interfere with intensive aquaculture (Johnson et al., 1971; Shelton, Shelton and Edwards, 1975). Inert animal surfaces, such as shells, are colonized in a manner similar to that of inanimate surfaces (Sieburth, 1975).

The microbiology and biology of surfaces still offers many challenges. There is a need for basic information on how to interfere with the fouling sequence in a manner that could be utilized to prevent the fouling of

PLATE 2-7. The microbial colonization of seaweed surfaces varies with the species and growing season, among other factors. (*A*) The usual type of bacterial attachment is perpendicular, as seen in a thin section of the green alga *Ulva lactuca* (Narragansett Bay, R.I.). (*B*) A similar green alga, with but a single layer of cells, *Monostroma* appears to select only one type of bacteria that show intense growth along its edge (Pigeon Key, Fla.). (*C*) Another green alga, *Cladophora*, is heavily colonized by the pennate diatom *Cocconeis*, which appears to compete with bacteria for the surface (Pigeon Key). (*D*) The dominant bacterial form on many intertidal algae, including *Fucus vesiculosis*, is the filamentous *Leucothrix mucor* (Narragansett Bay). (*A*) courtesy of Paul W. Johnson; (*B,C*) from Sieburth, 1975, *Microbial Seascapes*, University Park Press, Baltimore, Md.; (*D*) courtesy of J. Lawton Tootle.

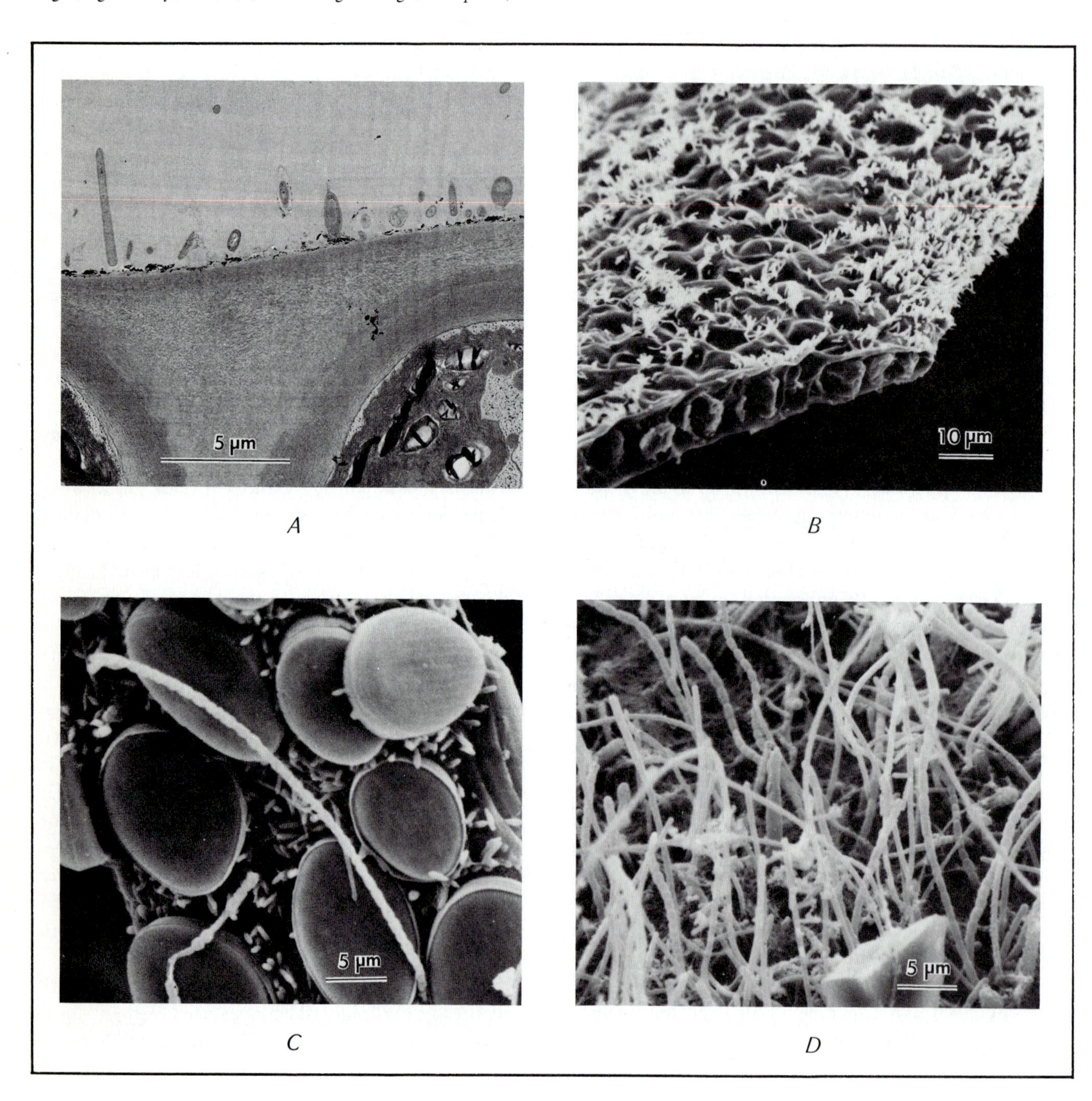

PLATE 2-8. The surface of marine animals may repel colonization (*A*), permit the rapid colonization of new surfaces (*C*), or permit the denser colonization of protected areas (*B,D*). (*A*) The skin of the cunner fish (*Tautogolabrus adspersus* Walbaum), with its patterned proteinaceous surface and pores, appears to repel bacteria. (*B*) The crevice between segments of a caprellid amphipod supports pennate diatoms and filamentous bacteria. (*C*) The surface of a 3-day-old lobster (*Homarus americanus*) supports microcolonies of the filamentous bacterium *Leucothrix mucor* as well as the more usual rod forms. (*D*) The ribbed chitinous surface of the tube-dwelling amphipod *Jassa falcata* supports bacteria in its protected folds. (*A*) courtesy of Kenneth R. Hinga; (*B-D*) from Sieburth, 1975, *Microbial Seascapes*, University Park Press, Baltimore, Md.

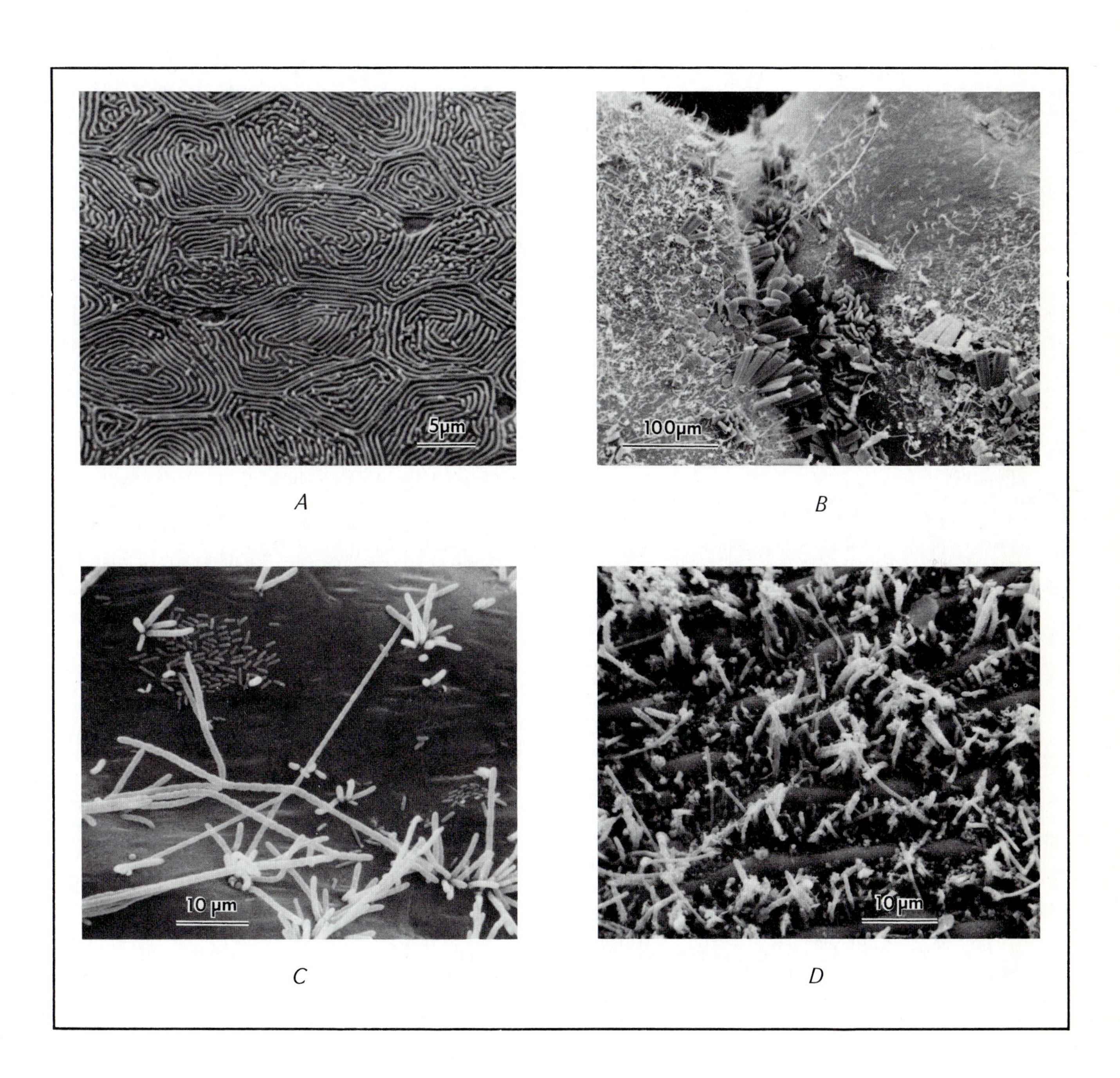

ships' hulls and other submerged surfaces. From the opposite direction, we should learn how to better utilize the normal fouling process. In addition to artificial reefs, we could farm surfaces for both edible algae and sedentary invertebrate animals. Edible species of barnacles, which are really sessile shrimp that kick food into their mouths with their feet, are an example. The fouling within pipes through which seawater is pumped into aquaria removes much of the suspended particles and markedly changes the chemistry of the water. This process could be reversed, with polluted effluents from land mixed with seawater and processed by passing them over heavily fouled surfaces before discharging the treated water into the sea.

REFERENCES

Acker, R.F., B.F. Brown, J.R. DePalma, and W.P. Iverson (eds.). 1972. *Proc. 3rd Intl. Congress Mar. Corrosion & Fouling,* Natl. Bur. Standards, Gaithersburg, Md., 1031 p.

Anderson, G.C., and R.P. Zeutschel. 1970. Release of dissolved organic matter by marine phytoplankton in coastal and offshore areas of the northeast Pacific Ocean. *Limnol. Oceanogr.* 15:402–7.

Anderson, J.W., and G.C. Stephens. 1969. Uptake of organic material by aquatic invertebrates. VI. Role of epiflora in apparent uptake of glycine by marine crustaceans. *Mar. Biol.* 4:243–49.

Antia, N.J., C.D. McAllister, T.R. Parsons, K. Stephens, and J.D.H. Strickland. 1963. Further measurements of primary production using a large-volume plastic sphere. *Limnol. Oceanogr.* 8:166–83.

Baier, R.E. 1970. Surface properties influencing biological adhesion. In: *Adhesion in Biological Systems* (R.S. Manly, ed.). Academic Press, New York and London, pp. 15–48.

Baier, R.E. 1972a. Influence of the initial surface condition of materials on bioadhesion. In: *Proc. 3rd Intl. Congr. Mar. Corr. & Fouling,* Natl. Bur. Standards, Gaithersburg, Md., pp. 633–39.

Baier, R.E. 1972b. Organic films on natural waters: Their retrieval, identification and modes of elimination. *J. Geophys. Res.* 77:5062–75.

Baier, R.E., D.W. Goupil, S. Perlmutter, and R. King. 1974. Dominant chemical composition of sea-surface films, natural slicks, and foams. *J. Rech. Atmosph.* 8:571–600.

Banse, K. 1974. On the role of bacterioplankton in the tropical ocean. *Mar. Biol.* 24:1–5.

Beebe, C.W. 1926. *The Arcturus Adventure.* Putnam, New York, 439 p.

Beers, J.R., and G.L. Stewart, 1969. The vertical distribution of micro-zooplankton and some ecological observations. *J. Cons. int. Explor. Mer* 33:30–44.

Beers, J.R., and G.L. Stewart. 1971. Micro-zooplankters in the plankton communities of the upper waters of the eastern tropical Pacific. *Deep-Sea Research* 18:861–83.

Bell, W., and R. Mitchell. 1972. Chemotactic and growth responses of marine bacteria to algal extracellular products. *Biol. Bull.* 143:265–77.

Bell, W.H., J.M. Lang and R. Mitchell. 1974. Selective stimulation of marine bacteria by algal extracellular products. *Limnol. Oceanogr.* 19:833–39.

Bezdek, H.F., and A.F. Carlucci. 1972. Surface concentration of marine bacteria. *Limnol. Oceanogr.* 17:566–69.

Bezdek, H.F., and A.F. Carlucci. 1974. Concentration and removal of liquid microlayers from a seawater surface by bursting bubbles. *Limnol. Oceanogr.* 19:126–32.

Blanchard, D.C., and L. Syzdek. 1970. Mechanism for the water-to-air transfer and concentration of bacteria. *Science* 170:626–28.

Blanchard, D.C., and L. Syzdek. 1972. Concentration of bacteria in jet drops from bursting bubbles. *J. Geophys. Res.* 77:5087–99.

Blanchard, D.C., and L. Syzdek. 1974a. Bubble tube: Apparatus for determining rate of collection of bacteria by an air bubble rising in water. *Limnol. Oceanogr.* 19:133–38.

Blanchard, D.C., and L. Syzdek. 1974b. Importance of bubble scavenging in the water-to-air transfer of organic material and bacteria. *J. Rech. Atmosph.* 8:529–40.

Blanchard, D.C., and L. Syzdek. 1975. Electrostatic collection of jet and film drops. *Limnol. Oceanogr.* 20:762–74.

Bonneau, E.R. 1977. Polymorphic behavior of *Ulva lactuca* (Chlorophyta) in axenic culture. *J. Phycol.* 13:133–40.

Burney, C.M., and J.McN. Sieburth. 1977. Dissolved carbohydrates in seawater. II. A spectrophotometric procedure for total carbohydrate analysis and polysaccharide estimation. *Mar. Chem.* 5:15–28.

Burney, C.M., K.M. Johnson, D.M. Lavoie, and J.McN. Sieburth. (in press). Dissolved carbohydrate concentrations and their microbial cycling in the North Atlantic. *Deep-Sea Res.*

Carlucci, A.F., and S.L. Shimp. 1974. Isolation and growth of a marine bacterium in low concentrations of substrate. In: *Effect of the Ocean Environment on Microbial Activities* (R.R. Colwell and R.Y. Morita, eds.). University Park Press, Baltimore, Md., pp. 363–67.

Carlucci, A.F., and S.B. Silbernagel. 1966. Bioassay of seawater. I. A ^{14}C uptake method for the determination of concentrations of vitamin B_{12} in seawater. *Can. J. Microbiol.* 12:175–83.

Carlucci, A.F., and P.M. Williams. 1965. Concentration of bacteria from seawater by bubble scavenging. *J. Cons. perm. int. Explor. Mer* 30:28–33.

Caron, D.M., and J.McN. Sieburth. (in prep.) The microbial fouling sequence and its mechanical disruption.

Cheng, L. 1975. Marine pleuston—animals at the sea–air interface. *Oceanogr. Mar. Biol. Ann. Rev.* 13:181–212.

Chesselet, R., R.A. Duce, and D.C. Blanchard (eds.). 1974. International Symposium on the Chemistry of Sea/Air Particulate Exchange Processes. *J. Rech. Atmosph.* 8(3–4):510–1006.

Colunga, L., and R. Stone. 1974. *Proc. Intl. Conf. Artificial Reefs*, Texas A&M University, College Station, Tex., 152 p.

Conover, J.T., and J.McN. Sieburth. 1964. Effect of *Sargassum* distribution on its epibiota and antibacterial activity. *Bot. Mar.* 6:147–57.

Corpe, W.A. 1970. Attachment of marine bacteria to solid surfaces. In: *Adhesion in Biological Systems* (R.S. Manly, ed.). Academic Press, New York and London, pp. 73–87.

Corpe, W.A. 1972. Microfouling: the role of primary film forming marine bacteria. *Proc. 3rd Intl. Congr. Mar. Corr. & Fouling*, Natl. Bur. Standards, Gaithersburg, Md., pp. 598–609.

Costerton, J.W., G.G. Geesey and K.J. Cheng. 1978. How bacteria stick. *Sci. American* 238:86–95.

Crisp, D.J. 1972. The role of the biologist in anti-fouling research. *Proc. 3rd Intl. Congr. Mar. Corr. & Fouling.* Natl. Bur. Standards, Gaithersburg, Md., pp. 88–93.

Crisp, D.J., and J.S. Ryland. 1960. Influence of filming and of surface texture on the settlement of marine organisms. *Nature* 180:119.

Crow, S.A., D.G. Ahearn, and W.L. Cook. 1975. Densities of bacteria and fungi as determined by a membrane-adsorption procedure. *Limnol. Oceanogr.* 20:644–46.

Cushing, D.H., and J.J. Walsh (eds.). 1976. *The Ecology of the Seas.* W.B. Saunders Co., Philadelphia, 467 p.

David, P.M. 1965a. The neuston net. A device for sampling the surface fauna of the ocean. *J. mar. biol. Assoc. U.K.* 45:313–20.

David, P.M. 1965b. The surface fauna of the ocean. *Endeavour* 24:95–100.

Davis, P.G., D.M. Caron, and J.McN. Sieburth. 1978. Oceanic amoebae from the North Atlantic: Culture, distribution, and taxonomy. *Trans. Amer. Micros. Soc.* 97:73–88.

Degens, E.T. 1970. Molecular nature of nitrogenous compounds in seawater and recent marine sediments. In: *Organic Matter in Natural Waters* (D.W. Hood, ed.). Inst. Mar. Sci. University of Alaska, pp. 77–106.

Dexter, S.C., J.D. Sullivan, Jr., J. Williams III, and S.W. Watson, 1975. Influence of substrate wettability on the attachment of marine bacteria to various surfaces. *Appl. Microbiol.* 30:298–308.

Dietz, A.S., L.J. Albright, and T. Tuominen. 1976. Heterotrophic activities of bacterioneuston and bacterioplankton. *Can. J. Microbiol.* 22:1699–709.

Dietz, R.S., and E.C. LaFond. 1950. Natural slicks on the ocean. *J. Mar. Res.* 9:69–76.

DiSalvo, L. H. 1973. Contamination of surfaces by bacterial neuston. *Limnol. Oceanogr.* 18:165–68.

Duce, R.A., J.G. Quinn, C.E. Olney, S.R. Piotrowicz, B.J. Ray, and T.L. Wade. 1972. Enrichment of heavy metals and organic compounds in the surface microlayer of Narragansett Bay, Rhode Island. *Science* 176:161–63.

Eriksson, S., C. Sellei, and K. Wallström. 1977. The structure of the plankton community of the Öregrundsgrepen (Southwest Bothnian Sea). *Helgoländer wiss. Meeresunters.* 30:582–97.

Ewing, G. 1950. Slicks, surface films and internal waves. *J. Mar. Res.* 9:161–87.

Faller, A.G. 1971. Oceanic turbulence and the Langmuir circulations. *Ann. Rev. Ecol. Syst.* 2:201–36.

Fauré-Fremiet, E. 1967. Documents et observations écologiques et pratiques sur la culture des infusoires cilíes. *Hydrobiologia* 18:300–20.

Fenchel, T. 1970. Studies on the decomposition of organic detritus derived from the turtle grass *Thalassia testudinum. Limnol. Oceanogr.* 15:14–20.

Fliermans, C.B., and R.W. Gorden. 1977. Modification of membrane diffusion chambers for deep-water studies. *Appl. Environ. Microbiol.* 33:207–10.

Fries, L. 1975. Some observations on the morphology of *Enteromorpha linza* (L.) J. Ag. and *Enteromorpha compressa* (L.) Grev. in axenic culture. *Bot. Mar.* 18:251–53.

Gallagher, J.L. 1975. The significance of the surface film in salt marsh plankton metabolism. *Limnol. Oceanogr.* 20:120–23.

Garrett, W.D. 1965. Collection of slick-forming materials from the sea surface. *Limnol. Oceanogr.* 10:602–5.

Garrett, W.D. 1967. The organic chemical composition of the ocean surface. *Deep-Sea Res.* 14:221–27.

Gessner, R.V., R.D. Goos, and J.McN. Sieburth. 1972. The fungal microcosm of the internodes of *Spartina alterniflora. Mar. Biol.* 16:269–73.

Goering, J.J., and D.W. Menzel. 1965. The nutrient chemistry of the sea surface. *Deep-Sea Res.* 12:839–43.

Gonye, E.R., and E.J. Carpenter. 1974. Production of iron-binding compounds by marine microorganisms. *Limnol. Oceanogr.* 19:840–41.

Graham, J.J., and J.McN. Sieburth. 1973. Survival of *Salmonella typhimurium* in artificial and coastal seawater. *Rev. Intern. Océanogr. Méd.* 29:5–29.

Hanic, L.A., and J.S. Craigie. 1969. Studies on the algal cuticle. *J. Phycol.* 5:89–102.

Hardy, J.T. 1973. Phytoneuston ecology of a temperate marine lagoon. *Limnol. Oceanogr.* 18:525–33.

Harrison, P.G., and K.H. Mann. 1975. Detritus formation from eelgrass (*Zostera marina* L.): The relative effects of fragmentation, leaching, and decay. *Limnol. Oceanogr.* 20:924–34.

Hedgpeth, J.W. 1957. Classification of marine environments. In: *Treatise on Marine Ecology and Paleoecology, Vol. 1, Ecology.* Natl. Res. Council, Washington D.C., Mem. 67, Geol. Soc. America, pp. 17–28.

Hellebust, J.A. 1965. Excretion of some organic compounds by marine phytoplankton. *Limnol. Oceanogr.* 10:192–206.

Herbert, D., R. Elsworth, and R.C. Telling. 1956. The continuous culture of bacteria; a theoretical and experimental study. *J. gen. Microbiol.* 14:601–22.

Herman, S.S., and J.A. Mihursky. 1964. Infestation of the copepod *Acartia tonsa* with the stalked ciliate *Zoothamnium*. *Science* 146:543–44.

Hirche, H-J. 1974. Die Copepoden *Eurytemora affinis* POPPE und *Acartia tonsa* DANA und ihre Besiedlung durch *Myoschiston centropagidarum* PRECHT (Peritricha) in der Schlei. *Kieler Meeresforsch.* 30:43–64.

Holm-Hansen, O., C.J. Lorenzen, R.W. Holmes, and J.D.H. Strickland. 1965. Fluorometric determination of chlorophyll. *J. Cons. perm. int. Explor. Mer* 30:3–15.

Horne, A.J., G.E. Fogg, and D.J. Eagle. 1969. Studies *in situ* of the primary production of an area of inshore Antarctic sea. *J. mar. biol. Assoc. U.K.* 49:393–405.

Ignatiades, L. 1973. Studies on the factors affecting the release of organic matter by *Skeletonema costatum* (Greville) Cleve in field conditions. *J. mar. biol. Assoc. U.K.* 53:923–35.

Jannasch, H.W. 1968a. Growth characteristics of heterotrophic bacteria in seawater. *J. Bacteriol.* 95:722–23.

Jannasch, H.W. 1968b. Competitive elimination of *Enterobacteriaceae* from seawater. *Appl. Microbiol.* 16:1616–18.

Jannasch, H.W. 1970. Threshold concentrations of carbon sources limiting bacterial growth in seawater. In: *Organic Matter in Natural Waters* (D.W. Hood, ed.). Inst. Mar. Sci., University of Alaska, pp. 321–28.

Jarvis, N.L., W.D. Garrett, M.S. Scheiman, and C.O. Timmons. 1967. Surface chemical characterization of surface-active material in seawater. *Limnol. Oceanogr.* 12:88–96.

Jensen, A., B. Rystad, and L. Skoglund. 1972. The use of dialysis culture in phytoplankton studies. *J. exp. mar. Biol. Ecol.* 8:241–48.

Johnson, K.M., and J.McN. Sieburth. 1977. Dissolved carbohydrates in seawater. I. A precise spectrophotometric analysis for monosaccharides. *Mar. Chem.* 5:1–13.

Johnson, P.W., and J.McN. Sieburth. 1976. In situ morphology of nitrifying-like bacteria in aquaculture systems. *Appl. Environ. Microbiol.* 31:423–32.

Johnson, P.W., J.McN. Sieburth, A. Sastry, C.R. Arnold, and M.S. Doty. 1971. *Leucothrix mucor* infestation of benthic crustacea, fish eggs and tropical algae. *Limnol. Oceanogr.* 16:962–69.

Kapraun, D.F. 1970. Field and cultural studies of *Ulva* and *Enteromorpha* in the vicinity of Port Aransas, Texas. *Contr. Mar. Sci.* 15:205–85.

Kinne, O. (ed.). 1972–75. *Marine Ecology. A comprehensive, integrated treatise on life in oceans and coastal waters. Vol. I. Environmental Factors* (1972). *Vol. II. Physiological Mechanisms* (1975). Wiley-Interscience, London and New York.

Koblents-Mishke, O.I., and V.I. Vedernikov. 1973. Tentative comparison of primary production and quantity of phytoplankton on the ocean surface. *Oceanology* 13:55–62.

Kropach, C. 1975. The yellow-bellied sea snake, *Pelamis*, in the Eastern Pacific. In: *The Biology of Sea Snakes* (W.A. Dunson, ed.). University Park Press, Baltimore, Md., pp. 185–213.

Lavoie, D.M. 1975. Application of diffusion culture to ecological observations on marine microorganisms. M.S. Thesis, University of Rhode Island, 91 p.

Laycock, R.A. 1974. The detrital food chain based on seaweeds. I. Bacteria associated with the surface of *Laminaria* fronds. *Mar. Biol.* 25:223–31.

Lorenzen, C.J. 1967. Vertical distribution of chlorophyll and phaeo-pigments: Baja California. *Deep-Sea Res.* 14:735–45.

MacIntyre, F. 1970. Geochemical fractionation during mass transfer from sea to air by breaking bubbles. *Tellus* 22:451–61.

MacIntyre, F. 1974a. The top millimeter of the ocean. *Sci. American* 230(5):62–77.

MacIntyre, F. 1974b. Non-lipid-related possibilities for chemical fractionation in bubble film caps. *J. Rech. Atmosph.* 8:515–27.

MacIntyre, F. 1974c. Chemical fractionation and sea-surface microlayer processes. In: *The Sea, Vol. 5, Marine Chemistry* (E.D. Goldberg, ed.). Wiley, New York, pp. 245–99.

Mann, K. 1973. Seaweeds: Their productivity and strategy for growth. *Science* 182:975–81.

Marshall, K.C. 1976. *Interfaces in Microbial Ecology*. Harvard University Press, Cambridge, Mass., 156 p.

Marshall, K.C., R. Stout, and R. Mitchell. 1971a. Selective sorption of bacteria from seawater. *Can. J. Microbiol.* 17:1413–16.

Marshall, K.C., R. Stout, and R. Mitchell. 1971b. Mechanism of the initial events in the sorption of marine bacteria to surfaces. *J. Gen. Microbiol.* 68:337–48.

Marumo, R., N. Taga, and T. Nakai. 1971. Neustonic bacteria and phytoplankton in surface microlayers of the equatorial waters. *Bull. Plankton. Soc. Japan* 18:36–41.

McCutchen, C.W. 1970. Surface films compacted by moving water: Demarcation lines reveal film edges. *Science* 170:61–64.

McDowell, R.S., and C.W. McCutchen. 1971. The Thoreau-Reynolds Ridge, a lost and found phenomenon. *Science* 172:973.

McFeters, G.A., and D.G. Stuart. 1972. Survival of coliform bacteria in natural waters: Field and laboratory studies with membrane-filter chambers. *Appl. Microbiol.* 24:805–11.

Meade, T.L. 1974. *The Technology of Closed System Culture of Salmonids.* University of Rhode Island Mar. Tech. Rept. 30 (Sea Grant), 30 p.

Meadows, P.S., and G.B. Williams. 1963. Settlement of *Spirorbis borealis* Daudin larvae on surfaces bearing films of microorganisms. *Nature* 198:610–11.

Menzel, D.W., and J.H. Ryther. 1970. Distribution and cycling of organic matter in the oceans. In: *Organic Matter in Natural Waters* (D.W. Hood, ed.). Inst. Mar. Sci. University of Alaska, pp. 31–54.

Mills, E.L. 1975. Benthic organisms and the structure of marine ecosystems. *J. Fish. Res. Bd. Canada* 32:1657–63.

Monod, J. 1942. *Récherches sur la croissance des cultures bactériennes.* Herman & Cie., Paris.

Myklestad, S. 1974. Production of carbohydrates by marine planktonic diatoms. I. Comparison of nine different species in culture. *J. exp. mar. Biol. Ecol.* 15:261–74.

Myklestad, S., and A. Haug. 1972. Production of carbohydrates by the marine diatom *Chaetoceros affinis* var. *willei* (Gran) Hustedt. I. Effect of the concentration of nutrients in the culture medium. *J. exp. mar. Biol. Ecol.* 9:125–36.

Myklestad, S., A. Haug, and B. Larsen. 1972. Production of carbohydrates by the marine diatom *Chaetoceros affinis* var. *willei* (Gran) Hustedt. II. Preliminary investigation of the extracellular polysaccharide. *J. exp. mar. Biol. Ecol.* 9:137–44.

Naumann, E. 1917. Beitrag zur Kenntnis des Teichnannoplankton. II. Uber das Neuston des Süsserwassers. *Biol. Zentralbl.* 37:98–106.

Nixon, S.W., C.A. Oviatt, and S.L. Northby. 1973. *Ecology of Small Boat Marinas.* University of Rhode Island Mar. Tech. Rept. Ser., No. 5, 20 p.

Norris, R.E. 1965. Neustonic marine Craspedomonadales (Choanoflagellates) from Washington and California. *J. Protozool.* 12:589–602.

Ogura, N. 1974. Molecular weight fractionation of dissolved organic matter in coastal seawater by ultrafiltration. *Mar. Biol.* 24:305–12.

O'Neill, T.B., and G.L. Wilcox. 1971. The formation of a "primary film" on materials submerged in the sea at Port Hueneme California. *Pacific Science* 25:1–12.

Parker, B., and G. Barsom. 1970. Biological and chemical significance of surface microlayers in aquatic ecosystems. *Bioscience* 20:87–93.

Parsons, T.R., M. Takahashi, and B.T. Hargrave (1977). *Biological Oceanographic Processes.* 2nd Ed., Pergamon Press, Oxford, 332 p.

Persoone, G. 1968. Ecologie des infusoires dans les salissures de substrats immerges dans un port de mer. I. Le film primaire et le recouvrement primaire. *Protistologica* 4:187–94.

Prakash, A., L. Skoglund, B. Rystad, and A. Jensen. 1973. Growth and cell-size distribution of marine planktonic algae in batch and dialysis cultures. *J. Fish. Res. Bd. Canada* 30:143–55.

Provasoli, L., and I.J. Pintner. 1964. Symbiotic relationships between microorganisms and seaweeds. Amer. J. Bot. 51:681.

Provasoli, L., and I.J. Pintner (in press). Bacteria-induced polymorphism in axenic laboratory strains of *Ulva. Proc. 8th Intl. Seaweed Symp. 1975.*

Ranwell, D.S., and B.M. Downing. 1959. Brent goose (*Branta bernicla* (L.)) winter feeding pattern and *Zostera* resources at Scolt Head Island, Norfolk. *Animal Behaviour* 7:42–56.

Raymont, J.E.G. 1963. *Plankton and Productivity in the Oceans.* Pergamon Press, Oxford, 672 p.

Rhoads, D.C. 1973. The influence of deposit-feeding benthos on water turbidity and nutrient recycling. *Amer. J. Sci.* 273:1–22.

Ryther, J.H. 1969. Photosynthesis and fish production in the sea. *Science* 166:72–76.

Salonen, K. 1974. Effectiveness of cellulose ester and perforated polycarbonate membrane filters in separating bacteria and phytoplankton. *Ann. Bot. Fennici* 11:133–35.

Schultz, J.S., and P. Gerhardt. 1969. Dialysis culture of microorganisms: Design, theory, and results. *Bact. Reviews* 33:1–47.

Sechler, G.E., and K. Gunderson. 1972. Role of surface chemical composition on the microbial contribution to primary films. *Proc. 3rd Intl. Congr. Mar. Corr. & Fouling,* Natl. Bur. Standards, Gaithersburg, Md., pp. 610–16.

Seshadri, R., and J.McN. Sieburth. 1975. Seaweeds as a reservoir of *Candida* yeasts in inshore waters. *Mar. Biol.* 30:105–17.

Sheldon, R.W. 1972. Size separation of marine seston by membrane and glass-fiber filters. *Limnol. Oceanogr.* 17:494–98.

Shelton, R.G.J., P.M.J. Shelton, and A.S. Edwards. 1975. Observations with the scanning electron microscope on a filamentous bacterium present on the aesthetasc setae of the brown shrimp *Crangon crangon* (L.). *J. mar biol. Assoc. U.K.* 55:795–800.

Sieburth, J.McN. 1960. Acrylic acid, an "antibiotic" principle in *Phaeocystis* blooms in Antarctic waters. *Science* 132:676–77.

Sieburth, J.McN. 1965. Bacteriological samplers for air–water and water–sediment interfaces. *Ocean Sci. & Ocean Eng., Trans. Joint Conf. Mar. Tech. Soc. and ASLO,* Washington D.C., pp. 1064–68.

Sieburth, J.McN. 1967. Seasonal selection of estuarine bacteria by water temperature. *J. exp. mar. Biol. Ecol.* 1:98–121.

Sieburth, J.McN. 1968a. The influence of algal antibiosis on the ecology of marine microorganisms. In: *Advances in Microbiology of the Sea* (M.R. Droop and E.J.F. Wood, eds.). Academic Press, London and New York, pp. 63–94.

Sieburth, J.McN. 1968b. Observations on bacteria planktonic in Narragansett Bay, Rhode Island; a résumé. *Bull. Misaki Mar. Biol. Inst., Kyoto University,* 12:49–64.

Sieburth, J.McN. 1971a. Distribution and activity of oceanic bacteria. *Deep-Sea Res.* 18:1111–21.

Sieburth, J.McN. 1971b. An instance of bacterial inhibition in oceanic surface water. *Mar. Biol.* 11:98–100.

Sieburth, J.McN. 1975. *Microbial Seascapes.* University Park Press, Baltimore, Md., 248 p.

Sieburth, J.McN. 1976. Bacterial substrates and productivity in marine ecosystems. *Ann. Rev. Ecol. Syst.* 7:259–85.

Sieburth, J.McN. 1977. Biomass and productivity of microorganisms in planktonic ecosystems. *Helgoländer wiss. Meeresunters.* 30:697–704.

Sieburth, J.McN., and J.T. Conover. 1965. Sargassum tannin, an antibiotic that retards fouling. *Nature* 208:52–53.

Sieburth, J.McN., and C.D. Thomas. 1973. Fouling on eelgrass (*Zostera marina* L.). *J. Phycol.* 9:46–50.

Sieburth, J.McN., R.D. Brooks, R.V. Gessner, C.D. Thomas, and J.L. Tootle. 1974. Microbial colonization of marine plant surfaces as observed by scanning electron microscopy. In: *Effect of the Ocean Environment on Microbial Activities* (R.R. Colwell and R.Y. Morita, eds.). University Park Press, Baltimore, Md., pp. 318–26.

Sieburth, J.McN., P–J. Willis, K.M. Johnson, C.M. Burney, D.M. Lavoie, K.R. Hinga, D.A. Caron, F.W. French III, P.W. Johnson, and P.G. Davis. 1976. Dissolved organic matter and heterotrophic microneuston in the surface microlayers of the North Atlantic. *Science* 194:1415–18.

Sieburth, J.McN., K.M. Johnson, C.M. Burney, and D.M. Lavoie. 1977. Estimation of *in situ* rates of heterotrophy using diurnal changes in dissolved organic matter and growth rates of picoplankton in diffusion culture. *Helgolander wiss. Meeresunters.* 30:565–74.

Skerman, T.M. 1956. The nature and development of primary films on surfaces submerged in the sea. *New Zealand J. Sci. Technol.* 38(B):44–57.

Slanetz, L.W., and C.H. Bartley. 1965. Survival of fecal streptococci in sea water. *Health Lab. Sci.* 2:142–48.

Smestad, B., A. Haug, and S. Myklestad. 1974. Production of carbohydrate by the marine diatom *Chaetoceros affinis* var. *willei* (Gran) Hustedt. III. Structural studies of the extracellular polysaccharide. *Acta Chem. Scand.* B 28:662–66.

Smestad, B., A. Haug, and S. Myklestad. 1975. Structural studies of the extracellular polysaccharide produced by the diatom *Chaetoceros curvisetus* Cleve. *Acta Chem. Scand.* B 29:337–40.

Smith, D.F., and W.J. Wiebe. 1976. Constant release of photosynthate from marine phytoplankton. *Appl. Environ. Microbiol.* 32:75–79.

Sorokin, Ju.I. 1971. On the role of bacteria in the productivity of tropical oceanic waters. *Int. Revue ges. Hydrobiol.* 56:1–48.

Steele, J.H. 1974. *The Structure of Marine Ecosystems.* Harvard University Press, Cambridge, Mass., 128 p.

Steimle, F., and R.B. Stone. 1973. Bibliography on artificial reefs. *Coastal Plains Ctr. for Mar. Devel. Ser.*, Wilmington, N.C., 129 p.

Strickland, J.D.H. 1968a. Continuous measurement of *in vivo* chlorophyll; a precautionary note. *Deep-Sea Res.* 15:225–27.

Strickland, J.D.H. 1968b. A comparison of profiles of nutrient and chlorophyll concentrations taken from discrete depths and by continuous recording. *Limnol. Oceanogr.* 13:388–91.

Strickland, J.D.H., and T.R. Parsons. 1972. *A Practical Handbook of Seawater Analysis.* Bull. 167 (2nd Ed.). Fish. Res. Bd. Canada, Ottawa, 310 p.

Sutcliffe, W.H., Jr., E.R. Baylor, and D.W. Menzel. 1963. Sea surface chemistry and Langmuir circulation. *Deep-Sea Res.* 10:233–43.

Swift, D.G. 1973. Vitamin B_{12} as an ecological factor for centric diatoms in the Gulf of Maine. Ph.D. Thesis, Johns Hopkins University, 182 p.

Taguchi, S., and K. Nakajima. 1971. Plankton and seston in the sea surface of three inlets of Japan. *Bull. Plankton Soc. Japan* 18:29–36.

Tootle, J.L. 1974. The fouling microflora of intertidal seaweeds: Seasonal, species and cuticular regulation. M.S. Thesis, University of Rhode Island, 64 p.

Tsyban, A.V. 1971. Marine bacterioneuston. *J. Oceanogr. Soc. Japan* 27:56–66.

Unger, I. 1966. Artificial reefs—a review. Spec. Pub. No. 4, Amer. Littoral Soc., Highland, N.J., 74 p.

Vargo, G.A., P.E. Hargraves, and P.W. Johnson. 1975. Scanning electron microscopy of dialysis tubes incubated in flowing seawater. *Mar. Biol.* 31:113–20.

Vollenweider, R.A. (ed.). 1969. *A Manual on Methods for Measuring Primary Production in Aquatic Environments,* IBP Handbook No. 12 (reprinted 1971). Blackwell Scientific, Oxford and Edinburgh, 213 p.

Wagner, F.S., Jr. 1969. Composition of the dissolved organic compounds in seawater: A review. *Contr. Mar. Sci.* 14:115–53.

Wangersky, P.J. 1976. The surface film as a physical environment. *Ann. Rev. Ecol. Syst.* 7:161–76.

Watson, S.W., and J.B. Waterbury. 1969. The sterile hot brines of the Red Sea. In: *Hot Brines and Recent Heavy Metal Deposits in the Red Sea* (E.T. Degens and D.A. Ross, eds.). Springer-Verlag, New York, pp. 272–81.

Welsh, B.L. 1975. The role of grass shrimp, *Palaemonetes pugio*, in a tidal marsh ecosystem. *Ecology* 56:513–30.

Williams, P.M. 1967. Sea surface chemistry: Organic carbon and organic and inorganic nitrogen and phosphorus in surface films and subsurface waters. *Deep-Sea Res.* 14:791–800.

Williams, P.J.LeB., and C.S. Yentsch. 1976. An examination of photosynthetic production, excretion of photosynthetic products, and heterotrophic utilization of dissolved organic compounds with reference to results from a coastal subtropical sea. *Mar. Biol.* 35:31–40.

Young, L.Y., and R. Mitchell. 1972. The role of chemotactic responses in primary microbial film formation. *Proc. 3rd Intl. Congr. Mar. Corr. and Fouling.* Natl. Bur. Standards, Gaithersburg, Md., pp. 617–24.

Zaitsev, Yu. P. 1970. *Marine Neustonology* (K.A. Vinogradov, ed.). Israel Prog. Sci. Transl., Jerusalem, 1971, 207 p.

ZoBell, C.E. 1943. The effect of solid surfaces upon bacterial activity. *J. Bacteriol.* 46:39–59.

CHAPTER 3

Protected Microbial Habitats

The role of hidden microorganisms is as important as that of their more visible counterparts in the exposed habitats just described. These hidden microorganisms live in enclosed or protected habitats that can be divided into three groups: the enteric–fecal–sestonic habitats, the benthic habitats, and the endobiotic habitats. The enteric–fecal–sestonic and benthic habitats have much in common, since both are based on organic debris and the latter is largely dependent upon the sedimentation of the former. The endobiotic habitat of microorganisms is within the cell of other microorganisms or within the thallus or body of multicellular organisms. Endobiotic microorganisms, which may be beneficial symbionts, innocuous parasites, or dangerous pathogens, can be very important in affecting the ecosystem of their host and, in turn, that of predators of their hosts.

A. ENTERIC–FECAL–SESTONIC HABITATS

A central theme of oceanic biology is the trophic relationship of species, food chains, and ecosystems. Microorganisms are involved with food before and during its ingestion and digestion, and with excrement and fecal debris that enter the food chain. Microorganisms involved in dissimilative processes are part of an enteric–fecal–sestonic ecosystem that links the synthetic biomass formed in the euphotic zone with the dissimilative processes and biomass on the sea floor. Despite the obvious existence and importance of these ecosystems, they have not received the attention they deserve. It is this particulate material and the soluble waste materials that connect the different trophic levels and marine habitats, which tie together the components of an overall marine ecosystem.

Microorganisms in planktonic and epibiotic ecosystems nurture ecosystem-specific invertebrate fauna. The carnivorous swimming animals or nekton, including squid, fish, and mammals, live upon these invertebrates and each other. The nekton populates the mid- and deep waters (Isaacs and Schwartzlose, 1975) as well as the rich surface waters. In the process of harvesting the primary production of the photic zone (either directly or indirectly), the invertebrates and the vertebrates excrete one-quarter of their food material as excrement. The enteric–fecal–sestonic habitat includes the ingesta within the animal, the waste material eliminated as feces, and the sedimenting and suspended debris resulting from this process. A parallel input of organic debris results from the normal, seasonal, and catastrophic die-offs of the components of plant, animal, and microbial populations. The sedimenting organic debris, from carcasses to fecal pellets, provides the energy for the benthic organisms on the sea floor (Sokolova, 1959; Schrader, 1971; Dayton and Hessler, 1972; Grassle and Sanders, 1973; Bishop et al. 1977). Seston may also form a significant portion of the food of coral reef animals (Hickel, 1974).

Two examples of the enteric–fecal–sestonic habitat are shown in Plates 3-1 and 3-2. The suspended organic debris that serves as food for suspension-feeding molluscs is shown in Plate 3-1. The presence of bacteria on diatom fragments is clearly seen in thin section (as well as by epifluorescence microscopy, Plate 14-2). A classic example of a habitat-specific bacterium is the spirochete *Cristispira*, which is associated with the gut of lamellibranch molluscs and can be clearly seen in the crystalline style of oysters where it becomes concentrated. The great amount of waste matter deposited on the sea floor (biodeposition) by the many molluscs of inshore waters is indicated by the amount seen in the underwater photograph of mussels, which are completely surrounded by fecal pellets. Another example of the enteric–fecal–sestonic habitat is the fecal pellets of calanoid copepods, which feed upon the flagellates and diatoms of rich coastal water overlying continental shelves (Plate 3-2). The only indication that prasinophycean and haptophycean flagellates have been ingested is the presence of as many as six types of species-specific body scales in the pellets, whereas all that remains of the diatoms are their empty frustules. The only microorganisms surviving in the fecal pellets are the apparently undigestible chroococcoid cyanobacteria and phagotrophic flagellates containing ingested bacteria.

1. Enteric Microflora

The evolution of the gut in animals has provided heterotrophic bacteria and protozoa with a culture vessel whose efficiency varies with the type of animal (Luckey, 1972). The enteric, or gut, microflora is actually two microfloras. One is the microbiota of the ingesta, which can originate from and reflect the microbiota of the food (Plates 3-1,*B* and 3-3,*A*). Other examples are the chitin-digesting bacteria associated with the copepod integument (Plate 2-8,*D*), which grow in the ingesta of copepod-consuming populations (Lear, 1963; Seki and Taga, 1963; Seki, 1965a) and the marine enterobacteria in the genera *Beneckea* and *Photobacterium* (Baumann and Baumann, 1977; Chapter 15,A). The other is the microbiota of indigenous (autochthonous) microorganisms that are attached to or intimately associated with specific areas of the intestinal tract, especially in such areas as the villous surface (villi) (Savage, 1972; Erlandsen, Thomas, and Wendelschafer, 1973). A marine example is the bacterial flora of the crystalline style of lamellibranch molluscs, the pestle-like grinding organ contained in a sac adjoining the intestinal tract that triturates food particles against the gastric shield. Intertidal species, such as the eastern oyster (*Ostrea virginiana*), have a very distinctive bacterial flora consisting of the large spirochete *Cristispira* (Plate 3-1,*C*) and a small spirillum, *Spirillum ostreae* (Noguchi, 1921; Dimitroff, 1926a, b; Kuhn, 1974; Chapter 16, D,E). The role of these bacteria in the style and in digestion is unknown. The ubiquitous coccobacillary forms, however, are not present in the oyster style.

In the buccal and the anal cavity, at either end of the gut, ciliates have adapted to the particle-rich microenvironment. Particulate matter arising from the tearing and chewing of food as well as from the increased flow of water due to ingestion nurtures a variety of ciliates that attach to the mouth parts, mantle, gill, or feeding appendages. In the anal cavity of many invertebrates, a virtually host-specific ciliate population feeds upon fecal particulates and microorganisms. These highly evolved forms are discussed in Chapter 24.

2. Modifying Factors

The dual microflora system in the gut is also affected by the age of the animal and the nature of its food.

PLATE 3-1. The ingesta, gut bacteria, and fecal pellets of suspension-feeding molluscs in turbid nearshore waters rich in suspended organic debris. (*A*) Cloud of suspended particles provides a background for a colony of hydroid *Tubularia crocea* attached to the piling (actual size). (*B*) Electron micrograph of a thin section of suspended debris, including diatomaceous and other fragments colonized by bacteria. (*C*) Phase-contrast photomicrograph of the crystalline style (digestive organ) from the oyster *Ostrea virginiana*, showing the habitat-specific spirochete *Cristispira*. (*D*) Fecal matter litters the area between juvenile mussels (*Mytilus edulis*) (actual size). (*A*, *D*) underwater photographs courtesy of H.L. "Wes" Pratt; (*B*,*C*) courtesy of Paul W. Johnson.

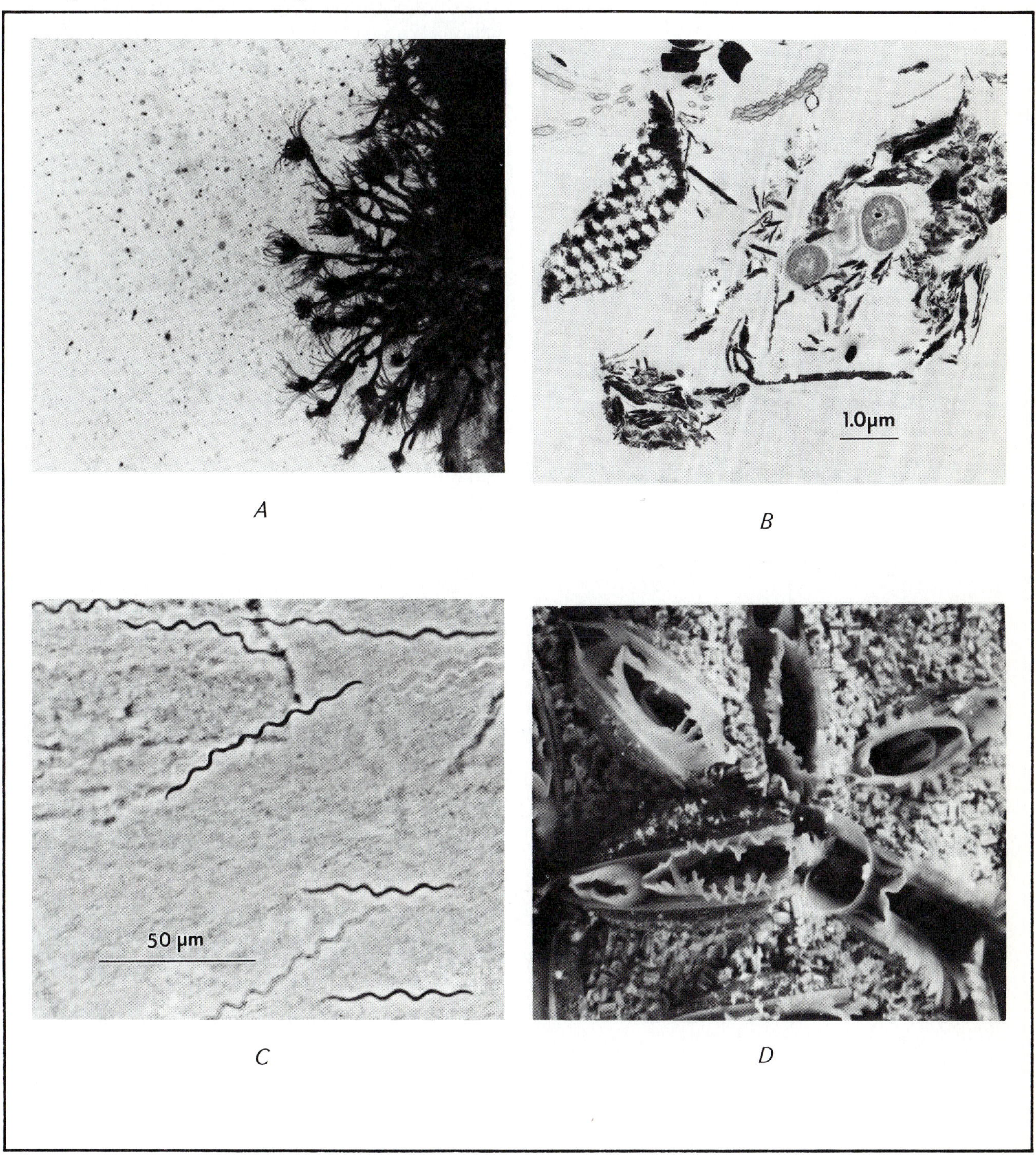

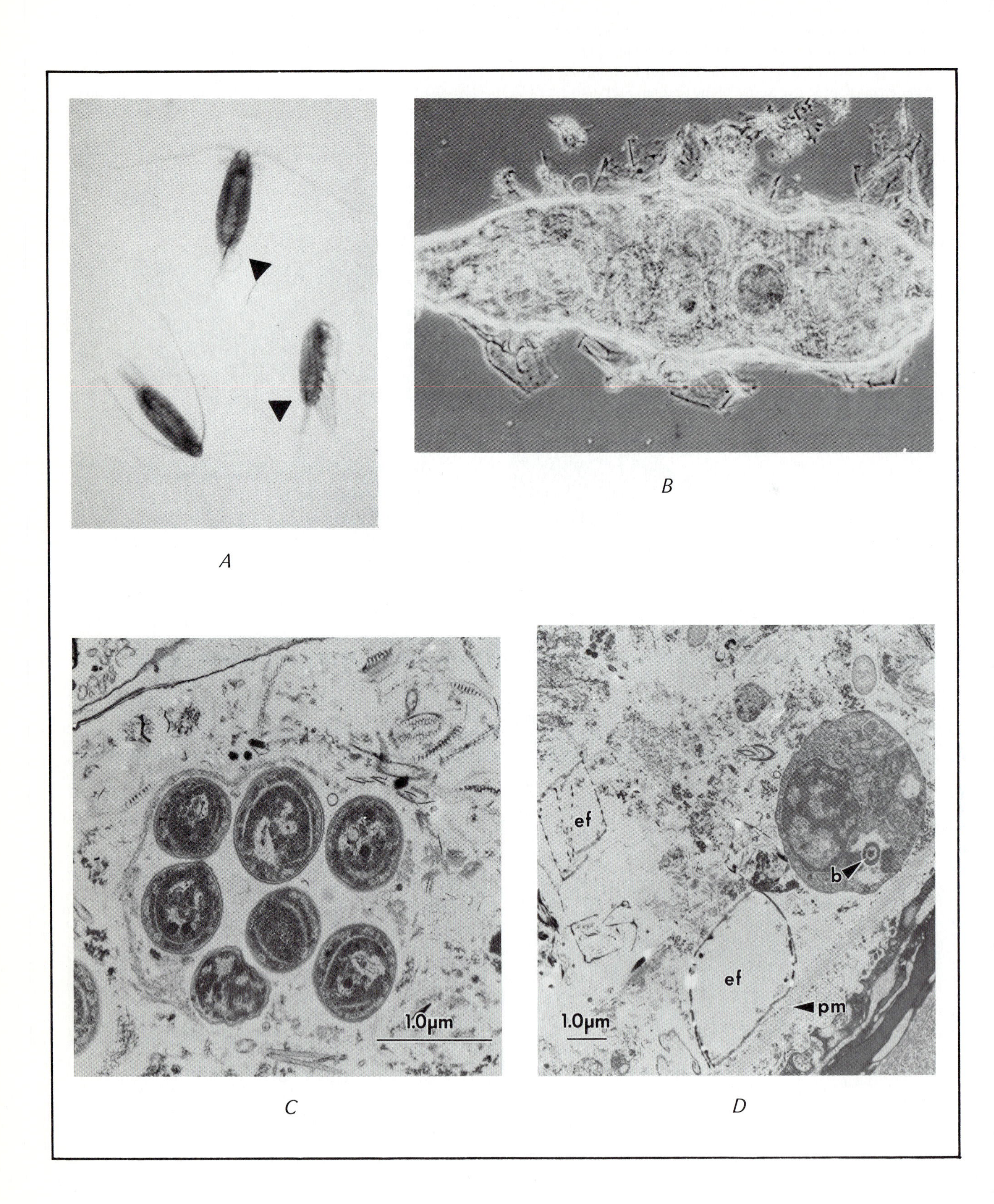
A
B
C
1.0µm
D
ef
b
ef
pm
1.0µm

PLATE 3-2. The enteric–fecal habitat of the calanoid copepod *Calanus finmarchicus*, which grazes upon the phytoplankton in continental shelf waters. (*A*) Copepods, with fecal pellets (arrows) within the gut. (*B*) Dissected fecal pellet, showing the individuals of the centric diatom *Coscinodiscus* in various states of digestion. (*C*) and (*D*) Transmission electron micrographs of thin sections of fecal pellets within the copepod gut, showing (*C*) prasinophycean scales, which are the telltale remains of a digested flagellate next to a microcolony of chroococcoid cyanobacteria, with their distinctive photosynthetic cytomembrane structure, which is apparently resistant to digestion, and (*D*) empty diatom frustules (ef) and a phagotrophic flagellate, with an engulfed bacterium (b) adjacent to the peritrophic membrane (pm) enclosing the fecal pellet. All micrographs courtesy of Paul W. Johnson.

As animals go from larval to adult forms, their diet, and therefore their microbiota, would be expected to change, as it does in terrestrial animals. Primary chemical substances in the food, such as sugars (Raibaud et al., 1972), can modify the microflora. The food of herbivorous and carnivorous animal species might be expected to select for carbohydrate-digesting (saccharolytic) and protein-digesting (proteolytic) bacteria, respectively.

Secondary chemical substances in the food may also modify the intestinal microflora. Predators with a gut content rich in the body fluids of their prey may have a reduced gut microbiota due to the inhibitory activity of a number of constituents of the blood of the prey, as suggested for the gut of predatory birds in Antarctica (Sieburth, 1959). Under certain conditions, phytoplankton blooms dominated by a species producing biologically active secondary substances may affect the intestinal microflora of animals ingesting it. The haptophycean microalga *Phaeocystis poucheti* (Guillard and Hellebust, 1971; Chapter 12,A) and other phytoplankters, such as the diatom *Skeletonema costatum*, contain the sulfonium compound dimethyl beta-propiothetin, which is hydrolyzed to release dimethyl sulfide and acrylic acid. When penguins ingest euphasids that have been feeding upon *Phaeocystis*, acrylic acid is released, which, in the acidic gut, can markedly inhibit the aerobic bacterial flora (Sieburth, 1960, 1961, 1968a; Soucek and Mushin, 1970).

3. Microorganisms as Food

The major transfer of living organic matter up the food chain occurs, in the microbial world, through the feeding of one microorganism upon another. The simple and classical food chain of phytoplankton to zooplankton and zooplankton to nekton must be modified to account for the many microbial components of the water column that are usually overlooked. These include the free planktobacteria and the epibacteria on debris; the microphagous (bacteria-eating) protozoa, such as amoebae, heliozoans, microflagellates, and ciliates; and the larger macrophagous protozoa that feed upon the phytoplankton and the protozooplankton, such as the planktonic foraminifera, radiolarians, and larger flagellates and ciliates. These are the microorganisms responsible for the active and rapid production, release, and uptake of particulate and soluble organic matter. This is where the drama of the sea unfolds on an hour by hour basis, controlled in large degree by diel periodicity, the light/dark cycle, which turns photosynthesis on and off and effectively coordinates the activities of the osmotrophic and phagotrophic microorganisms that are nurtured by the phototrophs.

The major, or normal, source of food for herbivorous or omnivorous copepods is obviously the primary producers in the phytoplankton. The cell contents of diatoms are quickly released within the gut as indicated by fluorescence microscopy (Johannes and Satomi, 1966) and in thin section (Plate 3-2,*D*), leaving many whole but empty diatom frustules in the voided fecal pellets (Schrader, 1971, 1972; Plate 3-2,*B*). Phytoflagellates quickly disappear in the gut of copepods, leaving only their scales as evidence of their ingestion (see Plate 3-2,*C*). Bacteria, however, can account for 30 to 40% of the microbial biomass (Sieburth et al., 1977). Bacteria can serve as food for the zooplankton (Khailov and Finenko, 1970; Sorokin, 1971) and even replace the normal phytoplankton diet of zooplankters in an aphotic basin (Sibert and Brown, 1975). Some species of fish apparently digest bacteria (Odum, 1970; Moriarty, 1975), and this digestion may be the cause of the reported sterility of the empty gut in fish. The microbiota on the detritus of sea and marsh grasses, in the sequence bacteria, flagellates, and ciliates, is seen as the true source of food for deposit feeders (Newell, 1965; Fenchel, 1969, 1972). In the aphotic zone of the sea heterotrophic microorganisms, osmotrophic bacteria and osmotrophic and phagotrophic protozoa, become the secondary producers upon which aphotic life is based.

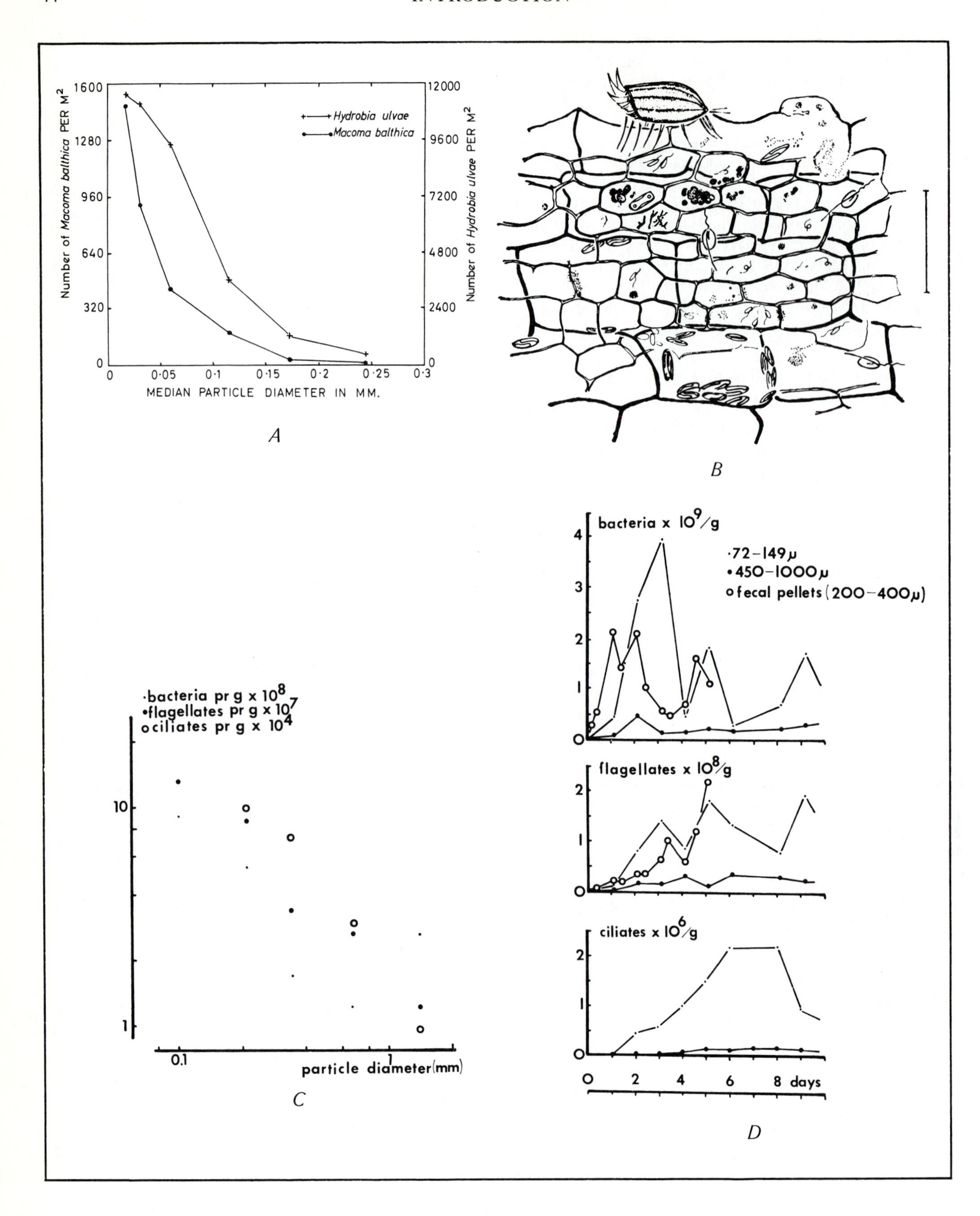
Number of Macoma balthica PER M²
Number of Hydrobia ulvae PER M²
Hydrobia ulvae
Macoma balthica
MEDIAN PARTICLE DIAMETER IN MM.
A
B
·bacteria pr g x 10⁸
•flagellates pr g x 10⁷
o ciliates pr g x 10⁴
particle diameter(mm)
C
bacteria x 10⁹/g
·72–149μ
•450–1000μ
o fecal pellets (200–400μ)
flagellates x 10⁸/g
ciliates x 10⁶/g
days
D

PLATE 3-3. The association of deposit-feeders, such as the prosobranch *Hydrobia ulvae* and the bivalve *Macoma balthica*, with the finer sediment particles (*A*), is dependent upon the microbial communities that develop on organic debris, such as that from the turtlegrass *Thalassia testudinum* (*B*) (scale = 0.1 mm). The populations of bacteria, flagellates, and ciliates increase with the decrease in size of the *Thalassia* debris (*C*) and show a definite sequence of development and population density on fresh amphipod fecal pellets and fractions of *Thalassia* debris smaller and larger than the fecal pellets (*D*). The ciliates on the fecal pellets were not counted. (*A*) from R. Newell, 1965, *Proc. Zool. Soc. London* 144:25–45; (*B,C,D*) from T. Fenchel, 1970, *Limnol. Oceanogr.* 15:14–20.

4. Bacterial Digestion and Enrichment

Even in the copepods with a food passage time as short as 30 minutes, enteric bacteria apparently grow and are then eaten by flagellates (Plate 3-2,*D*). Larger animals, such as squid, fish, birds, and mammals, have comparatively longer intestinal tracts, and food passage takes hours to days. Microbial activities would soon deplete the oxygen content of the ingesta as it passes through the gut, thus selecting for anaerobic bacteria, which require anaerobic techniques for their culture (Chapter 17,A). Intestinal microflora studies have not been reported for marine animals with different types of diets, e.g., phytoplankton, zooplankton, squid, and fish eating forms. Such studies might lead to an understanding of the role of the intestinal microflora in marine animals as similar studies have in terrestrial herbivores (Hungate, 1966; McBee, 1971). But the major role of the fecal microflora may be in a post-defecation stage. Snorkelers or SCUBA divers swimming over the shallow sea floor, if they know what to look for, cannot help but be impressed by the amount and variety of fecal deposits. Even the photographs of the deep-sea floor in geological atlases clearly show the presence of feces of the larger animals. Feces vary with the species and are quite distinctive (Moore, 1931; Arakawa, 1970; Kraeuter and Haven, 1970), ranging in size from the membrane-covered, copepod fecal pellets (Plate 3-2,*B*) that sink to the sea floor (Schrader, 1971, 1972; Honjo, 1976) to the large coiled deposits of benthic sea hares. Nitrogen-poor fecal pellets are apparently enriched with nitrogen by bacteria as they age (Newell, 1965). Coprophagy, the eating of feces, may constitute a major source of food for a number of species that consume the feces of other species as well as their own (Johannes and Satomi, 1966; Frankenberg and Smith, 1967).

5. Fecal Debris

It is obvious that much of the amorphous and flake types of debris (Hobbie et al., 1972) suspended throughout the water column must be the debris resulting from feeding and is not synthesized *de novo* from dissolved organic carbon. But observations by Riley (1963) and Riley, van Hemert, and Wangersky (1965) on the increase in suspended organic debris following phytoplankton blooms led them to a concept of the *de novo* synthesis of organic aggregates in which the large reserves of dissolved organic carbon (Wangersky, 1965) formed flakes and amorphous particles through a foam-tower or bubble-scavenging mechanism (Sutcliffe, Baylor, and Menzel, 1963; Riley, Wangersky, and van Hemert, 1964; Riley, van Hemert, and Wangersky, 1965). But the particles larger than 2 μm are in the 8 to 44 μm size range (Hobson, 1967). Particle formation (organic aggregation) in seawater occurs mainly in the less than 4 μm size range and is due to bacterial growth (Sheldon, Evelyn and Parsons, 1967). A re-evaluation of bubble-scavenging by Menzel (1966) indicated that it could be a laboratory artifact. Although some flocculent material is formed by precipitation of colloidal material (Sieburth, 1968b) or by film collapse (Wheeler, 1975), these processes are minor compared to the feeding process, which produces large quantities of debris. The smaller, non-sedimenting fragments (McCave, 1975), which could be largely the remnants of cells from disintegrating fecal pellets (Plates 3-1 and 3-2), presumably decay *in situ* (as illustrated by Honjo, 1976) to release dissolved organic carbon above threshold levels to nurture the bacterioplankton.

Not all feces disintegrate or are consumed in coprophagy. A large amount of the feces (Bishop et al., 1977) and larger debris will not be consumed and will sediment on the bottom, since there is a threshold concentration for particulate matter that must be exceeded before filter-feeding organisms will feed (Parsons and Seki, 1970), just as there is a threshold concentration for dissolved organic carbon utilization. This ensures a more or less steady-state deposition of larger debris on the sea floor. This material undergoes biodeterioration on the sea floor, which constitutes "secondary production" for the development of benthic ecosystems.

B. BENTHIC HABITAT

The benthos is life on the sea floor. The food input to organisms on the sea floor is directly related to primary productivity in the euphotic zone, but the amount of food reaching the sea floor is inversely related to its depth from the mixed layer (Hargrave, 1973). This is due to the *in-situ* recycling of most of the easily utilizable organic matter in the mixed layer and the deposition of more refractory particulate matter, such as chitin, to the sediment. Chitin in the upper aerobic layer of sediment is broken down by bacteria of the genus *Beneckea*, whereas in the deeper anaerobic zone break down is apparently due to the sulfate-reducing bacteria of the genus *Desulfovibrio* (Seki, 1965b). Break down at the sediment surface apparently reflects the supply of organic matter not oxidized in the water column. In the estuarine environment of a Scottish sea loch, the use of sediment traps and measurements of *in-situ* respiration indicated that almost one-third of primary (photosynthetic) biomass produced reached the bottom where most of it is utilized (Davies, 1975). Except where there is a terrigenous input, organic carbon accumulation is less than 10% of the total annual supply of sedimented matter (Hargrave, 1973), and sediments are actively broken down in winter as well as in summer (Seki, Skelding, and Parsons, 1968).

1. Sediment Enrichment

An examination of particulate matter at the sediment–water interface in shallow waters in the vicinity of Woods Hole (Johnson, 1974) indicated that some 61% of the particles were potential food, being 43% organic mineral aggregates, 3.8% fecal matter, 2.3% skeletal elements, 1.9% microalgae and fungi, and 0.7% non-algal plant fragments. The organic mineral aggregates, which are rich in carbohydrates, are produced by deposit-feeding animals, such as bivalves and polychaetes, as they feed, and during which process they rework the sediment. Extensive reworking of the sediments makes them unstable so that organic mineral aggregates and fecal pellets are easily resuspended by a weak tidal flow to produce permanently turbid or "upwelled" estuarine waters, which are important in nutrient cycling and molluscan nutrition (Rhoads, 1973; Bisagni, 1975). In shallow estuaries, the zooplankton is relatively unimportant and the phytoplankton is consumed mainly by the benthic invertebrates and phytophagous fish (Odum, 1970). The feeding on resuspended seston (organic debris) and phytoplankton by the bivalve molluscs and other invertebrates produces a copious biodeposition (Haven and Morales-Alamo, 1966) of pseudofeces and feces, the latter being distinctive to the family level (Arakawa, 1970) and even to genus and species (Kraeuter and Haven, 1970). It is the process of biodeposition that coats mineral fragments to produce organic mineral aggregates, which are heavily colonized by microorganisms. The non-algal debris may come from stands of cordgrass, such as *Spartina*, at the edge of the marsh (Welsh, 1975) and from the beds of seagrasses, such as the eelgrass *Zostera* and the turtlegrass *Thalassia testudinum* (Fenchel, 1970), as well as from mangroves (Newell, 1976). As these plants break down to become organic debris, they are reworked to smaller fragments by deposit feeders, which can support an increased microbiota (see Plate 3-3).

In the interstitial ecosystem of sandy beaches (Plate 3-4), the particles are caught in the upper 5 cm and give rise to the usual bacterial–protozoan community. But below this, a bacterial flora attached to sand grains (Meadows and Anderson, 1968) removes some 1 to 4 mg of dissolved organic carbon/liter and maintains a meiofauna of nematodes, turbellarians, and copepods with a daily carbon requirement of 21 μg of carbon/kg of sand. (McIntyre, Munroe and Steele, 1970). The community metabolism of these intertidal sandflats (Pamatmat, 1968) is augmented by the production of organic matter by benthic diatoms. Productivity increases with the depth of the overlying water and is minimized by wave action (Steele and Baird, 1968). These benthic diatoms migrate vertically with the tides (Palmer, 1975), and their biomass and activity increase as the sands give way to the silts and clays of mud flats (Leach, 1970).

2. Deep-sea Sediments

In contrast to the margins of the sea, the mean depth of the sea is 4 km, and some 85% of the sea floor is in the abyssal environment, at depths exceeding 2 km, where it is lightless and cold (some 2 °C), and where pressures reach 200 to 1000 atm. The slow sinking rates of phytoplankton (Smayda, 1970) compared to the comparatively high sinking rates of copepod fecal pellets (Smayda, 1969; Strickland, 1970; Mills, 1975; Honjo, 1976), which could reach the sea floor in as short a time as 1 month, indicate that the residues of zooplankton feeding may be a main energy source in abyssal waters and deep-sea sediments (Bishop et al. 1977). Copepods in the euphotic zone reportedly feed selectively on the larger diatoms (Mullin, 1963) and

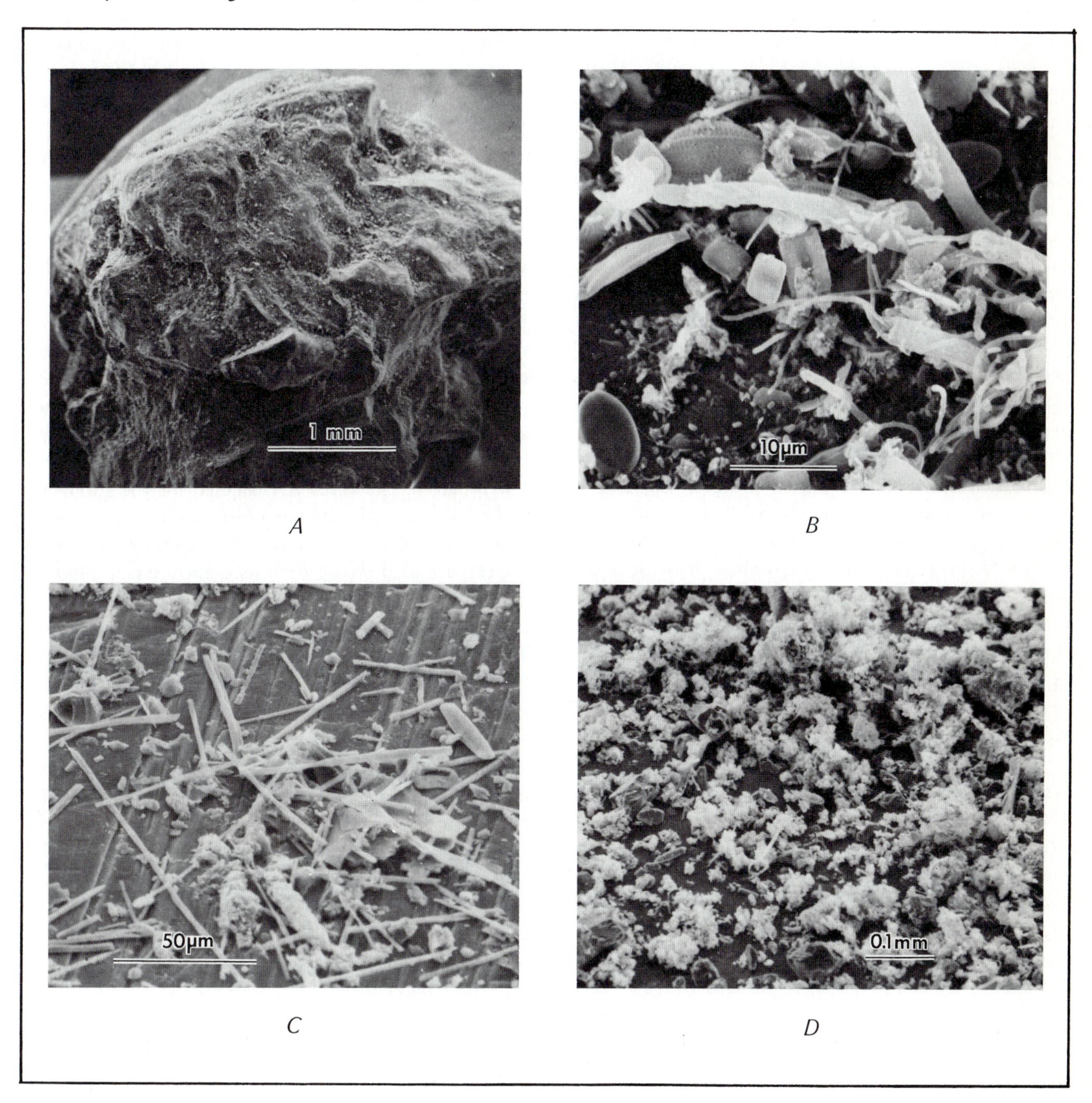

PLATE 3-4. The environment and microbiota of sand and silt from Narragansett Bay, R.I. (*A*) Sand grain, showing non-colonized abraded areas and microbial growth in crevices. (*B*) Diatoms and bacteria colonize a protected crevice on a sand grain. (*C*) The prisms of mussel shells and fecal fragments washed from the pores of intertidal sand. (*D*) The large amount of organic debris or flocs colonized by bacteria and the small amount of fine mineral grains in subtidal mud. Electron micrographs from Sieburth, 1975, *Microbial Seascapes*, University Park Press, Baltimore, Md.

other microorganisms in the plankton (Mullin, 1966). In most species, the fragments and frustules passing the pyloric filter press of the foregut are enveloped in a chitinous (peritrophic) membrane in the anterior part of the intestine; this presumably protects the intestinal epithelium from the abrasive action of the food fragments (Forster, 1953).

Only fecal pellets that retain their integrity through their chitinous membranes (Schrader, 1971, 1972) and are not consumed by coprophagous zooplankton in abyssal waters (Vinogradov, 1962) would sink to the sea floor in a reasonable length of time (Smayda, 1969). The more finely divided debris from fragmented pellets would sink very slowly and would not reach the sea floor. In addition to the slow rain of fecal pellets and zooplankton remains, the deep-sea sediments are enriched by the carcasses of the larger organisms in the nekton. These carcasses of organisms killed by predation, disease, and old age would sediment at high rates and would arrive in the deep sea with much flesh intact. Although there would be far fewer carcasses than zooplankton debris, their biomass would be considerable because of their size. The rarer occurrence of this considerable biomass would be compensated for by the deep-sea amphipods and fish (Isaacs and Schwartzlose, 1975), who would quickly consume it and scatter its organic remains in the form of feces over a wide area to be used by benthic microorganisms. The prevalence of deep-sea scavengers that consume the baits used to lure them to deep-sea cameras indicates that there are available consumers for the carcasses of larger organisms and that their feces could be an important source of energy. But regardless of whether it is the feces of copepods or of deep-sea scavengers, sedimented fecal material is the primary energy source for microorganisms on the surface of the deep-sea floor.

The energy input to the sea floor below depths of 2 km is not known, although it is believed to be less than 10% of the primary productivity in the euphotic zone (Riley, 1970; Hargrave, 1973). Sediment traps that measure the rates of sedimentation into the abyssal zone and onto the sea floor are currently being used in a number of laboratories, including the laboratory of the writer. A sediment trap deployed at 2000 m in Tongue of the Ocean, the Bahamas (Wiebe, Boyd, and Winget, 1976) collected a daily average of 650 fecal pellets/m^2, which amounts to 36.5 mg of carbon/m^2. Several attempts have been made to measure the benthic respiration supported by this sedimenting debris. Pamatmat (1973) measured the respiration of decompressed deep-sea sediments aboard ship. In order to obtain more valid data, a number of procedures and apparatus have been used to make *in-situ* measurements. A few observations have been made with chambers placed and recovered by submersibles (Smith and Teal, 1973; Smith, 1974). A free vehicle that uses oxygen electrodes to record oxygen uptake in the chamber was described by Smith et al. (1976). A less expensive free vehicle recovers replicate water samples (Hinga, 1974), which can be used to measure nutrient fluxes as well as oxygen consumption by the enclosed sediment. The number of deep-sea measurements obtained with any of these instruments is very few, due to the time and money involved in deploying and recovering them. Total oxygen uptake can be fractionated by inhibiting the respiration of all biological forms with formaldehyde and estimating anaerobic metabolism by measuring the rate of inorganic chemical oxidation of the reduced end products of metabolism (Pamatmat, 1971; Smith and Teal, 1973; Smith et al., 1976). Methods better than antibiotic inhibition are needed to accurately differentiate animal from microbial respiration.

The deep-sea coring programs conducted by marine geologists (paleontologists) offer a potentially unique opportunity to marine microbiologists. Usually only the lower part of the cores is examined for the surviving skeletons of such fossil microorganisms as foraminifera, radiolaria, silicoflagellates, and hystotrichospheres. The upper part of the core with its contemporary biota is usually ignored or discarded. Microbiologists are remiss in not developing programs, in not riding these cruises, and in not taking advantage of opportunities to do meaningful studies in deep-sea microbiology. Apparatus similar to the *in-situ* pressure water samplers and culture vessels of Jannasch and Wirsen (1977) could be developed to obtain cores and their overlying water, keep them at pressure, and yet allow subsamples to be obtained at timed intervals so that rates of oxygen uptake and nutrient flux could be determined as well as the nature of and changes in the microbiota.

3. Aerobic Zone

In the aerobic film of sedimenting debris, a sequence of bacteria, flagellate, and ciliate development occurs (Fenchel, 1970; Plate 3-3,*D*). Bacteria are believed to be the primary decomposers in the sea, and rates of decomposition are increased by the triturating activities of deposit-feeding animals (Fenchel, 1972). The aerobic decomposition of the debris in sediments is a function both of their organic content and of the surface area of the debris particles (Hargrave, 1972). The ease with which this aerobic decomposition can be inhibited is indicated by the preservation and fossilization of animals in sediments (Schäfer, 1972) and, in one instance, by the paradoxical preservation of the box lunch aboard the research submersible *Alvin* during its accidental submergence on the sea floor for 10 months (Jannasch et al., 1971). The inhibition of the biodeterioration of the labile solid and liquid substrates in the lunch has led to observations on the low rates of uptake of soluble substrates in bottle experiments (Jannasch et al., 1971; Jannasch and Wirsen, 1973), presumably due to low temperature and high hydrostatic pressure inhibition (Wirsen and Jannasch, 1975). The low rates also obtained with shallow water samples from 50 m, with a 50% reduction over atmospheric controls (Wirsen and Jannasch, 1975), may be due to methodology, such as inadequate controls and the choice of substrate concentration (Azam and Holm-Hansen, 1974) and observation intervals, since hydrostatic pressures up to 18 atm have no effect on the total oxygen uptake of cores from a depth of 180 m (Pamatmat, 1971). But the main problem with this approach is trying to explain the lack of biodeterioration of solid substrates with observations on soluble substrates isolated in closed bottles. Preliminary observations with box-lunch type substrates indicated that these materials, when held in perforated containers near the sea floor at 5000 m, were consumed within 2 months, but were intact when held in sealed lunch boxes, presumably because of the exclusion of the passage of water and of such scavenging animals as lyssianasid amphipods (Sieburth and Dietz, 1974). The concluding remarks of Jannasch and Wirsen (1977) indicate that they have now come to a similar conclusion.

4. Anaerobic Zone

Below the oxidized layer with its oxidative bacterial processes, a redox-potential discontinuity point (Fenchel and Riedl, 1970) marks the beginning of a strongly reducing environment (Plate 3-5). The aerobic decomposition of organic debris incorporated into the sediment quickly uses up the oxygen that diffuses into the pore water and helps form the reduced or sulfide zone. About 50% of the carbon mineralized in the anaerobic sediments can be due to sulfate reduction by *Desulfovibrio* (Chapter 17,C), which obtains its oxygen from sulfate ions and releases hydrogen sulfide to maintain the required anaerobic conditions (Jørgensen and Fenchel, 1974). This zone and its biotic community underlies the oxidized layer of all nearshore sediments except those with high energy windows in the intertidal and surge zones (Fenchel and Riedl, 1970). The bacteriological processes that go on in such zones have been discussed in general terms (Oppenheimer, 1960). In areas of finer sediments, black sulfide muds give rise to a profuse population of colorless and photosynthetic sulfur bacteria, which are grazed by a ciliate microfauna. The structure and function of this ecosystem in nearshore systems have been well illustrated in the monograph by Fenchel (1969; Plate 3-6). Sulfide systems, however, can also occur offshore under certain conditions. Along the shelf and slopes of Chile and Peru, where a large organic input to the sediment impinges on an extensive deoxygenated water layer, a considerable sulfide system is formed (Gallardo, 1977). The gliding bacterium *Thioploca* (Chapter 16,A) which is dominant, forms bundles of filaments that are so large and numerous that they can be seen with the unaided eye; they are well known to the local fishermen. This procaryotic biomass may play a role in the nutrition of the commercial quantities of shrimp that live in this area, which is largely devoid of the usual benthic macrofauna.

The author speculates that the scarcity of bottom fish over areas of Georges Bank where foreign factory ships processed large volumes of fish and discarded organic refuse before the 200-mile limit went into effect in 1977, may possibly have been due, in part, to exclusion by localized areas of sulfide accumulation, rather than to overfishing alone. The rapid improvement in catch in the months following the enforcement of the 200-mile limit tends to support this hypothesis.

5. Bacterial–Protozoan Partnership

The bacterial decomposition of organic debris is really a bacterial–protozoan partnership. It is possible

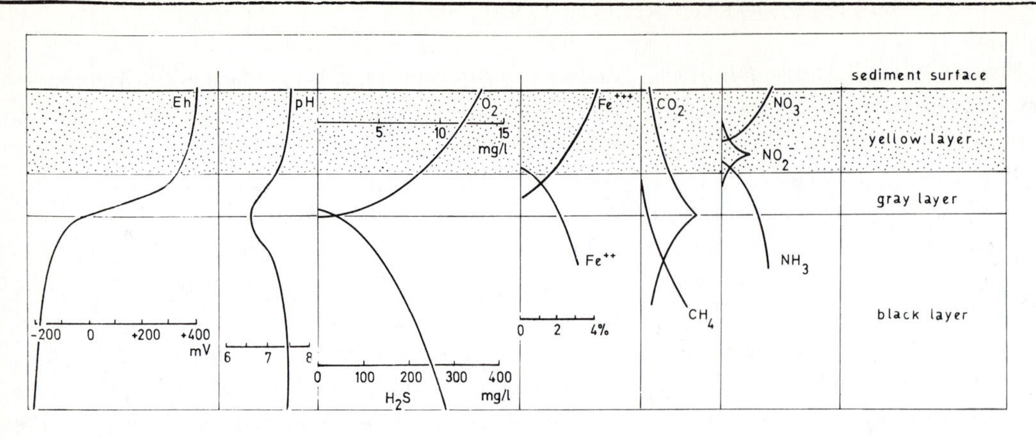
sediment surface
yellow layer
gray layer
black layer
Eh
pH
O2
Fe+++
CO2
NO3-
NO2-
Fe++
NH3
CH4
mg/l
mV
H2S
mg/l
4%

A

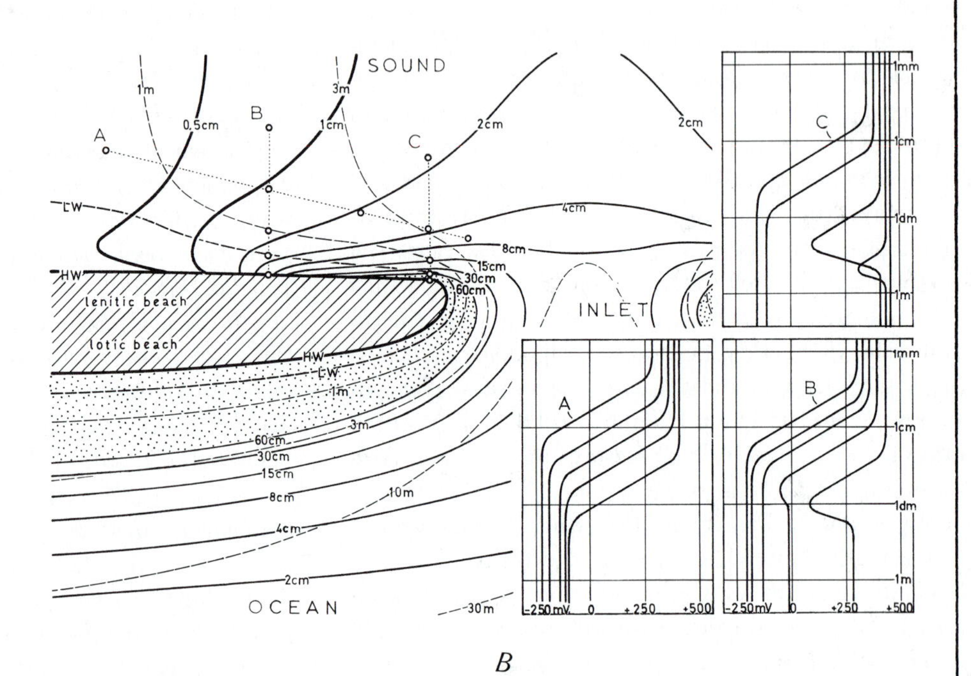
SOUND
INLET
OCEAN
lenitic beach
lotic beach
HW
LW
A
B
C
0.5cm
1cm
2cm
4cm
8cm
15cm
30cm
60cm
1m
3m
10m
30m
1mm
1cm
1dm
-250mV
0
+250
+500

B

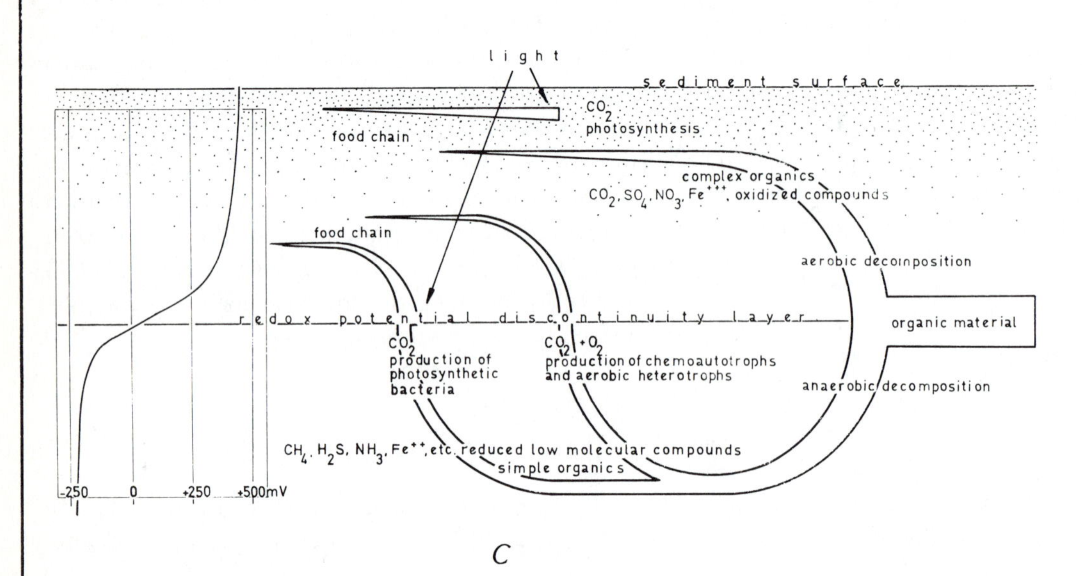
light
sediment surface
CO2
photosynthesis
food chain
complex organics
CO2, SO4, NO3, Fe+++, oxidized compounds
food chain
aerobic decomposition
redox potential discontinuity layer
organic material
CO2
production of photosynthetic bacteria
CO2 + O2
production of chemoautotrophs and aerobic heterotrophs
anaerobic decomposition
CH4, H2S, NH3, Fe++, etc. reduced low molecular compounds
simple organics
-250
0
+250
+500mV

C

PLATE 3-5. The nature, position, and energy flow in sulfide sediments below the oxidized and redox-potential discontinuity layers. (*A*) Schematic representation of profiles of Eh (redox potential), pH, oxygen, hydrogen sulfide, iron, carbon dioxide, methane, ammonia, nitrite, and nitrate through the oxidized, redox-potential discontinuity, and sulfide layers. (*B*) Position of the sulfide sediments (white zone) in relation to land area (cross-hatched) between the ocean and the sound and the high energy surf zone (dotted), which is mixed and irrigated and thus prevents the accumulation and decomposition of the organic matter required for sulfide sediments. (*C*) A schematic representation of energy flow, showing major microbial processes and products produced in the oxidized and reduced zones. All from T.M. Fenchel and R.J. Riedl, 1970, *Mar. Biol.* 7:255–68.

that the apochlorotic flagellates associated with debris act to supplement the osmotrophic activities of bacteria (Odum, 1970). A cultural study has shown that bacteria-eating flagellates are the dominant forms in sediments (Lighthart, 1969). It was possible to detect amoebae only in nearshore samples, but not in those from 800 to 1,500 m, but a direct microscopic examination of sediment from 1200 m in the San Diego trough indicated that naked amoeboid forms were the predominant eucaryotes in the microfauna (Burnett, 1973). The testate forms of amoebae, the benthic foraminifera, are also important. Although selective feeding on a few algal species meets most of their energy requirement (Muller, 1975), bacteria are apparently indispensable for their nutrition (Muller and Lee, 1969). In transects from coastal to deep aphotic waters, the benthic foraminifera became increasingly important with depth, outnumbering the nematodes (Tietjen, 1971), and bacteria and protozoa play an increasingly important role in the nutrition of these benthic foraminifera. In the abyssal communities of the Pacific, benthic foraminifera also appeared to be an important part of the community, equaling the nematodes in abundance, and exceeding the metazoans in quantity (Hessler and Jumars, 1974).

In addition to the dominant agglutinating species of foraminifera, related organisms with anastomosing organic tubules are also present (Hessler, 1974) and may be involved in manganese nodule formation (Greenslate, 1974a). It is of interest that manganese nodules (Horn, 1972) appear to originate in the microcavities of skeletal debris, and especially diatom frustules (Greenslate, 1974b). Bacteria growing on this debris change the oxidation state of iron and manganese, which they precipitate (Ehrlich, 1972), and apparently nurture benthic foraminifera whose tests are incorporated in the nodule (Greenslate, 1974a; Wendt, 1974). The abundance of bacteria-ingesting amoeboid forms in the deep sea indicates that substantial populations of active bacteria are present in abyssal sediments.

6. Benthic Animals

The abundance and nature of deposit feeders on the sea floor (Hessler and Jumars, 1974) are further evidence of the existence of significant populations of active bacteria and protozoa in bottom sediments. With increasing distance from shore, the predominant filter feeders are replaced by debris- and mud-feeders (Zatsepin, 1970). Carnivorous species of starfish are replaced by deposit-feeding species (Carey, 1972). Over a decade ago, it was recognized that animals of the abyssal plain, as characterized by isopods, obtained over 90% of their nutrients from finely divided detritus (Menzies, 1962), and that the nutrient value was not associated with the organic debris but with the bacteria and other microorganisms it contained. Observations on a prosobranch and a bivalve indicated that the populations of these deposit-feeders increased as finer grained deposits did, because of a greater surface area for bacterial growth rather than more accumulation of organic matter (Newell, 1965; Plate 3-3,*A*). The role of bacteria in decomposer food chains and their disappearance from the ingesta as it passes through the gut of deposit-feeders is now well established (Fenchel, 1972). Even in the intertidal zone, deposit-feeding bivalves selectively feed on the top millimeter of the surface and grow better on the material in this layer than on the debris falling from the water column (Hylleberg and Gallucci, 1975). There is a high degree of sorting in the mantle cavity before some 97% of the sediment is rejected as pseudofeces. Because of this sorting process, and because of microbial digestion, the feces of these bivalves are richer than the components of the sediment, and microorganisms are more abundant in the stomach contents than in either the sediment or the feces. This selectivity for food particles is the reason why different species of deposit-feeders can coexist. An amphipod coexisting with a diatom-preferring prosobranch could use not only bacteria on particles in the 4 to 63 μm range, but also bacteria suspended in water with silt and clay particles that were pumped by the animal through its burrow

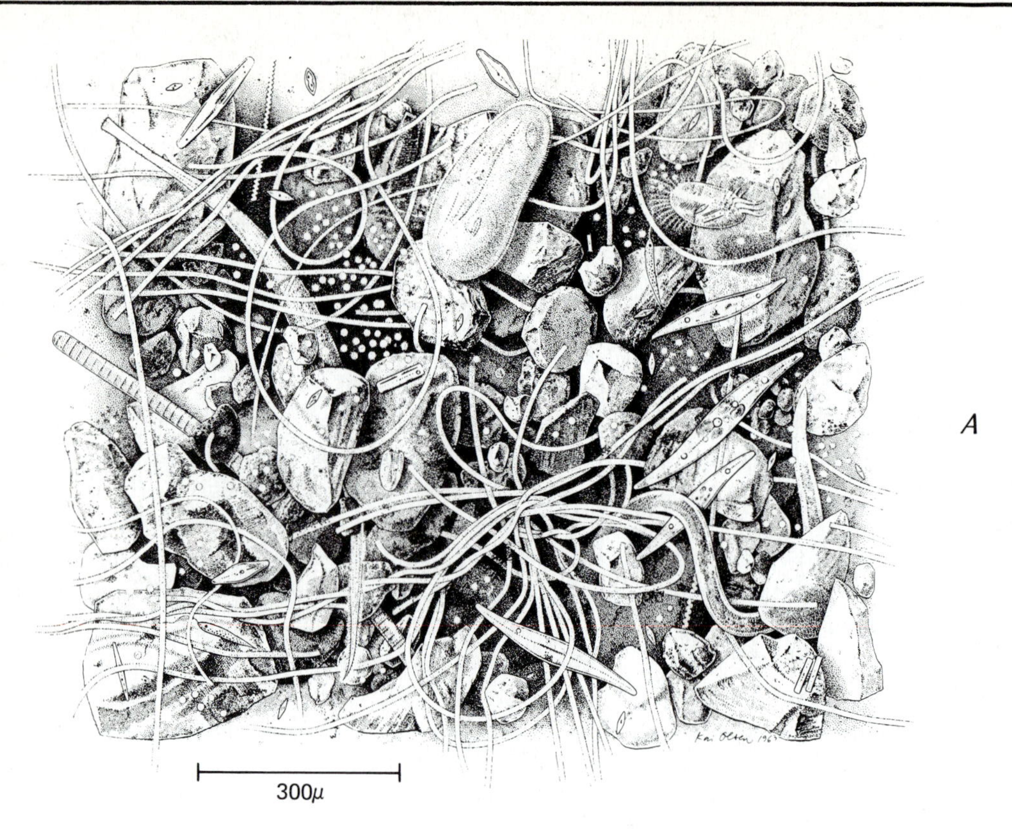

A

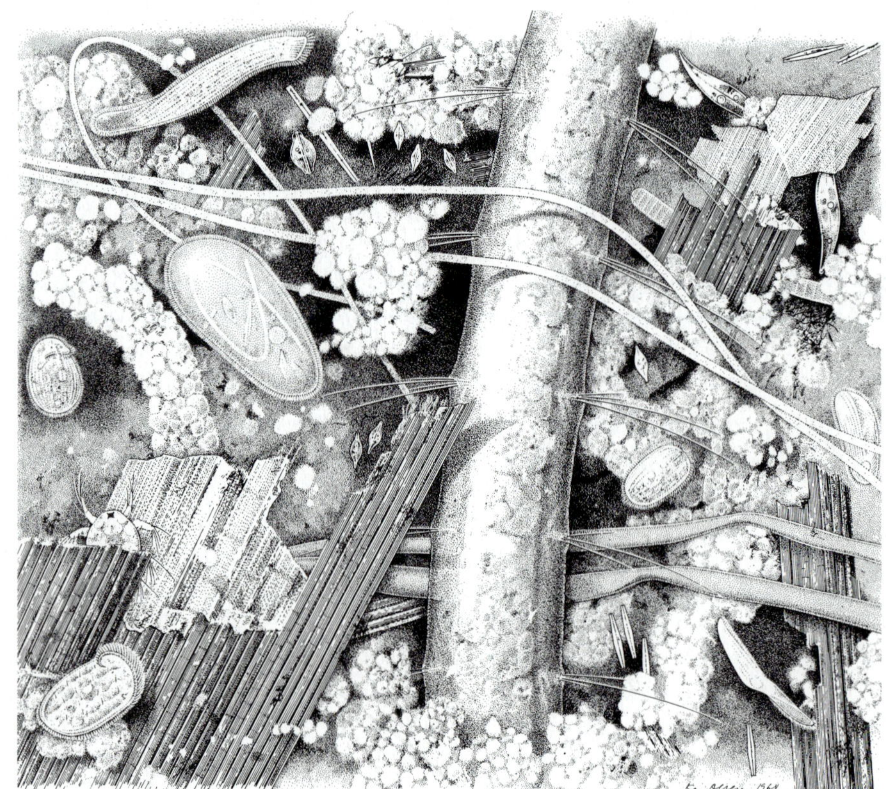

B

PLATE 3-6. Microbial populations of sulfide sediments. (*A*) Patch of sulfur bacteria, containing *Beggiatoa* filaments, the globular *Thiovulum* with an *Oscillatoria* (trichomes), pennate diatoms, euglenids, a nematode, and four species of ciliates. (*B*) Surface of sulfuretum, with granular colonies of the sulfur bacterium *Lamprocystis*, filaments of *Beggiatoa*, an *Oscillatoria*, pennate diatoms, large harpacticoid copepod nauplii, and eight species of ciliates. From T.M. Fenchel, 1969, *Ophelia* 6:1–182.

(Fenchel, Kofoed, and Lappalainen, 1975). Details of the microbial–deposit-feeder relationship and an energy budget have been worked out for a freshwater amphipod. All test algae and bacteria except the cyanobacteria, when added to natural sediments, were digested with more than a 50% efficiency. But sediments, with their naturally associated microorganisms, which were only assimilated with a 6 to 15% efficiency, still supplied 0.0033 to 0.0081 calorie per amphipod hourly, which compared to a total hourly requirement of 0.005 calorie per amphipod (Hargrave, 1970a). The amphipod-stimulated microbial activities, at natural population densities of less than 10% of the microflora production, supplied all the required energy of the deposit-feeders (Hargrave, 1970b). Of the calories assimilated, 49% were used for respiration, 36% were lost as soluble excretory products, while only 15% were retained to support growth, egg production, and molt (Hargrave, 1971).

The growing evidence for an organic debris–microbial economy for deposit-feeders down to the abyssal depths of the sea does not support the conclusions drawn by some bacteriologists that bacterial activity on the sea floor is extremely slow (Jannasch et al., 1971; Jannasch and Wirsen, 1973; Seki et al., 1974; Schwarz and Colwell, 1975; Wirsen and Jannasch, 1975). It is difficult to believe that bacteria, which are highly adaptable, have not adjusted to the food, temperatures, and pressures in the deep sea, whereas animals have (Hochachka, 1974). The bacterial populations that are cultured in bottles on soluble substrates must vary significantly from those that decompose particulate organic debris in an open system *in situ*.

Bacteria that decompose organic debris on the sea floor release dissolved organic matter into the sediments (Degens, Reuter, and Shaw, 1964). In addition to the dissolved organic carbon that is used by the bacteria (Litchfield et al., 1974), it is possible that some soluble organic matter is also used directly by some benthic animals. There is a growing literature on the uptake of soluble organic matter by pogonophores, polychaetes, and echinoderms, among others (Southward and Southward, 1970, 1972a,b; Ferguson, 1971). But uptake of dissolved organic carbon by an epibiota (Sieburth, 1975) rather than by the epidermal cells of the host is a possibility. Further work is needed.

C. ENDOBIOTIC HABITAT

The endobiotic environment consists of the cells of microorganisms as well as the tissues of plants and animals that serve as hosts to a spectrum of microbial forms in a variety of symbiotic (living together) associations. These relationships include mutualism (organisms of different species live together for the benefit of both), parasitism (only the parasite derives nourishment from the host, but does not necessarily cause disease), and pathogenesis (the pathogen can cause a disease in the host). Mutualistic symbiosis and disease are very important factors that affect the ecology of both microorganisms and macroorganisms in the sea. Host-dependent microorganisms, however, are beyond the scope of this text, which concerns free-living forms only. But the occasional mutualistic symbiont, parasite, or pathogen will be discussed, when appropriate, in the following chapters. A few examples and references are given here to characterize this specialized habitat for microorganisms in the sea. The four main areas of study to date have been plant pathogens, the mutualistic phototrophic symbionts of protozoa and invertebrate animals, animal pathogens, and bacteria as endosymbionts and hosts.

1. Plant Pathogens

Diseases of the phototrophic microorganisms and macroscopic plants responsible for primary production in the sea can be locally important in the ecology not only of the infected populations but of those organisms that live upon them. The prasinophycean *Halosphaera viridis* is a unicellular green alga that commonly occurs in the plankton of the Mediterranean Sea. During the summer of 1967, the *Halosphaera* in the vicinity of Villefranche-sur-Mer, France, were infected by a peridinean dinoflagellate described as *Myxodinium pipiens* by Cachon, Cachon, and Bouquaheux (1969; Plate 3-7,*A*). This bubble-like parasite remains on the outside of the alga and completely devours its contents. A more transient infection, which affected only a single diatom species, *Coscinodiscus centralis*, in a bloom in Hecate Strait, British Columbia, was caused by a phycomycetous fungus that could not be detected over a wider area 3 weeks later (Parsons, 1962). A similar phycomycete, *Lagenisma coscinodisci*, which affects *Coscinodiscus concinnus*, is shown in Plate 3-7,*C*. Other species of phycomycetes and their phytoplankton hosts are discussed in Chapter 21. The diatom *Skeletonema costatum*, which is a dominant species in temperate coastal waters, is reportedly lysed by a pseudomonad in the indigenous microbial population (Mitchell, 1971).

When seaweeds are examined closely, they are often found to have galls. A gall that typically occurs on the pelagic brown seaweed *Sargassum natans* is caused by the ascomycetous fungus *Haloguignardia oceanica*, described by Kohlmeyer (1972) (Plate 3-7,*D*,*E*). Other galls are less noticeable but quite regular in occurrence, especially on such red seaweeds as *Chondrus crispus*. These galls are apparently caused by a large unclassified oval bacterium (Lagerheim, 1901; Cantacuzene, 1930), the latter investigator having conducted a very thorough study employing Koch's postulates to verify the etiology. A number of fungi associated with marine plants are listed by Johnson and Sparrow (1961). They range from saprophytes to obligate parasites. Some appear to be weak parasites, growing on living tissues without marked pathogenic effects. Some species of fungi associated with the turtlegrass *Thalassia testudinum* were only seen on necrotic leaves, whereas others were present on both healthy and necrotic leaves (Meyers et al., 1965); this also is seen to occur on the cordgrass *Spartina alterniflora* (Gessner, Goos, and Sieburth, 1972). Spores of the brown seaweed *Chorda tomentosa* cultured in unsterilized seawater underwent severe lysis, failed to germinate or develop cell walls, and were found to contain virus-like particles (Plate 3-7,*C*), whereas spores cultured in sterile seawater developed normally and were free of these particles (Toth and Wilce, 1972). Other virus-like particle associations with algae are discussed by Chapman and Lang (1973). The seaweeds can also house endosymbionts. A classic example is the occurrence of the fungus *Mycosphaera ascophylli* in the brown seaweed *Ascophyllum nodosum* (see Chapter 19).

PLATE 3-7. Infections of algal cells, spores, and tissues. (*A*) The prasinophycean alga *Halosphaera viridis* parasitized by a dinoflagellate, *Myxodinium pipiens* (1,400 X). (*B*) The centric diatom *Coscinodiscus concinnus* infected by the phycomycetous fungus *Lagenisma coscinodisci*. (*C*) A thin section of a spore of the brown seaweed *Chorda tomentosa* undergoing lysis; it contains virus-like particles. The mitochondria (M) are swollen, and the vesicles (V) are irregularly shaped; (S) is presumed to be storage carbohydrate. (*D*) The pelagic brown seaweed *Sargassum natans*, with galls caused by the ascomycetous fungus *Haloguignardia oceanica*. (*E*) Longitudinal section of a gall of *Sargassum natans*, with a large ascocarp of *Haloguignardia oceanica* and an open acervulus of a deuteromycetous fungus *Sphaceloma cecidii*, top. (*A*) from J. Cachon, M. Cachon, and F. Bouquaheux, 1969, *Phycologia* 8:157–64; (*B*) from G. Drebes, 1974, *Marine Phytoplankton*, G. Thieme Verlag, Stuttgart; (*C*) from R. Toth and R.T. Wilce, 1972, *J. Phycol.* 8:126–30; (*D*,*E*) from J. Kohlmeyer, 1972, *J. Elisha Mitchell Scientific Soc.* 88:255–59.

Information on the diseases of seaweeds and phytoplankton is widely scattered in the literature and has usually been obtained incidentally during phycological, mycological, and ecological studies. To focus attention on this long neglected area of marine biology, Andrews (1976) has reviewed the diseases of marine algae, including the non-infectious diseases caused by natural and man-made changes in the environment.

2. Phototrophic Symbionts

The phototrophic microalgae and photobacteria that live symbiotically with their protozoan and coelenterate hosts form a very interesting and challenging microcosm in the marine environment. Droop (1963) has reviewed these symbioses and compiled an impressive list of hosts. The symbiotic phototrophs classically have been divided into three groups: the golden-brown zooxanthellae containing the dinoflagellates (Dinophyceae), the greenish zoochlorellae containing species from both the Chlorophyceae and the Prasinophyceae, and the bluish-green Cyanellae containing the cyanobacteria. In addition, a few diatoms can shed their frustules and become endosymbionts. The phototrophic symbionts are usually associated with the digestive cells of their hosts. Two texts deal with general symbiotic associations (Henry, 1966; Gotto, 1969). The phototrophic symbionts of invertebrates have been reviewed by Taylor (1973), and those of the foraminifera are discussed by Lee (1974).

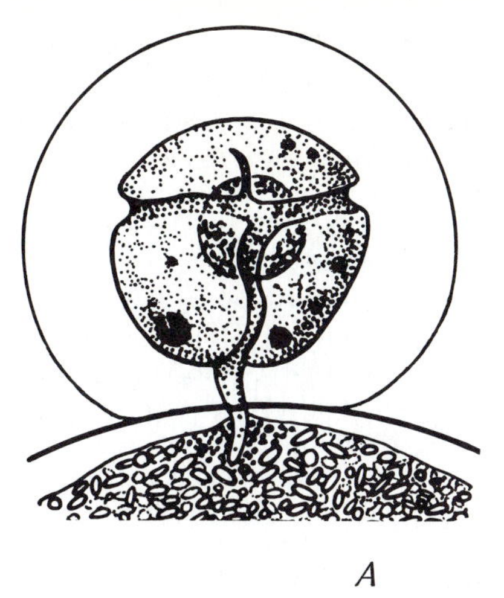

A

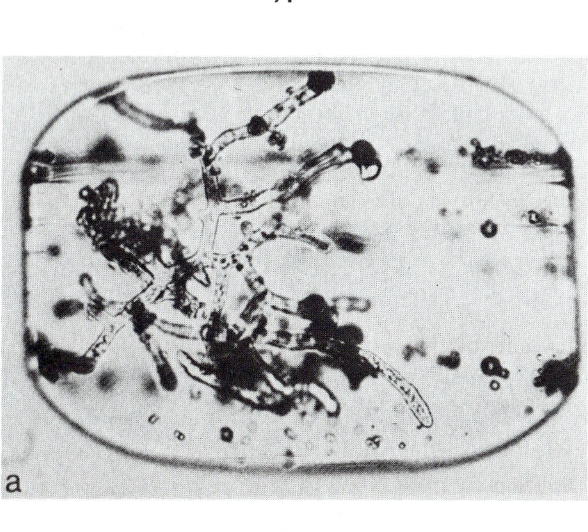
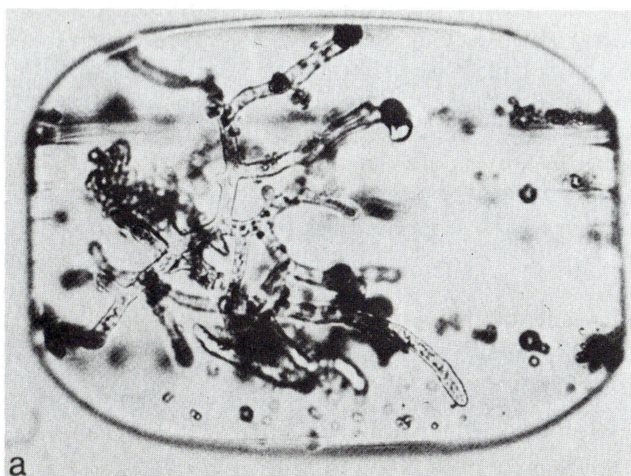

B

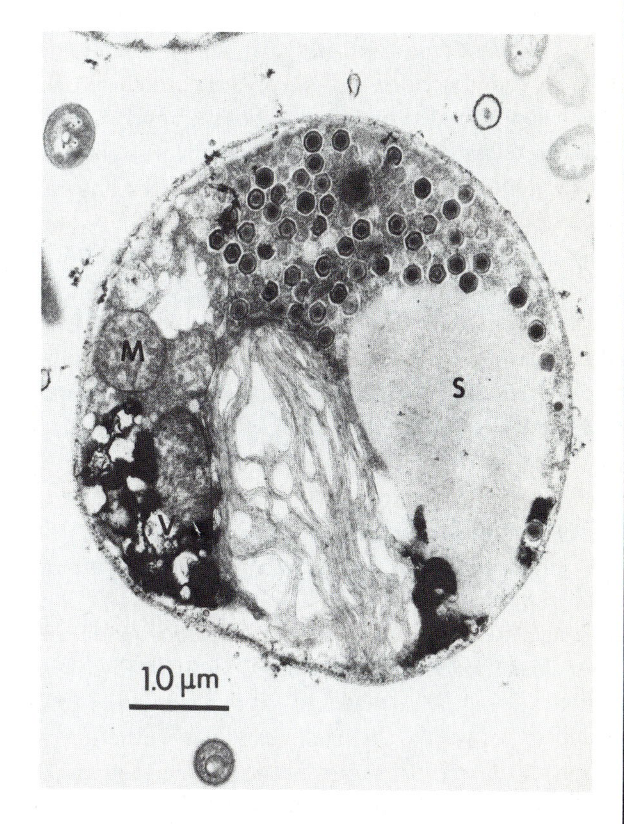

C

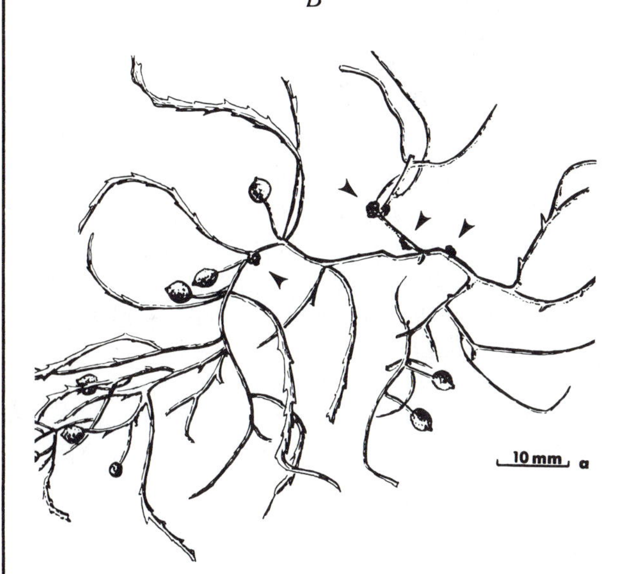

D

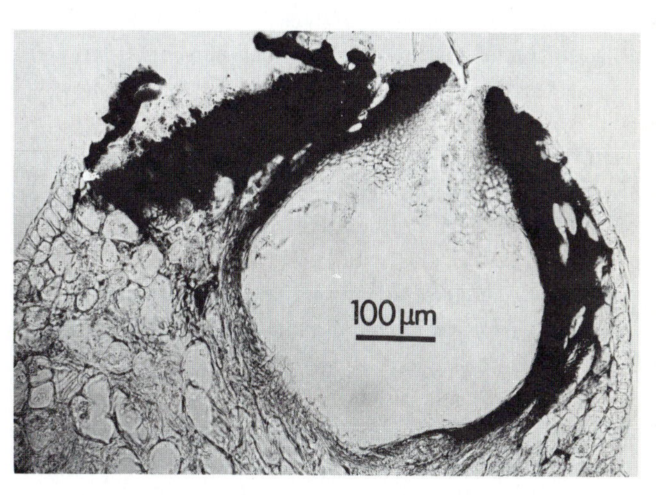

E

A major breakthrough in identifying the zooxanthellae (the main group of symbionts) as dinoflagellates was the culture of the symbionts of the scyphozoan jellyfish *Cassiopeia* and the sea anemone *Condylactis* (McLaughlin and Zahl, 1957). The *Cassiopeia* isolate was named *Symbiodinium microadriaticum* by Freudenthal (1962), who described its morphology and life cycle. The taxonomic description has been emended on the basis of ultrastructural studies, and the zooxanthellae from six species of tridacnid clams have been shown to be identical (Kevin et al., 1969; Taylor, 1969). A previously described zooxanthella from the "by-the-wind sailor" *Velella velella*, *Endodinium chattoni*, can be distinguished from the above symbiont (Taylor, 1971). Another *Endodinium*, *E. nutricola*, is symbiotic in the sphaerocollid radiolarians (Hollande and Carré, 1974). Continuing studies are needed to identify and catalog the zooxanthellae in different hosts including the unidentified zooxanthella in the sea anemone studied by Trench (1971a) (see Plate 3-8).

There are a number of well-described symbionts from a variety of microalgal classes. The free-living dinoflagellate *Amphidinium klebsii* isolated from the flatworm *Amphiscolops langerhansi* by Taylor (1971) is shown in Plates 3-8,*A*,*B*. A model of similar symbiosis in the turbellarian (worm) *Convoluta roscoffensis* with the prasinophycean alga *Platymonas convolutae* has been studied with homologous and heterologous species (Provasoli, Yamasu, and Manton, 1968). The symbiosis of another turbellarian, *Convoluta convoluta*, with the pennate diatom *Licmophora* is most interesting (Apelt, 1969). The newly hatched *C. convoluta*, which are free of symbionts, acquire the microorganisms that become symbionts from their diatom and algal spore diet. *Licmophora hyalina* and *Licmophora communis* fed to the newly hatched host species slip out of their silica shells to produce high population densities in the peripheral parenchyma. The algal symbiont in the foraminifera *Heterostegina depressa* is also a diatom (Schmaljohann and Röttger, 1976), whereas the symbiont of *Archaias angulatus* is the chlorophyte *Chlamydomonas hedleyi* (Lee et al., 1974).

A number of relatively unmodified endosymbiotic phototrophs are readily isolated and can grow without their hosts. There are, however, two examples of highly modified endosymbionts. The best studied is the photosynthetic eucaryote within the planktonic ciliate *Mesodinium rubrum*, which often forms large, nontoxic red tides; its occurrence has been reported since the times of Leeuwenhoek and Darwin. Taylor, Blackbourn, and Blackbourn (1971) reported that this widely distributed ciliate has an "incomplete symbiont," which consists of numerous functional chloroplasts associated with distinctive, non-ciliate mitochondria within a single membrane but which apparently lacks a nucleus, endoplasmic reticulum, and a Golgi body. Although the pigments are characteristic of cryptomonads, the lamellae are not. A further study of preserved material by Hibberd (1977) confirmed that the endosymbiont is surrounded by a single membrane, but indicated that it was not as "incomplete" as was previously thought, since there is a large nucleus, an endoplasmic reticulum, and even a Golgi body at certain stages in the life cycle. Further ultrastructural features confirmed that the single symbiont has cryptomonad affinities and that the symbiosis is permanent and obligate for both the symbiont and the host. The only other example of a highly modified symbiont surrounded by a single membrane is of that occurring in the dinoflagellate *Peridinium balticum*, which has a eucaryotic nucleus (Tomas, Cox, and Steidinger, 1973; Tomas and Cox, 1973).

The value of these symbionts to their hosts range from minimal to maximal, depending, in part, upon the observer. Beyers (1966) concluded from a study of the

PLATE 3-8. Algal symbionts in marine invertebrates. (*A*) Optical micrograph, transverse section, of the flatworm *Amphiscolops langerhansi*, with a single layer of the free-living dinoflagellate *Amphidinium klebsii* as an endosymbiont. Algae (a), dorsal surface (d), ventral surface (v), peripheral parenchyma (pp), central parenchyma (p), and epithelium (e) (75 X). (*B*) Reconstruction of the above association based on optical and electron micrographs: cilia (c), cilia rootlets (r), mitochondrion (m), lipid (l), nucleus (n), endoplasmic reticulum (er), and longitudinal muscle (lm) and circular muscle (cm) of the host and the flagellum (f), chloroplast (cp), pyrenoid (p), starch (s), golgi body (g), pusule (pu), and accumulation body (ac) of the dinoflagellate (6000 X). (*C*) Thick-section optical micrograph through the body wall of the sea anemone *Anthopleura elegantissima*, showing the relative positions of the epidermis (epi), thin mesogloea (m), and the endoderm (en) containing the zooxanthellae (zx). (*D*) Transmission electron micrograph of a thin section of the above association in the tentacle, showing a detail of the zooxanthella: assimilation product (as), chromosome (ch), chloroplast (cp), mitochondrion (m), nucleus (n), periplast membrane (p), and pyrenoid (py) (13,000 X). (*A*,*B*) from D.L. Taylor, 1971, *J. Mar. Biol. Ass. UK* 51:301–13; (*C*,*D*) from R.K. Trench, 1971a, *Proc. Roy. Soc. London* Ser. B 177:225–35.

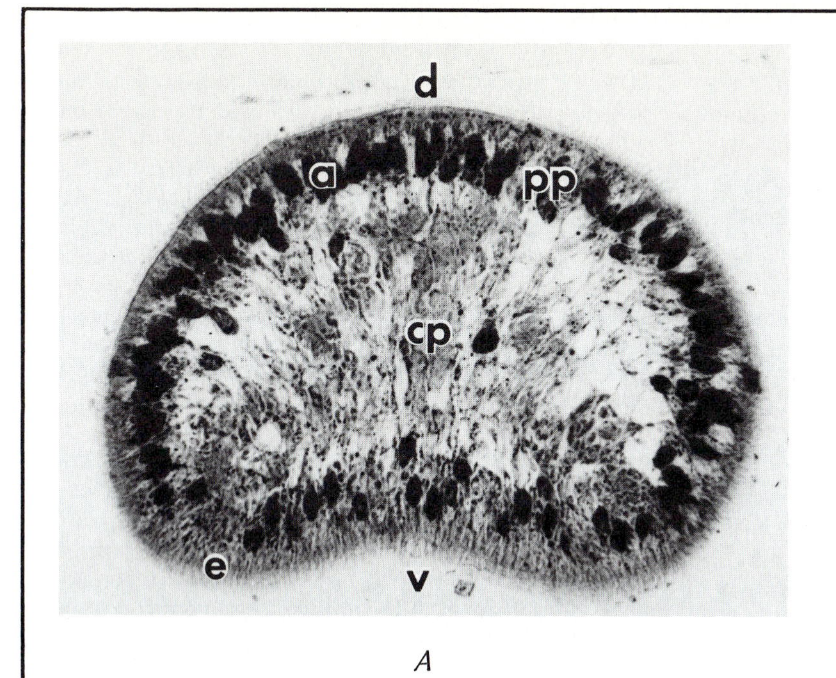
d
a
pp
cp
e
v

A

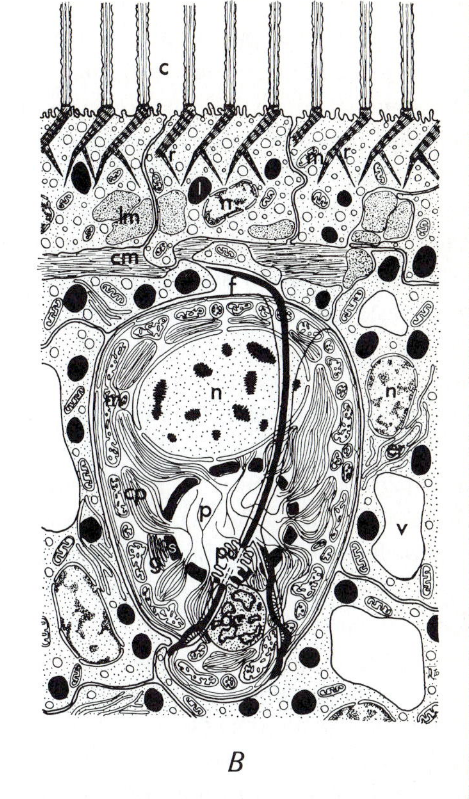
c
lm
cm
f
n
n
er
cp
p
v

B

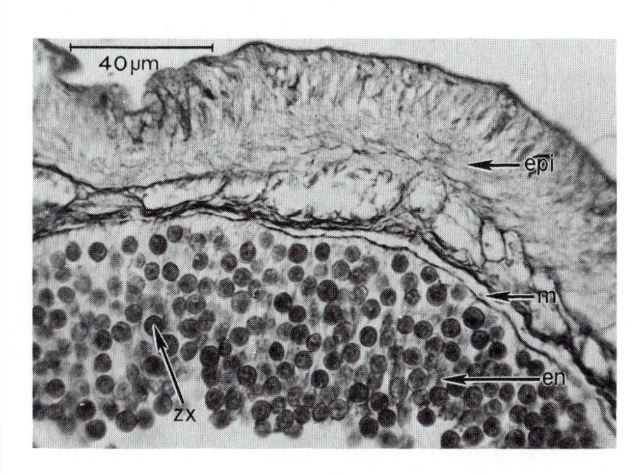
40μm
epi
m
en
zx

C

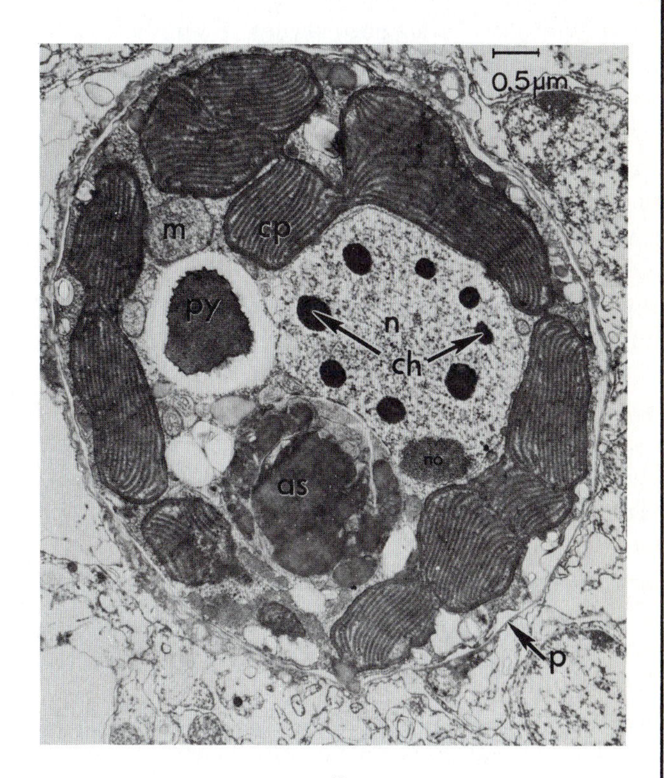
0.5μm
m
cp
py
n
ch
no
as
p

D

diel pattern of net photosynthesis and night respiration of a green hydra and of small coral heads that these plant–animal symbioses had the same plant–animal relationships found in larger ecosystems. Lewis and Smith (1971) have reviewed the literature on the movement of photosynthetic products (photosynthate) between symbionts and studied the effect of certain primary products of photosynthesis on increasing the release rate of these products. Trench (1971a) has shown that 25 to 50% of the photosynthate of the zooxanthellae, a symbiont of the sea anemone *Anthopleura elegantissima* (Plate 3-8,*C*,*D*), is incorporated into the host animal's tissue. The products of algal release were determined by Trench (1971b) and production was shown to be stimulated by the host's tissues (Trench, 1971c). On prolonged isolation, the zooxanthellae produced progessively less photosynthate, and although production was restored to normal levels upon the return of the symbiont to the host, the qualitative and quantitative distribution was quite different.

A phenomenon distinct from phototrophic endosymbiosis but apparently serving the same ecological purpose is the maintenance of chloroplasts from whole algal cells ingested by the host, a kind of temporary "algal symbiosis." This phenomenon in opisthobranch molluscs has been discussed for various species by Taylor (1970), Greene (1970), and Trench (1975). One example is given in the fascinating three-paper series of Trench, Boyle, and Smith (1973a,b, 1974), which showed that the chloroplasts from the siphonaceous green alga *Codium fragile* remain robust and continue functioning with a selective release of photosynthate after ingestion by the mollusc *Elysia viridis*. A similar retention of chloroplasts also occurs in planktonic ciliates. Two unidentified species similar to those in the genera *Phorodon* and *Strombidium* had intact and apparently functional chloroplasts, whose thylakoids and pyrenoids were similar to those of the Dinophyceae, Haptophyceae, and Chrysophyceae (Blackbourn, Taylor and Blackbourn, 1973).

3. Animal Pathogens

The diseases of marine animals comprise a study unto itself. This broad area has been reviewed by Sindermann (1970). Other sources of more selective information are the symposium on fish diseases edited by Mawdesley-Thomas (1972) and the review on oyster diseases by Sprague (1971). Such works indicate that all manner of microorganisms have become pathogenic and infest a great variety of animal life in the sea. A few examples are illustrated in Plate 3-9 and are briefly discussed to show the diversity of pathogens and hosts. The most poorly represented group among the infectious microorganisms is the microalgae. The epizoic diatoms already mentioned and further discussed at the end of Chapter 10 may kill animals under conditions of extreme stress. Some dinoflagellates have become apochlorotic and, in addition to being algal pathogens as previously discussed, also attack animal hosts. An example is the parasitic dinoflagellate *Dissodinium pseudocalani*, which lives on the eggs of the copepod *Pseudocalanus elongatus* (Drebes, 1969; Plate 3-9,*D*).

Plate 3-9. Infections of animal tissues. (*A*) Blood smear from the lobster *Homarus americanus* infected with the micrococcus *Aerococcus viridans* (*Gaffkya homari*), which causes a septicemia with a high mortality rate (890 X). (*B*) Three ova of the pea-crab *Pinnotheres pisum* attached to the setae of a pleopod; all are in different stages of infection by the phycomycetous fungus *Leptolegnia marina*. (*C*) The polychaete *Ophryotrocha puerilus*, with its body cavity infected with the oocysts of the sporozoan parasite *Eucoccidium ophryotrochae* (magnification not given). (*D*) The egg mass of the copepod *Pseudocalanus elongatus*, with two eggs infected with the ectoparasitic dinoflagellate *Dissodinium pseudocalani*. (*E*) Chlamydia (Rickettsia) from a hard-shelled clam inclusion, showing all the developmental stages including the large initial body (1), the paired daughter cells (1D), the intermediate stages (2E,2M,2L), and a small, fully condensed elementary body (3). (*A*) from D.G. Wilder and D.W. McLeese, 1961, *J. Fish. Res. Bd. Canada* 18:367–75; (*B*) from D. Atkins, 1954, *J. Mar. Biol. Ass. UK* 33:613–25; (*C*) from K.G. Grell, 1960, *Naturwissenschaften* 47:47–48; (*D*) from G. Drebes, 1969, *Helgoländer wiss. Meeresunters.* 19:58–67; (*E*) from J.C. Harshbarger, S.C. Chang, and S.V. Otto, 1977, *Science* 196:666–68, copyright 1977 American Association for Advancement of Science.

a. Bacteria. The bacteria are probably the best represented group among the infectious microorganisms. Severe localized outbreaks of contact dermatitis among swimmers in Oahu, Hawaii have been caused by a toxic substance produced by the cyanobacterium *Lyngbya majuscula* (Moikeha, Chu, and Berger, 1971). Another skin infection of man, a chronic granuloma that occasionally resembles a sporotrichosis infection, is caused by an acid-fast bacterium, *Mycobacterium marinum* (Silcox and

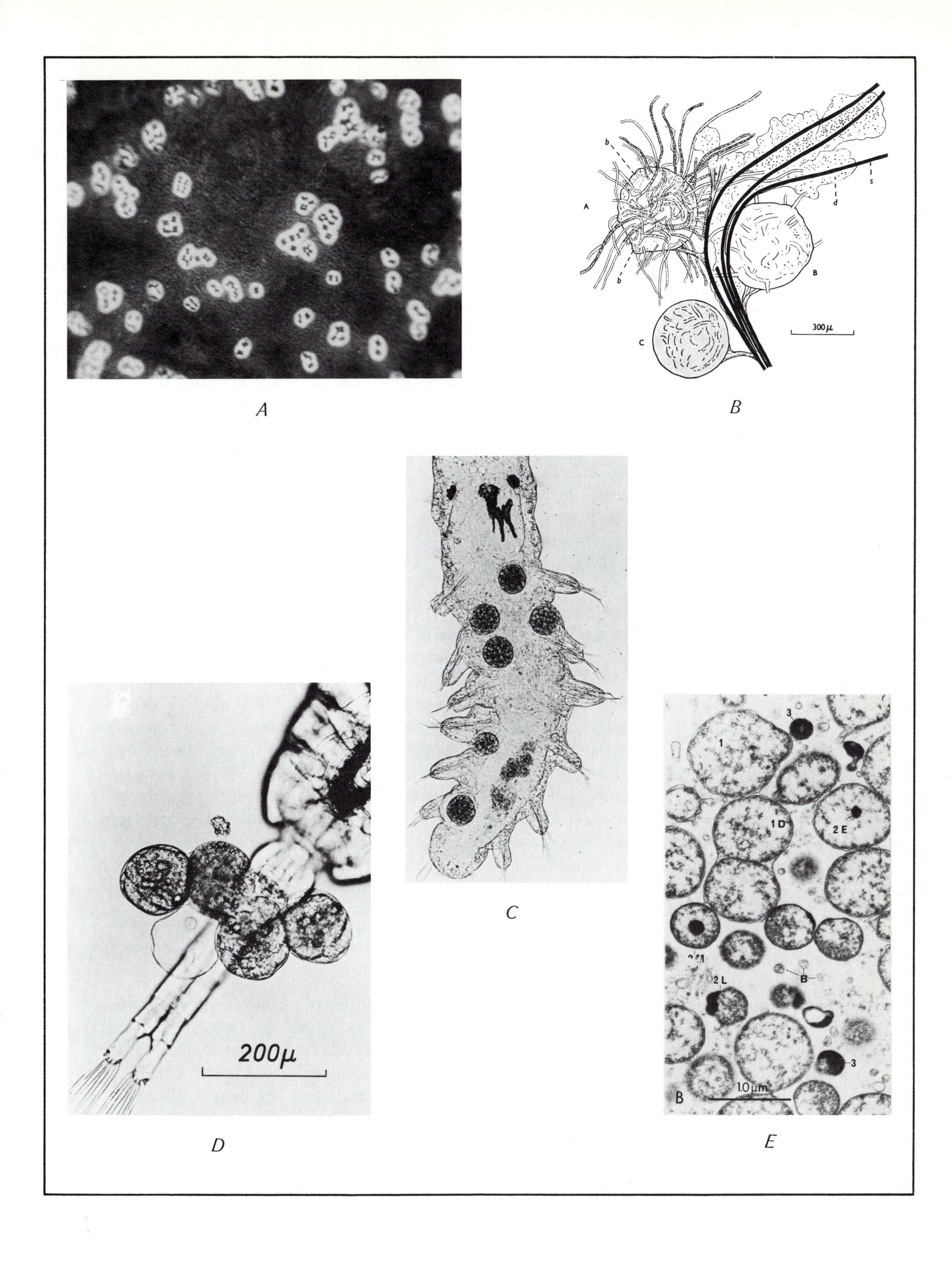

b
A
s
d
b
B
C
300μ
200μ
1
3
1 D
2 E
2 L
B
1.0 μm
A
B
C
D
E

David, 1971). The causative agent was originally described from dead tropical marine fish in Philadelphia (Aronson, 1926) and is now recognized as an occupational hazard of aquarists (Adams et al., 1970). Other bacterial pathogens of marine origin can also affect man and other mammals. *Vibrio (Beneckea) parahaemolytica* (Baumann, Baumann, and Reichelt, 1973), which was originally obtained from an outbreak of food poisoning in Japan caused by the fish dish "shirasu," is now known to be widespread in temperate waters. It is apparently harbored as an epibiotic microorganism on zooplankton (Kaneko and Colwell, 1973) and can cause morbidity in the blue crab *Callinectes sapidus* (Krantz, Colwell, and Lovelace, 1969). A nephritic disease of the California sea lion *Zalophus californianus* causes fever, paralysis of the hindlimbs, thirst, and eventual death. The isolation of *Leptospira*, the clinical picture, and the high titers to four species of *Leptospira* indicated a leptospiral etiology in affected animals (Vedros et al., 1971). The same species of sea lion also has a high incidence of abortion, caused by a virus indistinguishable from the virus of vesicular exanthema of swine, and these sea lions may serve as a reservoir of this infection for California swine herds raised along the coast and exposed to aborted sea lion pups (Smith et al., 1973).

A septicemia of the lobster *Homarus americanus* (Plate 3-9,*A*) is an example of how the incidence of a disease can be increased by crowding and by the practices of man. Heavy mortalities in Maine lobster pounds attracted attention to the pathogenic *Aerococcus viridans* (*Gaffkya homari*) (Snieszko and Taylor, 1947). The immobilization of lobster claws to prevent cannibalism has traditionally been performed by inserting plugs between the claws. This disrupts the integument and allows entry by the pathogen. A practice that has come into general use in recent years is claw immobilization with bands, which avoids this injury and lowers the incidence of fatal aerococcus septicemia (Wilder and McLeese, 1961). A few species of pseudomonads and vibrios, which are among the more common genera of bacteria in seawater (Chapter 15), have been incriminated in a number of infections. The mortality of the larvae of the hard-shelled clam *Mercenaria mercenaria* in culture was caused by the growth of either a pseudomonad or a vibrio isolated from moribund larvae by Guillard (1959). Mortality could be controlled by antibiotic treatment. A gaping disease of the oyster *Ostrea gigas* was attributed to *Pseudomonas enalia* by Colwell and Sparks (1967). Another pseudomonad, *P. (Flavobacterium) piscicida*, is implicated in the mortality of fish associated with red tides (Meyers et al., 1959). A number of bacteria, including pseudomonads and vibrios, apparently attack and break down the softer parts of nematodes in culture (Hopper and Meyers, 1966). A freshly caught plaice with an infective lesion, extensive necrosis, cavity formation, and a pustular exudate also yielded a pseudomonad (Hodgkiss and Shewan, 1950). Another pseudomonad was implicated in the tail-rot of the codfish *Gadus callarias* (Oppenheimer, 1958). *Vibrio anguillarum* and its synonyms (Hendrie, Hodgkiss, and Shewan, 1971) cause several diseases in a number of marine fish; these include juvenile mortality in the Chinook salmon *Oncorhynchus tshawytscha* (Cisar and Fryer, 1969) and a fin disease in the winter flounder *Pseudopleuronectes americanus* (Levin, Wolke, and Cabelli, 1972). Vibriosis due to *V. anguillarum* is seen in coalfish, cod, flounder, plaice, and salmon in Norwegian waters (Haastein and Holt, 1972).

The Chlamydiae (Rickettsia) are gram-negative, coccoid, pathogenic bacteria that multiply only within the cytoplasm of host cells, have a unique developmental cycle, and cause a variety of diseases, ranging from psittacosis to lymphogranuloma venereum, in man, other vertebrates, and birds. Until recently, only vertebrate hosts were known, but chlamydia-like microorganisms have been found in inclusions from the quahaug or hard-shelled clam *Mercenaria mercenaria* (Harshbarger, Chang and Otto, 1977). All the developmental stages are shown in Plate 3-9,*E*. In some of these initial bodies of the chlamydia-like forms, icosahedral virus particles (bacteriophage) in crystalline arrays could be seen. If these chlamydia-like forms are transmissible to man, then the gourmet treat of raw quahaugs on the half shell, a practice passed down from the native New Englanders, will be threatened.

b. Fungi. There are numerous fungal pathogens that attack shellfish, crustaceans, and fish, among other marine animals. Most of the reports concern cultured populations or economically important species. Oysters in the Gulf and the waters off the southeastern states are affected by a disease caused by the phycomycete *Dermocystidium marinum* (Mackin, 1961; Chapter 21). It is somewhat ironic that a predator of the oyster, *Urosalpinx cinerea*, also is susceptible to infection by the phycomycete

Haliphthoros milfordensis (Vishniac, 1958). Another phycomycete, *Leptolegnia marina*, which attacks the eggs of the gravid pea-crab *Pinnotheres pisum* (Atkins, 1954) is shown in Plate 3-9,*B*. Problems encountered with fungal pathogens during crustacean culture have been discussed by Anderson and Conroy (1968). A fungus of doubtful position, *Ichthyophonus hoferi*, causes a systemic infection in the sea herring *Clupea harengus* (Daniel, 1933a,b) and is one of the most serious fungal pathogens for a number of freshwater and marine species including the mackerel (Sindermann, 1958, 1963).

c. Protozoa. The protozoan parasites of marine animals belong mainly to two parasite classes, the Sporozoa and the Cnidospora, which are not fully discussed in this text. The Sporozoa are characterized by spores or cysts that lack polar filaments and contain one or more sporozoites. The Sporozoa, unlike most protozoa, have a single nucleus and only the microgametes of some groups have flagella. These protozoans show an alternation of generations, with sexual reproduction by gamogony and asexual reproduction by multiple fission (sporogamy). A further asexual multiple fission (schizogony) occurs in some Sporozoa and can disrupt the alternation of generations. A sporozoan infection of a polychaete described by Grell (1960, 1973) is shown in Plate 3-9,*C*. The Cnidospora are distinguished from the Sporozoa by spores or cysts that contain one or more polar filaments coiled up inside the spore; they are extruded for transmission to a host. These spores contain one or several amoeboid cells (amoebula) rather than the sporozoites of the Sporozoa. There are two groups within the Cnidospora: the Myxosporidia and the Microsporidia. The former are mainly fish parasites that inhabit the intercellular spaces of such hollow organs as the gallbladder or urinary bladder, and such tissues as those of the gills, muscles, liver, kidney, and spleen. The Microsporidia, by contrast, are much smaller intracellular parasites of fish and invertebrates. The amoebula penetrate the host cells to reproduce by repeated binary or multiple fission, as in the microsporidian infection of the nematode *Metoncholaimus scissus* (Hopper, Meyers, and Cefalu, 1970).

d. Virus. The first isolation of a fish virus was from infectious pancreatic necrosis in the eastern brook trout *Salvelinus fontinalis* (Wolf et al., 1960). The viral origin of a disease in the sockeye salmon *Oncorhynchus nerka* was reported by Rucker et al. (1953) in guarded terms, although it has now been cultured (Wingfield, Fryer, and Pilcher, 1969). The methodology of plaquing fish viruses is described by Wolf and Quimby (1973), but despite the success in growing a number of fish viruses, other viral-like diseases, such as ulcerative dermal necrosis of salmonids, remain an enigma (Roberts et al., 1970). Snieszko (1974) has written a very timely review on the effects of environmental stress on the outbreak of infectious diseases in fish including those of viral etiology.

4. Bacteria as Endosymbionts and Hosts

Endosymbiotic bacteria are apparently quite common, and their occurrence in both phototrophic and heterotrophic marine dinoflagellates are reviewed by Gold and Pollingher (1971). The symbiotic and parasitic bacteria of protozoa in general are reviewed by Preer, Preer, and Jurand (1974). A number of other examples of bacterial–protozoan associations are pointed out as they are encountered in the microorganisms discussed in later chapters.

Marine bacteria also are hosts to infectious agents. Early investigators of bacteriophage in the sea used terrestrial isolates for test microorganisms, but Spencer (1955, 1960) demonstrated that bacteria indigenous to inshore and offshore waters are also attacked by bacteriophage from marine waters. Studies on the bacteriophages of marine bacteria have been summarized by Ahrens (1971) and Zachary (1974), who studied the distribution and abundance of phages for agrobacteria and *Beneckea natriegens*, respectively. They concluded that such phage were abundant and widely distributed in the Baltic Sea and the coastal marshes of the United States. The phage that attack the blue-green photobacteria are discussed by Safferman (1973). The fairly common occurrence of several types of bacteriophage in negatively stained preparations of natural populations of bacterioplankton from Narragansett Bay, Rhode Island (Plate 3-10), the continental shelf, and the open sea further attest to the widespread occurrence of bacteriophage and lysogeny in natural populations. Studies on the seasonal incidence of bacteriophage for the main genera of epibacteria (Chapters 15 and 16), made in conjunction with

PLATE 3-10. The readily detectable occurrence of bacteriophage with and on natural populations of bacterioplankton from Narragansett Bay, R.I., as observed by electron microscopy. (*A*) Free phage particle (type A), with a contractile tail. (*B*) Phage particle (type B) attached to a bacterial cell, with a long noncontractile tail (*A* and *B*, negatively stained preparations). (*C*) Thin section of a phage, with a contracted tail after its DNA had been injected into the host cell. (*D*) Thin section of phage particles inside a bacterial cell. Electron micrographs courtesy of Paul W. Johnson.

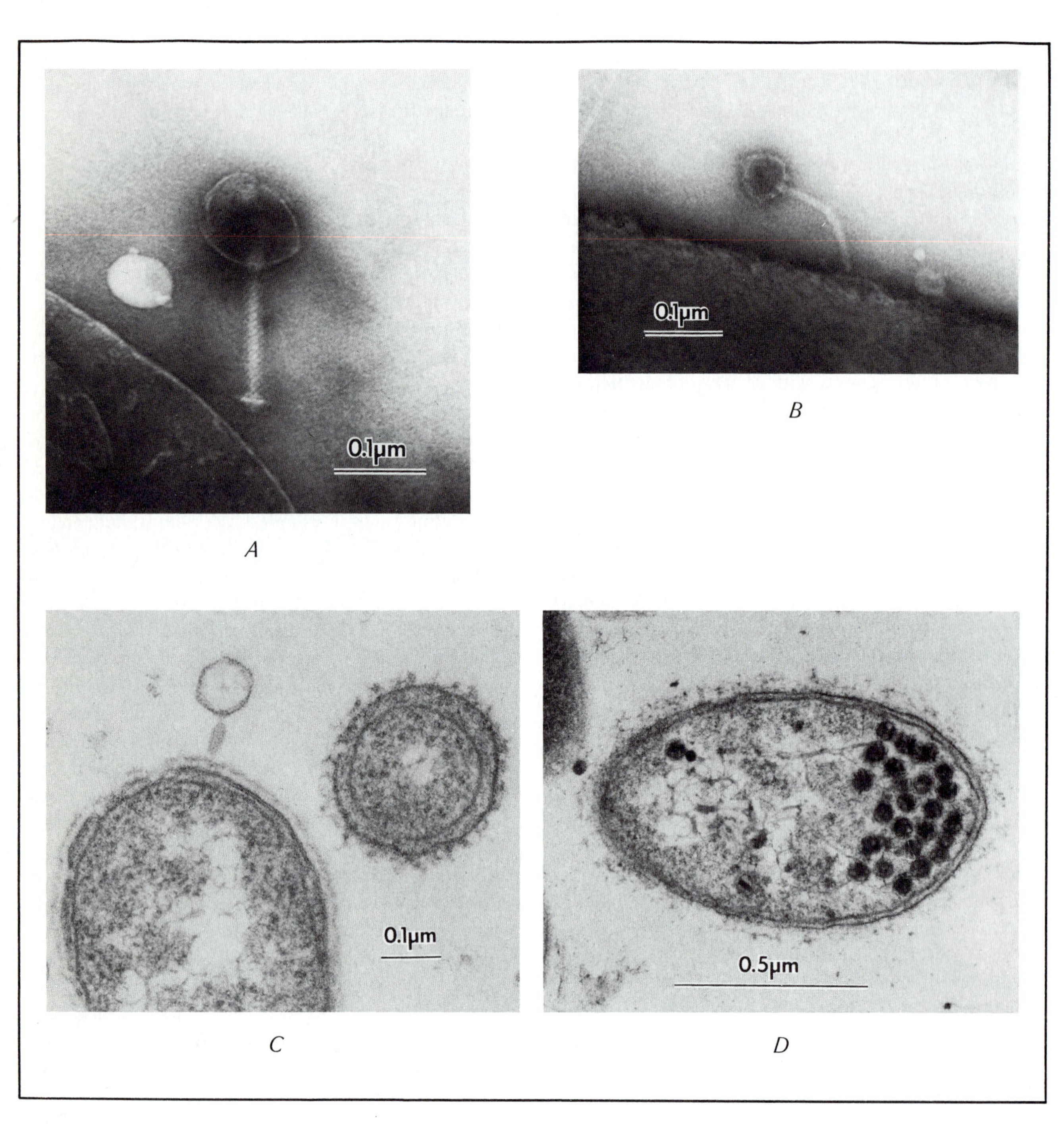

observations on the seasonal occurrence of the hosts (Sieburth, 1967, 1968a,b), are indicated. The intriguing bacteria in the genus *Bdellovibrio* (Stolp and Starr, 1963), which attack, parasitize, and lyse other bacteria, have also been shown to have salt-dependent members that attack marine bacteria preferentially over terrestrial strains (see Chapter 16,C).

REFERENCES

Adams, R.M., J.S. Remington, J. Steinberg, and J.S. Seibert. 1970. Tropical fish aquariums: A source of *Mycobacterium marinum* infections resembling sporotrichosis. *J. Amer. Med. Assn.* 211:457–61.

Ahrens, R.R. 1971. Untersuchungen zur Verbeitung von Phagen der Gattung *Agrobacterium* in der Ostsee. *Kiel. Meeresforsch.* 27:102–12.

Anderson, J.I.W., and D.A. Conroy. 1968. The significance of disease in preliminary attempts to raise crustacea in sea water. *Bull. Off. Int. Epiz.* 69:1239–47.

Andrews, J.H. 1976. The pathology of marine algae. *Biol. Rev.* 51:211–53.

Apelt, G. 1969. Die Symbiose zwischen dem acoelen Turbellar *Convoluta convoluta* und Diatomeen den Gattung *Licmophora*. *Mar. Biol.* 3:165–87.

Arakawa, K.Y. 1970. Scatological studies of the Bivalvia (Mollusca). *Adv. Mar. Biol.* 8:307–436.

Aronson, J.D. 1926. Spontaneous tuberculosis in salt water fish. *J. Infect. Dis.* 39:315–20.

Atkins, D. 1954. Further notes on a marine member of the Saprolegniaceae, *Leptolegnia marina* n. sp., infecting certain invertebrates. *J. mar. biol. Assoc. U.K.* 33:613–25.

Azam, F., and O. Holm-Hansen. 1974. Microbial degradation of organic matter in the deep sea. In: *Research on the Marine Food Chain, Progr. Rept. July 1973–June 1974.* Inst. Mar. Res. Univ. California, San Diego, Ca., pp. 349–62.

Baumann, P., and L. Baumann. 1977. Biology of the marine enterobacteria: Genera *Beneckea* and *Photobacterium*. *Ann. Rev. Microbiol.* 31:39–61.

Baumann, P., L. Baumann, and J.L. Reichelt. 1973. Taxonomy of marine bacteria: *Beneckea parahaemolytica* and *Beneckea alginolytica*. *J. Bacteriol.* 113:1144–55.

Beyers, R.J. 1966. Metabolic similarities between symbiotic coelenterates and aquatic ecosystems. *Arch. Hydrobiol.* 62:273–84.

Bisagni, J.J. 1975. The origin and development of turbidity structure in the lower Providence River and upper Narragansett Bay, Rhode Island. M.S. Thesis, University of Rhode Island.

Bishop, J.K.B., J.M. Edmond, D.R. Ketten, M.P. Bacon, and W.B. Silker. 1977. The chemistry, biology, and vertical flux of particulate matter from the upper 400 m of the equatorial Atlantic Ocean. *Deep-Sea Res.* 24:511–48.

Blackbourn, D.J., F.J.R. Taylor, and J. Blackbourn. 1973. Foreign organelle retention by ciliates. *J. Protozool.* 20:286–88.

Burnett, B.R. 1973. Observations of the microfauna of the deep-sea benthos using light and scanning electron microscopy. *Deep-Sea Res.* 20:413–17.

Cachon, J., M. Cachon, and F. Bouquaheux. 1969. *Myxodinium pipiens* gen. nov., sp. nov., péridinien parasite d'Halosphaera. *Phycologia* 8:157–64.

Cantacuzène, A. 1930. Contribution a l'étude des tumeurs bactériennes chez les algues marines. Doctoral Thesis, University of Paris, Masson & Cie., Paris, 86 p.

Carey, A.G., Jr. 1972. Food sources of sublittoral, bathyal and abyssal asteroids in the northeast Pacific Ocean. *Ophelia* 10:35–47.

Chapman, R.L., and N.J. Lang. 1973. Virus-like particles and nuclear inclusions in the red alga *Porphyridium purpureum* (Bory) Drew et Ross. *J. Phycol.* 9:117–22.

Cisar, J.O., and J.L. Fryer. 1969. An epizootic of vibriosis in chinook salmon. *Bull. Wildlife Dis. Assoc.* 5:73–76.

Colwell, R.R., and A.K. Sparks. 1967. Properties of *Pseudomonas enalia*, a marine bacterium pathogenic for the invertebrate *Crassostrea gigas* (Thunberg). *Appl. Microbiol.* 15:980–86.

Daniel, G. 1933a. Studies on *Ichthyophonus hoferi*, a parasitic fungus of the herring *Clupea harengus*. I. The parasite as it is found in the herring. *Amer. J. Hygiene* 17:262–76.

Daniel, G. 1933b. Studies on *Ichthyophonus hoferi*, a parasitic fungus of the herring *Clupea harengus*. II. The gross and microscopic lesions produced by the parasite. *Amer. J. Hygiene* 17:491–501.

Davies, J.M. 1975. Energy flow through the benthos in a Scottish Sea loch. *Mar. Biol.* 31:353–62.

Dayton, P.K., and R.R. Hessler. 1972. Role of biological disturbance in maintaining diversity in the deep sea. *Deep-Sea Res.* 19:199–208.

Degens, E.T., J.H. Reuter, and K.N.F. Shaw. 1964. Biochemical compounds in offshore California sediments and sea waters. *Geochim. Cosmochim. Acta* 28:45–66.

Dimitroff, V.T. 1926a. *Spirillum virginianum* nov. spec. *J. Bacteriol.* 12:19–49.

Dimitroff, V.T. 1926b. Spirochaetes in Baltimore market oysters. *J. Bacteriol.* 12:135–65.

Drebes, G. 1969. *Dissodinium pseudocalani* sp. nov., ein parasitischer Dinoflagellat auf Copepodeneiern. *Helgol. wiss. Meeres.* 19:58–67.

Drebes, G. 1974. *Marines Phytoplankton*. G. Thieme Verlag, Stuttgart, p. 167.

Droop, M.R. 1963. Algae and invertebrates in symbiosis. *Symp. Soc. Gen. Microbiol.* 13:171–99.

Ehrlich, H.L. 1972. The role of microbes in manganese nodule genesis and degradation. In: *Ferromanganese Deposits on the Ocean Floor* (D.R. Horn, ed.). Office IDOE, NSF, Washington, D.C., pp. 63–70.

Erlandsen, S.L., A. Thomas, and G. Wendelschafer. 1973. A simple technique for correlating SEM with TEM on biological tissue originally embedded in epoxy resin for TEM. In: *Scanning Electron Microscopy* (O. Johari and I. Corvin, eds.). IIT Res. Inst., Chicago, pp. 349–56.

Fenchel, T.M. 1969. The ecology of marine microbenthos. IV. Structure and function of the benthic ecosystem, its chemical and physical factors and the microfauna communities with special reference to the ciliated protozoa. *Ophelia* 6:1–182.

Fenchel, T. 1970. Studies on the decomposition of organic detritus derived from the turtle grass *Thalassia testudinum*. *Limnol. Oceanogr.* 15:14–20.

Fenchel, T. 1972. Aspects of decomposer food chains in marine benthos. *Verh. Deutsch. Zool. Ges.* 65:14–22.

Fenchel, T., and R.J. Riedl. 1970. The sulfide system: A new biotic community underneath the oxidized layer of marine sand bottoms. *Mar. Biol.* 7:255–68.

Fenchel, T., L.H. Kofoed, and A. Lappalainen. 1975. Particle size-selection of two deposit feeders: The amphipod *Corophium volutator* and the prosobranch *Hydrobia ulvae*. *Mar. Biol.* 30:119–28.

Ferguson, J.C. 1971. Uptake and release of free amino acids by starfishes. *Biol. Bull.* 141:122–29.

Forster, G.R. 1953. Peritrophic membranes in the Caridea (Crustacea Decapoda). *J. mar. biol. Assoc. U.K.* 32:315–18.

Frankenberg, D., and K.L. Smith, Jr. 1967. Coprophagy in marine animals. *Limnol. Oceanogr.* 12:443–50.

Freudenthal, H.D. 1962. *Symbiodinium* gen. nov. and *Symbiodinium microadriaticum* sp. nov., a zooxanthella: Taxonomy, life cycle and morphology. *J. Protozool.* 9:45–52.

Gallardo, V.A. 1977. On the discovery of large benthic microbial communities in the sulfide biota under the low-oxygen water-mass of the Peru-Chile subsurface countercurrent. *Nature* 268:331–32.

Gessner, R.V., R.D. Goos and J.McN. Sieburth. 1972. The fungal microcosm of the internodes of *Spartina alterniflora*. *Mar. Biol.* 16:269–73.

Gold, K., and U. Pollingher. 1971. Occurrence of endosymbiotic bacteria in marine dinoflagellates. *J. Phycol.* 7:264–65.

Gotto, R.V. 1969. *Marine Animals. Partnerships and Other Associations.* American Elsevier Publ. Co., New York, 96 p.

Grassle, J.F., and H.L. Sanders. 1973. Life histories and the role of disturbance. *Deep-Sea Res.* 20:643–59.

Greene, R.W. 1970. Symbiosis in sacoglossan opisthobranchs: Functional capacity of symbiotic chloroplasts. *Mar Biol.* 7:138–42.

Greenslate, J. 1974a. Microorganisms participate in the construction of manganese nodules. *Nature* 249:181–83.

Greenslate, J. 1974b. Manganese and biotic debris associations in some deep-sea sediments. *Science* 186:529–31.

Grell, K.G. 1960. Reziproke infektion mit Eucoccidien aus vershieden Wirten. *Naturwissenschaft.* 47:47–48.

Grell, K.G. 1973. *Protozoology.* Springer-Verlag, New York and Heidelberg, 554 p.

Guillard, R.R.L. 1959. Further evidence of the destruction of bivalve larvae by bacteria. *Biol. Bull.* 117:258–66.

Guillard, R.R.L., and J.A. Hellebust. 1971. Growth and the production of extracellular substances by two strains of *Phaeocystis poucheti*. *J. Phycol.* 7:330–38.

Haastein, T., and G. Holt. 1972. The occurrence of vibrio disease in wild Norwegian fish. *J. Fish. Biol.* 4:33–37.

Hargrave, B.T. 1970a. The utilization of benthic microflora by *Hyalella azteca* (Amphipoda). *J. Animal Ecol.* 39:427–37.

Hargrave, B.T. 1970b. The effect of a deposit-feeding amphipod on the metabolism of benthic microflora. *Limnol. Oceanogr.* 15:21–30.

Hargrave, B.T. 1971. An energy budget for a deposit-feeding amphipod. *Limnol. Oceanogr.* 16:99–103.

Hargrave, B.T. 1972. Aerobic decomposition of sediment and detritus as a function of particle surface area and organic content. *Limnol. Oceanogr.* 17:583–96.

Hargrave, B.T. 1973. Coupling carbon flow through some pelagic and benthic communities. *J. Fish. Res. Bd. Canada* 30:1317–26.

Harshbarger, J.C., S.C. Chang, and S.V. Otto. 1977. Chlamydiae (with phages), mycoplasmas, and Rickettsiae in Chesapeake Bay bivalves. *Science* 196:666–68.

Haven, D.S., and R. Morales-Alamo. 1966. Aspects of biodeposition by oysters and other invertebrate filter feeders. *Limnol. Oceanogr.* 11:487–98.

Hendrie, M.S., W. Hodgkiss, and J.M. Shewan. 1971. Proposal that the species *Vibrio anguillarum* Bergman 1909, *Vibrio piscium* David 1927, and *Vibrio ichthyodermis* (Wells and ZoBell) Shewan Hobbs and Hodgkiss 1960 be combined as a single species, *Vibrio anguillarum*. *Int. J. Syst. Bacteriol.* 21:64–68.

Henry, S.M. (ed.). 1966. *Symbiosis. Vol. 1. Associations of Microorganisms, Plants and Marine organisms.* Academic Press, New York and London, 478 p.

Hessler, R.R. 1974. The structure of deep benthic communities from central oceanic waters. In: *The Biology of the Oceanic Pacific* (C.B. Miller, ed.). Oregon State University Press, Corvallis, pp. 79–93.

Hessler, R.R., and P.A. Jumars. 1974. Abyssal community analysis from replicate box cores in the central North Pacific. *Deep-Sea Res.* 21:185–209.

Hibberd, D.J. 1977. Observations on the ultrastructure of the cryptomonad endosymbiont of the red-water ciliate *Mesodinium rubrum*. *J. mar. biol. Assoc. U.K.* 57:45–61.

Hickel, W. 1974. Seston composition of the bottom waters of Great Lameshur Bay, St. John, U.S. Virgin Islands. *Mar. Biol.* 24:125–30.

Hinga, K.R. 1974. *In-situ* benthic respirometry based on micro-Winkler. M.S. Thesis, University of Rhode Island, 52 p.

Hobbie, J.E., O. Holm-Hansen, T.T. Packard, L.R. Pomeroy, R.W. Sheldon, J.P. Thomas, and W.J. Wiebe. 1972. A study of the distribution and activity of microorganisms in ocean water. *Limnol. Oceanogr.* 17:544–55.

Hobson, L.A. 1967. The seasonal and vertical distribution of suspended particulate matter in an area of the northeast Pacific Ocean. *Limnol. Oceanogr.* 12:642–49.

Hochachka, P.W. 1974. Enzymatic adaptations to deep sea life. In: *The Biology of the Oceanic Pacific* (C.B. Miller, ed.). Oregon State University Press, Corvallis, pp. 107–36.

Hodgkiss, W., and J.M. Shewan. 1950. *Pseudomonas* infection in a plaice. *J. Pathol. Bacteriol.* 62:655–57.

Hollande, A., and D. Carré. 1974. Les xanthelles des Radiolaires Sphaerocolloides, des Acanthaires et de *Velella velella:* infrastructure—cytochimie—taxonomie. *Protistologica* 10:573–601.

Honjo, S. 1976. Coccoliths: Production, transportation and sedimentation. *Marine Micropaleontology* 1:65–79.

Hopper, B.E., and S.P. Meyers. 1966. Observations on the bionomics of the marine nematode *Metoncholaimus* sp. *Nature* 209:899–900.

Hopper, B.E., S.P. Meyers, and R. Cefalu. 1970. Microsporidian infection of a marine nematode, *Metoncholaimus scissus. J. Invert. Pathol.* 16:371–77.

Horn, D.R. (ed.). 1972. *Ferromanganese deposits on the ocean floor.* Office Int. Decade Ocean Explor., National Science Foundation, Washington, D.C., 293 p.

Hungate, R.E. 1966. *The Rumen and its Microbes.* Academic Press, New York and London, 533 p.

Hylleberg, J., and V.F. Gallucci. 1975. Selectivity in feeding by the deposit-feeding bivalve *Macoma nasuta. Mar. Biol.* 32:167–78.

Isaacs, J.D., and R.A. Schwartzlose. 1975. Active animals of the deep-sea floor. *Sci. American* 233(4):84–91, 139.

Jannasch, H.W., and C.O. Wirsen. 1973. Deep-sea microorganisms: *In situ* response to nutrient enrichment. *Science* 180:641–43.

Jannasch, H.W., and C.O. Wirsen. 1977. Microbial life in the deep sea. *Sci. American* 236(6):42–52.

Jannasch, H.W., K. Eimhjellen, C.O. Wirsen, and A. Farmanfarmaian. 1971. Microbial degradation of organic matter in the deep sea. *Science* 171:672–75.

Jannasch, H.W., C.O. Wirsen, and C.L. Winget. 1973. A bacteriological, pressure-retaining, deep-sea sampler and culture vessel. *Deep-Sea Res.* 20:661–64.

Johannes, R.E., and M. Satomi. 1966. Composition and nutritive value of fecal pellets of a marine crustacean. *Limnol. Oceanogr.* 11:191–97.

Johnson, P.W., J.McN. Sieburth, A. Sastry, C.R. Arnold, and M.S. Doty. 1971. *Leucothrix mucor* infestation of benthic crustacea, fish eggs and tropical algae. *Limnol. Oceanogr.* 16:962–69.

Johnson, R.G. 1974. Particulate matter at the sediment-water interface in coastal environments. *J. Mar. Res.* 32:313–30.

Johnson, T.W., Jr., and F.K. Sparrow, Jr. 1961. *Fungi in Oceans and Estuaries.* J. Cramer, Weinheim, and Hafner Publ., New York, 668 p.

Jørgensen, B.B., and T. Fenchel. 1974. The sulfur cycle of a marine sediment model system. *Mar. Biol.* 24:189–201.

Kaneko, T., and R.R. Colwell. 1973. Ecology of *Vibrio parahaemolyticus* in Chesapeake Bay. *J. Bacteriol.* 113:24–32.

Kevin, M.J., W.T. Hall, J.J.A. McLaughlin, and P.A. Zahl. 1969. *Symbiodinium microadriaticum* Freudenthal, a revised taxonomic description, ultrastructure. *J. Phycol.* 5:341–50.

Khailov, K.M., and Z.Z. Finenko. 1970. Organic macromolecular compounds dissolved in sea-water and their inclusion into food chains. In: *Marine Food Chains* (J.H. Steele, ed.). University of California Press, Berkeley and Los Angeles, pp. 6–18.

Kohlmeyer, J. 1972. Parasitic *Haloguignardia oceanica* (Ascomycetes) and hyperparasitic *Sphaceloma cecidii* sp. nov. (Deuteromycetes) in drift *Sargassum* in North Carolina. *J. Elisha Mitchell Soc.* 88:255–59.

Kraeuter, J., and S. Haven. 1970. Fecal pellets of common invertebrates of lower York River and Lower Chesapeake Bay, Virginia. *Chesapeake Science* 11:159–73.

Krantz, G.E., R.R. Colwell, and E. Lovelace. 1969. *Vibrio parahaemolyticus* from the blue crab *Callinectes sapidus* in Chesapeake Bay. *Science* 164:1286–87.

Kuhn, D. 1974. *Cristispira* Gross 1910. In: *Bergey's Manual of Determinative Bacteriology* (8th Ed.) (R.E. Buchanan and N.E. Gibbons, eds.). Williams & Wilkins Co., Baltimore, Md., pp. 171–74.

Lagerheim, G. 1901. Mykologische Studien. III. Beitrage zur Kenntnis der parasitischen Bakterien und der bakteroiden Pilze. *Centralbl. Bakteriol. Abt 2*, 7:248–50.

Leach, J.H. 1970. Epibenthic algal production in an intertidal mudflat. *Limnol. Oceanogr.* 15:514–21.

Lear, D.W., Jr. 1963. Occurrence and significance of chitinoclastic bacteria in pelagic waters and zooplankton. In: *Marine Microbiology* (C.H. Oppenheimer, ed.). C.C. Thomas, Springfield, Ill., pp. 594–610.

Lee, J.J. 1974. Towards understanding the niche of Foraminifera. In: *Foraminifera* (R.H. Hedley and C.G. Adams, eds.), Vol. 1. Academic Press, London and New York, pp. 207–60.

Lee, J.J., L.J. Crockett, J. Hagen, and R.J. Stone. 1974. The taxonomic identity and physiological ecology of *Chlamydomonas hedleyi* sp. nov., algal flagellate symbiont from the foraminifer *Archais angulatus. Br. phycol. J.* 9:407–22.

Levin, M.A., R.E. Wolke, and V.J. Cabelli. 1972. *Vibrio anguillarum* as a cause of disease in winter flounder (*Pseudopleuronectes americanus*). *Can. J. Microbiol.* 18:1585–92.

Lewis, D.H., and D.C. Smith. 1971. The autotrophic nutrition of symbiotic marine coelenterates with special reference to hermatypic corals. I. Movement of photosynthetic products between the symbionts. *Proc. Royal Soc. London Ser. B* 178:111–29.

Lighthart, B. 1969. Planktonic and benthic bacteriovorous Protozoa at eleven stations in Puget Sound and adjacent Pacific Ocean. *J. Fish. Res. Bd. Canada* 26:299–304.

Litchfield, C.D., A.L.S. Munro, L.C. Massie, and G.D. Floodgate. 1974. Biochemistry and microbiology of some Irish Sea sediments: I. Amino-acid analyses. *Mar. Biol.* 26:249–60.

Luckey, T.D. 1972. Introduction to intestinal microecology. *Amer. J. Clin. Nutr.* 25:1292–94.

Mackin, J.G. 1961. Oyster disease caused by *Dermocystidium marinum* and other microorganisms in Louisiana. *Pub. Inst. Mar. Sci., Contr. Mar. Sci.* 7:133–229.

Mawdesley-Thomas, L.E. (ed.). 1972. *Diseases of Fish. Symp. Zool. Soc. London No. 30.* Academic Press, London, 380 p.

McBee, R.H. 1971. Significance of intestinal microflora in herbivory. *Ann Rev. Ecol. Syst.* 2:165–76.

McCave, I.N. 1975. Vertical flux of particles in the ocean. *Deep-Sea Res.* 22:491–502.

McIntyre, A.D., A.L.S. Munro, and J.H. Steele. 1970. Energy flow in a sand ecosystem. In: *Marine Food Chains* (J.H. Steele, ed.). University of California Press, Berkeley and Los Angeles, pp. 19–31.

McLaughlin, J.J.A., and P.A. Zahl. 1957. Studies in marine biology. II. *In vitro* culture of Zooxanthellae. *Proc. Soc. Exp. Biol. Med.* 95:115–20.

Meadows, P.S., and J.G. Anderson. 1968. Micro-organisms attached to marine sand grains. *J. mar. biol. Assoc. U.K.* 48:161–75.

Menzel, D.W. 1966. Bubbling of seawater and the production of organic particles: A re-evaluation. *Deep-Sea Res.* 13:963–66.

Menzies, R.J. 1962. On the food and feeding habits of abyssal organisms as exemplified by the Isopoda. *Int. Revue ges. Hydrobiol.* 47:339–58.

Meyers, S.P., P.A. Orpurt, J. Simms, and L.L. Boral. 1965. Thalassiomycetes. VII. Observations on fungal infestation of turtle grass, *Thalassia testudinum* König. *Bull. Mar. Sci.* 15:548–64.

Meyers, S.P., M.H. Baslow, S.J. Bein, and C.E. Marks. 1959. Studies on *Flavobacterium piscicidia* Bein. *J. Bacteriol.* 78:225–30.

Mills, E.L. 1975. Benthic organisms and the structure of marine ecosystems. *J. Fish. Res. Bd. Canada* 32:1657–63.

Mitchell, R. 1971. Role of predators in the reversal of imbalances in microbial ecosystems. *Nature* 230:257–58.

Moikeha, S.N., G.W. Chu, and L.R. Berger. 1971. Dermatitis-producing alga *Lyngbya majuscula* Gomont in Hawaii. I. Isolation and characterization of the toxic factor. *J. Phycol.* 7:4–8.

Moore, H.B. 1931. The specific identification of faecal pellets. *J. mar. biol. Assoc. U.K.* 17:325–58.

Moriarty, D.J.W. 1975. A method for estimating the biomass of bacteria in aquatic sediments and its application to trophic studies. *Oecologia* 20:219–30.

Muller, W.A. 1975. Competition for food and other niche related studies of three species of salt-marsh foraminifera. *Mar. Biol.* 31:339–51.

Muller, W.A., and J.J. Lee. 1969. Apparent indispensability of bacteria in foraminiferan nutrition. *J. Protozool.* 16:471–78.

Mullin, M.M. 1963. Some factors affecting the feeding of marine copepods of the genus *Calanus. Limnol. Oceanogr.* 8:239–50.

Mullin, M.M. 1966. Selective feeding by Calanoid copepods from the Indian Ocean. In: *Some Contemporary Studies in Marine Science* (H. Barnes, ed.). George Allen & Unwin, London, pp. 545–54.

Newell, R. 1965. The role of detritus in the nutrition of two marine deposit feeders, the prosobranch *Hydrobia ulvae* and the bivalve *Macoma balthica. Proc. Zool. Soc. London* 144:25–45.

Newell, S.Y. 1976. Mangrove fungi: the succession in the mycoflora of red mangrove (*Rhizophora mangle* L.) seedlings. In: *Recent Advances in Aquatic Mycology* (E.B.G. Jones, ed.). Wiley, New York, pp. 51–91.

Noguchi, H. 1921. *Cristispira* in North American shellfish. A note on a spirillum found in oysters. *J. exp. Med.* 34:295–315.

Odum, W.E. 1970. Utilization of the direct grazing and plant detritus food chains by the striped mullet *Mugil cephalus.* In: *Marine Food Chains* (J.H. Steele, ed.). University of California Press, Berkeley and Los Angeles, pp. 222–40.

Oppenheimer, C.H. 1958. A bacterium causing tail rot in the Norwegian codfish. *Inst. Mar. Sci.* 5:160–62.

Oppenheimer, C.H. 1960. Bacterial activity in sediments of shallow marine bays. *Geochim. Cosmochim. Acta* 19:244–60.

Palmer, J.D. 1975. Biological clocks of the tidal zone. *Sci. American* 232(2):70–79.

Pamatmat, M.M. 1968. Ecology and metabolism of a benthic community on an intertidal sandflat. *Int. Revue ges. Hydrobiol.* 53:211–98.

Pamatmat, M.M. 1971. Oxygen consumption by the seabed. VI. Seasonal cycle of chemical oxidation and respiration in Puget Sound. *Int. Revue ges. Hydrobiol.* 56:769–93.

Pamatmat, M.M. 1973. Benthic community metabolism on the continental terrace and in the deep sea in the North Pacific. *Int. Revue ges. Hydrobiol.* 58:345–68.

Parsons, T.R. 1962. Infection of a marine diatom by *Lagenidium* sp. *Can. J. Bot.* 40:523.

Parsons, T.R., and H. Seki. 1970. Importance and general implications of organic matter in aquatic environments. In: *Organic Matter in Natural Waters* (D.W. Hood, ed.). University of Alaska Inst. Mar. Science Publ. No. 1, pp. 1–27.

Preer, J.R., L.B. Preer, and A. Jurand. 1974. Kappa and other endosymbionts in *Paramecium aurelia. Bacteriol. Rev.* 38:113–63.

Provasoli, L., T. Yamasu, and I. Manton. 1968. Experiments on the resynthesis of symbiosis in *Convoluta roscoffensis* with different flagellate cultures. *J. mar. biol. Assoc. U.K.* 48:465–79.

Raibaud, P., R. Ducluzeau, M-C. Muller, and G.D. Abrams. 1972. Diet and the equilibrium between bacteria and yeast implanted in gnotobiotic rats. *Amer. J. Clin. Nutr.* 25:1467–74.

Rhoads, D.C. 1973. The influence of deposit-feeding benthos on water turbidity and nutrient recycling. *Amer. J. Sci.* 273:1–22.

Riley, G.A. 1963. Organic aggregates in seawater and the dynamics of their formation and utilization. *Limnol. Oceanogr.* 8:372–81.

Riley, G.A. 1970. Particulate organic matter in sea water. *Adv. Mar. Biol.* 8:1–118.

Riley, G.A., P.J. Wangersky, and D. Van Hemert. 1964. Organic aggregates in tropical and subtropical surface waters of the North Atlantic Ocean. *Limnol. Oceanogr.* 9:546–50.

Riley, G.A., D. Van Hemert, and P.J. Wangersky. 1965. Organic aggregates in surface and deep waters of the Sargasso Sea. *Limnol. Oceanogr.* 10:354–63.

Roberts, R.J., W.M. Shearer, A.L.S. Munro, and K.G.R. Elson. 1970. Studies on ulcerative dermal necrosis of salmonids. II. The sequential pathology of the lesions. *J. Fish. Biol.* 2:373–78.

Rucker, R.R., W.J. Whipple, J.R. Parvin, and C.A. Evans. 1953. A contagious disease of salmon possibly of virus origin. *Fish. Bull. Fish. Wildl. Serv.* 54(76):35–46.

Safferman, R.S. 1973. Phycoviruses. In: *The Biology of Blue-Green Algae* (N.G. Carr and B.A. Whitton, eds.). Blackwell Scientific Publ. and University of California Press, Berkeley, pp. 214–37.

Savage, D.C. 1972. Associations and physiological interactions of indigenous microorganisms and gastrointestinal epithelia. *Amer. J. Clin. Nutr.* 25:1372–79.

Schäfer, W. 1972. *Ecology and Palaeoecology of Marine Environments* (G.Y. Craig, ed.). Oliver & Boyd, Edinburgh, 568 p.

Schmaljohann, R., and R. Röttger. 1976. Die Symbionten der Grossforaminifere *Heterostegina depressa* sind Diatomeen. *Naturwissenschaften* 63:486.

Schrader, H-J. 1971. Fecal pellets: Role in sedimentation of pelagic diatoms. *Science* 174:55–57.

Schrader, H-J. 1972. Anlösung und Konservation von Diatomeenschalen beim Absinken am Beispiel des Landsort-Tiefs in der Ostsee. *Nova Hedwigia Beiheft* 39:191–216.

Schwarz, J.R., and R.R. Colwell. 1975. Heterotrophic activity of deep-sea sediment bacteria. *Appl. Microbiol.* 30:639–49.

Seki, H. 1965a. Microbiological studies on the decomposition of chitin in marine environment. IX. Rough estimation on chitin decomposition in the ocean. *J. Oceanogr. Soc. Japan* 21:253–60.

Seki, H. 1965b. Microbiological studies on the decomposition of chitin in marine environment. X. Decomposition of chitin in marine sediments. *J. Oceanogr. Soc. Japan* 21:261–69.

Seki, H., and N. Taga. 1963. Microbiological studies on the decomposition of chitin in marine environment. I. Occurrence of chitinoclastic bacteria in the neritic region. *J. Oceanogr. Soc. Japan* 19:101–8.

Seki, H., J. Skelding, and T.R. Parsons. 1968. Observations on the decomposition of a marine sediment. *Limnol. Oceanogr.* 13:440–47.

Seki, H., E. Wada, I. Koike, and A. Hattori. 1974. Evidence of high organotrophic potentiality of bacteria in the deep ocean. *Mar. Biol.* 26:1–4.

Sheldon, R.W., T.P.T. Evelyn, and T.R. Parsons. 1967. On the occurrence and formation of small particles in sea water. *Limnol. Oceanogr.* 12:367–76.

Sibert, J., and T.J. Brown. 1975. Characteristics and potential significance of heterotrophic activity in a polluted fjord estuary. *J. exp. mar. Biol. Ecol.* 19:97–104.

Sieburth, J.McN. 1959. Gastrointestinal microflora of Antarctic birds. *J. Bacteriol.* 77:521–31.

Sieburth, J.McN. 1960. Acrylic acid, an "antibiotic" principle in *Phaeocystis* blooms in Antarctic waters. *Science* 132:676–77.

Sieburth, J.McN. 1961. Antibiotic properties of acrylic acid, a factor in the gastrointestinal antibiosis of polar marine animals. *J. Bacteriol.* 82:72–79.

Sieburth, J.McN. 1967. Seasonal selection of estuarine bacteria by water temperature. *J. exp. mar. Biol. Ecol.* 1:98–121.

Sieburth, J.McN. 1968a. The influence of algal antibiosis on the ecology of marine microorganisms. In: *Advances in Microbiology of the Sea* (M.R. Droop and E.J.F. Wood, eds.). Academic Press, London and New York, pp. 63–94.

Sieburth, J.McN. 1968b. Observations on bacteria planktonic in Narragansett Bay, Rhode Island; a résumé. *Bull. Misaki Mar. Biol. Inst., Kyoto University* 12:49–64.

Sieburth, J.McN. 1975. *Microbial Seascapes.* University Park Press, Baltimore, Md., 248 p.

Sieburth, J.McN., and A.S. Dietz. 1974. Biodeterioration in the sea and its inhibition. In: *Effect of the Ocean Environment on Microbial Activities* (R.R. Colwell and R.Y. Morita, eds.). University Park Press, Baltimore, Md., pp. 318–26.

Sieburth, J.McN., K.M. Johnson, C.M. Burney, and D.M. Lavoie 1977. Estimation of in-situ rates of heterotrophy using diurnal changes in dissolved organic matter and growth rates of picoplankton in diffusion culture. *Helgoländer wiss. Meeresunters.* 30:565–74.

Silcox, V.A., and H.L. David. 1971. Differential identification of *Mycobacterium kansasii* and *Mycobacterium marinum. Appl. Microbiol.* 21:327–34.

Sindermann, C.J. 1958. An epizootic in Gulf of St. Lawrence fishes. *Trans. North Amer. Wildl. Conf.* 23:349–60.

Sindermann, C.J. 1963. Disease in marine populations. *Trans. North Amer. Wildl. Conf.* 28:336–56.

Sindermann, C.J. 1970. *Principal Diseases of Marine Fish and Shellfish.* Academic Press, London and New York, 369 p.

Smayda, T.J. 1969. Some measurements of the sinking rate of fecal pellets. *Limnol. Oceanogr.* 14:621–25.

Smayda, T.J. 1970. The suspension and sinking of phytoplankton in the sea. *Oceanogr. Mar. Biol. Ann. Rev.* 8:353–414.

Smith, A.W., T.G. Akers, S.H. Madin, and N.A. Vedros. 1973. San Miguel sea lion virus isolation, preliminary characterization and relationship to vesicular exanthema of swine virus. *Nature* 244:108–10.

Smith, K.L., Jr. 1974. Oxygen demands of San Diego Trough sediments: An *in-situ* study. *Limnol. Oceanogr.* 19:939–44.

Smith, K.L., Jr., and J.M. Teal. 1973. Deep-sea benthic community respiration, an *in-situ* study at 1850 meters. *Science* 179:282–83.

Smith, K.L., Jr., C.H. Clifford, A.H. Eliason, B. Walden, G.T. Rowe, and J.M. Teal. 1976. A free vehicle for measuring benthic community metabolism. *Limnol. Oceanogr.* 21:164–70.

Snieszko, S.F. 1974. The effects of environmental stress on outbreaks of infectious diseases of fish. *J. Fish. Biol.* 6:197–208.

Snieszko, S.F., and C.C. Taylor. 1947. A bacterial disease of the lobster (*Homarus americanus*). *Science* 105:500.

Sokolova, M.N. 1959. On the distribution of deep-water bottom animals in relation to their feeding habits and the character of sedimentation. *Deep-Sea Res.* 6:1–4.

Sorokin, Yu. I. 1971. On the role of bacteria in the productivity of tropical oceanic waters. *Int. Revue ges. Hydrobiol.* 56:1–48.

Soucek, Z., and R. Mushin. 1970. Gastrointestinal bacteria of certain Antarctic birds and mammals. *Appl. Microbiol.* 20:561–66.

Southward, A.J., and E.C. Southward. 1970. Observations on the role of dissolved organic compounds in the nutrition of benthic invertebrates. Experiments on three species of Pogonophora. *Sarsia* 45:69–96.

Southward, A.J., and E.C. Southward. 1972a. Observations on the role of dissolved organic compounds in the nutrition of benthic invertebrates. II. Uptake by other animals living in the same habitat as Pogonophores, and by some littoral Polychaeta. *Sarsia* 48:61–70.

Southward, A.J., and E.C. Southward. 1972b. Observations on the role of dissolved organic compounds in the nutrition of benthic invertebrates. III. Uptake in relation to organic content of the habitat. *Sarsia* 50:29–46.

Spencer, R. 1955. A marine bacteriophage. *Nature* 175:690–91.

Spencer, R. 1960. Indigenous marine bacteriophages. *J. Bacteriol.* 79:614.

Sprague, V. 1971. Diseases of oysters. *Ann. Rev. Microbiol.* 25:211–30.

Steele, J.H., and I.E. Baird. 1968. Production ecology of a sandy beach. *Limnol. Oceanogr.* 13:14–25.

Stolp, H., and M.P. Starr. 1963. *Bdellovibrio bacteriovorus* gen. et sp. n., predatory, ectoparasitic and bacteriolytic microorganism. *Antonie van Leeuwenhoek* 29:217–48.

Strickland, J.D.H. 1970. Introduction to recycling of organic matter. In: *Marine Food Chains* (J.H. Steele, ed.). University of California Press, Berkeley and Los Angeles, pp. 3–5.

Sutcliffe, W.H., Jr., E.R. Baylor, and D.W. Menzel. 1963. Sea surface chemistry and Langmuir circulation. *Deep-Sea Res.* 10:233–43.

Taylor, D.L. 1969. Identity of zooxanthellae isolated from some Pacific Tridacnidae. *J. Phycol.* 5:336–40.

Taylor, D.L. 1970. Chloroplasts as symbiotic organelles. *Intern. Rev. Cytol.* 27:29–64.

Taylor, D.L. 1971. On the symbiosis between *Amphidinium klebsii* (Dinophyceae) and *Amphiscolops langerhansi* (Turbellaria: Acoela). *J. mar. biol. Assoc. U.K.* 51:301–13.

Taylor, D.L. 1973. Algal symbionts of invertebrates. *Ann. Rev. Microbiol.* 27:171–87.

Taylor, F.J.R., D.J. Blackbourn, and J. Blackbourn. 1971. The red-water ciliate *Mesodinium rubrum* and its "incomplete symbionts": A review containing new ultrastructural observations. *J. Fish. Res. Bd. Canada* 28:391–407.

Tietjen, J.H. 1971. Ecology and distribution of deep-sea meiobenthos off North Carolina. *Deep-Sea Res.* 18:941–57.

Tomas, R.N., and E.R. Cox. 1973. Observations on the symbiosis of *Peridinium balticum* and its intracellular alga. I. Ultrastructure. *J. Phycol.* 9:304–23.

Tomas, R.N., E.R. Cox, and K.A. Steidinger. 1973. *Peridinium balticum* (Levander) Lemmerman, an unusual dinoflagellate with a mesocaryotic and an eucaryotic nucleus. *J. Phycol.* 9:91–98.

Toth, R., and R.T. Wilce. 1972. Viruslike particles in the marine alga *Chorda tomentosa* Lyngbye (Phaeophyceae). *J. Phycol.* 8:126–30.

Trench, R.K. 1971a. The physiology and biochemistry of zooxanthellae symbiotic with marine coelenterates. I. The assimilation of photosynthetic products of zooxanthellae by two marine coelenterates. *Proc. Royal Soc. London Ser. B* 177:225–35.

Trench, R.K. 1971b. The physiology and biochemistry of zooxanthellae symbiotic with marine coelenterates. II. Liberation of fixed ^{14}C by zooxanthellae *in vitro*. *Proc. Royal Soc. London Ser. B* 177:237–50.

Trench, R.K. 1971c. The physiology and biochemistry of zooxanthellae symbiotic with marine coelenterates. III. The effect of homogenates of host tissues on the excretion of photosynthetic products *in vitro* by zooxanthellae from two marine coelenterates. *Proc. Royal Soc. London Ser. B* 177:251–64.

Trench, R.K. 1975. Of "leaves that crawl": Functional chloroplasts in animal cells. *Symp. Soc. Exp. Biol.* 29:229–65.

Trench, R.K., J.E. Boyle, and D.C. Smith. 1973a. The association between chloroplasts of *Codium fragile* and the mollusc *Elysia viridis*. I. Characteristics of isolated *Codium* chloroplasts. *Proc. Royal Soc. London Ser. B* 184:51–61.

Trench, R.K., J.E. Boyle, and D.C. Smith. 1973b. The association between chloroplasts of *Codium fragile* and the mollusc *Elysia viridis*. II. Chloroplast ultrastructure and photosynthetic carbon fixation in *E. viridis*. *Proc. Royal Soc. London Ser. B* 184:63–81.

Trench, R.K., J.E. Boyle, and D.C. Smith. 1974. The association between chloroplasts of *Codium fragile* and the mollusc *Elysia viridis*. III. Movement of photosynthetically fixed ^{14}C in tissues of intact living *E. viridis* and in *Tridachia crispata*. *Proc. Royal Soc. London Ser. B* 184:453–64.

Vedros, N.A., A.W. Smith, J. Schonewald, G. Migaki, and R.C. Hubbard. 1971. Leptospirosis epizootic among California sea lions. *Science* 172:1250–51.

Vinogradov, M.W. 1962. Feeding of the deep-sea zooplankton. *Rapp. Procès-verb. Cons. Int. Explor. Mer* 153:114–20.

Vishniac, H.S. 1958. A new marine phycomycete. *Mycologia* 50:66–79.

Wangersky, P.J. 1965. The organic chemistry of sea water. *Amer. Scientist* 53:358–74.

Welsh, B.L. 1975. The role of grass shrimp, *Palaemonetes pugio*, in a tidal marsh ecosystem. *Ecology* 56:513–30.

Wendt, J. 1974. Encrusting organisms in deep-sea manganese nodules. In: *Pelagic sediments: On Land and under the Sea* (K.J. Hsü and H.C. Jenkyns, eds.). Spec. Publ. Int. Ass. Sediment. 1:437–47.

Wheeler, J.R. 1975. Formation and collapse of surface films. *Limnol. Oceanogr.* 20:338–42.

Wiebe, P.H., S.H. Boyd, and C. Winget. 1976. Particulate matter sinking to the deep-sea floor at 2000 m in the Tongue of the Ocean, Bahamas, with a description of a new sedimentation trap. *J. Mar. Res.* 34:341–54.

Wilder, D.G., and D.W. McLeese. 1961. A comparison of three methods of inactivating lobster claws. *J. Fish Res. Bd. Canada* 18:367–375.

Wingfield, W.H., J.L. Fryer, and K.S. Pilcher. 1969. Properties of the sockeye salmon virus (Oregon strain) (33719). *Proc. Soc. Exp. Biol. Med.* 130:1055–59.

Wirsen, C.O., and H.W. Jannasch. 1975. Activity of marine psychrophilic bacteria at elevated hydrostatic pressures and low temperatures. *Mar. Biol.* 31:201–8.

Wolf, K., and M.C. Quimby. 1973. Fish viruses: Buffers and methods for plaquing eight agents under normal atmosphere. *Appl. Microbiol.* 25:659–64.

Wolf, K., S.F. Snieszko, C.E. Dunbar, and E. Pyle. 1960. Virus nature of infectious pancreatic necrosis in trout. *Proc. Soc. Exp. Biol. Med.* 104:105–8.

Zachary, A. 1974. Isolation of bacteriophages of the marine bacterium *Beneckea natriegens* from coastal salt marshes. *Appl. Microbiol.* 27:980–82.

Zatsepin, V.I. 1970. On the significance of various ecological groups of animals in the bottom communities of the Greenland, Norwegian, and the Barents Seas. In: *Marine Food Chains* (J.H. Steele, ed.). University of California Press, Berkeley and Los Angeles, pp. 207–21.

PART II

Methodology

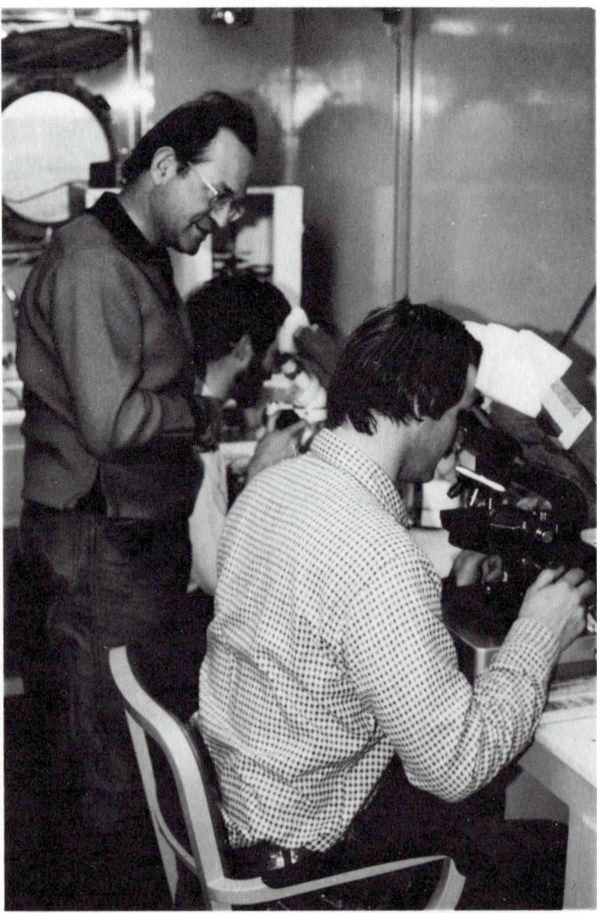
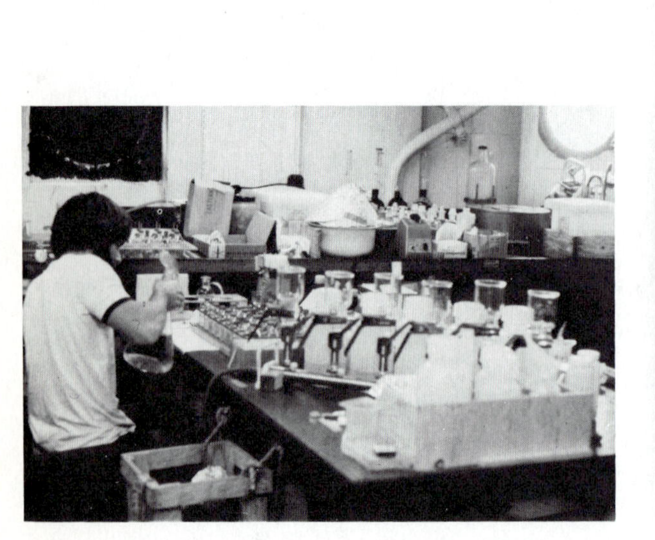

Methodologies prerequisite for studies in marine microbiology: sampling, microscopic examination, quantitative characterization, and culture. Clockwise from upper left: Marine Technician James Hannon supervises a plankton tow from the "hero platform" during the R/V *Endeavor* maiden cruise EN-001, November, 1976, in the Gulf of St. Lawrence; Chemical Marine Research Specialist Kenneth Johnson prepares carbohydrate analyses during the R/V *Trident* cruise TR-170 enroute to Spain, August, 1975; the author observes as student James Fontaine and Professor Paul Hargraves examine plankton with microscopes fitted with forehead braces, during Cruise EN-001; student Peter Yorgey prepares cultures during Cruise TR-170.
Photographs courtesy of Kenneth R. Hinga (TR-170) and Paul W. Johnson (EN-001).

The microbiology of natural marine environments is studied in order to characterize specific microbiotas from specific habitats and thus gain an understanding of how microorganisms affect the biological, chemical, and physical nature of that habitat. The chemical and physical characterization of the sampling area and the samples (Schlieper, 1968, 1972; Strickland and Parsons, 1972) is important collateral information necessary for interpreting microbiological observations. Physical and chemical parameters are beyond the scope of this text, as is the effect of these parameters on the microorganisms (but see Kinne, 1972, 1975). Here we are only concerned with the microbiology of the samples. Any studies on natural populations depend upon accurate sampling. It is often in sampling that the investigator can either alter or selectively obtain microorganisms in such a way as to bias the results. Sample treatment is the key to the microbiological study of natural populations. The degree of success in unraveling the nature of the microorganisms and the processes that occur in natural microbial assemblages will depend in large measure on how well one chooses and masters the basic techniques of sample treatment.

Nothing is more frustrating than a poor preparation, an unaligned microscope that gives a suboptimal image, a procedure for quantification that generates unreliable values, or a culture technique that fails to grow the desired microorganism. Conversely, nothing gives more pure pleasure and excitement than to see a new microbial seascape unfold with superior resolution, to obtain reliable and significant data on these populations, and to grow the desired species in culture. The mastery of techniques and an acquaintance with the microorganisms they reveal will provide the tools necessary for studying microbial life in the sea.

References

Kinne, O. 1972, 1975. (ed.). *Marine Ecology, a comprehensive. integrated treatise on life in oceans and coastal waters. Vol. I. Environmental Factors* (1972). *Vol. II. Physiological Mechanisms* (1975). Wiley-Interscience, New York.

Schlieper, C. (ed.). 1968. *Methoden der Meeresbiologischen Forschung.* Gustav Fischer Verlag, Jena, 322 p.

Schlieper, C. (ed.). 1972. *Research Methods in Marine Biology* (English translation). University of Washington Press, Seattle, 356 p.

Strickland, J.D.H., and T.R. Parsons. 1972. *A Practical Handbook of Seawater Analysis.* (2nd Ed,) Fish. Res. Bd. Canada, Ottawa, Bull. No. 167, 301 p.

CHAPTER 4

Sampling

In order to study microorganisms as they occur in nature, samples that will contain representative microorganisms in a usable condition must be obtained. Before embarking on a collecting expedition, the researcher should be adequately prepared not only for the sampling but also for the preparation, examination, and analysis or preservation of the samples before significant deterioration takes place. One must determine what type of sample is required, choose the type of habitat to be visited, and have a working knowledge of the samplers to be used. In addition, one must have the equipment and expertise needed to make the collateral observations that characterize the environment and the sample. When environmental observations and sampling have been completed, precautions must be taken to maintain the sample in good condition during transportation and examination.

As we have seen in the last two chapters, microorganisms are ubiquitous throughout the water column but tend to be concentrated where organic matter is produced and where organic matter accumulates at the boundaries of the sea. There are three basic types of microbiological samples: water from the sea surface containing the surface microlayer and the microneuston, subsurface water and the organisms suspended and in motion in the water column, and the organisms living in association with the sea floor. Sampling for these three types of samples can be more or less complex, depending upon the sampling platform used.

A. SAMPLING PLATFORMS

1. Onshore Sites

Sampling to obtain organisms for classroom work and for developing and testing methods need not entail putting a boat out to sea or the purchase or fabrication of elaborate samplers. Even samples for a synoptic study can often be obtained from judiciously selected onshore sites. A quick and rich source of material is from aquaria. There are usually a number of aquaria set up for various purposes at marine stations, oceanographic laboratories, and similar facilities, which can supply a spectrum of microenvironments and microorganisms. Films on the surface and sides of tanks as well as sediments in tanks (Lackey, 1961) offer rich microcosms for study. In addition to existing aquaria, special aquaria can be set up for enriching certain sequences of microbiological events. One problem encountered with aquaria is that most seawater systems have water pipes that are usually so fouled with attached and free-living invertebrates and mussels that the type and amount of suspended particles at the outflow are determined by them, as is the chemical composition of the water. Some seawater systems are constructed with piping that can be easily removed, scraped, and flushed regularly to keep them relatively clean. These provide a better quality water. Aquaria located on piers can be supplied with seawater raised by peristaltic pumps through short lengths of plastic tubing, which can be cleaned or replaced regularly. Such seawater will have a more normal microbiota and chemical composition than that in the usual aquarium. Pumped samples both from shore (Jensen and Sakshaug, 1970; Lavoie, 1975) and aboard ship (Sieburth et al., 1977) offer a means for monitoring changes in the microbiology and chemistry of the water with time or tide as well as providing large volumes of water for filtering out particles of a specific size.

Such structures as piers that jut out into the sea and bridges across estuaries and to and between islands can offer adequate and convenient sampling sites for both class and research purposes. The main disadvantages of sampling from such structures are the large biomass of fouling and benthic forms on the supporting structures and the usually increased turbulence and turbidity of the waters passing such structures. But the convenient use of these structures outweighs their disadvantages for many purposes. Water samples can be easily obtained with buckets, sterile sample jars, and other simple water samplers. Plankton nets of porosities down to 10 μm can be streamed during the ebb and flow of the tide. Simple gravity cores can also be taken from such structures by using a portable hand winch with hydrographic wire. The wire and winch also facilitate taking simultaneous water samples from a series of depths when the wire is held vertical by an end weight.

The molluscs, seaweeds, and invertebrate fauna from pilings are a rich source of material for studying microorganisms. These fouling organisms attract baitfish suitable for microbiological study, which can be easily collected in good condition by setting out baited minnow traps overnight. The collection of materials from docks and rocky shores is greatly facilitated by snorkeling and SCUBA diving. Such sampling should be made only after adequate instruction in YMCA or NAUI programs or their equivalent, and should be made in the company of a diving buddy or two, for safety. The swimming and diving microbiologist soon learns how to examine surfaces and suspended particles with a submersible magnifier ("MacroSnooper," Pratt, 1973); to collect plants and animals in Whirl-Pak bags; to collect water samples by righting sterile tubes or jars after removing the caps, or with Vacutainers; to obtain cores with short lengths of plastic core liner and rubber stoppers; and even to obtain plankton samples by pulling small nets. Skindiving as a means of observing the physical and biological nature of shallow waters has been described (Woods and Lythgoe, 1971) as has useful hardware for surveying shallow sea bottoms (Fager et al., 1966). Snorkeling in shallow water yields much information, and it should also be conducted at night to observe nocturnal activities. Most of the material used in *Microbial Seascapes* (Sieburth, 1975) was obtained while snorkeling the rocky subtidal water at the Narragansett Bay Campus and along the coral reefs and keys in the vicinity of the Pigeon Key Biological Station in Florida. Habitats other than rocky coasts and coral reefs worth studying from shore include the stands of such seagrasses as *Zostera* (Sieburth and Thomas, 1973) that occur in shallow waters, the cordgrass *Spartina* that grows with its roots in the intertidal zone (Gessner, Goos, and Sieburth, 1972), and mangroves, the trees that walk into the sea, in tropical environments (Gore and Lavies, 1977).

It is not necessary to immerse oneself to look at and learn about life in the intertidal zone. At low tide, seaweeds and other biota are exposed, although in the absence of a suspending fluid the more delicate forms collapse and are difficult to observe. Such material should be placed in seawater to restore the organisms to their normal configuration for examination. Low tides also leave tide pools that have a resident flora and fauna, with their associated microorganisms (Langlois, 1975). The surface skin of tide pool water can also develop a significant microbiota (Norris, 1965). Some of the phenomena of the open ocean, such as algal antibiosis, (Conover and Sieburth, 1964; Sieburth and Conover, 1965) also occur in tide pools (Conover and Sieburth, 1966; Langlois, 1975).

2. Nearshore Sites

To get away from the direct influence of land, the fouling organisms at the sea's edge and the excessive turbulence and turbidity experienced from beaches, piers, and bridges, it is necessary to use small boats for sampling. These can run the gamut from canoes, inflatables like the Zodiac, which are a wise choice in rocky areas, through a variety of recreational boats, such as Jon boats and Boston Whalers that can be outfitted with outboard, fathometer, davit, winch and wire for lowering bottles and corers, and streaming plankton nets. At most marine institutions there are also a variety of small research vessels, such as converted fishing vessels outfitted for trawling and hauling heavier gear as well as the usual water and plankton samplers. They are often run by experienced ex-fishermen who know the waters and the bottoms and often much about the biology of the area. It is wise for the newcomer to utilize such expertise.

In order to study the microbiology of the more exotic habitats, such as coral reefs and open tropical waters, it is not necessary to go on deep-sea expeditions. Many small marine stations scattered throughout the world can be reached by air (Hiatt, 1963; Webb, 1974) and have living, laboratory, and small boat facilities available at very reasonable rates compared to the daily cost of operating large research vessels. Facilities such as the Bermuda Biological Station offer quick access to the open sea, where planktonic foraminifera can be obtained (Anderson and Bé, 1976; Bé and Anderson, 1976), as well as access to the seagrass beds and coral reefs, all within a 5-mile radius. Even when such facilities are not available at a desired location, sleeping and makeshift laboratory quarters can be rented, as can the boat and the services of a local fisherman. Such a temporary laboratory set up at Majuro Atoll in the Marshall Islands by the author provided material for the papers by Johnson et al. (1971) and Sorokin (1974). Ocean islands are relatively inexpensive platforms for studying both neritic and oceanic problems in a number of different geographic locations.

3. Offshore Sites

Some research requires the use of open-ocean research vessels. There is no other way to study life away from the margins of the sea or to probe the abyssal waters, plains, and trenches of the deep sea with their specialized problems associated with pressure and temperature (Jannasch and Wirsen, 1977a,b). Offshore research vessels fall into three categories. Vessels under 100 feet are good for several hundred to a thousand miles' range and up to two weeks at sea with minimal work and equipment space. Then there are the intermediate-size vessels between 100 and 200 feet, which have both the range and the facilities for staying at sea for many weeks and for making ocean crossings. In the American fleet, new vessels, such as those of the *Oceanus* class owned by the National Science Foundation, are replacing the surplus World War II vessels, such as *Chain* and *Trident*, which were converted to research vessels and served their purpose very well. The larger ships, such as the *Agor* class owned by the Navy, are over 200 feet long, take a large crew to operate, and are too expensive for all but a few institutions who operate them for the larger, multi-investigator, multi-nation programs. Some special purpose research vessels are the shallow draft *Alpha Helix*, *Lulu* the mother ship of the research submersible *Alvin*, and the floating spar *Flip*, which can assume a vertical position and allow one to descend some 200 feet to make observations and obtain samples. Most research vessels are nothing more than floating hotels and empty laboratory space to be fitted out before each cruise. The permanent apparatus consists of A, J, and U frames for getting wire over the side; winches and wire used for sampling; water samplers; and equipment for the routine measurement of position, depth, temperature, salinity, and dissolved oxygen to characterize water masses. Specialized equipment for sampling as well as all the equipment for the examination and estimation of biological, microbiological, and chemical samples and parameters must be purchased, designed, fabricated, prepared, and operated by the scientists in cooperation with the marine technicians who do the routine sampling and analysis. The logistic, economic, and human problems, not to mention the vagaries of the sea and the rolling bitches called research vessels, are enough to discourage all but the most resolute or the demented from taking part in such endeavors. But the challenging problems discussed in Chapters 2 and 3 are much too intriguing to ignore.

A problem with oceanic biology practiced from a ship is that the oceanographer is sampling blind and cannot tell what is being missed. In the upper euphotic zone, the blue water plankton that is too fragile to survive in plankton nets must be observed, captured, and studied by SCUBA divers (Hamner, 1974; Hamner et al., 1975). For deeper waters, down to the sea floor, observations and sampling can be made from such research submersibles as the *Alvin* and the *Archimede* (Ballard and Kristof, 1975; Heirtzler and Kristof, 1975). Most biological studies, however, must be conducted from the usual research vessel, using traditional methods to monitor the biology of the water. The distribution of organisms vertically and horizontally in marine basins (Platt and Filion, 1973) and open waters (Mullin and Brooks, 1976; Riley, 1976) is often too patchy and variable to use standard oceanographic depths, which can straddle zones of aggregation. To avoid blind sampling, detailed profiling finds strata of interest. Phytoplankton maxima can be detected by using a submarine photometer to show light attenuation with depth and to calculate the compensation point or 1% light level (isolumen) where phytoplankters tend to aggregate. A more precise method is to obtain a vertical profile of fluorescent particles (containing plant pigments) by monitoring water pumped in vertical and horizontal transects with a fluorometer (Holm-Hansen et al., 1965). Parameters other than phytoplankton with maxima worth detecting appear to be those for oxygen, biomass as indicated by adenosine triphosphate (ATP), and components of the dissolved organic carbon such as carbohydrates (Sieburth et al., 1977). A water mass is first marked by a drogued buoy (Volkmann, Knauss, and Vine, 1956; Shields and McFall, 1974; Vachon, 1975), and then a vertical profile is analyzed for a number of parameters to detect the water strata of interest. By determining the temperature of each strata of interest, the oceanographer can relocate the depth of the same strata (isotherm) at later time intervals by dropping an XBT. An XBT is a disposable bathythermograph, which draws a depth–temperature curve (to depths of 450 or 750 m) electronically by sensing water temperature with a quick-response thermistor that is weighted and sinks but is linked to the recorder by an excess length of thin copper wire. The results of such profiling and time sequence studies provide a rational basis for the selection of sampling depths at each new station and for following the same strata with time despite an ever-changing depth due to internal waves that change with the sea state.

Plate 4-1. Samplers for the sea–air interface. (*A*) A bulb used to skim the surface water. (*B*) The fly-screen sampler, which lifts off the upper 250 μm layer in the mesh and drains when held vertically. (*C*) Collection of a 27 μm thick layer by adsorption onto glass. (*D*) A continuous large volume sampler based on the above principle. (*E*) The bubble microtome, which ejects the concentrated microlayer as a jet drop is used as the mechanism in (*F*) the bubble interfacial microlayer sampler, which collects dried sea salt. (*A*) from A.V. Tsyban, 1967, *Hydrobiol. J.* 3:84–86; (*B*) photograph courtesy of Paul W. Johnson; (*C*) from G.W. Harvey and L.A. Burzell, 1972, *Limnol. Oceanogr.* 17:156–57; (*D*) from G.W. Harvey, 1966, *Limnol. Oceanogr.* 11:608–13; (*E*) reprinted with permission from F. MacIntyre, 1968, *J. Phys. Chem.* 72:589–92, copyright by the American Chemical Society; (*F*) from J.L. Fasching et al., *J. Récherches Atmosphériques* 3–4:650–52, 1974.

One does not have to be a member of an oceanographic institution to initiate or to participate in deep-sea studies. The University National Oceanographic Laboratory System (UNOLS) is a clearinghouse coordinated by the Woods Hole Oceanographic Institution for some 30 university-run ships sponsored by the U.S. Government through a number of agencies. Arrangements can be made by anyone holding federal grants to be accommodated on one of these vessels for their grant-related work if a year is allowed to work the program in with the ship schedules. Other investigators and their students can often be accommodated on a space-available basis. But the incidental costs of marine and electronics technicians, the costs of the replacement of lost equipment, and the cost of shipping equipment and samples in addition to the cost of training, equipping, and transporting personnel is a sobering experience second only to obtaining adequate funding for such a costly endeavor in the first place. This, plus the difficulties of doing microbiological and chemical work from a rolling platform with a team, each member having to contribute to the overall program, despite *mal-de-mer* and fatigue, are among the reasons why only a hearty few participate in this sport.

B. SAMPLERS AND SAMPLING

1. The Surface Microlayer

To obtain a valid surface sample, it is important to avoid disturbing or contaminating the surface film be-

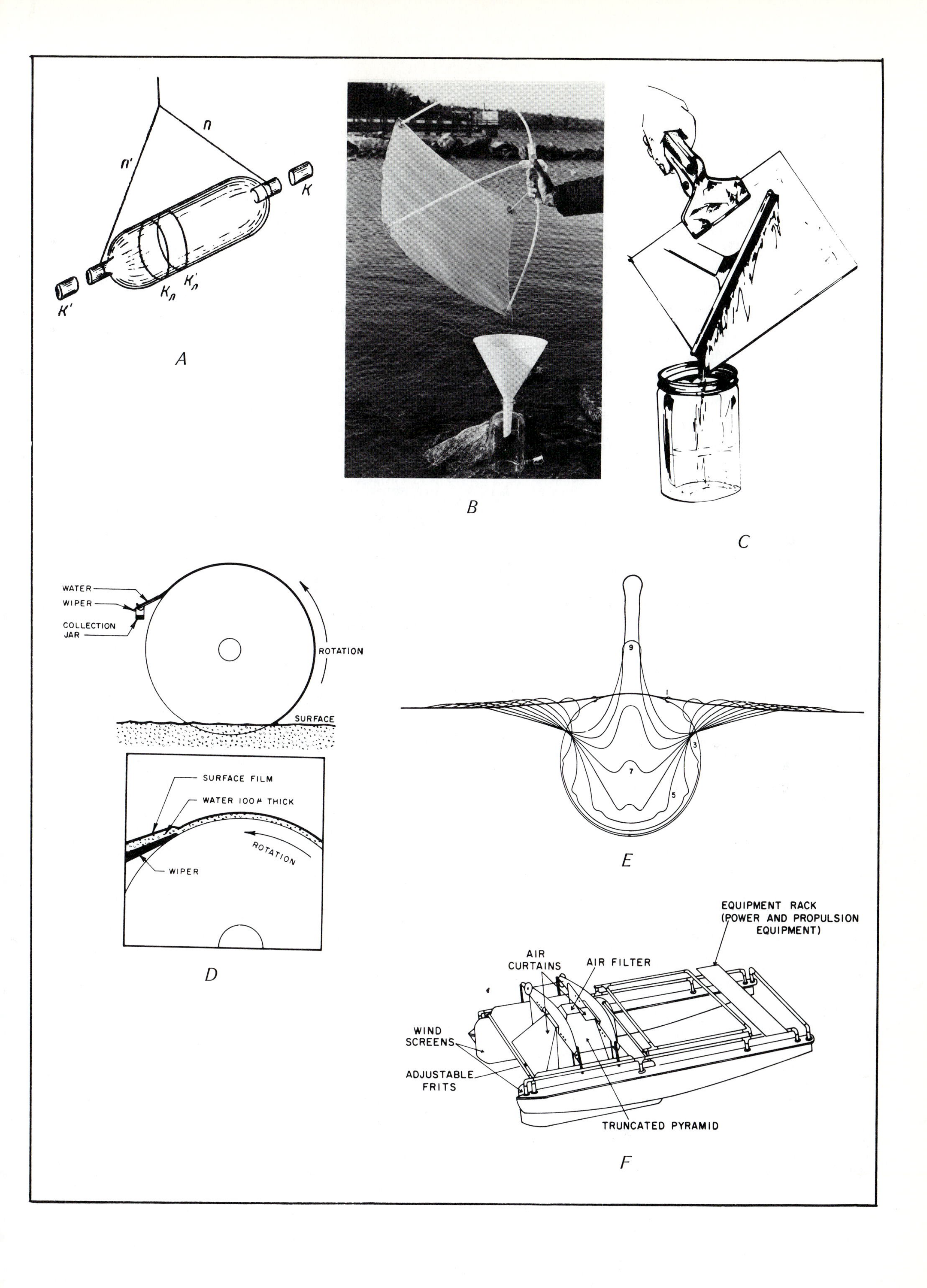
n
n'
K
K'
K_n
K'_n
A
B
C
WATER
WIPER
COLLECTION JAR
ROTATION
SURFACE
SURFACE FILM
WATER 100 µ THICK
ROTATION
WIPER
D
9
1
3
7
5
E
EQUIPMENT RACK (POWER AND PROPULSION EQUIPMENT)
AIR CURTAINS
AIR FILTER
WIND SCREENS
ADJUSTABLE FRITS
TRUNCATED PYRAMID
F

fore sampling. Although such sampling from a research vessel can be done from the bow (Sieburth, 1965, 1971a), bow pulpit, or walking boom (Beebe, 1926), sampling from a small boat that puts the operator next to the water is preferable when the sea state permits. Inflatables, such as the Zodiac, are ideal, since they survive the trauma of launch and retrieval and have a low freeboard. Surface samplers and sampling techniques have been summarized by MacIntyre (1974). A variety of types are shown in Plate 4-1. Collapsed plastic bottles with floating capillaries as an inlet (Goering and Menzel, 1965) and the Volumetric Pipette sampler of Tsyban (1967, 1971) are ill-advised, as they take in a large and variable quantity of subsurface water that dilutes the surface microlayer. Continuous samplers have been devised that scrape the microlayer from rotating surfaces (Harvey, 1966; Schlieper, 1972), but so far their design limits them to protected waters and very mild sea states. The screen sampler devised by Garrett (1965) and applied by Sieburth (1965, 1971a,b), Jarvis et al. (1967), Duce et al. (1972), Duce, Stumm, and Prospero (1972) and Sieburth et al. (1976), among others, is probably the most practical sampler to date (Parker, 1978). It can be used in fairly rough seas, and although time-consuming, 5 liters can be harvested in 80 minutes, some 90 ml at a dip, with three people and two samplers. The screen sampler has defined limitations of a 70% sampling efficiency of the surface film, and the approximately 250 μm thick sample dilutes the 0.1 μm surface microlayer by a factor of 1:2500.

The glass plate sampler that uses a squeegee to remove the sample (Harvey and Burzell, 1972) has been used for studying the bacterioneuston by Dietz, Albright, and Tuominen (1976). Although in the original report, a 60 to 100 μm thick film was obtained, Hatcher and Parker (1974) describe a 20 μm film, whereas in the author's lab, 27 μm films have been obtained. The concentrations of dissolved and particulate materials obtained with a glass plate sampler can be double that obtained with a nylon screen yielding 270 μm thick samples; but the glass plate appears to either exclude or fail to release a major part of the surface film. When the sample concentrations are corrected for sample thickness (Sieburth et al., 1976), screen samplers yield values some five times that of glass samplers. To obtain samples of equal volume takes twice as long with a glass plate sampler (both sides of a 30 by 30 cm dipping area) than with a screen sampler (some 65 by 65 cm in area). The use of Teflon plates (Larsson, Odham, and Sodergren, 1974; Miget et al., 1974), polycarbonate filters (Crow, Ahearn, and Cook, 1975), and a nonperforated polycarbonate membrane offer interesting alternatives to the screen and glass plate samplers and require further study. The bubble microtome (MacIntyre, 1968; Fasching et al., 1974) is an interesting approach for sampling trace metals and other chemicals, but the dry sea-salt samples have not been found to be useful for microbiological sampling.

2. Water Column

a. Nekton. The water column is sampled to obtain a cross section of the organisms contained in a specific stratum of water. Nekton or swimming forms are obtained by gill-nets, seines, trawls, and baited lines: gill-nets and purse seines for the species occurring in the upper layers and midwater and bottom trawls for those occurring at mid-depth and on the bottom (Kändler, 1972). For microbiological studies, fish caught with lines are superior to those whose surfaces are abraded by gill-netting or purse-seining as well as to those fish caught in otter trawls, which are tumbled along the bottom where their gills become choked with bottom sediment. A variety of traps are used to capture nekton and benthic animals in good shape for microbiological studies. Baited minnow traps work well with the smaller marine species.

b. Plankton size fractions. Plankton consists of all the floating forms from bacteria to the giant jellyfish and includes the zooplankton, protozooplankton, phytoplankton, mycoplankton, and bacterioplankton. In most studies on plankton, only specific categories of organisms are sampled. Although these categories are based on morphology, an explicit trophic mode is also implicated (Lenz, 1972). Identification and enumeration is by direct microscopic examination, which is difficult, tedious, and time-consuming. But size is used as a means of concentrating and fractionating the plankton from its suspending water when we use plankton nets, screens, or filters with specific mesh or pore sizes. The application of the Coulter Counter for estimating size fractions of the plankton (Sheldon and Parsons, 1967a,b; Parsons,

1969) has revived the use of terms that describe the plankton on the basis of size (Dussart, 1965) rather than taxonomic groupings or trophic mode.

A proposed scheme that includes all size fractions and uses a non-ambiguous nomenclature based on 10^{-3} (milli) units of the International (Metric) System of measurement as prefixes for the smaller units has been developed by Sieburth, Smetacek, and Lenz (1978). These are included with the plankton categories and suggested size fractions for the nekton in Table 4-1. This scheme is not open-ended as are the older schemes, and it includes everything from bacteriophage (virio-plankton) to whales (meter-neton). This scheme shows the size overlap between the plankton categories, especially the

TABLE 4-1. A proposed size classification scheme for components of pelagic ecosystems. The classes of plankton are compared with plankton size fractions whose terminology is based on traditional usage. Since a compatible size classification scheme does not seem to exist for the nekton, one is proposed whose terminology is based on the metric system of measurement. From Sieburth, Smetacek, and Lenz (1978).

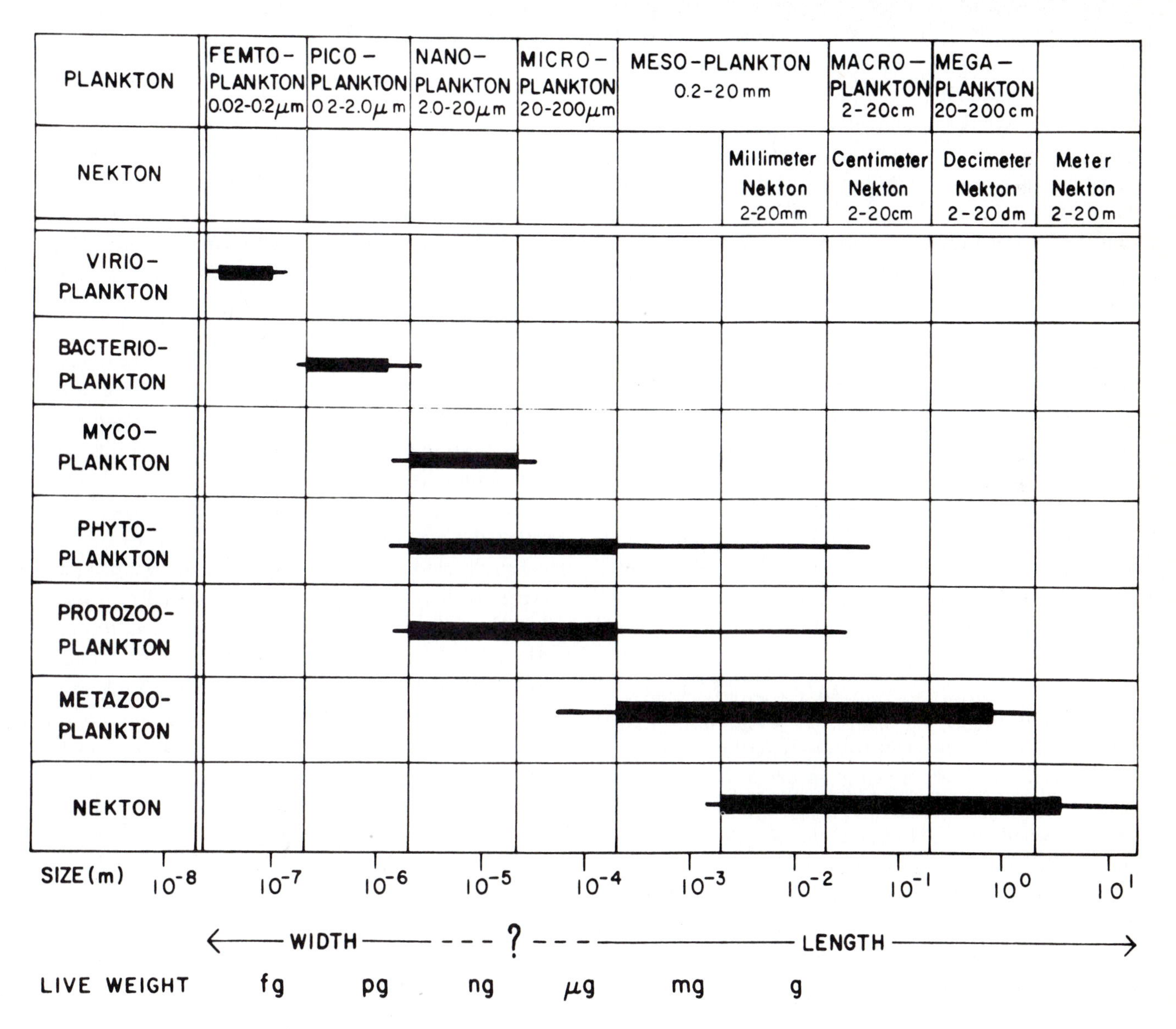

larger size fractions. It should be noted that in discussions of size fractions, width or length is not always clearly stated. The length of the larger animals in the nekton, such as amphipods, squid, fish, and mammals, is measured. Even the forms in the plankton as small as copepods and chains of diatoms that are caught in nets are thus measured. But, for the even smaller forms in the plankton (nanoplankton and picoplankton) that pass the standard mesh nets, length does not determine retention. Water flowing through nets and filters forms streamlines from the water mass through the pores. The nanoplankters and picoplankters caught up in the flow of water apparently line up longitudinally with the streamlines and only their width determines whether they will pass through or be retained. An example is the large spirochete *Cristispira*, some 90 μm in length and just under 2 μm in width, which easily passes through a 3 μm porosity Nuclepore membrane, with very few individuals being caught. The transition area between width and length is indicated in Table 4-1. For the two smallest plankton size fractions, the femtoplankton and picoplankton, the sizes are almost identical to the virio-plankton and the bacterioplankton, respectively. The implications of this are discussed in Chapters 6,C, 7,C, and 14. Although a 0.2 to 2.0 μm fraction will be mainly bacterioplankton, from a practical viewpoint a 0.1 to 1.0 μm fraction is preferred, since it will include the narrowest minibacteria and most of the larger bacterioplankton while excluding the smallest flagellates. For the plankton, the only serious overlap involves the protists. Cultural procedures are required to enumerate the mycoplankton, whereas the phytoplankton can be distinguished from the protozooplankters and differential counts obtained using the autofluorescence of chloroplasts in epifluorescence microscopy.

PLATE 4-2. Sampling for net plankton. (*A*) The conventional plankton net. (*B*) Release mechanism for closing the mouth of a net before hauling it to the surface. (*C*) The intermediate speed Clarke–Bumpus net, with an opening and closing mechanism. (*D*) Gulf Stream samplers towed at high speeds are used to catch the faster zooplankton, including older fish larvae that escape conventional nets. (*E*) The unique Hardy Plankton Recorder that traps plankton on a strip of gauze, which is rolled into formalin as a preservative by a propeller action dependent upon speed. (*F*) Deck mount unit used to collect plankton from pumped water. (*A*) from J. Lenz, 1972, in *Research Methods in Marine Biology* (C. Schlieper, ed.), University of Washington Press, Seattle; (*B*) courtesy of Sir Alister Hardy, 1956, from *The Open Sea*, Houghton-Mifflin, Boston; (*C, D*) from J.H. Wickstead, 1965, *An Introduction to the Study of Tropical Plankton*, Hutchinson Publ., London; (*E*) from J. Fraser, 1962, *Nature Adrift*, G.T. Foulis, Yeovil, England; (*F*) from J.R. Beers, G.L. Stewart, and J.D.H. Strickland, 1967, *J. Fish. Res. Bd. Canada* 24:1811–18.

c. Net plankton. Plankton is usually captured in nets that come in the standard configuration shown in Plate 4-2,*A*, which are fished at speeds of 2 to 4 knots. A net bucket that collects the sample is screwed or tied onto the small end (cod end) of the net. Plankton nets are usually attached to a wire above a terminal weight or depressor and towed horizontally at a known depth and speed for a period of time; the amount of water that passes through is estimated from a flow meter in the mouth of the net. After the tow, the plankton adhering to the sides of the net are washed into the sample jar and then removed to a tray for examination. These tows sample at mostly one depth, although some material is caught while the net is being lowered and raised. In a tow that samples more precisely, the nets are fitted with a mechanism for opening and closing; these stream the net at one specific depth (Plates 4-2,*B*,*C*). The opening and closing mechanisms are operated by a messenger, a weight that slides down a hydrographic wire and operates a release mechanism on a sampling apparatus. The Clarke–Bumpus net (Plate 4-2*C*) has a special throat mechanism before the net, which opens the net at the start of the tow and closes it at the end of the tow. A mechanism similar to that shown in Plate 4-2,*B*, which attaches to the hydrographic wire, can deploy a rolled net at a specific depth and then close it before recovery with a bridle sewed around the net at mid-length, which collapses the net when the towing bridle is released. The faster zooplankton and even some of the smaller nekton are captured by high-speed plankton samplers, such as the Gulf Stream (Plate 4-2,*D*), which are towed at speeds of 10 knots. The Charles J. Fish modification (Fish and Snodgrass, 1962) was an attempt to use an electrically operated rosette mechanism to deploy sample buckets at different depths in order to obtain a series of samples at each deployment. A similar series of samples is obtained at the usual tow speeds with the Bé Multiple Net system (Bé, 1962) in which a

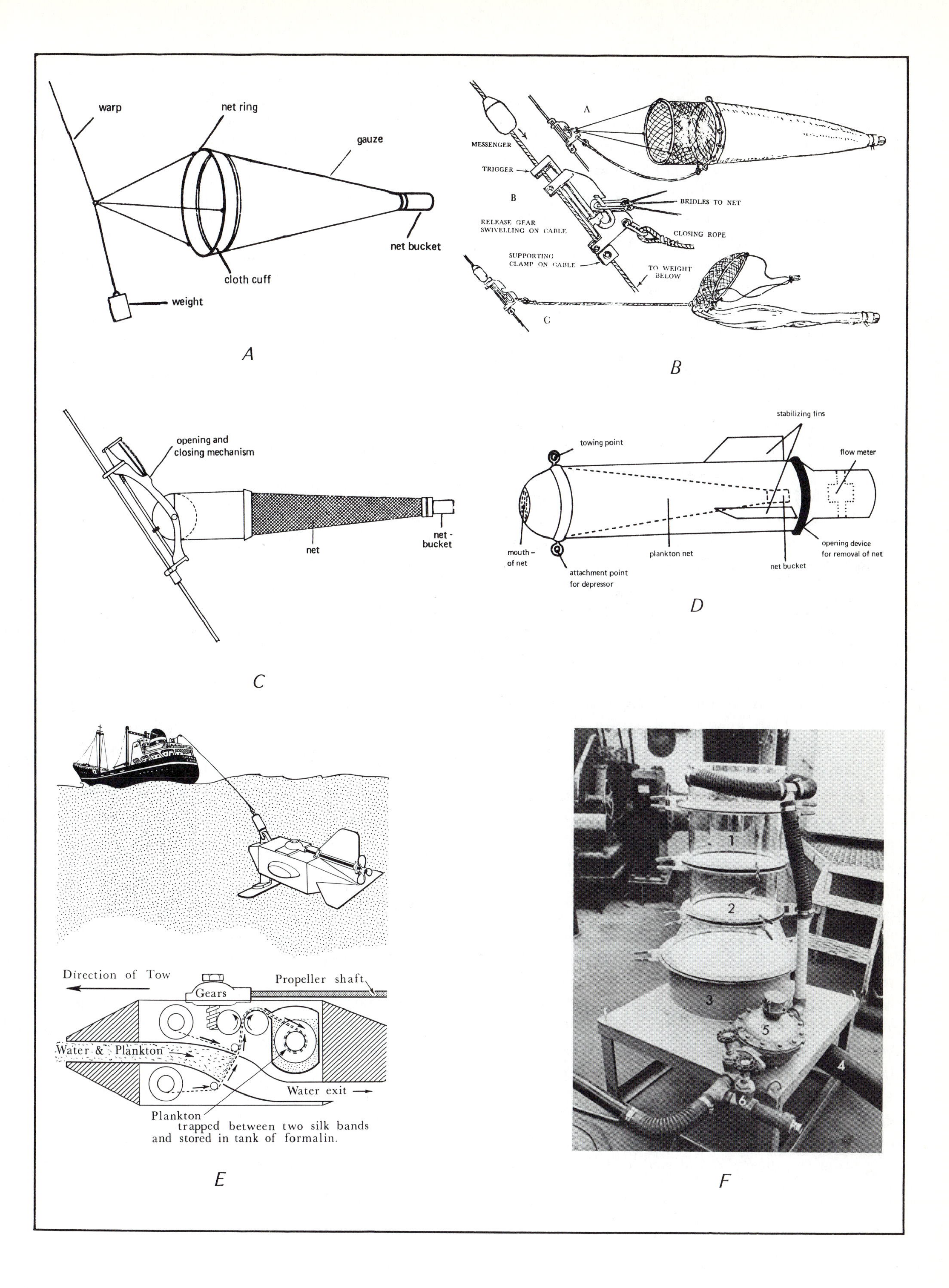

warp
net ring
gauze
net bucket
cloth cuff
weight
A
MESSENGER
TRIGGER
B
RELEASE GEAR SWIVELLING ON CABLE
BRIDLES TO NET
CLOSING ROPE
SUPPORTING CLAMP ON CABLE
TO WEIGHT BELOW
C
B
opening and closing mechanism
net
net-bucket
C
stabilizing fins
towing point
flow meter
mouth of net
plankton net
net bucket
opening device for removal of net
attachment point for depressor
D
Direction of Tow
Propeller shaft
Gears
Water & Plankton
Water exit
Plankton trapped between two silk bands and stored in tank of formalin.
E
1
2
3
4
5
6
F

series of nested nets are fished one at a time by sequentially exposing the throat of each individual net. Another useful, discrete depth plankton sampler was described by Aron et al. (1964). The quantity of plankton caught can be doubled by fishing two nets of similar mesh size at the same time with a "Bongo Net" system, as used by the MARMAP program of the U.S. National Marine Fisheries Service. A rather unique type of plankton net is the Hardy Plankton Recorder (Hardy, 1956; Fraser, 1962) shown in Plate 4-2,*E*. This device is deployed on such "ships of convenience" as British trawlers as they travel to and from the fishery grounds, so that plankton surveys can be made over areas larger than those possible with a limited number of research vessels. As water passes through the device and plankton is caught on one strip of sampling silk that is slowly advanced by a propellor-driven transmission, a second silk strip forms a continuous plankton "sandwich," which is rolled up tightly and stored in a chamber containing formalin as a preservative. In addition to their use in plankton surveys, the stored plankton records are available to investigators searching for the distribution patterns of newly recognized species. The Hardy Plankton Recorder has been repackaged with an Undulating Oceanographic Recorder (UOR) (Bruce and Aiken, 1975), which records the major parameters of salinity and temperature with depth in the euphotic zone so that it can be towed by "ships of convenience," at speeds of 7 to 15 knots (Aiken, Bruce, and Lindley, 1977).

Plankton nets are useful for collecting specimens and determining what types of organisms are present, but as quantitative samplers, they have definite limitations. Flagellates in the nanoplankton, which sometimes dominate the phytoplankton, especially in the open sea (Pomeroy and Johannes, 1968), pass through conventional phytoplankton nets. Even special 10 μm nanoplankton nets fished at half-knot speeds pass much of the smaller flagellates. Conventional phytoplankton nets with 30 to 75 μm pores retain the larger diatoms and flagellates along with the slower swimming and smaller zooplankton. Zooplankton are usually fished for with 200 to 900 μm pore nets. Such nets pass the earlier nauplier stages while catching fish larvae and an occasional small fish. An even bigger disadvantage than not getting all the members of a trophic group is net clogging. Not only does the mesh size change, but as clogging increases, water is diverted and does not pass through the net.

PLATE 4-3. Nonsterile sampling for sub-net plankton. (*A*) The standard Nansen bottle, the standby of physical and chemical oceanographers, was often used by biologists, despite its dirty metallic interior. (*B*) The Van Dorn Bottle, a larger volume plastic sampler. (*C*) the 5-liter Niskin bottle, which has largely replaced the Nansen bottle. (*D*) the Go-Flo bottle, with ball-type end closures (all plastic) that permit the cleaned bottle to go through surface layers closed but to open automatically at a pre-set depth until it is closed at depth by a messenger or an electronic release on a rosette frame. (*A*) from M. Grant Gross, 1972, *Oceanography: A View of the Earth*, Prentice-Hall, Englewood Cliffs, N. J., p. 190; (*B*) from W.G. Van Dorn, 1956, *Trans. Amer. Geophys. Union* 37:682–84, copyrighted by American Geophysical Union; (*C*) from S. Niskin, 1964, *Limnol. Oceanogr.* 9:591–94; (*D*) courtesy of S. Niskin, General Oceanics, Miami, Fla.

d. Shipboard processing. An alternate method of obtaining plankton is to pump water through tubing from discrete depths and pass it through stacked sieves to obtain different size fractions (Beers, Stewart, and Strickland, 1967; Plate 4-2,F). The smallest living organisms that can be concentrated by plankton nets and pumping through screens is usually above 30 μm in size. The most practical way of obtaining concentrated samples of all the microbial plankton without losing the smaller fractions, is to collect water samples and process them aboard ship (see Plate 6-1). The larger and sparser the size fraction desired, the larger the water sample required. One way of obtaining large volumes of water is to tape a polyethylene tube to a weighted hydrographic wire and lower it to the desired depth. The water is raised with a peristaltic pump placed within the ship as close to the water line as possible, or it can be raised at a faster rate using a vacuum system. But samples obtained by the latter system are partially deoxygenated. For the small microorganisms, water volumes sampled from acid-cleaned 5-liter Niskin samplers (Niskin, 1964) or Go-Flo bottles are sufficient (see Plate 4-3). The largest water samplers that are practical to handle are the 30-liter version of the Niskin sampler, which weighs some 90 lb when filled, and the 60-liter Go-Flo bottles. Deployment of ten 30-liter samplers obtains nearly 300 liters of sample, which yields over 250 ml of an 1000-fold concentrate with a continuous reverse flow concentrator (see Chapter 6). Such particulate preparations from the open ocean exceed the normal concentration in rich estuarine samples and can be examined microscopically, assayed chemically, used for culture, or preserved for taxonomic and ultrastructural

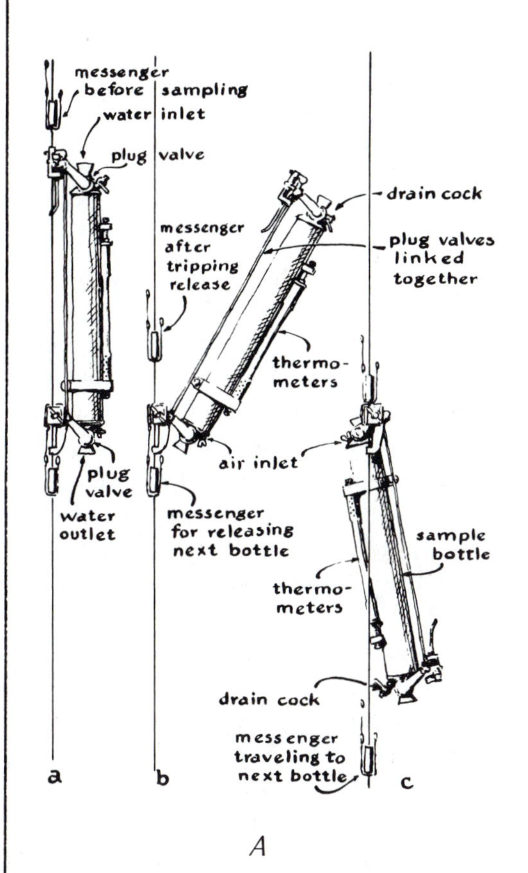
messenger
before
sampling
water inlet
plug valve
drain cock
messenger
after
tripping
release
plug valves
linked
together
thermo-
meters
air inlet
plug
valve
water
outlet
messenger
for releasing
next bottle
sample
bottle
thermo-
meters
drain cock
messenger
traveling to
next bottle
a
b
c

A

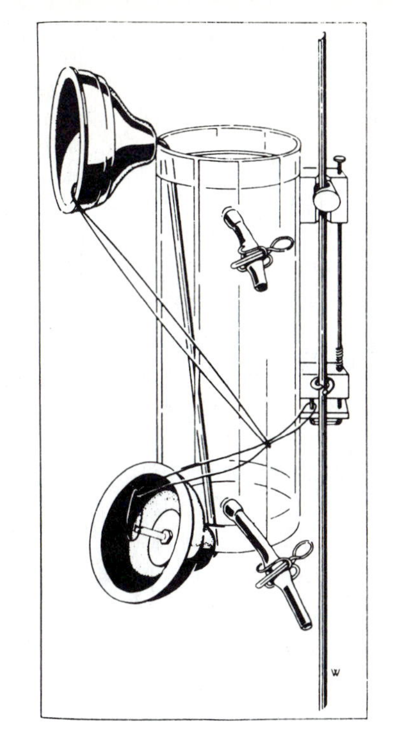

B

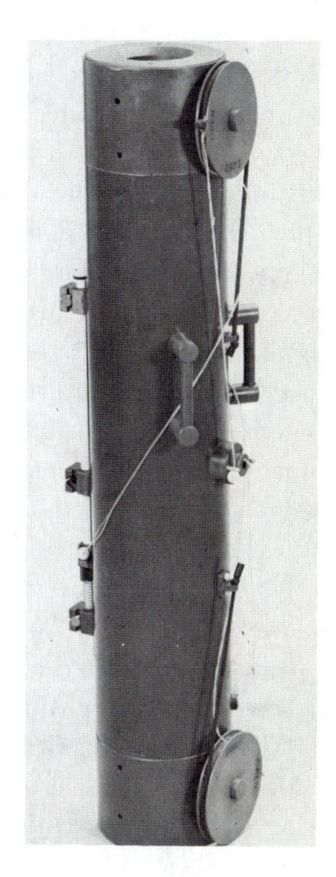

D

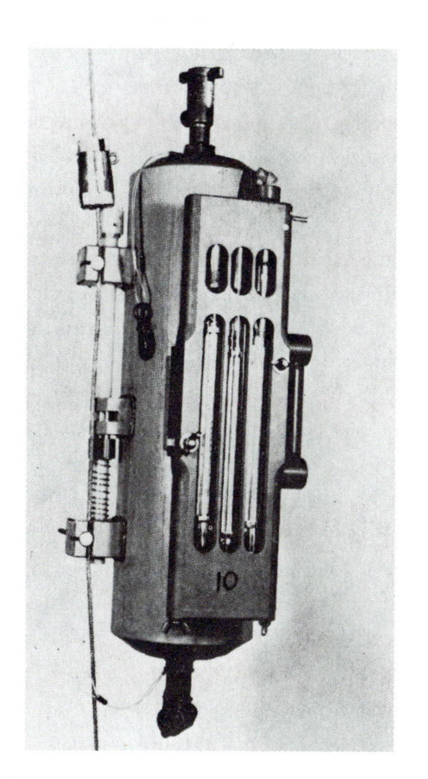
10

C

studies. Reliable samples of picoplankton, nanoplankton, and microplankton can be obtained in this manner without the problems inherent in net sampling.

e. *Sterile samplers.* Bacteriologists have traditionally used some type of sterile water sampler (Plate 4-4) to obtain their samples, and tradition shuns the use of open bottles (ZoBell, 1946; Sorokin, 1962; Sieburth, 1971a). But when it comes to the practicality of obtaining many large (5- to 300-liter) samples at sea, practicing oceanographic microbiologists have had to modify tradition (Sorokin, 1971; Sieburth et al., 1976) and use freshly acid-washed samplers, which apparently destroy the microbiota in the surface layer but are rinsed sufficiently after 20 minutes at depth to not destroy the desired samples. This may horrify purists and medically trained bacteriologists who believe that a sample is only valid if it is aseptically removed from the sea. This obsession with the need for aseptic samples, when relatively uncontaminated ones will suffice, is just one example of misplaced priorities that have delayed the development of marine bacteriology. The use of nonsterile samplers has received a bad press due to the poor selection of samplers and the absence of sanitary precautions. It was only natural that the early bacteriologist Fischer (1887, 1894) would use the narrow-necked brass sampler already perfected by oceanographers. But one objection to its use is the bactericidal effect of the metal ions in brass, as noted by Drew (1914) and ZoBell (1941). A more important objection is that these narrow-necked bottles cannot be cleaned properly. The use of such bottles by Kriss (1963) has been criticized by Sorokin (1962), Bogoyavlenski (1964), and Sieburth (1971a). It is just as well that the narrow-necked, metallic Nansen bottle, encrusted with copper sulfate and having an indigenous microbiota, has passed from the scene and been replaced by wide-necked, easily washed, inert plastic samplers, such as the Niskin bottle (Niskin, 1964).

When nonsterile samplers are adequately washed before use they appear to yield valid results (Kriss, Lebedeva, and Tsyban, 1966; Sorokin, 1971; Sieburth et al., 1976). The only real reservation to using open-on-descent bottles is that they must pass through the surface layer with its rich organic content and microbiota. The acid on freshly scrubbed bottles apparently prevents surface contamination, as noted earlier. But the problem of surface contamination can also be avoided by using the Go-Flo version of the Niskin bottle (General Oceanics), which can be Teflon-lined and fitted with Go-Flo end closures that close the bottle on descent to a depth of 10 m, or another adjustable depth. At a given depth, they open automatically and continue to flush until sampling depth, where they are closed by a messenger on a wire or electronically when several bottles in a rosette are being used.

PLATE 4-4. Sterile sampling for bacterioplankton. (*A*) The J-Z sampler for obtaining water from different depths in evacuated bottles and collapsed rubber bulbs. (*B*) A simple form of the J-Z sampler for nearshore work. (*C*) A system for piggy-backing a collapsed rubber bulb sampler on a Nansen bottle. (*D*) A spring-activated metal frame for expanding the detachable 2-liter, sterile plastic bag on the Niskin bacteriological sampler. (*E*) A deep-sea bacteriological sampler that uses a falling piston to replace the sterile freshwater in the sample chamber. (*F*) A syringe sampler for a small volume, with a protected inlet tube. (*G*) A sampler for maintaining the water sample at the *in-situ* pressure. (*A*) from C.E. ZoBell, 1941, *J. Mar. Res.* 4:173–88; (*B*) from J. Sieburth, 1963, *Limnol. Oceanogr.* 8:489–92; (*C*) from J. Sieburth, J. Frey, and J. Conover, 1963, *Deep-Sea Res.* 10:757–58; (*D*) from S. Niskin, 1962, *Deep-Sea Res.* 9:501–3; (*E*) from Y. Sorokin, 1964, *J. Conseil* 29:25–40; (*F*) from H. Jannasch and W. Maddux, 1967, *J. Mar. Res.* 25:185–89; (*G*) from H. Jannasch, C. Wirsen, and C. Winget, 1973, *Deep-Sea Res.* 20:661–64.

Sterile water samples may still be appropriate for nearshore samples from one or two depths and when volumes less than a liter are required for a coliform count or an ATP assay. The larger surface to volume ratio in small-volume samplers makes the avoidance of contamination of the outer and inner surfaces more critical. The corked jug as described by Johnston (1892) and simplified by Whipple (1927) is a poor choice, since, as ZoBell (1946, p. 26) correctly points out, the external surface contaminates the sample, whereas the hydrostatic pressure below a few meters prevents the removal of the cork. All-glass samplers with a thin-walled neck that is evacuated, sealed, and opened at depth with a messenger were used by Russell (1892) and Issatchenko (1914). The evacuated bottle is good for sampling at several hundred meters and its subsequent development as the J-Z sampler (ZoBell, 1941; Plate 4-4,*A*) has been described by ZoBell (1946). Below several hundred meters, collapsed rubber bulbs can be used, but care should be taken to wash and sterilize them after use to prevent microbial growth (Sorokin, 1962; Willingham and Buck, 1965). A simple form of the J-Z sampler was described by Sieburth (1963; Plate

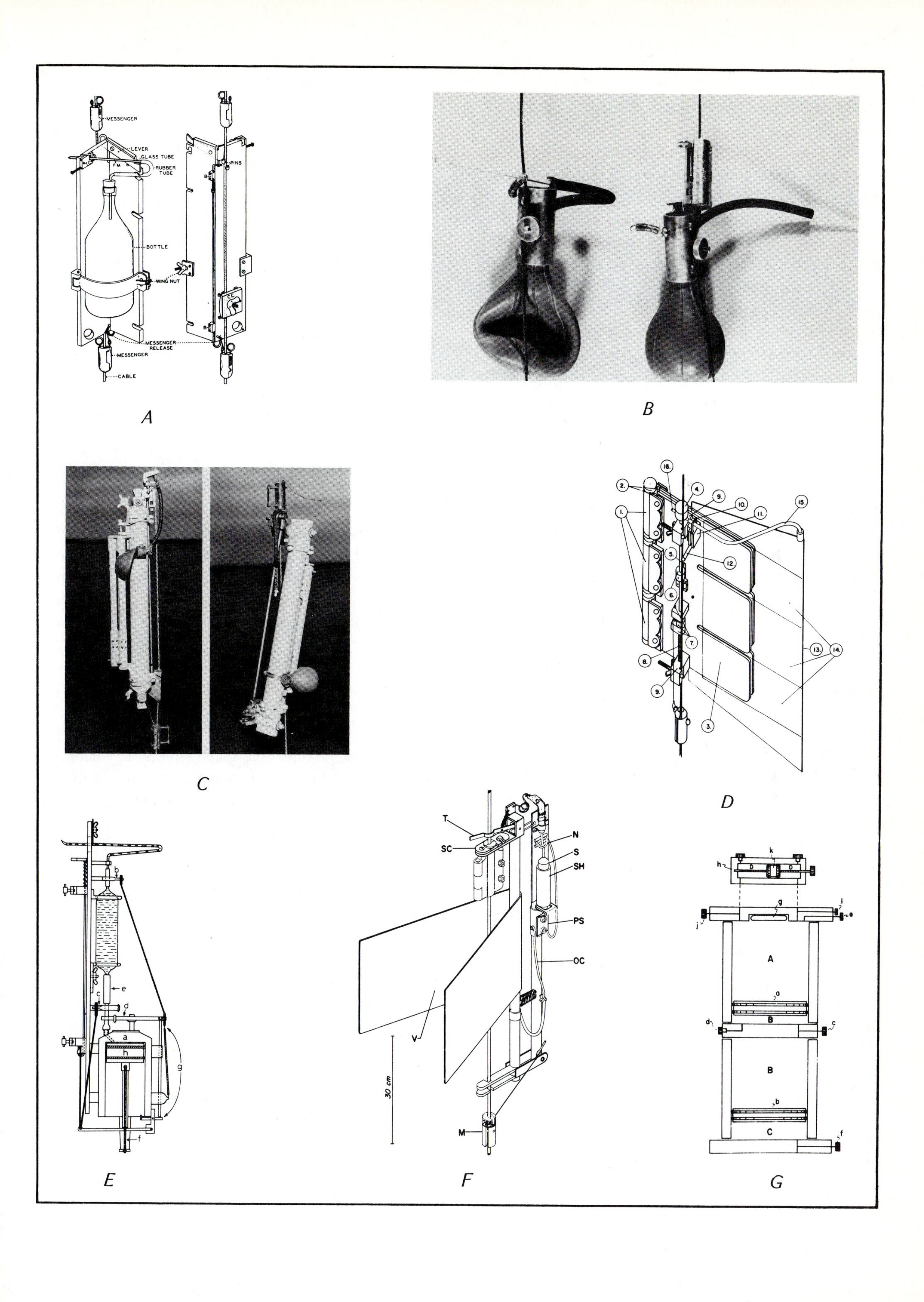
MESSENGER
LEVER
GLASS TUBE
RUBBER TUBE
PINS
BOTTLE
WING NUT
MESSENGER RELEASE
MESSENGER
CABLE
A
B
C
D
T
SC
N
S
SH
PS
OC
V
30 cm
M
E
F
A
B
C
G

4-4,*B*) and a form for anoxic waters by Schegg (1970; Plate 4-5,*F*).

Piggyback devices have been used to take advantage of the reliable tripping mechanisms of the metallic water samplers (Gee, 1932; Sieburth, Frey and Conover, 1963; Plate 4-4,*C*). A well-designed version of the syringe or piston sampler for very small volumes is described by Jannasch and Maddux (1967; Plate 4-4,*F*). Samples of up to 2 liters are obtained by a spring and hinge mechanism that spreads a plastic bag in the Niskin bacteriological sampler (Niskin, 1962; Plate 4-4,*D*). Problems in operation are discussed by Jannasch and Maddux (1967) and in the volume edited by Oppenheimer (1968); this writer objects to the leakage of organic carbon from the sample bags and its possible toxicity. The displacement type sampler, which uses a sterile liquid to keep out contamination and equalize pressure, is probably the most advanced sampler, but it is hardly practical for routine use. Matthews (1913) described an early form and Sorokin (1962, 1964; Plate 4-4,*E*) a more modern version. A novel device for subsurface water sampling from a helicopter was developed by Pinon and Pijck (1975). Jannasch, Wirsen, and Winget (1973) have described a sampler/incubator for studying natural deep-sea populations without decompression (Plate 4-4,*G*), and the application of this apparatus and similar equipment is described by Jannasch and Wirsen (1977b).

3. Sea Floor

The biology of the sea floor largely depends on the microbiology of the sedimenting organic debris. This "secondary productivity" nurtures both an infauna in the sediment as well as an epifauna that roams over the sea floor and lives on this infauna. To study these microbe-based processes on the sea floor one can examine the recently deposited sediment at the sea–mud interface where decomposition is taking place, the overlying water that shows the flux of nutrients released back to the water by these microbial activities, as well as the respiration occurring in the sediments to get an idea of the combined metabolism of the fauna and of the microorganisms that support it.

a. Bottom water. Water at the interface with the sea floor can have very different chemical and thermal properties from the water above it (Bruneau, Jerlov, and Koczy, 1953; Koczy, 1953). A variety of devices used to sample bottom water are shown in Plate 4-5. Edwards and Richards (1956; Plate 4-5,*A*) developed a sampler based on a circular Nansen bottle rack from which a released bottom weight rotated a bar that tripped the bottles shut as they ascended from 1 to 60 m above the bottom. A more modern version by Sholkovitz (1970; Plate 4-5,*B*) is a free vehicle that incorporates ten plastic Van Dorn bottles mounted horizontally 20 cm apart on a frame with a releasable weight that is dropped overboard. A magnesium link that dissolves in 3 hours allows time for the bottom to settle and the bottles to flush with undisturbed water before it triggers the bottles to close and releases the weight, allowing the floats to bring the sampler back to the surface.

The samples, as collected above, show the vertical distribution of dissolved nutrients in the water some distance above the sediment but do not collect a sediment sample with the overlying water. Hjørt and Ruud (1938; Plate 4-5,*D*) devised a hinged sampler that collected both the sediment and the overlying water. A multi-depth pipette was used to withdraw the water samples. A modern version on a larger scale was devised by Schink, Fanning, and Piety (1966; Plate 4-5,*C*). An elongated shaft on a gravity corer housed six plastic Frautschey bottles (modified Van Dorns) that would take water samples at 1-m intervals above the sea floor while obtaining a core of sediment. Bottom water was also collected with sediment in the Carey and Paul

PLATE 4-5. Sampling devices for obtaining bottom water. (*A*) A rack for tripping Nansen bottles at specific levels from the bottom. (*B*) A free-fall vehicle, with a dissolving link that trips a series of bottles and releases an anchor so floats can return the sampler to the surface. (*C*) A sediment sampler that collects water at 1-m intervals above the sediment. (*D*) A hinged sediment sampler that collects overlying water, which can be drawn from 1-cm levels above the mud line by glass tubes (II). (*E*) An ooze-sucker for obtaining surface debris and bottom water in a collapsed rubber bulb. (*F*) A modified J-Z sampler for obtaining and maintaining anoxic water.

(*A*) from R.S. Edwards and F.A. Richards, 1956, *Deep-Sea Res.* 4:65–66; (*B*) from E.R. Sholkovitz, 1970, *Limnol. Oceanogr.* 15:641–44; (*C*) from D. Schink, K. Fanning, and J. Piety, 1966, *J. Mar. Res.* 24:365–73; (*D*) from J. Hjort and J. Ruud, 1938, *Hvalradets Skrifter* 17:145–51; (*E*) from J.McN. Sieburth, 1965, *Ocean Science and Ocean Engineering*, Marine Technology Society, Fig. 3; (*F*) from E. Schegg, 1970, *Limnol. Oceanogr.* 15:820–22.

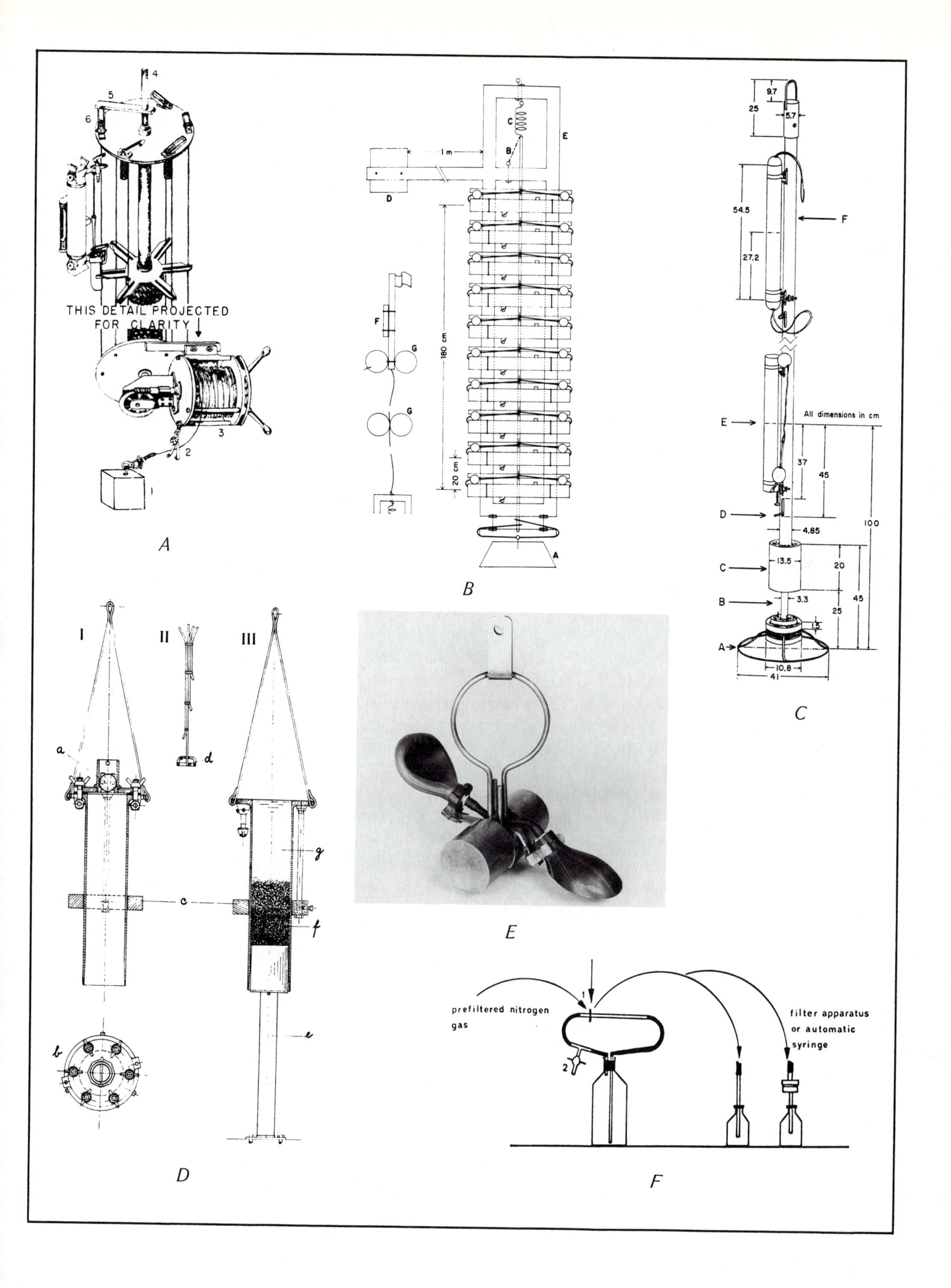
THIS DETAIL PROJECTED FOR CLARITY
A
1 m
180 cm
20 cm
B
All dimensions in cm
C
I
II
III
D
E
prefiltered nitrogen gas
filter apparatus or automatic syringe
F

(1968) modification of the Smith-McIntyre grab sampler (Plate 4-7,*B*).

Specialized samplers have been devised for taking samples of water associated with the benthic microbiota. Waters behind sills such as fjords and trenches can become anoxic and thus maintain an oxygen-sensitive anaerobic microbiota. A useful variation of the J-Z sampler for such water bodies is described by Schegg (1970; Plate 4-5,*F*); here the water is collected in sample bottles preflushed with nitrogen and is then removed in a closed system. In order to study the bacterial flora in the loose flocculent material from the sea–sediment interface (Sieburth, 1967), a modern version of Moore's ooze-sucker was devised (Sieburth, 1965; Plate 4-5,*E*). A connecting bar between balanced weights closes the rubber tubing connecting one-half of a rubber bulb that covers the sampling area to a collapsed bulb that receives the sediment–water slurry. Upon reaching the bottom, the weight touches first and is raised, thereby opening the tubing that permits the collapsed bulb to aspirate a sample. When the sampler is retrieved, the weighted bar closes the tubing again to retain the sample.

b. Sediments. The type of sediment to be obtained will depend upon the purpose of the study. The area to be sampled can be narrowed down with the aid of published data and the consultation of charts and sailing directions. Once the area is reached, final selection can be made using a precision depth profiler and even an underway sampler, such as the "scoopfish" (Emery and Champion, 1948). The type of bottom sought and the ultimate use of the sample will dictate the type of sampler used. Methods for obtaining the benthos have been summarized by Holme (1964) and Holme and McIntyre (1971).

Silts and clays constitute the bulk of the sediments found on the ocean floor. Microbiologically, only the upper layers with microbial and animal activity are of interest, and therefore only relatively short cores are needed. Gravity corers are commonly used to obtain short cores in these sediments. The Emery and Dietz (1941) corer, a conventional type of gravity corer, is often carried aboard research vessels. Its stainless steel coring pipe is fitted with a one-way valve on top that lets the pipe flush during descent but seals it on ascent to help prevent core washout. Lead weights encircle the pipe at the top to provide enough weight for the sharp cutting head to penetrate into the sediment while the sediment is being forced into a plastic lining, which facilitates the removal of the core and its storage. A circle of flexible fingers of metal or plastic between the core liner and the cutting head acts to retain the core. The fingers of the core retainer are pushed up against the liner by the sediment during penetration but are bent down by the weight and bulk of the sediment on retrieval to block the opening and help retain the core. These internal core retainers can distort the core and, for microbiological purposes, they should not be used, if possible, or they should be replaced by the external core retainer proposed by Mills (1961; Plate 4-6,*C*) or by Fenchel (1967; Plate 4-6,*F*).

PLATE 4-6. Coring devices for sampling bottom sediments. (*A*) A sampler for taking short, undisturbed cores that are retained by a one-way valve system. (*B*) A corer for soft estuarine muds. (*C*) A corer for microbiological purposes fitted with an external core retainer that samples both fluid and semifluid sediments and minimizes agitation of the top few centimeters of the bottom sediment. (*D*) A multiple corer for sampling microenvironmental variation in the distribution of microorganisms. (*E*) A free-fall sediment sampler that dispenses with the need for a wire and winch. (*F*) A simple ball core retainer for shallow water coring for microorganisms. (*A*) from J.S. Craib, 1965, *J. Conseil* 30:34–39; (*B*) from N.A. Holme and A.D. McIntyre, 1971, *IBP Handbook No. 16*, Blackwell Sci. Publ., Oxford; (*C*) from A.A. Mills, 1961, *Deep-Sea Res.* 7:294–95; (*D*) from G. Fowler and L. Kulm, 1966, *Limnol. Oceanogr.* 11:630–33; (*E*) from P. Sachs and S. Raymond, 1965, *J. Mar. Res.* 23:44–53; (*F*) from T. Fenchel, 1967, *Ophelia* 4:121–37.

When longer cores are required, additional penetration can be obtained by releasing the corer with a pilot weight, which allows a 7-m free fall (Dietz, 1952). For even longer cores, up to 20 m in length, piston corers are used. A piston within the coring tube rests on the sediment and is held in place with the wire as the coring pipe penetrates the sediment. The vacuum created by the piston above the core reduces the friction that limits core length. Sachs and Raymond (1965) developed a free-falling corer that returns the core to the surface by a float (Plate 4-6,*E*).

Unlike silts and clays, sandy bottoms are usually difficult to sample. Gravity corers often fail in sands, and repeated casts may be needed to obtain a short core. The large box sampler of Bouma (1969) removes a large rectangular sample (30 cm × 30 cm × 45 cm) and per-

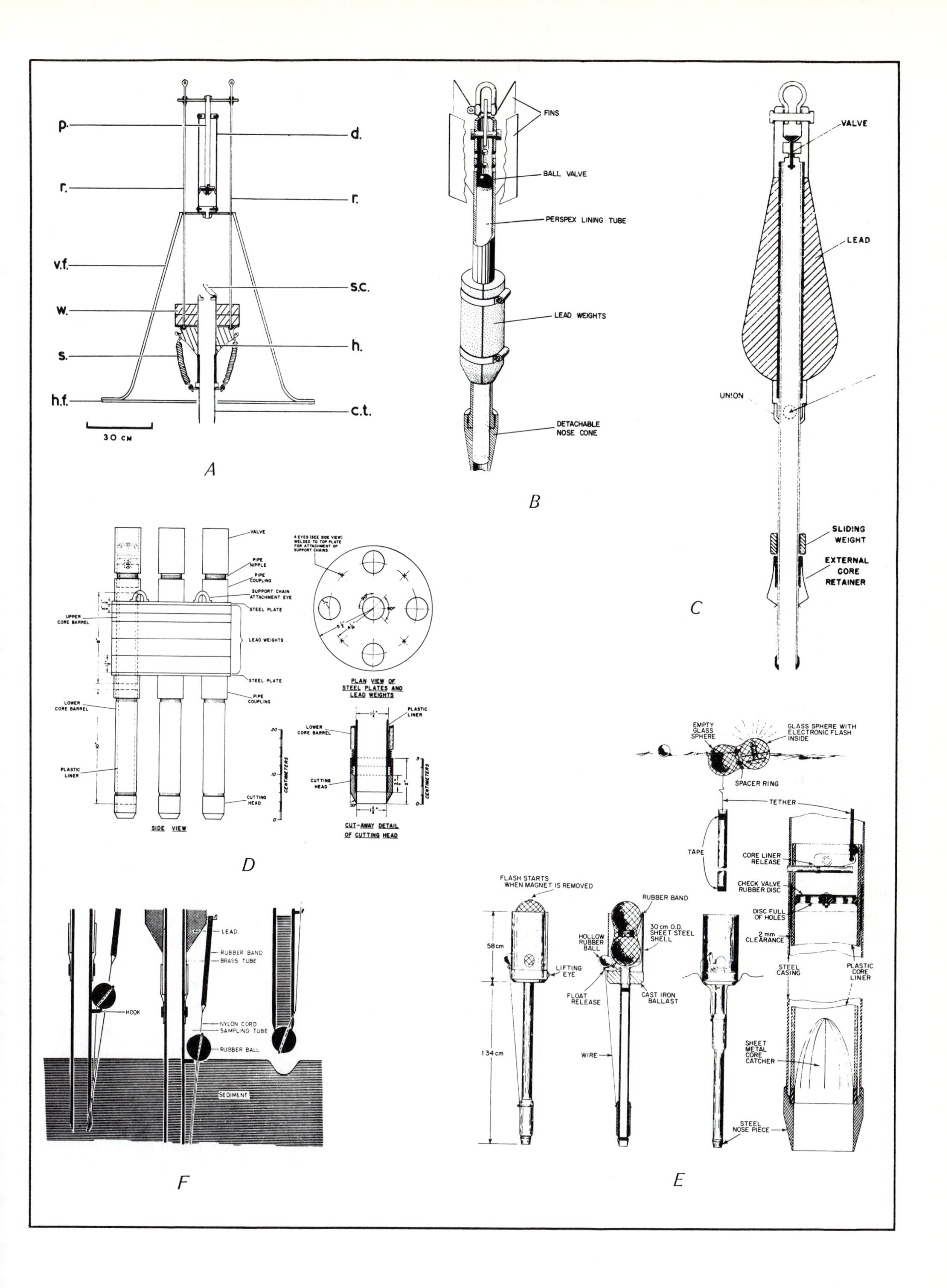

p.
d.
r.
r.
v.f.
s.c.
w.
s.
h.
h.f.
c.t.
30 CM
A
FINS
BALL VALVE
PERSPEX LINING TUBE
LEAD WEIGHTS
DETACHABLE NOSE CONE
B
VALVE
LEAD
UNION
SLIDING WEIGHT
EXTERNAL CORE RETAINER
C
VALVE
PIPE NIPPLE
PIPE COUPLING
SUPPORT CHAIN ATTACHMENT EYE
STEEL PLATE
UPPER CORE BARREL
LEAD WEIGHTS
STEEL PLATE
PIPE COUPLING
LOWER CORE BARREL
PLASTIC LINER
CUTTING HEAD
SIDE VIEW
4 EYES (SEE SIDE VIEW) WELDED TO TOP PLATE FOR ATTACHMENT OF SUPPORT CHAINS
PLAN VIEW OF STEEL PLATES AND LEAD WEIGHTS
PLASTIC LINER
LOWER CORE BARREL
CUTTING HEAD
CENTIMETERS
CUT-AWAY DETAIL OF CUTTING HEAD
D
EMPTY GLASS SPHERE
GLASS SPHERE WITH ELECTRONIC FLASH INSIDE
SPACER RING
TETHER
TAPE
CORE LINER RELEASE
CHECK VALVE RUBBER DISC
DISC FULL OF HOLES
2 mm CLEARANCE
STEEL CASING
PLASTIC CORE LINER
SHEET METAL CORE CATCHER
STEEL NOSE PIECE
FLASH STARTS WHEN MAGNET IS REMOVED
58 cm
134 cm
LIFTING EYE
RUBBER BAND
30 cm O.D. SHEET STEEL SHELL
HOLLOW RUBBER BALL
FLOAT RELEASE
CAST IRON BALLAST
WIRE
E
LEAD
RUBBER BAND
BRASS TUBE
HOOK
NYLON CORD
SAMPLING TUBE
RUBBER BALL
SEDIMENT
F

mits a three-dimensional analysis of sand-inhabiting organisms (the psammon). Longer cores up to 4.5 m can be obtained in sandy bottoms with a vibrating piston corer. Shipboard electricity conducted through a two-conductor–insulated hydrographic cable powers a sealed vibrator motor, which usually restricts such samplers to a depth of 2000 m.

Other coring devices may be more appropriate for the type of samples needed for microbiological observations. A hydrostatic sampler that takes a large diameter short core is the Knutsen Suction Corer (Holme, 1964). A water pump creates a vacuum that sucks up a sample, which is retained by inversion as the sampler is raised. The intersample variation between cores (Anthony, 1963) can be determined and minimized by the use of multiple corers, such as that described by Fowler and Kulm (1966; Plate 4-6,*D*). In shallow waters, the disturbance of the top few millimeters can be minimized by the use of a frame-mounted corer, such as that designed by Craib (1965; Plate 4-6,*A*), which looks like a very easy to use corer. The bottom sampler designed by Emery (1958) for bacteriological studies does not appear to have been used and seems little more than a curiosity.

c. Animals, rocks, and manganese nodules. To obtain specimens of animals and minerals for microbiological studies, a surface area larger than that obtained by corers must be sampled. A number of grabs and dredges have been devised for this purpose (Plate 4-7). The weight of the grab allows penetration of the sediment before its jaws bite and grab a sample. Grabs can be modified for a short distance free fall (as are gravity corers) to allow deeper penetration. The amount of sediment sample obtained with a given sampler depends upon the nature of the sediment. A hard object like a rock or a clam caught between the jaws prevents closure and allows the sample to wash out. These samplers are discussed by Holme (1964). The basic samplers are the clamshell type, such as the Peterson bottom grab (Plate 4-7,*A*), which has two weighted jaws that close as the sampler is withdrawn, to obtain a 0.1 to 0.2 m^2 sample. Other grabs of this type have a pressure plate mechanism to release the jaws. A modification is the Okean grab, designed for deep-sea work, that has the lids of the "clamshells" open on descent to minimize water resistance. Better leverage for closure is achieved by the long arms of the Van Veen grab, but water resistance during ascent can cause them to trip and lose the sample. The Aberdeen grab sampler (Smith and McIntyre, 1954; Plate 4-7,*B*), a much improved version, is held within a framework that ensures a good grab even with a drifting ship in rough weather. Only when the sampler rests squarely on the bottom does a trigger plate allow a pair of large springs to force the jaws into the sediment and to release the levers that close them. This sampler was modified by Carey and Paul (1968) as a benthic environmental sampler, which takes water samples before the sediment is disturbed by the grab. Very popular grabs in the United States are variations of the orange-peel grab (Plate 4-7,*C*). Three or four curved jaws take a hemispherical sample, and a modification involves a canvas sleeve over the top one-half to minimize washout.

PLATE 4-7. Large object sampling, with grab and dredge. (*A*) The Peterson grab obtaining a sample from the sea bed. (*B*) A spring-loaded grab that is only triggered when it is in the correct position for sampling. (*C*) An orange peel sampler fitted with a canvas hood to minimize washout. (*D*) A naturalist's rectangular dredge for marine animals. (*E*) A deep-sea rock dredge. (*A*) from N.A. Holme, 1964, *Adv. Mar. Biol.* 2:171–260; (*B*) from W. Smith and A.D. McIntyre, 1954, *J. Mar. Biolo. Ass. UK* 33:257–64; (*C*) from O. Hartmann, 1955, *Allan Hancock Pacific Exp.* 19; (*D,E*) from N.A. Holme and A.D. McIntyre, 1971, *IBP Handbook No. 16*, Blackwell Sci. Publ., Oxford.

When the objects to be sampled occur too infrequently to be obtained by grab samples, a larger area can be covered by dragging a dredge along the bottom (Plate 4-7,*D,E*). A dredge is a rectangular metal frame that keeps the mouth of a chain bag open while a toothed or smooth edge digs into the bottom and scoops benthic material into the attached chain bag. The mesh of this bag, or a net insert, determines the size of the objects retained. For harvesting oysters, rocks, and manganese nodules, a large mesh is used to let the sediment pass through while the appropriate size object is retained. For catching the infauna and more sedentary epifauna, a sack dredge with a finer net inside stirs up the sediment and fauna which are then caught in either the inside net or a bottom egg net that follows behind the dredge.

An alternate method for obtaining the deep fauna is a bucket or conical dredge with a circular rim to dig into the sediment and a finer mesh bag, which quickly fills with sediment. Anchor dredges are designed for packed sediments and are deployed like an anchor with considerable scope to the line (three or more times the depth).

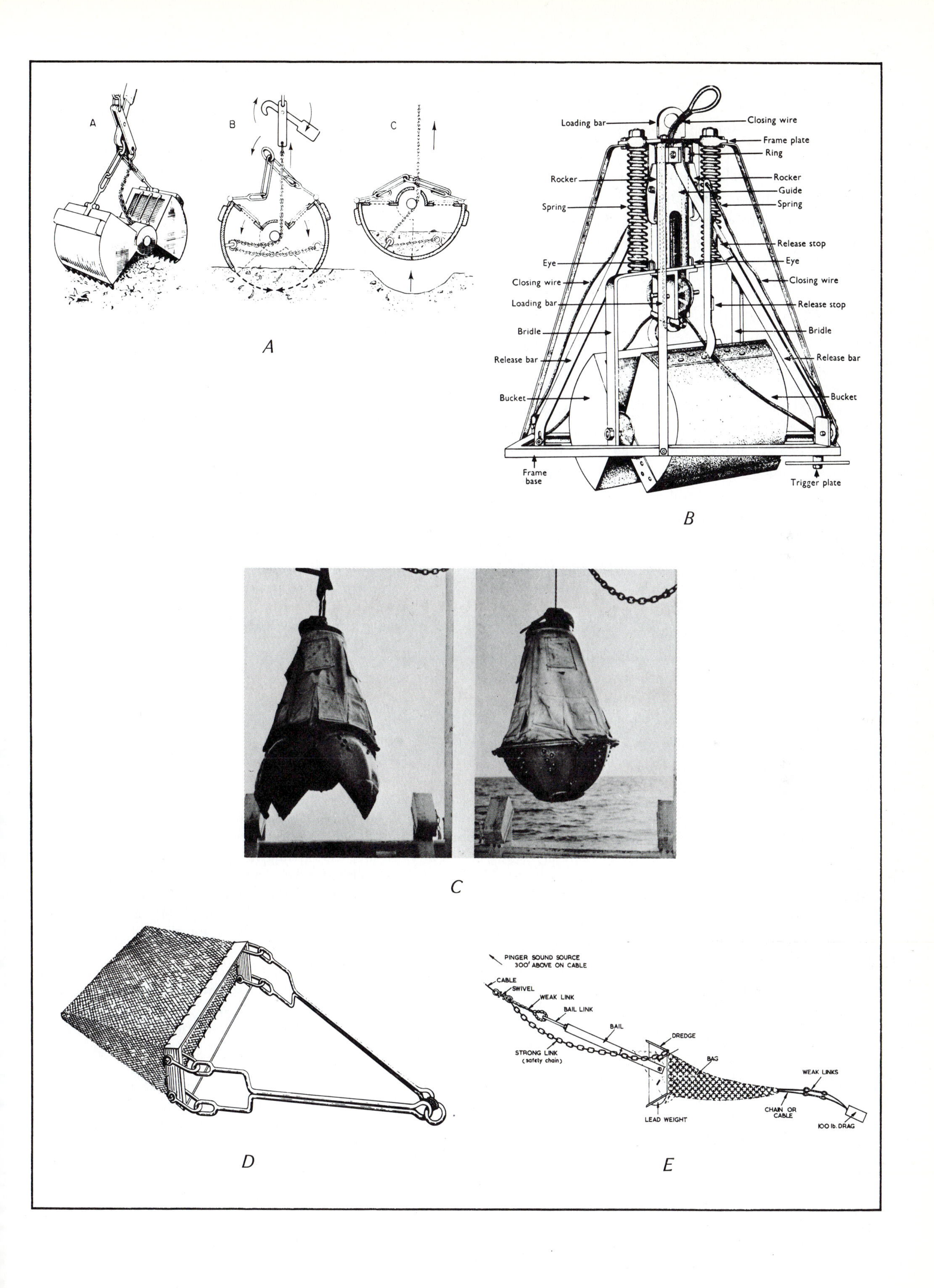
A
B
C
Loading bar
Closing wire
Frame plate
Ring
Rocker
Rocker
Guide
Spring
Spring
Release stop
Eye
Eye
Closing wire
Closing wire
Loading bar
Release stop
Bridle
Bridle
Release bar
Release bar
Bucket
Bucket
Frame base
Trigger plate
PINGER SOUND SOURCE 300' ABOVE ON CABLE
CABLE
SWIVEL
WEAK LINK
BAIL LINK
BAIL
DREDGE
STRONG LINK (safety chain)
BAG
WEAK LINKS
LEAD WEIGHT
CHAIN OR CABLE
100 lb. DRAG

A

B

C

D

E

A strain on the line makes the dredge dig in and take a sample, while pulling on the dredge from the reverse direction retrieves the sampler. It is obvious that sediments, rocks, animals, and manganese nodules retrieved by dredge from the deep sea may be in less than ideal condition for microbiological studies. Samples selected by articulated arms of research submersibles, such as *Alvin*, is another method of sampling dispersed objects.

d. SCUBA and snorkeling. The benthos of shallow waters is ideally sampled by divers who can control the selectivity or randomness of the sampling as well as making "eyeball observations" on environmental conditions, animal behavior and population distribution. A great advantage is the precision and lack of damage with which sampling can be accomplished. Fager et al. (1966) describe equipment and techniques for such studies. The multilayer corer is one good example. This corer has a pint jar screwed on top and an area of 45 cm^2 is pushed into the sand until the top plate touches the sand. Sand in front of the corer is dug out and plates are worked into place into slots which divide the core into 1 cm thick layers. The sampler is inverted, the plates removed, and the layers are put into pint jars, one at a time. However, usable cores can be obtained simply by driving short lengths of plastic core liners into the sediment, stoppering the top, and withdrawing the liner with core that is retained with a vacuum but is sealed on the bottom with another stopper for transport. The cutting wedge sampler (Schlieper, 1972) is also useful. SCUBA diving and snorkeling, which are useful for underwater observation and sampling, are indispensable for testing prototype sampling gear under field conditions.

REFERENCES

Aiken, J., R.H. Bruce, and J.A. Lindley. 1977. Ecological investigations with the Undulating Oceanographic Recorder: The hydrography and plankton of the waters adjacent to the Orkney and Shetland Islands. *Mar. Biol.* 39:77–91.

Anderson, O.R., and A.W.H. Bé. 1976. A cytochemical fine structure study of phagotrophy in a planktonic foraminifer, *Hastigerina pelagica* (D'Orbigny). *Biol. Bull.* 151:437–49.

Anthony, E.H. 1963. Regional variability of bacteria in North Atlantic sediments. In: *Marine Microbiology* (C.H. Oppenheimer, ed.). C.C. Thomas, Springfield, Ill., pp. 522–32.

Aron, W., N. Raxter, R. Noel, and W. Andrews. 1964. A description of a discrete depth plankton sampler with some notes on the towing behavior of a 6-foot Isaacs-Kidd mid-water trawl and a one-meter ring net. *Limnol. Oceanogr.* 9:324–33.

Ballard, R.D., and E. Kristof. 1975. Project Famous II. Dive into the great rift. *Natl. Geogr.* 147(5):604–15.

Bé, A.W.H. 1962. Quantitative multiple opening-and-closing plankton samplers. *Deep-Sea Res.* 9:144–51.

Bé, A.W.H., and O.R. Anderson. 1976. Gametogenesis in planktonic foraminifera. *Science* 192:890–92.

Beebe, C.W. 1926. *The Arcturus Adventure.* Putnam, New York, 439 p.

Beers, J.R., G.L. Stewart, and J.D.H. Strickland. 1967. A pumping system for sampling small plankton. *J. Fish. Res. Bd. Can.* 24:1811–18.

Bogoyavlenskii, A.N. 1964. On the distribution of heterotrophic microorganisms in the Indian Ocean and in Antarctic waters. *Deep-Sea Res.* 11:105–8.

Bouma, A.H. 1969. Large box sampler. In: *Methods for the Study of Sedimentary Structures.* Wiley, New York, pp. 332–45.

Bruce, R.H., and J. Aiken. 1975. The Undulating Oceanographic Recorder—a new instrument system for sampling plankton and recording physical variables in the euphotic zone from a ship underway. *Mar. Biol.* 32:85–97.

Bruneau, L., J.G. Jerlov, and F.F. Koczy. 1953. Physical and chemical methods. *Rept. Swedish Deep-Sea Exped.* 3:99–112.

Carey, A.G., Jr., and R.R. Paul. 1968. A modification of the Smith-McIntyre grab for simultaneous collection of sediment and bottom water. *Limnol. Oceanogr.* 13:545–49.

Conover, J.T., with J.McN. Sieburth. 1964. Effect of Sargassum distribution on its epibiota and antibacterial activity. *Bot. Mar.* 6:147–57.

Conover, J.T., and J.McN. Sieburth. 1966. Effect of tannins excreted from Phaeophyta on planktonic animal survival in tide pools. *Proc. 5th Intl. Seaweed Symp., Halifax, N.S.*, Pergamon Press, London, pp. 99–100.

Craib, J.S. 1965. A sampler for taking short undisturbed marine cores. *J. Cons. Perm. Int. Explor. Mer* 30:34–39.

Crow, S.A., D.G. Ahearn, and W.L. Cook. 1975. Densities of bacteria and fungi in coastal surface films as determined by a membrane-adsorption procedure. *Limnol. Oceanogr.* 20:644–46.

Dietz, A.S., L.J. Albright, and T. Tuominen. 1976. Heterotrophic activities of bacterioneuston and bacterioplankton. *Can. J. Microbiol.* 22:1699–709.

Dietz, R.S. 1952. Methods of exploring the ocean floor. In: *Symposium on Oceanographic Instrumentation.* Rancho Santa Fe, Ca., pp. 194–209.

Drew, G.H. 1914. On the precipitation of calcium carbonate in the sea by marine bacteria, and on the action of denitrifying bacteria in tropical and temperate seas. *Papers from Tortugas Lab., Carnegie Inst., Washington D.C.* 5:7–45.

Duce, R.A., J.G. Quinn, C.E. Olney, S.R. Piotrowicz, B.J.

Ray, and T.L. Wade. 1972. Enrichment of heavy metals and organic compounds in the surface microlayer of Narragansett Bay, Rhode Island. *Science* 176:161–63.

Duce, R.A., W. Stumm, and J.M. Prospero (eds.). 1972. Working symposium on sea–air chemistry. *J. Geophys. Res.* 77(27):5059–349.

Dussart, B.M. 1965. Les différentes catégories de plancton. *Hydrobiologia* 26:72–74.

Edwards, R.S., and F.A. Richards. 1956. A bottom-water sampler. *Deep-Sea Res.* 4:65–66.

Emery, K.O. 1958. Bacterial bottom sampler. *Limnol. Oceanogr.* 3:109–11.

Emery, K.O., and A.R. Champion. 1948. Underway bottom sampler. *J. Sed. Petrol.* 18:30–33.

Emery, K.O., and R.S. Dietz. 1941. Gravity coring instrument and mechanics of sediment coring. *Bull. Geol. Soc. Amer.* 52:1685–714.

Fager, E.W., A.O. Flechsig, R.F. Ford, R.I. Clutter, and R.J. Ghelardi. 1966. Equipment for use in ecological studies using SCUBA. *Limnol. Oceanogr.* 11:503–9.

Fasching, J.L., R.A. Courant, R.A. Duce, and S.R. Piotrowicz. 1974. A new surface microlayer sampler utilizing the bubble microtome. *J. Rech. Atmos.* 8:650–52.

Fenchel, T. 1967. The ecology of marine microbenthos. I. The quantitative importance of ciliates as compared with metazoans in various types of sediments. *Ophelia* 4:121–37.

Fischer, B. 1887. Bacteriologische Untersuchungen auf einer Reise nach Westindien. II. Ueber einen lichtentwickelunden im Meerwasser gefundenen Spaltpilz. *Zeits. Hyg.* 2:54–95.

Fischer, B. 1894. Die Bakterien des Meeres nach den Untersuchungen der Plankton-Expedition unter gleichzeitiger Berucksichtigung einiger alterer und neuerer Untersuchungen. *Ergeb. Plankton-Exped. Humboldt-Stiftung.* 4:1–83.

Fish, C.J., and J.M. Snodgrass. 1962. The Scripps-Narragansett high-speed multiple plankton sampler. *Rapp. Proc.-Verb. Cons. Int. Explor. Mer* 153:23–24.

Fowler, G.A., and L.D. Kulm. 1966. A multiple corer. *Limnol. Oceanogr.* 11:630–33.

Fraser, J. 1962. *Nature Adrift.* G.T. Foulis, London, 178 p. (Reissued by Xerox University Microfilms, Ann Arbor Michigan.)

Garrett, W.D. 1965. Collection of slick-forming materials from the sea surface. *Limnol. Oceanogr.* 10:602–5.

Gee, H. 1932. Bacteriological water sampler. *Bull. Scripps Inst. Oceanogr., Tech. Ser.* 3:191–200.

Gessner, R.V., R.D. Goos, and J.McN. Sieburth. 1972. The fungal microcosm of the internodes of *Spartina alterniflora. Mar. Biol.* 16:269–73.

Goering, J.J., and D.W. Menzel. 1965. The nutrient chemistry of the sea surface. *Deep-Sea Res.* 12:839–43.

Gore, R., and B. Lavies. 1977. The tree nobody liked. *Natl. Geogr.* 151:669–89.

Gross, M.G. 1972. *Oceanography—A View of the Earth.* Prentice-Hall, Englewood Cliffs, N.J., p. 190.

Hamner, W.M. 1974. Blue-water plankton. *Natl. Geogr.* 146:530–545.

Hamner, W.M., L.P. Madin, A.L. Alldredge, R.W. Gilmer, and P.P. Hamner. 1975. Underwater observations of gelatinous zooplankton: Sampling problems, feeding biology, and behavior. *Limnol. Oceanogr.* 20:907–917.

Hardy, A.C. 1956. *The Open Sea.* Collins, London and Houghton-Mifflin, Boston, pp. 76–79.

Hartmann, O. 1955. *Quantitative Survey of the Benthos of San Pedro Basin, Southern California. Part I. Preliminary Results* (Allan Hancock Pacific Exped. 19). Allan Hancock Foundation, University of Southern California, Los Angeles, Plates 3 & 4.

Harvey, G.W. 1966. Microlayer collection from the sea surface: A new method and initial results. *Limnol. Oceanogr.* 11:603–18.

Harvey, G.W., and L.A. Burzell. 1972. A simple microlayer method for small samples. *Limnol. Oceanogr.* 17:156–57.

Hatcher, R.F., and B.C. Parker. 1974. Laboratory comparisons of four surface microlayer samplers. *Limnol. Oceanogr.* 19:162–65.

Heirtzler, J.R., and E. Kristof. 1975. Project Famous. I. Man's first voyage down to the mid-Atlantic ridge. *Natl. Geogr.* 147:586–603.

Hiatt, R.W. 1963. *World Directory of Hydrobiological and Fisheries Institutions.* Amer. Inst. Biol. Sci., Washington, D.C., 320 p.

Hjørt, J., and J.T. Ruud. 1938. A bottom-sampler for the mud line. *Hvalradets Skrifter* 17:145–51.

Holm-Hansen, O., C.J. Lorenzen, R.W. Holmes, and J.D.H. Strickland, 1965. Fluorometric determination of chlorophyll. *J. Cons. perm. int. explor. Mer* 30:3–15.

Holme, N.A. 1964. Methods of sampling the benthos. *Adv. Mar. Biol.* 2:171–260.

Holme, N.A., and A.D. McIntyre (eds.). 1971. *Methods for the Study of Marine Benthos.* IBP Handbook No. 16, Blackwell Scientific Publ. Oxford, 323 p.

Issatchenko, B.L. 1914. *Investigation on the Microbes of the Icy Arctic Ocean* (Monograph in Russian). Petrograd, 247 p.

Jannasch, H.W., and W.S. Maddux. 1967. A note on bacteriological sampling in seawater. *J. Mar. Res.* 25:185–89.

Jannasch, H.W., and C.O. Wirsen. 1977a. Retrieval of concentrated and undecompressed microbial populations from the deep sea. *Appl. Environ. Microbiol.* 33:642–46.

Jannasch, H.W., and C.O. Wirsen. 1977b. Microbial life in the deep sea. *Sci. American* 236(6):42–52.

Jannasch, H.W., C.O. Wirsen, and C.L. Winget. 1973. A bacteriological pressure-retaining, deep-sea sampler and culture vessel. *Deep-Sea Res.* 20:661–64.

Jarvis, N.L., W.D. Garrett, M.A. Scheiman, and C.O. Timmons. 1967. Surface characterization of surface-active material in seawater. *Limnol. Oceanogr.* 12:88–96.

Jensen, A., and E. Sakshaug. 1970. Producer-consumer relationships in the sea. II. Correlation between *Mytilus* pigmentation and the density and composition of phytoplanktonic populations in inshore waters. *J. exp. mar. Biol. Ecol.* 5:246–53.

Johnson, P.W., J.McN. Sieburth, A. Sastry, C.R. Arnold, and M.S. Doty. 1971. *Leucothrix mucor* infestation of benthic crustacea, fish eggs and tropical algae. *Limnol. Oceanogr.* 16:962–69.

Johnston, W. 1892. On the collection of samples of water for bacteriological analysis. *Can. Rec. Sci.* 5:19–28.

Kändler, R. 1972. Analysis of the Stock. B. Nekton. In: *Research Methods in Marine Biology* (C. Schlieper, ed.). University of Washington Press, Seattle, pp. 64–88.

Koczy, F.F. 1953. Chemistry and hydrography of the water layers next to the ocean floor. *Statens (Sweden) Naturvet. Forskn. Aarsl.* (1951–52) 7:97–102.

Kriss, A.E. 1963. *Marine Microbiology (Deep Sea).* (J.M. Shewan and Z. Kabata, transls.). Oliver and Boyd, Edinburgh, 536 p.

Kriss, A.E., M.N. Lebedeva, and A.V. Tsyban. 1966. Comparative estimate of a Nansen and microbiological water bottle for sterile collection of water samples from depths of seas and oceans. *Deep-Sea Res.* 13:205–12.

Lackey, J.B. 1961. Bottom sampling and environmental niches. *Limnol. Oceanogr.* 6:271–79.

Langlois, G.A. 1975. Effect of algal exudates on substratum selection by motile telotrochs of the marine peritrich ciliate *Vorticella marina*. *J. Protozool.* 22:115–23.

Larsson, K., G. Odham, and A. Sodergren. 1974. On lipid surface films on the sea. I. A simple method for sampling and studies of composition. *Mar. Chem.* 2:49–57.

Lavoie, D.M. 1975. Application of diffusion culture to ecological observations on marine microorganisms. M.S. Thesis, University of Rhode Island, 91 p.

Lenz, J. 1972. Analysis of the Stock. A. Plankton. In: *Research Methods in Marine Biology* (C. Schlieper, ed.). University of Washington Press, Seattle, pp. 46–63.

MacIntyre, F. 1968. Bubbles: A boundary-layer "microtome" for micron-thick samples of a liquid surface. *J. Phys. Chem.* 72:589–92.

MacIntyre, F. 1974. The top millimeter of the ocean. *Sci. American* 230(5):62–77.

Matthews, D.J. 1913. A deep-sea bacteriological water-bottle. *J. Mar. Biol. Assoc. U.K.* 9:525–29.

Miget, R., H. Kator, C. Oppenheimer, J.L. Laseter, and E.J. Ledet. 1974. New sampling device for the recovery of petroleum hydrocarbons and fatty acids from aqueous surface films. *Analyt. Chem.* 46:1154–57.

Mills, A.A. 1961. An external core-retainer. *Deep-Sea Res.* 7:294–95.

Mullin, M.M., and E.R. Brooks. 1976. Some consequences of distributional heterogeneity of phytoplankton and zooplankton. *Limnol. Oceanogr.* 21:784–96.

Niskin, S.J. 1962. A water sampler for microbiological studies. *Deep-Sea Res.* 9:501–3.

Niskin, S.J. 1964. A reversing-thermometer mechanism for attachment to oceanographic devices. *Limnol. Oceanogr.* 9:591–94.

Norris, R.E. 1965. Neustonic marine Craspedomonadales (Choanoflagellates) from Washington and California. *J. Protozool.* 12:589–94.

Oppenheimer, C.H. (ed.). 1968. *Marine Biology IV. Unresolved Problems in Marine Microbiology.* New York Academy of Science, New York, 485 p.

Parker, B. 1978. Neuston Sampling. In: *A Phytoplankton Manual* (A. Sournia, ed.). UNESCO, Paris, Ch. 3.5.

Parsons, T.R. 1969. The use of particle size spectra in determining the structure of a plankton community. *J. Oceanogr. Soc. Japan* 25:172–81.

Pinon, J., and J. Pijck. 1975. Water sampling by helicopter. *Rev. Int. Oceanogr. Med.* 37–38:153–58.

Platt, T., and C. Filion. 1973. Spatial variability of the productivity biomass ratio for phytoplankton in a small marine basin. *Limnol. Oceanogr.* 18:743–49.

Pomeroy, L.R., and R.E. Johannes. 1968. Occurrence and respiration of ultraplankton in the upper 500 meters of the ocean. *Deep-Sea Res.* 15:381–91.

Pratt, H.L. 1973. MacroSnooping. *Skin Diver,* Sept., 38–39.

Riley, G.A. 1976. A model of plankton patchiness. *Limnol. Oceanogr.* 21:873–80.

Russell, H.L. 1892. Bacterial investigation of the sea and its floor. *Bot. Gaz.* 17:312–21.

Sachs, P.L., and S.O. Raymond. 1965. A new unattached sediment sampler. *J. Mar. Res.* 23:44–53.

Schegg, E. 1970. A new bacteriological sampling bottle. *Limnol. Oceanogr.* 15:820–22.

Schink, D.R., K.A. Fanning, and J. Piety. 1966. A sea-bottom sampler that collects both water and sediment simultaneously. *J. Mar. Res.* 24:365–73.

Schlieper, C. (ed.). 1972. *Research Methods in Marine Biology* (Engl. transl.), University of Washington Press, Seattle, 356 p.

Sheldon, R.W., and T.R. Parsons. 1967a. *A Practical Manual on the Use of the Coulter Counter in Marine Science.* Coulter Electronics, Toronto, 66 p.

Sheldon, R.W., and T.R. Parsons. 1967b. A continuous size spectrum for particulate matter in the sea. *J. Fish. Res. Bd. Canada* 24:909–915.

Shields, B., and J. McFall (eds.). 1974. *Free Drifting Buoys.* NASA Conf. Publications, Springfield, Va., 371 p.

Sholkowitz, E.R. 1970. A free vehicle bottom-water sampler. *Limnol. Oceanogr.* 15:641–44.

Sieburth, J.McN. 1963. A simple form of the ZoBell bacteriological sampler for shallow water. *Limnol. Oceanogr.* 8:489–92.

Sieburth, J.McN. 1965. Bacteriological samplers for air-water and water-sediment interfaces. In: *Ocean Science and Ocean Engineering* (Trans. Joint Conf. Mar. Tech. Soc./ASLO). Washington, D.C., pp. 1064–68.

Sieburth, J.McN. 1967. Seasonal selection of estuarine bacteria by water temperature. *J. exp. mar. Biol. Ecol.* 1:98–121.

Sieburth, J.McN. 1971a. Distribution and activity of oceanic bacteria. *Deep-Sea Res.* 18:1111–21.

Sieburth, J.McN. 1971b. An instance of bacterial inhibition in oceanic surface water. *Mar. Biol.* 11:98–100.

Sieburth, J.McN. 1975. *Microbial Seascapes.* University Park Press, Baltimore, Md., 248 p.

Sieburth, J.McN., and J.T. Conover. 1965. Sargassum tannin, an antibiotic that retards fouling. *Nature* 108:52–53.

Sieburth, J.McN., and C.D. Thomas. 1973. Fouling on eelgrass (*Zostera marina* L.). *J. Phycol.* 9:46–50.

Sieburth, J.McN., J.A. Frey, and J.T. Conover. 1963. Micro biological sampling with a piggy-back device during routine Nansen bottle casts. *Deep-Sea Res.* 10:757–58.

Sieburth, J.McN., P-J. Willis, K.M. Johnson, C.M. Burney, D.M. Lavoie, K.R. Hinga, D.A. Caron, F.W. French, P.W. Johnson, and P.G. Davis. 1976. Dissolved organic carbon and living particulates in the surface microlayer of the North Atlantic. *Science* 194:1415–18.

Sieburth, J.McN., K.M. Johnson, C.M. Burney, and D.M. Lavoie. 1977. Estimation of *in-situ* rates of heterotrophy using diurnal changes in dissolved organic matter and growth rates of picoplankton in diffusion culture. *Helgoländ. wiss. Meeresunters.* 30:565–74.

Sieburth, J.McN., V. Smetacek, and J. Lenz. 1978. Pelagic ecosystem structure: Heterotrophic components of the plankton and their relationship to plankton size-fractions. *Limnol. Oceanogr.* 23:1256–63.

Smith, W., and A.D. McIntyre. 1954. A spring-loaded bottom-sampler. *J. mar. biol. Ass. U.K.* 33:257–64.

Sorokin, Yu. I. 1962. Problems of the technique for collecting samples for investigating marine microorganisms. *Okeanologiya* 2:888–97.

Sorokin, Yu.I. 1964. A quantitative study of the microflora in the central Pacific Ocean. *J. Cons. perm. int. Explor. Mer* 29:25–40.

Sorokin, Yu.I. 1971. On the role of bacteria in the productivity of tropical oceanic waters. *Int. Revue ges. Hydrobiol.* 56:1–48.

Sorokin, Yu.I. 1974. Bacteria as a component of the coral reef community. *Proc. 2nd Int. Coral Reef Symp.* 1:3–10.

Tsyban, A.V. 1967. On an apparatus for the collection of microbiological samples in the near-surface micro-horizon of the sea. *Hydrobiologia* 3:84–86.

Tsyban, A.V. 1971. Marine bacterioneuston. *J. Oceanogr. Soc. Japan* 27:56–66.

Vachon, W.A. 1975. *Instrumented Full-scale Tests of a Drifting Buoy and Drogue.* Charles Stark Draper Lab., Inc., Cambridge, Mass. 163 p.

Van Dorn, W.G. 1956. Large-volume water samplers. *Trans. Amer. Geophys. Union* 37:682–84.

Volkmann, G., J. Knauss, and A. Vine. 1956. The use of parachute drogues in the measurement of subsurface ocean currents. *Trans. Amer. Geophys. Union* 37:573–77.

Webb, J.E. 1974. *Guide to the Marine Stations of the North Atlantic Waters. Part 1. North European and the East Atlantic Coast.* Royal Society, London, 262 p.

Whipple, G.C. 1927. *The Microscopy of Drinking Water.* Wiley, New York, 586 p.

Wickstead, J.H. 1965. *An Introduction to the Study of Tropical Plankton.* Hutchinson Publications, London, pp. 37, 41.

Willingham, C.A., and J.D. Buck. 1965. A preliminary comparative study of fungal contamination in non-sterile water samplers. *Deep-Sea Res.* 12:693–95.

Woods, J.D., and J.N. Lythgoe (eds.). 1971. *Underwater Science.* Oxford University Press, London, 330 p.

ZoBell, C.E. 1941. Apparatus for collecting water samples from different depths for bacteriological analysis. *J. Mar. Res.* 4:173–88.

ZoBell, C.E. 1946. *Marine Microbiology.* Chronica Botanica Co., Waltham, Mass., 240 p.

CHAPTER 5

Microscopic Examination

The availability of functional microscopes in the 19th century (Padgitt, 1975) permitted the study of the natural history of aquatic microorganisms. An outstanding example of such a study is the superbly illustrated atlas and its accompanying monograph by Ehrenberg (1838), which was completed under the patronage of Friedrich Wilhelm, Crown Prince of Prussia. Bacteria, microalgae, and protozoa, as well as some metazoa, are presented in a systematic but interdisciplinary manner since they are shown as they occur in nature. Unfortunately this simplistic viewpoint was also his undoing (Jahn, 1971). Another example of the keen observation of microorganisms is the monograph by Leidy (1879), which resulted from his work on a U.S. Geological Survey expedition in the area around Fort Bridger in the then Wyoming Territory. Leidy used a number of microscopes, including a 50-dollar model to study the aquatic microbiota, to produce a most remarkable monograph on the taxonomy and natural history of the amoeboid protozoa, especially the testaceans, the amoebae that live in houses. From the exquisite colored plates of these two volumes, it is awesome to see how these two individuals pushed their equipment to the utmost. Although there are better microscopes today, all too often they are mistreated, in poor shape, or used improperly, especially in teaching labs. Even a common instrument, when used well by such professionals as a Douglas P. Wilson or a James B. Lackey, will reveal more than the most sophisticated of today's instruments improperly used. Select the proper microscope for the job you have to do, read the instruction manual, put the microscope (or have it put) in good mechanical shape, and then learn the few principles required for optimal results.

A. STEREOMICROSCOPY

The stereomicroscope gives a three-dimensional image of the sample by having two separate objective–eyepiece systems mounted at the approximate angle of binocular vision (15°). It is ideal for the low power (2 to 200 X) viewing of everything from net plankton to fouled surfaces. In the better models, such as the versatile Wild-Heerbrugg M-5, one can work on the sample while looking at it, since there is great depth of focus and a large working distance, and the image is unreversed. The modular nature of this microscope permits a number of ways the stages and stands can be used. Different stands are used for bright-field, dark-field, and polarization work. By changing stands, settings, and light sources, different degrees of illumination between incident and transmitted light can be obtained. Incident light is ideal for top and side lighting, whereas light transmitted by a mirrored surface is useful for visualizing colonies on agar plates. The dark-field mode shows plankton as fragile luminous forms. The muted light reflected from the white reverse side of the mirror will demonstrate the presence of fecal pellets in copepods. Fiber optics are particularly useful for obtaining adequate light on submerged surfaces, such as well-fouled clumps of the blue mussel *Mytilus edulis*. The three-dimensional drama that can be observed in such microcosms shows the microbial interplay that occurs between the large bacteria and algae, and the larger protozoa and metazoa that prey upon them. It is fascinating to watch the dinoflagellates churning along, the large diatoms like *Bacillaria* gliding by, the tintinnids zipping back and forth, and the vorticellids springing up and down on their coiled stalks. By taking the time to experiment with lighting and different eyepiece and objective combinations, one can learn how and when to use this versatile microscope for viewing the larger microorganisms that have sufficient contrast. It is a necessary tool for examining the small colonies that develop on seawater agar media (at a magnification of 5 to 50 X) and for preparing blocks and stubs for electron microscopy. The only time a good stereomicroscope is disappointing is when one attempts to photograph a particularly pleasing preparation. Since only one of the body tubes is customarily used, the depth and resolution obtained with the stereomicroscope cannot be filmed. Furthermore, the low numerical aperture (NA) of most objectives limits the resolution of the photomicrograph.

B. VISIBLE LIGHT MICROSCOPY

Except for the specialized stereomicroscope, most visible light microscopy is accomplished with the compound microscope, the workhorse of the microbiologist. A most readable description of how the compound microscope works is included in "Photography through the Microscope" by the Eastman Kodak Company. This and the manual for the microscope are required reading *before* sitting down at the microscope. LEARN THE FUNCTION OF EACH PART OF THE COMPOUND MICROSCOPE. Before examining specimens, LEARN THE FEW ESSENTIAL STEPS IN SETTING UP KÖHLER ILLUMINATION AND USE THEM EVERY TIME YOU START YOUR MICROSCOPIC WORK. Unless conditions are optimal, the resolution, contrast, and quality of the images will suffer. The common laboratory microscope is fitted for bright field. At little cost, it can be converted to dark field for the examination of living specimens with maximal contrast. Improved resolution of structure can be achieved with a variety of optics for contrast microscopy, including amplitude, modulation phase-contrast, and interference-contrast microscopy. The student and the researcher should also be familiar with blue and ultraviolet fluorescence microscopy. Descriptions of the modern optical microscopes and the emerging equipment for automatic image analysis are given by Tribe, Eraut, and Snook (1975) and Bradbury (1976).

1. Illumination

For bright-field, phase-contrast, and polarized light microscopy, as well as photomicrography, a good tungsten filament microscope lamp with corrected condenser and iris diaphragm is set up for the Köhler illumination required. Some microscopes are fitted with self-contained illumination systems, which minimize misalignment and are more convenient for routine work. With both separate and self-contained light sources, the enlarged image of the lamp filament is focused upon the microscope mirror and focal plane of the condenser. The microscope condenser then focuses the image of the lamp condenser into the object plane to yield a field with even illumination. The iris diaphragm is used to limit the area of illumination to the field of view, to control glare and image quality. In addition to low voltage tungsten lamps, quartz halogen lamps, which prevent blackening; high pressure mercury arcs, with their monochromatic light; and xenon high pressure arcs, with their great intensity and high color temperature, are used for photomicrography, fluorescence microscopy, and color photography, respectively.

As one looks at increasingly smaller and less pigmented organisms under the compound microscope, visualization of the organisms and their structures becomes more difficult due to the minimal contrast between their often transparent bodies and the suspending medium (Plate 5-2,*A*). To visualize the microorganisms and their structures, there must be contrast between organisms and their suspending medium.

2. Dark-field Microscopy

Contrast between microorganisms and their suspending fluid can be achieved economically and easily by the use of dark-field or dark-ground microscopy. Light passes through the object but outside the acceptance angle of the objective. The illuminated objects refract the light and stand out as luminous bodies against a black background. In a way, the dark-field is an extension of incident light illumination as used in stereomicroscopy. For objectives of 25 X and under, this is achieved simply by inserting a dark disk called a patch stop centrally in the focal plane of the condenser and critically focusing the light into a hollow cone whose apex is in the specimen plane. These stops can be made easily for microscopes that lack them (Bradbury, 1970, 1976). For objectives of 40 X and over, special reflecting dark-field condensers are used for immersion oil contact with the underside of the microscope slide. The thickness of the slide is critical. Immersion oil objectives for dark-field are fitted with an iris diaphragm to make their numerical aperture smaller than that of the condenser. When the larger motile microorganisms, such as spirochetes, flagellates, and zooplankton, are viewed in this manner, their fragile beauty and behavior can be observed. Excellent examples of this are the micrographs by D.P. Wilson that appear in his many fine papers in the Journal of the Marine Biological Association of the United Kingdom as well as in Sir Alister Hardy's monograph, *The World of Plankton* (1956). This technique is a very useful but much overlooked one in marine microbiology. Some phase-contrast and interference-contrast condensers are fitted with dark-field stops. Dark-field resolves the microorganisms but not their structures (see Plate 5-2,*B*).

3. Bright-field Microscopy

An alternate way of increasing the contrast between the specimen and the suspending fluid is to decrease the intensity or amplitude of the vibrating light waves passing through it. This is achieved by destructive interference of the light waves. Destructive interference occurs when the light waves are deflected 180°; that is, their direction is reversed. The specimen itself deflects the light waves 90°. What is needed is some means to add another 90° of deflection. One way this is achieved is by killing the organisms and mastering the chemistry and art of staining. Some stains, such as the protargol stain for the basal bodies of cilia in ciliates and the Gram stain for the cell envelope of bacteria, are basic procedures upon which the taxonomy of these groups are based. Another way that the contrast needed to visualize living organisms and their larger organelles can be obtained is by optically deflecting the light waves (phase-contrast microscopy) the additional 90° required for destructive interference. This phase contrast is useful in studying living organisms because much can be learned from their motion, behavior, cell development, and life cycle.

4. Phase-Contrast Microscopy

As we have seen in bright-field microscopy, the change in phase (angle) is not detected by the human eye until it is deflected 180° so that destructive interference affects the amplitude or intensity, which is detected. In the phase-contrast microscope, there are two light paths, direct light and diffracted light, which is out of phase by one-quarter wavelength or 90°. The addition of the specimen causes a further one-quarter wavelength phase change, which additively makes a half arc or 180° change to cause destructive interference that shows up as contrast in the image. If the diffracted light is slowed with respect to the direct light, positive contrast is obtained. Here the specimen is darker than the background. Conversely, the direct light can be slowed with respect to the diffracted light, to give negative contrast in which the specimen is bright on a dark background. Most microscopes offer positive contrast (see Plate 4-2,*C*), whereas the recently introduced amplitude modulated microscope from the American Optical

Company (Buffalo, N.Y.) permits a continuous spectrum from positive to negative contrast (Casida, 1976).

The phase change in conventional phase-contrast microscopes is obtained by using a phase annulus in the condenser that can be aligned with the annular phase plate of the objective (as shown in Plate 5-1,*C*). This is accomplished with the aid of a phase telescope either built into the microscope or temporarily inserted in place of one eyepiece. The disadvantage of phase-contrast microscopy is that halos of light occur at the edges of structures and from objects that are not in focus, and these become increasingly more bothersome at higher magnifications of smaller objects, such as the bacteria.

5. Polarized Microscopy

In addition to phase and amplitude, the specimen can also affect the plane of light passing through it. The vibration plane of light waves assumes any angle with regard to the direction of propagation. The insertion of one piece of an inexpensive plastic Polaroid sheet between the condenser and the light source (polarizer) and of another above the objective (analyzer) can simply and immediately change any microscope into a polarizing microscope useful for detecting and identifying birefringent materials. For more careful work, specially designed polarizing microscopes with convenient refinements are available.

When the two Polaroid sheets are in the same direction of vibration, maximal light passes through. As the angle is increased, the light decreases until, at 90°, no light passes through. Birefringent material, such as some crystalline substances, splits a plane-polarized light ray into two rays. When viewed at a 45° angle to the crossed polarizer and analyzer, light passes the analyzer and appears bright on a dark field. When this principle is applied in a polarizing microscope, siliceous and calcareous materials in microorganisms, suspended particles, and sediments can be detected and distinguished. Thus, the calcareous spines of a planktonic foraminifera can be distinguished from the siliceous spines of radiolaria. Regular arrays of micelles embedded in a matrix of a different refractive index also split a plane-polarized light ray into two rays. This birefringence due to regular arrays of microstructures is common in biological material and can be used to identify cellulose fibrils, starch grains, and phospholipid droplets.

6. Interference Microscopy

The phase-contrast microscope uses the specimen to produce two light beams, and contrast is caused by interference between the two images. In interference contrast, two beams are produced optically; one carries the image, while the reference beam is used to cause interference patterns. In the commonly used Nomarski system, a polarizer below the condenser produces a beam of plane-polarized light that is split by a birefringent Wollaston prism, which also rotates the beam so that it is also perpendicular. This beam passes through the condenser, the slide with the specimen, and the objective before being recombined by another Wollaston prism with oppositely cut faces (Padawer, 1968). A diagonally oriented polarizing analyzer before the eyepieces polarizes the two beams circularly to cause an interference-contrast image. Although this image does not enhance the contrast, as in phase contrast, interference microscopy eliminates the halo effect, resolves finer structural details, and is less sensitive to depth of field, since it "cuts" a thin optical section. There is also a marked three-dimensional or pseudo-relief effect (see Plate 5-2,*D*). With slight changes in phase, Newtonian colors (false colors) are produced; these are not only aesthetically pleasing but are useful in studying microstructure.

A recent technique, called modulation contrast (Hoffman and Gross, 1975a,b), is quite simple and can be used with any microscope. It optically shadows a specimen to give a three-dimensional effect similar to Nomarski interference contrast.

C. FLUORESCENCE MICROSCOPY

Certain substances can absorb short wavelength radiation in the blue to ultraviolet region of the spectrum and re-emit the energy as light of a longer wavelength. If the light is emitted for an appreciable time after excitation, it is called phosphorescence. If the light is only emitted during excitation, it is called fluorescence. The fluorescence of chlorophyll-containing chloroplasts of phytoplankton in the natural state is called autofluorescence or primary fluorescence. Other cells, which do not possess autofluorescence, such as heterotrophic bacteria, can take up a fluorochrome, such as acridine orange or a fluorescent tagged anti-

PLATE 5-1. Comparison of different modes of optical and electron microscopy. (*A*) Optical path in the bright-field compound microscope, showing the magnification of an object P-Q to its reversed image (Q X–P X). (*B*) Optical path in the incident light fluorescent microscope, showing the bending of the horizontal light beam by the dichroic mirror. (*C*) Optical path in the phase-contrast microscope, showing the phase-altering phase plate in the objective and the annular phase diaphragm in the condenser. (*D*) Schematic diagram of the similarities of the light and the transmission electron microscopes, which penetrate thin specimens and have a shallow depth of field, and the basic differences of the scanning electron microscope, which only looks at the surface but has a great depth of field. (*A*) from M.A. Tribe, M.R. Eraut, and R.K. Snook, 1975, *Microscopy and its Application to Biology*, Cambridge University Press; (*B*) Anon., *Image-forming and Illuminating Systems of the Microscope*, Ernst Leitz, Wetzlar, Germany (512/99 Engl); (*C*) Anon., *Phase Contrast Equipment, Wild-Heerbrugg Ltd. Manual*, Heerbrugg, Switzerland; (*D*) from P.E. Mee, 1972 *Microstructures* October/November.

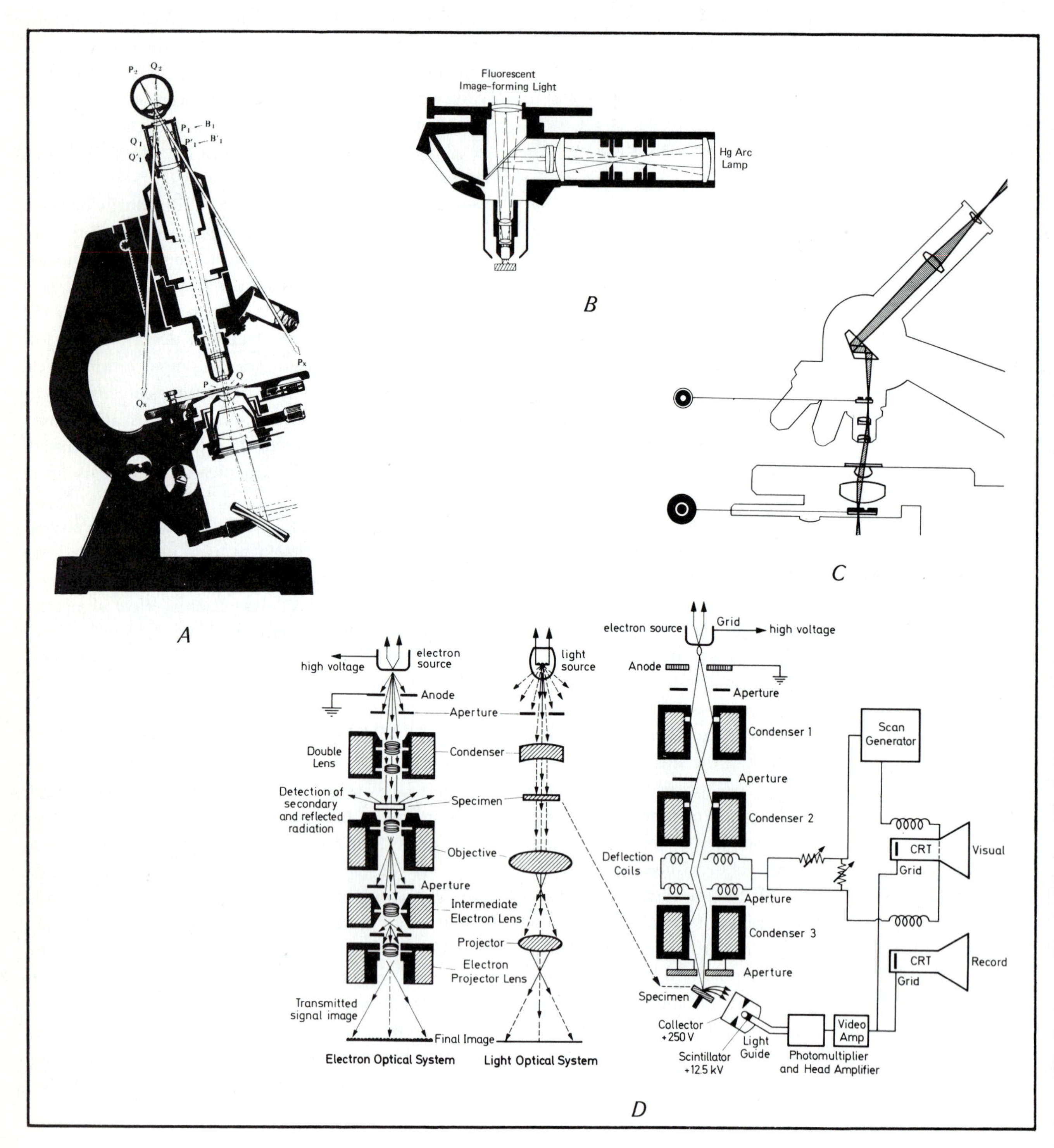

body, and show secondary fluorescence. Secondary fluorescence is strongly affected by pH, quenching, desiccation, and many other factors, so care should be exercised in the preparation of specimens. A conventional optical microscope can be operated in either transmitted light or the bright-field or dark-field modes of fluorescence microscopy but it is best to use a special microscope with incident light optics (Plate 5-1,*B*). A high intensity lamp that emits radiation in the required region of the light spectrum is usually used, although low voltage or low intensity lamps can be used in some applications, but with minimal results. An exciting filter placed between the light source and the microscope condenser allows a specific wavelength to pass through the specimen to induce fluorescence. The unwanted exciting radiation, which is harmful to the eyes, is stopped by a barrier filter in the optical system between the objective and eyepieces while the longer wavelength fluorescent light is transmitted to form a bright, self-luminous image of the specimen on a dark background.

1. Blue Light Fluorescence

An optimum wavelength of 400 nm in the short-wave region of the visible spectrum is passed by a blue exciting filter that causes the fluorescence. This exciting light is stopped by an orange–yellow barrier filter before the eyepieces. Therefore, all blue and blue–green fluorescence is cut out while yellow and green fluorescence is seen. A fluorochrome, such as acridine orange, is used to stain bacterial cells (Francisco, Mah, and Rabin, 1973; Zimmerman and Meyer-Reil, 1974; Daley and Hobbie, 1975) and other nucleic acid-containing materials. Here, RNA stains orange–red, whereas DNA stains yellow–green (Plate 5-2,*E*).

Another form of blue fluorescence uses fluorescein isothiocyanate (FITC) as a fluorochrome to tag specific antibodies. Such tagged antibodies are used to detect specific microorganisms through the specificity of the antigen–antibody reaction. The FITC absorbs visible blue light at 495 nm and re-emits energy as a green fluorescence at 520 nm; this is achieved with special filters. The technique is valuable in ecological studies because it permits detection and enumeration of specific microorganisms in natural populations. It must be kept in mind that, because of the filters used, there is much falsification of color in blue light fluorescence.

2. Ultraviolet Fluorescence

Only ultraviolet light is used in this mode. Mercury vapor lamps are a suitable source. The exciting filter, which passes the optimal wavelength of 360 nm, also passes red, which is absorbed by a red filter. After the almost pure ultraviolet light passes the specimen, it is completely absorbed with a nearly colorless barrier filter. All the fluorescence can be seen in this mode, since there is no falsification of color. Ultraviolet fluorescence works well with autofluorescing material, such as the chlorophyll-containing microorganisms. It is possible to use primary fluorescence with secondary fluorescence to distinguish phototrophic and heterotrophic members of a population. A total count can be made of the nanoplankton using an acridine orange-stained preparation: a count of the phototrophs is then made using the autofluorescence of the chloroplasts in a non-stained preparation, and the heterotrophic nanoplankters are then estimated by difference (Davis and Sieburth, unpublished data). Ultraviolet fluorescence can also be used with secondary fluorescence produced by the brightener Calcofluor (Darken, 1962; Harrington and Raper, 1968; Weaver and Zibilske, 1975; Hughes and McCully, 1975). This fluorochrome stains cellulose, among other substances, in living organisms, and microorganisms with a high content of cellulosic fibrils, such as the dinoflagellates, with their cellulose plates, are vividly visualized. When these brighteners are used with europium chelates (Scaff, Dyer, and Mori, 1969), they complex with nucleic acids and perhaps other vital components of viable cells. Anderson and Slinger (1975a,b) have used this phenomenon to develop a technique for enumerating microorganisms on soil organic matter. This technique may have advantages over the acridine orange technique and may be useful for studying plankton microorganisms concentrated on Nuclepore filters as well as microorganisms attached to suspended debris particles. Similarly, fluorescamine, which forms a fluorescent complex with amine groups, can be used as a convenient label for the native forms of

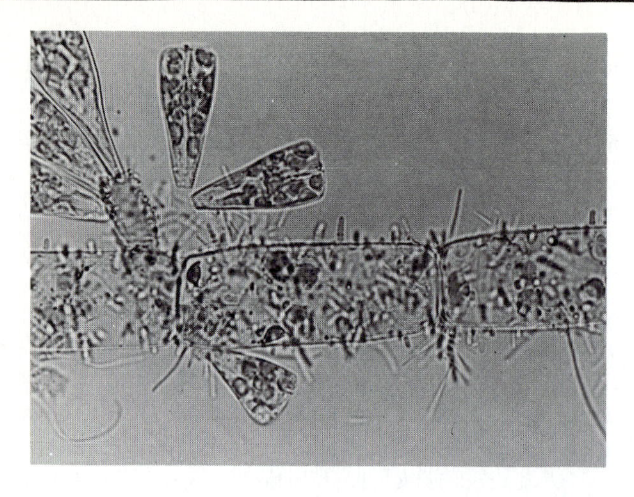

A

B

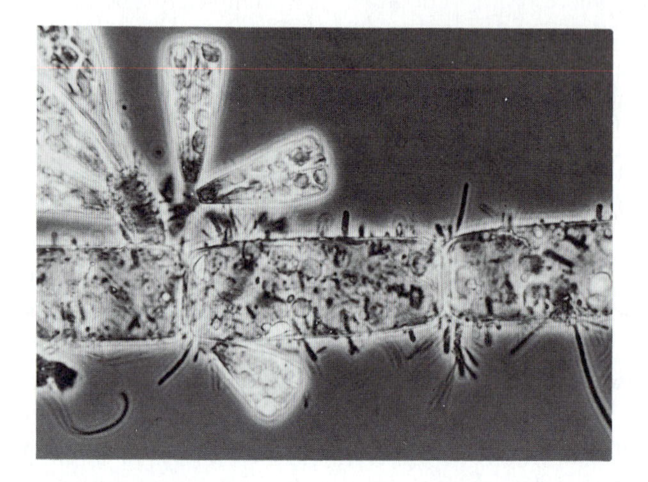

C

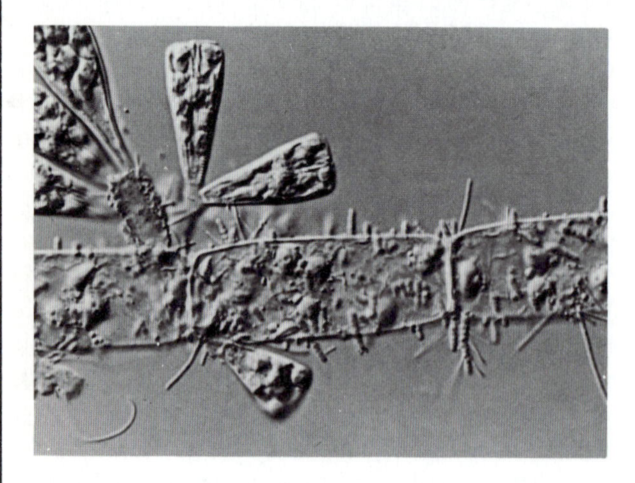

D

E

PLATE 5-2. A comparison of different modes of light microscopy, showing the filamentous alga *Pylaiella littoralis* and its epiphytic population of the pennate diatom *Licmophora* sp. and bacteria, including filaments of *Leucothrix mucor*. (*A*) Bright field. (*B*) Dark field. (*C*) Phase contrast. (*D*) Differential interference contrast (Nomarski). (*E*) Epifluorescence (blue light excitation with acridine orange stain) (all magnifications, 560 X). Photomicrographs courtesy of Paul W. Johnson.

chitin that are so prevalent and so important in the marine environment.

3. Optical Arrangements

When transmitted light is used, the microscope is usually set up for the dark-field type of illumination so that light is concentrated on the specimen, against a dark background. In recent years, the incident mode of illumination has been preferred for fluorescence microscopy. The exciting radiation is focused on the object from above, with the objective serving as the condenser. Not only is the light focused just on the field of view, it is not dissipated by passing through the specimen. Fluorescence microscopy is also ideal for examining and estimating microorganisms concentrated or growing on opaque surfaces. At the lower powers, many irregular surfaces from copepods to pebbles can be examined. When microorganisms originally suspended in water, are concentrated on a membrane for enumeration, it is preferable to stain them and use Nuclepore filters (Watson et al., 1977; Hobbie, Daley, and Jasper, 1977) rather than use pre-stained Sartorius filters (Daley and Hobbie, 1975), which apparently give low counts because the cells are obscured in the spongelike matrix of the filter (Bowden, 1977).

True phase contrast cannot be achieved with fluorescence, since the specimen must be observed as a self-luminous object and it is not possible to put a phase plate in or above the objective. But when the transmitted light mode is used, fluorescence can be viewed in bright-field, and simultaneous phase-contrast observations can be made from a second light source. In this mixed light mode, the specimen is first viewed for fluorescing areas and then the white light is slowly mixed in so that the non-fluorescing parts can be seen in phase contrast. With incident light fluorescence microscopy, a similar effect can be obtained in a dark room by side-lighting the specimen with fiber optics to see such larger objects as debris and diatom frustules.

D. TRANSMISSION ELECTRON MICROSCOPY

The principles involved in transmission electron microscopy are similar to those of light microscopy (Plate 5-1,*D*). The electron microscope was developed to extend the limit of resolution of the optical microscope. With the shortest wavelength of visible light (about 400 nm), the limit of resolution is 0.2 μm. The effective wavelength of an electron beam is about 10,000 X less than that of a light wave and offers a corresponding increase in the theoretical resolving power. The principle of operation of transmission electron microscopy is succinctly covered by Grimstone (1968); some of its biological applications are covered in more detail in the volumes edited by Hayat (1970, 1972) and Dawes (1971).

Each method of microscopy, optical and electron, has depended upon the development of suitable methods to prepare and visualize the specimen before it became useful. Many of the techniques for electron microscopy were first developed for transmission electron microscopy and later applied to scanning electron microscopy. The most critical problem in electron microscopy has been how to preserve morphology and ultrastructure. This was largely solved by the methods for fixation, dehydration, and embedding that have evolved over the years. The unique problem in transmission electron microscopy was the selection of a suitable support for the specimen, since the glass slides used in optical microscopy are opaque to electrons. Circular grids of copper, 2 to 3 mm in diameter, with holes of various sizes are used; they may be covered with a thin coat of electron-transparent material such as nitrocellulose, which is sometimes reinforced by coating it with carbon. A drop of this electron-transparent material in a volatile solvent is allowed to spread on the surface of water and a thin coat is lifted onto the grid to act as the specimen support. These coated grids are usually used to study specimens by transmission electron microscopy using the techniques of shadowing, negative-staining, and thin sections, as follows.

1. Shadowing

Shadowing is the depositing of a metallic coating on a specimen at an angle so that the otherwise transparent architecture can be seen in relief. A suspension of the microorganisms to be examined is dried on a coated grid. The specimens on the grid are then shadowed by a technique similar to that used for specimens on scanning electron microscopy stubs (Chapter 5,D). A metal wire heated to incandescence in a vacuum chamber vaporizes to coat the specimen, which is rotated to ensure uniform coating. The specimen can then be placed at a specific angle to the metal source so that it will be shadowed for more surface relief. On the photographic negative, the specimen appears white, whereas the shadows are dark. Negative prints are often used for micrographs, since positive prints show a dark object with light shadows and are a little harder to read. In addition to the specimen itself, replicas of specimen surfaces can be made with electron-transparent plastics or molecular carbon and then shadowed. In this technique, internal structures are not revealed but precise data on external shape and dimensions are obtained; these data are useful for determining the size of bacteria and virus particles. Shadow-cast preparations of bacteria (Plate 5-3,*B*) show the external architecture of the cells, including pili and flagella, to a greater useful magnification than is possible with the current generation of scanning electron microscopes.

2. Negative Staining

Negative staining is used to visualize transparent cells and their organelles by turning all non-cellular areas opaque. Stains for electron microscopy are salts of heavy metals, such as tungsten, uranium, molybdenum, and lead, which have high atomic numbers. In negative staining, the stain is used to form a thick opaque layer around the cells or virus particles so that the specimen stands out as a light object against a dark background (Plate 5-3,*A*). The particles to be examined are suspended in a solution of sodium or potassium phosphotungstate or uranyl acetate and sprayed on or dropped onto the coated grids. In a good preparation, the stain surrounds the particles or organisms, filling all its crevices so that as it dries it leaves a structureless solid.

PLATE 5-3. A comparison of bacteria examined by different modes of transmission electron microscopy (*A–D*) and by scanning electron microscopy (*E*). (*A*) Negatively stained marine bacterium, showing a single polar flagellum. (*B*) Shadow-cast preparation of *Bdellovibrio* cells, showing the cell surface as well as the flagella. (*C*) Thin section of *Nitroscococcus oceanus*, showing the characteristic internal cytomembranes in this dividing cell. (*D*) Shadowed replica of a frozen-etched preparation, also of *N. oceanus*, showing the cytomembranes within the cell. (*E*) Scanning electron micrograph of a critical-point-dried squid pen (chitinous surface), after exposure in seawater, showing an assortment of attached bacteria. (*A*) courtesy of Paul W. Johnson; (*B*) from D. Abram and B.K. Davis, 1970, *J. Bacteriol.* 104:948–65; (*C,D*) from C.C. Remsen, F.W. Valois, and S.W. Watson, 1967, *J. Bacteriol.* 94:422–33; (*E*) courtesy of Dawn L. Lavoie.

3. Thin Sections

In order to observe the internal structures of cells, more elaborate preparative procedures are required. The paraffin blocks and sections used for histology are too thick (2 μm) and are thus suitable only for optical microscopy. Thinner sections are required for transmission electron microscopy. After fixation and dehydration, the specimen is embedded in a methylacrylate or an epoxy plastic, which is polymerized and hardened by heating. The embedded specimens are then trimmed and sectioned. A microtome with either a glass or a diamond knife is first used to prepare thick sections (0.5 to 1.0 μm), which are examined by light microscopy to find the section of interest. Then thin sections (600 to 900 A or 0.06 to 0.09 μm thick) are floated on water and picked up on grids. To enhance the low intrinsic electron-scattering power of the specimen, structures within the specimen are stained. Uranyl salts are taken up by proteins and lead salts by lipids. The embedding plastic does not take up the stains, but it is permeable to aqueous solutions of the stains.

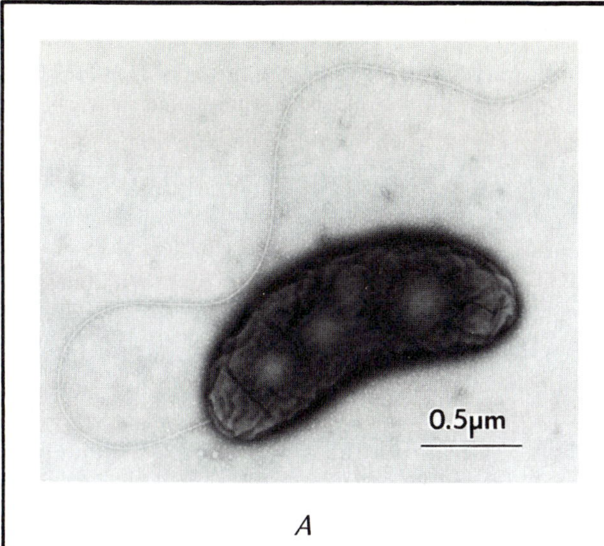

A

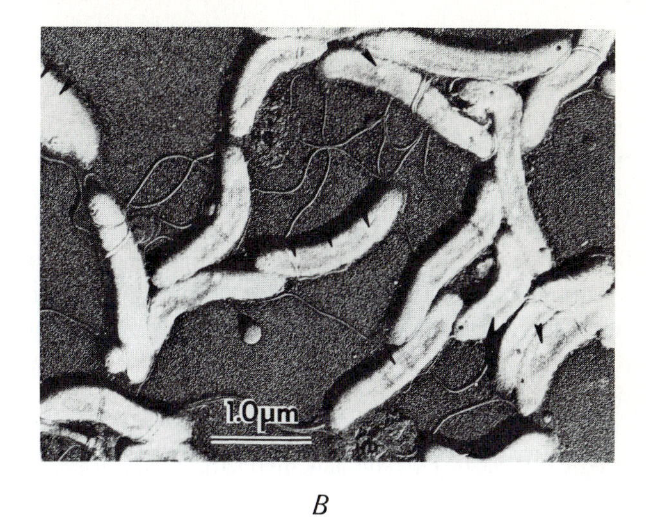

B

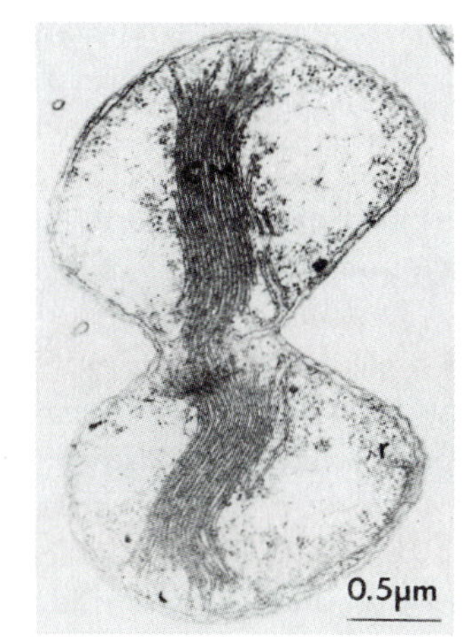

C

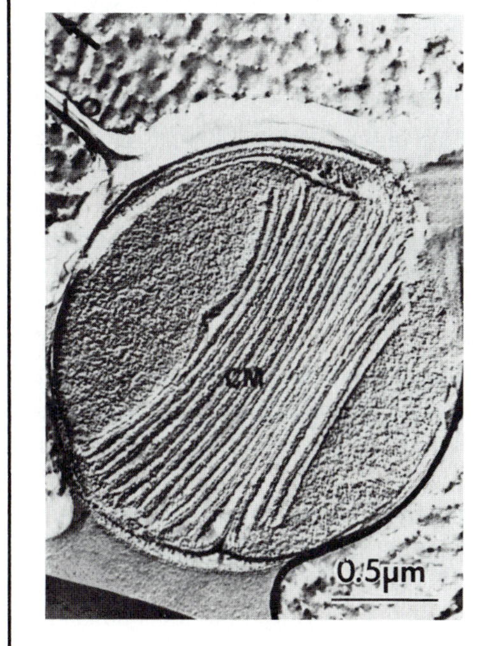

D

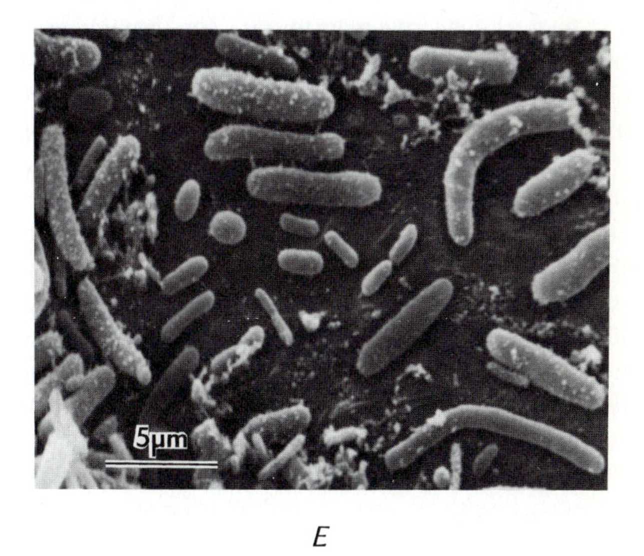

E

Such materials as microbial films or membranes (Plate 2-6,*B–D*), algal tissue (Plate 2-7,*A*), and animal tissue (Plate 3-2,*C*) are simply embedded before thin sectioning. But loose assemblages, such as cell suspensions (Plate 14-5) and suspended organic debris (Plate 3-1,*B*), must be centrifuged to form a pellet, which is then embedded. Such materials from natural environments not only show the concentrations and spatial relationships of natural assemblages of microorganisms, but through their distinctive fine structure and size permit the identification of cell types, which is often not possible by light or scanning microscopy (Caldwell and Tiedje, 1975a,b; Johnson and Sieburth, 1976).

E. SCANNING ELECTRON MICROSCOPY

The scanning electron microscope operates in a unique mode by looking only at the surface of objects made more reflective with a metallic coating. The transmitted light microscope and the transmission electron microscope (Plate 5-1,*D*) operate in the opposite mode, in which light and electrons, respectively, pass through a thin specimen and form an image for the eye or a viewing screen. The scanning electron microscope operates more in the manner of the stereomicroscope and incident–light microscope. As just seen, by using the latter mode in fluorescence microscopy, irregular and opaque surfaces can be examined for autofluorescing and fluorochromed microorganisms. The scanning electron microscope does this with electrons. A beam of electrons, focused by magnetic lenses similar to those used in the transmission electron microscope scan the surface of the specimen. Secondary electrons emitted from the specimen surface are detected electronically, amplified, and displayed on a cathode-ray tube. Unlike the transmitted light and transmission electron microscopes, which can detect internal structures, the scanning electron microscope sees only the outer three-dimensional shape. The internal structures are hidden by the outer surface. Although the inability to see internal structures in scanning electron microscopy is limiting at times, this limitation is heavily outweighed by the ability to observe thick, opaque, and irregular surfaces and to have a 400- to 500-fold increase in depth of field over that of both the light microscopes and the transmission electron microscope. The scanning electron microscope is an ideal tool for studying the architecture of the larger microorganisms in the Protista and for examining microscopic forms where they occur on surfaces, as shown in *Microbial Seascapes* (Sieburth, 1975).

Material to be examined by scanning electron microscopy is preserved by fixatives in much the same way as is material for transmission electron microscopy. At lower filament voltages and especially with the lanthanum hexaboride (rather than the tungsten filament) electron gun, wet uncoated specimens can be observed directly. More often the specimens are dehydrated. Water is replaced with a volatile solvent by passing the specimens through a series of solvent baths of increasing concentrations of either acetone alone or ethanol and amyl acetate. The samples are then dried at a critical point in a pressure housing in which liquid carbon dioxide, Freon, or nitrous oxide is used to replace the amyl acetate or acetone and, in turn, is rapidly changed to the gas at a critical pressure obtained by heating the pressure housing. The dry specimens are then glued onto scanning electron microscope stubs that fit into the specimen holder. Before examination, the stub-mounted specimens are coated with a thin film of gold or a gold–palladium alloy, which releases the secondary electrons that make the specimen visible. Coating is achieved with either the older type thermal vacuum evaporator or with the more controllable sputter coater (Draggan, 1976). The stubs are then examined at low or high operating voltages with different apertures, depending upon the degree of resolution desired. Useful magnifications range from 20 to 20,000 X. The unique property of the scanning electron microscope, with its very large depth of field, is that only this apparatus permits a single image in which details at all depths of focus can be seen at one time. Details seen by light microscopy or transmission electron microscopy one layer or depth of focus at a time, and reconstructed into a whole by a composite drawing, can now be seen at one time. The ability to scan large areas and then to enlarge an area of interest while retaining a three-dimensional perspective makes the scanning electron microscope unique and versatile. Scanning electron microscopy can be used to study attached microorganisms as they occur on their natural substrates and to observe the architecture of the larger protists and the smaller metazoa. Energy-dispersive analytical modules, which can be added to a scanning electron microscope, yield qualitative and quantitative analyses on a variety of elements, which can be correlated with the organisms or their organs or organelles, thereby greatly increasing the usefulness of the microscope.

No one mode of microscopy can fulfill all the requirements for a thorough examination of suspended microorganisms and colonized substrates. Each mode, or its variation, is better adapted for one purpose than another. Microscopes are important and potent tools for the microbiologist. Microscopic procedures are the primary ones used for the estimation of populations, for the determination of species composition of the protists, and for the calculation of estimated biomass of all trophic types in natural, enriched, and experimental populations.

REFERENCES

Abram, D., and B.K. Davis. 1970. Structural properties and features of parasitic *Bdellovibrio bacteriovorus*. *J. Bacteriol.* 104:948–65.

Anderson, J.R., and J.M. Slinger. 1975a. Europium chelate and fluorescent brightener staining of soil propagules and their photographic counting. I. Methods. *Soil Biol. Biochem.* 7:206–9.

Anderson, J.R., and J.M. Slinger. 1975b. Europium chelate and fluorescent brightener staining of soil propagules and their photographic counting. II. Efficiency. *Soil Biol. Biochem.* 7:211–15.

Bowden, W.B. 1977. Comparison of two direct-count techniques for enumerating aquatic bacteria. *Appl. Environ. Microbiol.* 33:1229–32.

Bradbury, S. 1970. *Additional Notes on the use of the Microscope.* Royal Microscopical Society, Oxford, 17 p.

Bradbury, S. 1976. *The Optical Microscope in Biology, Studies in Biology No. 59.* Edward Arnold Ltd., London, 76 p.

Caldwell, D.E., and J.M. Tiedje. 1975a. A morphological study of anaerobic bacteria from the hypolimnia of two Michigan lakes. *Can. J. Microbiol.* 21:362–76.

Caldwell, D.E., and J.M. Tiedje. 1975b. The structure of anaerobic bacterial communities in the hypolimnia of several Michigan lakes. *Can. J. Microbiol.* 21:377–85.

Casida, L.E., Jr. 1976. Continuously variable amplitude contrast microscopy for the detection and study of microorganisms in soil. *Appl. Environ. Microbiol.* 31:605–8.

Daley, R.J., and J.E. Hobbie. 1975. Direct counts of aquatic bacteria by a modified epifluorescence technique. *Limnol. Oceanogr.* 20:875–82.

Darken, M.A. 1962. Absorption and transport of fluorescent brighteners by microorganisms. *Appl. Microbiol.* 10:387–93.

Dawes, C.J. 1971. *Biological Techniques in Electron Microscopy.* Barnes & Noble, New York, 193 p.

Draggan, S. 1976. Improved specimen coating technique for scanning electron microscope observation of decomposer microorganisms. *Appl. Environ. Microbiol.* 31:313–15.

Eastman Kodak. 1974. *Photography through the Microscope* (6th Ed.). Eastman Kodak Co., Rochester, New York, 76 p.

Ehrenberg, C.G. 1838. *Die Infusionsthierchen als vollkommene Organismen.* Verlag Leopold Voss, Leipzig, 547 p; Atlas of 64 color plates.

Francisco, D.E., R.A. Mah, and A.C. Rabin. 1973. Acridine orange-epifluorescence technique for counting bacteria in natural waters. *Trans. Amer. Micros. Soc.* 92:416–21.

Grimstone, A.V. 1968. *The Electron Microscope in Biology, Studies in Biology No. 9.* Edward Arnold Ltd., London, 54 p.

Hardy, A.C. 1956. *The Open Sea. Its Natural History: The World of Plankton.* Collins, London and Houghton-Mifflin, Boston, 335 p.

Harrington, B.J., and K.B. Raper. 1968. Use of a fluorescent brightener to demonstrate cellulose in the cellular slime molds. *Appl. Microbiol.* 16:106–13.

Hayat, M.A. 1970–72. *Principles and Techniques of Electron Microscopy. Biological Applications.* Van Nostrand Reinhold, New York, Vol. 1, 1970, 412 p., Vol. 2, 1972, 268 p.

Hobbie, J.E., R.J. Daley, and S. Jasper. 1977. Use of Nuclepore filters for counting bacteria by fluorescence microscopy. *Appl. Environ. Microbiol.* 33:1225–28.

Hoffman, R., and L. Gross. 1975a. The modulation contrast microscope. *Nature* 254:586–88.

Hoffman, R., and L. Gross. 1975b. Modulation contrast microscope. *Appl. Optics* 14:1169–76.

Hughes, J., and M.E. McCully. 1975. The use of an optical brightener in the study of plant structure. *Stain Technol.* 50:319–29.

Jahn, I. 1971. Ehrenberg, Christian Gottfried. In: *Dictionary of Scientific Biography*, Vol. 4 (Gillispie, C.C., ed.). Scribner's, New York, pp. 288–292.

Johnson, P.W., and J.McN. Sieburth. 1976. *In-situ* morphology of nitrifying-like bacteria in aquaculture systems. *Appl. Environ. Microbiol.* 31:423–32.

Leidy, J. 1879. *Fresh-water rhizopods of North America*, Vol. 12. *U.S. Geological Survey of the Territories.* U.S. Government Printing Office, Washington, D.C., 234 p.

Mee, P.E. 1972. Microscopy and its contribution to computer technology. *Microstructures*, October/November 1972.

Padawer, J. 1968. The Nomarski interference-contrast microscope. An experimental basis for image interpretation. *J. Royal Micros. Soc.* 88:305–49.

Padgitt, D.D. 1975. *A Short History of the Early American Microscopes.* Microscopical Publications, Chicago, Ill., 147 p.

Remsen, C.C., F.W. Valois, and S.W. Watson. 1967. Fine structure of the cytomembranes of *Nitrosocystis oceanus*. *J. Bacteriol.* 94:422–33.

Scaff, W.L., Jr., D.L. Dyer, and K. Mori. 1969. Fluorescent europium chelate stain. *J. Bacteriol.* 98:246–48.

Sieburth, J.McN. 1975. *Microbial Seascapes.* University Park Press, Baltimore, Md., 248 p.

Tribe, M.A., M.R. Eraut, and R.K. Snook. 1975. *Basic Biology Course. Unit I. Microscopy and its Application to Biology. Book 1. Light Microscopy. Book 2. Electron Microscopy.* Cambridge University Press, London and New York.

Watson, S.W., T.J. Novitsky, H.L. Quinby, and F.W. Valois. 1977. Determination of bacterial number and biomass in the marine environment. *Appl. Environ. Microbiol.* 33:940–54.

Weaver, R.W., and L. Zibilske. 1975. Affinity of cellular constituents of two bacteria for fluorescent brighteners. *Appl. Microbiol.* 29:287–92.

Wild-Heerbrugg, Ltd. 1974. *Phase Contrast Equipment.* Heerbrugg, Switzerland, No. M1 410E-V.74.

Zimmerman, R., and L-A. Meyer-Reil. 1974. A new method for fluorescence staining of bacterial populations on membrane filters. *Kiel Meeresforsch.* 30:24–27.

CHAPTER 6

Quantitative Characterization

Marine microbiology is approaching a stage of maturity in which the descriptive phase is being supplemented by studies on rates and processes. The foundation of such studies is the reliable, quantitative estimation of populations occurring in samples. Microbial populations can be estimated in four ways: by direct microscopic counts, physical and chemical analysis, and by culture.

A. MICROSCOPIC ESTIMATION

In rich coastal waters, the larger and more common protists are sufficiently concentrated that 1-ml portions of freshly collected sample held in the well of a Sedgwick–Rafter cell can be estimated at a magnification of 100X, preferably with phase-contrast optics or dark field. For the smaller microorganisms that are sufficiently concentrated, such as phytoplankton cultures, the hemocytometer with its chamber of specified depth and calibrated grids can be used for examination and enumeration with high-dry optics to magnifications of 600X. More dilute samples with even smaller microorganisms can be concentrated by the Utermöhl technique (Utermöhl, 1927, 1931, 1958; Hasle, 1959). Here, a sample of preserved microorganisms is placed in a calibrated glass cylinder with a removable bottom, which itself is a shallow chamber with an optical glass bottom. After the microorganisms have settled (in about 2 days) the supernatant water and the cylinder are carefully removed, a coverslip is placed over the sedimented microorganisms in the shallow well, and the microorganisms are examined and enumerated from the bottom of the slide with an inverted microscope, using phase-contrast optics. With objectives up to oil immersion, the microorganisms can be examined to magnifications of 1500X. This is a very useful tool for the larger microorganisms, but unfortunately, some of the smaller microorganisms in the nanoplankton adhere to the walls of the sample bottles and sedimenting chambers and erroneously low counts are obtained. Dilute samples can also be stained, concentrated on a membrane filter with grids, dried, clarified with immersion oil, and the microorganisms counted while the preparation is examined with bright field. Contrast optics cannot be used with clarified membranes. An opaque cellulose acetate membrane filter can be rendered essentially transparent because the membrane consists of 85% air space, which when filled with an immersion oil of suitable numerical index, takes on the optical properties of glass. This technique is applicable to bacteria (Jannasch, 1958; Jannasch and Jones, 1959) as well as to the larger protists with rigid cell walls (APHA, 1975). The use of black-stained Nuclepore or cellulose-acetate filters, suitable fluorochromes, and epifluorescent microscopy (Chapter 5,C) makes possible a fluorescent modification of this technique (Francisco, Mah, and Rabin, 1973; Zimmerman and Meyer-Reil, 1974; Daley and Hobbie, 1975), which is being rapidly adopted.

When it is desirable to look at dilute samples with phase-contrast optics, the suspended microorganisms can be concentrated by filtration without sedimenting the microorganisms on a filter. In Cholodny-Bachman filtration (Jannasch and Jones, 1959; Plate 6-1,*A*), the microorganisms are kept in suspension with a rotating blade as the sample volume is reduced to yield a concentration sufficient for examination with a wet mount or in a counting chamber. Larger organisms are gently concentrated by reverse-flow filtration (Dodson and Thomas, 1964; Plate 6-1,*B*), in which a cylinder with a bottom screen is allowed to sink within a slightly wider sample cylinder. As the filtrate rises within the screened cylinder it is removed, and the excluded organisms are gently concentrated in the decreasing volume of the outer sample cylinder. An apparatus for the continuous reverse-flow concentration of particulates larger than 2 μm in volumes as large as several hundred liters of open ocean water combines the best features of both the Cholodny-Bachman and Dodson-Thomas filtration techniques (Hinga, Davis, and Sieburth, in press; Plate 6-1,*C*). In the 2-liter chamber version, the sample flows with gravity into the bottom half of a two-compartment chamber in which 35 × 35 cm squares of supported 2μm Nuclepore membrane separate the two chambers. The filtrate fills the upper chamber and leaves through the overflow while the accumulating particles greater than 2 μm in diameter sediment away from the membrane or are kept from plugging it by a gently rotating

PLATE 6-1. Apparatus for concentrating suspended particles from seawater. (*A*) The Cholodny–Bachman method of filtration: the ultrafine sintered glass plate (B) with a lower filtration speed than the membrane filter (A) and the stirring loop (C) keep the concentrating particles in suspension as the volume of sample is reduced to several milliliters. (*B*) The batch method of reverse-flow filtration for the gentle concentration of plankton. Nylon netting stretched over the Lucite tube supports paper or membrane filters, and the filtrate that passes into the tube is removed to concentrate particles in the reduced sample volume. (*C*) The enclosed and continuous reverse-flow concentrator, using a top supported, 2 μm Nuclepore membrane. Sample volumes of 200 liters can be reduced to a 1,000-fold concentrate at a controlled temperature in 2 hours, using a 1,250 cm^2 membrane surface and a flow rate of 1 liter/minute controlled by the level of the outflow tube. (*A*) from H.W. Jannasch and G.E. Jones, 1959, *Limnol. Oceanogr.* 4:128–39; (*B*) courtesy of A.N. Dodson and W.H. Thomas; (*C*) K.R. Hinga, P.G. Davis, and J.McN. Sieburth (in press).

bar. The fresh concentrates are suitable for culture and microscopic examination aboard ship (Sieburth et al., 1976, 1977), whereas the preserved concentrates can be used for chemical analysis and microscopy, including electron microscopy, ashore. Plankton netting can be substituted for Nuclepore membrane for the larger size fractions. Simpler and smaller configurations of 10 × 10 cm squares of membrane or screen, or standard 142 mm diameter Nuclepore membranes between an inlet and an outlet chamber (and with an optional magnetic mixing bar in the inlet chamber) can be used to concentrate nearshore samples and neuston samples, and for the final concentration of the residues from the larger concentrator. An even smaller 7-ml volume reverse flow chamber with standard 47-mm filter membranes can be used to remove the protist plankton from the bacterioplankton or to gently concentrate the fragile nanoplankton for ATP assays.

To obtain counts from a specific microscopic preparation, the area of the microscope field at that magnification must be determined to obtain a factor to convert the count to a population per standard volume. The diameter and area of a microscope field are so small that a number of fields must be counted for statistical significance. With an objective of 40X, the diameter is about 350 μm, and the area of the field 1×10^5 μm^2. This area is very small. When one looks at such preparations and sees only one or two microorganisms, it is a natural tendency to conclude that the sample is virtually devoid of microorganisms. But, in a wet mount with a thickness of 10 μm, which is used to examine both bacteria

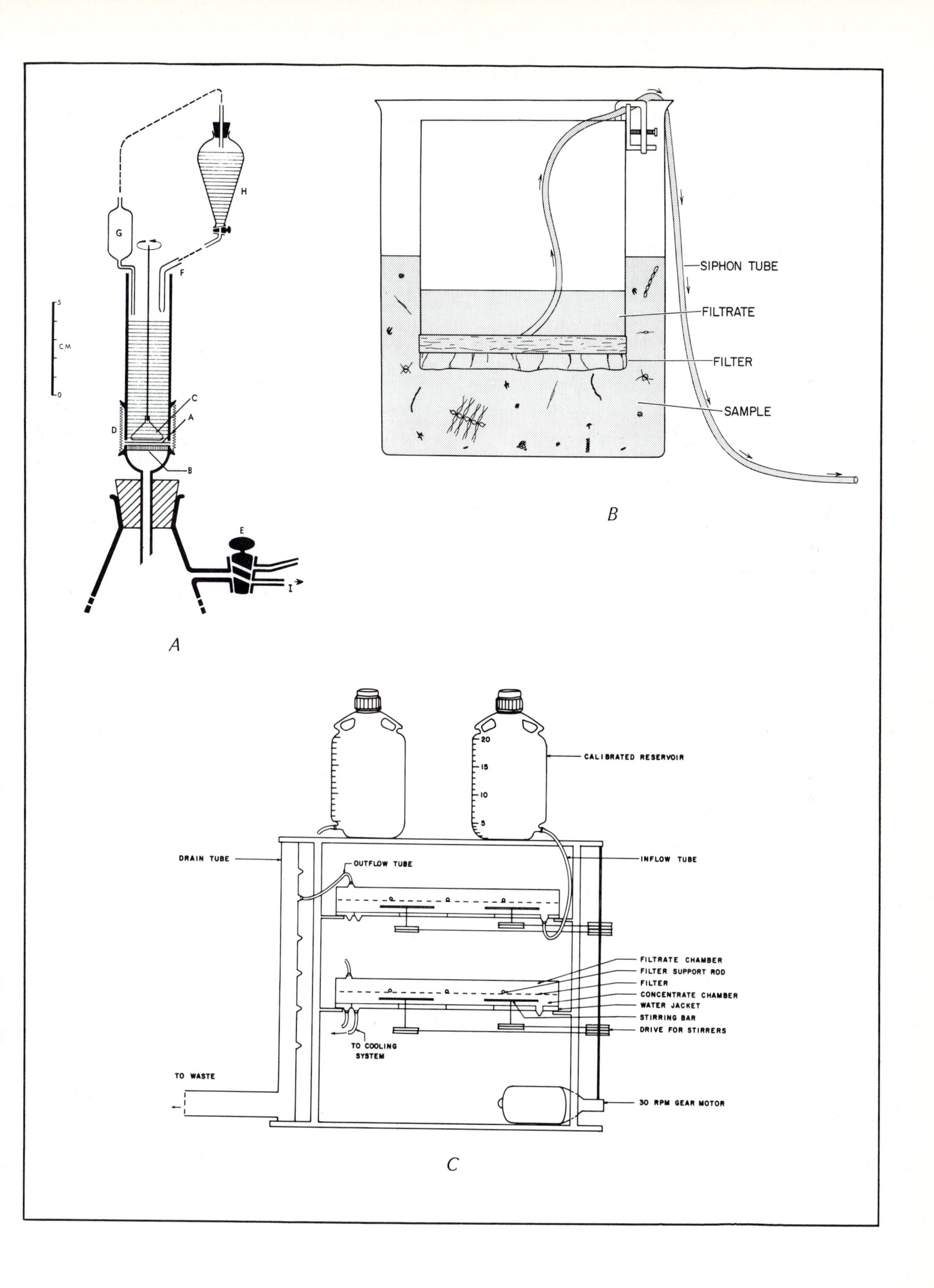
H
G
F
5
CM
0
C
A
D
B
E
I
SIPHON TUBE
FILTRATE
FILTER
SAMPLE
B
A
20
15
10
5
CALIBRATED RESERVOIR
DRAIN TUBE
OUTFLOW TUBE
INFLOW TUBE
FILTRATE CHAMBER
FILTER SUPPORT ROD
FILTER
CONCENTRATE CHAMBER
WATER JACKET
STIRRING BAR
DRIVE FOR STIRRERS
TO COOLING SYSTEM
TO WASTE
30 RPM GEAR MOTOR
C

and protists with a 40X objective, each field contains a volume of only 1×10^{-6} ml. Every microorganism seen in one field of such a preparation represents 1 million organisms/ml. This is a sizable number, especially for natural populations from the sea. For a 2 μm thick sample, each microorganism seen would represent 5 million organisms/ml. When examined with a 100X objective under oil immersion, the area observed is smaller, representing some 5×10^8 organisms/ml for each microorganism seen.

There are two alternatives for the quantitative estimation of populations on opaque and irregular surfaces. For the larger microorganisms or for specimens with relatively thin bodies, epifluorescence microscopy can be used for counting, using either autofluorescence or secondary fluorescence. When a greater depth of field and higher magnifications are desirable, scanning electron microscopy would be preferable. For data handling, the micrographs taken with these and other microscopes can be simply and quickly analyzed by using the test point set method of stereological analysis (Weibel, 1973). Transparent sheets of grids or patterns are laid over the micrographs to obtain the number of coincident points, which, with the area of view and an appropriate formula, are used to calculate probable populations. The scanning electron micrograph must be taken at a 0° angle or the calculation must take into account the usual 45° angle of view. For the well-funded laboratory that does much microscopic estimation, equipment now available for examining either microscopic fields or micrographs can automatically and selectively count the numbers and volumes of particles of a specific size and record other data, which is tallied, stored, printed, or displayed (Bradbury, 1976).

B. PHYSICAL ESTIMATION

Apparatus for the physical estimation of particles are produced by a number of companies, but all are based on the Coulter Counter, which both enumerates and estimates the volumes of particles. The Coulter Counter is an electronic particle counter developed for the automatic counting of blood cells. It has been used in much microbiological and some marine research (Sheldon and Parsons, 1966, 1967a; Kubitschek, 1969; Drake and Tsuchiya, 1973). To use the Coulter Counter, a specific volume of sample containing particles in an electrolyte is drawn into a glass tube through an aperture of specific size across which an electric field is maintained. For each particle, there is a change in electrical resistance equal to the volume of electrolyte displaced. The volume of each particle in a specific solute volume is estimated and counted, while the upper and lower thresholds of particle size and sensitivity are controlled. Linear relationships occur between 2 and 40% of the aperture size. Particles ranging from about 1.0 to 1000 μm can be accurately counted. The use of the Coulter Counter for bacterial cultures has been discussed by Toennies et al. (1961). But for the significant populations of minibacteria with a subthreshold volume that occur in the natural populations from the marine environment (Chapter 14), the Coulter Counter is of little use.

PLATE 6-2. The use of particle size fractions in plankton characterization. (*A*) Variation in particle size with time in the surface water of Sannich Inlet, in which nanoplankton with an associated peak in ultrananoplankton (picoplankton) occurred before and after a massive bloom of microplankton. (*B*) Continuous size spectrum of particulate matter from the above waters, showing that unidentified flagellates in the nanoplankton dominate the sample and that peaks of increasing size were associated with species of diatoms, copepod eggs, and a copepod. (*A*) from T.R. Parsons, 1969, *J. Oceanogr. Soc. Japan* 25:172–81; (*B*) from R.W. Sheldon and T.R. Parsons, 1967b, *J. Fish. Res. Bd. Canada* 24:909–15.

The growth constant of phytoplankton cultures has been determined by Maloney, Donovan, and Robinson (1962) and El-Sayed and Lee (1963) by Coulter-counting the number of cells at the beginning and end of a specified time interval. Using the cell volumes, as well as the numbers obtained with the Counter, and plotting these arithmetically with time, Parsons (1965) determined the growth rate of chain-forming phytoplankton more precisely. Such procedures could be used to determine the generation time of other microorganisms. The growth rate of natural phytoplankton populations has been determined by following the changes in volume of particles of different sizes with time, in light and dark bottle experiments (Vollenweider, 1969).

Size frequency distributions obtained with the Coulter Counter can be used to identify indicator species in mixtures when they differ in size by more than 3 μm (Parsons, 1969). A complete particle spectrum from 3 to 1000 μm can be obtained when a 300 μm zooplankton net sample is used to supplement water samples (Plate 6-2). This spectrum includes the nano-

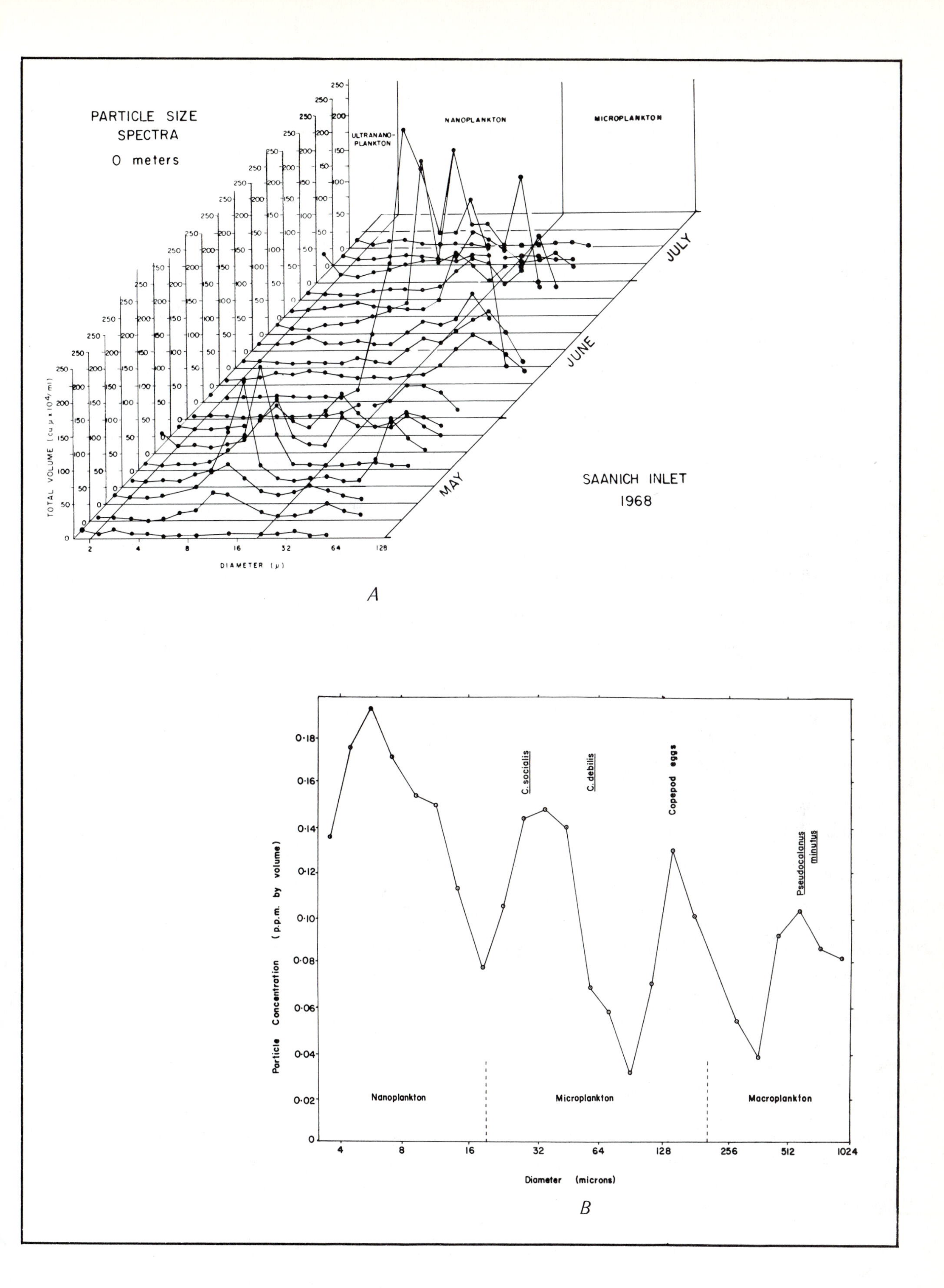

PARTICLE SIZE
SPECTRA
O meters
ULTRANANO-PLANKTON
NANOPLANKTON
MICROPLANKTON
JULY
JUNE
MAY
SAANICH INLET
1968
TOTAL VOLUME (cu μ x 10⁴/ml)
DIAMETER (μ)
A
C. socialis
C. debilis
Copepod eggs
Pseudocalanus minutus
Nanoplankton
Microplankton
Macroplankton
Particle Concentration (p.p.m. by volume)
Diameter (microns)
B

plankton (2 to 20 μm), microplankton (20 to 200 μm), and mesoplankton (0.2 to 2.0 mm). Peaks in particle diameter can be identified with phytoplankton species, copepods, and their eggs. The Coulter Counter has also been used to measure the number of phytoplankton grazed by zooplankton (Sheldon and Parsons, 1966). Despite its many applications, however, there are definite limits to the application of the Coulter Counter, since it cannot discriminate between such similar-size particles, which vary greatly in nature, as organic debris and phytoplankton (Vollenweider, 1969), phototrophs, phagotrophs, etc.

A prototype of a cell separator utilizing the Coulter Counter in which particles, both living and nonliving, are physically separated on the basis of electronically measured volume has been described (Fulwyler, 1965). After the cell volume has been sensed by a Coulter aperture within a piezoelectric droplet generator, an electric pulse proportional to the volume charges the emerging droplet, which falls through an electrostatic field that deflects the droplet into an appropriate collection vessel. A mixture of mouse and human red blood cells have been separated by a simple version of this device. The new generation of particle analyzers, such as spectrofluorometers, might make use of this mechanism to distinguish chemically between the physically similar particles to yield useful automatic counts and to produce real-time data ashore and at sea.

C. CHEMICAL ESTIMATION

Another approach to estimating the biomass of microorganisms is the quantitative determination of certain cell constituents that are not found to any degree in organic detritus (Holm-Hansen, 1970). One of the most promising analytical methods so far is the measurement of adenosine triphosphate (ATP). It is based upon the luciferin–luciferase reaction (Strehler and McElroy, 1957), which is responsible for the light emitted by the firefly. The amount of light produced is linearly proportional to the amount of ATP present. This principle has been applied to a method for estimating the biomass of marine microorganisms suspended in water samples (Holm-Hansen and Booth, 1966).

To make an ATP extract for analysis, membrane filters (0.2 μm) that are used to concentrate the suspended cells from the water samples are extracted in boiling Tris buffer. The ATP of sediment samples is best recovered by acidic extraction (Lee, Harris, and Williams, 1971; Karl and LaRock, 1975). To estimate the ATP concentration in the ATP extracts from organisms, a volume of extract is injected into a test tube in a dark chamber containing a freshly hydrated luciferin–luciferase mixture. The amount of light produced, which is proportional to the amount of ATP present, is measured by a photomultiplier tube and associated amplifying and recording equipment. One quantum of light is emitted for each molecule of ATP hydrolyzed by the luciferin–luciferase enzyme–substrate system (Seliger and McElroy, 1960). The ATP content of marine bacteria ranges from 0.5 to 6.5×10^{-9} μg ATP/cell or 0.3 to 1.1% of the cell carbon, with an average of 0.4% (Hamilton and Holm-Hansen, 1967). Similar values have been obtained with microalgae (Syrett, 1958; Holm-Hansen and Booth, 1966). When cells other than exponential growth phase are used, however, a much wider range of chlorophyll *a*/ATP and cell carbon/ATP ratios are obtained (Sakshaug and Holm-Hansen, 1977). When converting from cell carbon to number of cells, especially in oceanic samples, the order of magnitude smaller cell volume of the minibacteria (Hoppe, 1976; Watson et al., 1977) must be taken into consideration.

Profiles of particulate ATP from water samples throughout the water column without size fractionation showed high levels in the euphotic zone and thermocline, presumably due to phytoplankton and decomposing debris, respectively (Holm-Hansen and Booth, 1966; Plate 6-3,*A*). Since there is an appreciable biomass of phototrophs in the euphotic zone, the estimation of heterotrophic biomass by ATP assay has been restricted to deep waters (Holm-Hansen, 1970). An attempt to estimate the bacterial biomass of water samples indirectly, by correcting total ATP for phytoplankton bio-

PLATE 6-3. Distribution of particulate adenosine triphosphate (ATP) with depth. (*A*) The total particulate fraction off the California coast. (*B*) The distribution of procaryote and protist fractions, with environmental factors and chemical analyses in the mid-North Atlantic. (*C*) A comparison of ATP and adenylate energy charge in the western North Atlantic. (*A*) from O. Holm-Hansen and C.R. Booth, 1966, *Limnol. Oceanogr.* 11:510–19; (*B*) from J.McN. Sieburth, K.M. Johnson, C.M. Burney, and D.M. Lavoie, 1977, *Helgolander wiss. Meeresunters.* 30:565–74; (*C*) from W.J. Wiebe and K. Bancroft, 1975, *Proc. Natl. Acad. Sci.* 72:2112–15.

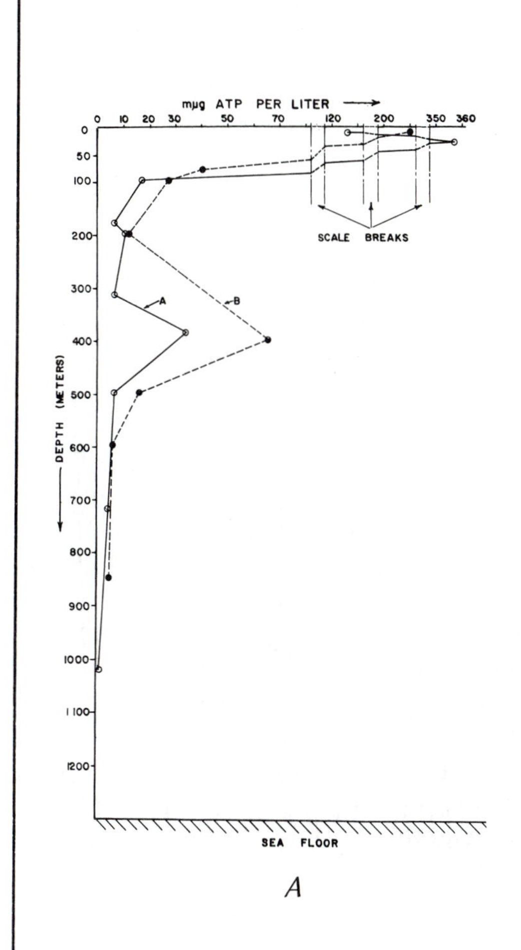
mμg ATP PER LITER
0 10 20 30 50 70 120 200 350 360
SCALE BREAKS
DEPTH (METERS)
A
B
SEA FLOOR

A

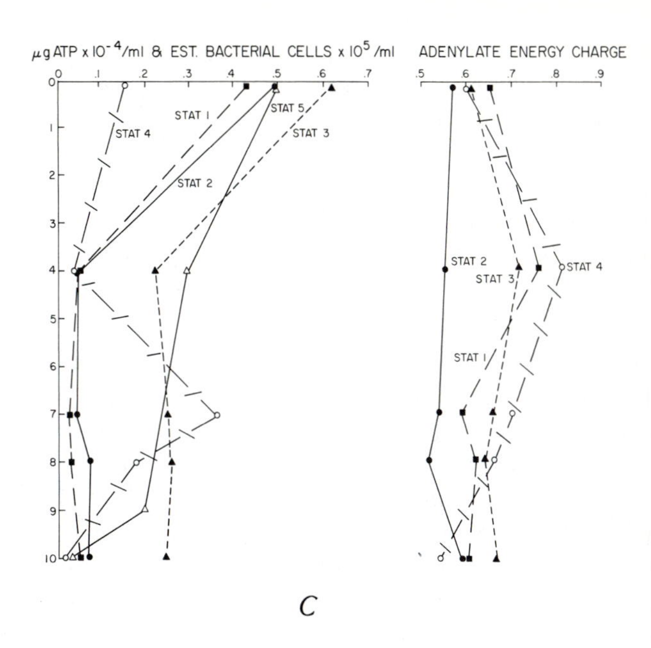
μg ATP x 10^-4/ml & EST. BACTERIAL CELLS x 10^5/ml
ADENYLATE ENERGY CHARGE
STAT 1
STAT 2
STAT 3
STAT 4
STAT 5

C

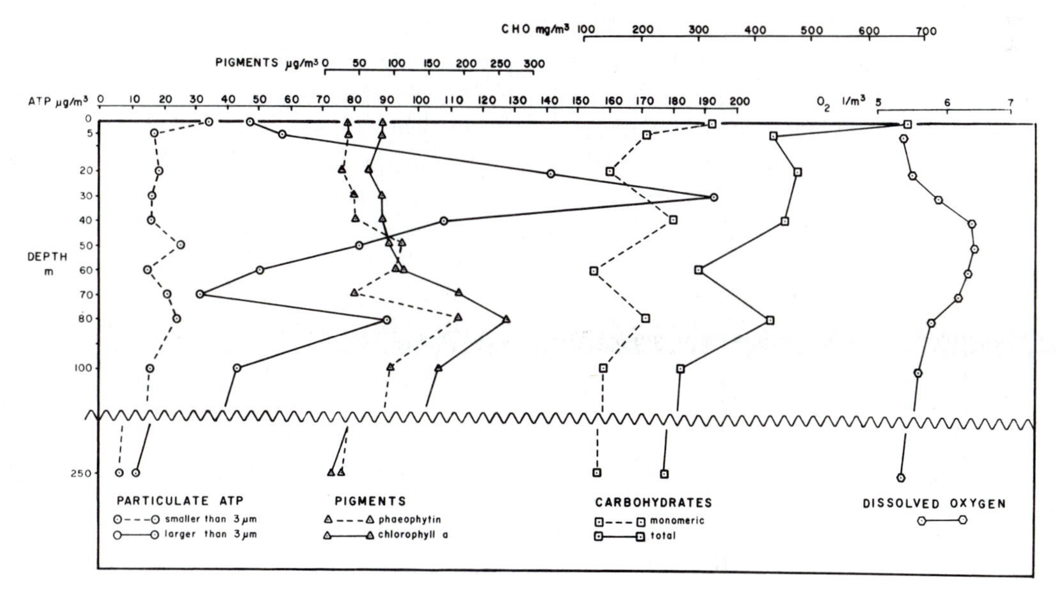
CHO mg/m³
PIGMENTS μg/m³
ATP μg/m³
O2 l/m³
DEPTH m
PARTICULATE ATP
smaller than 3 μm
larger than 3 μm
PIGMENTS
phaeophytin
chlorophyll a
CARBOHYDRATES
monomeric
total
DISSOLVED OXYGEN

B

mass estimated with chlorophyll values, has shown that systematic errors are too large for most samples (Jassby, 1975). An obvious approach to separating the bacterial biomass from the protist biomass is to use Nuclepore filters, which act as sieves (Sheldon, 1972), to remove the larger protists from the smaller procaryotes and to measure the ATP in both fractions. Bacteria on the larger organic debris would be excluded from the free bacteria and might therefore lead to an underestimate of the total bacterial biomass by no more than 5% (Hoppe, 1976) to 10%, since free bacteria passing a 1 μm Nuclepore membrane apparently account for 90% of the heterotrophic activity (Azam and Hodson, 1977). The values of 25% (Sorokin, 1971a,b) or more (Jannasch and Jones, 1959) appear to be overestimates due to an underestimation of total cell numbers. For a selectively filtered fraction between 0.2 and 3 μm, values on the order of 1 pg/ml have been obtained throughout the water column of deep-oceanic waters, whereas maxima of 10 to 100 pg ATP/ml occur in the subsurface euphotic zone (Sieburth et al., 1977), and values to 1000 pg ATP/ml occur in the surface microlayer (Sieburth et al., 1976). In addition to the bacterioplankton, such preparations can also include some of the smaller flagellates. The use of 1 μm Nuclepore filters in subsequent work has excluded the smaller flagellates passing the 2 and 3 μm porosities and has yielded only bacterioplankton, with similar ATP values. Concomitant ATP assays of the 3 to 1000 μm size fraction indicate the depths at which protists aggregate, whereas pigment determinations indicate where healthy phytoplankton accumulates (Plate 6-3,*B*). In subsequent work, this protist fraction has been decreased to 1 to 153 μm to include the smaller nanoplankton and exclude most of the metazooplankton. Assays of ATP have also been used to study terrestrial decomposition. (Ausmus, 1973).

A portion of the cell extracts obtained for ATP analysis can also be used to determine adenylate energy charge (AEC) (Atkinson, 1968, 1969, 1971; Chapman, Fall, and Atkinson, 1971). Adenosine monophosphate (AMP) and adenosine disphosphate (ADP) are converted to ATP with the appropriate enzyme, substrate, and incubation time; the ATP is then measured and added to the original ATP concentration. Energy charge is calculated from the equation:

$$AEC = ATP + \frac{1}{2}\,ADP \,/\, ATP + ADP + AMP$$

Since organisms maintain a certain ratio of ATP to AMP and ADP, the ratio that is actually observed is an indicator of physiological state. Dying cells have values between 0 and 0.6, cells in stationary growth have values between 0.7 and 0.8, and intact metabolizing cells have values between 0.85 and 1.0. The variation of ATP content with physiological state, which can yield a wide range of ATP:C ratios, in contradiction to the premise of the test, is one of the disadvantages of the ATP assay method. A preliminary study of the adenylate energy charge in marine populations indicates that this parameter can also be applied to natural populations (Wiebe and Bancroft, 1975; Plate 6-3,*C*). In addition to adenylate energy charge, the ATP:C ratio in the size fractions may be another index of physiological state.

The determination of chlorophyll *a* and its breakdown products (phaeopigments) (Holm-Hansen et al., 1965) would permit an estimate of the extent and state of the phytoplankton in the protist size fraction. The size differences between bacteria, protists, and copepods, which fall mainly into the picoplankton, nanoplankton–microplankton, and mesoplankton fractions, respectively (Chapter 4), might be used to study the dynamics of nutrition with a minimum of tedious microscopic sorting and counting of microorganisms. The big problem with all ATP assays is the high cost of the ATP kit (purified luciferin–luciferase). In an extensive program, one student can use some 3,000 to 5,000 dollars' worth of reagent a year. Such prohibitive costs may ultimately limit its use to calibration, while routine counts are made with epifluorescence measurements.

A more recent, but promising procedure for estimating bacterial biomass and numbers in water samples and on surfaces is based on an assay described by Levin and Bang (1964), which determines the amount of lipopolysaccharide (LPS) from the cell walls of the dominant gram-negative bacteria with a purified lysate of the amoebocytes (blood cells) of the horseshoe crab *Limulus* (Limulus Amoebocyte Lysate, LAL) (Sullivan and Watson, 1974). The basic assumptions for the LAL assay of LPS as an index of bacterial populations in seawater are (1) LAL is specific for LPS, a compound that only occurs naturally in the cell walls of gram-negative bacteria, (2) the LPS of these bacterial cells is relatively constant, and (3) the percentage of gram-positive bacterial cells in the sea is so small that it can be ignored (Watson et al., 1977). Purified LAL reacts specifically with LPS when incubated at 37 °C for 1 hour to form either a clot at higher concentrations (Sullivan and Watson, 1974;

Dexter et al., 1975) or a turbid solution at lower concentrations (Watson et al., 1977). The ubiquitous nature of bacteria, and therefore of LPS, makes the fastidious use of noncontaminated, single-use plastic dishes and glassware, the precombustion of salt used in the dilution blanks as well as all reusable glassware at 180 °C for 3 hours, and the use of special, LPS-free distilled water essential for successful analysis.

Water samples and bacterial cultures contain both free and bound LPS. The total LPS is extracted by boiling the sample for 1 minute, and its concentration determined. The free LPS is determined in cell-free supernatants obtained by centrifugation. The cell-bound LPS is estimated by difference. In addition to water samples (Watson et al., 1977), particulate matter and surfaces (Dexter et al., 1975) can also be extracted and assayed for LPS. For the coagulation procedure, semi-decimal dilutions in 10 × 75 mm disposable glass tubes are mixed with an equal volume of LAL and, after incubation, are slowly inverted by 180° to determine the presence of clots that do not break. For the turbidometric procedure, one volume of LAL is added to six volumes of a dilution series of the sample and the turbidity formed during incubation, which occurs in the linear part of the curve, is measured at 360 nm in a spectrophotometer. For both procedures, the LAL sensitivity for each run is determined with LPS standards (about 10^{-10} to 10^{-11} g/ml). The amount of LPS determined can be converted into bacterial numbers by using a factor of 2.78 ± 1.42 fg of LPS/bacterial cell (Watson et al., 1977), obtained by epifluorescent counts, carbon analyses, and cell volumes estimated by transmission electron microscopy. The major advantage of this procedure is that bacteria can be determined in the presence of other microorganisms without interference. In conjunction with selective filtration, it should be useful for distinguishing between bacteria free in the water (bacterioplankton) and those colonizing organic debris (epibacteria). One possible disadvantage is that the LAL assay also determines LPS from non-viable cells. More important disadvantages may be the high cost of the LAL reagent, which is approaching that of the ATP reagents, its short half-life in the lyophilized form, and its 8-hour shelf life when rehydrated and chilled.

Bacterial biomass can also be estimated through the chemical analysis of muramic acid (3-O-carboxyethyl-D-glucosamine), which has only been detected so far in the muramyl peptide cell walls of the procaryotes (all bacteria including the cyanobacteria except the halophiles, which occur only in concentrated brines at the edge of the sea). Tipper (1968) reported that mild alkaline treatment of muramic acid resulted in the quantitative release of D-lactic acid and that the D-lactate liberated could be determined with D-lactate dehydrogenase. Millar and Casida (1970) developed a procedure to purify the amino compounds released from the acid hydrolysates of soil samples and bacterial cultures, but they used colorimetric analysis to determine the amount of lactic acid released from muramic acid. The values, which ranged from 0 to 150 μg of muramic acid/g dry weight of soil agreed with the values for the total microscopic counts of bacteria and were two to three orders of magnitude greater than those indicated by the colony forming units on agar plates, as one might expect. Values of 3.44 ± 0.5 and 9.6 ± 1.9 μg of muramic acid/mg dry weight were obtained for gram-negative and gram-positive bacteria, respectively. Moriarty (1975) used the enzymatic procedure of Tipper (1968) to determine the lactic acid released from non-purified acid hydrolysates in an attempt to measure the bacterial biomass of sediments, especially those ingested by deposit-feeding animals. For pure cultures, Moriarty (1975) obtained values of 20 μg and 100 μg of muramic acid/mg of cell carbon (equal to 9 and 45 μg of muramic acid/g dry weight) for gram-negative and gram-positive bacteria, respectively. These values are three and five times the values obtained for gram-negative and gram-positive bacteria, respectively, by Millar and Casida (1970). King and White (1972) found that a sensitive automated assay for lactic acid gave probably erroneous results for similar environmental samples from an estuary which were 10 to 20 × greater than those obtained with a procedure similar to that of Millar and Casida (1970), which involved chromatographic separation and colorimetric analysis of the hydrolysate. These procedures indicated a bacterial content of approximately 0.5% and 10% for estuarine mud and organic debris, respectively. Ausmus (1973) has shown that the ratio of ATP to muramic acid in bacterial cultures is 1:4 to 1:5. The assay of muramic acid and ATP along with scanning electron microscopy has been used to document the succession of a procaryote by a protist biomass on oak leaf litter in an estuarine environment (Morrison et al., 1977).

The assay of muramic acid by measurement of released lactate could not be used to determine the bacterioplankton content of water, due to the sample size required. Sample sizes required for reliable determinations are 50 mg for bacterial cells, 0.2 to 1.0 g for organic debris, and 1 to 10 g for sediment. As pointed out by Moriarty (1975), King and White (1975), and Morrison et al. (1977), the assay of muramic acid, which is specific for bacteria, is useful for determining the content of bacteria in the presence of eucaryotic cells in organic debris, sediment, fouling films and the gut contents of detritus-feeders. The effect of environmental changes on the muramic acid content of gram-positive bacteria only seem to involve twofold differences in concentration (Ellwood and Tempest, 1972). But more serious limitations may be the fourfold concentrations that occur in bacterial spores (Millar and Casida, 1970) and the five hundred-fold concentrations in cyanobacteria (Moriarty, 1975).

An alternate chemical approach to determining microbial biomass in the sea has been the measurement of deoxyribonucleic acid (DNA). The simple and sensitive fluorometric method of Kissane and Robins (1958), which was used by Sutcliffe (1965) to estimate growth in small marine organisms, has been adapted by Holm-Hansen, Sutcliffe, and Sharp (1968) to measure the DNA suspended in seawater. The fluorescence of a DNA-3,5 diaminobenzoic acid complex is measured with a fluorometer (which can also be used to measure plant pigments). A reliable range of 0.2 to 40 μg of DNA is obtained from 0.1 to 4 liters of sample. The DNA amounts to some 1% of the cell carbon of microalgae. Unfortunately, DNA is also present in debris so it cannot be used as an index of viable biomass. This test, however, might be valuable for the detection of DNA-containing viral particles in the 0.02 to 0.2 μm femtoplankton size fraction (see Chapter 4,B).

D. CULTURAL ESTIMATION

The primary purpose of cultivation is to obtain enriched, unispecies, and axenic cultures for experimental, taxonomic, physiological, and ultrastructural studies, and it is discussed in the next chapter. For most groups of microorganisms, the microscopic, physical, and chemical methods just described are more useful than are cultural methods for estimation. A case in point is the use of cultures for estimating bacteria. Before epifluorescence, the LAL test for LPS, or ATP determinations of the picoplankton size fraction, the only means for bacterial estimation were direct counts of stained preparations (Jannasch and Jones, 1959; Sorokin, 1971a,b), in which one did not know for sure what to count or what the counts indicated, or cultural estimation, which only detected high nutrient bacteria transient in the plankton. For many of the chemotrophic, phototrophic, and low-nutrient bacteria, cultural estimation appears to be of little value. Only epibacteria requiring high levels of dissolved nutrients or highly nutrient surfaces seem suitable for cultural estimation today. An example would be the estimation of bacterial populations capable of utilizing cellulosic and chitinous debris throughout the water column. But realistic cultural procedures for these bacteria remain to be worked out. Other organisms for which cultural estimations may be suitable are such phycomycetous fungi as *Lagenidium*, and *Thraustochytrium*, which are facultative parasites of phytoplankton and crustaceans. Selected hosts, such as diatoms and brine shrimp, may prove to be suitable substrates that could be used in dilution extinction procedures in addition to pollen grains.

In the case of phagotrophs, cultural estimation is useful for microorganisms that are too small, fragile, or dilute to be enumerated microscopically. The small and fragile choanoflagellates are a good example. Throndsen (1969) has used dilution extinction culture to estimate the populations of these and similar flagellates in Norwegian waters. Amoebae, which are naked and small but occur in appreciable numbers in the open sea, are another example (Sieburth et al., 1976; Davis, Caron, and Sieburth, 1978). Serial dilutions of reverse-flow concentrates were supplemented with a small plug of nutrient agar or a rice grain to support the low bacterial populations required by these bacteria-eating forms. In dilution extinction and most probable number estimates, dilutions of a sample or its concentrate are used to inoculate replicate tubes of a liquid medium containing nutrients that select for a specific physiological or taxonomic group of microorganisms (APHA, 1975). After a suitable incubation period and examination, the number of positive cultures in the replicates (3, 5, or 10) is used, along with most probable number (MPN) tables (Fisher and Yates, 1963) and the concentration or dilution factor, to estimate the population originally present in the sample.

The methods and equipment developed in the last decade to yield more reliable quantitative procedures that can be applied to studies on marine microoganisms are just beginning to have a major impact. The semi-quantitative and highly speculative papers of the last decades are giving way to more sophisticated and exact studies. With each passing year, the techniques of yesterday are no longer sufficient. Tough problems, thought insoluble a decade ago, are being attacked and partially resolved with each issue of a pertinent journal. The active researcher must stay abreast of these developments as they occur.

REFERENCES

American Public Health Assn. 1975. *Standard Methods for the Examination of Water and Wastewater* (14th Ed.). APHA, Washington, D.C., pp. 1018–19.

Atkinson, D.E. 1968. The energy charge of the adenylate pool as a regulatory parameter. Interaction with feedback modifiers. *Biochemistry* 7:4030–34.

Atkinson, D.E. 1969. Regulation of enzyme function. *Ann. Rev. Microbiol.* 23:47–68.

Atkinson, D.E. 1971. Adenine nucleotides as stoichiometric coupling agents in metabolism and as regulatory modifiers: The adenylate energy charge. In: *Metabolic Pathways* (D.M. Greenberg, ed.), Vol. 5 (3rd Ed.). Academic Press, New York, pp. 1–21.

Ausmus, B.S. 1973. The use of the ATP assay in terrestrial decomposition studies. In: *Modern Methods in the Study of Microbial Ecology* (T. Rosswall, ed.). NSR, Stockholm, p. 223.

Azam, F., and R.E. Hodson. 1977. Size distribution and activity of marine microheterotrophs *Limnol. Oceanogr.* 22:492–501.

Bradbury, S. 1976. *The Optical Microscope in Biology. Studies in Biology No. 59.* Edward Arnold Ltd., London, 76 p.

Chapman, A.G., L. Fall, and D.E. Atkinson. 1971. Adenylate energy charge in *Escherichia coli* during growth and starvation. *J. Bacteriol.* 108:1072–86.

Daley, R.J., and J.E. Hobbie. 1975. Direct counts of aquatic bacteria by a modified epifluorescence technique. *Limnol. Oceanogr.* 20:875–82.

Davis, P.G., D.A. Caron, and J.McN. Sieburth. 1978. Oceanic amoebae from the North Atlantic: Culture, distribution, and taxonomy. *Trans. Am. Micros. Soc.* 97:73–88.

Dexter, S.C., J.D. Sullivan, Jr., J. Williams III, and S.W. Watson. 1975. Influence of substrate wettability on the attachment of marine bacteria to various surfaces. *Appl. Microbiol.* 30:298–308.

Dodson, A.N., and W.H. Thomas. 1964. Concentrating plankton in a gentle fashion. *Limnol. Oceanogr.* 9:455–56.

Drake, J.F., and H.M. Tsuchiya. 1973. Differential counting in mixed cultures with Coulter Counters. *Appl. Microbiol.* 26:9–13.

El-Sayed, S.Z., and B.D. Lee. 1963. Evaluation of an automatic technique for counting unicellular organisms. *J. Mar. Res.* 21:59–73.

Ellwood, D.C., and D.W. Tempest. 1972. Effects of environment on bacterial wall content and composition. *Adv. Microb. Physiol.* 7:83–117.

Fisher, R.A., and F. Yates. 1963. *Statistical Tables for Biological Agricultural and Medical Research.* Hafner Publ. Co., Darien, Conn.

Francisco, D.E., R.A. Mah, and A.C. Rabin. 1973. Acridine orange-epifluorescence technique for counting bacteria in natural waters. *Trans. Amer. Micros. Soc.* 92:416–21.

Fulwyler, M.J. 1965. Electronic separation of biological cells by volume. *Science* 150:910–11.

Hamilton, R.D., and O. Holm-Hansen. 1967. Adenosine triphosphate content of marine bacteria. *Limnol. Oceanogr.* 12:319–24.

Hasle, G.R. 1959. A quantitative study of phytoplankton from the equatorial Pacific. *Deep-Sea Res.* 6:38–59.

Hinga, K.R., P.G. Davis, and J.McN. Sieburth. (in press). Enclosed chambers for the reverse flow concentration and selective filtration of particles. *Limnol. Oceanogr.*

Holm-Hansen, O. 1970. Determination of microbial biomass in deep ocean water. In: *Organic Matter in Natural Waters* (D.W. Hood, ed.). Inst. Mar. Sci., University of Alaska, pp. 287–300.

Holm-Hansen, O., and C.R. Booth. 1966. The measurement of adenosine triphosphate in the ocean and its ecological significance. *Limnol. Oceanogr.* 11:510–19.

Holm-Hansen, O., C.J. Lorenzen, R.W. Holmes, and J.D.H. Strickland. 1965. Fluorometric determination of chlorophyll. *J. Cons. perm. int. Explor. Mer* 30:3–15.

Holm-Hansen, O., W.H. Sutcliffe, Jr., and J. Sharp. 1968. Measurement of deoxyribonucleic acid in the ocean and its ecological significance. *Limnol. Oceanogr.* 13:507–14.

Hoppe, H-G. 1976. Determination and properties of actively metabolizing heterotrophic bacteria in the sea, investigated by means of micro-autoradiography. *Mar. Biol.* 36:291–302.

Jannasch, H.W. 1958. Studies on planktonic bacteria by means of a direct membrane filter method. *J. gen Microbiol.* 18:609–20.

Jannasch, H.W., and G.E. Jones. 1959. Bacterial populations in sea water as determined by different methods of enumeration. *Limnol. Oceanogr.* 4:128–39.

Jassby, A.D. 1975. An evaluation of ATP estimations of bacterial biomass in the presence of phytoplankton. *Limnol. Oceanogr.* 20:646–48.

Karl, D.R., and P.A. LaRock. 1975. Adenosine triphosphate measurements in soil and marine sediments. *J. Fish. Res. Bd. Canada* 32:599–607.

King, J.D., and D.C. White. 1977. Muramic acid as a measure of microbial biomass in estuarine and marine samples. *Appl. Environ. Microbiol.* 33:777–83.

Kissane, J.M., and E. Robins. 1958. The fluorometric measurement of deoxyribonucleic acid in animal tissues with special reference to the central nervous system. *J. Biol. Chem.* 233:184–88.

Kubitschek, H.E. 1969. Counting and sizing microorganisms with the Coulter Counter. In: *Methods in Microbiology, Vol. 1* (R. Norris and D.W. Ribbons, eds.). Academic Press, London and New York, pp. 593–610.

Lee, C.C., R.F. Harris, and J.D.H. Williams. 1971. Adenosine triphosphate in lake sediments. I. Determination. II. Origin and significance. *Soil Sci. Soc. Amer. Proc.* 35:82–86(I); 86–91(II).

Levin, J., and F.B. Bang. 1964. A description of cellular coagulation in the *Limulus*. *Bull. Johns Hopkins Hosp.* 115:337.

Maloney, T.E., E.J. Donovan, Jr., and E.L. Robinson. 1962. Determination of numbers and sizes of algal cells with an electronic particle counter. *Phycologia* 2:1–8.

Millar, W.N., and L.E. Casida, Jr. 1970. Evidence for muramic acid in soil. *Can. J. Microbiol.* 16:299–304.

Moriarty, D.J.W. 1975. A method for establishing the biomass of bacteria in aquatic sediments and its application to trophic studies. *Oecologia* 20:219–29.

Morrison, S.J., J.D. King, R.J. Bobbie, R.E. Bechtold, and D.C. White. 1977. Evidence for microfloral succession on allochthonous plant litter in Apalachicola Bay, Florida, USA. *Mar. Biol.* 41:229–40.

Parsons, T.R. 1965. An automated technique for determining the growth rate of chain-forming phytoplankton. *Limnol. Oceanogr.* 10:598–602.

Parsons, T.R. 1969. The use of particle size spectra in determining the structure of a plankton community. *J. Oceanogr. Soc. Japan* 25:172–81.

Sakshaug, E., and O. Holm-Hansen. 1977. Chemical composition of *Skeletonema costatum* (Grev.) Cleve and *Pavlova (Monochrysis) lutheri*: (Droop) Green as a function of nitrate-, phosphate-, and iron limited growth. *J. exp. mar. Biol. Ecol.* 29:1–34.

Seliger, H.H., and W.D. McElroy. 1960. Spectral emission and quantum yield of firefly bioluminescence. *Arch. Biochem. Biophys.* 88:136–41.

Sheldon, R.W. 1972. Size separation of marine seston by membrane and glass-fiber filters. *Limnol. Oceanogr.* 17:494–98.

Sheldon, R.W., and T.R. Parsons. 1966. *On Some Applications of the Coulter Counter to Marine Research*. Fish. Res. Bd. Canada, Rept. No. 214, 36 p.

Sheldon, R.W., and T.R. Parsons, 1967a. *A Practical Manual on the Use of the Coulter Counter in Marine Science*. Coulter Electronic Sales Co., Toronto, 66 p.

Sheldon, R.W., and T.R. Parsons. 1967b. A continuous size spectrum for particulate matter in the sea. *J. Fish. Res. Bd. Canada* 24:909–15.

Sieburth, J.McN., P-J. Willis, K.M. Johnson, C.M. Burney, D.M. Lavoie, K.R. Hinga, D.A. Caron, F.W. French III, P.W. Johnson, and P.G. Davis. 1976. Dissolved organic matter and heterotrophic microneuston in the surface microlayers of the North Atlantic. *Science* 194:1415–18.

Sieburth, J.McN., K.M. Johnson, C.M. Burney, and D.M. Lavoie. 1977. Estimation of *in-situ* rates of heterotrophy using diurnal changes in dissolved organic matter and growth rates of picoplankton in diffusion culture. *Helgoländer wiss. Meeresunters.* 30:565–74.

Sorokin, Yu. I. 1971a. On the role of bacteria in the productivity of tropical oceanic waters. *Int. revue ges. Hydrobiol.* 56:1–48.

Sorokin, Yu.I. 1971b. Abundance and production of bacteria in the open water of the Central Pacific. *Oceanology* 11(1):85–94.

Strehler, B.L., and W.D. McElroy. 1957. Assay of adenosine triphosphate. In: *Methods in Enzymology* (Colowick and Kaplan, eds.), Vol. 3. Academic Press, New York and London, pp. 871–73.

Sullivan, J.D., and S.W. Watson. 1974. Factors affecting the sensitivity of *Limulus* lysate. *Appl. Microbiol.* 28:1023–26.

Sutcliffe, W.H., Jr. 1965. Growth estimates from ribonucleic acid content in some small organisms. *Limnol. Oceanogr.* 10(Suppl.):R253–R258.

Syrett, P.J. 1958. Respiration rate and internal adenosine triphosphate concentration in *Chlorella*. *Arch. Biochem. Biophys.* 75:117–24.

Throndsen, J. 1969. Flagellates of Norwegian coastal waters. *Nytt. Mag. Bot (Norw. J. Bot.)* 16:161–216.

Tipper, D.J. 1968. Alkali-catalyzed elimination of D-lactic acid from muramic acid and its derivatives and the determination of muramic acid. *Biochemistry* 7:1441–49.

Toennies, G., L. Iszard, N.B. Rogers, and G.D. Shockman. 1961. Cell multiplication studied with an electronic particle counter. *J. Bacteriol.* 82:857–66.

Utermöhl, H. 1927. Untersuchungen über den Gesamtplanktongehalt des Kanarenstromes. *Arch. Hydrobiol.* 18:464–525.

Utermöhl, H. 1931. Neue Wege in der quantitativen Erfassung des Planktons. *Verh. int. Ver. Limnol.* 5:567–96.

Utermöhl, H. 1958. Zur Vervollkommung der quantitativen Phytoplankton-Methodik. *Mitt. int. Ver. Limnol.* 9:1–38.

Vollenweider, R.A. (ed.). 1969. *A Manual on Methods for Measuring Primary Production in Aquatic Environments*. IBP Handbook No. 12, Blackwell Scientific, Oxford, 213 p.

Watson, S.W., T.J. Novitsky, H.L. Quinby, and F.W. Valois. 1977. Determinations of bacterial number and biomass in the marine environment. *Appl. Environ. Microbiol.* 33:940–46.

Weibel, E.R. 1973. Stereological techniques for electron microscopic morphometry. In: *Principles and Techniques of Electron Microscopy, Biological Applications, Vol. 3* (M.A. Hayat, ed.). Van Nostrand Reinhold, New York, pp. 237–96.

Wiebe, W.J., and K. Bancroft. 1975. Use of the adenylate energy charge ratio to measure growth state of natural microbial communities. *Proc. Nat. Acad. Sci. USA* 72(6):2112–15.

Zimmerman, R., and L-A. Meyer-Reil. 1974. A new method for fluorescence staining of bacterial populations on membrane filters. *Kieler. Meeresforsch.* 30:24–27.

CHAPTER 7

Culture

Every viable microorganism in a non-resting state has an endogenous metabolism. When the nutrient intake or reserve materials fail to satisfy its nutrient requirements, the microorganism either encysts or perishes. When the nutrients assimilated exceed those required for maintenance, there is an enlargement and replication known as growth. When a microorganism flourishes in nature, enrichment culture, or pure culture, its growth requirements are being satisfied. Some microorganisms are not very fussy in their requirements for growth and are easily cultured in the laboratory. For others, much careful observation and imaginative experimentation is required to determine the unique conditions to make them yield to artificial cultivation. The purpose of this chapter is to look briefly at the fundamentals of the culture of marine microorganisms. Some examples of special cultural requirements are included in the discussions of particular microorganisms in the following chapters.

The one growth requirement common to all trophic modes of marine microorganisms is that for seawater. This can be met by using either natural or artificial seawater. Once the ionic requirements provided by seawater are met, as well as requirements for trace metals and organic micronutrients, then the macronutrients required for one of the trophic modes must be considered. These are plant nutrients and light for the phototrophs, dissolved organic matter for the osmotrophs, and particulate food for the phagotrophs. The way in which these macronutrients are supplied depends upon the culture mode, which may be batch culture, continuous culture, or diffusion culture. The mode chosen will depend upon whether the objective is to isolate, to maintain, to enumerate, to determine nutrient requirements, to observe the selective process in multispecies competition, or to observe single or multispecies growth rates under conditions approaching those *in situ*. Such environmental variables as oxygen, pH, light, and temperature must also be maintained within the limits required by the microorganisms under study.

A. SEAWATER

Marine microorganisms are intermediate in their salt requirements between freshwater species and the halophilic species growing in saturated brines. The salinity of seawater is approximately three times that of physiological saline and one-eighth that of saturated brine. Most marine species will not grow in freshwater and require the presence of salts, although they can often survive or grow at reduced salinities. Conversely, marine species are inhibited by the salinities of brines approaching saturation. The salts in seawater are essentially the steady-state concentrations of ions resulting from an input from eroded soil and losses due to evaporation and precipitation. The ionic species occur in a very constant proportion one to the other, the absolute amount depending upon the salinity, which varies from sea to sea and environment to environment. Oceanic seawater with 35 o/oo salinity contains the following ionic species at concentrations in grams per kilogram: Cl^-, 19.35; Na^+, 10.76; SO_4^{2-}, 2.71; Mg^{2+}, 1.29; Ca^{2+}, 0.41; K^+, 0.39; and HCO_3^-, 0.14 (Kester et al., 1967). Other ionic species are bromine, strontium, boron, and fluoride in addition to the trace metals and such inorganic nutrients as phosphate, iron, and nitrate. The total concentration of salts was once measured as chlorinity, using silver nitrate titration, which was then converted to total salinity in parts per thousand (S o/oo). Precise determinations to at least three decimals for physical–chemical studies are now made with electronic salinometers, which measure conductance. For biological work, an optical refractometer (American Optical Company) is sufficiently precise for most purposes, and is a useful instrument for nearshore field studies.

Both natural and artificial seawaters are used for culture. During the early days of culture when vitamins and growth factors were not well understood, natural seawater was preferred. To minimize the inhibitory effects of "poor waters" (Wilson and Armstrong, 1952, 1954, 1958, 1961), seawater was obtained in large quantities, preferably from a poorly productive area like the Sargasso Sea. Failing this, local waters were obtained during lulls in phytoplankton blooms. After coarse filtration to remove the larger suspended particles, carboys of water were held in a dark, cold room to "age" (ZoBell, 1941). The purpose of aging was to give the indigenous bacteria time to mineralize the organic matter, and thus minimize the variability of the organic content of the water. The local variation in the quality of natural seawater, which enhances or inhibits the survival and development of invertebrate larvae, is well documented (Wilson and Armstrong, 1952, 1954, 1958, 1961). The chemical nature of enhancing and inhibiting substances has been discussed by Lucas (1947, 1949, 1955, 1958), Nigrelli (1958, 1962), Sieburth (1964, 1968), and Johnston (1963a,b, 1964). Superimposed upon this natural variability is the ever increasing occurrence, concentration, and variety of such man-made pollutants as polychlorinated biphenyls, insecticides and herbicides that find their way into the sea. Since natural seawater only furnishes the necessary salts for marine bacteria and is not required for unknown growth factors, MacLeod and Onofrey (1956) concluded that there seems to be no compelling reason to use natural seawater. Some phycologists will disagree. One must decide whether to accept the biological variability and contaminants of freshly collected seawater and the tedious chore of collecting, filtering, aging, and storing it, or to incur the expense of preparing stock solutions of a reproducible artificial seawater. The current trend seems to be toward artificial seawaters.

There are many formulations for artificial seawater. The classic and most widely used one is that of Lyman and Fleming (1940). For physical–chemical purposes, a seawater should reproduce natural seawater as closely as possible and contain all the principal ions and trace elements. But for biological purposes, only the inorganic constitutents that are required for the synthesis and functioning of the cell need be included. Many formulations appear to be unnecessarily complex for their objectives (Segedi and Kelley, 1964; Agalides, 1966; LaRoche, Eisler and Tarzwell, 1970; Marshall, Stout, and Mitchell, 1971). Others have been designed for specific applications (Provasoli, McLaughlin, and Droop, 1957; McLachlan and Gorham, 1961; Courtright, Breese, and Krueger, 1971; Simidu, 1974) and contain plant nutrients, inorganic ions, buffers, and chelators that are unnecessary or counterindicated for uses other than those originally intended. Despite this, substances such as buffers and chelators required for phytoplankton have found their way into general purpose, commercial artificial seawaters. There are solubility and purity problems with a number of these preparations. To avoid these problems, artificial seawater should be made from scratch.

Kester et al. (1967) comment that little attention has been given to the techniques for preparing artificial seawater. They have developed a method that yields reproducible and satisfying results for physical–chemical studies. At the request of the author, Kester has simplified the formulation for artificial seawater further, to yield one for biological applications (Balderston, 1974; Table 7-1). Our goal was to keep trace metal contaminants to a minimum by using as few salts as possible while adding the major ions and to have the option of maintaining or reducing these levels of trace metals by autoclaving the salt solutions before or after they are combined (Jones, 1967a,b; Graham and Sieburth, 1973). This artificial seawater has become the standard seawater in the author's laboratory and has been used by Seshadri and Sieburth (1971, 1975), Graham and Sieburth (1973), Lavoie (1975), and Gessner (1975), among others. To the precautions of Kester et al. (1967) a word of caution should be added concerning the organic content of artificial seawater. Among the dry ingredients, the magnesium salt is the major source of organic carbon contamination. The source and the handling of the distilled and deionized water can also be very critical. Freshly distilled water has a dissolved organic carbon content of up to 3 mg of carbon/liter, which falls to below 1 mg after being held several days in storage containers. Apparently bacteria such as *Caulobacter* and *Hyphomicrobium*, which are indigenous to the walls of distilled water reservoirs, rapidly mineralize the labile dissolved organic carbon. These small amounts of dissolved organic carbon in freshly distilled water are apparently enough to chelate toxic heavy metals and prolong the survival time of the pathogen *Salmonella typhimurium* (Graham and Sieburth, 1973). Ion exchange cartridges used in water purification systems (as a substitute for distilled water systems) can also be a major source of organic carbon contamination. With care, a seawater with a dissolved organic carbon concentration of 1 to 2 mg of carbon/liter can be prepared to reproduce the organic matter content of natural seawater. A basal artificial seawater free of added inorganic and organic nutrients, buffers, and chelators has a number of possible applications. Such an artificial seawater can be used to maintain or rear animals in aquaria and to dilute solutions in chemical and cultural work. With supplementation, it can be used to grow phototrophic, osmotrophic, and phagotrophic microorganisms.

TABLE 7-1. The Kester–Sieburth formulation of artificial seawater for general biological use (30 o/oo S).
"Baker Analyzed Reagent Grade" chemicals are used throughout

FOR APPROXIMATELY 10 LITERS, THE FOLLOWING PROCEDURE IS USED:

1. Dry the following salts at 125 °C (257 °F) for at least 4 hours and cool in a desiccator: NaCl, ≃250 g; Na_2SO_4, ≃50 g; KCl, ≃10 g.
2. Solution A: Dissolve the following salts in 7 liters of distilled water at room temperature (20–26 °C): NaCl (dried), 205.1 g; KCl (dried), 5.8 g; $MgCl_2 \cdot 6H_2O$, 92.8 g; $CaCl_2 \cdot 2H_20$, 13.0 g *or* 9.8 g anhydrous $CaCl_2$.
3. Solution B: Dissolve 34.4 g dry Na_2SO_4 in 2.58 liters of distilled water at room temperature.
4. Solution C: Dissolve the following salts in 1 liter of distilled water: $NaHCO_3$, 16.8 g; Na_2HPO_4, 0.014 g. Filter-sterilize using a 0.2 μm Gelman, Sartorius, or Nuclepore sterile membrane filter in a sterile membrane filter unit or a plastic disposable unit, such as Nalgene No. 120.
5. For 1 liter of nonsterile seawater solution, combine 730 ml of Solution A and 260 ml of Solution B with 10 ml of Solution C and use as natural seawater. Autoclaving the combined solutions will precipitate excess heavy metals and yield a seawater with an ionic content approaching that of natural seawater.
6. For 1 liter of microbiological medium in which it is desired to prevent heavy metal precipitation, add the necessary nutrient ingredients including agar for 1 liter to 730 ml Solution A in one flask. In a smaller flask add 260 ml of Solution B. Autoclave Solutions A and B separately. Add 10 ml of Solution C to Solution B before combining and mixing Solutions A and B.
7. Permit sufficient oxygenation between preparation and use so that the pH may equilibrate and the dissolved oxygen will approach saturation.

B. SUPPLEMENTS

Batch culture is commonly used for the routine isolation, examination, and maintenance of marine microorganisms. The basal seawater, either natural or artificial, must be enriched with nutrients to support appreciable populations. The macronutrients and even the micronutrients at times, however, must be added in a manner such that they are released slowly so that their presence or more rapid utilization does not cause conditions that overwhelm the microorganisms to be cultured. The procedure is different for each of the principal trophic modes.

1. Phototrophs

The basal seawater is enriched with the "plant nutrients" nitrogen and phosphorus in the form of nitrate, ammonium, and phosphate salts. Silicate is added when diatoms and silicoflagellates are cultured. A buffer such as Tris [Tris (hydroxymethyl) methane] is used to control pH changes produced by the addition and utilization of the inorganic plant nutrients. Metals added as a trace metal mix include zinc, manganese, molybdenum, cobalt, copper, iron, and boron. A chelator such as EDTA (ethylenediaminetetra-acetic acid) is added to bind excess heavy metals so that their available concentrations are nontoxic. Vitamins required by many species are added as a vitamin mix and include at least cyanocobalamin, biotin, and thiamin. Since Tris and EDTA are resistant to bacterial attack, they are widely used to buffer macro- and micronutrients. In fact, EDTA can actually inhibit bacterial growth and probably does, to some degree, in many algal media.

A simple medium to prepare is *Erdschreiber* (Gross, 1937). This medium of Schreiber, containing natural seawater and nitrate and phosphate salts, is enriched with an extract of soil (*Erde*) that provides trace metals and humic substances that act as chelators and "vitamins." Although this medium has been effectively used for the isolation and maintenance of many diatoms and flagellates (it is the stock medium used by Mary Parke at the Plymouth Laboratory, England), the trend is toward more defined media. A semi-defined medium is the "f" medium of Guillard and Ryther (1962). Natural seawater sterilized with the plant nutrients is supplemented with sterile trace metal and vitamin mixes. When the concentration of the nutrient added is halved, the medium becomes the widely used "f/2" medium, whereas dilutions as great as "f/200" are indicated for some applications.

Artificial seawaters can be and are substituted for natural seawater in the above media. Other more defined media start with the salts of seawater and include macro- and micronutrients, buffers, chelators, and growth factors. Examples of widely used media are those that were developed at the old Haskins and Millport laboratories (before their relocations) and described by Provasoli, McLaughlin, and Droop (1957), Droop (1961), and Provasoli (1964). A more detailed but succinct survey of media (McLachlan, 1973) appears in the *Handbook of Phycological Methods* edited by Stein (1973), which contains much useful information for phototroph culture.

2. Osmotrophs

The probable abundance of the osmotrophs in the sea in descending order is bacteria, microflagellates, and microfungi. Even though the requirements for the bacteria that dominate the bacterioplankton have not been determined in artificial media, it appears that marine bacteria, in general, require dissolved organic carbon at concentrations from 100 to 1,000 μg of carbon/liter. Some bacteria prefer solid substrates but all appear capable of growth on soluble organic matter. The microfungi appear to require much higher concentrations of substrate than the bacteria, preferring a carbohydrate substrate with a concentration of 1,000 to 10,000 mg of carbon/liter. The microflagellates requiring dissolved organic carbon for their macronutrients are often colorless forms of the phytoflagellates. These forms sometimes have odd micronutrient requirements, as discussed in Chapter 22. From cultural studies and their standing crops in the sea, the organic carbon requirements of the osmotrophic microflagellates appear to be intermediate between those of the bacteria and of the microfungi.

All three types of osmotrophs can be grown in liquid media. The problem with this type of culture is that when the soluble nutrients are added in sufficient concentration to yield a population of microorganisms that turn the medium turbid, the abundant growth also induces senescence and death. In mixed cultures, there is also species competition and selection. To avoid these problems, microbiologists have developed solid media upon which colony forming units develop; they can then be detected, enumerated, and subcultured for further study.

These solid media are usually prepared by hydrating 15 g of bacteriological agar in 1 liter of boiling seawater. Commercial agar melts around 90 °C and forms a gel as the medium cools below 42 °C. The gelling agent in bacteriological agar produced from hot water extracts of certain species of red algae (such agarophytes as *Gelidium*, *Gracilaria*, *Phyllophora*, *Acanthopeltis*, and *Ahnfeltia*) is the phycocolloid and polysaccharide agarose (Percival and McDowell, 1967). Only some 1% or less of the colony forming units in rich coastal waters can utilize agarose and cause a concavity that surrounds the sinking colony. For bacteria of medical significance, such nutrients as peptone, meat extract, yeast extract, blood, and carbohydrates are added at a concentration of 5 to 25 g/liter to the agar gels. Because most of the older marine microbiologists have had medical training, virtually all the published media also contain such nitrogenous nutrients (at a level of 2 to 6 g/liter), in addition to the agar-gelled seawater, in order to obtain the maximal number of colony forming units (ZoBell,, 1941; Oppenheimer and ZoBell 1952; Carlucci and Pramer, 1959; Sieburth, 1967).

Bacteriological agar is a mixture of polysaccharides. In addition to the gelling agent agarose, which is a neutral galactan, there are charged galactans, which, at one time, were thought to be a single contaminant material, "agaropectin" (Percival and McDowell, 1967). It is now known that agar is a complex mixture of polysaccharides with the same galactan backbone, but with charged groups that are substituted to a variable degree (Duckworth and Yaphe, 1971a). There are three extremes in structure in this spectrum of polysaccharides. In addition to the cold-water–insoluble, neutral galactans, there are two cold-water–soluble fractions, pyruvated galactans and sulfated galactans. The structure of these extreme fractions has been characterized by Duckworth and Yaphe (1971b). Agars high in sulfated galactans are apparently low in pyruvated galactans and vice versa (Young, Duckworth, and Yaphe, 1971). The presence of these cold-water–soluble polysaccharides, which can be readily utilized by marine bacteria plated on seawater agars, however, is ignored by most marine bacteriologists.

The ability of the soluble, charged polysaccharides of unsupplemented seawater agar to support the growth of bacteria has not been adequately studied. Carlucci and Shimp (1974) isolated a bacterium from an unpurified seawater agar medium and assumed that the colony forming units were growing on the dissolved organic carbon in the seawater and not on the polysaccharide contaminants in the medium. The growth of these isolates to populations of 5×10^6 organisms/ml in unsupplemented natural seawater can be accounted for by the some 200 μg of carbon/liter of total polysaccharide in seawater (Burney and Sieburth, 1977). Rather than being the hypothetical "low nutrient" bacteria that may dominate the bacterioplankton but do not form colonies, these bacteria isolated by Carlucci and Shimp (1974) appear to be polysaccharide-utilizing bacteria of the "high nutrient" type that can form colonies on solid media. In Narragansett Bay, Rhode Island, the maximal population of colony forming units supported by a seawater agar medium supplemented with 1 g/liter of peptone is only 20% higher than that population supported by a seawater agar without supplementation. If this seawater–agar gel is soaked in several changes of sterile artificial seawater at 5 °C for several days to permit the charged polysaccharide contaminants to diffuse out, the seawater–agar medium can no longer support the abundant polysaccharide-utilizing populations occurring in natural seawater. Such crudely purified agar can be used to make a semi-defined agarose media for specific bacteria, such as *Leucothrix mucor* (Chapter 16,A,2). The use of this and other semi-defined media in addition to the charged polysaccharide (seawater–agar) medium should permit the enumeration and isolation of some of the substrate-specific epibacteria. The low-nutrient bacteria that dominate the sea, however, will require a medium much lower in available nutrients than dialyzed agar, since growth inhibition of minibacteria at 100 mg of dialyzable galactan/liter gives way to growth at concentrations of 1 to 10 mg/liter.

In contrast to the media for phototrophs, in which supplements added in excess are bound by chelation, the excess nutrients in bacteriological media are immediately available and apparently select for high nutrient bacteria. The closest that bacteriological media come to slowly releasing soluble organic nutrients for the low nutrient bacterioplankton are the seldom used liquid media with particulate substrates that are utilized at a low rate. These include such substrates as cellulose, chitin, starch grains, and casein, which are depolymerized slowly by appropriate species and therefore release the readily used monomers slowly. It is ironic that protozoologists use grains of rice and wheat to grow indigenous populations of bacteria that nurture protozoa but do not overwhelm them, whereas bacteriologists have apparently missed the significance of this procedure. Serial decimal dilutions of seawater slowly enriched by a grain of rice or wheat, which would slowly release its carbohydrates, might be a far more reasonable medium for growing low nutrient bacteria than any used to date. A more controlled method of releasing such important substrates as glucose, glycolate, mannitol, cellobiose, and glucosamine at ecological levels would be to use a non-nutrient gel in slide culture and to add these substrates to a central well at a concentration of 0.5 μg/ml such as occurs in strata of microbiological activity in the sea. Such procedures may eventually lead to the isolation and maintenance of microcolonies of minibacteria that dominate the open ocean (chapter 14).

3. Phagotrophs

Most of the protozoa take in solid food particles by phagocytosis, whether or not special organelles for ingestion are present. Exceptions are the smaller flagellates or microflagellates, which may be entirely osmotrophic and which function like bacteria. Others, being mixotrophic, may obtain some nutrients from soluble organic substances by active cell transport through the cell wall or by the abscission of vesicles from the unit membrane to the interior of the cell (pinocytosis). For a more complete discussion on nutrition, see Grell (1973) and Hall (1965). Some of the ingesting protozoa, such as *Tetrahymena*, have been weaned onto soluble substrates in axenic (bacteria-free) culture (Elliott, 1958; Holz, 1964), but microorganisms, such as bacteria, flagellates, diatoms, and ciliates, are the normal particulate food for the ingesting marine protozoa.

The bacteria-ingesting (microphagous) species, including amoeboid, flagellated, and ciliated forms, can be cultured on media containing plant infusions and seeds like rice and wheat grains. As just discussed, these substrates slowly release nutrients and maintain a more or less controlled bacterial population that does not overwhelm the culture system when bacteria-ingesting protozoa are present. An alternate method is that of Singh (1946), in which washed suspensions of resting bacterial cells of a suitable species are added to dilutions of cultures or samples in a non-nutrient diluent. The protist-ingesting (macrophagous) species can be sustained on a variety of microorganisms, including phytoflagellates, diatoms, and other protozoa. When phytoflagellates and diatoms are supplied, the growth of the cultured protozoa can be regulated by controlling the growth conditions of these microalgae.

C. CULTURE MODES

A culture vessel for the microorganisms and a mechanism for providing nutrients under optimal conditions are the basic requirements for culture. They are also the limitations. There are three primary culture modes: batch culture, continuous culture, and diffusion culture. Each has its own applications. Improvements in culture techniques are an ever present challenge to microbiologists trying to maintain and observe microorganisms in culture. Multicultures or microcosms are one approach and are well suited for observing the types and associations of microorganisms that can occur under different cultural conditions. Pure or unispecies cultures, however, are required for taxonomic, physiological, nutritional, and even some ultrastructural studies.

1. Batch Cultures

In batch culture, after inoculation, the nutrient medium in the culture vessel is not renewed or supplemented subsequently. Most culture work is in the batch mode. For liquid culture either nothing more is done to suitable tubes, flasks, carboys, and fermentors or else they are aerated, sparged with gases, shaken, rotated, illuminated, warmed, cooled, or pressurized to provide some of the environmental parameters required for the growth of specific microorganisms. For culture on solid surfaces, nutrients are included in a suitable solidifying agent such as agar, dispensed into Petri dishes or tubes, and surface-inoculated. An alternative method is to collect microorganisms on a suitable membrane filter, which is placed on a cellulose pad soaked with nutrients or on a nutrient agar gel in a Petri dish. Batch culture is simple and convenient and can be used for handling large numbers of cultures. Among its many applications are:

1. Selection of specific microorganisms by the suppression of competing microorganisms. Examples are flagellate selection by inhibiting diatoms with germanium dioxide, and yeast and mold selection by inhibiting bacteria with antibacterial agents or low pH.
2. Enrichment of specific groups by the addition of specific substrates. Examples are nitrifying bacteria enrichment by ammonium and one carbon-utilizing bacteria enrichment by methanol.
3. The isolation of bacterial colonies and strains on solid media by streaking the inoculum and picking out individual colony forming units.
4. Establishment of clones of protists from single cells and the preparation of axenic cultures by the passage of single cells through successive antibiotic washes.
5. The maintenance of strains and clones of unispecies and axenic cultures.
6. Laboratory studies on the effect of salts, nutrients, temperature, and other parameters on the growth and life cycles of microorganisms.

It is also well to remember the limitations of batch culture when pure cultures are used. The main points are:

1. The substrate concentration is constantly decreasing as the culture grows.
2. As the substrate is utilized, by-products, including autoinhibitory substances, accumulate.
3. Exponential growth occurs for only a very short time after lag phase before the factors discussed in 1 and 2 cause the stationary and death phases.

When environmental levels of substrate are used in batch culture, the amount of growth is so small that the usual methods for estimating populations may not be sensitive enough to show significant differences. This has been overcome by using ^{14}C counting procedures to follow the uptake of $^{14}CO_3$ by phototrophs (Vollenweider, 1969) and heterotrophs (Romanenko, 1964; Sorokin, 1971a,b; Overbeck and Daley, 1973) and the uptake of organic carbon by heterotrophs (Antia et al., 1963; Hobbie and Wright, 1965; Wright and Hobbie, 1965; Vaccaro and Jannasch, 1966, 1967; Azam and Holm-Hansen, 1973). Although it is possible to measure radiocarbon uptake precisely, the basic shortcomings of batch culture cannot be corrected for. Unlike the natural environment, where nutrients will diffuse from sources of released nutrients toward the microorganisms that produce the nutrient gradients, the closed system with its finite amount of nutrients precludes this and therefore may determine the results of these productivity and heterotrophic potential measurements.

2. Continuous Culture

In continuous culture, the culture container is fitted with a medium reservoir having a constant rate of inflow and an effluent port with a leveling device (see Plate 7-1,*C*). The individual cells are suspended in a nearly constant volume and are cultured at a nearly steady state of growth, which is maintained by the addition of fresh growth medium and the removal of an equivalent volume of the culture at rates commensurate with growth. The total number of cells and the rate of cell division become nearly constant under specific conditions of growth (see Plate 7-1,*B*). When nutrients are supplied in abundance so that they are above the critical levels that control growth rate, the culture device is a turbidostat and the growth rate is limited by internal mechanisms within the cell (Kubitschek, 1970). To maintain a constant population under these conditions, a photoelectric device uses cell density (turbidity) to control the mechanisms for withdrawing a portion of the culture and replacing it with new medium. Turbidostats are useful for producing large volumes of cells for chemical analysis and for the nutrition of protozoa, metazoa, and mollusks. When critical substrates are at concentrations that affect the growth rate, growth is under external control (Plate 7-1,*B*). Continuous culture systems that control growth externally by limiting the supply of critical growth factors are called chemostats. The basis of the chemostat is that the concentration of a limiting substrate (s) controls the specific growth rate (μ) of the culture. This dependence is shown by the equation of Monod: $\mu = \mu_{max}\, s/(Ks + s)$, where μ_{max} is the maximum specific growth rate and Ks is a saturation constant equal to the substrate concentration where $\mu = \frac{1}{2}\mu_{max}$. The steady-state concentration of the growth-limiting substrate (s) depends upon the dilution rate (D) and is represented by the equation of Herbert, Elsworth, and Telling (1956): $s = KsD/(\mu_{max} - D)$. This equation applies for all dilution rates up to the point where D equals the growth rate. When the rate of dilution exceeds the growth rate, the culture population becomes progressively sparser until all the cells are washed out of the system. The exception is wall-growing species, which are not suitable for conventional chemostat studies. But attaching microorganisms serve a function in flow-through systems, such as aquaculture facilities (Johnson and Sieburth, 1976; Balderston and Sieburth, 1976), and should not be ignored. The theory of the chemostat has been given by Herbert, Elsworth, and Telling (1956), Tempest (1970), and Kubitschek (1970). Among the advantages of the chemostat over batch culture for unispecies culture are:

1. The easier control of growth and division rates over long time periods.
2. The constancy of the chemical environment.
3. The determination and maintenance of cell concentrations that are independent of growth rate.
4. The maintenance of very low concentrations of factors that control the growth of the culture.
5. The control of cell size and composition through rate of growth.

The chemostat can be used in three modes: unispecies, multispecies, and multistage. In the unispecies

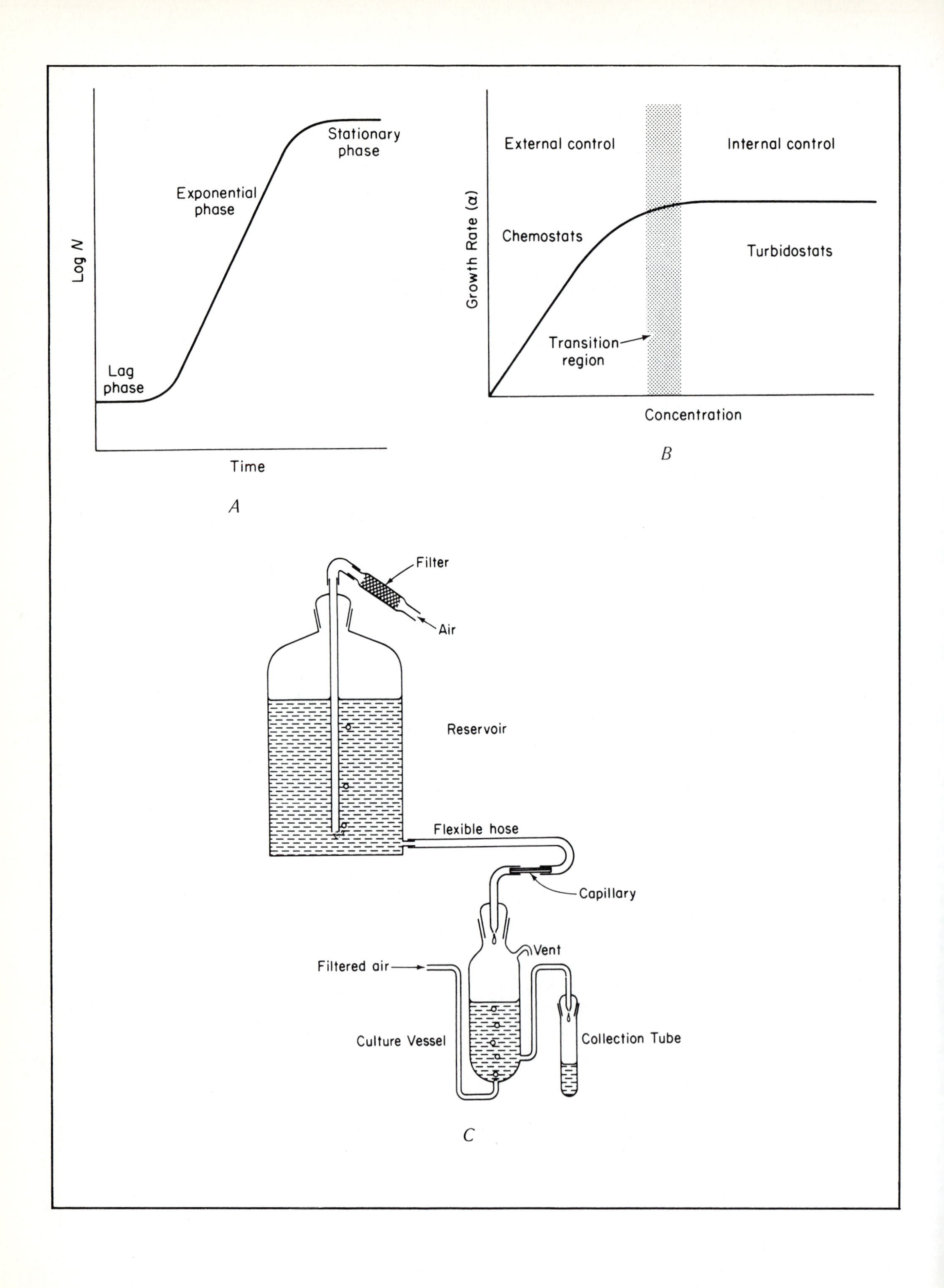
Log N
Stationary phase
Exponential phase
Lag phase
Time
A
Growth Rate (α)
External control
Internal control
Chemostats
Turbidostats
Transition region
Concentration
B
Filter
Air
Reservoir
Flexible hose
Capillary
Vent
Filtered air
Culture Vessel
Collection Tube
C

PLATE 7-1. A comparison of batch and continuous culture. (*A*) Simplified growth curve in batch culture showing an initial lag in number while the cell size increases, a brief exponential growth phase, and the stationary population when death balances growth. Cell size, composition, and physiological state vary markedly throughout the curve. (*B*) Dependence of growth rate on the concentration of a limiting nutrient in continuous culture. Chemostats control the growth rate by changing the concentration of a limiting nutrient; the turbidostats operate at non-limiting concentrations when the growth rate is controlled internally by cells. Cell size, composition, and physiological state are stabilized in both regions of the curve. (*C*) A simple form of chemostat, in which a constant flow rate of the medium is controlled by a capillary and a constant culture volume is maintained by a leveling siphon. All from H.E. Kubitschek, *Introduction to Research with Continuous Cultures*, 1970, reprinted by permission of Prentice-Hall, Englewood Cliffs, N.J.

mode, the concentrations at which nutrients limit growth (threshold concentration) can be determined. Examples are the limiting concentrations of various carbon sources on different species of marine bacteria (Jannasch, 1963, 1965, 1970) and the cobalamine requirements of marine phytoplankton (Droop, 1973). Since the threshold values observed depend upon the dilution rate of the culture system, the values obtained are not absolute but relative and cannot be directly applied to the natural environment. Unlike natural ecosystems, which are multispecies systems, chemostats become unispecies systems by washing out the slower growing species. This characteristic of the chemostat has been utilized to study the factors that affect interspecies competition (Veldkamp and Jannasch, 1972; Veldkamp and Kuenen, 1973). Specific applications have been the competition between obligate and facultative psychrophiles (microorganisms requiring and being able to grow at temperatures below 20 °C, respectively) (Harder and Veldkamp, 1971); the differentiation of species that are either active or inactive at low substrate concentrations (Jannasch, 1967); and the competitive elimination of enteric bacteria from seawater (Jannasch, 1968). Multistage culture, when two or more chemostats are hooked up in series (Herbert, 1964), has been used to observe the dependence of one trophic level on another. Bell and Mitchell (1972) have used algal cultures to determine the effect of extracellular algal products on the enhancement and inhibition of bacteria. The effect of such environmental factors as light, nutrients, and flow rates on primary producers and their ciliate and crustacean consumers has been studied by Taub and Dollar (1968) and Taub and MacKenzie (1973).

For the environmentally oriented microbiologist, the eventual goal of culture is to obtain growth rates of natural populations in natural waters. The observed doubling rates of 20 to 200 hours for bacteria in natural waters obtained with the chemostat (Jannasch, 1969) seem excessively long. An alternate culture mode is needed to maintain a multispecies culture free of predators in as natural a milieu as possible. A promising mode is that of dialysis or diffusion culture.

3. Diffusion Culture

In diffusion culture, a vessel in which the wall surface incorporates a semipermeable membrane holds a population of cells for growth studies while exposing it to diffusing nutrients and other environmental parameters except predators (see Plates 7-2 and 7-3). The purpose of diffusion culture is to avoid the problems of batch culture by allowing the replacement of low ecological levels of naturally occurring nutrients as they are used and allowing the removal of metabolic by-products as they are produced, while avoiding the major problems of continuous culture, the washout of microorganisms and the selection toward a unispecies system. The volume of the culture vessel can be of sufficient size (200 ml to 15 liters) so that several methods of estimation, including carbon, chlorophyll, and ATP assays, and cell enumeration, using epifluorescence microscopy can be used to detect changes occurring in natural populations. There are two modes of diffusion culture, passive and active.

In passive diffusion, a bag made of dialysis tubing (Beard and Meadowcraft, 1935; Slanetz and Bartley, 1965; Hendricks and Morrison, 1967; Baskett and Lulves, 1974) or a vessel with membrane filter walls (McFeters and Stuart, 1972) is inoculated with a culture that is monitored over a period of time to observe growth and death patterns. When dialysis bags and membranes are made of cellulose esters, the cellulose-digesting bacteria penetrate them in 1 to 3 days depending upon water temperature, and the membrane itself is a source of nutrients (regenerated cellulose and its breakdown products). The McFeters chamber, when fitted with a Millipore membrane (which has a textured sur-

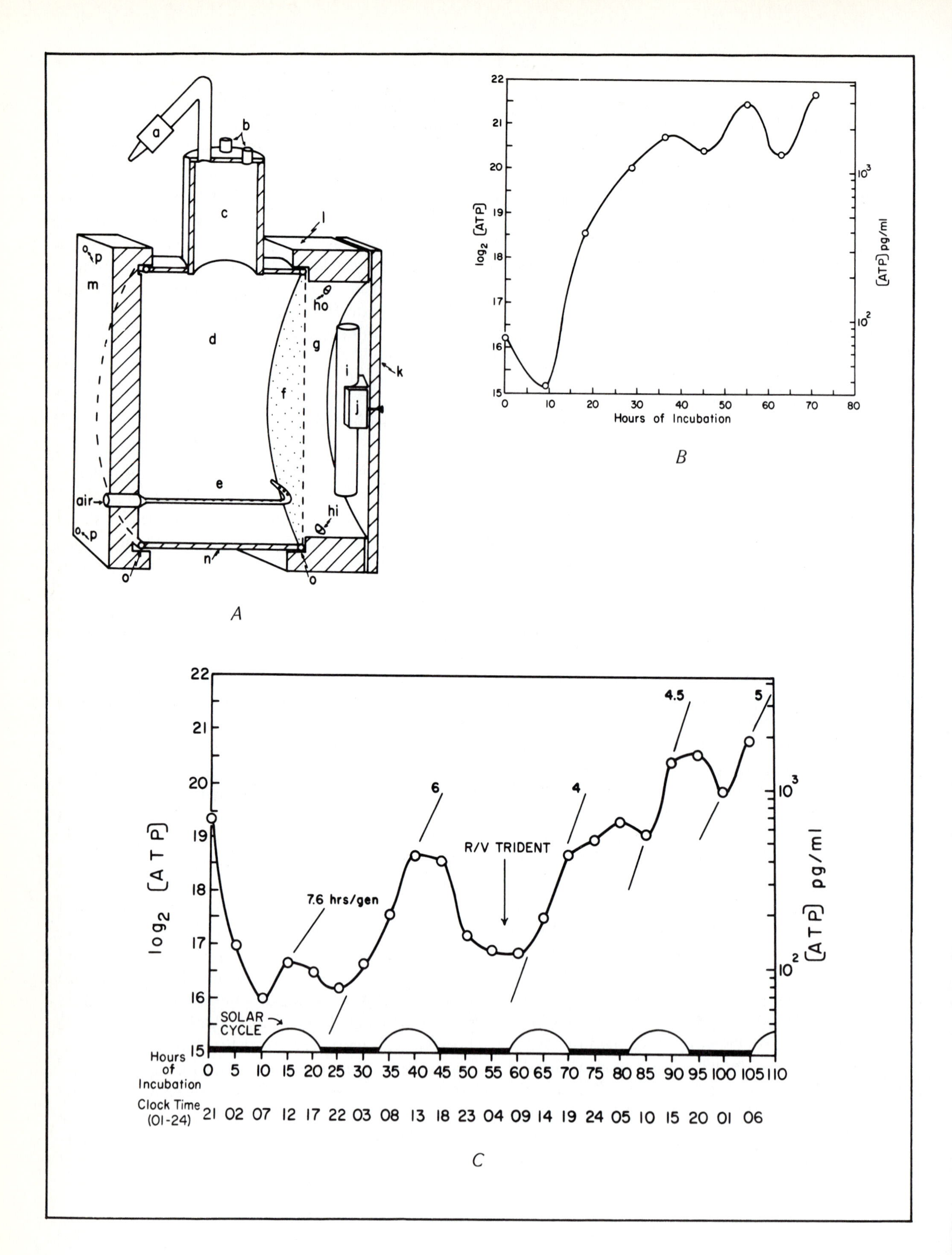

a
b
c
l
p
m
d
ho
g
i
k
f
j
e
air
hi
p
n
o
o
log₂ [ATP]
[ATP] pg/ml
Hours of Incubation
R/V TRIDENT
7.6 hrs/gen
6
4
4.5
5
SOLAR CYCLE
Hours of Incubation
Clock Time (01-24)

A

B

C

PLATE 7-2. Diffusion culture chamber and representative growth patterns of natural populations of bacterioplankton in the picoplankton size fraction. (*A*) Cutaway view of a 180-ml polycarbonate chamber, showing a 0.1 μm Nuclepore filter (f), an aerator tube (e) and a mixing bar (i) for membrane agitation, and a one-way valve (a) for air exhaust. (*B*) Growth curve of a pure culture on a homogeneous batch of unenriched, filter-sterilized seawater; note the maximum carrying capacity of the system, which is reached after 36 hours. (*C*) Diel growth patterns of a multispecies natural population on *in-situ* water at a dock and its perturbation by the docking of a research vessel (ATP, adenosine triphosphate). All from Sieburth et al., 1977, *Helgoländer wiss. Meeresunters.* 30:565–74.

face), can be seriously modified or plugged within a day or two by fouling microorganisms. The use of Nuclepore membranes is an improvement, but they will still foul and plug, thereby isolating the contained population in batch mode, as demonstrated in the growth curves of Fliermans and Gorden (1977), where growth occurred for 2 days but declined until the 4th day, when a minimal flora was maintained with an apparent interspecies fluctuation.

In active diffusion, the membrane surfaces are agitated to minimize fouling and to maximize the diffusion of soluble nutrients through the membrane. An early active diffusion chamber consisted of a dialysis bag containing an air bubble and glass beads, the incline of which was changed during rotation by wave action (Jensen, Rystad, and Skoglund, 1972) or by a small electric motor (Graham and Sieburth, 1973) to agitate both the outside and inside surfaces, but external fouling was still a problem (Vargo, Hargraves, and Johnson, 1975). An improvement for lab culture was the use of a membrane filter with a stirring mechanism (Jensen, Rystad, and Skoglund, 1972). For the culture of bacterioplankton populations in the picoplankton size fraction of seawater, a diffusion chamber of inert, strong, sieve-like Nuclepore membranes has been developed. The principles of diffusion culture laid down by Schultz and Gerhardt (1969) apply here. The data obtained with active diffusion culture chambers will depend largely upon the choice of membrane (which affects design) and the configuration of the culture chamber (which affects performance) (Lavoie, 1975).

In both natural waters and in diffusion culture, the increase in the biomass of microorganisms will depend upon the diffusion rate and the amount of substrate available to the microorganisms by diffusion. An increase in concentration above that required for maintenance will result in growth, whereas a decrease will result in starvation and death. The primary requirement is an adequate rate of diffusion with a response time that is neither too long nor too short. The response time should be sufficiently shorter than the intrinsic growth rate of the microorganisms so that true exponential growth rates are measured; these will reflect rapid changes in external nutrient concentrations. This depends upon the permeability of the membrane, the area of the membrane, and the volume of the chamber as well as the degree to which the membrane is agitated. This is indicated by the formula

$$T_p = \frac{-\ln(1-p)}{P_m(A/V)}$$

where T_p = the time (hrs) required to reach a certain population (p) of the external concentration; P_m = the permeability coefficient (cm/hr), A = the area (cm²) of the diffusion membrane, and V = the volume (cm³) of the diffusion chamber. Chambers with a response time of 50% in about half the doubling time of the microbial populations appear to give reasonable data. Chambers with a response time that approaches or exceeds the doubling rate would restrict growth, to cause an artifact.

Doubling rates are not affected by short response times, but as the response time decreases, the maximum cell density or carrying capacity of the chamber increases. This is indicated by the formula $X_{max} = (P_m \times A/V) \times (\Delta S/Y_e)$, where X_{max} = the maximum cell density (g/cm³), ΔS = the difference in nutrient concentration between the chamber and the external medium (g/cm³), and Y_e = the maintenance requirements per hour. If the response time is too low, cell populations greater than those possible in the natural environment will be obtained, to cause a second artifact. Preliminary results with the diffusion culture of picoplankton in Narragansett Bay (Plate 7-2,*C*), which indicate values of 84 mg of carbon/m³ daily appear to correlate well with data on phytoplankton production, dissolved organic carbon release rates, and bacterial efficiency (Lavoie, 1975). For the open North Atlantic Ocean, however, a hourly rate of 5.1 μg of carbon/liter greatly exceeds the accepted values of phytoplankton productivity. This value would be suspect but it agrees with the 9.8 μg of carbon/liter for the hourly dissolved organic carbon uptake at the depths of aggregation of chlorophyll impoverished

A

C

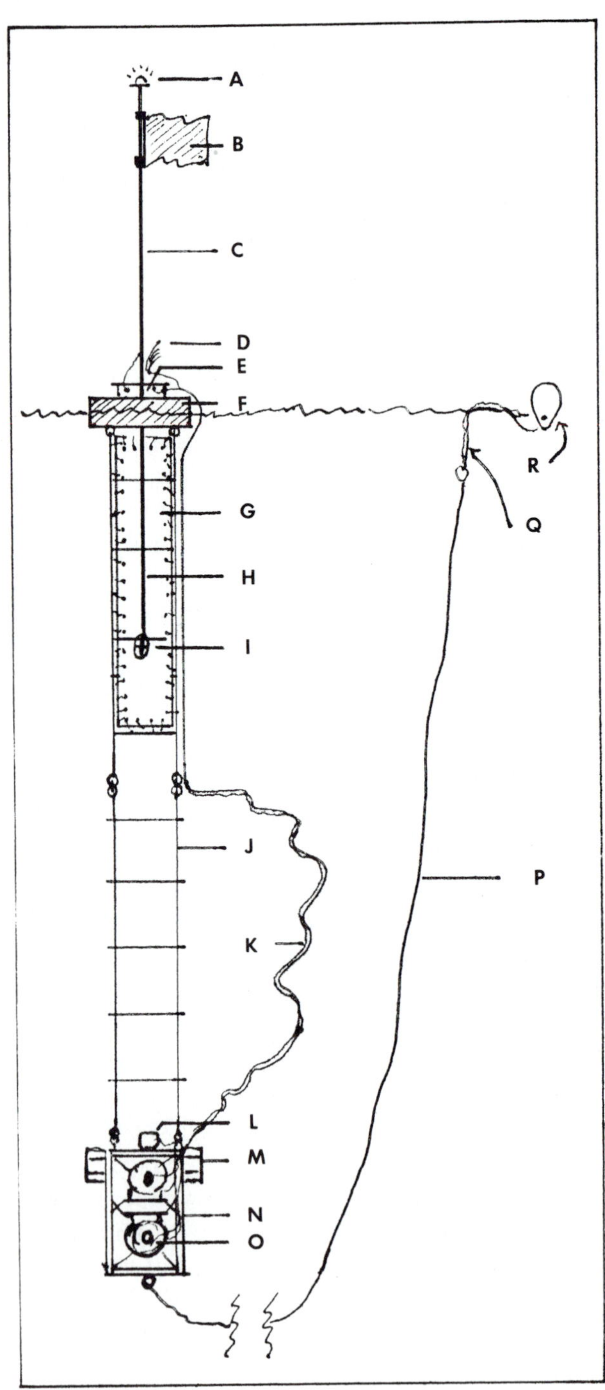

B

PLATE 7-3. A 5-liter active diffusion chamber for environmental monitoring and its drogued buoy array and harness for stable suspension in the open sea. (*A*) Exploded view of the chamber, showing (A) end bell without motor (with filling ports B, C, four screened holes—sealed with tape during filling and deployment); (D) Nuclepore membrane (0.1 μm porosity and 48 cm diameter, mounted on frame); (E) 5-liter chamber, with sampling port (F) and central wiper blade (G); (H) external wiper blade; (I) end bell with motor; (J) 1 rpm motor in pressure housing; and (K) motor cable, with underwater connector to 12-V battery. (*B*) The array for the stable suspension of the diffusion chamber: (A) strobe light for night-time tracking; (B) swiveling flag for daytime tracking; (C) 5-m mast; (D) Impolene tubing for sampling chambers; (E) waterproof battery box, with cords to strobe and chambers; (F) low-freeboard float; (G) drogue panel tied to harness; (H) 3-m counterbalance, with added weight (I); (J) shock cord assembly; (K) harness, with safety wire, sample tubing, and cord to motor; (L) batch culture control; (M) carboy partly filled with air for buoyancy; (N) shock cords; (O) diffusion chamber; (P) tagline for chamber—first recovery; (Q) floating propylene line; (R) tagline buoy. (*C*) Underwater photograph of diffusion chamber deployed in frame at 15 m on Cruise EN-015 in the western Sargasso Sea. (*A,B*) J.McN. Sieburth, unpubl.; (*C*) courtesy of Phillip Sharkey.

protist plankton (Sieburth et al., 1977). These preliminary results on diffusion culture, which indicate that the bacteria in the picoplankton size fraction are closely linked to the diel activities of the plankton (Plate 7-2,*C*), are very encouraging.

But the small volume (180-ml) chambers shown in Plate 7-2,*A* and the subsequent models with two membranes and four magnetic stirring bars agitating the liquid on both surfaces of the membranes were fraught with problems. A 5-liter chamber (shown in Plate 7-3) was designed to take the maximum width Nuclepore membrane (48 cm). The volume of the central chamber can be increased to 15 liters by the addition of two 5-liter spacers on either side of the central 5-liter chamber and by the use of a new central wiper assembly. A 1 rpm motor was selected to reduce the excessive agitation caused by the high speed magnetic mixer. The membrane was fitted with hubs and O-rings so that the motor shaft would pass through the membrane assemblies and directly turn the soft rubber wiper blades in contact with both sides of the membrane as well as the walls of the central growth chamber. This slow and effective wiping and agitation prevents fouling and sedimentation and promotes diffusion through the membrane. Several different suspending arrays have been designed for deploying these diffusion chambers. For shallow waters, the chambers on a frame are positively buoyant and are held at a specific height above a bottom mooring by a line. For open waters, a freely floating drogued buoy has a shock cord harness, which stably suspends the chambers at the desired depth (Plate 7-3,*B*). The latter array has maintained a pair of chambers (Plate 7-3,*C*) in the western Sargasso Sea in late October in 20-ft seas with wind speeds of Beaufort 8 for some 3½ days. The membranes remained intact and the chambers yielded useful data on the decay rates of enteric bacteria exposed to the open sea. Plastic tubing with a 1.5 mm i.d. was used to obtain samples at the surface by using a hand-operated vacuum pump to aspirate two tubing volumes (which were discarded), and then the sample. The use of a four-man diving team was indispensable in monitoring the deployment of the chambers and in modifying the design of the suspending array to make the chambers stable regardless of the movement of the buoy.

Such a diffusion culture apparatus may be used to measure certain phototrophic fractions (nanoplankton) as well as heterotrophic productivity, to monitor pathogen survival, and to determine the perturbations caused by pollutants. Only further work will show the validity of the preliminary work and the applications and limits of this culture technique.

REFERENCES

Agalides, E. 1966. Synthetic seawater for sharks. *Geo-Marine Tech.* 2:V249–V253.

Antia, N.J., C.D. McAllister, T.R. Parsons, K. Stephens, and J.D.H. Strickland. 1963. Further measurements of primary production using a large-volume plastic sphere. *Limnol. Oceanogr.* 8;166–83.

Azam, F., and O. Holm-Hansen. 1973. Use of tritiated substrates in the study of heterotrophy in sea water. *Mar. Biol.* 23:191–96.

Balderston, W.L. 1974. Denitrification in closed-system aquaculture. Ph.D. Thesis, University of Rhode Island, 125 p.

Balderston, W.L., and J.McN. Sieburth, 1976. Nitrate removal in closed-system aquaculture by columnar denitrification. *Appl. Environ. Microbiol.* 32:808–18.

Baskett, R.C., and W.J. Lulves. 1974. A method of measuring bacterial growth in aquatic environments using dialysis culture. *J. Fish. Res. Bd. Canada* 31:372–74.

Beard, P.J., and N.F. Meadowcraft. 1935. Survival and rate of death of intestinal bacteria in sea water. *Amer. J. Publ. Health* 25:1023–26.

Bell, W., and R. Mitchell. 1972. Chemotactic and growth responses of marine bacteria to algal extracellular products. *Biol. Bull.* 143:265–77.

Burney, C.M., and J.McN. Sieburth. 1977. Dissolved carbohydrates in seawater. II. A spectrophotometric procedure for total carbohydrate analysis and polysaccharide estimation. *Mar. Chem.* 5:15–28.

Carlucci, A.F., and D. Pramer. 1959. Factors affecting the survival of bacteria in sea water. *Appl. Microbiol.* 7:388–92.

Carlucci, A.F., and S.L. Shimp. 1974. Isolation and growth of a marine bacterium in low concentrations of substrate. In: *Effect of the Ocean Environment on Microbial Activities* (R.R. Colwell and R.Y. Morita, eds.). University Park Press, Baltimore, Md., pp. 363–67.

Courtright, R.C., W.P. Breese, and H. Krueger. 1971. Formulation of a synthetic seawater for bioassays with *Mytilus edulis* embryos. *Water Res.* 5:877–88.

Droop, M.R. 1961. Some chemical considerations in the design of synthetic culture media for marine algae. *Bot. Mar.* 2:231–46.

Droop, M.R. 1973. Nutrient limitation in osmotrophic protista. *Amer. Zool.* 13:209–14.

Duckworth, M., and W. Yaphe. 1971a. The structure of agar. Part I. Fractionation of a complex mixture of polysaccharides. *Carbohydr. Res.* 16:189–97.

Duckworth, M., and W. Yaphe. 1971b. The structure of agar. Part II. The use of a bacterial agarase to elucidate structural features of the charged polysaccharides in agar. *Carbohydr. Res.* 16:435–45.

Elliott, A.M. 1959. Biology of *Tetrahymena*. *Ann. Rev. Microbiol.* 13:79–96.

Fliermans, C.B., and R.W. Gorden. 1977. Modification of membrane diffusion chambers for deep-water studies. *Appl. Environ. Microbiol.* 33:207–10.

Gessner, R.V. 1975. The seasonal succession and distribution of filamentous fungi associated with *Spartina alterniflora* and *in vitro* studies on the growth, nutrition and taxonomic position of *Sphaerulina pedicellata*. Ph.D. Thesis, University of Rhode Island.

Graham, J.J., and J.McN. Sieburth. 1973. Survival of *Salmonella typhimurium* in artificial and coastal sea water. *Rev. Intern. Oceanogr. Méd.* 29:5–29.

Grell, K.G. 1973. *Protozoology*. Springer-Verlag, New York. 554 p.

Gross, F. 1937. Notes on the culture of some marine plankton organisms. *J. mar. biol. Assoc. U.K.* 21:753–68.

Guillard, R.R.L., and J.H. Ryther. 1962. Studies of marine planktonic diatoms. I. *Cyclotella nana* Hustedt, and *Detonula confervacea* (Cleve) Gran. *Can. J. Microbiol.* 8:229–39.

Hall, R.P. 1965. *Protozoan Nutrition.* Blaisdell Publ., New York, 90 p.

Harder, W., and H. Veldkamp. 1971. Competition of marine psychrophilic bacteria at low temperatures. *Antonie van Leeuwenhoek* 37:51–63.

Hendricks, C.W., and S.M. Morrison. 1967. Multiplication and growth of selected enteric bacteria in clear mountain stream water. *Water Res.* 1:567–76.

Herbert, D. 1964. Multi-stage continuous culture. In: *Continuous Cultivation of Microorganisms* (I. Malek, K. Beran, and J. Hospodka, eds.). Publ. Czech. Acad. Sci., Prague, pp. 23–44.

Herbert, D., R. Elsworth, and R.C. Telling. 1956. The continuous culture of bacteria: A theoretical and experimental study. *J. gen. Microbiol.* 14:601–22.

Hobbie, J.E., and R.T. Wright. 1965. Bioassay with bacterial uptake kinetics: Glucose in freshwater. *Limnol. Oceanogr.* 10:471–74.

Holz, G.G., Jr. 1964. Nutrition and metabolism of ciliates. In: *Biochemistry and Physiology of Protozoa* (S.H. Hutner, ed.), Vol. 3. Academic Press, New York and London, pp. 199–242.

Jannasch, H.W. 1963. Studies on the ecology of a marine spirillum in the chemostat. In: *Marine Microbiology* (C.H. Oppenheimer, ed.). C.C. Thomas, Springfield, Ill., pp. 558–66.

Jannasch, H.W. 1965. Starter populations as determined under steady state conditions. *Biotech. Bioeng.* 7:279–83.

Jannasch, H.W. 1967. Growth of marine bacteria at limiting concentrations of organic carbon in seawater. *Limnol. Oceanogr.* 12:264–71.

Jannasch, H.W. 1968. Competitive elimination of *Enterobacteriaceae* from seawater. *Appl. Microbiol.* 16:1616–18.

Jannasch, H.W. 1969. Estimations of bacterial growth rates in natural waters. *J. Bacteriol.* 99:156–60.

Jannasch, H.W. 1970. Threshold concentrations of carbon sources limiting bacterial growth in sea water. In: *Organic Matter in Natural Waters* (D.W. Hood, ed.). Inst. Mar. Sci. University of Alaska, pp. 321–28.

Jensen, A., B. Rystad, and L. Skoglund. 1972. The use of dialysis culture in phytoplankton studies. *J. exp. mar. Biol. Ecol.* 8:241–48.

Johnson, P.W., and J.McN. Sieburth. 1976. *In-situ* morphology of nitrifying-like bacteria in aquaculture systems. *Appl. Environ. Microbiol.* 31:423–32.

Johnston, R. 1963a. Antimetabolites as an aid to the study of phytoplankton nutrition. *J. mar. biol. Assoc. U.K.* 43:409–25.

Johnston, R. 1963b. Sea water, the natural medium of phytoplankton. I. General features. *J. mar. biol. Assoc. U.K.* 43:409–25.

Johnston, R. 1964. Sea water, the natural medium of phytoplankton. II. Trace metals, chelation, and general discussion. *J. mar. biol. Assoc. U.K.* 44:87–109.

Jones, G.E. 1967a. Precipitates from autoclaved sea water. *Limnol. Oceanogr.* 12:165–67.

Jones, G.E. 1967b. Growth of *Escherichia coli* in heat- and copper-treated synthetic sea water. *Limnol. Oceanogr.* 12:167–72.

Kester, D.R., I.W. Duedall, D.N. Connors, and R.M. Pytkowicz. 1967. Preparation of artificial seawater. *Limnol. Oceanogr.* 12:176–79.

Kriss, A.E. 1963. *Marine Microbiology (Deep-Sea)* (J.M. Shewan and Z. Kabata, trs.). Oliver & Boyd, Edinburgh and London, 536 p.

Kubitschek, H.E. 1970. *Introduction to Research with Continuous Cultures.* Prentice-Hall, Englewood Cliffs, N.J., 195 p.

LaRoche, G., R. Eisler, and C.M. Tarzwell. 1970. Bioassay procedures for oil and oil dispersant toxicity evaluation. *J. Water Poll. Contr. Fed.* 42:1982–89.

Lavoie, D.M. 1975. Application of diffusion culture to ecological observations on marine microorganisms. M.S. Thesis, University of Rhode Island, 91 p.

Lucas, C.E. 1947. The ecological effects of external metabolites. *Biol. Rev.* 22:270–95.

Lucas, C.E. 1949. External metabolites and ecological adaptations. *Symp. Soc. Expl. Biol.* 3:336–56.

Lucas, C.E. 1955. External metabolites in the sea. *Deep-Sea Res.* 3(Suppl.):139–48.

Lucas, C.E. 1958. External metabolites and productivity. *J. Cons. int. Explor. Mer* 144:155–58.

Lyman, J., and R.H. Fleming. 1940. Composition of seawater. *J. Mar. Res.* 3:134–46.

MacLeod, R.A., and E. Onofrey. 1956. Nutrition and metabolism of marine bacteria. II. Observations on the relation of sea water to the growth of marine bacteria. *J. Bacteriol.* 7:661–67.

Marshall, K.C., R. Stout, and R. Mitchell. 1971. Mechanism of the initial events in the sorption of marine bacteria to surfaces. *J. gen. Microbiol.* 68:337–48.

McFeters, G.A., and D.G. Stuart. 1972. Survival of coliform bacteria in natural waters: Field and laboratory studies with membrane-filter chambers. *Appl. Microbiol.* 24:805–11.

McLachlan, J. 1973. Growth media—marine. In: *Handbook of Phycological Methods* (J.R. Stein, ed.). Cambridge University Press, New York, pp. 25–51.

McLachlan, J., and P.R. Gorham. 1961. Growth of *Microcystis aeruginosa* Kütz in a precipitate-free medium buffered with Tris. *Can. J. Microbiol.* 7:869–82.

Nigrelli, R.F. 1958. Dutchman's Baccy Juice, or growth-promoting and growth-inhibiting substances of marine origin. *Trans. N.Y. Acad. Sci.* 20:248–62.

Nigrelli, R.F. 1962. Antimicrobial substances from marine organisms. Introduction: The role of antibiosis in the sea. *Trans. N.Y. Acad. Sci. Ser. II* 25:496–97.

Oppenheimer, C.H., and C.E. ZoBell. 1952. The growth and viability of sixty-three species of marine bacteria as influenced by hydrostatic pressure. *J. Mar. Res.* 11:10–18.

Overbeck, J., and R.J. Daley. 1973. Some precautionary comments on the Romanenko technique for estimating heterotrophic bacterial production. *Bull. Ecol. Res. Comm. Stockholm* 17:342–44.

Percival, E., and R.H. McDowell. 1967. *Chemistry and Enzymology of Marine Algal Polysaccharides.* Academic Press, London and New York, 219 p.

Provasoli, L. 1964. Growing marine seaweeds. *Proc. 4th Intl. Seaweed Symp.* Pergamon, Oxford, pp. 9–17.

Provasoli, L., J.J.A. McLaughlin, and M.R. Droop. 1957. The development of artificial media for marine algae. *Arch. Mikrobiol.* 25:392–428.

Romanenko, V.I. 1964. Heterotroph CO_2 assimilation by bacterial flora of water. *Mikrobiologika* 33:679–83.

Schultz, J.S., and P. Gerhardt. 1969. Dialysis culture of microorganisms: Design, theory, and results. *Bacteriol. Rev.* 33:1–47.

Segedi, R., and W.E. Kelley. 1964. A new formula for artificial seawater. *U.S. Fish. Wildl. Serv. Res. Rept.* 63:17–19.

Seshadri, R., and J.McN. Sieburth. 1971. Cultural estimation of yeasts on seaweeds. *Appl. Microbiol.* 22:507–12.

Seshadri, R., and J.McN. Sieburth. 1975. Seaweeds as a reservoir of *Candida* yeasts in inshore waters. *Mar. Biol.* 30:105–17.

Sieburth, J.McN. 1964. Antibacterial substances produced by marine algae. *Devel. Ind. Microbiol.* 5:124–34.

Sieburth, J.McN. 1965. Bacteriological samplers for air-water and water-sediment interfaces. In: *Ocean Science & Ocean Engineering, Trans. Joint Conf. Mar. Tech. Soc. & ASLO.* Washington D.C., pp. 1064–68.

Sieburth, J.McN. 1967. Seasonal selection of estuarine bacteria by water temperature. *J. exp. mar. Biol. Ecol.* 1:98–21.

Sieburth, J.McN. 1968. The influence of algal antibiosis on the ecology of marine microorganisms. In: *Advances in Microbiology of the Sea* (M.R. Droop and E.J.F. Wood, eds.). Academic Press, London and New York, pp. 63–94.

Sieburth, J.McN. 1971. Distribution and activity of oceanic bacteria. *Deep-Sea Res.* 18:1111–21.

Sieburth, J.McN., P-J. Willis, K.M. Johnson, C.M. Burney, D.M. Lavoie, K.R. Hinga, D.A. Caron, F.W. French III, P.W. Johnson, and P.G. Davis. 1976. Dissolved organic matter and heterotrophic microneuston in the surface microlayers of the North Atlantic. *Science* 194:1415–18.

Sieburth, J.McN., K.M. Johnson, C.M. Burney, and D.M. Lavoie. 1977. Estimation of *in situ* rates of heterotrophy using diurnal changes in dissolved organic matter and growth rates of picoplankton in diffusion culture. *Helgoländer wiss. Meeresunters.* 30:565–74.

Simidu, U. 1974. Improvement of media for enumeration and isolation of heterotrophic bacteria in seawater. In: *Effect of the Ocean Environment on Microbial Activities* (R.R. Colwell and R.Y. Morita, eds.). University Park Press, Baltimore, Md., pp. 249–57.

Singh, B.N. 1946. A method of estimating the numbers of soil protozoa especially amoebae, based on their differential feeding on bacteria. *Ann. Appl. Biol.* 33:112–19.

Slanetz, L.W., and C.H. Bartley. 1965. Survival of fecal streptococci in sea water. *Health Lab. Sci.* 2:142–48.

Sorokin, Yu. I. 1971a. On the role of bacteria in the productivity of tropical oceanic waters. *Int. Revue ges. Hydrobiol.* 56:1–48.

Sorokin, Yu. I. 1971b. Abundance and production of bacteria in the open water of the Central Pacific. *Oceanology* 11(1):85–94.

Stein, J.R. (ed.). 1973. *Handbook of Phycological Methods*. Cambridge University Press, New York, 448 p.

Taub, F.B., and A.M. Dollar. 1968. The nutritional inadequacy of *Chlorella* and *Chlamydomonas* as food for *Daphnia pulex*. *Limnol. Oceanogr.* 10(Suppl):R253–R258.

Taub, F.B., and D.H. McKenzie. 1973. Continuous cultures of an alga and its grazer. *Bull. Ecol. Res. Comm. Stockholm* 17:371–77.

Tempest, D.W. 1970. The continuous cultivation of microorganisms. I. Theory of the chemostat. In: *Methods in Microbiology* (J.R. Norris and D.W. Ribbons, eds.), Vol. 2. Academic Press, New York and London, pp. 259–76.

Vaccaro, R.F., and H.W. Jannasch. 1966. Studies on heterotrophic activity in seawater based on glucose assimilation. *Limnol. Oceanogr.* 11:596–607.

Vaccaro, R.F., and H.W. Jannasch. 1967. Variations in uptake kinetics for glucose by natural populations in seawater. *Limnol. Oceanogr.* 12:540–42.

Vargo, G.A., P.E. Hargraves, and P.W. Johnson. 1975. Scanning electron microscopy of dialysis tubes incubated in flowing seawater. *Mar. Biol.* 31:113–20.

Veldkamp, H., and H.W. Jannasch. 1972. Mixed culture studies with the chemostat. *J. Appl. Chem. Biotechnol.* 22:105–23.

Veldkamp, H., and J.G. Kuenen. 1973. The chemostat as a model system for ecological studies. *Bull. Ecol. Res. Comm. Stockholm* 17:347–55.

Vollenweider, R.A. (ed.). 1969. *A Manual on Methods for Measuring Primary Production in Aquatic Environments* (IBP Handbook No. 12). Blackwell Scientific, Oxford, 213 p.

Wilson, D.P., and F.A.J. Armstrong. 1952. Further experiments on biological differences between natural seawaters. *J. mar. biol. Assoc. U.K.* 31:335–49.

Wilson, D.P., and F.A.J. Armstrong. 1954. Biological differences between sea waters: Experiments in 1953. *J. mar. biol. Assoc. U.K.* 33:347–60.

Wilson, D.P., and F.A.J. Armstrong. 1958. Biological differences between sea waters: Experiments in 1954 and 1955. *J. mar biol. Assoc. U.K.* 37:331–48.

Wilson, D.P., and F.A.J. Armstrong. 1961. Biological differences between sea waters: Experiments in 1960. *J. mar. biol. Assoc. U.K.* 41:663–81.

Wright, R.T., and J.E. Hobbie. 1965. The uptake of organic solutes in lake water. *Limnol. Oceanogr.* 10:22–28.

Young, K., M. Duckworth, and W. Yaphe. 1971. The structure of agar. Part III. Pyruvic acid, a common feature of agars from different agarophytes. *Carbohydr. Res.* 16:446–48.

ZoBell, C.E. 1941. Studies on marine bacteria. I. The cultural requirements of heterotrophic aerobes. *J. Mar. Res.* 4:42–75.

PART III

Phototrophic and Chemolithotrophic Procaryotes

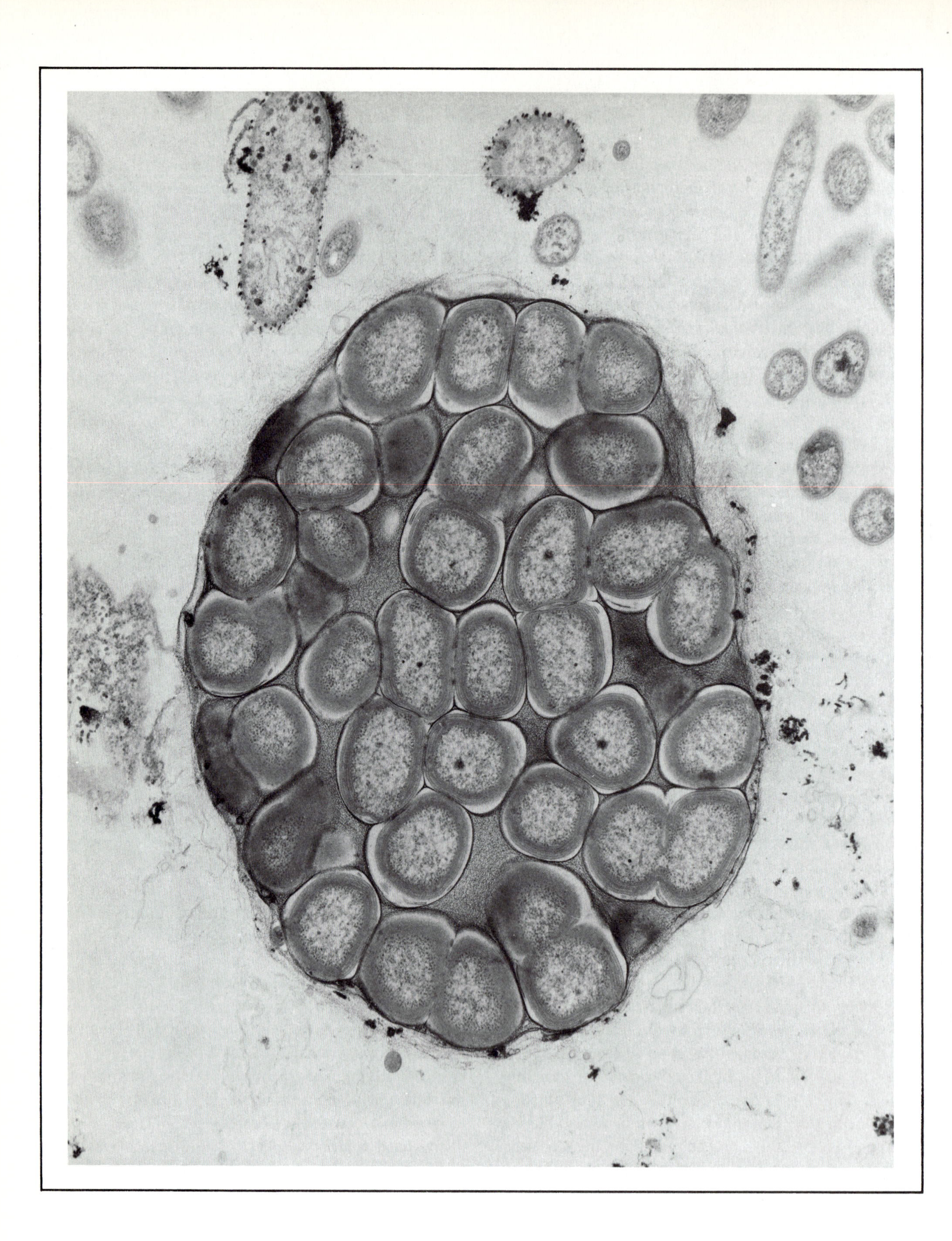

Thin section of bacteria suspended in the water column of a closed-system, salmon-rearing tank, showing a species of *Nitrosomonas* with its characteristic peripheral lamellae and cyst-type colony, with dividing cells (30,000 X). From P.W. Johnson and J.McN. Sieburth, 1976, *Appl. Environ. Microbiol.* 31:423–32).

There is every indication that the first forms of life were procaryotic, with an osmotrophic or a chemoorganotrophic nutritional mode that allowed them to utilize the prebiological accumulation of dissolved organic matter. While the atmosphere was still anaerobic, but as the supply of organic matter dwindled, the phototrophic mode of nutrition probably developed in the anoxyphotobacteria. The later appearance of the oxyphotobacteria not only introduced aerobic phototrophism and the evolution of oxygen that would be carried into the eucaryotes by endosymbiosis (Chapter 1), but these photobacteria were apparently the progenitors of a number of apochlorotic gliding forms with both chemoorganotrophic (Chapter 16,A) and chemolithotrophic modes of nutrition. The chemolithotrophic mode of nutrition is by oxidation of reduced inorganic substances, which are often the end products of the metabolism of other organisms, to provide energy with the bulk of the cell carbon coming from carbon dioxide. The release of ammonia during metabolism and decay can enrich for the nitrifying bacteria, some of which can oxidize the toxic ammonia to nitrite, while other nitrifying bacteria can oxidize the equally toxic nitrite to the less toxic nitrate. The reduced forms of sulfur, such as sulfur itself and the sulfides, can be oxidized by a number of forms ranging from the thiobacilli, which can be thought of as autotrophic pseudomonads, to the larger, sulfur-granule-containing forms, some of which are gliding trichomes related to the blue-green photobacteria (cyanobacteria).

CHAPTER 8

Phototrophic Bacteria

The phototrophic bacteria were observed in the 18th century by botanists and naturalists drawn to their colored accumulations, but it was not until the first half of the 19th century that genera and species were described by such taxonomists as Ehrenberg (1838) and the "blue-green algae" (cyanobacteria) were distinguished from the green and red algae by Näegli (1849). Both the anoxyphotobacteria and the oxyphotobacteria can be grown in enrichment culture in the laboratory under suitable conditions. The exacting anaerobic culture conditions required by the anoxyphotobacteria, including such reduced substances as sulfide, as well as their unique photosynthesis without oxygen evolution, apparently discouraged most botanists from taking over these forms, putting them in herbaria, and using classical morphological methods for classification. The soil bacteriologist Winogradsky (1888), however, not only learned how to enrich for these bacteria, but worked out some of the physiology of anaerobic bacterial photosynthesis. Molisch (1907) differentiated the two physiological groups among the purple photobacteria and described a number of species. Over the years, a number of species have been characterized, with those refractory to isolation being described morphologically from naturally occurring and enriched multicultures. The requirements for growth have been slowly worked out, especially by Pfennig (1961, 1965). The use of controlled conditions for enrichment, has made possible the pure culture of most described species (Pfennig and Trüper, 1973). These pure cultures have been characterized by biochemical analysis. The correlation of biochemical activity with cellular morphology and ultrastructure has put the taxonomy of this group on a firm basis.

Unlike the anoxyphotobacteria, the cyanobacteria, which are usually considered as cyanophytes or blue-green algae, are aerobes and evolve oxygen. They were naturally studied by botanists, who were intrigued by these microscopic, uniquely pigmented forms. Some of these botanists used mixed enrichment cultures to augment the descriptions of fresh material and herbarium specimens. The remarkable property of the cyanobacteria to survive for many decades in herbaria and then to grow upon the addition of water, nutrients, and illumination has also led to the tremendous taxonomic problems of this group, as discussed by Desikachary (1973) and Fogg et al. (1973). Since different cultural conditions affect morphology, there has been a proliferation of descriptions of species over the 150 years these microorganisms have been described by classical botanical taxonomy. This proliferation is compounded by the multispecies nature of natural and enriched assemblages, in which a few cells of one species may be included in the description of the more dominant species. Drouet (Drouet and Daily, 1956; Drouet, 1968, 1973) has spent a lifetime examining not only his collections but collections in herbaria the world over in the monumental task of trying to simplify the classical taxonomy of the main groups of blue-green photobacteria. Although this simplification has been in the right direction, there is no firm basis upon which to build it and it actually hinders the development of a rational system based on biochemistry, physiology, and ultrastructure, in which the photobacteria are studied under standardized conditions and evaluated by numerical taxonomy (Whitton, 1969). Both the filamentous and the coccoid cyanobacteria are slowly being brought into pure culture, a necessary step before the analyses needed for current bacterial taxonomy can be conducted. Such pioneering studies have already been undertaken by Stanier and his colleagues (Stanier et al., 1971; Kenyon, Rippka, and Stanier, 1972). When this study has progressed to a significant degree, then the cyanobacteria and perhaps other oxyphotobacteria (Lewin, 1977) will have the same firm taxonomic criteria as the anoxyphotobacteria, and the Division One of the Kingdom Procaryotae, the cyanobacteria, will not be the void it is in the 8th edition of *Bergey's Manual of Determinative Bacteriology*, edited by Buchanan and Gibbons (1974) (called hereafter *Bergey's 8th*). Division One has now been modified and the position of the photobacteria among the higher taxa of bacteria has been clarified by Gibbons and Murray (1978), whose proposal has been adopted here despite a reluctance by some to synonymize the words "procaryota" and "bacteria" (Lewin, 1976).

A. ANOXYPHOTOBACTERIA

These photosynthetic bacteria use reduced substances (sulfide, molecular hydrogen, and small organic carbon compounds produced during fermentation by anaerobic organotrophs) as external electron donors to photoassimilate carbon dioxide anaerobically without evolving oxygen. They are the bacterial prey for the protozoa, nematodes, and other grazers in oxygen-poor environments. The order Rhodospirillales consists of three families of phototrophic bacteria differentiated by pigments and the ability to use different sulfur compounds and to photoassimilate simple organic compounds. They are the purple sulfur bacteria (Chromatiaceae), the purple non-sulfur bacteria (Rhodospirillaceae), and the green sulfur bacteria (Chlorobiaceae) as defined by Pfennig and Trüper in *Bergey's 8th*.

The purple non-sulfur bacteria, found in inshore waters, are associated with decomposing plant materials and bottom sediments. Their numbers increase with the amount of organic matter, but they rarely accumulate to visible concentrations. Most species are inhibited by higher concentrations of hydrogen sulfide, and at such concentrations, they are overgrown by the sulfur bacteria. In addition to anoxygenic photosynthesis, many purple non-sulfur bacteria can grow heterotrophically in the light or dark at atmospheric oxygen levels. It was believed that purple non-sulfur bacteria were unable to utilize hydrogen sulfide as a photosynthetic electron donor and hence their name. Hansen and Veldkamp (1972), however, reported the isolation of a photosynthetic purple bacterium from marine mud flats that had a high sulfide tolerance and was able to convert sulfide and thiosulfate into sulfate without passing through the intermediate step present in purple sulfur bacteria where elemental sulfur is accumulated. Further studies by Hansen and Veldkamp (1973) indicated that all strains showed excellent photolithotrophic growth on hydrogen, hydrogen sulfide, and thiosulfate, whereas a wide range of organic compounds could be utilized anaerobically in the dark and that growth on organic compounds aerobically in the dark was possible. This very versatile purple photobacterium, named *Rhodopseudomonas sulfidophila*, prompted Hansen and van Gemerden (1972) to re-evaluate the ability of other purple non-sulfur bacteria to utilize sulfide as a photosynthetic electron donor. They found that *Rhodospirillum rubrum*, *Rhodopseudomonas capsulata*, and *Rhodopseudomonas sphaeroides* oxidized sulfide to extracellular elemental sulfur only, whereas *Rhodopseudomonas palustris* converted sulfide into sulfate, bypassing the elemental sulfur step. To demonstrate sulfide utilization in sulfide-sensitive species, batch cultures were unsuccessful, and chemostat culture had to be used.

The purple and green sulfur bacteria are restricted to anaerobic and sulfide-containing environments where sufficient light for growth can penetrate. There are three distinct habitats for these bacteria: sulfur springs containing hydrogen sulfide and other reduced sulfur compounds, the sulfide-containing sands and muds of salt marshes, and the chemocline of anoxic, sulfide-containing water basins. The sulfide in the last two habitits is produced by the reduction of the sulfate in seawater by *Desulfovibrio*, as discussed in Chapter 17,C. Volcanic areas with fissures that permit sulfurous spring waters and gases (fumaroles) to percolate into shallow marine waters are quite rare; they occur in such places as Surtsey Island in Iceland and Deception Island in the South Shetland Archipelago. The mass accumulations of purple and green photobacteria in the muds of estuaries of the salt marsh type were first described by Warming (1875) on the Danish coast. Since then, similar accumulations have been observed in the adjacent Schleswig–Holstein area of Germany and along much of the east coast of the United States. The vertical distribution of photosynthetic pigments and the penetration of light into such marine sediments were studied by Fenchel and Straarup (1971) (Plate 8-1). Long wavelength light penetrated furthest, which in part explains why the purple and green photobacteria that utilize near-infrared for photosynthesis can live deeper in the sediment than the eucaryotic algae and the cyanobacteria. The chemistry and biology of the sulfide layer and its community under the oxidized layer of sand was studied by Fenchel and Riedl (1970) and is discussed in Chapter 3,B. The purple photobacteria of these rich sediments play a significant role in the nutrition of the microbenthos (Fenchel, 1968, 1969).

Purple and green sulfur bacteria are often conspicuous in the chemocline of saline water in meromictic ponds, lagoons, and lakes adjacent to the sea but with no flow to the sea; these include those on Cape Cod, Massachusetts, and in southern Rhode Island, as well as those of anoxic fjords, such as Lake Pettaquamscutt in Rhode Island, Lake Nitinat in British Columbia, and Lake Faro near Messina Italy (Plate 8-1), which are connected to the sea. Studies on anoxic freshwaters and seawater have shown how the photobacteria accumulate below the thermocline in a chemocline where oxygen disappears, where hydrogen sulfide is present, but where there is still light for growth (Genovese, 1963; Kondrat'eva, 1965; Pfennig, 1967; Takahashi and Ichimura, 1968). The study by Trüper and Genovese (1968) on the saline Lake Faro described species of purple and green photobacteria that typically occur in such bodies of water. There is an indication that the open areas of nearshore waters, which can become anoxic temporarily, may also support the growth of purple sulfur bacteria. The New York Bight area, enriched by the metropolitan sewage coming from Raritan Bay, apparently sparked a marked bloom of the dinoflagellate *Ceratium* in the spring of 1976. The senescence of such a bloom may help produce anoxic and sulfide-containing layers in the photic zone from May to October, which, on occasion, yielded red-colored filtrates suggestive of purple sulfur bacteria. The presence of purple photobacteria under such conditions can be easily verified by using the method of Trüper and Yentsch (1967) for determining the *in vivo* absorption spectra.

Most studies on the culture, morphology, ultrastructure, and taxonomy of the purple and green photobacteria have been made with freshwater isolates. Many species, however, may live in both freshwater and marine environments and therefore all forms are discussed here. The purple and green photobacteria are gram negative as are most of the bacteria in the sea. The purple non-sulfur bacteria (Rhodospirillaceae) are relatively small rods, spheres, vibrios, spirals, and budding forms motile by means of a polar flagellum or a flagellar tuft. The photosynthetic membranes are one of four characteristic forms and vary with genus and species (Plate 8-2). Vesicles, the numerous tubular invaginations that occur over the cytoplasmic membrane and extend inward as bulged and connected tubes, are present in both *Rhodopseudomonas* and *Rhodospirillum* species. Stacks of short lamellae near the periphery are present in the strictly anaerobic species of *Rhodospirillum*. Polar stacks, the many-layered peripheral lamellae that run parallel to the long axis of the cell, are present in certain species of *Rhodopseudomonas* and the budding genus *Rhodomicrobium*. Tubes occurring as simple invaginations from the periphery are present in several other species. As the light intensity decreases and the oxygen level increases, which force these microorganisms into a heterotrophic metabolic phase, both pigment development and membrane system development are markedly depressed (Cohen-Bazire and Sistrom, 1966).

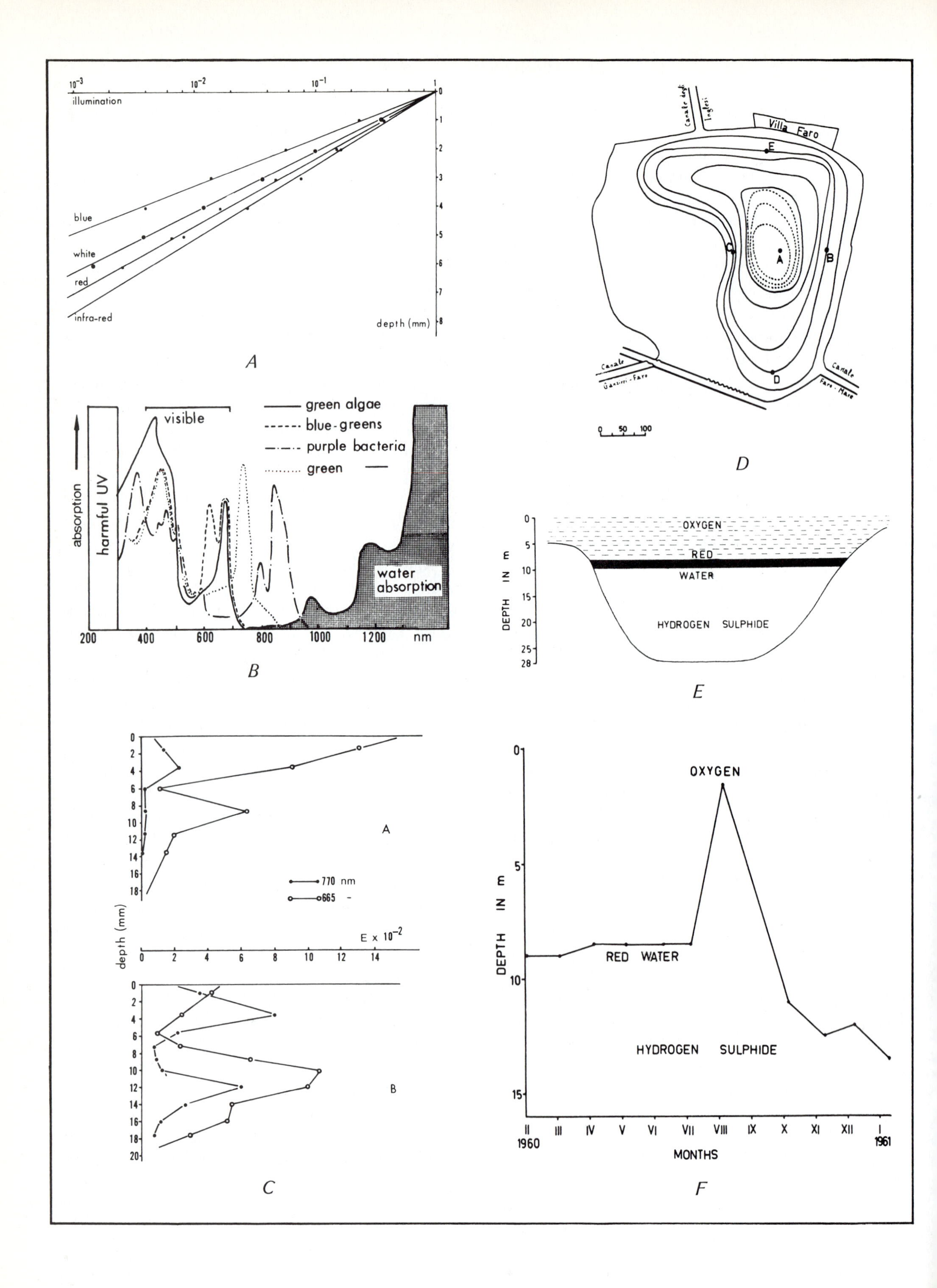

illumination
blue
white
red
infra-red
depth (mm)
A
visible
green algae
blue-greens
purple bacteria
green
absorption
harmful UV
water absorption
nm
B
Villa Faro
Canale degli Inglesi
Canale Ganzirri-Faro
Canale Faro-Mare
D
OXYGEN
RED
WATER
HYDROGEN SULPHIDE
DEPTH IN m
E
770 nm
665 -
depth (mm)
E x 10^{-2}
C
OXYGEN
RED WATER
HYDROGEN SULPHIDE
DEPTH IN m
MONTHS
F

PLATE 8-1. The two habitats of anoxyphotobacteria. In sulfide-containing sediments (*A*), penetration of light of different wavelengths (colors) to different depths in quartz sand and (*B*) the different *in vivo* absorption curves for the phototrophic organisms present in sediments control (*C*) the distribution of chlorophyll *a* and bacteriochlorophyll *a*. In anoxic basins, such as (*D*) Lake Faro (Sicily), seawater in communication over a shallow sill permits (*E*) a red layer of purple sulfur bacteria to accumulate at the chemocline (interface of oxygen and sulfide-containing water) in the photic zone, but (*F*) its occurrence and depth can vary seasonally.
(*A,B,C*) from T. Fenchel and B.J. Straarup, 1971, *Oikos* 22:172–82; (*D,E,F*) from S. Genovese, 1963, *Marine Microbiology* (C.H. Oppenheimer, ed.), courtesy of C.C. Thomas, Publisher, Springfield, Ill., Ch. 20.

The purple sulfur bacteria (Chromatiaceae) are small and large spheres, ovoids, short and long rods, vibrios, and spirals, which have a polar flagellum or a flagellar tuft when motile. If hydrogen sulfide is the electron donor, sulfur accumulates as an intermediate oxidation product either inside or outside the cell (Plate 8-3). These droplets or globules of sulfur are regularly shaped and highly refractile bodies, which are easily seen by bright-field or phase-contrast microscopy. Buoyant species can contain gas vacuoles, which appear as a single refractile or reddish empty area by light microscopy. Such species have been selectively enriched for in freshwater, but so far they have not been reported from the marine environment. Electron microscopy shows that the gas vacuoles are parallel arrays of vesicles, 700 Å in diameter, of varying lengths, with conical ends that are bounded by a 20 Å membrane. They appear identical to those in the green sulfur bacteria, the cyanobacteria, and a number of the organotrophic bacteria. All the purple sulfur bacteria have the vesicle-type photosynthetic membrane system, except for *Ectothiorodospira mobilis*, which has a stacked membrane, and *Thiocapsa pfennigii*, which has a tubular type.

The green sulfur bacteria (Chlorobiaceae), usually small cells of varying shapes, are nonmotile and develop in high sulfide concentrations, where they deposit sulfur outside the cell. Some species form gas vacuoles. Unlike the purple bacteria, with their photosynthetic membrane system continuous with the cytoplasmic membrane, the green sulfur bacteria have a unique photosynthetic structure, the chlorobium vesicle. These large oblong bodies (300 to 400 Å by 1,000 to 1,500 Å) contain the pigments and cover the entire cortical region of the cell just under the cytoplasmic membrane (Echlin and Morris, 1965; Cohen-Bazire and Sistrom, 1966; Pfennig, 1967; Pfennig and Trüper, 1973).

The Winogradsky column (Winogradsky, 1888) is a very useful tool for the demonstration (Hutner, 1962) and the enrichment (Larsen, 1952) of the purple and green sulfur bacteria. Marine mud is mixed with calcium sulfate and shredded paper or seagrasses, such as *Zostera* or *Spartina;* this material is then placed in the bottom of a glass cylinder, which is then filled almost to the top with black mud and overlaid, to the top, with seawater containing 0.1% ammonium chloride and continually incubated with incandescent light. Organotrophic bacteria attacking the cellulose in the paper or plants consume any dissolved oxygen and produce the organic acids used by sulfate-reducing bacteria (*Desulfovibrio;* Chapter 17,C). The anaerobic sulfate-reducers produce hydrogen sulfide, which not only maintains anaerobiosis but enriches for the phototrophic sulfur bacteria. The latter appear as enlarging spots at the mud–tube interface, ranging in color from the greens of the Chlorobiaceae lower in the column to the mauves, magentas, purples, and browns of the Chromatiaceae higher in the column. Extracellular sulfur deposited by the cells forms rings around some colonies. As the column ages and the hydrogen sulfide concentration decreases, cyanobacteria and various algae are seen in the upper levels of the overlying water.

Van Niel (1932), in his benchmark monograph, emphasized the importance of pH and sulfide concentration for the selective enrichment of purple and green sulfur bacteria, but the variety of organisms obtained by his procedures were much more limited than those occurring in nature (*Bergey's 7th*). The physical–chemical limits of growth in the natural environment for microorganisms in the sulfur cycle have been documented by Baas-Becking and Wood (1955), whose figure showing this is widely reproduced in textbooks. A species of green sulfur bacterium enriched with the Winogradsky column and purified in sulfide shake agar culture (van Niel, 1932) developed poorly on subsequent transfers in this medium (Larsen, 1952). Faster and normal growth was obtained in a synthetic medium containing ferric chloride and trace metals. It was correctly pointed out that the early descriptions of such microorganisms as *Chlorobium limicola* were based on unhealthy cultures grown on inadequate media (Larsen, 1952).

The enriching procedure described by van Niel

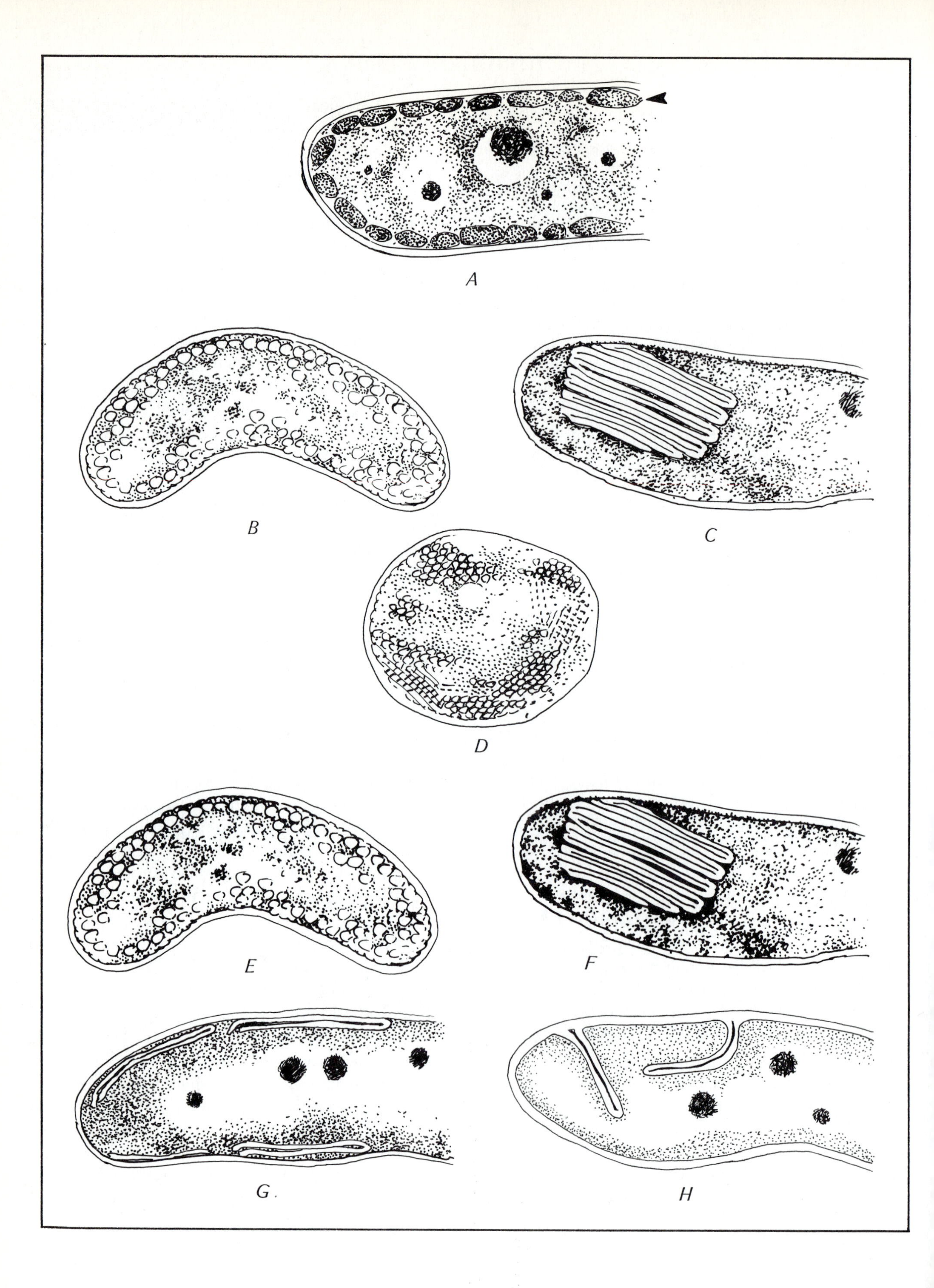
A
B
C
D
E
F
G
H

PLATE 8-2. The form of intracytoplasmic membranes and *Chlorobium* vesicles that occur in the families of the purple and green photobacteria. (*A*) The *Chlorobium* vesicles (arrow) of the family Chlorobiaceae (also present in *Prosthecochloris*). The membrane systems of the family Chromatiaceae showing (*B*) type found in *Chromatium*, *Thiocapsa*, and *Thiocystis;* (*C*) type found in *Ectothiorhodospira;* and (*D*) type found in *Thiocapsa pfennigii.* The membrane systems of the family Rhodospirillaceae showing (*E*) type found in *Rhodopseudomonas*, (*E, and F*) types found in *Rhodospirilla*, (*G*) type found in *Rhodomicrobium*, and (*H*) type found in *Rhodocyclus.* After N. Pfennig and H. Trüper, 1973, in *Handbook of Microbiology, Vol. I. Organismic Microbiology* (A.I. Laskin and H.A. Lechevalier, eds.). CRC Press, Cleveland, Ohio, Table 3, p. 19.

(1944) in his monograph on the purple non-sulfur bacteria permitted the isolation of the then-described species. In the process of obtaining the large purple sulfur bacteria (*Chromatium okenii* and *Thiospirillum jenense*) in pure culture, Pfennig (1961, 1965) developed a new synthetic medium containing vitamin B^{12}. With this medium it has been possible to grow in pure culture purple and green sulfur bacteria that had never been cultivated before. For the enrichment of the purple non-sulfur bacteria, sulfate has to be almost completely replaced by chloride in order to prevent the growth of sulfate-reducing bacteria, which would produce sulfide and enrich for the purple and green sulfur bacteria (van Niel, 1944).

The cultural conditions devised by Pfennig (1961) for the enrichment of the anaerobic phototrophic bacteria are summarized in Table 8-1. In addition to variations in pH and sulfide concentration, the wavelength and intensity of light and the temperature of incubation were varied to selectively enrich for various subgroups within the major groups. These cultural procedures have permitted the isolation in pure culture of species previously described in natural populations as well as some new species. Pfennig and Trüper (1971a,b,c,d, 1973; *Bergey's 8th*) have used observations made on these pure cultures to straighten out the confused taxonomy of these bacteria. Techniques for the enrichment, isolation, and maintenance of the purple and green photobacteria have been reviewed by van Niel (1971).

The numerous papers describing the culture of the anoxyphotobacteria fail to mention generation times. A yield of 2.5 g wet weight of green sulfur bacteria/liter in

TABLE 8-1. Cultural conditions for enriching different groups of anaerobic phototrophic bacteria by the system of Pfennig.[a]

GROUP	$Na_2S \cdot 9H_2O$ (%)	pH RANGE	LIGHT QUALITY AND INTENSITY	TEMPERATURE (°C)
Green sulfur bacteria:			350–760 nm	
Species without gas vacuoles	0.05–0.1	6.6–6.9	20–100 ft-c	20–30
Species with gas vacuoles	0.02–0.05	6.6–6.9	10–20 ft-c	10
Purple sulfur bacteria:			800–900 nm	
Species without gas vacuoles	0.05–0.1		20–100 ft-c	20–30
Species with gas vacuoles	0.02–0.05	7.3–7.6	10–20 ft-c	10
Species producing external elemental sulfur	0.05–0.1	7.3–7.6	20–100 ft-c	30–40
Purple non-sulfur bacteria:[b]				
Species with Bchl *a*	No sulfide[c,d]	7.0–7.5	Visible 20–100 ft-c	25–30
Species with Bchl *b*	No sulfide[c,d]	7.0–7.5	Infrared radiation 1,000–1,050 nm	25–30

(Pfennig, 1967, after Brock, 1970)

[a] Mineral salts medium (Pfennig, 1965) of 10 to 30 o/oo sodium chloride; bicarbonate for carbon dioxide, vitamin B_{12}, anaerobic incubation in sealed bottles (see also van Niel, 1971).
[b] Bchl, bacteriochlorophyll.
[c] Hydrogen or organic carbon as electron donors.
[d] Plus growth factors.

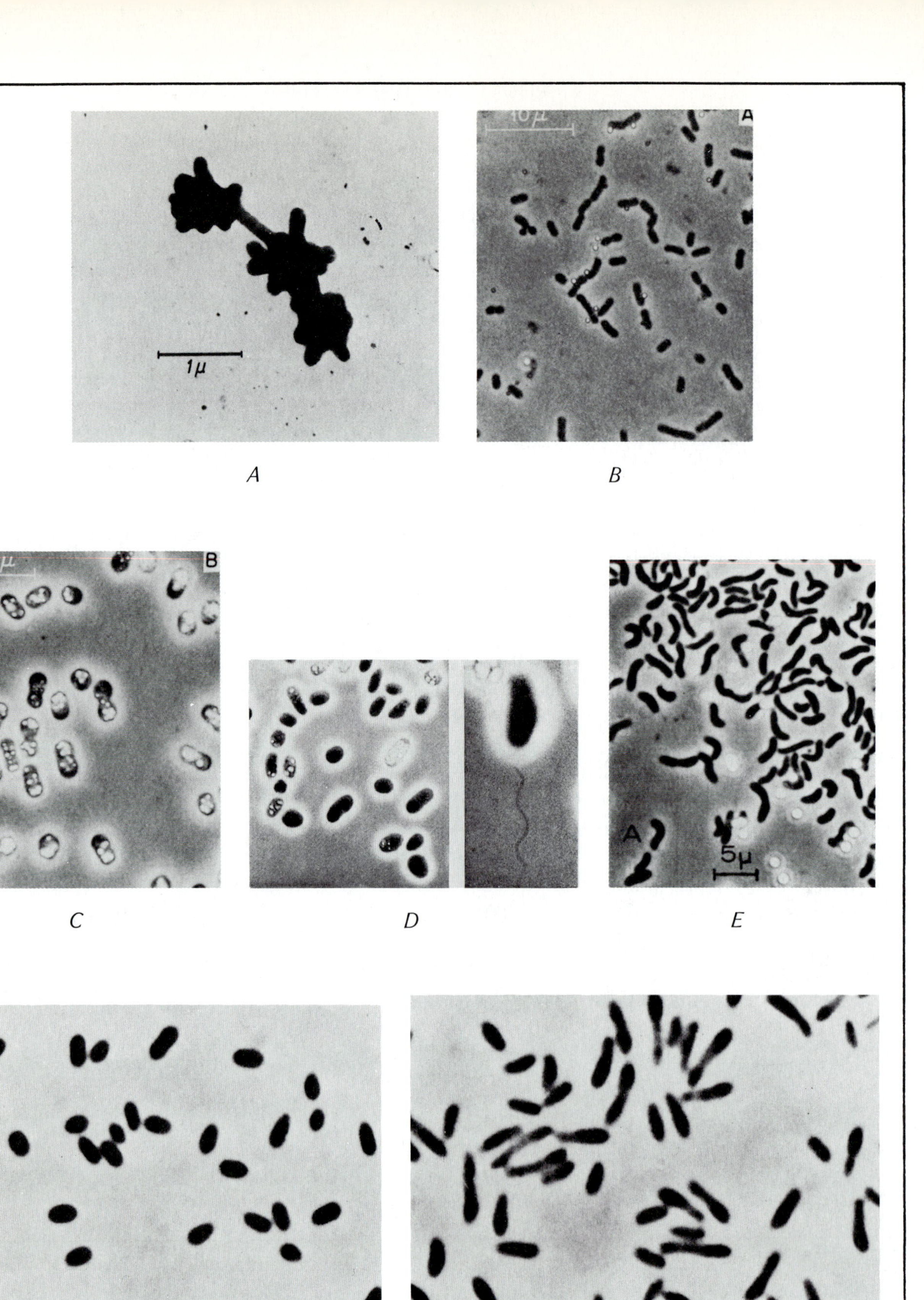
1μ
10μ
A
10μ
B
A
5μ
A
B
C
D
E
F
G

PLATE 8-3. Marine isolates representative of the families of purple and green photobacteria. Members of the family Chlorobiaceae: (*A*) The green species *Prosthecochloris aestuarii* from the sulfide mud of a brackish lagoon in the Crimea, showing the appendaged cells. (*B*) The brown species *Chlorobium phaeobacteriodes* (with its external sulfur granules), which is the predominant bacterium in the red zone of Lake Faro (see Plate 8-1). Members of the family Chromatiaceae: (*C*) a small species of *Chromatium* from Lake Faro; (*D*) a large species of *Chromatium*, *C. buderi;* (*E*) the short spirals of *Ectothiorhodospira mobilis* grown on sulfide, with sulfur globules deposited outside the cells. Members of the purple, non-sulfur family Rhodospirilliales, with no sulfur granules: (*F*) the versatile, sulfide-tolerant *Rhodopseudomonas sulfidophila*, (*G*) compared with the sulfide-sensitive *Rhodopseudomonas palustris*. (*A*) from V.M. Gorlenko, 1970, *Zeits. Allg. Mikrobiol.* 10:147–149; (*B, C*) from H.G. Trüper and S. Genovese, 1968, *Limnol. Oceanogr.* 13:225–32; (*D*) from H.G. Trüper and H.W. Jannasch, 1968, *Arch. Mikrobiol.* 61:363–72; (*E*) from H.G. Trüper, 1968, *J. Bacteriol.* 95:1910–20; (*F, G*) from T.A. Hansen and H. Veldkamp, 1973, *Arch. Mikrobiol.* 92:45–58.

a 36- to 48-hour period (Larsen, 1952) suggests an appreciable growth rate, whereas the dry weight yields extrapolated from *Chromatium okenii* growth data (Trüper and Schlegel, 1964) suggest a generation time of 80 to 160 hours in very dim light, depending upon the presence of acetate. For *Chromatium vinosum*, a generation time of 7 hours was determined under optimal conditions by van Gemerden (1968).

In the anoxyphotobacteria, only photosystem I (in the far-infrared) (San Pietro, 1971) occurs. These bacteria use reduced sulfur compounds, molecular hydrogen, or simple organic carbon compounds as external electron donors. Since water is not used as an electron donor, as it is by the oxyphotobacteria and the phototrophic eucaryotes, oxygen is not evolved. Carbon dioxide photoassimilation involves ferrodoxin as the electron donor and the principal enzymes ribulose 1-5 diphosphate carboxylase and ribulose 5-phosphate kinase of the reductive pentose phosphate cycle. Molecular nitrogen fixation is also a common property of these bacteria (Gest and Kamen, 1960). During dark periods under anaerobic conditions, the anoxyphotobacteria maintain a fermentative metabolism; both the sulfur and the nonsulfur bacteria produce carbon dioxide and organic acids, especially acetic acid, whereas the sulfur bacteria also use elemental sulfur to produce sulfide (Gest, San Pietro, and Vernon, 1963; Pfennig, 1967).

The flagellated purple sulfur and purple non-sulfur bacteria accumulate in areas of optimal light intensity as a result of phototaxis; this phototaxis is unlike that in the eucaryotes, since both the carotenoids and bacteriochlorophylls are active. A 1 to 3% decrease in the intensity of illumination will reverse the swimming direction of the cells. The strict anaerobes also show a negative phototaxis to dissolved oxygen with increasing light intensity. The facultative aerobic, purple non-sulfur bacteria have a positive phototactic response under anaerobic conditions in the light and a positive aerotactic response in the dark (Clayton, 1959; Pfennig, 1967).

The establishment of order out of chaos in the anoxyphotobacteria is a textbook example of what can be done when microorganisms are brought into pure culture and their taxonomy based on their morphology, fine structure, biochemical composition, and physiology. As a result of the improved culture methods developed for this group, most of the described species have been obtained in pure culture, and non-valid family names and species have been eliminated (Trüper and Pfennig, 1971) or amended or new nomenclatural combinations have been formed (Pfennig and Trüper, 1971a,b). The 41 type or neotype strains used to establish the new families and genera have been described and maintained in established collections (Pfennig and Trüper, 1971d). The taxonomy of these microorganisms in *Bergey's 7th* is now completely obsolete and with it the long familiar family names Thiorhodaceae and Athiorhodaceae. The new hierarchy of the higher taxa are as follows (Pfennig and Trüper, 1971c; *Bergey's 8th*):

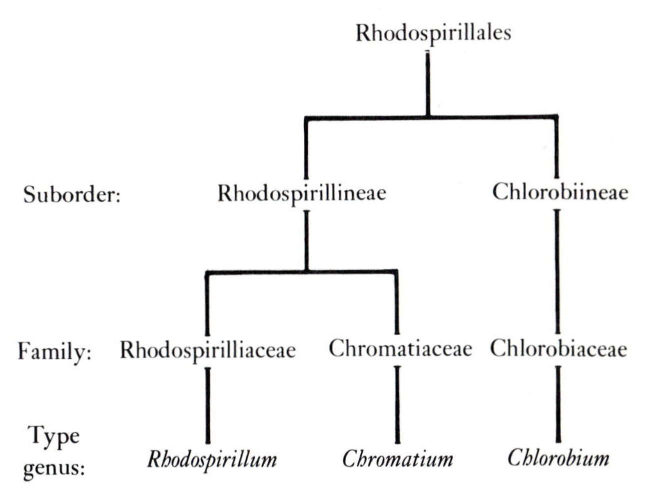

Thus, the new order Rhodospirillales is divided into two suborders: the Chlorobiineae, which have bacteriochlorophyll *c*, *d*, or the recently characterized *e* (Gloe et al., 1975) as major pigments, and the Rhodospirillineae, which have bacteriochlorophyll *a* or *b* as major pigments. The Rhodospirillineae are further divided into the families Rhodospirilliaceae (purple non-sulfur bacteria) and Chromatiaceae (purple sulfur bacteria). The genera and species within each of these families are classified on the basis of morphology, size, membrane system, photoassimilation electron donors, major bacteriochlorophylls and carotenoids, DNA base compositions, which range from 45 to 73 mole % G+C, motility, form of aggregation, presence or absence of sulfur granules, growth factor requirements, hydrogenase activity, and mode of multiplication, all observed with pure cultures (Pfennig and Trüper, 1973; *Bergey's 8th*). A comparative systematic survey of these bacteria has been made by Pfennig (1977).

Now that adequate cultural procedures for isolation and taxonomic criteria for characterization have been developed, the observations made mainly on freshwater species are being extended into the marine environment. With this has come the first pure culture of some long recognized species as well as the recognition of new species (Plate 8-3). The very versatile, purple non-sulfur bacterium *Rhodopseudomonas sulfidophila*, which can tolerate and use sulfides, has been isolated from marine mud flats in the Netherlands (Hansen and van Gemerden, 1972). Among the purple and green sulfur bacteria are a number of new isolates. The large *Chromatium* seen frequently in the marine environment has been isolated from both the Galapagos salt flats and Barnstable Beach on Cape Cod, compared with other species of *Chromatium*, and named *Chromatium buderi* by Trüper and Jannasch (1968). In addition to large, sulfur globule-containing, purple sulfur bacteria in the genera *Chromatium* and *Thiocystis*, there is a smaller, rod-shaped type, which deposits sulfur outside the cell when grown on sulfide; it was described but not named by van Niel (1932). The valid description of this short spiral microorganism by Pelsh (1937) as *Ectothiorhodospira mobilis* (*mobile*) was ignored by new isolators and two editions of *Bergey's Manual* until two strains from the Galapagos salt flats found Trüper (1968) who recognized what and who they were. This species is motile by a polar tuft of flagella and has several large stacks of lamellar membranes (Remsen et al., 1968). When some organic compounds are utilized as hydrogen donors instead of sulfur compounds, sulfate can serve as a source of sulfur. An unusual prosthecate green sulfur bacterium with some 20 appendages and chlorobium vesicles has been isolated from the mud of Lake Sivash, a brackish lagoon in communication with the sea of Azov (Gorlenko and Zhilina, 1968). This microorganism has been appropriately named *Prosthecochloris aestuarii* by Gorlenko (1970).

Three surveys have been conducted to study the distribution and occurrence of photosynthetic sulfur bacteria in mud flats and lagoon-type water basins. Gorlenko (1968) studied two brackish lakes connected to each other and the Sea of Azov in the Crimea and isolated one purple non-sulfur, one green sulfur, and seven purple sulfur species. Trüper (1970) has summarized the results of his large scale isolations from the Galapagos Islands, South Africa, Italy, and the New England coast, in which he used a modified version of the medium of Pfennig (1965). Forty-seven strains comprising thirteen species were obtained, many in pure culture for the first time. Strains in the Chromatiaceae, genus *Chromatium*, were *C. buderi*, *C. gracile*, *C. vinosum*, and *C. violascens; Thiocystis violacea;* genus *Thiocapsa*, *T. pfennigii* and *T. roseopersicina;* and *Ectothiorhodospira mobilis*, whereas strains in the Chlorobiaceae, genus *Chlorobium*, were *C. limicola*, *C. limicola* f. *thiosulfatophilum*, *C. vibrioforme*, and *C. phaeobacterioides;* and *Prosthecochloris aestuarii*. Matheron and Baulaigue (1972) isolated fifteen strains of photosynthetic sulfur bacteria from various sulfide-containing marine habitats in the Marseillaise region of France. These included six *Chromatium*, three of *Thiocapsa*, two of *Ectothiorhodospira*, two of *Chlorobium*, and two of *Prosthecochloris*. The anoxyphotobacteria in the marine habitat are apparently a diverse community, and while some of the terrestrial species, such as those with gas vacuoles, are not seen, there are some species that may be unique to the marine environment.

B. OXYPHOTOBACTERIA

The oxyphotobacteria are phototrophic procaryotic cells that lack the eucaryotic cellular organization of the algae but possess photosystems I and II, use water as an electron donor, and therefore produce oxygen in the light as do the phototrophic eucaryotes. Almost all these forms are cyanobacteria, possessing only one chlorophyll, *a*, and carotenoids usually located in membranous structures known as thylakoids internal to the cytoplasmic membrane. At least three molecular classes of phycobiliproteins are located in disc-shaped granules. The cyanobacteria are placed in the order Cyanobacteriales, while a single genus and species of green procaryote that contains chlorophyll *b* as well as *a* but lacks the phycobiliproteins, is placed in a new order, the Prochlorales.

The association of cyanobacteria-like cells with ascidians (sea squirts) has been described and reviewed by Lewin and Cheng (1975) and Newcomb and Pugh (1975). A prevalent species has been described as *Synechocystis didemni* by Lewin (1975). Despite a pigment complement characteristic of "chlorophytes" rather than "cyanophytes," Schulz-Baldes and Lewin (1976) concluded that this microorganism was a typical cyanophyte. Reversing himself, Lewin (1977) concluded that this green procaryotic microorganism could not be assigned to either the Cyanophyta or the Chlorophyta and created a new division (Prochlorophyta), Class (Prochlorophyceae), Order (Prochlorales), Family (Prochloraceae) and genus (*Prochloron*), for *Prochloron didemni* (= *Synechocystis didemni*). Despite the objection to calling the blue-greens bacteria (Lewin, 1976), Gibbons and Murray (1978) include both the orders Cyanobacteriales and Prochlorales in the subclass Oxyphotobacteriae parallel to the subclass Anoxyphotobacteriae; both of which are gram-negative photobacteria.

The Cyanobacteria, (Stanier and Cohen-Bazire, 1977), which are referred to as the Myxophyceae (Wallroth, 1833) and Cyanophyceae (Sachs, 1874) in the botanical literature, get their characteristic color from the phycobiliproteins (OhEocha, 1965) allophycocyanin, allophycocyanin B and phycocyanin and sometimes phycoerythrin and phycoerythrincyanin, which mask the green of the only chlorophyllous pigment chlorophyll *a*. Aggregates of the phycobiliprotein pigments occur as granules (phycobilisomes) in the outer surfaces of the paired photosynthetic lamellae (thylakoids) that are extensive in the cytoplasm (Gray, Lipschultz, and Gantt, 1973). The thylakoids have an ultrastructure (Cohen-Bazire and Sistrom, 1966; Lang and Whitton, 1973) not unlike the extensive cytomembranes of the purple and green photobacteria (Chapter 8,A) and the nitrifying bacteria (Chapter 9,A). A rigid, multilayered cell wall with an inner peptidoglycan layer always encloses the cells, which are sometimes covered with a fibrous or gelatinous sheath. The general ultrastructure of the cyanobacteria (Echlin and Morris, 1965; Lang, 1968; Drews, 1973) is similar to that of the gram-negative chemoorganotrophs (see Wolk, 1973, for a review of the cytological chemistry and physiology of the filamentous forms). Some filamentous cyanobacteria have a gliding motility at some stage of their development, which is dependent upon surface contact. The mucilage that holds the unicellular forms in colonies and forms the sheaths that hold the trichomes and filaments together is thought to be responsible for their gliding motility (Walsby, 1968). Alternatively, the helical fibrillar array between the cell and the sheath could produce unidirectional traveling waves and a constant rotation to move the trichomes along a solid surface (Halfen and Castenholtz, 1970). The filamentous planktonic species of cyanobacteria, such as *Trichodesmium*, have membrane-bound gas vacuoles that are similar in structure to those in the anoxyphotobacteria and the heterotrophic bacteria (Cohen-Bazire, Kunisawa, and Pfennig, 1969). These gas vacuoles (Walsby, 1973) are extensive and essentially surround the photosynthetic membranes. They apparently provide buoyancy and function as a light shield by reducing the amount of light adsorbed by the photosynthetic pigments of cells on or near the surface where high light intensity and shorter wavelengths might damage the cells (van Baalen and Brown, 1969; Waaland, Waaland, and Branton, 1971).

The cyanobacteria have three basic forms: unicellular coccoid, filamentous, and filamentous with heterocysts. The unicellular coccoid forms reproduce by binary fission, multiple fission, or the serial release of the apical cells (exospores) from sessile individuals. The filamentous forms are chains of cells (trichomes) with no branching, false branching, or true branching. These forms grow by repeated intercalary ("in-between") cell division and reproduce by either the terminal release of short motile chains of cells (hormogonia) or a random fragmentation of the filaments. In the filamentous forms with heterocysts, the heterocysts or akinetes occur as specialized cells with refractive polar granules that are larger than the adjacent vegetative cells; they also have a thicker outer wall (Fay et al., 1968; Lang and Fay, 1971). These non-reproductive cells are the resting stage that subsequently germinate to release a hormogonium that is quite motile and forms new trichomes. As the heterocyst develops, it loses its photosynthetic (carbon dioxide reduction) ability, becomes more reductive, and apparently is the center of nitrogenase activity (Stewart, Haystead, and Pearson, 1969). Intercellular connections between the heterocysts and the vegetative cells in the filaments apparently allow the exchange of the products of nitrogen fixation and photosynthesis between the cells. Other species that lack heterocysts can also fix nitrogen under low oxygen tensions.

Although Stanier et al. (1971) state that the unicellular chroococcoid forms are apparently a minor component of natural algal populations, van Baalen (1962) has stated that they are numerous in estuarine muds and sands. The occurrence of chroococcoid cyanobacteria in the plankton has not been studied until now. Paul Johnson and the author are examining concentrates of bacterioplankton and greater than 3 μm plankton in thin section by transmission electron microscopy. Cells that were missed with light microscopy and even with electron microscopy of negatively stained preparations are detected in thin section due to the presence of thylakoids; and they are ubiquitous in the bacterioplankton of estuarine, shelf, and oceanic waters. In oceanic waters, they can exceed the biomass of chemoorganotrophic bacteria. These cyanobacteria occur widely as single cells and as microcolonies. Their diversity and abundance in the plankton of the North Atlantic are shown in Plate 8-4. The blue-green community developing in undisturbed sand on the Baltic Island of Fehmarn was described by Anagnostidis and Schwabe (1966). There was a sharp zonation controlled by water, light, and oxygen. Below the layer of sand that dried out at low tide was a layer in which a great variety of unicellular as well as filamentous cyanobacteria lived; immediately below was a layer with purple and green photobacteria, and deeper still a layer of sulfide-containing anaerobic mud. The diversity of unicellular coccoid forms on European coasts is illustrated in the plates of Frémy (1929–33).

A pure culture study of 40 isolates of unicellular cyanobacteria (Stanier et al., 1971) is a pioneering step in characterization using current procedures for bacterial taxonomy and is also the beginning of a resolution of typological groups with the traditional form genera. This is just the first step, but it shows the way this group can and should be studied. The typological groups defined by Stanier et al. (1971) are shown in Plate 8-5 and include three marine isolates of van Baalen (1962) that were all type IA. The filamentous forms without heterocysts belong to the Oscillatoriaceae (Plate 8-6). They can occur along the edges of the sea in marshes and muddy shorelines, where they form mats. These mats can become quite thick along warmer shores, such as the Gulf of Mexico, and have been described by Sorensen and Conover (1962), who suggested that these communities are dominated by *Lyngbya conferroides*. Wood (1965) stated that such mat communities stabilize shorelines and serve as food for grazing animals, including juvenile fish. A planktonic member of this group is the genus *Trichodesmium*, in which at least three species are recognized: *T. erythraeum*, *T. thiebautii*, and *T. hildenbrandii* (Desikachary, 1959). In his monograph on this group, Drouet (1968) reclassifies *Trichodesmium* as *Oscillatoria* and puts all the species into *Oscillatoria* (*Trichodesmium*) *erythrea*, where they probably belong. *Trichodesmium* occurs as bundles of filaments that look like sheaves of wheat under a stereomicroscope, but radial colonies are also seen (Taylor, Lee, and Bunt, 1973; Marumo, 1975; Plate 8-5).

PLATE 8-4. The widespread occurrence, diversity, and abundance of chroococcoid cyanobacteria in the plankton of estuarine, shelf, and open ocean waters of the North Atlantic Ocean as seen in transmission electron micrographs of thin sections. (*A*) A single cyanobacterium from the bacterioplankton of Narragansett Bay, R.I. (*B*) A single cyanobacterium in a fecal pellet within the gut of the copepod *Calanus finmarchicus* from the waters of the Nova Scotian shelf. (*C*) An ensheathed microcolony of cyanobacteria from water at a depth of 80 m in the Iberian Basin, showing an extra cell wall layer (arrow). (*D*) A single cyanobacterium from the above water sample showing fine strands of mucilage on the cell wall. (*E*) Abundance of the biomass of larger single cells of cyanobacteria among the bacterioplankton in a pellet concentrated from the picoplankton size fraction from water at a depth of 100 m in the Sargasso Sea. Note the numerically predominant minibacteria (arrows). (*F*) A higher magnification of a single cyanobacterium in the above preparation showing peripheral thylakoids. Micrographs courtesy of Paul W. Johnson.

Carpenter and Price (1977) have determined the distribution and productivity of *Trichodesmium*, and its apparent ability to fix nitrogen, in the Sargasso Sea and the Caribbean Sea. Proof of nitrogen fixation will have to wait until *Trichodesmium* is brought into pure culture. *Trichodesmium*, apparently of little importance in the Sargasso Sea, is a major component of the Caribbean, accounting for about 60% of the total chlorophyll in the upper 50 m and about 20% of all productivity. The peak in *Trichodesmium* concentration is much shallower than that for the other primary producers, 75% of the *Trichodesmium* production occurring in the upper 15 m. Colonies coming to the surface are visible to the naked eye as reddish particles, which have been called "sea sawdust." These colonies in the surface microlayer are concentrated in surface slicks, which occur in windrows induced by Langmuir circulations (Sieburth and Conover, 1965; Carpenter and Price, 1977). Such surface slicks can cover vast areas of tropical seas (Baas-Becking, 1951; Wood, 1965; Sieburth and Conover, 1965; Sournia, 1968).

Steven and Glombitza (1972) reported that *Trichodes-*

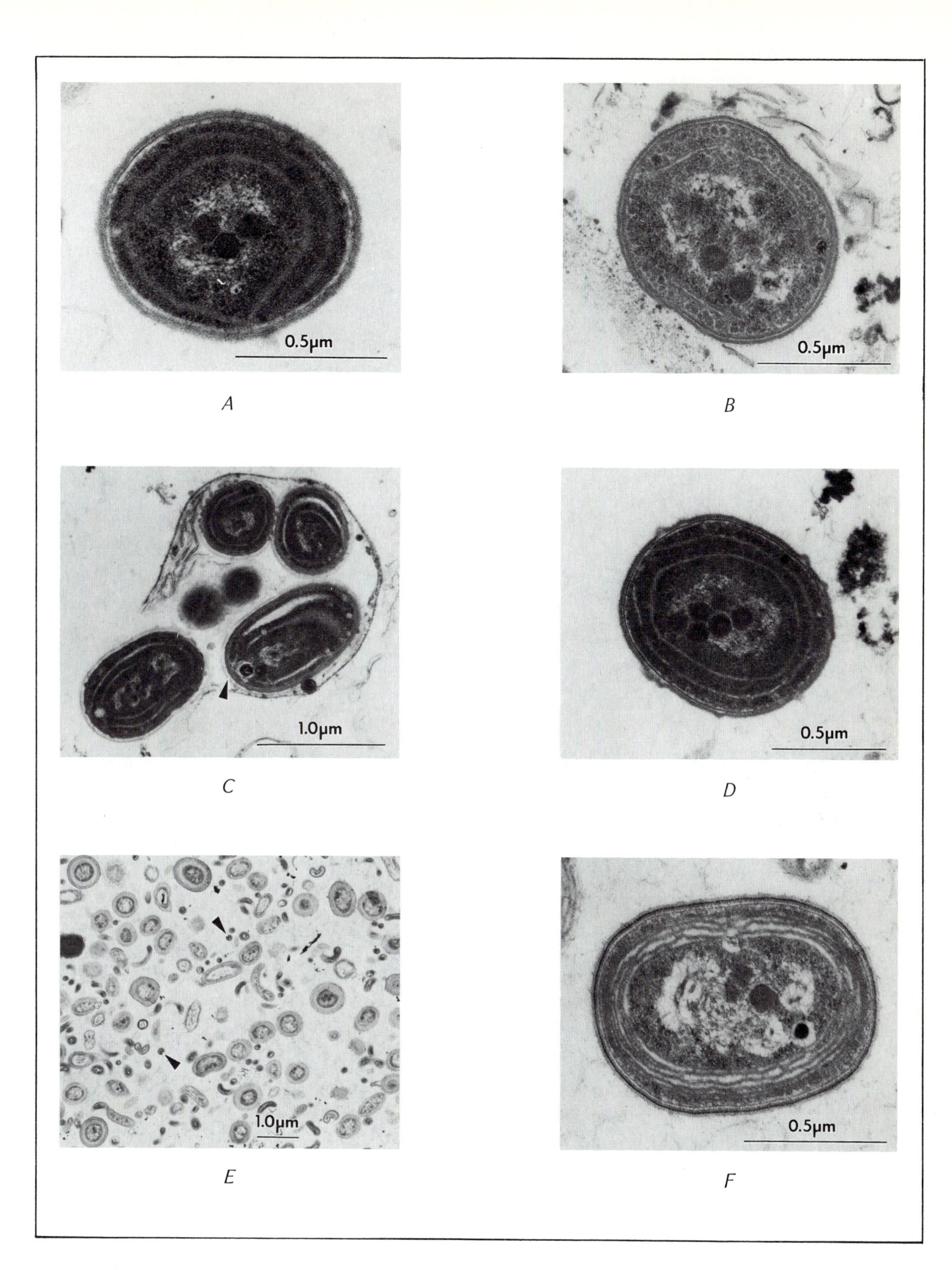

0.5μm
A
0.5μm
B
1.0μm
C
0.5μm
D
1.0μm
E
0.5μm
F

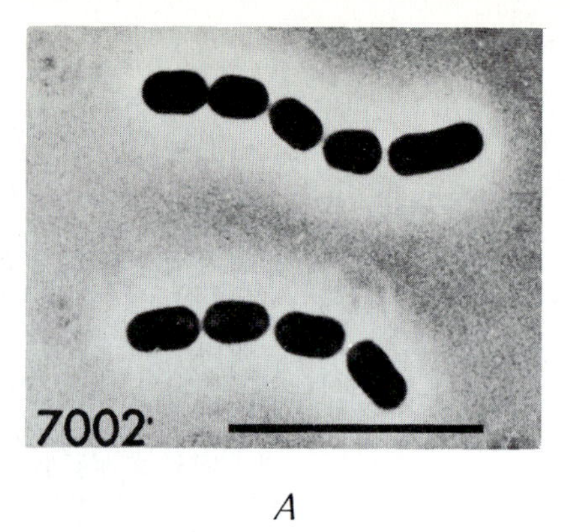

A

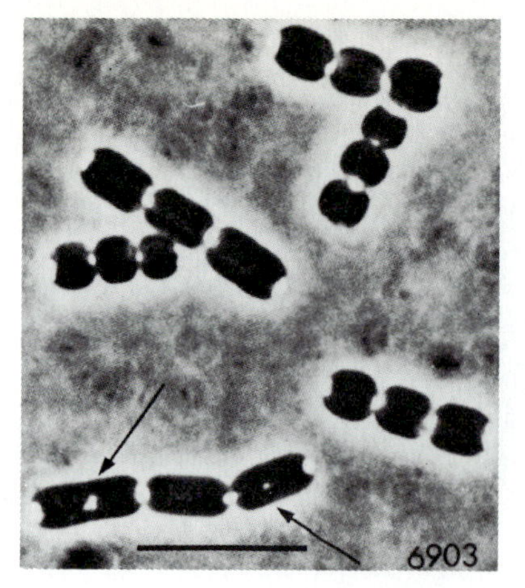

B

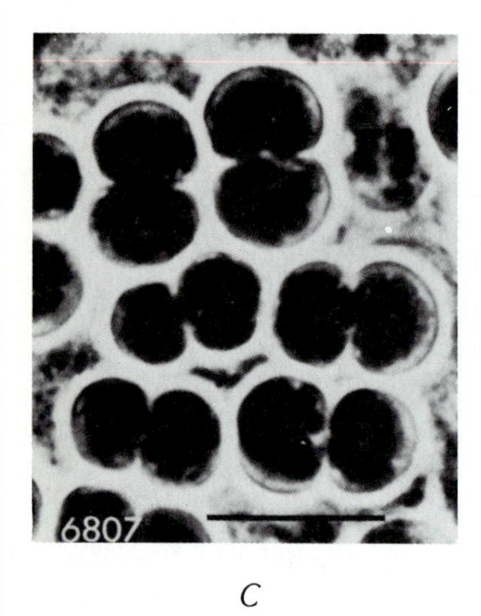

C

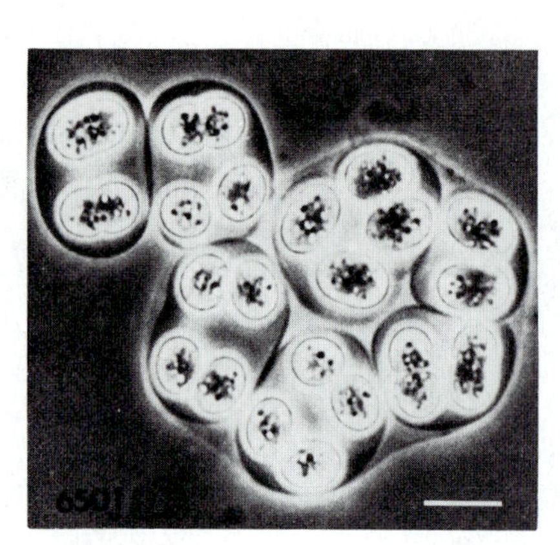

D

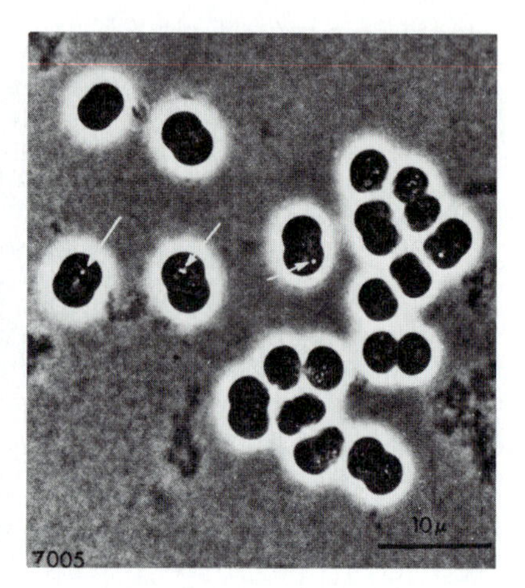

E

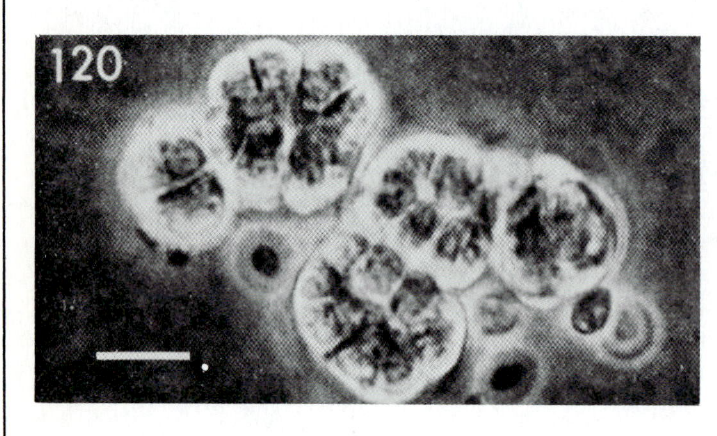

F

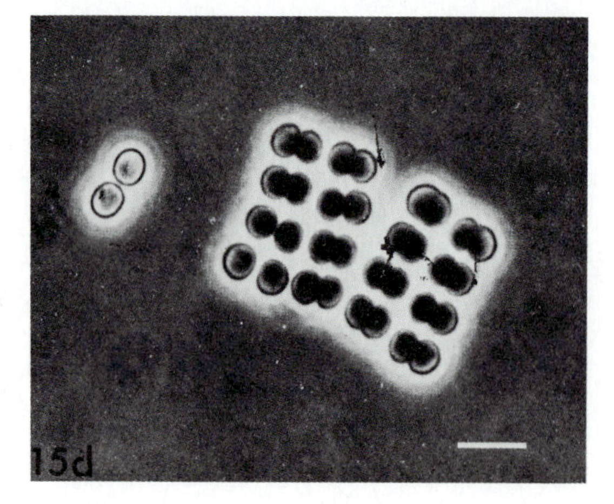

G

PLATE 8-5. The division of the Berkeley Collection of axenic unicellular cyanobacteria into typological groups. Group I, cylindrical or ellipsoidal cells reproducing by repeated binary fission in a single plane to form short chains of cells: (*A*) group IA, no refractile polar granules; (*B*) group IB, refractile polar granules (arrows, newly formed polar granules develop at dividing cell septa). Group II, spherical cells reproducing by binary fission in two or three successive planes at right angles to one another: (*C*) group IIA, unsheathed cells, with no gas vacuoles; (*D*) group IIB, sheathed cells, with no gas vacuoles; (*E*) group IIC, unsheathed cells, with gas vacuoles (arrows). Group III: (*F*) cells in which irregular growth leads to parenchymous, tightly packed masses of cells; (*G*) regular, 16-celled colonies of *Merismopedia*-like aggregates, in an old culture of a Group IIA strain (bars = 10 μm). From R.Y. Stanier, R. Kunisawa, M. Mandel, and G. Cohen-Bazire, 1971, *Bact. Rev.* 35:171–205.

mium thiebautii has an apparent periodicity of about 120 days between surface blooms; it is unrelated to the solar cycle. They believe that *Trichodesmium* is brought to the surface by gas vacuoles and that it then decays and disappears for up to a month before a new crop is recruited from depths of 50 to 100 m. Barth (1967), however, observed that *Trichodesmium* blooms appeared after a rise in water temperature, which, Fogg (1973) points out, also indicates a calm sea condition conducive to accumulation at the surface. The reported periodicity is also questioned by Carpenter and McCarthy (1975), who found that the presence of *Trichodesmium* blooms correlates with sunny calm days and the periods of its absence with windy overcast days. Nitrogen fixation also apparently correlates with sea state. Stewart and Lex (1970) have found that forms without heterocysts can fix nitrogen under such microaerophilic conditions as occur in salt marshes. These are apparently the conditions under which the non-heterocystous *Trichodesmium* fixes nitrogen (Carpenter and McCarthy, 1975). The central areas of the filaments, which are held in bundles, lack photosynthetic activity and thus only fix nitrogen under reduced oxygen tension. Aeration of suspensions of *Trichodesmium* inhibits nitrogen fixation. The degree of nitrogen fixation in the Sargasso Sea during *Trichodesmium* blooms correlates well with sea state and any resultant agitation and aeration of the bundles of *Trichodesmium*.

The filamentous forms with heterocysts are included in the Families Nostocaceae, Rivulariaceae, and Scytonemataceae (Plate 8-7). The encrusting vegetation that discolors the rocks near the high tide line to form the "black zone" noted by Stephenson and Stephenson (1949, 1972) is caused by a mixture of dusky lichens, such as *Verrucaria*, and the heterocystous filaments of *Calothrix* (Allen, 1963; Mälmeström, 1972) of the Rivulariaceae family. The *Calothrix* extends into the sublittoral zone, where it is obscured by macroalgae. The *Calothrix* community in algal reef flats fixes atmospheric nitrogen at rates comparable to those of managed agriculture and apparently contributes to the high productivity of adjacent coral reefs and atoll lagoons (Wiebe, Johannes, and Webb, 1975). But filaments of *Calothrix*, presumably from such algal reefs, have been seen in nearshore plankton (Allen, 1963). Tufts of epiphytic blue-green bacteria, such as *Dichothrix* and *Calothrix*, often dominate the surface of the large floating patches of gulfweed (*Sargassum natans* and *Sargassum fluitans* originating in the Sargasso Sea) (Carpenter, 1972; Ryland, 1974) as well as benthic *Sargassum* (Feldmann, 1958). A small planktonic *Nostoc* has been reported in sizable numbers in the aphotic zone (beneath the illuminated depths) in upwelling areas off Monaco, off Algeria, and in the Indian Ocean, for example (Bernard and Lecal, 1960; Bernard, 1963). Planktonic marine diatoms in the genera *Rhizosolenia* and *Pleurosigma* can contain a blue-green form known as *Richelia intracellularis* (Plate 8-7,*A*), which is distinguished by its wide terminal heterocysts (Pascher, 1929).

The distribution of blue-green photobacteria in the marine environment is further discussed by Frémy (1929–1933), Fogg (1973), and Fogg et al. (1973). From a geological point of view, these photosynthetic bacteria have been involved in calcium carbonate deposition from the early Precambrian time, and they constitute the only fossil record for more than two and a half billion years of the earth's early history (Chapter 1; Golubic, 1973). These deposits occur as stromatolites, which are layered rocks and which, in the case of the cyanobacteria deposits, are carbonates; deposits produced by other microorganisms can be silicates, such as the minerals being formed in Yellowstone Park (Brock, 1973). As in Yellowstone, stromatolites are still being formed in certain marine areas, including Shark Bay, Eastern Australia (Davies, 1970). But the cyanobacteria, in addition to depositing carbonates, also bore through corals and carbonate rocks (Golubic, 1973). Pia (1937) has provided an extensive review of the early work on rock-destroying algae and fungi. Golubic, Perkins, and Lukas (1975) have updated the literature on these endolithic microorganisms. An ecological study on the distribution and relative importance of endolithic green algae,

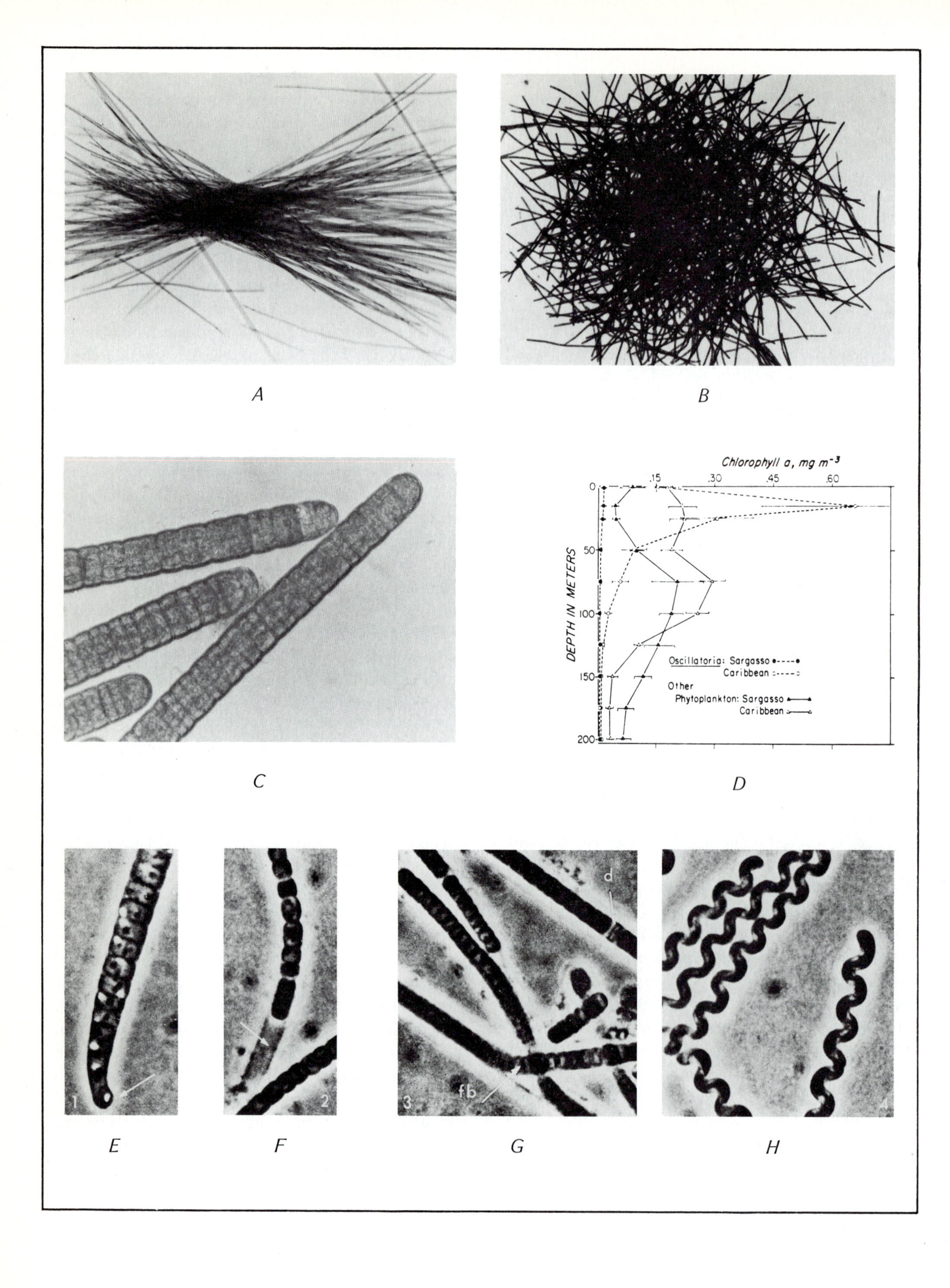
A
B
Chlorophyll a, mg m⁻³
.15
.30
.45
.60
0
50
100
150
200
DEPTH IN METERS
Oscillatoria: Sargasso
Caribbean
Other
Phytoplankton: Sargasso
Caribbean
C
D
1
2
d
3
fb
4
E
F
G
H

PLATE 8-6. Filamentous, non-heterocystous cyanobacteria. (*A*) Confluent and (*B*) radial trichomes in colonies of *Trichodesmium thiebautii* Gomont and (*C*) the trichomes of *Trichodesmium erythraeum* Ehrenberg (magnifications not given). (*D*) The distribution of *Trichodesmium* and other phototrophic biomass with depth in the Sargasso and Caribbean seas, as indicated by chlorophyll *a* concentration. The filaments of axenic non-heterocystous strains characteristic of the (*E*) *Oscillatoria*, (*F*) *Lyngbya*, (*G*) *Plectonema*, and (*H*) *Spirulina* types (*E–H*, 2,000 X). (*A,B,C*) courtesy of Ryuzo Marumo; *D* from E.J. Carpenter and C.C. Price, 1977, *Limnol. Oceanogr.* 22:60–72; (*E,H*) from C.N. Kenyon, R. Rippka, and R.Y. Stanier, 1972, *Arch. Mikrobiol.* 83:216–36.

cyanobacteria, and fungi in Atlantic and Pacific corals has been conducted by Lukas (1973).

Enrichment culture and plating on solid agar media have been used successfully to isolate cyanobacteria for many decades. The techniques described by Mary Belle Allen (1952) were the basis for procedures used to obtain an assortment of unicellular and filamentous forms from the estuarine environment by van Baalen (1962). The eucaryotic microalgae, which also grow in the nutritionally nonselective media used, are greatly inhibited by elevating the temperature of incubation to the upper temperate limits of the cyanobacteria (Allen and Stanier, 1968). A most important observation was that a marked inhibition on agar media was completely overcome by separately sterilizing double-strength mineral salts and agar solutions and keeping agar concentration of the media between 1 and 1.5% (Allen, 1968). These media permit the quantitative estimation of chroococcoid blue-green populations by plate counts. Procedures for the enrichment, isolation, and purification of the cyanobacteria have been summarized by Allen (1973). Many grow better in the presence of other bacteria, and a species of *Nostoc* has been shown to be dependent upon a species of the stalked bacterium *Caulobacter* [presumably due to the latters' production of an auxin-like hormone (Bunt, 1961)]. This indicates that we still have much to learn about the culture of these bacteria, including the finding of adequate agents to suppress the chemoorganotrophic bacteria (Tchan and Gould, 1961). The coccoid forms with generation times approaching 2 hours have been the main type isolated from marine environments (van Baalen, 1962), whereas *Trichodesmium* has resisted many attempts at culture, although there is one report that it has been obtained in pure culture (Ramamurthy, 1972).

The blue-green photobacteria have the same two photosystems and noncyclic phosphorylation scheme seen in the unicellular and multicellular eucaryotic phototrophs. Photosystem II has been considered essential for the growth of cyanobacteria in the presence of hydrogen sulfide; it can provide electrons from hydrogen sulfide when reduction from the photolysis of water is greatly curtailed (Stewart and Pearson, 1970). The work of Cohen, Padan, and Shilo (1975) has shown that "anoxygenic" photosynthesis with Photosystem I does occur in the cyanobacteria, which also partially explains why these bacteria are found in reducing habitats, such as sulfide-containing muds, as well as in aerobic environments.

The classification of the cyanobacteria is at a crossroads. Current taxonomic criteria for bacteria are just now being applied to characterize the form genera as classified by classical botanical morphology (Stanier et al., 1971; Kenyon, Rippka, and Stanier, 1972). According to the International Code of Botanical Nomenclature, the starting point for the taxonomy of non-heterocystous filamentous forms is the treatise by Gomont (1892), whereas the starting point for the nomenclature of heterocystous filamentous forms is the review of Bornet and Flahault (1886–1888). No starting point has been designated for the coccoid forms in the Chroococcales. Desikachary (1959) has proposed the large monograph by Lemmermann (1907) that covers all three forms, whereas Stanier et al. (1971) want to go back to the earlier, shorter treatise by Näegli (1849). Whitton (1969) states that the nomenclatural system frequently proves unsatisfactory and may be a positive hindrance to advancing understanding of the general biology of these microorganisms. He suggests a twofold approach of using a binomial when desired (based on the monograph by Geitler, 1932) while standardized criteria are used for numerical taxonomy. Except for the starting point, this is exactly what Stanier et al. (1971) and Kenyon, Rippka, and Stanier (1972) are now doing with the coccoid and filamentous forms, respectively. These studies have shown that, as with most other bacteria, morphology alone is insufficient for characterization. Other characteristics used were DNA base composition, fatty acid composition, pigmentation, vitamin requirements, and the ability to fix atmospheric nitrogen. Analyses and observations cannot be made on "unialgal" clones contaminated with other bacteria.

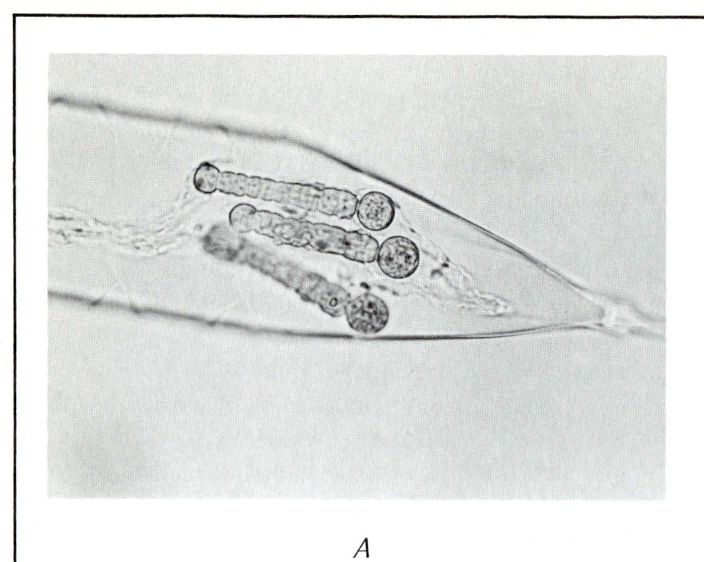

A

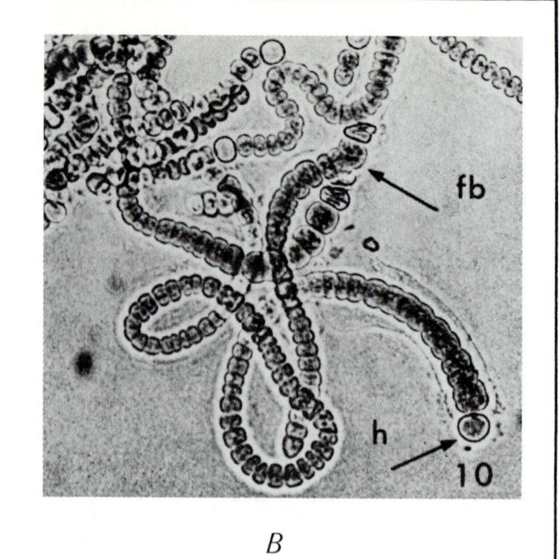

B

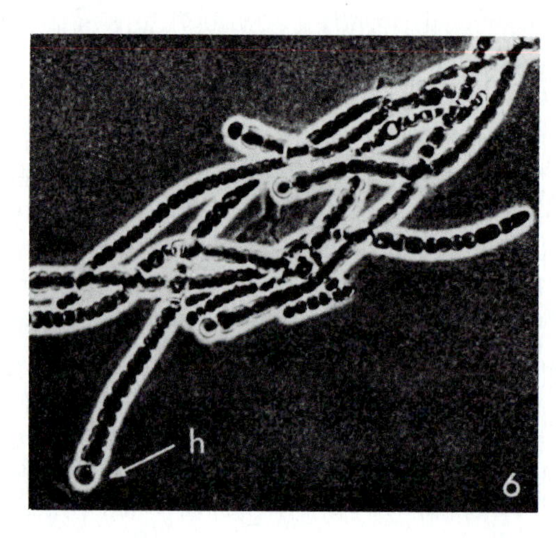

C

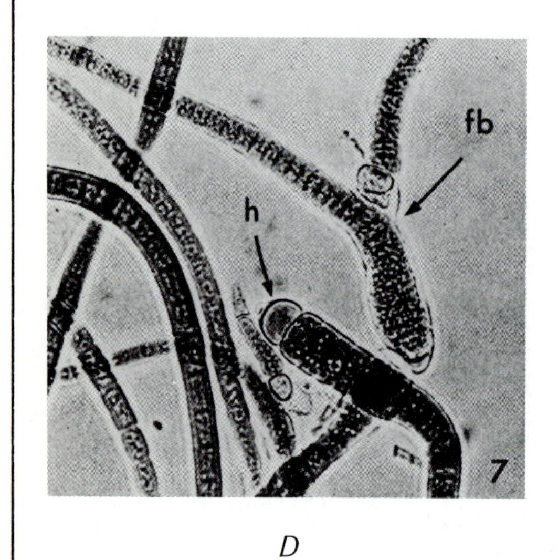

D

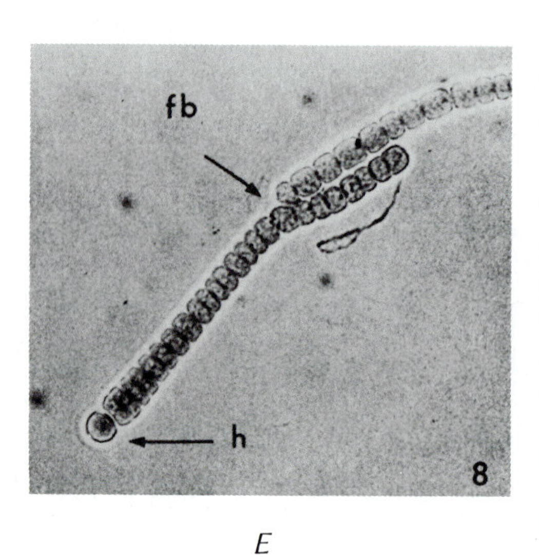

E

PLATE 8-7. Filamentous, heterocystous cyanobacteria. (*A*) *Richelia intracellularis* in the diatom *Rhizosolenia* from Caribbean plankton. Terminal heterocysts are 10 μm. The filaments of axenic heterocystous strains characteristic of (*B*) *Microchaete* and (*C*), (*D*), and (*E*) *Calothrix* types (fb, false branching; h, heterocyst) (*B,D,E* 600 X; *C* 320 X). (*A*) courtesy of Edward Carpenter. (*B–E*) from C.N. Kenyon, R. Rippka and R.Y. Stanier, 1972, *Arch. Mikrobiol.* 83:216–36.

REFERENCES

Allen, M.B. 1952. The cultivation of Myxophyceae. *Arch. Mikrobiol.* 17:34–53.

Allen, M.B. 1963. Nitrogen fixing organisms in the sea. In: *Marine Microbiology* (C.H. Oppenheimer, ed.). C.C. Thomas, Springfield, Ill., pp. 85–92.

Allen, M.M. 1968. Simple conditions for growth of unicellular blue-green algae on plates. *J. Phycol.* 4:1–4.

Allen, M.M. 1973. Methods for Cyanophyceae. In: *Handbook of Phycological Methods* (J.R. Stein, ed.). Cambridge University Press, pp. 127–38.

Allen, M.M., and R.Y. Stanier. 1968. Selective isolation of blue-green algae from water and soil. *J. gen. Microbiol.* 51:203–9.

Anagnostidis, K., and G.H. Schwabe, 1966. Über artenreiche Bestände von Cyanophyten und Bacteriophyten in einem Farbstreifensandwatt sowie über das Auftreten *Gomontiella*-Artig deformierter *Oscillatoria* trichome. *Nova Hedwigia* 11:417–41.

Baas-Becking, L.G.M. 1951. Notes on some Cyanophyceae of the Pacific region. *Proc. Kon. Ned. Akad. Wetenschap. Ser. C.* 54:213–25.

Baas-Becking, L.G.M., and E.J. Ferguson Wood. 1955. Biological processes in the estuarine environment. I,II. Ecology of the sulfur cycle. *Koninkl. Ned. Akad. Wetenschap. Proc. Ser. B.* 58:160–81.

Barth, R. 1967. Observaçoes sobre ocorrência em massa de Cyanophycea. *Publ. Inst. Pesq. Mar.* 6:1–8.

Bernard, F. 1963. Density of flagellates and myxophyceae in the heterotrophic layers related to environment. In: *Marine Microbiology* (C.H. Oppenheimer, ed.). C.C. Thomas, Springfield, Ill., pp. 215–28.

Bernard, F., and J. Lecal. 1960. Plancton unicellulaire récolté dans l'océan Indien par le Charcot (1950) et le Norsel (1955–56). *Bull. Inst. Océan.* 1166:1–59.

Bornet, E., and C. Flahault. 1886–88. Révision des Nostocacées Heterocystées. *Ann. Sci. Nat. Bot. VII,* 3:323–381; 4:343–373; 5:51–129; 7:177–262. (Reprinted by J. Cramer, Weinheimer, 1959.)

Brock, T.D. 1970. *Biology of Microorganisms* (1st Ed.). Prentice-Hall, Englewood Cliffs, N.J., 737 p.

Brock, T.D. 1973. Evolutionary and ecological aspects of the cyanophytes. In: *The Biology of Blue-green Algae* (N.G. Carr and B.A. Whitton, eds.). University of California Press, Berkeley and Los Angeles, pp. 487–500.

Buchanan, R.E., and N.E. Gibbons (eds.). 1974. *Bergey's Manual of Determinative Bacteriology* (8th Ed.). Williams & Wilkins, Baltimore, Md., 1246 p.

Bunt, J.S. 1961. Blue-green algae—growth. *Nature* 192:1274–75.

Carpenter, E.J. 1972. Nitrogen fixation by a blue-green epiphyte on pelagic *Sargassum. Science* 178:1207–9.

Carpenter, E.J. 1973. Nitrogen fixation by *Oscillatoria* (*Trichodesmium*) *thiebautii* in the southwestern Sargasso Sea. *Deep-Sea Res.* 20:285–88.

Carpenter, E.J., and J.J. McCarthy. 1975. Nitrogen fixation and uptake of combined nitrogenous nutrients by *Oscillatoria* (*Trichodesmium*) *thiebautii* in the western Sargasso Sea. *Limnol. Oceanogr.* 20:389–401.

Carpenter, E.J., and C.C. Price, IV. 1977. Nitrogen fixation, distribution, and production of *Oscillatoria* (*Trichodesmium*) spp. in the western Sargasso and Caribbean Seas. *Limnol. Oceanogr.* 22:60–72.

Clayton, R.K. 1959. Phototaxis of purple bacteria. In: *Handbuch der Pflanzenphysiologie* (W. Ruhland, ed.), Vol. 17. Springer-Verlag, Berlin, pp. 371–87.

Cohen, Y., E. Padan, and M. Shilo. 1975. Facultative anoxygenic photosynthesis in the cyanobacterium *Oscillatoria limnetica. J. Bacteriol.* 123:855–61.

Cohen-Bazire, G., and W.R. Sistrom. 1966. The procaryotic photosynthetic apparatus. In: *The Chlorophylls* (L.P. Vernon and G.R. Seely, eds.). Academic Press, London and New York, pp. 313–41.

Cohen-Bazire, G., R. Kunisawa, and N. Pfennig. 1969. Comparative study of the structure of gas vacuoles. *J. Bacteriol.* 100:1049–61.

Davies, G.R. 1970. Algal–laminated sediments, Gladstone Embayment, Shark Bay, Western Australia. In: *Carbonate Sedimentation and Environments, Shark Bay, Western Australia* (B.W. Logan, G.R. Davies, J.F. Read, and D.E. Cebulski, eds.), vol. 13. Amer. Assn. Petrol. Geol. Mem., Tulsa, Okla., pp. 169–205.

Desikachary, T.V. 1959. *Cyanophyta.* Indian Council of Agr. Res., New Delhi, 686 p.

Desikachary, T.V. 1973. Status of classical taxonomy. In: *The Biology of Blue-green Algae* (N.G. Carr and B.A. Whitton, eds.). University of California Press, Berkeley and Los Angeles, pp. 473–81.

Drews, G. 1973. Fine structure and chemical composition of the cell envelopes. In: *The Biology of Blue-green Algae* (N.G. Carr and B.A. Whitton, ed.). University of California Press, Berkeley and Los Angeles, pp. 99–116.

Drouet, F. 1968. *Revision of the Classification of the Oscillatoriaceae,* Monograph 15. Academy of Natural Sciences, Philadelphia. 370 p.

Drouet, F. 1973. *Revision of the Nostocaceae with Cylindrical Trichomes.* Hafner Press, New York, 292 p.

Drouet, F., and W.A. Daily. 1956. *Revision of the coccoid myxophyceae. Butler University Botanical Studies, Vol. 12,* 218 p.

(Revisions and Corrections, Vol. 26(2), 1957; facsimile by Hafner Press, 1973.)

Echlin, P., and I. Morris. 1965. The relationship between blue-green algae and bacteria. *Biol. Rev.* 40:143–87.

Ehrenberg, C.G. 1838. *Die Infusionsthierchen als volkommene Organismen.* Verlag Leopold Voss, Leipzig, 547 p.

Fay, P., W.D.P. Stewart, A.F. Walsby, and G.E. Fogg. 1968. Is the heterocyst the site of nitrogen fixation in blue-green algae? *Nature* 220:810–12.

Feldmann, J. 1958. Les Cyanophycées marine de la Guadeloupe. *Rev. Algol. Nov. Ser.* 4:25–40.

Fenchel, T. 1968. The ecology of marine microbenthos. II. The food of marine benthic ciliates. *Ophelia* 5:73–121.

Fenchel, T. 1969. The ecology of marine microbenthos. IV. Structure and function of the benthic ecosystem, its chemical and physical factors and the microfauna communities with special reference to the ciliated protozoa. *Ophelia* 6:1–82.

Fenchel, T.M., and R.J. Riedl. 1970. The sulfide system: A new biotic community underneath the oxidized layer of marine sand bottom. *Mar. Biol.* 7:255–68.

Fenchel, T., and B.J. Straarup. 1971. Vertical distribution of photosynthetic pigments and the penetration of light in marine sediments. *Oikos* 22:172–82.

Fogg, G.E. 1973. Physiology and ecology of marine blue-green algae. In: *The Biology of Blue-green Algae* (N.G. Carr and B.A. Whitton, eds.). University of California Press, Berkeley, pp. 368–78.

Fogg, G.E., W.D.P. Stewart, P. Fay, and A.E. Walsby. 1973. *The Blue-green Algae.* Academic Press, London and New York. 459 p.

Frémy, P. 1929–33. *Cyanophycées des côtes d'Europe.* (Saint-Lo Edition reprinted by A. Asher, Amsterdam, 1975, 232 p.)

Geitler, L. 1932. Cyanophyceae. In: *Kryptogamenflora von Deutschland, Österreich und der Schweiz* (L. Rabenhorst, ed.), Vol. 14. Akad. Verlagsgesellschaft, Leipzig.

Genovese, S. 1963. The distribution of the H_2S in the Lake of Faro (Messina) with particular regard to the presence of "red water." In: *Marine Microbiology* (C.H. Oppenheimer, ed.). C.C. Thomas, Springfield, Ill., pp. 194–228.

Gest, H., and M.C. Kamen. 1960. The photosynthetic bacteria. In: *Handbuch der Pflanzenphysiologie* (W. Ruhland, ed.), Vol. 5. Springer-Verlag, Berlin, pp. 568–612.

Gest, H., A. San Pietro, and L.P. Vernon. 1963. *Bacterial Photosynthesis.* Antioch Press, Yellow Springs, Ohio, 523 p.

Gibbons, N.E., and R.G.E. Murray. 1978. Proposals concerning the higher taxa of bacteria. *Int. J. Syst. Bacteriol.* 28:1–6.

Gloe, A., N. Pfennig, H. Brockmann, Jr., and W. Trowitzsch. 1975. A new bacteriochlorophyll from brown-colored Chlorobiaceae. *Arch. Mikrobiol.* 102:103–9.

Golubic, S. 1973. The relationship between blue-green algae and carbonate deposition. In: *The Biology of Blue-green Algae* (N.G. Carr and B.A. Whitton, eds.). University of California Press, Berkeley and Los Angeles, pp. 434–72.

Golubic, S., R.D. Perkins, and K.J. Lukas. 1975. Boring microorganisms and microborings in carbonate substrates. In: *The Study of Trace Fossils; a Synthesis of Principles, Problems and Procedures in Ichnology* (R.W. Frey, ed.). Springer-Verlag, New York, pp. 229–59.

Gomont, M. 1892. Monographie des Oscillariées. *Ann. Sci. Nat. (Bot.)* 15:263–368; 16:91–264.

Gorlenko, V.M. 1968. Photosynthetizing sulfur bacteria from south Crimean basins. *Microbiology* 37:617–20.

Gorlenko, V.M. 1970. A new phototrophic green sulphur bacterium—*Prosthecochloris aestuarii* nov. gen. nov. spec. *Zeits. Allg. Mikrobiol.* 10:147–49.

Gorlenko, V.M., and T.N. Zhilina. 1968. Study of the ultrastructure of green sulfur bacteria, strain SK-413. *Microbiology* 37:892–97.

Gray, B.H., C.A. Lipschultz, and E. Gantt. 1973. Phycobilisomes from a blue-green alga *Nostoc* species. *J. Bacteriol.* 116:471–78.

Halfen, L.N., and R.W. Castenholz. 1970. Gliding in a blue-green alga: A possible mechanism. *Nature* 225:1163–64.

Hansen, T.A., and H. van Gemerden. 1972. Sulfide utilization by purple nonsulfur bacteria. *Arch. Mikrobiol.* 86:49–56.

Hansen, T.A., and H. Veldkamp. 1972. A new type of photosynthetic purple bacterium. *Antonie van Leeuwenhoek* 38:629.

Hansen, T.A., and H. Veldkamp. 1973. *Rhodopseudomonas sulfidophila* nov. spec., a new species of the purple nonsulfur bacteria. *Arch. Mikrobiol.* 92:45–58.

Hutner, S.H. 1962. Nutrition of protists. In: *This is Life* (W.H. Johnson and W.C. Steere, eds.). Holt, Rinehart and Winston, New York, pp. 109–37.

Johnson, P.W., and J.McN. Sieburth. 1976. *In-situ* morphology of nitrifying-like bacteria in aquaculture systems. *Appl. Environ. Microbiol.* 31:423–32.

Kenyon, C.N., R. Rippka, and R.Y. Stanier. 1972. Fatty acid composition and physiological properties of some filamentous blue-green algae. *Arch. Mikrobiol.* 83:216–36.

Kondrat'eva, E.N. 1963/65. *Photosynthetic bacteria.* Akad. Nauk. SSSR, Moskva (in Russian, 1963, English transl. 1965, Israel Progr. Sci. Transl., Jerusalem, 1965), 243 p.

Lang, N.J. 1968. The fine structure of blue-green algae. *Ann. Rev. Microbiol.* 22:15–46.

Lang, N.J., and P. Fay. 1971. The heterocysts of blue-green algae. II. Details of ultrastructure. *Proc. Royal Soc. London Ser. B* 178:193–203.

Lang, N.J., and B.A. Whitton. 1973. Arrangement and structure of thylakoids. In: *The Biology of Blue-green Algae* (N.G. Carr and B.A. Whitton, eds.). University of California Press, Berkeley and Los Angeles, pp. 66–79.

Larsen, H. 1952. On the culture and general physiology of the green sulfur bacteria. *J. Bacteriol.* 64:187–96.

Lemmermann, E. 1907. Algen I. In: *Kryptogamenflora der Mark Brandenberg, Vol. 3.* Borntraeger, Leipzig, 712 p.

Lewin, R.A. 1975. A marine *Synechocystis* (Cyanophyta, Chroococcales) epizoic on ascidians. *Phycologia,* 14:153–160.

Lewin, R.A. 1976. Naming the blue-greens. *Nature* 259:360.

Lewin, R.A. 1977. *Prochloron*, type genus of the Prochlorophyta. *Phycologia*, 16:217.

Lewin, R.A., and L. Cheng. 1975. Associations of microscopic algae with didemnid ascidians. *Phycologia*, 14:149–52.

Lukas, K.J. 1973. Taxonomy and ecology of the endolithic microflora of reef corals with a review of the literature on endolithic microphytes. Ph.D. Thesis, University of Rhode Island, 159 p.

Malmeström, B. 1972. The genus *Calothrix* in the black zone. *Bot. Mar.* 15:87–90.

Marumo, R. (ed.). 1975. *Report of Studies on the Community of Marine Pelagic Blue-green Algae, 1972–1974*. Ocean Research Inst., University of Tokyo, Tokyo, 77 p.

Matheron, R., and R. Baulaigue. 1972. Bacteriés photosynthetiques sulfureuses marines. *Arch. Mikrobiol.* 86:291–304.

Molisch, H. 1907. *Die Purpurbakterien nach neuen Untersuchungen*. G. Fischer, Jena, I–VII, pp. 1–95.

Näegli, C. 1849. Gattungen einzelligen Algen, physiologische und systematisch bearbeit. *Neue Denkschr. allg. Schweiz.-ges. Naturw.* 10:139.

Newcomb, E.H., and T.D. Pugh. 1975. Blue-green algae associated with ascidians of the Great Barrier Reef. *Nature*, 253:533–34.

OhEocha, C. 1965. Phycobilins. In: *Chemistry and Biochemistry of Plant Pigments* (T.W. Goodwin, ed.). Academic Press, New York and London, pp. 175–96.

Pascher, A. 1929. Studien über Symbiosen. Uber einige Endosymbiosen von Blaualgen in Einzellern. *Jahrb. wiss. Bot.* 71:386–462.

Pelsh, A.D. 1937. Photosynthetic sulfur bacteria of the eastern reservoir of Lake Sakskoe. *Mikrobiologiya* 6:1090–1100.

Pfennig, N. 1961. Eine vollsynthetische Nahrlosung zur selektiven Anreicherung einiger Schwefelpurpurbakterien. *Naturwissenschaften* 48:136.

Pfennig, N. 1965. Anreicherungs kulturen für rote und grüne Schwefelbakterien. *Zentr. Bakteriol. Parasitenk. Abt. I.* Suppl. 1, 179–89, 503–4.

Pfennig, N. 1967. Photosynthetic bacteria. *Ann. Rev. Microbiol.* 21:285–324.

Pfennig, N. 1977. Phototrophic green and purple bacteria: A comparative, systematic survey. *Ann. Rev. Microbiol.* 31:275–90.

Pfennig, N., and H.G. Trüper. 1971a. New nomenclatural combinations in the phototrophic bacteria. *Int. J. Syst. Bacteriol.* 21:11–14.

Pfennig, N., and H.G. Trüper. 1971b. Conservation of the family name Chromatiaceae Bavendamm 1924 with the type genus *Chromatium* Perty 1852. *Int. J. Syst. Bacteriol.* 21:15–16.

Pfennig, N., and H.G. Trüper. 1971c. Higher taxa of the phototrophic bacteria. *Int. J. Syst. Bacteriol.* 21:17–18.

Pfennig, N., and H.G. Trüper. 1971d. Type and neotype strains of the species of phototrophic bacteria maintained in pure culture. *Int. J. Syst. Bacteriol.* 21:19–24.

Pfennig, N., and H.G. Trüper. 1973. The Rhodospirillales (phototrophic or photosynthetic bacteria). In: *Handbook of Microbiology Vol. I. Organismic Microbiology* (A.I. Laskin and H.A. Lechevalier, eds.). CRC Press, Cleveland, Ohio, pp. 17–27.

Pia, J. 1937. Die kalklösenden Thallophyten. *Arch. Hydrobiol.* 31:264–328.

Ramamurthy, V.D. 1972. On the culture of the marine blue-green alga, *Trichodesmium erythraeum* Ehr. In: *Taxonomy and Biology of Blue-green Algae* (T.V. Desikachary, ed.). University of Madras, India, pp. 425–27.

Remsen, C.C., S.W. Watson, J.B. Waterbury, and H.G. Trüper. 1968. Fine structure of *Ectothiorhodospira mobilis* Pelsh. *J. Bacteriol.* 95:2374–92.

Ryland, J.S. 1974. Observations on some epibionts of gulf-weed, *Sargassum natans* (L.) Meyen. *J. exp. mar. Biol. Ecol.* 14:17–25.

Sachs, J. 1874. *Lehrbuch der Botanik*, 4 Aufl. W. Engelmann, Leipzig, pp. 1–928.

San Pietro, A. (ed.). 1971. Photosynthesis. Part A. In: *Methods in Enzymology* (S.P. Colowick and N.O. Kaplan, eds.-in-chief), Vol. 23. Academic Press, New York and London, 743 p.

Schulz-Baldes, M., and R.A. Lewin. 1976. Fine structure of *Synechocystis didemni* (Cyanophyta: Chroococcales). *Phycologia*, 15:1–6.

Sieburth, J.McN., and J.T. Conover. 1965. Slicks associated with *Trichodesmium* blooms in the Sargasso Sea. *Nature* 205:830–31.

Sorensen, L.O., with J.T. Conover. 1962. Algal mat communities of *Lyngbya confervoides* (C. Agardh) Gomont. *Pub. Inst. Mar. Sci.* 8:61–74.

Sournia, A. 1968. La cyanophycée *Oscillatoria* (= *Trichodesmium*) dans le plancton marin: Taxonomie, et observations dans le Canal de Mozambique. *Nova Hedwigia* 15:1–12.

Stanier, R.Y., and G. Cohen-Bazire. 1977. Phototrophic procaryotes: The Cyanobacteria. *Ann. Rev. Microbiol.* 31:225–74.

Stanier, R.Y., R. Kunisawa, M. Mandel, and G. Cohen-Bazire. 1971. Purification and properties of unicellular blue-green algae (order Chroococcales). *Bacteriol. Rev.* 35:171–205.

Stephenson, T.A., and A. Stephenson. 1949. The universal features of zonation between tide-marks on rocky coasts. *J. Ecology* 37:289–305.

Stephenson, T.A., and A. Stephenson. 1972. *Life between Tidemarks on Rocky Shores*. W.H. Freeman, San Francisco, Ca., 425 p.

Steven, D.M., and R. Glombitza. 1972. Oscillatory variation of a phytoplankton in a tropical ocean. *Nature* 237:105–7.

Stewart, W.D.P., and M. Lex. 1970. Nitrogenase activity in the blue-green alga *Plectonema boryanum* Strain 594. *Arch. Mikrobiol.* 73:250–60.

Stewart, W.D.P., and H.W. Pearson. 1970. Effects of aerobic and anaerobic conditions on growth and metabolism of blue-green algae. *Proc. Royal Soc. London Ser. B* 175:293–311.

Stewart, W.D.P., A. Haystead, and H.W. Pearson. 1969. Nitrogenase activity in heterocysts of blue-green algae. *Nature* 224:226–28.

Takahashi, M., and S. Ichimura. 1968. Vertical distribution and organic matter production of photosynthetic sulfur bacteria in Japanese lakes. *Limnol. Oceanogr.* 13:644–55.

Taylor, B.G., C.C. Lee, and J.S. Bunt. 1973. Nitrogen-fixation associated with the marine blue-green alga, *Trichodesmium* as measured by the acetylene-reduction technique. *Arch. Mikrobiol.* 88:205–12.

Tchan, Y.T., and J. Gould. 1961. Use of antibiotics to purify blue-green algae. *Nature* 192:1276.

Trüper, H.G. 1968. *Ectothiorhodospira mobilis* Pelsh, a photosynthetic bacterium depositing sulfur outside the cells. *J. Bacteriol.* 95:1910–20.

Trüper, H.G. 1970. Culture and isolation of phototrophic sulfur bacteria from the marine environment. *Helgoländ. wiss. Meeresunters.* 20:6–16.

Trüper, H.G., and S. Genovese. 1968. Characterization of photosynthetic sulfur bacteria causing red water in Lake Faro (Messina, Sicily). *Limnol. Oceanogr.* 13:225–32.

Trüper, H.G., and H.W. Jannasch. 1968. *Chromatium buderi* nov. spec., eine neue Art der "grossen" Thiorhodaceae. *Arch. Mikrobiol.* 61:363–72.

Trüper, H.G., and N. Pfennig. 1971. Family of phototrophic green sulfur bacteria: Chlorobiaceae Copeland, the correct family name; rejection of Chlorobacterium Lauterborn; and the taxonomic situation of the consortium-forming species. *Int. J. Syst. Bacteriol.* 21:8–10.

Trüper, H.G., and H.G. Schlegel. 1964. Sulphur metabolism in Thiorhodaceae. I. Quantitative measurements on growing cells of *Chromatium okenii. Antonie van Leeuwenhoek* 30:225–38.

Trüper, H.G., and C.S. Yentsch. 1967. Use of glass fiber filters for the rapid preparation of *in vivo* absorption spectra of photosynthetic bacteria. *J. Bacteriol.* 94:1255–56.

van Baalen, C. 1962. Studies on marine blue-green algae. *Bot. Mar.* 4:129–39.

van Baalen, C., and R.M. Brown, Jr. 1969. The ultrastructure of the marine blue-green alga, *Trichodesmium erythraeum*, with special reference to the cell wall, gas vacuoles and cylindrical bodies. *Arch. Mikrobiol.* 69:79–91.

van Gemerden, H. 1968. Growth measurements of *Chromatium* cultures. *Arch. Mikrobiol.* 4:103–10.

van Niel, C.B. 1932. On the morphology and physiology of the purple and green sulphur bacteria. *Arch. Mikrobiol.* 3:1–112.

van Niel, C.B. 1944. The culture, general physiology, morphology and classification of the non-sulphur purple and brown bacteria. *Bacteriol. Rev.* 8:1–118.

van Niel, C.B. 1971. Techniques for the enrichment, isolation and maintenance of the photosynthetic bacteria. In: *Methods in Enzymology* (A. San Pietro, ed.), Vol. 23, Part A. Academic Press, New York and London, pp. 3–28.

Waaland, J.R., S.D. Waaland, and D. Branton. 1971. Gas vacuoles. Light shielding in blue-green algae. *J. Cell Biol.* 48:212–15.

Wallroth, F.G. 1833. *Flora Cryptogamica Germaniae Algues et Champignona* 2:182.

Walsby, A.E. 1968. Mucilage secretion and the movements of blue-green algae. *Protoplasma* 65:223–38.

Walsby, A.E. 1973. Gas vacuoles. In: *The Biology of Blue-green Algae* (N.G. Carr and B.A. Whitton, eds.). University of California Press, Berkeley and Los Angeles, pp. 340–52.

Warming, E. 1875. Om nogle ved Danmarks kyster levende baketerier. *Vidensk. Meddr. dansk naturh. Foren.* 20/28:3–116.

Whitton, B.A. 1969. The taxonomy of blue-green algae. *Br. phycol. J.* 4:121–23.

Wiebe, W.J., R.E. Johannes, and K.L. Webb. 1975. Nitrogen fixation in a coral reef community. *Science* 188:257–59.

Winogradsky, S. 1888. *Beiträge zur Morphologie und Physiologie der Bacterien. Heft I. Zur Morphologie und Physiologie der Schwefelbakterien.* Arthur Felix, Leipzig, 120 p.

Wolk, C.P. 1973. Physiology and cytological chemistry of blue-green algae. *Bacteriol. Rev.* 37:32–101.

Wood, E.J.F. 1965. *Marine Microbial Ecology.* Reinhold, New York, 243 p.

CHAPTER 9

Chemolithotrophs

The existence of bacteria that derive their energy from the oxidation of ammonia, nitrite, and the sulfur compounds was demonstrated by the pioneering studies of Winogradsky in the 1880s (Winogradsky, 1888, 1890). From this work evolved the concept of obligate autotrophy, a nutritional mode in which these *inorgoxidants* apparently derived all their nutrients from inorganic substances. Today we realize that all the autotrophic microorganisms that have been thoroughly tested both assimilate and metabolize organic substances to some degree, i.e., they are mixotrophic (Rittenberg, 1969, 1972; Kelly, 1971). The importance of chemolithotrophic bacteria lies in their ability to detoxify and transform some of the end products of metabolism of other organisms and to produce bacterial biomass for phagotrophic microorganisms in their habitats. The microorganisms and the mechanisms involved in chemolitho-autotrophy, which include nitrifying, sulfur-oxidizing, iron-oxidizing, and hydrogen bacteria, have been reviewed by Schlegel (1975). Little is known about the iron-oxidizing and hydrogen bacteria in the sea, and these forms are not discussed here. Most of our knowledge of these bacteria concerns those that transform inorganic compounds of nitrogen and sulfur. It is relatively easy to change the valence of these elements from one oxidation state to another, an ability a number of bacteria possess.

A. NITRIFYING BACTERIA

Nitrification is the process whereby ammonia, the end product of animal metabolism and the decay of proteinaceous materials, is oxidized through nitrite to nitrate. The nitrifying bacteria occur in two physiological groups, which coexist in the nitrifying community. The ammonia oxidizers carry out the reaction:

NH_3	$\rightarrow NH_2OH \rightarrow$	$H_2N_2O_2$ [or $2(NOH)$]	$\rightarrow NO_2'$
ammonia	hydroxylamine	? intermediate	nitrite

in which ammonia is first converted to hydroxylamine with the reduction of cytochrome C and then to a labile intermediate before nitrite is formed. The nitrite oxidizers may also use the cytochrome C system in forming nitrate:

$$\begin{array}{ccccc} NO_2^- & \rightarrow & NO_2 + \frac{1}{2}O_2 & \rightarrow & NO^-_3 \\ Fe^{3+}Cyt_{551} & \curvearrowright & Fe^{2+}Cyt_{551} & \curvearrowright & Fe^{3+}Cyt_{551} \end{array}$$

via another unknown intermediate (Wallace and Nicholas, 1969). The energy obtained from these reactions (which is low) is used to fix carbon dioxide and to synthesize organic carbon compounds. The nitrifying bacteria are recognized on the basis of the substrate oxidized, cell morphology, and ultrastructure. All are placed in the family Nitrobacteriaceae (Watson, 1971; *Bergey's 8th*). These gram-negative aerobic bacteria are sometimes flagellated, although they may occur more often under natural conditions as colonies in a slime layer. There are two types of colonies, a loose mass of zoogloeae or a tighter colony with the cells in a stainable polysaccharide matrix. The latter is called a cyst or a cyst-type colony, which contains actively dividing cells despite the connotation of its name.

A good discussion of the early studies on nitrifying bacteria in the sea is provided by Carey (1938). Nitrifying bacteria require a habitat where ammonia is produced and permitted to accumulate to levels that will enrich them. In the sea, the end products of animal metabolism must be a major source of this ammonia. Immediately following the ingestion of food, ammonia is rapidly released from the gills and excretory organs of marine animals. Ideal habitats for studying the nitrification of this ammonia are rearing facilities in which closed-system, high density aquaculture (mariculture) is being practiced. It has been assumed that the nitrifying bacteria in these systems are associated mainly with the large surface areas provided by "biological filters" (Meade, 1974). But an ultrastructural study of nitrifying bacteria in the microbiological film of the filter and suspended in the water column (Plate 9-1) indicated that the nitrifying bacteria in the film had an aberrant morphology, with invaginated lamellae (Plate 9-1,*B*), compared to the "normal" ultrastructure (Plate 9-1,*C*) of the suspended nitrifiers (Johnson and Sieburth, 1976). In experiments in which varying proportions of the suspended microflora were flushed out, most of the nitrification was by the suspended nitrifiers. The importance of the populations in the films may be as a reservoir or an inoculum for the water column. The low but consistent population in the open sea (Watson, 1965) indicates that some ammonia-rich surface must be acting as a reservoir. One might speculate that the presence of spheres the size of colonies of nitrifying bacteria on the surface of animals (Sieburth, 1975) may indicate that the gills and excretory areas of animals can act as such reservoirs for nitrifying bacteria.

Another major source of ammonia is the nitrogen-containing debris that sediments to the sea floor. This arises from fecal material and from the tissue of plants, animals, and microorganisms that are injured, senescent, or dead. During decomposition by the rich bacterial flora in the sediments, ammonia is released by deamination and, in turn, supports a nitrifying microflora. Although nitrifying bacteria are aerobic, and although they decrease in number with increasing depth into the sediment, they can occur in large numbers in sulfide-containing muds (Kawai, Yoshida, and Kimata, 1968) and in seawater at low oxygen concentrations (Sugahara, Sugiyama, and Kawai, 1974).

In addition to the ammonia used by nitrifying bacteria, many phototrophic and osmotrophic microorganisms use ammonia, as well as the nitrate formed by nitrification, as a direct source of nitrogen. Nitrate formed in the surface layers of sediment can be removed as nitrogen gas under anaerobic conditions through denitrification by such facultatively anaerobic organotrophs as *Pseudomonas stutzeri* and *Thiobacillus denitrificans.* This process requires an adequate amount of utilizable carbon (Balderston and Sieburth, 1976) and is inhibited by photosynthetic oxygen production (Jannasch, 1960). The ecological significance of inorganic nitrogen transformation has been discussed by Vaccaro (1965).

Oceanic nitrifying bacteria have been enriched for by adding either ammonia or nitrite (along with phosphate) to 5-liter seawater samples. Their presence is detected by following either nitrite or nitrite and nitrate concentrations (Watson, 1965; Watson and Waterbury, 1971). When there was either an exponential increase in nitrite indicating the growth of ammonia-oxidizing bacteria or when more than one-half of the nitrite had been oxidized to nitrate, the transfer and purification of the nitrifying bacteria was attempted with serial decimal dilutions, which were scaled up to 1-liter fermenters or larger chemostats in order to harvest enough cells for electron microscopy. Stock cultures of ammonia-oxidizing bacteria have been maintained in pH-stated, 1-liter fermenters, whereas 100-ml batch culture vessels were adequate for the nitrite-oxidizing bacteria. Fermenters

PLATE 9-1. The ultrastructure of a species of *Nitrosomonas* in a high-density, closed-system for salmon aquaculture, with an ammonia concentration of approximately 1 mg/liter. (*A*) Low power transmission electron micrograph of a microbial film on a filter surface, showing a mixture of heterotrophic forms as well as a colony of nitrifiers (arrow). (*B*) The "aberrant" forms of *Nitrosomonas* in the film, possessing invaginated peripheral cytomembranes. (*C*) The "normal" forms suspended in the water column, possessing smooth lamellae. (*D*) These cells deteriorated after 7 days' enrichment in 500 mg of ammonia/liter. Electron micrographs courtesy of Paul W. Johnson.

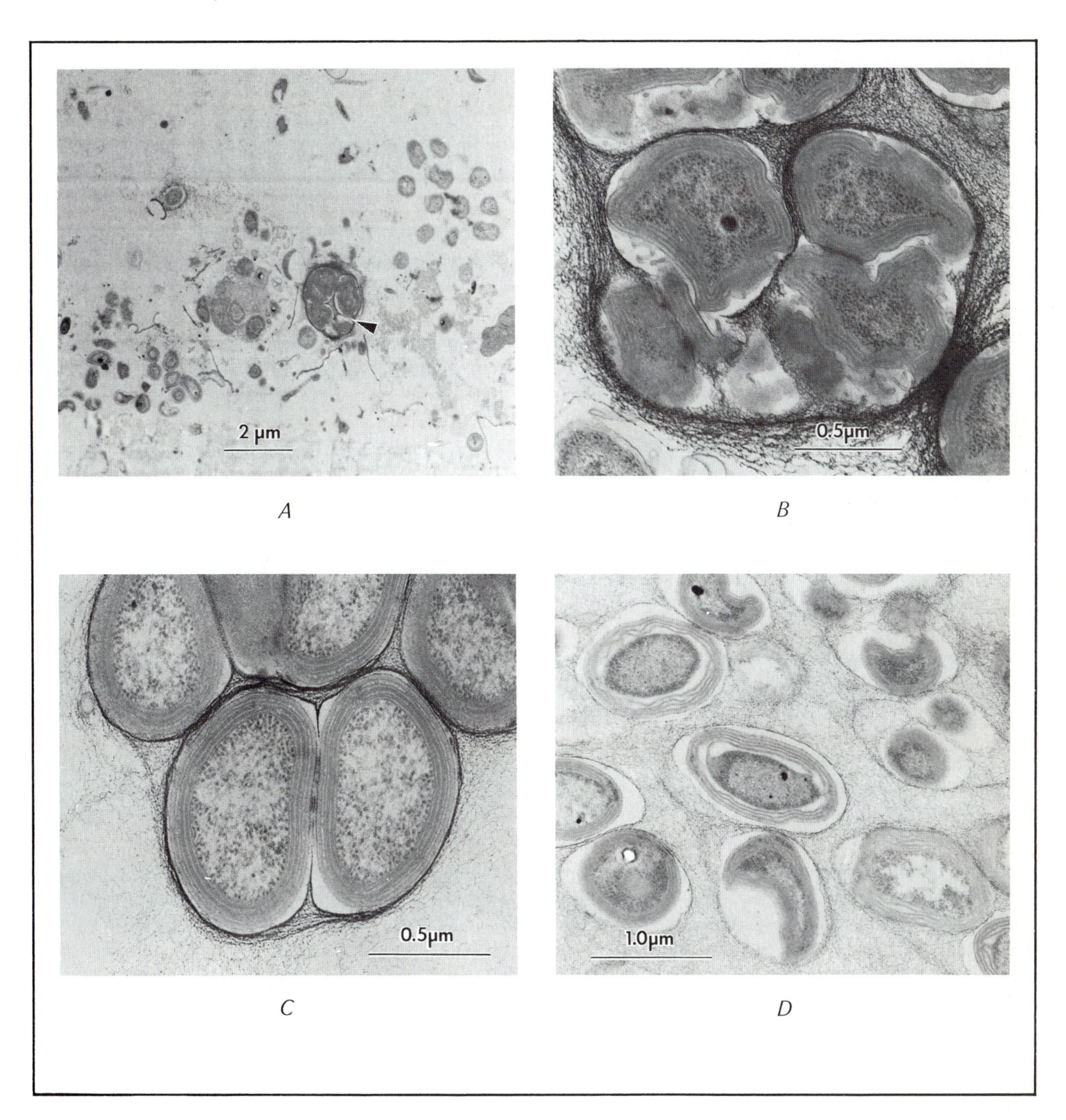

operated in a semicontinuous mode have maintained populations of 10^7 to 10^8 cells/ml with optimal generation times of 24 hours.

The ammonia-oxidizing bacteria have usually been enriched for with concentrations of ammonia exceeding 500 mg/liter. Attempts to enrich for ammonia-oxidizing bacteria at these concentrations from a high density aquaculture system, in which ammonia was maintained at around 1 mg/liter, failed (Johnson and Sieburth, 1976). An examination of nitrifying bacteria from nitrifying films maintained at low ammonia and at enriching (500 ml/liter) ammonia concentrations indicated that the cells that appeared healthy at the low ammonia concentrations became malformed and disintegrated at the ammonia concentration usually used for their enrichment (Plate 9-1,*D*). A diffusion culture system (Chapter 7-3,C), which would expose a captive population of nitrifying bacteria to low but constant levels of ammonia either in the laboratory or *in situ*, may be one way of enriching for and culturing such refractory strains.

In the nitrifying films and suspended flocs studied by Johnson and Sieburth (1976) where nitrite was being actively oxidized, bacteria with the morphology of the described nitrite-oxidizing bacteria could not be detected in thin sections of the films, although *Nitrobacter* was obtained in enrichment culture. One wonders how many of the cyst-type colonies that are present in the films but that lack distinctive cytomembrane structures are also nitrifying bacteria that have resisted culture and thus have not been described.

To obtain single colony isolates, Winogradsky (1890) used silica gel as the inorganic solidification agent, dialyzed out the unwanted acids, and dialyzed in the inorganic nutrients. Winogradsky wanted a completely inorganic medium in order to demonstrate obligate dependency on carbon dioxide as the sole carbon source. The concept of such a dependency has changed, and it is now known that all chemolithotrophic bacteria are able to assimilate bound organic carbon and that nitrifying bacteria are capable of mixotrophic and chemoorganotrophic growth. They are enriched for by the products of the heterotrophic bacteria that grow with them (Steinmüller and Bock, 1976; Bock, 1976). Agar has been used as a solidifying agent for growing isolated colonies of marine nitrifying bacteria (Watson, 1965; Koops, Harms, and Wehrman, 1976).

The papers by Watson (1965) and Watson and Mandel (1971) describe the taxonomic problems that have existed with the nitrifying bacteria. Watson and his colleagues at the Woods Hole Oceanographic Institution have rewritten the book on the taxonomy of nitrifying bacteria. Their studies included both the marine (Watson, 1965; Watson and Waterbury, 1971) and terrestrial (Watson, 1971; Watson et al., 1971) environments. This expertise permitted Watson (1971) to revise the family Nitrobacteriaceae, as can be seen in *Bergey's 8th*.

The nitrifying bacteria are gram negative and have a cell envelope that can differ from the classic gram-negative cell envelope of *Escherichia coli*. The seven-layer envelope of *Nitrosococcus* (*Nitrosocystis*) *oceanus* is an example of this difference (Watson and Remsen, 1970). Some nitrifiers also have an extensive cytomembrane system (Murray and Watson, 1965; Remsen, Valois, and Watson, 1967). Morphology, including the fine structure, of nitrifying bacteria is the only basis for placing these microorganisms in genera at the present time. Phase-contrast micrographs of whole cell suspensions and transmission electron micrographs of thin sections of the seven genera of nitrifying bacteria illustrated by Watson and Mandel (1971) are shown in Plate 9-2. Ammonia-oxidizing bacteria of the genus *Nitrosomonas* are nonmotile or single flagellated rods ($1-1.3 \times 2-2.5$ μm), with cytomembranes that form flattened lamellae at the periphery (Plate 9-2,*A*), whereas bacteria in the genus *Nitrosococcus* (*Nitrosocystis*) (Watson, 1965; Murray and Watson, 1965) are large cocci (1.8 to 2.2 μm), motile by a single flagellum or a single tuft of flagella, with cytomembranes that form flattened lamellae in the central region of the cell (Remsen, Valois, and Watson, 1967; Plate 9-2,*C*).

Koops, Harms, and Wehrmann (1976) recently isolated an ammonia-oxidizing bacterium from the North Sea with the characteristics of both the above genera—the peripheral membranes of *Nitrosomonas* and the spherical shape of *Nitrosococcus*. Since this microorganism also has a tuft of flagella, they placed it in the genus *Nitrosococcus* as *N. mobilis*, rather than form a new genus. These authors proposed that the genus *Nitrosococcus* be emended to include all coccoid ammonia-oxidizers regardless of fine structure. The other two genera of ammonia-oxidizing bacteria shown in Plate 9-2,*B*,*D* (*Nitrosospira* and *Nitrosolobus*, respectively) have only been isolated from soil, with no marine representatives so far.

All the genera of nitrite-oxidizing bacteria recognized in *Bergey's 8th* have marine representatives. The genus *Nitrobacter* (Plate 9-2,*E*) consists of usually nonmotile rod- to pear-shaped cells, which reproduce by

PLATE 9-2. A comparison of phase-contrast morphology and genera-specific ultrastructure in electron micrographs of ammonia-oxidizing (*A–D*) and nitrite-oxidizing (*E–G*) nitrifying bacteria. Marine species are recognized in all genera except those shown in (*B*) and (*D*). (*A*) *Nitrosomonas;* (*B*) *Nitrosospira;* (*C*) *Nitrosococcus;* (*D*) *Nitrosolobus;* (*E*) *Nitrobacter;* (*F*) *Nitrospina;* and (*G*) *Nitrococcus.* From S.W. Watson and M. Mandel, 1971, *J. Bacteriol.* 107:563–69.

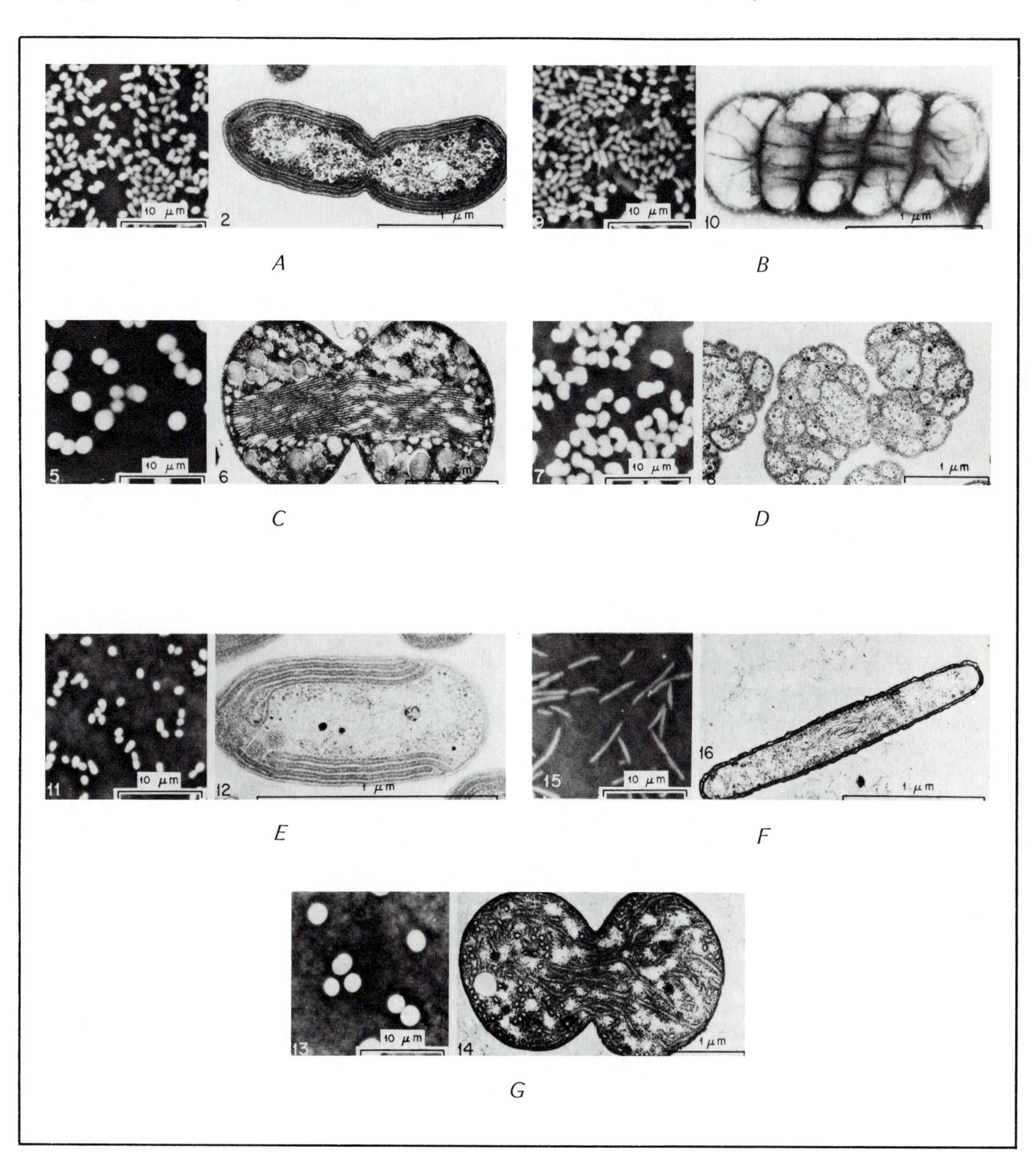

budding and possess a polar cap of cytomembranes. The marine strains are indistinguishable from *Nitrobacter winogradskyi* isolated from soil. Perhaps the genus *Nitrobacter* should really be called *Nitromicrobium*, since Hirsch (1970) has pointed out that Stutzer and Hartleb in 1899 had described a budding nitrite-oxidizing bacterium as *Nitromicrobium germinans*, which antedates the description of the type species by Winslow et al. in 1917. *Nitrosocystis*, the other genus of nitrifying bacteria recognized by *Bergey's 7th*, has now been combined with *Nitrobacter*. Two additional genera of nitrite-oxidizing bacteria isolated from the marine environment and described by Watson and Waterbury (1971) are also recognized by *Bergey's 8th*. *Nitrospina* (Plate 9-2,*F*) consists of nonmotile slender rods that become spherical in senescent cultures and, unlike the other recognized genera, have no extensive cytomembrane system. The type species *Nitrospina gracilis* was isolated from water collected in the Atlantic Ocean over 300 km from the mouth of the Amazon River. The genus *Nitrococcus* (Plate 9-2,*G*) is characterized by 1.5 to 1.8 μm spherical cells, which elongate to 3.5 μm just before division. The elongated cells have one or two subpolar flagella. This genus is characterized by an extensive cytomembrane system of tubes randomly arranged in the cytoplasm. Cultured cells are rich in cytochromes and turn cell suspensions yellow to reddish. The type species was isolated from a foamy slick less than 300 m from shore in the Galapagos Archipelago.

Nitrification in soil, wastes, and freshwater is receiving much current attention, but the estuarine and marsh environments have been largely overlooked. Aquaculture (mariculture) facilities seem to be ideal for studying the processes of nitrification (Johnson and Sieburth, 1976) and denitrification (Balderston and Sieburth, 1976) as well as the nitrifying bacteria that are enriched for at low substrate levels. The use of diffusion culture with low substrate concentration enrichment procedures will undoubtedly lead to the identification of nitrifying bacteria that were hitherto overlooked in estuarine and nearshore environments. Observations on the populations of nitrifying bacteria and their rates of activities in specific habitats would aid ecological studies in the marine environment. A good starting point would be the use of such techniques as most probable number (MPN) estimations (Matulewich, Strom, and Finstein, 1975), in conjunction with immunofluorescent (Fliermans, Bohlool, and Schmidt, 1974) and autoradiographic studies (Fliermans and Schmidt, 1975).

B. COLORLESS SULFUR BACTERIA

The colorless, sulfur-metabolizing bacteria fall into two disparate groups. Sulfur-oxidizing bacteria that oxidize free sulfur and thiosulfate to sulfate have been isolated in pure culture and have been found to oxidize sulfur both autotrophically and mixotrophically. The second group of colorless sulfur bacteria are large, morphologically distinct forms, usually with internal sulfur granules, which so far have proved resistant to pure culture. These bacteria occupy the same habitat and are therefore grouped together in *Bergey's 8th*.

1. Sulfur-oxidizing Bacteria

These bacteria form a spectrum ranging from obligate chemolithotrophy to heterotrophy. The obligate chemolithotrophs of the genus *Thiobacillus*, *T. thioparus*, *T. thiooxidans*, and *Thiomicrospira pelophila*, obtain energy from the oxidation of a number of reduced sulfur compounds, such as the sulfides produced by sulfate reduction (Chapter 17,C), elemental sulfur, and thiosulfate. The mixotroph *Thiobacillus novellus* also uses organic substrates for energy. The chemoorganotroph, *Thiobacillus trautweinii* is really a pseudomonad with thiosulfate-oxidizing enzymes.

Sulfide oxidation in nature is largely bacteriologically mediated. But in the absence of biological activity, sulfide can be slowly oxidized to sulfur, which then combines with the remaining sulfide to form polysulfides, as shown by Chen and Morris (1972) and diagrammed below.

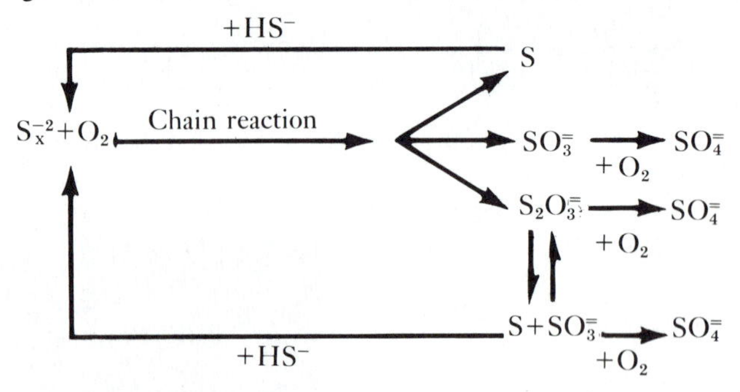

At acidic pH's below 6, the rate is very slow but becomes greatly increased at pH's of 7 to a maximum of 8.0, which falls into the pH range for seawater. The main role of sulfur-oxidizing bacteria is the oxidation of sulfide to sulfur, a rate-limiting step (Chen and Morris, 1972). Further oxidation is both bacterial and chemical.

Therefore the thiobacilli have a unique but precarious niche, where they contribute to, and compete with, an inorganic process. The final product of sulfur oxidation is sulfuric acid, which lowers the pH microzonally, with the formation of sulfate salts in the alkaline sea; these salts, in turn, can be reduced to hydrogen sulfide by the anaerobic sulfate-reducing bacteria in sediments (see Chapter 17,C).

Unlike terrestrial pseudomonads from the soil, which oxidize thiosulfate to tetrathionate but are unable to use this reaction to increase growth rate or yield (Trudinger, 1967), the growth rate and yield of three marine isolates increased when they were grown on organic compounds in the presence of thiosulfate, when the ratio of thiosulfate to organic substrate was high, and when the pH was kept constant (Tuttle, Holmes, and Jannasch, 1974).

The obligate chemolithotrophic sulfur-oxidizing bacteria are widespread in inshore marine muds, especially in eutrophic habitats (Baas Becking and Wood, 1955: Vishniac and Santer, 1957; Adair and Gunderson, 1969a). The frequency of *T. thioparus* occurrence may be due to its ability to utilize elemental sulfur, hydrogen sulfide, metal sulfides, and sulfite as energy sources. *Thiomicrospira pelophila* occupies a different niche in the same habitat with *T. thioparus* because it requires a concentration of sulfide that is inhibitory to *T. thioparus* (Kuenen and Veldkamp, 1972). The *T. thiooxidans* group that oxidizes elemental sulfur and sulfites is also common in estuarine muds (Baas Becking and Wood, 1955), and although rare in the open ocean, it has been isolated from large enrichment cultures (Tilton, Stewart, and Jones, 1967). *Thiobacillus novellus* has been isolated and identified from estuarine water (Adair and Gunderson, 1969b). The populations of thiobacilli in the open sea have been estimated on membrane filter cultures to range from 0 to 100/100 ml, the average being 5 cells/100 ml (Tilton, Cobet, and Jones, 1967), whereas estimates using enrichment culture ranged from 10 to 100 microorganisms/100 ml of seawater (Tuttle and Jannasch, 1972).

The thiobacilli are gram-negative rods with polar flagella; they can be thought of as autotrophic pseudomonads. The fine structure of their cell envelopes is similar if not identical to that of other gram-negative bacteria; the complex cytomembrane system found in the nitrifying bacteria is not present. *Thiobacillus thioparus* and *T. thiooxidans*, among other thiobacilli, have polyhedral inclusion bodies. These and other inclusions, as well as species-related differences in the middle layer of the cell envelope, make each of the recognized species of thiobacilli distinctive (Shively, Decker, and Greenawalt, 1970). *Thiomicrospira pelophila* is a very thin, comma-to-spiral–shaped bacterium (0.2×1 to $4\ \mu m$) with a polar flagellum and can only be seen by phase-contrast oil immersion microscopy.

Enrichment cultures for the thiobacilli are prepared by adding ammonium or nitrate salts as a source of nitrogen, potassium phosphate as a source of phosphate, ferric chloride as a source of iron, and a 1 to 2% sodium sulfate or other sulfur substrate as an energy source to seawater samples or their dilutions, which are aerated during incubation. Growth is indicated by turbidity and by the appearance of amorphous droplets of sulfur when thiosulfate is being used as an energy source. A similar medium enriches for thiobacilli from sediments. The same mineral seawater medium solidified with purified agar can also be used to isolate the organisms from enrichment cultures. Generation times as short as 4 to 5 hours make the thiobacilli easier to culture and to study than the nitrifying bacteria. The production of sulfuric acid and the neutral or alkaline pH required for the growth and viability of some thiobacilli must be taken into account when preparing media and making transfers (Vishniac and Santer, 1957; Tilton, Cobet, and Jones, 1967; Adair and Gunderson, 1969a; Tuttle and Jannasch, 1972). Higher concentrations of sulfide as well as the careful adjustment of pH (Kuenen and Veldkamp, 1972) are required for the enrichment of *Thiomicrospora pelophila*.

Information on the biochemical pathways of inorganic compounds of sulfur have been reviewed and summarized by Starkey (1956), Peck (1962), and Roy and Trudinger (1970). A general scheme of the pathways of sulfur oxidation, in which sulfite is central, is shown below.

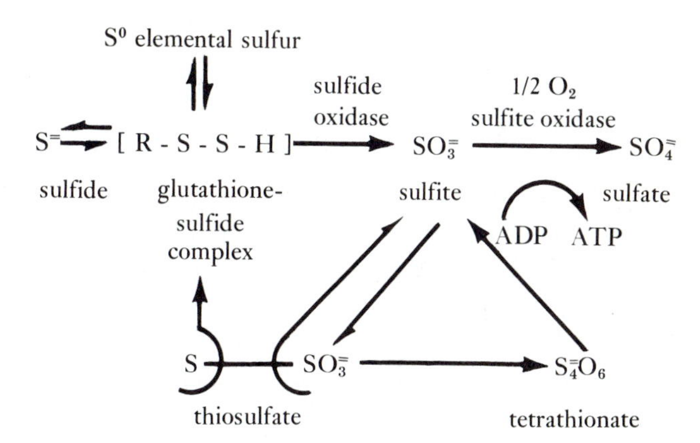

Thiosulfate is split into sulfite, which is then oxidized to sulfate with the production of ATP, while the sulfur ion is converted via the glutathione-sulfide complex to elemental sulfur (detected in culture) and sulfide. Elemental sulfur and sulfide react with glutathione to form the glutathione–sulfide complex before being oxidized to sulfite (Peck, 1968; Trudinger, 1969; Lyric and Suzuki, 1970). All thiobacilli are aerobic except *T. denitrificans*, which, like *Pseudomonas stutzeri*, can denitrify nitrate under anaerobic conditions to evolve gaseous nitrogen. It is of interest that *T. denitrificans* requires ammonia as a nitrogen source despite its denitrifying ability (Baalsrud and Baalsrud, 1954).

Although ultrastructure can apparently be used to differentiate known thiobacilli (Shively, Decker, and Greenawalt, 1970), the criteria used for classification are physiological, based on the utilization of usual and unusual substrates, the final pH obtained when growth is allowed to go to completion, and sensitivity to a spectrum of inhibitory agents. The absence of intermediate forms between the recognized species of thiobacilli (Parker and Temple, 1957) permits discrete groupings by numerical analysis (Hutchinson, Johnstone, and White, 1969). The DNA base composition (mole % G + C) for the species reported from the sea are *T. thiooxidans*, 51 to 52%; *T. thioparus*, 62 to 66%; *T. trautweinii*, 66%; and *T. novellus*, 66 to 68% (Jackson, Moriarty, and Nicholas, 1968). Since many chemoorganotrophic bacteria can oxidize thiosulfate without deriving energy from the process, such species as *T. trautweinii* have been eliminated from the genus *Thiobacillus* in *Bergey's 8th*. They will probably find their way into the pseudomonads.

The historical background of studies on marine thiobacilli is summarized by Tilton, Cobet, and Jones (1967) and Tilton, Stewart, and Jones (1967). It has been suggested that, although thiosulfate is the usual substrate used to enrich for thiobacilli, many isolates can use sulfide as a substrate and, if this occurs in the sea, bacterial sulfide utilization and the bacterial biomass resulting from it may be of significant importance in sulfide-rich environments (Adair and Gunderson, 1969a; Tuttle and Jannasch, 1972). Estuarine isolates of *T. thiooxidans* and *T. denitrificans* grown on colloidal sulfur and examined by scanning electron microscopy were found to physically attach to the sulfur, apparently forming sulfide as a metabolite, which could be precipitated by lead acetate (Baldensperger, Guarraia, and Humphreys, 1974). It has also been shown that *Thiobacillus*-type bacteria can utilize the methyl sulfides from cellulose mill effluents discharged into the marine environment as well as those formed naturally by the decomposition of sulfur-containing amino acids (Sivelä and Sundman, 1975).

PLATE 9-3. Some of the colorless sulfur bacteria encountered in sulfide-containing marine environments: (*A*) *Thiovulum majus;* (B and C) *Thiospira winogradsky* (150 and 1,000 X); (*D*) *Thiospira bipunctata* (775 X); (*E*) *Thiobacterium bovista*, showing the gelatinous colony enlarged to 20 X (7) and cells enlarged to 460 X (8) and 775 X (9). (*F*) A comparison of *Achromatium oxaliferum* (8, 9, and 10), *Macromonas mobilis* (11), and *Thiovulum majus* (12, 13, and 14). Compare with (*A*) for relative sizes. (*A*) Phase-contrast photomicrograph courtesy of Paul W. Johnson; (*B,C*) from W. Omelianski, 1905, *Zentralblatt Bakteriol. Parsit. Infekt. Hyg.*, Abt II 14:769–72; (*D,E*) from H. Molisch, 1912, *Zentralblatt Bakteriol. Parasit. Infekt. Hyg.* Abt II 33:55–62; (*F*) from R. Lauterborn, 1916, *Verb. Naturhist. Med. Vereins., Heidelberg*, 13:395–481.

2. Sulfur Granule-containing Bacteria.

These bacteria are widespread in inshore marine environments and are enriched to visible masses where hydrogen sulfide is released from decomposing organic matter. The first indication of their presence is a whitish veil or film, often seen in association with the whitish mats of *Beggiatoa* (discussed with the other gliding bacteria in Chapter 16,A). Unlike *Beggiatoa*, which can be cultured, there has been only one report on the successful culture of these bacteria (La Rivière, 1963), and most of our knowledge has been gained from microscopic observations on natural materials. The diversity of morphology of these large microorganisms, which usually have refractile sulfur granules, is shown in Plate 9-3. The colorless sulfur bacteria are described by La Rivière in *Bergey's 8th*. The occurrence of a number of species of these bacteria at different geographical areas in the marine environment has been recorded by Lackey (1961), Lackey and Lackey (1961), and Lackey, Lackey, and Morgan (1965).

Thiovulum majus, a highly motile large cell (5 to 25 μm), lives in the interface between oxygen and hydrogen sulfide, forming a visible veil or membrane. It has been thoroughly discussed by Starr and Skerman (1965), and its morphology has been summarized by de

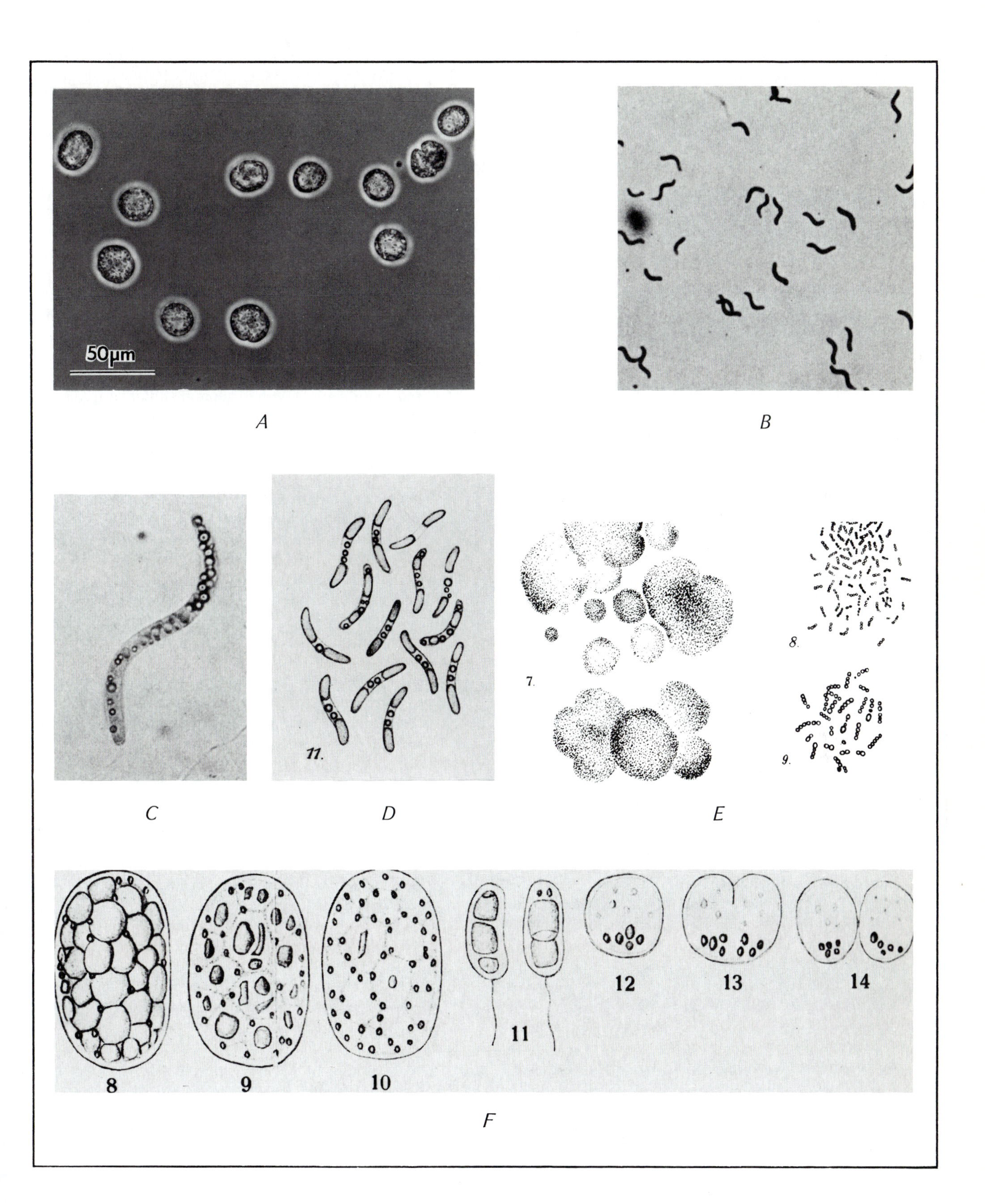
50µm
A
B
11.
7.
8.
9.
C
D
E
8
9
10
11
12
13
14
F

Boer, La Rivière, and Houwink (1961). It is usually enriched for from mud or seawater by allowing the green alga *Ulva lactuca* to decay as a source of hydrogen sulfide, with a trickle of seawater to supply the oxygen (van Niel, 1955). Enrichments that yield easily observed and much cleaner veils than those obtained with *Ulva* can be made by adding a few small dead fish to a thin layer of sediment in a jar, which is then submerged in a tank with running seawater. The delicate balance between oxygen and hydrogen sulfide was duplicated artificially in order to culture cells freed of contaminants by repeated migrations through sterile seawater (La Rivière, 1963). Attempts to repeat this work have met with failure so far.

The highly motile cells of *Macromonas mobilis* are distinguished by their single polar flagellum, which is visible without staining. The very large, spiral-shaped *Thiospira* species are motile by means of one or two non-visible flagella. Another large microorganism, *Achromatium*, which is 7 μm in diameter with some giant forms 35 to 100 μm in diameter, also has sulfur granules and is found in the same habitat. It is motile, moves with a rolling, rotating, jerky motion, and is included with the gliding bacteria in *Bergey's 8th* despite a report of the presence of peritrichous flagella (de Boer, La Rivière, and Schmidt, 1971). Among the other genera present in sulfide environments that can contain sulfur granules are *Beggiatoa*, *Thioploca*, and *Thiothrix*, which are discussed with the other gliding bacteria in Chapter 16,A.

REFERENCES

Adair, F.W., and K. Gundersen. 1969a. Chemoautotrophic sulfur bacteria in the marine environment. I. Isolation, cultivation and distribution. *Can. J. Microbiol.* 15:345–53.

Adair, F.W., and K. Gundersen. 1969b. Chemoautotrophic sulfur bacteria in the marine environment. II. Characterization of an obligately marine facultative autotroph. *Can. J. Microbiol.* 15:355–59.

Baalsrud, K., and K.S. Baalsrud. 1954. Studies on *Thiobacillus denitrificans*. *Arch. Mikrobiol.* 20:34–62.

Baas-Becking, L.G.M., and E.J. Ferguson Wood. 1955. Biological processes in the estuarine environment. I,II. Ecology of the sulfur cycle. *Koninkl. Ned. Akad. Wetenschap. Proc. Ser. B* 58:160–81.

Balderston, W.L., and J.McN. Sieburth. 1976. Nitrate removal in closed-system aquaculture by columnar denitrification. *Appl. Environ. Microbiol.* 32:808–18.

Baldensperger, J., L.S. Guarraia, and W.J. Humphreys. 1974. Scanning electron microscopy of thiobacilli grown on colloidal sulfur. *Arch. Mikrobiol.* 99:323–29.

Bock, E. 1976. Growth of *Nitrobacter* in the presence of organic matter. II. Chemoorganotrophic growth of *Nitrobacter agilis*. *Arch. Mikrobiol.* 108:305–12.

Carey, C. 1938. The occurrence and distribution of nitrifying bacteria in the sea. *J. Mar. Res.* 1:291–304.

Chen, K.Y., and J.C. Morris. 1972. Kinetics of oxidation of aqueous sulfide by O_2. *Environm. Sci. & Technol.* 6:529–37.

de Boer, W.E., J.W.M. La Rivière, and K. Schmidt. 1971. Some properties of *Achromatium oxaliferum*. *Antonie van Leeuwenhoek* 37:553–63.

de Boer, W.E., J.W.M. La Rivière, and A.L. Houwink. 1961. Observations on the morphology of *Thiovulum majus* Hinze. *Antonie van Leeuwenhoek* 27:447–56.

Fliermans, C.B., and E.L. Schmidt. 1975. Autoradiography and immunofluorescence combined for autecological study of single cell activity with *Nitrobacter* as a model system. *Appl. Microbiol.* 30:676–84.

Fliermans, C.B., B.B. Bohlool, and E.L. Schmidt. 1974. Autecological study of the chemoautotroph *Nitrobacter* by immunofluorescence. *Appl. Microbiol.* 27:124–29.

Hirsch, P. 1970. Budding nitrifying bacteria: The nomenclatural status of *Nitromicrobium germinans* Stutzer and Hartleb 1899 and *Nitrobacter winogradskyi* Winslow et al. 1917. *Int. J. Syst. Bacteriol.* 20:317–20.

Hutchinson, M., K.I. Johnstone, and D. White. 1969. Taxonomy of the genus *Thiobacillus:* The outcome of numerical taxonomy applied to the group as a whole. *J. gen. Microbiol.* 57:397–410.

Jackson, J.F., D.J.Moriarty, and D.J.D. Nicholas. 1968. Deoxyribonucleic acid base composition and taxonomy of thiobacilli and some nitrifying bacteria. *J. gen. Microbiol.* 53:53–60.

Jannasch, H.W. 1960. Denitrification as influenced by photosynthetic oxygen production. *J. gen. Microbiol.* 23:56–63.

Johnson, P.W., and J.McN. Sieburth. 1976. *In situ* morphology of nitrifying-like bacteria in aquaculture systems. *Appl. Environ. Microbiol.* 31:423–32.

Kawai, A., Y. Yoshida, and M. Kimata. 1968. Nitrifying bacteria in the coastal environments. *Bull. Misaki Mar. Biol. Inst. Kyoto Univ.* 12:181–94.

Kelly, D.P. 1971. Autotrophy: Concepts of lithotrophic bacteria and their organic metabolism. *Ann. Rev. Microbiol.* 25:177–210.

Koops, H-P., H. Harms, and H. Wehrmann. 1976. Isolation of a moderate halophilic ammonia-oxidizing bacterium, *Nitrosococcus mobilis* nov. sp. *Arch. Mikrobiol.* 107:277–82.

Kuenen, J.G., and H. Veldkamp. 1972. *Thiomicrospira pelophila*, gen. n., sp. n., a new obligately chemolithotrophic colourless sulfur bacterium. *Antonie van Leeuwenhoek* 38:241–56.

Lackey, J.B. 1961. Bottom sampling and environmental niches. *Limnol. Oceanogr.* 6:271–79.

Lackey, J.B., and E.W. Lackey. 1961. The habitat and description of a new genus of sulphur bacterium. *J. gen. Microbiol.* 26:29–39.

Lackey, J.B., E.W. Lackey, and G.B. Morgan. 1965. Taxonomy and ecology of the sulfur bacteria. *Eng. Progr. Univ. Florida Bull. Ser. No.* 119, 23 p.

La Rivière, J.W.M. 1963. Cultivation and properties of *Thiovulum majus* Hinze. In: *Marine Microbiology* (C.H. Oppenheimer, ed.). C.C. Thomas, Springfield, Ill., pp. 61–72.

Lavoie, D.M. 1975. Application of diffusion culture to ecological observations on marine microorganisms. M.S. Thesis, University of Rhode Island, 91 p.

Lauterborn, R. 1916. Die sapropelische Lebewelt. Ein Beitrag zur Biologie des Faulschlammes naturlicher Gewässer. *Verh. Naturhist. Med. Vereins, Heidelberg* 13:395–481.

Lyric, R.M., and I. Suzuki. 1970. Enzymes involved in the metabolism of thiosulfate by *Thiobacillus thioparus*. III. Properties of thiosulfate-oxidizing enzyme and proposed pathway of thiosulfate oxidation. *Can. J. Biochem.* 48:355–63.

Matulewich, V.A., P.F. Strom, and M.S. Finstein. 1975. Length of incubation for enumerating nitrifying bacteria present in various environments. *Appl. Microbiol.* 29:265–68.

Meade, T.L. 1974. The technology of closed system culture of salmonids. University of Rhode Island Mar. Tech. Rept. No. 30 (Sea Grant), 30 p.

Molisch, H. 1912. Neue farblose Schwefelbakterien. *Zentralbl. Bakteriol. Parasit. Infekt., Abt II* 33:55–62.

Murray, R.G.E., and S.W. Watson. 1965. Structure of *Nitrosocystis oceanus* and comparison with *Nitrosomonas* and *Nitrobacter*. *J. Bacteriol.* 89:1594–1609.

Omelianski, W. 1905. Ueber eine neue Art farbloser Thiospirillen. *Zentralbl. Bakteriol. Parasit. Infekt. Hyg., Abt II* 14:769–72.

Parker, C.D., and K.L. Temple. 1957. *Thiobacillus*. In: *Bergey's Manual of Determinative Bacteriology* (7th Ed.). Williams & Wilkins Co., Baltimore, Md., pp. 83–88.

Peck, H.D., Jr. 1962. Symposium on metabolism of inorganic compounds, V. Comparative metabolism of inorganic sulfur compounds in microorganisms. *Bacteriol. Rev.* 26:67–94.

Peck, H.D., Jr. 1968. Energy-coupling mechanisms in chemolithotrophic bacteria. *Ann. Rev. Microbiol.* 22:489–518.

Remsen, C.C., F.W. Valois, and S.W. Watson. 1967. Fine structure of the cytomembranes of *Nitrosocystis oceanus*. *J. Bacteriol.* 94:422–33.

Rittenberg, S.C. 1969. The roles of exogenous organic matter in the physiology of chemolithotrophic bacteria. *Adv. Microbial Phys.* 3:159–96.

Rittenberg, S.C. 1972. The obligate autotroph—the demise of a concept. *Antonie van Leeuwenhoek* 38:457–78.

Roy, A.B., and P.A. Trudinger. 1970. *The Biochemistry of Inorganic Compounds of Sulfur.* Cambridge University Press, London and New York, pp. 275–88.

Schlegel, H.G. 1975. Mechanisms of chemoautotrophy. In: *Marine Ecology, Vol. 2, Part 1* (O. Kinne, ed.). Wiley, London and New York, pp. 9–60.

Shively, J.M., G.L. Decker, and J.W. Greenawalt. 1970. Comparative ultrastructure of the *Thiobacilli*. *J. Bacteriol.* 101:618–27.

Sieburth, J.McN. 1975. *Microbial Seascapes.* University Park Press, Baltimore, Md., 248 p.

Sivelä, S., and V. Sundman. 1975. Demonstration of *Thiobacillus*-type bacteria which utilize methyl sulphides. *Arch. Mikrobiol.* 103:303–304.

Starkey, R.L. 1956. Transformations of sulfur by microorganisms. *Ind. Eng. Chem.* 48:1429–37.

Starr, M.P., and V.B.D. Skerman. 1965. Bacterial diversity: The natural history of selected morphologically unusual bacteria. *Ann. Rev. Microbiol.* 19:407–54.

Steinmüller, W., and E. Bock. 1976. Growth of *Nitrobacter* in the presence of organic matter. I. Mixotrophic growth. *Arch. Mikrobiol.* 108:299–304.

Sugahara, I., M. Sugiyama, and A. Kawai. 1974. Distribution and activity of nitrogen-cycle bacteria in water-sediment systems with different concentrations of oxygen. In: *Effect of the Ocean Environment on Microbial Activities* (R.R. Colwell and R.Y. Morita, eds.). University Park Press, Baltimore, Md., pp. 327–40.

Tilton, R.C., A.B. Cobet, and G.E. Jones. 1967. Marine thiobacilli. I. Isolation and distribution. *Can. J. Microbiol.* 13:1521–28.

Tilton, R.C., G.J. Stewart, and G.E. Jones. 1967. Marine thiobacilli. II. Culture and ultrastructure. *Can. J. Microbiol.* 13:1529–34.

Trudinger, P.A. 1967. Metabolism of thiosulfate and tetrathionate by heterotrophic bacteria from soil. *J. Bacteriol.* 93:550–59.

Trudinger, P.A. 1969. Assimilatory and dissimilatory metabolism of inorganic sulfur compounds by micro-organisms. *Adv. Microbiol. Physiol.* 3:111–58.

Tuttle, J.H., and H.W. Jannasch. 1972. Occurrence and types of Thiobacillus-like bacteria in the sea. *Limnol. Oceanogr.* 17:532–43.

Tuttle, J.H., P.E. Holmes, and H.W. Jannasch. 1974. Growth rate stimulation of marine pseudomonads by thiosulfate. *Arch. Mikrobiol.* 99:1–14.

Vaccaro, R.F. 1965. Inorganic nitrogen in sea water. In: *Chemical Oceanography* (J.P. Riley and G. Skirrow, eds.), Vol. 1 (1st Ed.). Academic Press, London and New York, pp. 365–408.

van Niel, C.B. 1955. Natural selection in the microbial world. *J. gen. Microbiol.* 13:201–17.

Vishniac, W., and M. Santer. 1957. The thiobacilli. *Bacteriol. Rev. 21:195–213.*

Wallace, W., and D.J.D. Nicholas. 1969. The biochemistry of nitrifying microorganisms. *Biol. Rev.* 44:359–91.

Watson, S.W. 1965. Characteristics of a marine nitrifying bacterium, *Nitrosocystis oceanus* sp. n. *Limnol. Oceanogr.* 10(Suppl.):R274–R289.

Watson, S.W. 1971. Taxonomic considerations of the family

Nitrobacteraceae Buchanan. Requests for opinions. *Int. J. Syst. Bacteriol.* 21:254–70.

Watson, S.W., and M. Mandel. 1971. Comparison of the morphology and deoxyribonucleic acid composition of 27 strains of nitrifying bacteria. *J. Bacteriol.* 107:563–69.

Watson, S.W., and C.C. Remsen. 1970. Cell envelope of *Nitrosocystis oceanus*. *J. Ultrastr. Res.* 33:148–60.

Watson, S.W., and J.B. Waterbury. 1971. Characteristics of two marine nitrite oxidizing bacteria, *Nitrospina gracilis* nov. gen. nov. sp. *Arch. Mikrobiol.* 77:203–30.

Watson, S.W., L.B. Graham, C.C. Remsen, and F.W. Valois. 1971. A lobular ammonia-oxidizing bacterium, *Nitrosolobus multiformis* nov. gen. nov. sp. *Arch. Mikrobiol.* 76:183–203.

Winogradsky, S. 1888. *Beiträge zur Morphologie und Physiologie der Bakterien. Heft I. Zur Morphologie und Physiologie der Schwefelbakterein.* Arthur Felix, Leipzig, 120 p.

Winogradsky, S. 1890. Sur les Organismes de la nitrification. *Ann. Inst. Pasteur (Paris)* 4:213–31.

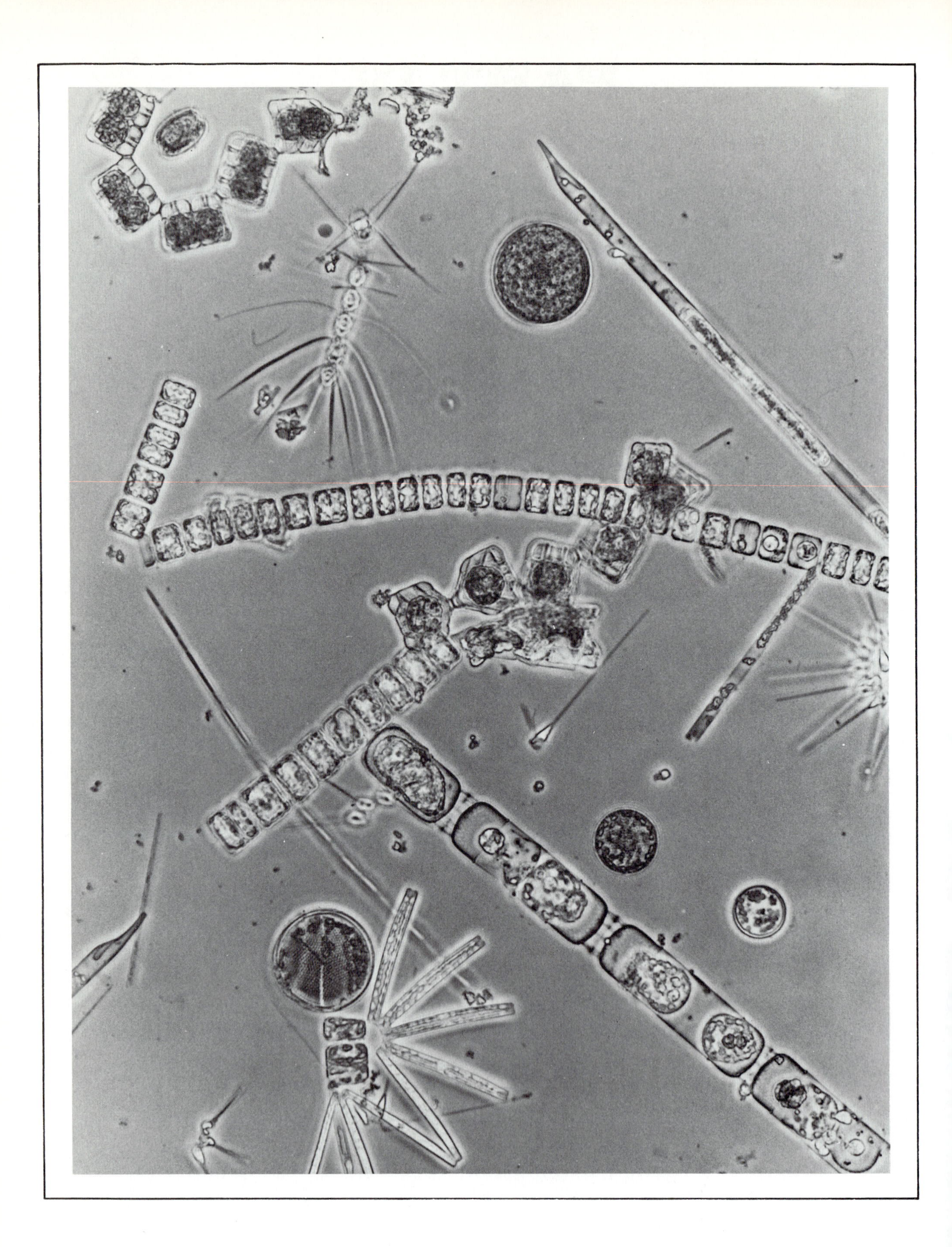

The diversity of the diatoms dominating a fall phytoplankton net tow from Rhode Island Sound (R/V *Endeavor* Cruise 001, November 1976) Phase-contrast photomicrograph courtesy of Paul W. Johnson. (magnification 300 X).

The first phototrophs in the fossil record are the cyanobacteria, which presumably gave rise to chlorophyll-containing phototrophic eucaryotes through endosymbiosis. The eucaryotic microalgae that then arose used carbon dioxide and solar energy to photosynthesize organic matter, with the evolution of gaseous oxygen. These forms helped turn the planet earth into the aerobic ecosystem we know today and in so doing also produce the organic matter upon which most other life forms in aquatic environments depend.

Marine species of microalgae are placed in twelve classes of algae as currently defined and modified (Christensen, 1962; Hibberd and Leedale, 1970; Dodge, 1973; Margulis, 1974; Parke and Dixon, 1976). For our purposes, we will discuss them in four chapters. The diatoms with their siliceous exoskeletons are discussed in Chapter 10, the equally well known dinoflagellates in Chapter 11. The haptophytes and prasinophytes covered in Chapter 12 have complex life cycles, which have only recently been clarified. Chapter 13 covers a miscellaneous collection of algal classes with few marine species. But some of these marine species, on occasion, compete successfully with the normally more dominant and better known algal groups.

REFERENCES

Christensen, T. 1962. Alger. In: *Botanik (Systematik Botanik)* (T.W. Böcher, M. Lange, and T. Sørensen, eds.), No. 2, Vdg. 2 (1st Ed.). Munksgaard, Copenhagen, 178 p.

Dodge, J.D. 1973. *The Fine Structure of Algal Cells.* Academic Press, London and New York, 261 p.

Hibberd, D.J., and G.F. Leedale. 1970. Eustigmatophyceae—a new algal class with unique organization of the motile cell. *Nature* 225:758–760.

Margulis, L. 1974. The classification and evolution of prokaryotes and eukaryotes. In: *Handbook of Genetics* (R.C. King, ed.), Vol. 1. Plenum, New York, pp. 1–41.

Parke, M., and P.S. Dixon. 1976. Checklist of British marine algae—third revision. *J. mar. biol. Assoc. U.K.* 56:527–594.

CHAPTER 10

Diatoms

The diatoms are characterized by elaborately ornamented frustules (bivalve shells of silica). Planktonic diatoms are often the major primary producers of microbial biomass in the photic zone of the sea, which is the foundation of the pelagic food chain, especially in nearshore waters. The diatoms are grazed by phagotrophic protozoans or filter-feeding herbivorous invertebrates, which, in turn, serve as prey for larger carnivores in the zooplankton (Raymont, 1963; Parsons, Takahashi, and Hargrave, 1977). Besides this biomass, from 1 to 50% of the daily photosynthate may be released as dissolved organic substances (Chapter 2,C). These released substances maintain a population of planktobacteria (Chapter 14), which apparently help maintain the phytoplankton in good condition by utilizing the algal auto-inhibitory by-products and by supplying growth factors. The littoral diatoms that live in the water, on surfaces, and in the illuminated bottom muds provide at least part of the food for such water-filtering animals as shellfish (Savage, 1925) and for the muck-feeders that burrow through the mud (Fenchel, 1969). The mucoid material produced by the littoral diatoms firmly binds the upper loose substrata and is instrumental in building up mud flats and banks (Linke, 1939; Grøntved, 1949; Hendey, 1964). The film-forming and attaching pennate diatoms (which are non-radially symmetrical) can initiate fouling on newly immersed surfaces (Skerman, 1956), although they usually seem to appear after the bacterial film is formed; they precede protozoa in photic environments.

Many thousands of marine diatom species are distributed throughout the photic zones of the sea. Some are found in the water as free-floating microorganisms in the plankton, while others can associate with any suitable illuminated surface as epiphytes. The planktonic species are considered to be either oceanic, spending their whole existence in the plankton of the open sea away from the influence of coasts, or neritic, being associated with coasts and reproducing there. The neritic species can also live a completely pelagic life, a mostly pelagic life spending some of the time on the bottom, or an essentially benthic life as part of the bottom microflora or attached to a fixed substrate and only entering the surface waters when forcibly removed from their natural habitats. Smayda (1958) has discussed the biogeography of diatoms.

Planktonic diatoms usually dominate the photic zone of temperate and polar seas and are seen as blooms in such nearshore waters as estuaries, bays, sounds, and continental shelves or in upwelling areas, where the submarine geology, winds, and currents force up nutrient-rich, deep-water masses. Since the light intensities of these uppermost layers of water inhibit many microalgae, the productive zone and peaks in population with depth vary with the nature of the water mass, being 2 to 20 m in turbid estuaries, 25 to 50 m in less turbid coastal waters, and 50 to 100 m in clear tropical waters (Dring, 1971). Distribution within the water column depends, in varying degrees, upon illumination, salinity, temperature, and nutrient supply, among other factors (Perkins, 1974). Populations at peak abundance run from hundreds of cells per liter in offshore waters to almost a billion cells per liter in eutrophic inshore waters. Biotic pressures on the planktonic diatom communities include depredations by the protozooplankton and metazooplankton, fungal epidemics of senescent blooms, and nutrient depletion. Nonbiotic pressures are extremes in light, nutrients, temperature, and the incursion of man-made pollution. In temperate and polar waters, the number of diatoms in the phytoplankton fluctuates from negligible numbers to very large blooms dominated by one or a few species. During and between blooms there is often a succession of species. The planktonic diatom populations (as shown in the Part IV title plate) are usually dominated by radially symmetrical forms known as centric diatoms (see Plate 10-1,*C*).

The pennate diatoms, which are named for the feather-like markings on their frustules, may have either the characteristic double-ended or canoe-like shape or a gonoid or trellisoid shape (Plate 10-1,*A*). They are essentially of the benthos, although sometimes they are found in the plankton. The motile forms with a raphe (slit) occur in films on a variety of surfaces from plastics to algae, and even on the skin of whales. The benthic diatoms have been reviewed by Round (1971). The stalked forms, such as species of *Achnanthes*, *Licmophora*, *Grammatophora*, and *Rhabdonema*, also occur on a variety of surfaces including animal surfaces. Both motile and

PLATE 10-1. Examples of the architectural patterns of diatoms. (*A*) Basic patterns of valves: (1) centric, radial about a central point; (2) gonoid, supported by angles; (3) pennate, usually symmetrical about an apical line; (4) trellisoid, arranged from margin to margin. (*B*) and (*C*) Photomicrographs of representative diatoms from British coastal waters, showing pennate and centric forms, respectively. From N.I. Hendey, 1964, *An Introductory Account of the Smaller Algae of British Waters*, Part V, Series IV. Fishery Investigations, London. Reproduced with permission of Controller of Her Britannic Majesty's Stationery Office.

attached species of pennate diatoms occur as lithophytes on calcareous and mineral rocks. Such tube-forming diatoms as *Amphipleura* form branched filaments, which, with the unaided eye, can be mistaken for fronds of the seaweed *Ectocarpus*. Pennates also occur on the mud flats, where species of *Navicula*, *Hantzschia*, and other biraphid species (with slit-like raphe on both the upper and the lower surface) freely move within the substratum, ascending upon exposure of the mud flat and descending upon flooding (Aleem, 1949, 1950; Round and Palmer, 1966). This rhythm is apparently controlled by a dual clock system, one being lunar day (i.e., tidal) clock and the other a solar day (light, dark) clock (Palmer and Round, 1967; Palmer, 1975).

Harper (1969) has discussed the migration of diatoms on sand grains; Harper and Harper (1967) measured diatomic adhesion and movement and have shown that the motility of the diatoms depends upon adhesion. The freely moving and stalked pennate diatoms are among the first microorganisms to colonize a newly immersed surface upon which they can glide or attach (Skerman, 1956). The bottom layers of pack ice in the polar seas are often discolored with species of diatoms that receive adequate illumination through the snow-covered ice. Unlike the planktonic diatoms, the populations of pennates in littoral habitats are more constant from season to season; there is, however, a marked periodicity, and different species dominate at specific times of the year (Hendey, 1964).

The distinguishing feature of all diatoms is the hydrated silica valve structure of two cups that fit one within the other to form a capsule enclosing the cell contents. The shape of the cells as viewed in the apical plane (the top view, which is really the end view in chains of cells) is used to divide diatoms into four basic patterns (Plate 10-1,*A*). The variations in form and pattern of some centric and pennate diatoms from British

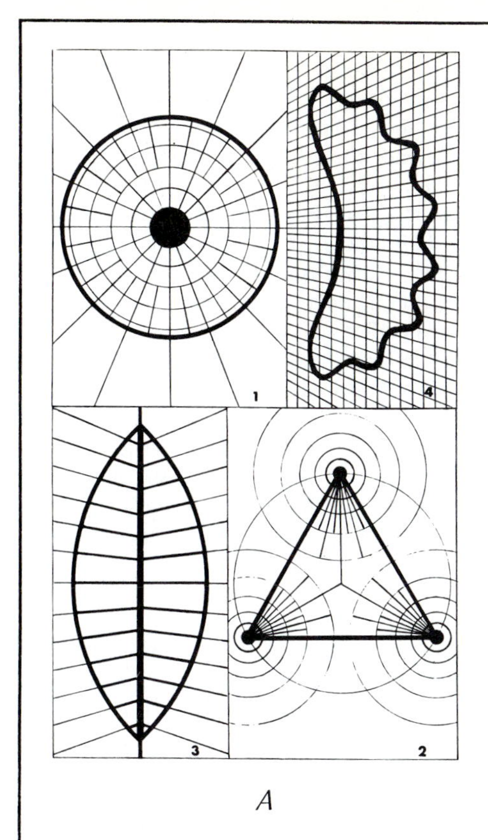

A

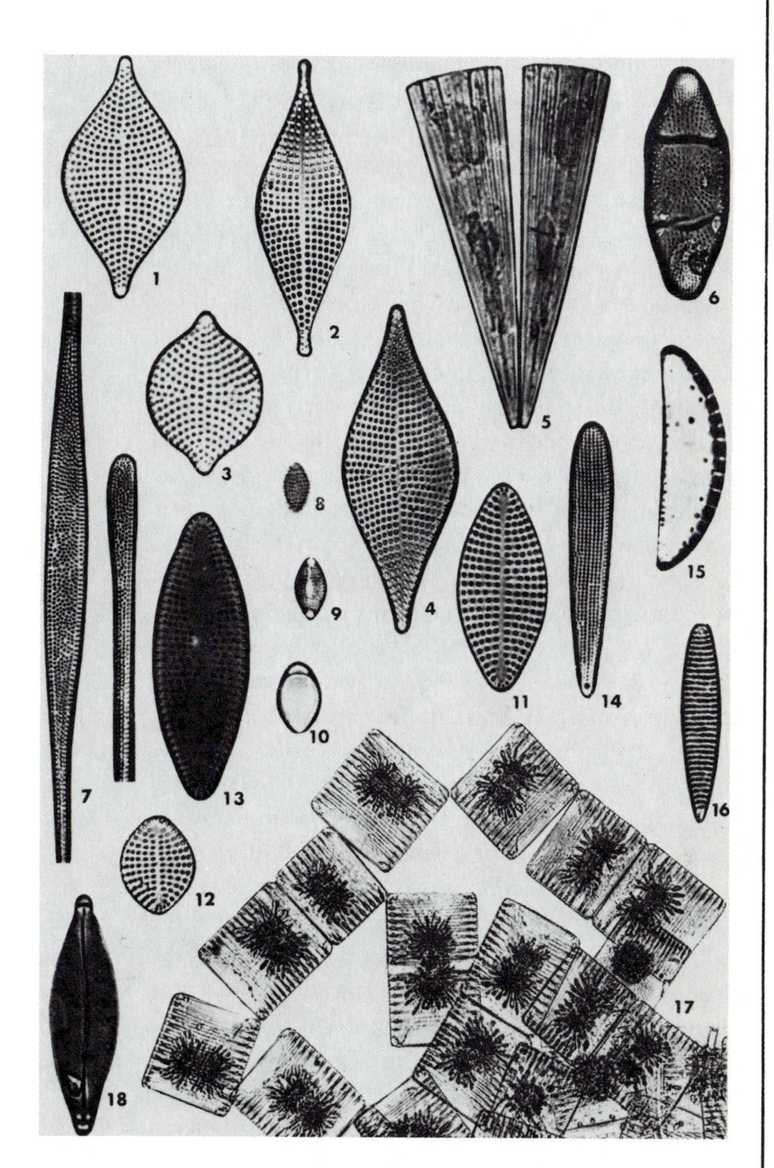

B

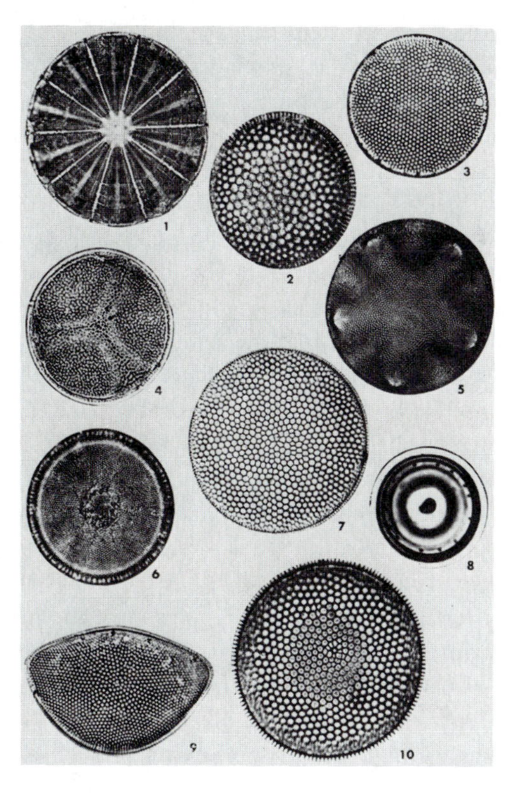

C

coastal waters are shown in Plate 10-1,*B*,*C*. When seen in girdle view (from the side), most pennate and some centric diatoms appear rectangular. Three examples of diatom frustules that span the range in shape from long capsule to box to "Petri dish" are seen in cross section in Plate 10-2,*B*. The two valves of the parent cell, which differ in size, have flanges or tape-like bands called cingula or girdle bands; these bands are somewhat flexible and join the two halves of the frustule together, fitting one inside the other like sleeves. Other details are shown in the scanning and transmission electron micrographs of both centric and pennate diatoms in Plates 10-3 and 10-4, respectively. Material diffuses in and out of the cell through the punctae (holes), approximately 0.5 μm or less in diameter, which occur on the valves and sometimes on the cingula. On the valves, they are often arranged in regular lines called striae; these may be straight, curved, transverse, or radiating in form. The areolae are larger honeycomb-like chambers within the cell wall, which may be open to the inside or the outside. They give cells like the large *Coscinodiscus* their characteristic appearance under the light microscope. The valves, which might be weakened by these chambers and the punctae, are reinforced by slight curvations in form, by double walls in some species, and especially, by thickened ribs called costae. The centric diatoms, which must be buoyant for their planktonic life (Smayda, 1970), have extensions or processes (Hendey, 1959; Hasle, 1972a) of many sorts, two of which are strutted tubules and setae. In addition to these structures in such diatoms as *Chaetoceros*, Gran (1912) recognized three other structural modifications that could aid flotation. These include the large, bladder-type cells, which have their protoplasm around the outer part of the cell and a thin fluid inside, as in *Porosira;* the ribbon-type or flattened colonial diatom, with a broad surface in one plane, as in *Fragilaria;* and the filamentous tubular diatom, in which the cells are greatly prolonged in one direction, as in *Rhizosolenia*. Other structural characteristics that may aid flotation are the thinner walls in the centric diatoms (Allen, 1941), and physiological changes in composition of long-living senescent cells (Riley, 1943) that include the production of large numbers of oil globules that would increase buoyancy (Hendey, 1937; Allen, 1941; Grøntved, 1949). The influence of such structures on buoyancy has been studied by Smayda and Boleyn (1966a,b,c), and the general topic of phytoplankton sinking has been reviewed by Smayda (1970).

PLATE 10-2. The morphology and ultrastructure of diatoms. (*A*) Thin section of a recently divided diatom, showing the newly formed theca inside the parental theca, the nucleus, and chloroplasts. (*B*) Schematic cross sections, showing the bivalve arrangement in diatoms as represented by the needle-shaped *Rhizosolenia*, the pillow-shaped *Biddulphia*, and the Petri-dish–shaped *Coscinodiscus*. (*C*) A diagram of binary fission in a diatom to demonstrate the MacDonald–Pfitzer hypothesis for diminution in cell size: girdle view, two generations. (*D*) Resting spore cycle in a centric diatom. (*A*) Electron micrograph courtesy of Paul W. Johnson; (*B,D*) from G. Drebes, 1974, *Marine Phytoplankton*, G. Thieme Verlag, Stuttgart; (*C*) from N.I. Hendey, 1964, *An Introductory Account of the Smaller Algae of British Waters*, Part V, Series IV. Fishery Investigations London. Reproduced with permission of Controller of Her Britannic Majesty's Stationery Office.

The pennate diatoms that live on surfaces are usually elongated cells. On the apical axis, they have a thickened axial area that reinforces the valves. If this area is structureless, it is called a pseudoraphe. Many species have a true raphe (on one or both valves), which is a slit with terminal polar nodules and a thickened central nodule that interrupts the slit. Motility is restricted to those species or to stages within a species that have a true raphe. Detailed observations have indicated that crystalloid bodies produce a mucoidal "locomotor secretion material," which leaves a trail on the substratum that flows in the opposite direction of the motion; that longitudinally oriented fibrillar bundles occur adjacent to all raphe systems; and that locomotion is inhibited by smooth muscle relaxants (Drum and Hopkins, 1966). In a novel and intriguing hypothesis, Gordon and Drum (1970) suggest that the pennate diatoms' gliding locomotion is due to the raphe fluid adhering to the substratum while the raphe acts as a parallel plate capillary and that capillary pressure draws out more fluid and pushes the diatom along.

Some of the cell contents of a diatom are shown in a transverse thin section in Plate 10-2,*A*. The nucleus of living diatoms can be stained with a 1:1,000,000 dilution of Methylene blue. The chloroplasts are the most visible structure in light microscopy, and their number, size, and distribution, which vary considerably from species to species, are used in classification (Aleem, 1950; Hendey, 1964). The color of the cells is due to the mixture of chlorophylls *a* and *c*, xanthophylls, and carotenoids present in the chloroplasts. Within this photosynthetic organelle is a proteinaceous body, the pyre-

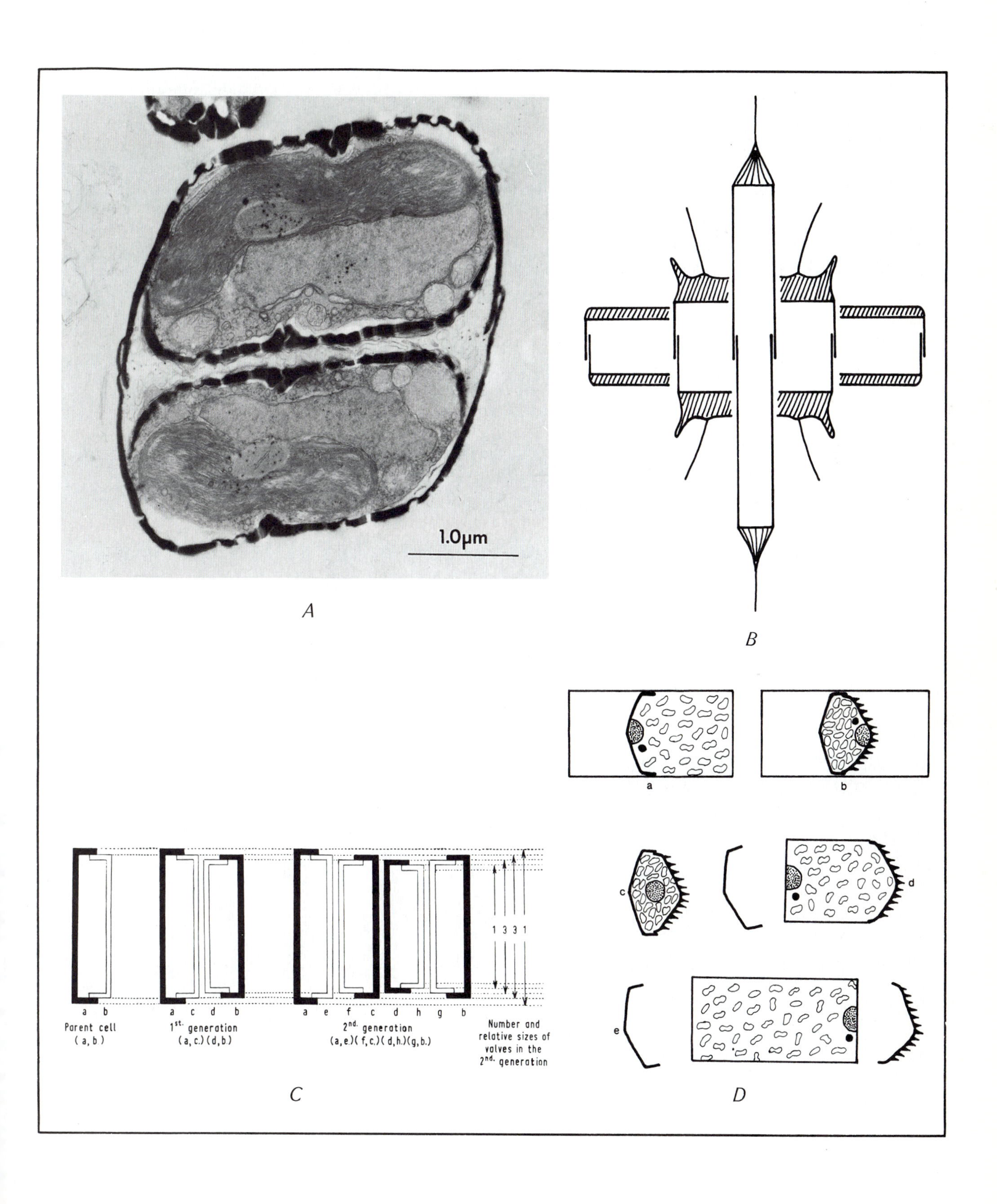
1.0μm
A
B
a
b
c
d
e
1 3 3 1
a b
Parent cell
(a, b)
a c d b
1st. generation
(a, c.) (d, b.)
a e f c d h g b
2nd. generation
(a, e.)(f, c.)(d, h.)(g, b.)
Number and
relative sizes of
valves in the
2nd. generation
C
D

A

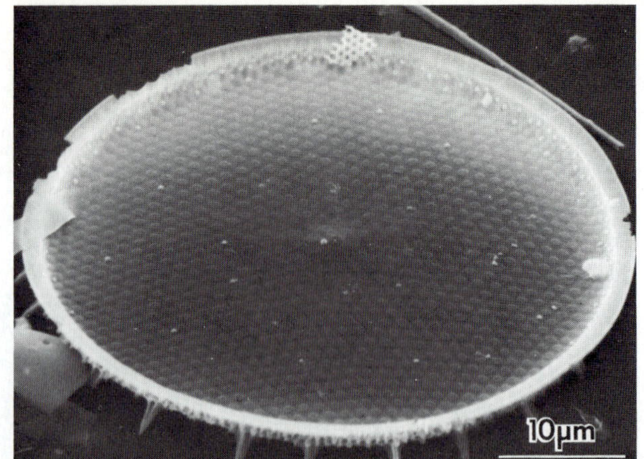

B

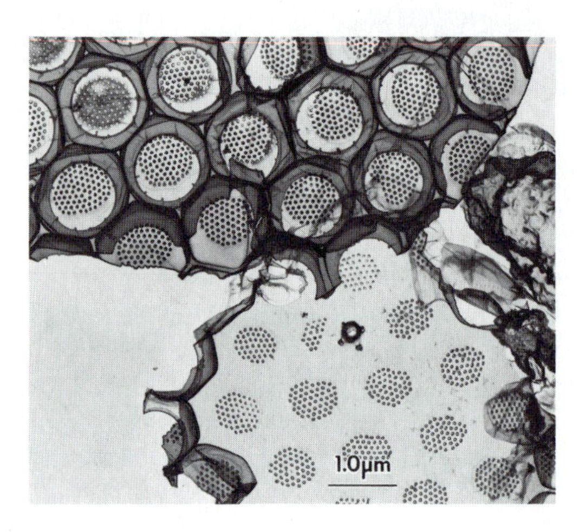

C

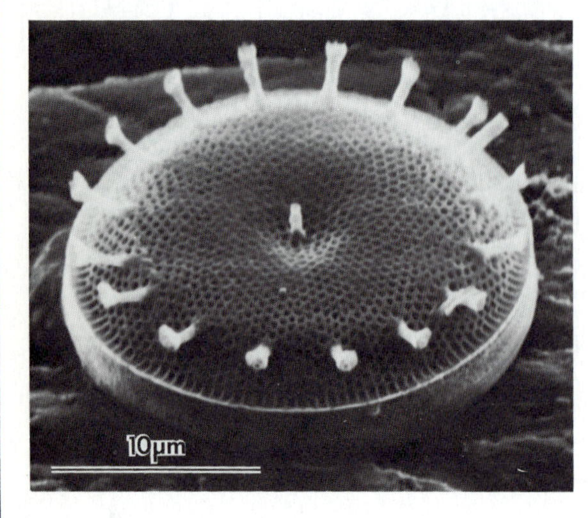

D

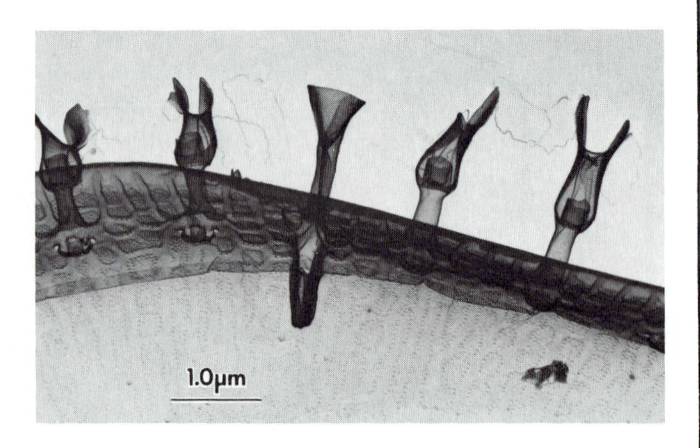

E

PLATE 10-3. Architectural details of valves of the centric diatom *Thalassiosira*. Scanning electron micrographs of (*A*) the outside and (*B*) the inside, compared with a transmission electron micrograph (*C*) of a carbon replica of the outside (top) and inside (bottom). The protuberances (marginal strutted tubuli and a labiate process) on the outside of another species is shown by a scanning (*D*) and, in detail, by a transmission electron micrograph (*E*) of a carbon replica. Electron micrographs courtesy of Paul E. Hargraves

noid, which converts the products of photosynthesis into reserve material (chrysolamarin, beta-1-3-glucan). The lamellar bands (thylakoids) of the chloroplast penetrate the central pyrenoid body.

The most common form of reproduction is by asexual cell division. In natural blooms, this appears to follow patterns of diel periodicity, with some species dividing at night and others dividing during the day (Smayda, 1975). During asexual reproduction by cell division of the diploid vegetative cells, the new valves for the dividing cells are formed internally (Stoermer, Pankratz, and Bowen, 1965; Coombs and Volcani, 1968), and therefore, one of the new valves should be smaller in diameter than the hypotheca (smaller valve) of the parent cell, if the valves have no plasticity. This expected decrease in valve size during fission is illustrated in Plate 10-2,*A*,*C*. At each division, the new valve would thus decrease in diameter by the thickness of the valve, according to the hypothesis of MacDonald and Pfitzer. Theoretically, the individuals would get smaller and smaller as shown in the following example from Rao and Desikachary (1970):

	Size of individuals a >b >c >d >e >f >	No. of cells
Mother cell	1	1
At the end of 1st division	1 + 1	2
At the end of 2nd division	1 + 2 + 1	4
At the end of 3rd division	1 + 3 + 3+ 1	8
At the end of 4th division	1 + 4 + 6+ 4+ 1	16
At the end of 5th division	1 + 5 + 10+ 10+ 5 + 1	32
At the end of 6th division	1 + 6 + 15+ 20+ 15 + 6 + 1	64

This expected decrease occurs in some species but not in others. Round (1972), in discussing this discrepancy between the textbook acceptance of the necessity for a regular and necessary diminution and observations to the contrary that such a phenomenon, although occurring in some species, is not generalized (Margalef, 1969), cites the paper by Geitler (1932), who showed a number of types of plasticity in many diatoms. The plasticity of the girdle bands suggested by Round (1972) would allow the replication of a constant valve size in at least some species. There is an eventual diminution in valve size in most species, however, and this is coupled with auxospore formation to restore the clone to its normal size. The sexual process of auxospore formation is shown in a stepwise series of micrographs in Plate 10-5,*B*. Cells in the same filament often form both oogonia and sperm, but not always (Hargraves, 1972). The sequence of events in one day in the life of a marine centric diatom undergoing mitosis, meiosis, and spermatogenesis is detailed in the four-paper study by Manton, Kowallik, and von Stosch (1969a,b; 1970a,b). Geitler (1969) has published an excellent review of auxospore formation in pennate diatoms.

A number of species undergo karyokinesis (nuclear division) and cytokinesis (division of cytoplasm after nuclear division) in a number of subdivisions to yield from 8 to 128 microspores. These microspores either die within the cell or are liberated to become active amoeboid or flagellated cells, which are male gametes (Plate 10-5,*A*). A type of spore that is more common than the sexually formed auxospore is the resting spore or endocyst, whose structure varies with the species (Hargraves, 1976). The resting spore cycle in a centric diatom is shown in Plate 10-2,*D*. Resting spores are apparently most often produced by asexual processes. The protoplasm within a cell contracts into a dark spherical mass, with very thick siliceous walls. Such cells presumably lose their buoyancy and fall to the bottom to await more favorable conditions (Hendey, 1964). Resting spores can be recognized not only in culture but in the plankton, in the gut and fecal pellets of copepods, and in sediments (Plate 10-6). A study by Hargraves and French (1975) on the survival of resting diatom spores indicates that these spores are only slightly more resistant than vegetative cells. These authors conclude that the seasonal increase in specific populations may not be due to the germination of resting spores but to a recruitment of diatoms from adjacent water masses.

Diatoms require silicon for frustule formation as well as phosphorus, nitrogen, iron, and a number of trace metals. Culture media may be based on enriched natural seawater or made up from artificial seawater. Soil extract, the indispensable ingredient of *Erdschreiber* media long used in diatom culture, can be replaced by mixtures of chelated trace metals and buffers. Vitamin requirements for diatoms include at least B_{12} (cyanoco-

A

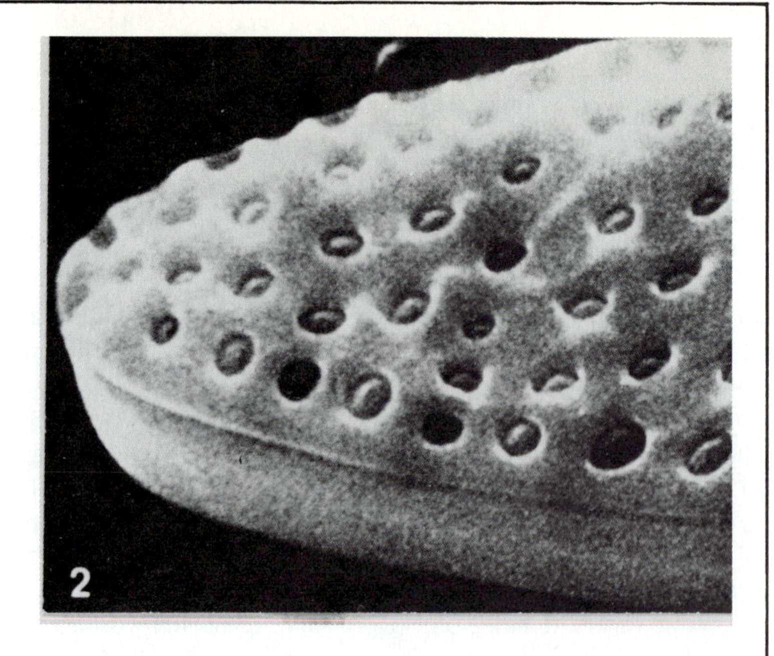

B

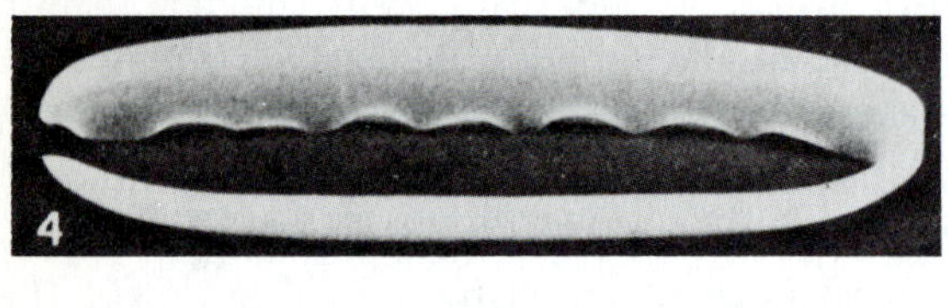

C

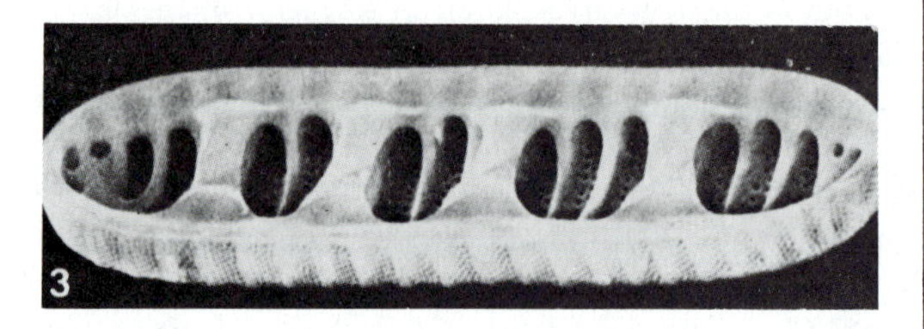

D

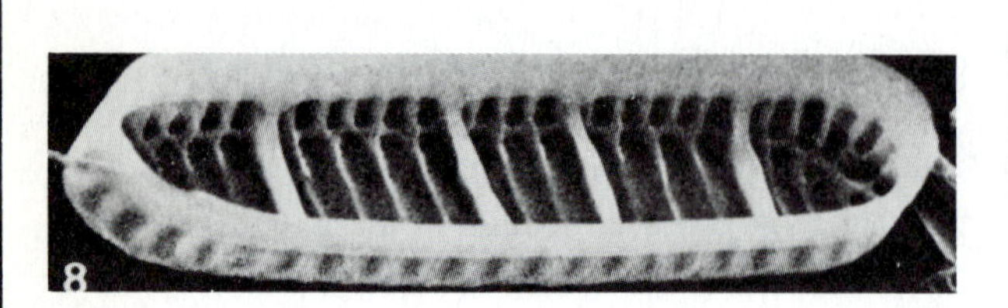

E

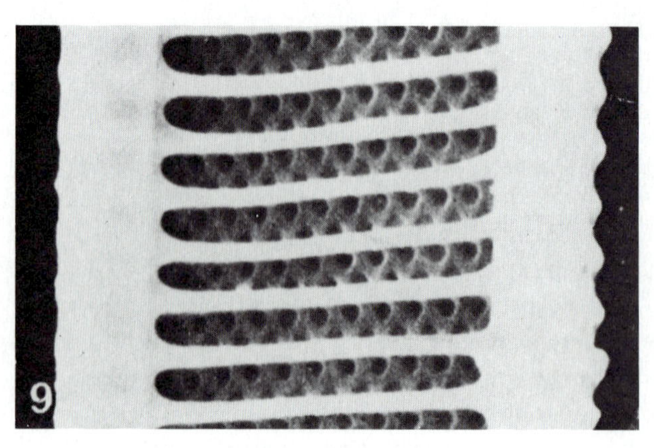

F

PLATE 10-4. Architectural details of valves of pennate diatoms as seen in scanning electron micrographs (except *A*). (*A*) Transmission electron micrograph of *Nitzschia kerguelensis*, showing valve and band (2,700 X). (*B*) Apical part of outside of valve of *N. kerguelensis* (7,500 X). (*C*), (*D*), and (*E*) *Denticula seminae*, showing band (2,400 X), inside view of valve with septa (3,700 X), and inside view of valve with pseudosepta (2,600 X), respectively. (*F*) Part of valve of *Pseudoeunotia doliolus*, showing two rows of poroids (5,600 X). From G.R. Hasle, 1972, *Nova Hedwigia Beiheft* 39:111–19.

balamin) and B_1 (thiamin). Detailed directions for formulating culture media, obtaining species in unialgal culture, and freeing them of bacteria are given in the handbook edited by Stein (1973). The art of successful diatom culture, such as choosing the correct time after autoclaving to inoculate the culture or to add reduced

PLATE 10-5. Development of male gametes and auxospore formation in the marine centric diatom *Bacteriastrum hyalinum*. (*A*) Spermatogonia form in a chain of six cells (a); 16 spermatogonia enclosed in each cell (b); a temporary swelling of the spermatogonia opens the mother cell (c); mature male gametes (sperms) inside the mother cell (d); meiotic stages in opened mother cells (e); two biflagellate cells derived from a spermatocyte after first meiotic division (f); four uniflagellate sperms developed after two meiotic divisions of a spermatocyte (g) (a–e, 100 μm scale; f–g, 25 μm scale). (*B*) During auxospore formation, two oogonia each have a sperm attached (a) (arrows); a young zygote, just leaving the oogonial shell (b); an auxospore formed laterally to the empty mother cells (c); a pre-perizonium secreted just inside the detached auxospore membrane (d); two detached auxospores attached to empty mother cells with spermatogonia forming in adjacent cells (e) (b–e, 100 μm scale; a, f, 50 μm). All from G. Drebes, 1972, *Nova Hedwigia Beiheft* 39:95–104 + 6 plates.

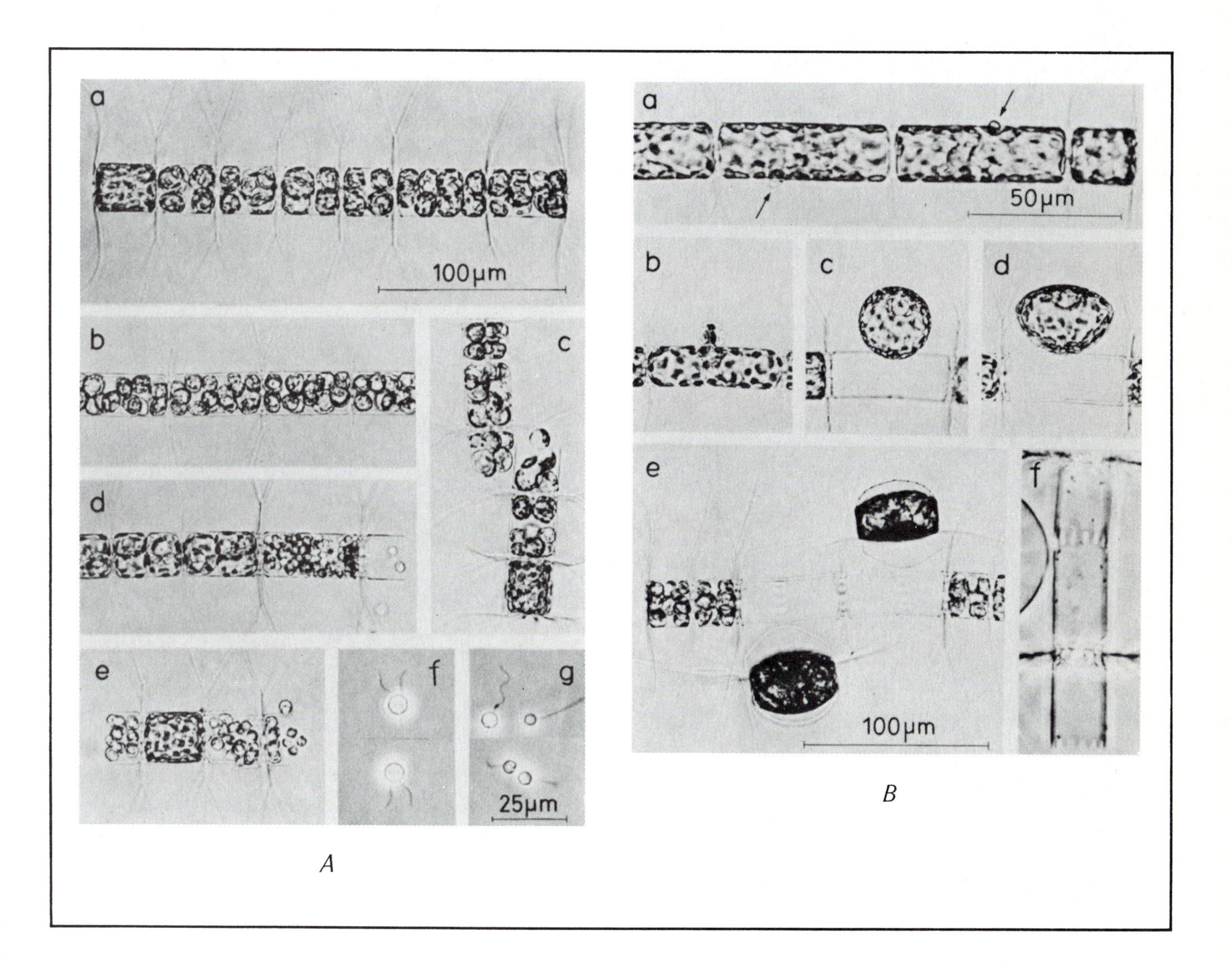

PLATE 10-6. The appearance of diatom resting spores (genus *Chaetocerus*): (*A*) in culture (phase contrast); (*B*) in the gut of a copepod (transmission electron micrograph of thin section); (*C*) in a copepod fecal pellet (arrows; phase contrast); (*D*) in sediment (arrow; phase contrast). Micrographs courtesy of Paul E. Hargraves.

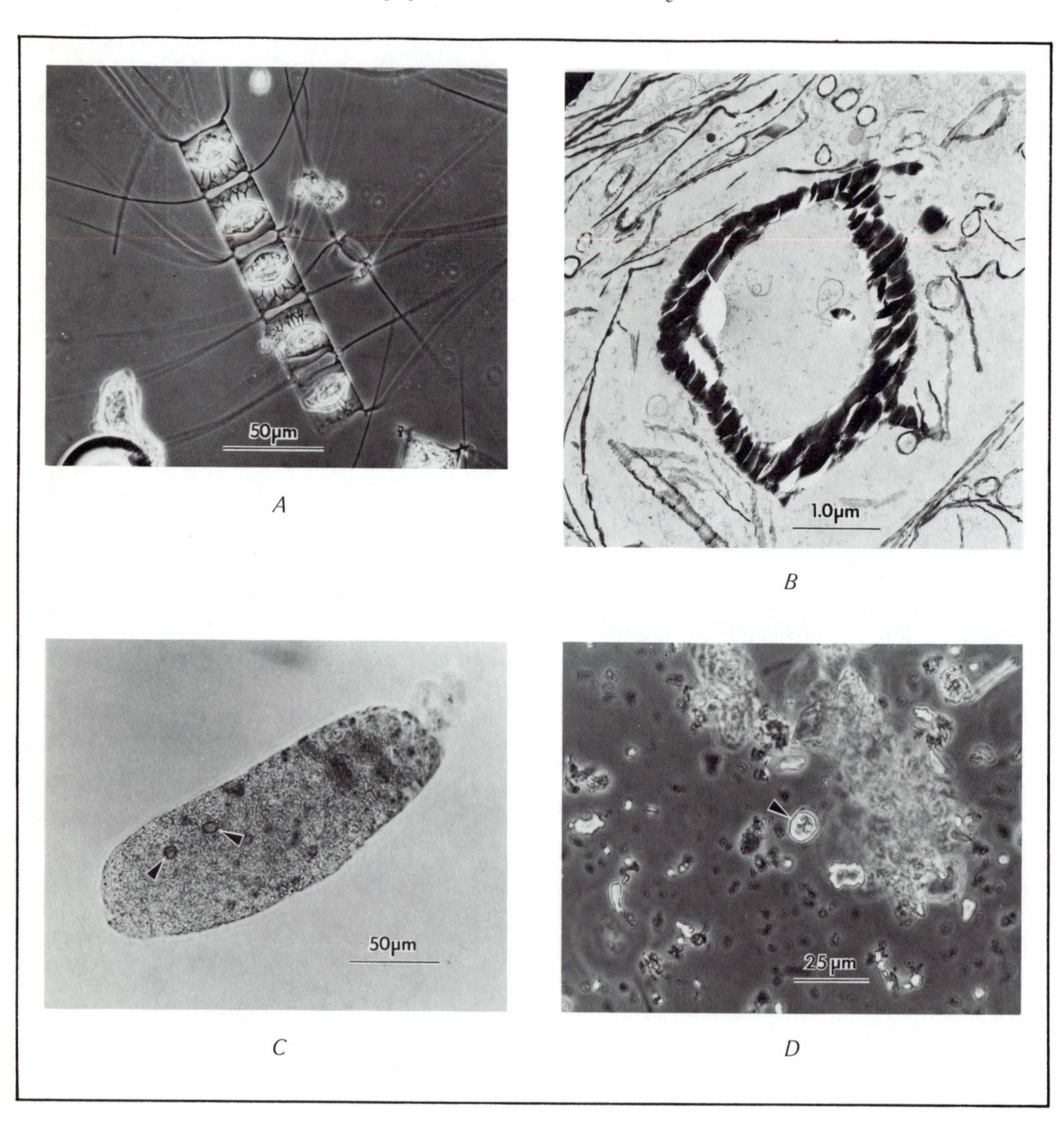

sulfur compounds to control oxygen tension (Droop, 1961), indicates that there is still much to learn. The maintenance of diatoms in culture depends also upon meeting the physical–chemical conditions required to induce auxospore formation (Lewin and Guillard, 1963). Although many species grow well in culture, it is the "fastidious" species that often provide us with the most insight into the physiology and ecology of diatoms. Dialysis culture (Jensen, Rystad, and Skoglund, 1972), although still in its infancy and fraught with problems, appears to be one way of obtaining and maintaining diatoms until their cultural conditions are met by an artificial culture medium. The maximum growth rate for marine diatoms is over three generations per day (Lewin and Guillard, 1963).

The general physiology of diatoms has been reviewed by Lewin and Guillard (1963). Recent developments in algal physiology and biochemistry are discussed in the multi-author text edited by Stewart (1974). The basic mechanism of the two photosystems is the same for all the eucaryotic phototrophic microalgae as well as the oxyphotobacteria discussed in Chapter 8,*b*. In Photosystem II, light of 400 to 500 nm wavelengths is absorbed by the accessory pigments and transferred to chlorophyll *a* in the photosynthetic membranes, where it causes charge separation in which an electron is transferred to the chlorophyll from the hydroxyl radical of water to evolve oxygen. The electron obtained at a negative redox potential is transported through an electron transport system of plastiquinone and cytochrome *b* to cytochrome *f* while ATP is formed at a positive redox potential. The cytochrome *f* is reoxidized in Photosystem I in the presence of chlorophyll and light near 700 nm in the far red. The electron is then transferred through ferrodoxin to NaDP, which is reduced and another ATP molecule is formed (Frenkel and Cost, 1966; Holm-Hansen, 1968; Brock, 1974). A critical part of diatom metabolism is, of course, the simultaneous uptake of silicon and the laying down of the valves (Lewin, 1962; Stoermer, Pankratz, and Bowen, 1965). Germanium has been found to be a specific inhibitor of silicon utilization and, therefore, of diatom growth (Lewin, 1966). Diatom DNA synthesis inhibited by silicon starvation but not by germanium inhibition can be reversed by the addition of some 50% of the silicon required for cell wall formation (Darley and Volcani, 1969). The process of germanium incorporation into diatom silica cell walls was studied by Azam, Hemmingsen, and Volcani (1973).

A number of centric and pennate diatoms have adapted to a partial or completely heterotrophic existence on rich organic surfaces. *Cyclotella cryptica*, a centric diatom from the littoral zone, was shown to be a facultative heterotroph in that it could grow on glucose in the dark (Lewin and Lewin, 1960). A study of heterotrophy at low substrate concentrations using the littoral diatom *C. cryptica* and the pelagic phytoplankters *Emiliania* (*Coccolithus*) *huxleyi*, *Skeletonema costatum*, and *Thalassiosira rotula* led Sloan and Strickland (1966) to conclude that heterotrophic survival would be impossible unless the algae were in contact with fairly labile particulate organic matter. Some centric diatoms can live benthically and could survive heterotrophically (Hellebust and Guillard, 1967). But the littoral marine diatoms are mainly pennates that do live in contact with rich organic surfaces. Of the forty-three pennate diatoms tested by Lewin and Lewin (1960), sixteen failed to grow on any substrate tested in the dark, two grew on lactate only, fifteen on glucose only, eight on either lactate or glucose, one on either acetate or glucose, and one on all three substrates.

The skin of whales becomes "infected" with yellow slime from the dominant diatom *Cocconeis ceticola* when the whales move into Antarctic waters (Bennett, 1920). This phenomenon has been well documented by Hart (1935), who found that a species of *Navicula* also penetrated the whale epidermis. *Cocconeis ceticola* and the penetration of pennates into the whale epidermis are shown in Plate 10-7,*A*. *Cocconeis scutellum* colonizes the surfaces of seagrasses (Sieburth and Thomas, 1973), algae, and terrestrial plant debris that finds its way into the sea (Sieburth, 1975; Plate 10-7,*B*). This diatom's widespread occurence in temperate and subtropical waters, its apparent attraction to organic-rich surfaces, and its penetration of the epithelium of plants and animals indicate that, in addition to being facultatively heterotrophic, it may even be a pathogen, as suggested by Guerrero (1958).

An even stronger case for the apparent dependence of some pennate diatoms on organic substances is the growth of endophytic pennates inside seaweeds (see Plate 10-7,*C*). *Navicula nidulans* was observed to live in quantities in the interior canal of the marine alga *Chylocladia* on the Australian coast (Cleve, 1892), here resembling the mucoid tubes of *Schizonema* (= *Navicula*) (Agardh, 1824). While studying the alginate gels in brown algae, Baardseth (1966) confirmed the presence

A

B

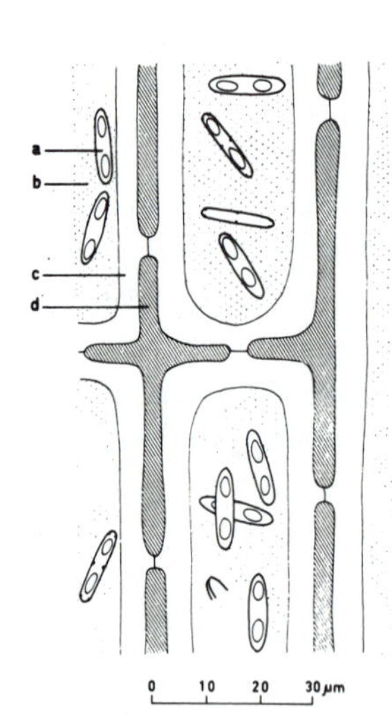

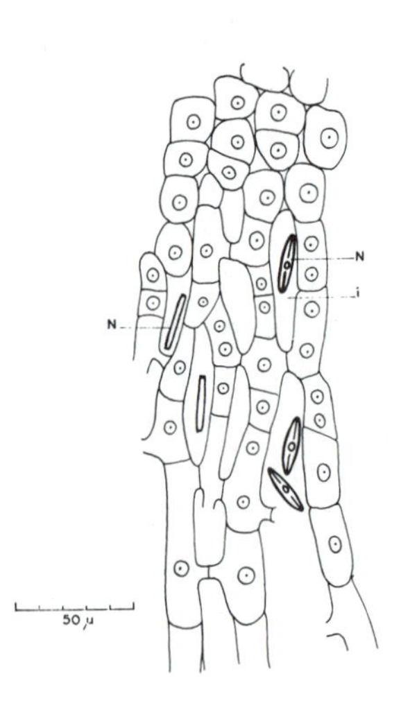

C

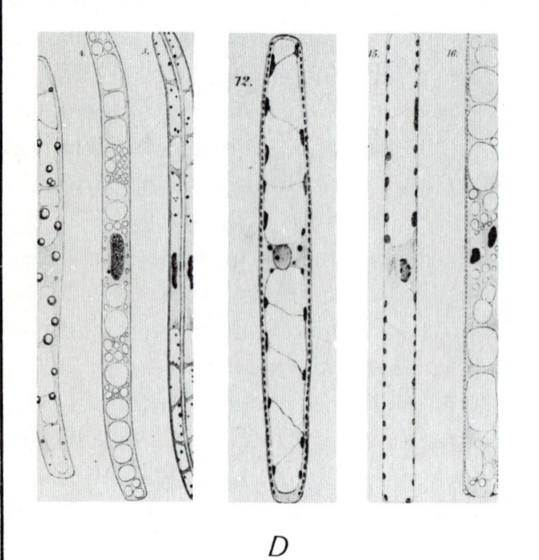

D

PLATE 10-7. Heterotrophy in diatoms. (*A*) *Cocconeis ceticola* in girdle (2) and ventral (3) views, and penetrating the epidermis of a blue whale (4); a species of *Navicula* (8), penetrating the epidermis of a sperm whale. (*B*) *Cocconeis scutellum*, infesting the surface of a green alga, *Cladophora*. (*C*) *Navicula dumontiae* endophytic in the red alga *Dumontiae incrassata* (left) and *N. endophytica* endophytic in the brown alga *Fucus vesiculosus* (right). (*D*) Colorless, obligate osmotrophic pennate diatoms from decomposing seaweeds. *Nitzschia leucosigma* (left three diatoms) and *Nitzschia putrida* (right), as drawn by Benecke (1900), in the original descriptions. (*A*) from T.J. Hart, 1935, *Discovery Repts.* 10:247–282; (*B*) scanning electron micrograph courtesy of Paul W. Johnson; (*C*) from E. Baardseth and J.P. Taasen, 1973, *Norweg. J. Botany* 20:79–87, and E. Baardseth, 1969, *Intl. Seaweed Symp. Proc.* 6:53–60; (*D*) from W. Benecke, 1900, *Jahrb. wiss. Botanis.* 35:535–72.

of dense concentrations of a solitary species of *Navicula* in the intercellular alginate gels in the receptacles and growing tissues of *Ascophyllum nodosum*. Hasle (1968) named this species *Navicula endophytica. Cocconeis scutellum* was also found to be abundant in the intercellular material of *A. nodosum*. The distribution and ecology of *N. endophytica* and *C. scutellum* as endophytes are discussed by Baardseth (1969) and, in more detail, by Taasen (1972). *Navicula dumontiae* has been reported from the interior mucilage of the red alga *Dumontia incrassata* by Baardseth and Taasen (1973). These authors also reported additional animal and algal finds of mucilage-dwelling diatoms and concluded that the endophytic diatoms are not trapped but thrive in the mucilage in high concentrations; usually involved are one, or more rarely two, species that are extremely rare or absent from the surrounding water.

A few diatom species are obligate heterotrophs. These colorless diatoms, which lack chloroplasts (i.e., they are apochlorotic), are shown in drawings from the original description by Benecke (1900) in Plate 10-7,*D*. Benecke described both *Nitzschia putrida* and *Nitzschia leucosigma*, which he considered to be common inhabitants of the littoral zone and associated with decaying seaweed. Pringsheim (1951) confirmed these observations. Lewin and Lewin (1967) isolated in pure culture Benecke's two species and a third, *Nitzschia alba*, from decaying seaweeds and other organic debris. Two of the species utilized glucose and glutamate as the sole carbon sources; all three utilized lactate and succinate. In addition to nitrate as a source of nitrogen, two species could also use glutamate; all three required both cobalamine and thiamin. Lewin and Lewin speculated that these diatoms originated as non-photosynthesizing mutants capable of rapid growth on organic substrates, which would be a requirement for successful competition with the bacteria and fungi found in the same habitat.

The diatoms are put in the class Bacillariophyceae, which is usually divided into the orders Pennales and Centrales because of their obvious differences in shape and symmetry. Some taxonomists believe, however, that this is an artificial separation and therefore group them together in one order, the Bacillariales (Hendey, 1974). The durable diatom skeletons of some species, whose intricate structural details persist in fossilized deposits of diatomaceous earth, provide an almost endless variety of elegant patterns that unfold in further detail with every increase in magnification up to the limit of resolution of the electron microscope. Taxonomy of the diatoms is based almost solely on the structure and pattern of these skeletons. The pennates are put into suborders according to the development of the raphe and the pseudoraphe, whereas the centric diatoms are put into suborders according to the shape and structure of the frustule and valves and any appendages present. On the generic and species levels, all diatoms are separated on the basis of detailed structure, such as the number of striae and the type, number, and arrangement of punctae and areolae.

As taxonomic studies advanced from light to electron microscopy (Helmcke and Krieger, 1953–1966; Hendey, 1959), some of the redundant taxa that existed because of inherent variations in cell size and different stages in the life cycle of some diatoms (Lewin and Guillard, 1963) have been eliminated. But other problems, particularly in morphology and phylogeny, have arisen. The fundamental basis for diatom taxonomy, which is the oldest among the microorganisms, seems to be withstanding technological developments in microscopy, although there will always be suggestions for improvements (Ross, 1972; Simonsen, 1972). References to the fundamental monographs can be found in Lewin and Guillard (1963), Hendey (1964, 1974), Van Landingham (1967–1975), and Hostetter and Stoermer (1971). The diatoms have undoubtedly received more attention than any other group of microorganisms in the sea. This is due in part to their obvious role in the food chain, to their beautiful anatomy, which has infatuated many through the years, and to their durability as fossil remains. A good starting point for the novice is the monographs by Cupp (1943) and Hendey (1964).

REFERENCES

Agardh, C.A. 1824. *Systema algarum*. Lundae, 312 p. (Reissued 1965 by A. Asher & Co., Amsterdam.)

Aleem, A.A. 1949. Distribution and ecology of marine littoral diatoms. (Consideration of the littoral diatom-flora with special reference to forms living in gelatinous tubes.) *Bot. Not., År* 1949:414–40.

Aleem, A.A. 1950. Distribution and ecology of British marine littoral diatoms. *J. Ecology* 38:75–106.

Allen, W.E. 1941. Depth relationships of plankton diatoms in sea water. *J. Mar. Res.* 4:107–11.

Azam, F., B.B. Hemmingsen, and B.E. Volcani. 1973. Germanium incorporation into silica diatom cell walls. *Arch. Mikrobiol.* 92:11–20.

Baardseth, E. 1966. Localization and structure of alginate gels. *Intl. Seaweed Symp. Proc.* 5:19–28.

Baardseth, E. 1969. Some aspects of the native intercellular substance in Fucaceae. *Intl. Seaweed Symp. Proc.* 6:53–60.

Baardseth, E., and J.P. Taasen. 1973. *Navicula dumontiae* sp. nov., an endophytic diatom inhabiting the mucilage of *Dumontia incrassata* (Rhodophyceae). *Norw. J. Bot.* 20:79–87.

Benecke, W. 1900. Ueber farblose Diatomeen der Kieler Föhrde. *Jahrb. wiss. Bot.* 35:535–72.

Bennett, A.G. (with appendix by E.W. Nelson). 1920. On the occurrence of diatoms on the skin of whales. *Proc. Royal Soc. London Ser. B* 91:352–57.

Brock, T.D. 1974. *Biology of Microorganisms* (2nd Ed.). Prentice-Hall, Englewood Cliffs, N.J., 852 p.

Cleve, P.T. 1892. Diatomées rares ou nouvelles. *Le Diatomiste* 1:75–78.

Coombs, J., and B.E. Volcani. 1968. Studies on the biochemistry and fine structure of silica-shell formation in diatoms. *Planta* 82:280–92.

Cupp, E.E. 1943. Marine plankton diatoms of the west coast of North America. *Bull. Scripps Inst. Oceanogr.* No. 5, 238 p.

Darley, W.M., and B.E. Volcani. 1969. Role of silicon in diatom metabolism. *Exp. Cell Res.* 58:334–42.

Drebes, G. 1972. The life history of the centric diatom *Bacteriastrum hyalinum* Lauder. *Nova Hedwigia Beiheft* 39:95–104.

Drebes, G. 1974. *Marines Phytoplankton*. Georg Thieme Verlag, Stuttgart, 186 p.

Dring, M.G. 1971. Light quality and the photomorphogenesis of algae in marine environments. *Fourth Eur. Mar. Biol. Symp.* (D.J. Crisp, ed.). Cambridge University Press, pp. 375–92.

Droop, M.R. 1961. Some chemical considerations in the design of synthetic culture media for marine algae. *Bot. Mar.* 2:231–46.

Drum, R.W., and J.T. Hopkins. 1966. Diatom locomotion: An explantion. *Protoplasma* 62:1–33.

Fenchel, T.M. 1969. The ecology of marine microbenthos. IV. Structure and function of the benthic ecosystem, its chemical and physical factors and the microfauna communities with special reference to the ciliated protozoa. *Ophelia* 6:1–182.

Frenkel, A.W., and K. Cost. 1966. Photosynthetic phosphorylation. In: *Comprehensive Biochemistry, Vol. 14. Biological Oxidations* (M. Florkin and E.H. Stotz, eds.). Elsevier, New York, pp. 397–423.

Geitler, L. 1932. Der Formwechsel der pennaten Diatomeen (Kieselalgen). *Arch. Protistenk.* 78:1–226.

Geitler, L. 1969. Comparative studies on the behavior of allogamous pennate diatoms in auxospore formation. *Amer. J. Bot.* 56:718–22.

Gordon, R., and R.W. Drum. 1970. A capillarity mechanism for diatom gliding locomotion. *Proc. Natl. Acad. Sci. USA* 67:338–44.

Gran, H.H. 1912. Pelagic plant life. In: *The Depths of the Ocean* (J. Murray and J. Hjort, eds.). MacMillan, London, pp. 307–86.

Grøntved, J. 1949. Investigations on the phytoplankton in the Danish Waddensea in July 1941. *Medd. Komm. Dan. Fisk. Havundersøgelser Kobenhaun, Ser. Plankton* 5(2):1–56.

Guerrero, P.G. 1958. La "Ficosis" enfermedad mortal en la hidrofitia. *Rev. Algologique* 4:161–69.

Hargraves, P.E. 1972. Studies on marine plankton diatoms. I. *Chaetoceros diadema* (Ehr.) Gran: Life cycle, structural morphology and regional distribution. *Phycologia* 11:247–57.

Hargraves, P.E. 1976. Studies on marine plankton diatoms. II. Resting spore morphology. *J. Phycol.* 12:118–28.

Hargraves, P.E., and F. French. 1975. Observations on the survival of diatom resting spores. *Nova Hedwigia Beiheft.* 53:229–38.

Harper, M.A. 1969. Movement and migrations of diatoms on sand grains. *Br. phycol. J.* 4:97–103.

Harper, M.A., and J.F. Harper. 1967. Measurements of diatom adhesion and their relationship with movement. *Br. phycol. J.* 3:195–207.

Hart, T.J. 1935. On the diatoms of the skin film of whales, and their possible bearing on problems of whale movements. *Discovery Reports* 10:247–82.

Hasle, G.R. 1968. *Navicula endophytica* sp. nov., a pennate diatom with an unusual existence. *Br. phycol. Bull.* 3:475–80.

Hasle, G.R. 1972a. Two types of valve processes in centric diatoms. *Nova Hedwigia Beiheft.* 39:55–78.

Hasle, G.R. 1972b. *Fragilariopsis* Hustedt as a section of the genus *Nitzschia* Hassall. *Nova Hedwigia Beiheft.* 39:111–19.

Hellebust, J.A., and R.R.L. Guillard. 1967. Uptake specificity for organic substrates by the marine diatom *Melosira mummuloides*. *J. Phycol.* 3:132–36.

Helmcke, J.G., and W. Krieger. 1953–66. *Diatomeenschalen im Elektronenmikroskopischen Bild.* J. Cramer, Lehre, Germany, 7 volumes, 613 plates.

Hendey, N.I. 1937. The plankton diatoms of the southern seas. *Discovery Reports* 16:151–364.

Hendey, N.I. 1959. The structure of the diatom cell wall as revealed by the electron microscope. *J. Quekett. Microscop. Club* 5:147–75.

Hendey, N.I. 1964. *An Introductory Account of the Smaller Algae of British waters. Part V: Bacillariophyceae (Diatoms).* Fishery Invest. Series IV, London, 317 p., 45 plates.

Hendey, N.I. 1974. A revised check-list of British marine diatoms. *J. mar. Biol. Assoc. U.K.* 54:277–300.

Holm-Hansen, O.H. 1968. Ecology, physiology and biochemistry of blue-green algae. *Ann. Rev. Microbiol.* 22:47–70.

Hostetter, H.P., and E.F. Stoermer. 1971. Bibliography on the Bacillariophyceae. In: *Selected Papers in Phycology* (J.R. Rosowski and B.C. Parker, eds.). University of Nebraska Botany Dept., Lincoln, Neb., pp. 784–90.

Jensen, A., B. Rystad, and L. Skoglund. 1972. The use of dialysis culture in phytoplankton studies. *J. exp. mar. Biol. Ecol.* 8:241–48.

Lewin, J.C. 1962. Silicification. In: *Physiology and Biochemistry of Algae* (R.A. Lewin, ed.). Academic Press, New York and London, pp. 445–55.

Lewin, J.C. 1966. Silicon metabolism in diatoms. V. Germanium dioxide, a specific inhibitor of diatom growth. *Phycologia* 6:1–12.

Lewin, J.C., and R.R.L. Guillard. 1963. Diatoms. *Ann. Rev. Microbiol.* 17:373–414.

Lewin, J.C., and R.A. Lewin. 1960. Auxotrophy and heterotrophy in marine littoral diatoms. *Can. J. Microbiol.* 6:127–34.

Lewin, J., and R.A. Lewin. 1967. Culture and nutrition of some apochlorotic diatoms of the genus *Nitzschia*. *J. Gen. Microbiol.* 4:361–67.

Linke, O. 1939. Die Biota des Jadebusenwattes. *Helgol. wiss. Meeresunters.* 1:201–348.

Margalef, R. 1969. Size of centric diatoms as an ecological indicator. *Mitt. int. Ver. Limnol.* 17:202–10.

Manton, I., K. Kowallik, and H.A. von Stosch. 1969a. Observations on the fine structure and development of the spindle at mitosis and meiosis in a marine centric diatom (*Lithodesmium undulatum*). I. Preliminary survey of mitosis in spermatogonia. *J. Microscopy* 89:295–320.

Manton, I., K. Kowallik, and H.A. von Stosch. 1969b. *Ibid.* II. The early meiotic stages in male gametogenesis. *J. Cell Sci.* 5:271–98.

Manton, I., K. Kowallik, and H.A. von Stosch. 1970a. *Ibid.* III. The later stages of meiosis I in male gametogenesis. *J. Cell Sci.* 6:131–57.

Manton, I., K. Kowallik, and H.A. von Stosch. 1970b. *Ibid.* IV. The second meiotic division and conclusion. *J. Cell. Sci.* 7:407–43.

Palmer, J.D. 1975. Biological clocks of the tidal zone. *Sci. American* 232(2):70–79.

Palmer, J.D., and F.E. Round. 1967. Persistent, vertical migration rhythms in benthic microflora. VI. The tidal and diurnal nature of the rhythm in the diatom *Hantzschia virgata*. *Biol. Bull.* 132:44–55.

Parsons, T.R., M. Takahashi, and B.T. Hargrave. 1977. *Biological Oceanographic Processes.* 2nd Ed. Pergamon, Oxford, 332 p.

Perkins, E.J. 1974. *The Biology of Estuaries and Coastal Waters.* Academic Press, London and New York, 678 p.

Pringsheim, E.G. 1951. Uber farblose Diatomeen. *Arch. Mikrobiol.* 16:18–27.

Rao, V.N.R., and T.V. Desikachary. 1970. MacDonald-Pfitzer hypothesis and cell size in diatoms. *Nova Hedwigia Beiheft.* 31:485–93.

Raymont, J.E.G. 1963. *Plankton and Productivity in the Oceans.* Pergamon, Oxford, 672 p.

Riley, G.A. 1943. Physiological aspects of spring diatom flowering. *Bull. Bingham Oceanogr. Coll.* 8, Art. 4, 53 p.

Ross, R. 1972. The current state of diatom taxonomy at the species level, with special reference to some species of *Navicula* Sect. *Lyratae*. *Nova Hedwigia Beiheft.* 39:1–36.

Round, F.E. 1971. Benthic marine diatoms. *Oceanogr. Mar. Biol. Ann. Rev.* 9:83–179.

Round, F.E. 1972. The problem of reduction of cell size during diatom cell division. *Nova Hedwigia* 23:291–303.

Round, F.E., and J.D. Palmer. 1966. Persistent, vertical-migration rhythms in benthic microflora. II. Field and laboratory studies on diatoms from the banks of the river Avon. *J. mar. biol. Assoc. U.K.* 46:191–214.

Savage, R.E. 1925. The food of the oyster. *Fish. Invest. London, Ser. 2* 8(1):1–50.

Sieburth, J.McN. 1975. *Microbial Seascapes.* University Park Press, Baltimore, Md., 248 p.

Sieburth, J.McN., and C.D. Thomas. 1973. Fouling on eelgrass (*Zostera marina* L.). *J. Phycol.* 9:46–50.

Simonsen, R. 1972. Ideas for a more natural system of the centric diatoms. *Nova Hedwigia Beiheft.* 39:37–54.

Skerman, T.M. 1956. The nature and development of primary films on surfaces submerged in the sea. *New Zeal. J. Sci. Tech. Ser. B* 38:44–57.

Sloan, P.R., and J.D.H. Strickland. 1966. Heterotrophy of four marine phytoplankters at low substrate concentrations. *J. Phycol.* 2:29–32.

Smayda, T.J. 1958. Biogeographical studies of marine phytoplankton. *Oikos* 9:158–91.

Smayda, T.J. 1970. The suspension and sinking of phytoplankton in the sea. *Oceanogr. Mar. Biol. Ann. Rev.* 8:353–414.

Smayda, T.J. 1975. Phased cell division in natural populations of the marine diatom *Ditylum brightwelli* and the potential significance of diel phytoplankton behavior in the sea. *Deep-Sea Res.* 22:151–65.

Smayda, T.J., and B.J. Boleyn. 1966a. Experimental observations on the flotation of marine diatoms. I. *Thalassiosira* cf. *nana Thalassiosira rotula*, and *Nitzschia seriata*. *Limnol. Oceanogr.* 10:499–509.

Smayda, T.J., and B.J. Boleyn. 1966b. *Ibid.* II. *Skeletonema costatum* and *Rhizosolenia setigera*. *Limnol. Oceanogr.* 11:18–34.

Smayda, T.J., and B.J. Boleyn. 1966c. *Ibid.* III. *Bacteriastrium hyalinum* and *Chaetoceros lauderi*. *Limnol. Oceanogr.* 11:35–43.

Stein, J.R. (ed.). 1973. *Handbook of Phycological Methods. Culture Methods and Growth Measurements.* Cambridge University Press, New York and London, 448 p.

Stewart, W.D.P. 1974. *Algal Physiology and Biochemistry.* University of California Press, Berkeley, 989 p.

Stoermer, E.F., H.S. Pankratz, and C.C. Bowen. 1965. Fine structure of the diatom *Amphipleura pellucida*. II. Cytoplasmic fine structure and frustule formation. *Amer. J. Bot.* 52:1067–78.

Taasen, J.P. 1972. Observations on *Navicula endophytica* Hasle (Bacillariophyceae). *Sarsia* 51:67–82.

Van Landingham, S.L. 1967–1975. *Catalogue of the Fossil and Recent Genera and Species of Diatoms and their Synonyms.* Parts 1–5. J. Cramer Verlag, Lehre, Germany.

CHAPTER 11

Dinoflagellates

The dinoflagellates are characterized by chromosomes that are usually condensed throughout the cell cycle and by two morphologically different flagella. The walls on certain forms are sufficiently thin to give them a "naked" or non-thecate appearance. In other forms, the cellulosic walls, or theca, in the form of plates of characteristic number and structure are sufficiently thick to give them an "armored" appearance. In addition to phototrophic forms, many dinoflagellates are apochlorotic and phagotrophic; these are discussed in Chapter 22,C. Dinoflagellates of both trophic modes may become major components of the plankton. Phototrophic dinoflagellates are second only to the diatoms as planktonic primary producers in the sea. Nonmotile stages of dinoflagellates, called zooxanthellae, are widely distributed as endosymbionts, especially in the cells of protozoans and the tissues of such invertebrates as corals, gorgonians, anemones, nudibranchs, and tridacnid clams, which they pigment. The zooxanthellae serve as an "internal phytoplankton" and provide their hosts with a substantial portion of their photosynthate. These forms are discussed in Chapter 3,C.

The free-living dinoflagellates are ubiquitous, occurring at low and high latitudes and in coastal and open ocean environments. Arctic forms that thrive in subzero waters have been described by Braarud (1935), although dinoflagellates can be thought of mainly as the warm water counterparts of diatoms (Ryther, 1954). In temperate waters, they may have a periodicity such that they increase in number in summer and decrease in winter (Allen, 1941a). After the winter–spring bloom of diatoms, the warming of the waters often leads to dinoflagellate blooms, which alternate with short spikes of diatom blooms from late spring through the fall. This sequence has been recorded for estuaries (Braarud, 1945a), slope water (Gran and Braarud, 1935), and the open sea (Herdman, 1922). There are fewer temperate species than tropical species, but the temperate species often occur as larger populations. Dinoflagellates often dominate the net plankton of the warmer waters of the subtropics and tropics, the populations usually being poor in numbers but rich in species. This is quite apparent where a warm water mass penetrates a usually colder mass (Balech, 1958). The features of the tropical dinoflagellate communities in the Indian Ocean have been discussed by Taylor (1973b).

Most of our knowledge of populations and distribution concerns the "armored" dinoflagellates, since many of the "naked" forms are either damaged by or pass through the usual phytoplankton nets in towing or are destroyed when the samples are prepared for examination. The use of unconcentrated water samples avoids some of the above problems, but limits the number of individuals that can be collected and counted, and many important species may be missed. Of some 5,000 water samples examined from the eastern North Pacific, dinoflagellate populations as high as 1,000 cells/liter occurred in some 500 samples only (Allen, 1941b). The larger populations were confined almost exclusively to the upper 30 m (most being in the upper 5 to 10 m) and to a zone within 40 km of the shore. Daily observations on samples from the Scripps Oceanographic Pier at La Jolla, California over a 20-year period (Allen, 1941b) indicated a streakiness and patchiness of surface distribution and a winter–summer periodicity, as well as 5-year periods of alternately higher and lower standing crops. Of the forty-six forms encountered, four were dominant and two of these formed red tides on occasion.

The motility and phototactic response of this group of phototrophs are major factors in their distribution and ecology. Observations on natural populations of dinoflagellates during blooms in the sewage-enriched, turbid (1% light at 10 to 12 m) and shallow (< 50 m) waters of the inner Oslo fjord (Hasle, 1950) indicated that maximum populations of 10^4 to 10^5 cells/liter occurred within the upper 2 m but not at 0 m. Different species had different phototactic responses. *Gonyaulax polyedra* and *Prorocentrum micans* rose to the surface by day and descended by night, while *Ceratium fusus* and *C. tripos* descended by day and rose to the surface by night. These regular migrations may account in part for the regular diurnal rhythm (diel periodicity) of cell division in the dinoflagellates (Sweeney and Hastings, 1962). Observations on the vertical migration of a natural bloom of *Ceratium furca* off the California coast were made by Eppley, Holm-Hansen, and Strickland (1968), using fluorometric analysis of chlorophyll as an index of *in-situ* biomass. The main layer of cells, in the upper 2 m, migrated to 5 m 2 hours after sunset and was dispersed between 5 and 16 m 4.5 hours after sunset. Similar observations by these authors on cultures of other dinoflagellates in a 3 × 10 m column indicated sinking rates of 1 to 2 m/hour; the upward and downward trends of these cultures preceded the turning on and off of the lights, respectively.

The concentration of dinoflagellate cells in coastal and estuarine environments can cause detectable surface discolorations known as "red water" or "red tides." Since these red tides are often accompanied by widespread kills of other marine organisms, they have received much attention (LoCicero, 1975). The reasons for their occurrence are not known. Some large tides are associated with the incursion of fresh waters from rivers, which not only lowers the salinity (Ryther, 1955) but may also supply iron, among other growth factors (Dragovich, Kelly, and Goodell, 1968). An investigation of why the dinoflagellate *Pyrodinium bahamense* persists endemically in large accumulations in Oyster Bay, Jamaica, where the flushing rate exceeds the division rate (Seliger et al., 1970), indicated two processes. The first was an avoidance of flushing by enrichment in the surface waters due to phototaxis during the morning, when prevailing winds drove the surface layer into the protected shallows; the second process was a further concentration on sunny days, when *P. bahamense*, which was originally concentrated in the lower saline layer, migrated into the warmer brackish layer to form a reddish-brown layer containing up to 10^7 cells/liter.

Some dinoflagellates have adapted to a life among the sand grains, as have some diatoms, appearing when the sand becomes exposed to discolor the sand and disappearing just before the sand is reimmersed. Such sand-inhabiting dinoflagellates are dominated by species of *Amphidinium* and *Hemidinium*, although species in the genera *Gymnodinium* and *Exuviaella* are also found (Herdman, 1920–1924.)

The most conspicuous members of the dinoflagellates are characterized by their unique flagellar arrangement and their cell walls of distinctive thecal plates of varying thickness, form, and pattern. Ultrastructural studies show that the most striking features of the dinoflagellates are the nucleus, with its prominent condensed chromosomes; a unique arrangement of flagella; trichocysts, pusules; and a spectrum of thecal forms.

The free-living dinoflagellates have three basic cell

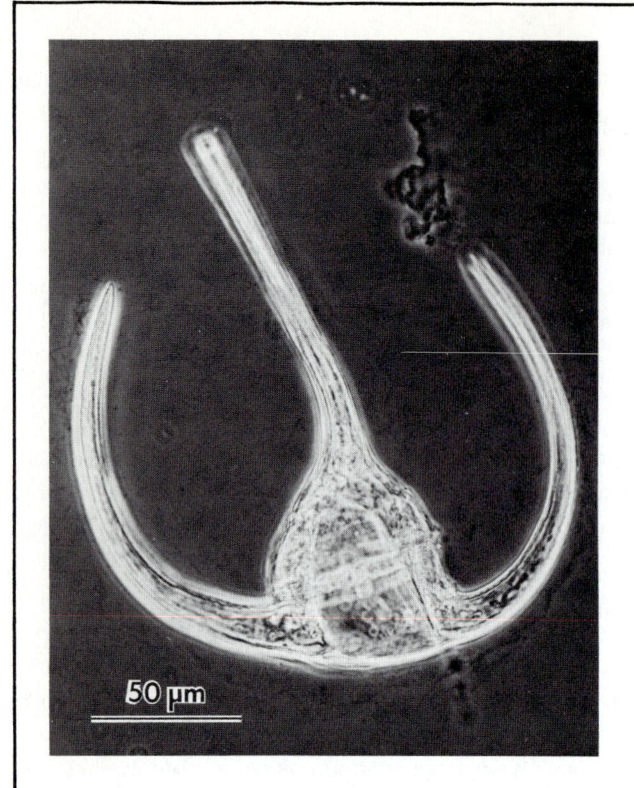

A

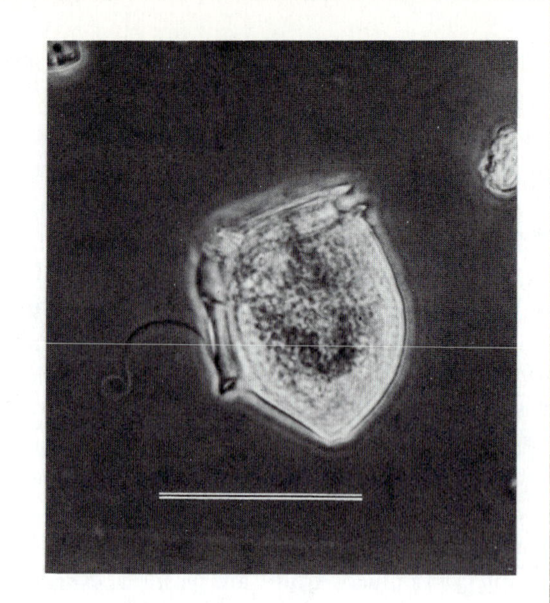

B

C

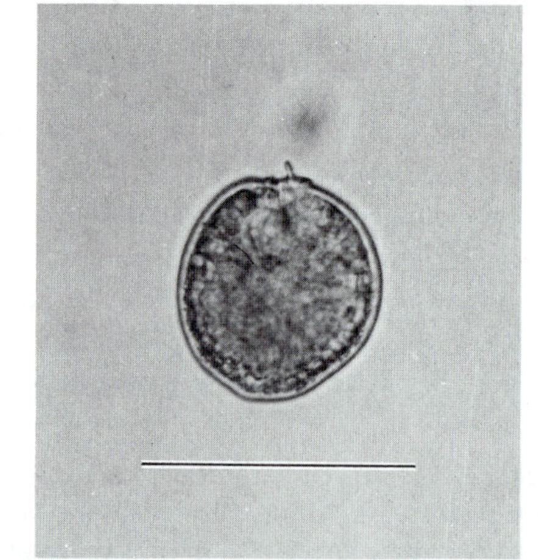

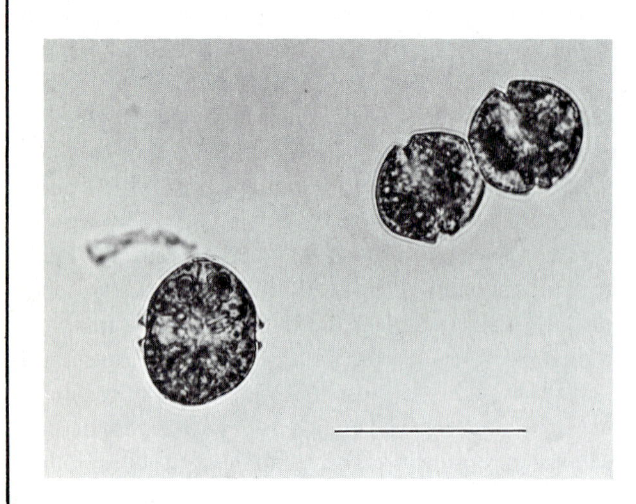

D

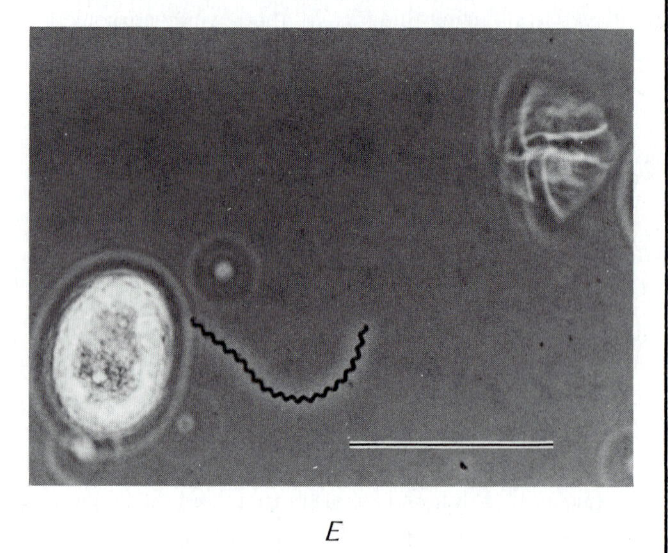

E

PLATE 11-1. The three basic cell morphologies of dinoflagellates, as seen by phase-contrast microscopy. (*A*) A *Ceratium* sp., with the peridinoid morphology of two hemispheres separated by a groove containing the transverse flagellum. (*B*) A species of *Dinophysis*, with the dinophysoid morphology of a small anterior hemisphere (epicone) and anterior groove containing the transverse flagellum. The plate arrangement is the real distinguishing feature, since some peridinoids look superficially similar. Note the longitudinal flagellum extending from the sulcus. (*C*) A *Prorocentrum* (*Exuviella*), with the prorocentrid morphology of a walnut-shell–like theca without a girdle and transverse flagellum. The two anterior flagella are not visible (*A*,*B*, and *C* from Narragansett Bay, R.I.). (*D*) *Gonyaulax tamarensis*, the red tide dinoflagellate from the Gulf of Maine, with peridinoid morphology, and (*E*) its spirochete-like transverse flagellum, which has become detached at its distal end to give the cell an aberrant motility. Phase-contrast micrographs courtesy of Paul E. Hargraves.

morphologies corresponding to the three orders recognized by Parke and Dixon (1968) (see Plate 11-1). Dinoflagellates in the order Peridiniales have cells in the form of two hemispheres (an anterior epitheca or epicone and a posterior hypotheca or hypocone) separated by a groove or girdle (cingulum). The two flagella differ in position and appearance: the longitudinal flagellum extending from the sulcus beats in the anterior plane, whereas the circular and ribbon-like, transverse flagellum undulates around the cingulum area. Dinoflagellates in the order Dinophysiales also have a well-developed girdle and sulcus in the anterior position, with a theca divided into subequal halves (Kofoid and Skogsberg, 1928), additional plates in the sulcus, and a pattern of more than 17 distinct plates, which bear no similarity to those of the order Peridiniales (Tai and Skogsberg, 1934). Dinoflagellates in the third order, the Prorocentrales, resemble a walnut and lack the two hemispheres and characteristic girdle of the above two orders. One flagellum beats in an anterior plane while the other coils against the cell. A number of taxonomic and systematic changes have been made in the Dinophyta since the second "Check-list of British Marine Algae" was compiled by Parke and Dixon (1968). This is reflected in the third revision of the "Check-list" (Parke and Dixon, 1976), in which the number of dinoflagellate orders has increased from three to nine to account for those forms that are apart from the main three groups, such as the phagotrophic *Noctiluca* and the parasitic species of *Dissodinium*. Recent revisions in this order have been discussed by Dodge (1975).

The flagella of the dinoflagellates, which have been described in detail by Leadbeater and Dodge (1967a,b), possess the normal axoneme structure with (9 + 2) filaments, the nine outer doublets forming triplets in the basal region. Unlike other flagella, they have two basal discs and two diaphragms in the transition region. The two flagella in the Peridiniales differ in both external morphology and extra-axonemal structure (see Plate 11-2). The longitudinal flagellum, which bears short fine hairs, consists of the axoneme only in the distal one-third and much packing material in the proximal two-thirds. The transverse flagellum is unique in that a striated strand running parallel to the axoneme gives it a ribbon-like form; its expanded sheath bears unilateral long fine hairs. The axoneme was thought to spiral around the striated strand, by Leadbeater and Dodge (1967a,b), but as a result of studies by scanning electron microscopy, this was shown not to be the case (Taylor, 1975), the axoneme remaining external to the strand. The mechanism of action of the flagella was observed by Jahn, Harmon, and Landman (1963), by a cinematic and photographic analysis of polystyrene bead patterns caused by flagellar action and a study of a model of the transverse flagellum. The longitudinal flagellum beats in a plane, in the form of a sine wave, while the transverse flagellum beat resembles a circular or an elliptical–helical wave. Their explanation for the forward thrust component of the transverse flagellum was based on the assumption that the flagellum was helical. LeBlond and Taylor (1976) have offered another explanation for the forward thrust associated with the transverse flagellum, which involves the shape of the axonemal wave and, secondarily, a form of "jet propulsion" in which water pushed around the girdle is redirected posteriorly by the sulcus. They noted that the axoneme attempts to beat in a spiral but must remain on the outer edge of the flagellar ribbon while doing so.

The classical distinction between "naked" or "unarmored" dinoflagellates and "armored" peridinoid dinoflagellates (which possess a cellulosic theca) is difficult to make. Many dinoflagellates earlier described as naked are now known to have very delicate plates as revealed by current microscopic studies. A comprehensive survey of the fine structure of the outer regions of the flagellated members of the dinoflagellates by transmission electron microscopy of thin sections (Dodge and Crawford, 1970) showed that all dinoflagellate forms had an outer region consisting of a continuous plasma

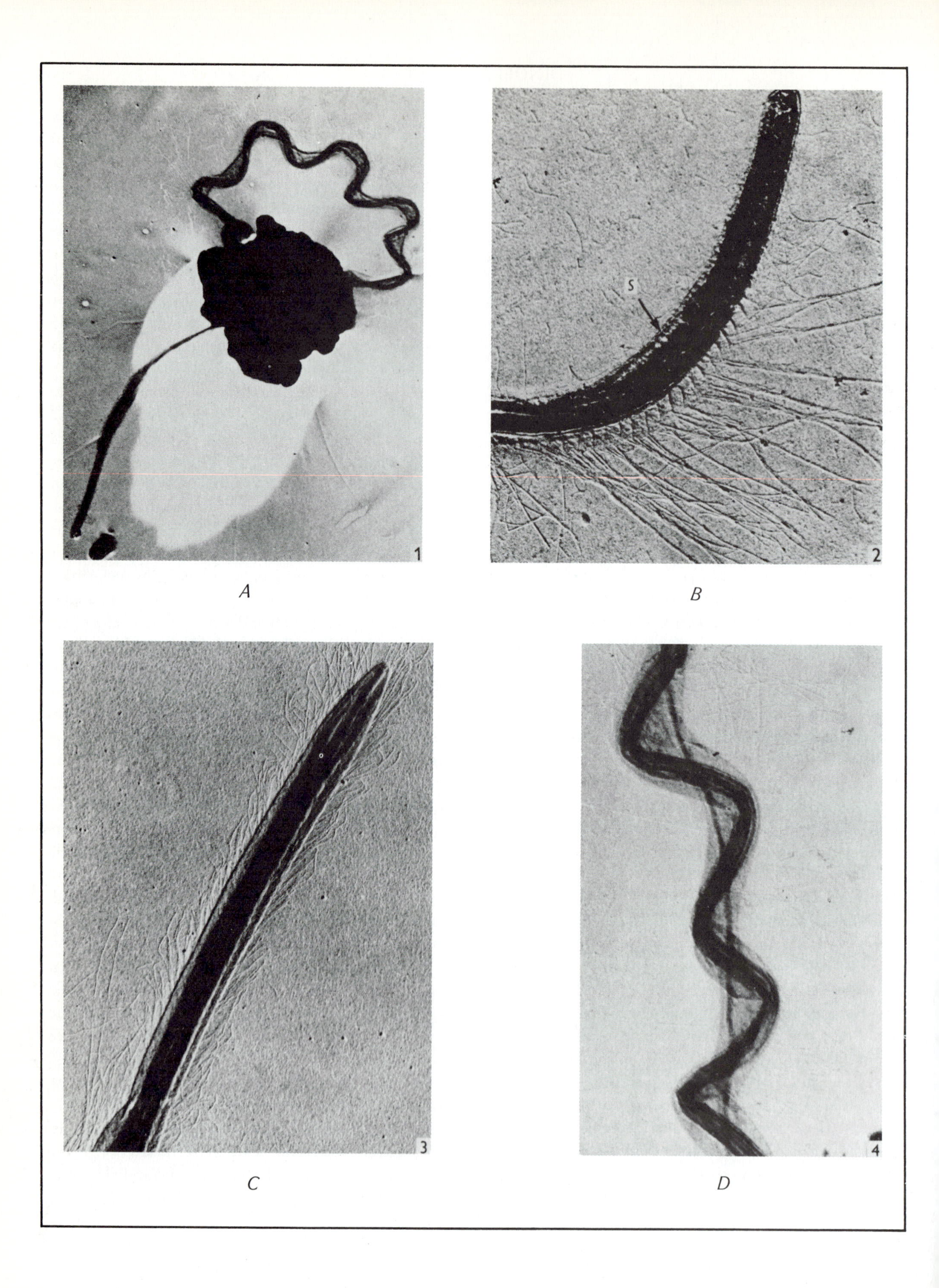

A
B
C
D

PLATE 11-2. Details of flagellar structure of the Peridiniales. (*A*) A species of *Gymnodinium*, showing the transverse flagellum displaced above the microorganism and the longitudinal flagellum in a posterior position (shadowed whole mount, 4,200 X). (*B*) The unilateral long hairs and striated strand terminating near the tip of the transverse flagellum (16,800 X). (*C*) The distal portion of the longitudinal flagellum of a species of *Woloszynskia*, showing the firm sheath around the axoneme and short fine hairs (19,200 X). (*D*) Small portion of transverse flagellum, showing the sheath bearing the unilateral array of long hairs and enclosing both the undulating axoneme and the straight striated strand (11,500 X). From B. Leadbeater and J.D. Dodge, 1967b, *J. gen. Microbiol.* 46:305–14.

membrane covering a single layer of flattened vesicles. Dodge and Crawford have expanded the term theca to include all the peripheral membranous components, but Loeblich (1976) and Taylor (1976) prefer to use the old name amphiesma for the entire complex, reserving theca for the plates only. The major differences among the dinoflagellates are the presence or absence of thecal plates in the amphiesmal vesicles and their thickness and form, when present, as well as the arrangement of amphiesmal microtubules, which fall into eight distinct categories (see Plate 11-3,*A*). As the plates become thicker, the number of vesicles decreases, from as many as one hundred and fifty to three or four. Plate 11-3,*C* shows some details of the thecal structure of a *Peridinium*, including epithecal and girdle plates and a thin section of the area of overlap. Electron microscopy shows how the plates fit together to form an entire theca and the nature of the ornamentation, and areas of growth zones may even be indicated (Cox and Arnott, 1971; Gocht and Netzel, 1974). The presence of small trichocyst pores before suture formation occurs indicates that the trichocysts play a role in pore formation (Kalley and Bisalputra, 1970). Although the encysted and coccoid dinoflagellates, like the endosymbionts or endoparasites or the forms derived from them, have a continuous wall, with interior and exterior membranes, their cell coverings differ greatly from those in flagellated forms (Dodge and Crawford, 1970). *Pyrocystis lunula* has a plasma membrane that is covered by a thin layer of small dispersed fibrils, which are themselves covered by about 24 layers of crossed parallel cellulose fibrils, with outer layers that are resistant to strong acids and bases (Swift and Remsen, 1970) (see Plate 11-3,*B*). Swift (1974) has pointed out that although some dinoflagellates called *Pyrocystis lunula* are in fact *P. lunula*, others should be considered a new species *Dissodinium pseudolunula* which is not closely related to *P. lunula*. Organisms such as *Dissodinium pseudolunula* may be the free-living stage of parasitic dinoflagellates like those described by Van Stosch, (1967), and Drebes, (1969).

The cell contents of a dinoflagellate without plates, *Amphidinium carterae* (named after Nellie Carter, hence the feminine ending), and a thecate dinoflagellate, *Prorocentrum micans*, are shown in Plate 11-4. The chloroplast is often a single peripheral structure with holes to permit the passage of flagella and trichocysts. Within the chloroplast envelope, the many lamellae that parallel the sides of the cell are grouped as three opposed discs (Dodge, 1968). The pyrenoids, which are seen only in photosynthetic species, synthesize reserve material from photosynthate produced by the chloroplasts. They occur as a single layer organelle in the center of *A. carterae* and as two large lateral structures in *P. micans* (Kowallik, 1969).

The nuclear organization in dinoflagellates is somewhat unique in being intermediate between that of the procaryotes and the eucaryotes (Dodge, 1966; Zingmark, 1970) (See Plate 11-4,*C*). Dinoflagellate chromosomes are visible with the light microscope, even at low magnifications, as discrete, condensed or elongate bodies during most stages of the life cycle. Chemical mutagenesis and renaturation studies indicate that, unlike the diatoms, which are diploid, the photosynthetic dinoflagellates are functionally haploid (Roberts et al., 1974). Cytological studies have shown little RNA, histone, or residual protein in the genophore of dinoflagellates (Kubai and Ris, 1969; Soyer and Haapala, 1974). For these reasons, it has been suggested that a third category, mesocaryota (Dodge, 1966), be added to the procaryota and the eucaryota. This might include the radiolaria and heliozoa as well as the dinoflagellates and other microorganisms with unusual nuclear characteristics. This proposal has been rejected by Allsopp (1969) and Loeblich (1976) and modified by Zingmark (1970).

The Golgi bodies lie between the nucleus and the pyrenoids. Pusules are organelles associated with each flagellar canal, which have two forms: (1) tubules or a sac or (2) a pusule vesicle opening directly into the flagellar canal. It has been suggested that pusules are excretory or osmoregulatory organelles (Dodge, 1972), but Taylor (1973b) has raised the question of a role in the

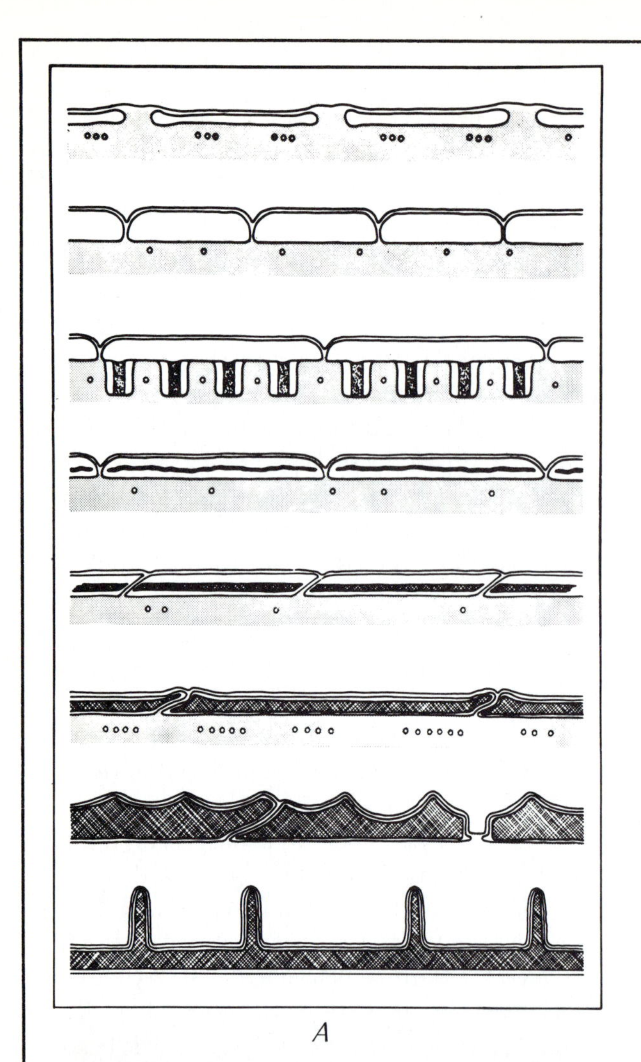

A

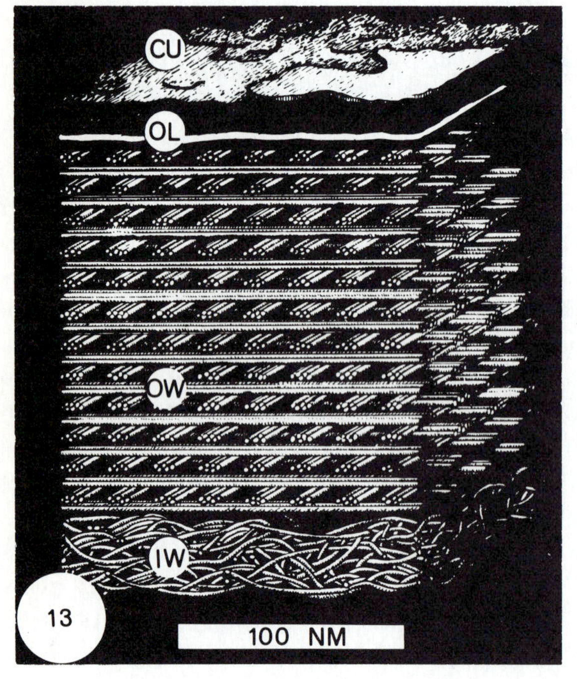

B

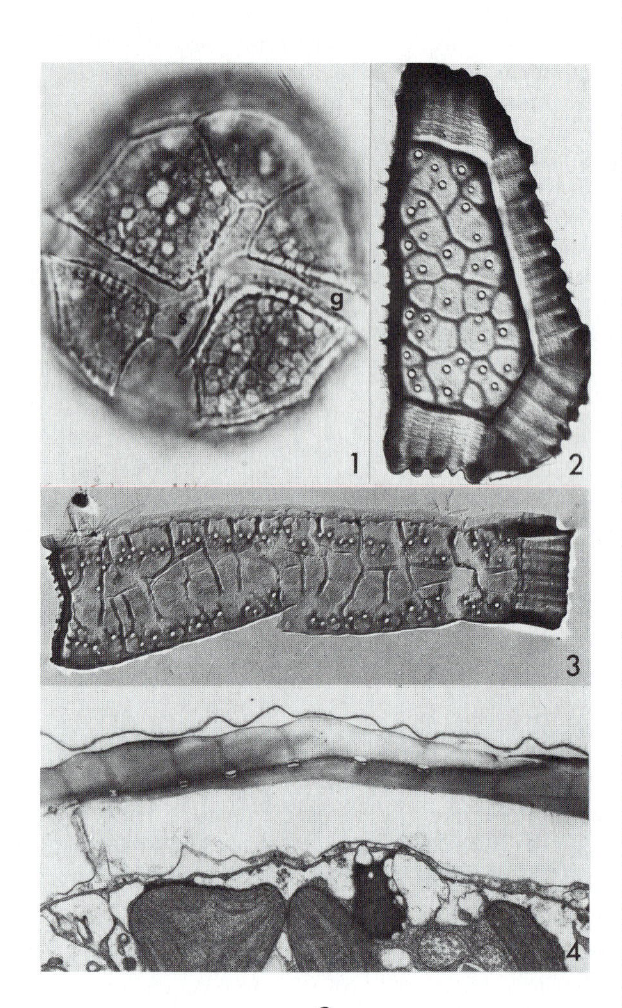

C

PLATE 11-3. The amphiesma of dinoflagellates. (*A*) Diagrams of the structure of the amphiesma of various dinoflagellates in vertical section. Thecal plates, when present, are cross-hatched; the upper line is the plasma membrane, and the cytoplasm is toward the bottom of the diagram. (*B*) A different arrangement of the cell wall of *Pyrocystis lunula*, with a resistant outer wall covering (CU), crossed parallel fibrils of the outer wall (OW), and the inner wall (IW), showing its dispersed textural structure. (*C*) Details of the thecal structure of a *Peridinium*, showing by (1) a light photomicrograph the ornate reticulations on the epicone (anterior one-half), girdle (g), sulcus (s), and hypocone (posterior one-half) (1,000 X). Transmission electron micrographs of a cleaned and shadowed epithecal (anterior) plate (2), showing reticulations on the plate, ridge, and projections on one side (2,900 X); the less elaborate girdle plate (3), with trichocyst pores (2,600 X); and (4) a vertical thin-section, showing the overlapping junction between two plates (1,000 X). (*A,C*) from J.D. Dodge and R.M. Crawford, 1970, *Bot. J. Linnaean Soc. London* 63:53–67; (*B*) from E. Swift and C.C. Remsen, 1970, *J. Phycol.* 6:79–86.

uptake of dissolved organic material. Trichocysts are cell organelles; they occur perpendicular to the wall, with a shaped neck protruding through the wall via the trichocyst opening, and are capable of ejecting rodlike projectiles (Bouck and Sweeney, 1966) (see Plate 11-4,*B*). Released trichocysts accumulate in culture media. The function of the trichocysts is unknown; they may be concerned with pore formation, osmoregulation, or protection by releasing toxic projectiles as do the ciliates. The eyespot, or stigma, is a prominent organelle in flagellated algae. It contains red or orange pigments in lipid globules (osmiophilic granules), and its position is constant for any given species. Although it is assumed to be a photosensitive organelle, its mechanism of operation is unknown. There are five types of algal eyespots, which are classified according to their spatial relations to the chloroplast and flagella and complexity (Dodge, 1969). Although each class of algae has a more or less specific type of eyespot, eyespots in dinoflagellates range from the most primitive, osmiophilic granules without a membrane to the largest and most complex, with a lens, retinoid, and pigment cup (Greuet, 1968; Dodge, 1971). This large, complex structure is far more than an eyespot; it is a unique aggregation of organelles and apparently functions visually.

Reproduction in the dinoflagellates is mostly asexual, that is, by simple cell division. In the few genera in which sexual reproduction has been documented, the haploid cells are potential gametes, and meiosis is postzygotic (von Stosch, 1965, 1967; Cao Vien, 1968). In some species without plates, division occurs only during a nonmotile phase while the cell is in a cyst that adheres easily to surfaces (Kubai and Ris, 1969). In cultures under optimal conditions, such dinoflagellates show a phasic period of growth where most cells are encysted at the plateau phase. During the late stages of cell division, before rupture of the cell wall, two, four, eight, or more newly formed cells become motile. In other dinoflagellates, the theca splits obliquely in two halves through the flagellar area, each daughter cell retaining one-half of the old wall and regenerating the missing one-half wall. In newly divided cells, the thecal horns are lopsided and easily recognized. This criterion of cell division was used early in the century to make the observation that, in plankton collected at intervals throughout the day and night, recently divided cells were absent during the day. Division peaked at 0300 for *Ceratium fusus* and between 0300 and 0700 for *C. tripos* (Gough, 1905; Jörgensen, 1911). This rhythmic cell division also occurs in culture. Division in cultures of *Peridinium* (= *Heterocapsa*) *triquetrum* occurred between 0500 and 1300 with a peak between 0900 and 1000 (Braarud and Pappas, 1951), whereas 85% of the division in *Gonyaulax polyedra*, which was grown in alternating 12-hour periods of light and dark, occurred during a 5-hour period that just followed the peak of the luminescent glow at the end of the dark period (Sweeney and Hastings, 1958). In other species, the thecae dissociate, permitting the naked protoplast to emerge and divide into two daughter cells, which then form new thecal plates (Buchanan, 1968).

Resting spore stages or cysts of dinoflagellates (Braarud, 1945b; Nordli, 1951) were virtually ignored until Evitt (1963) proposed that the hystrichospheres and dinoflagellates believed to be fossils by micropaleontologists (e.g., Eisenack, 1964) were really the cysts of modern dinoflagellates and not the thecae of fossil dinoflagellates. This hypothesis was confirmed by comparing hystrichospheres with modern dinoflagellate cysts (Evitt and Davidson, 1964). A method of culturing resting cysts to obtain unialgal dinoflagellate cultures (Wall, Guillard, and Dale, 1967) has permitted the comparison of the resting stages of over 30 species of modern-day dinoflagellates with fossil dinoflagellates and hystrichospheres (Wall and Dale, 1968; Wall et al., 1970). An example is the resting cyst of *Pyrodinium bahamense*, a thecate dinoflagellate well known for its biolumines-

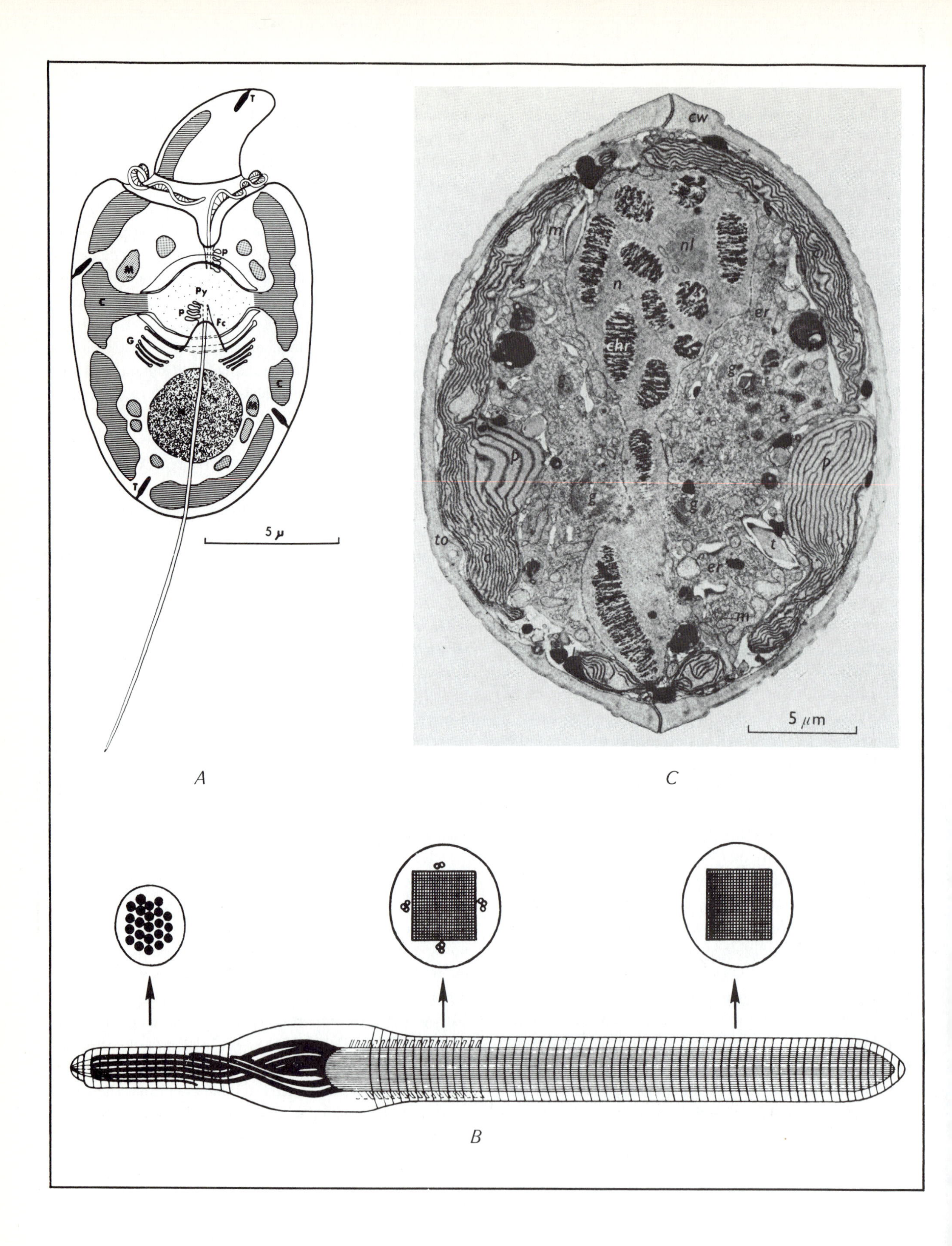
T
P
M
C
Py
Fc
G
5 μ
cw
nl
n
chr
m
s
er
g
p
to
t
5 μm
A
C
B

PLATE 11-4. The fine structure of dinoflagellates. (*A*) A diagram of *Amphidinium carterae* (Gymnodiniales), showing how the girdle containing the transverse flagellum separates the small anterior hemisphere (epicone) from the posterior hemisphere (hypocone) and the longitudinal flagellum in a posterior position. Only the peripheral chloroplasts (C) are shown so that the position of the nucleus (N), mitochondrion (M), Golgi body (G), pyrenoid (Py), and pusules (P) associated with the flagellar canals (Fc) and trichocysts (T) in the interior of the cell can be seen. (*B*) A diagram of the longitudinal section of a charged trichocyst from *Gonyaulax polyedra*. (*C*) A thin section through *Prorocentrum micans* (Prorocentrales), showing some of the main cell contents of the dinoflagellates: chloroplast (c), the unique condensed chromosomes (chr), cell wall (cs), endoplasmic reticulum (er), Golgi body (g), lipid body (l), mitochondrion (m), nucleus (n), nucleolus (nl), pyrenoid (p), starch (s), trichocyst (t), and trichocyst opening (to). (*A*) from J.D. Dodge and R.M. Crawford, 1968, *Protistologica* 4:231–41; (*B*) from G.B. Bouck and B.M. Sweeney, 1966, *Protoplasma* 61:205–23; (*C*) from K. Kowallik, 1969, *J. Cell Sci.* 5:251–69.

cence in tropical waters, such as the famed Bahia Fosforescente in Puerto Rico (Zahl, 1960), which is identical to the microfossil hystrichosphere *Hemicystodinium zoharyi* observed in marine sediments from southern Israel and in deep-sea cores from the Caribbean (Wall and Dale, 1969). Another example is the theca–cyst cycle of *Gonyaulax digitale*, shown in Plate 11-5. Factors controlling the distribution of gonyaulacean cysts around the British Isles are discussed by Reid (1974).

The media and procedures for culturing dinoflagellates are similar to those used for other oxyphototrophs. Methods for the isolation and purification of the microflagellates are summarized by Guillard (1973). Many of the media formulated by Provasoli and colleagues at the Haskins Laboratories (Provasoli, McLaughlin, and Droop, 1957) were developed to cultivate just such flagellated microorganisms (Provasoli and Pintner, 1953; Provasoli, McLaughlin, and Pintner, 1954). Other formulations appear to be equally successful and range from various modifications of Erdschreiber solution (Sweeney and Hastings, 1958; Dodge, 1963) used in the many studies by Dodge, Parke, Manton, and colleagues, in Great Britain, to modifications of the "f" medium of Guillard and Ryther (1962), which is used so successfully for the cultivation of many algae including the thecate dinoflagellates from their resting cysts (Wall and Dale, 1968). This simple procedure uses sedimented detritus obtained during cold water periods (0–3 °C), which is disaggregated by sonification and sieved to concentrate the 37 to 74 μm fraction from which spores are isolated by micropipette and cultured at 16–25 °C in microculture slides and tubes; the volume of medium is critical during cell development (Wall, Guillard, and Dale, 1967). To maintain the diel rhythm of luminescence in dinoflagellates, alternate periods of light and dark are required (Sweeney and Hastings, 1957). The generation time of *Gonyaulax polyedra* in constantly illuminated cultures was 12 hours versus 18 hours for cultures in an alternating 12-hour light and dark cycle (Sweeney and Hastings, 1958).

All the photosynthetic dinoflagellates possess chlorophyll *a* and *c* and β carotene. Among the xanthophylls unique to dinoflagellates are dinoxanthin, neodinoxanthin, peridinin, and neoperidinin. Carotenoids usually associated with other algal classes, such as fucoxanthin, have been reported from dinoflagellates (Mandelli, 1968), but this may be due to contamination by endosymbionts (Tomas, Cox, and Steidinger, 1973). Reserve food materials are starch and fat. Two distinctive physiological characteristics of dinoflagellates are the toxicity of the few species that cause red tides and the more general property of luminescence.

Blooms of dinoflagellates have been associated with massive fish kills and with paralytic shellfish poisoning, in which the toxin accumulated by such shellfish as clams and mussels seriously affects their predators, including man. The most commonly encountered toxic species of the genus *Gonyaulax* are *G.catenella*, *G. acatenella*, *G. tamarensis*, and *G. polyedra; Pyrodinium monilatum* (=*G. monilata*, see Taylor, 1976, for revision); and *Gymnodinium breve*. The symposium volume edited by LoCicero (1975) should be consulted for references. The first dinoflagellate toxin to be identified, saxitoxin, is a purine compound similar to tetrodoxin, the pufferfish poison, and one of the most potent, non-protein poisons known, having a mouse LD_{50} of 3.4 μg/kg (Bull and Pringle, 1968). Numerous other microorganisms can also cause red tides (Hart, 1966).

The 1972 and 1974 blooms of *Gonyaulax tamarensis* in the Gulf of Maine that closed down the large shellfish industry renewed interest in and research on paralytic shellfish toxins. Seven gonyautoxins have been isolated from *G. tamarensis* and characterized by Shimizu and colleagues at the University of Rhode Island. These differ from saxitoxin and the toxin of *Gymnodinium breve*. The isolation, chemistry, and pharmacology of dinoflagellates have been reviewed by Shimizu (1978).

Bioluminescence in the sea at night is always an intriguing phenomenon. The beauty of the display of light produced by the wake of a vessel or a swimming fish is only exceeded by the sight of these light flashes viewed from below during night diving. Much of the luminescence in the sea is caused by dinoflagellates (Harvey, 1952). The diel rhythm of bioluminescence, which occurs in the dark, is also observed in dinoflagellate culture. It can continue to occur as "night" periods for several days in the dark but this luminescence is inhibited by light during either the day or night. The amount of spontaneous light emitted at night depends directly upon the amount of illumination (and thus photosynthesis) during the day. Mechanically stimulated luminescence such as that caused by the movement of oars or the wake of a ship has a similar diel rhythm (Sweeney and Hastings, 1957). The well-defined peaks in spontaneous (unstimulated) intensity of glow and cell division as a function of photoperiodicity observed for *Gonyaulax polyedra* are not shared by either *Pyrodinium bahamense* or *Pyrocystis lunula* (Biggley et al., 1969), which indicates interspecies differences. The plastids in the luminescent lunate cysts of *P. lunula* show diel changes, moving to the center of the cell in intense light and thereby covering the refractile region that produces luminescence. In the dark, the plastids move away from the central region to permit brilliant luminescence (Swift and Taylor, 1967). The endogenous diel rhythm of bioluminescence has also been studied in a natural population of dinoflagellates from a temperate zone salt pond (Kelly and Katona, 1966). Diel periodicity also appears to apply to phototaxis. *Gyrodinium dorsum* grown in an alternate 12-hour light and dark cycle showed

PLATE 11-5. The theca–cyst cycle of *Gonyaulax digitale* (Pouchet) Kofoid, including a typical resting cyst (*Hystrichosphaera bentori* Rossignol). From Wall and Dale, 1968, *Micropaleontology* 14:265–304.

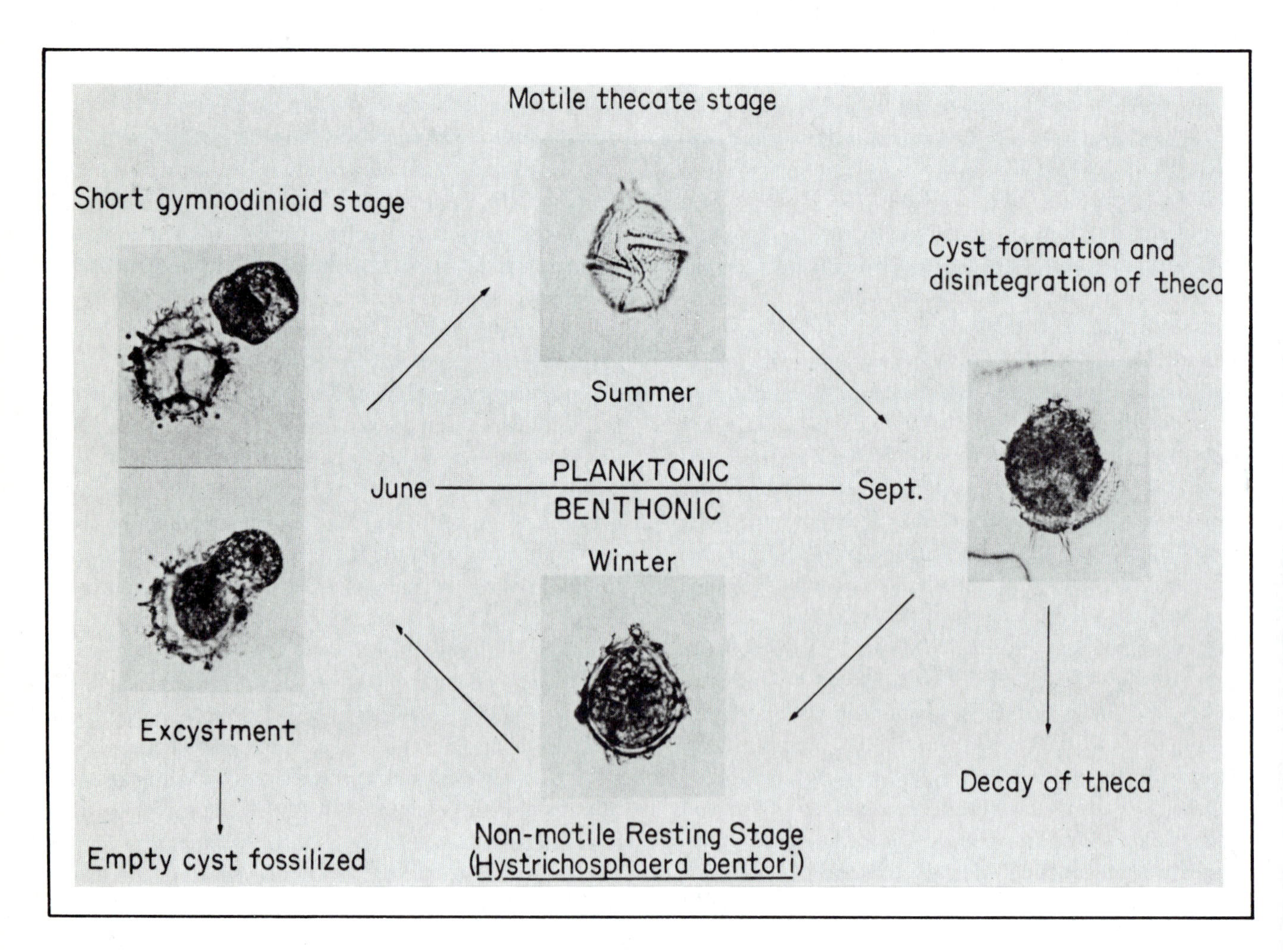

maximal phototaxis 1 hour before the expected light phase. This could not be measured by the stop-response (initial cessation of movement upon light stimulation) unless the cells were exposed to red light prior to blue light stimulation. Such behavior suggests a relationship between photatoxis and diel vertical migration (Forward and Davenport, 1970).

Ultrastructural studies during the past two decades on some 60 species have shown that the dinoflagellates are a clearly defined group (Dodge, 1971). The features that characterize them are the nucleus, which contains strongly condensed chromosomes of unique construction, the spectrum of thecal types, the transverse flagellum, the chloroplast, when present, and the pusules and trichocysts. The morphology and the structure of the theca were the basis of taxonomy by light microscopy, but ultrastructural studies with the electron microscope have shown that the theca is the only feature of taxonomic value (Dodge and Crawford, 1970; Dodge, 1971). Therefore, ultrastructural studies support rather than contradict current taxonomy.

Existing classification schemes can be found by consulting the monographs and floras including those by Kofoid and Swezy (1921), Lebour (1925), Kofoid and Skogsberg (1928), Schiller (1933), Tai and Skogsberg (1934), Graham (1942), and Taylor (1976). Recent studies on the Dinophysiales include those by Abé (1967a,b), Norris (1969), Norris and Berner (1970), Taylor (1971, 1973a), and Balech (1976). An alphabetical listing of the genera (Loeblich and Loeblich, 1966) and the "Check-list of British Marine Algae" (Parke and Dixon, 1976) are also useful. The division of the Pyrrophyta into the two classes Desmophyceae and Dinophyceae (Silva, 1962) is not justified because of the uniformity of their basic cell structure, according to Dodge (1971). Ultrastructural studies have shown that the lines of demarcation between orders and families are in need of revision (Dodge and Crawford, 1970; Dodge, 1971). That this has been accomplished to some degree is indicated by the present divisions suggested by Parke and Dodge in Parke and Dixon (1976).

REFERENCES

Abé, T.H. 1967a. The armoured dinoflagellata: II. Prorocentridae and Dinophysidae (A). *Publ. Seto Mar. Biol. Lab.* 14(5):369–89.

Abé, T.H. 1967b. The armoured dinoflagellata: II. Prorocentridae and Dinophysidae (B)—*Dinophysis* and its allied genera. *Publ. Seto Mar. Biol. Lab.* 15(1):37–78.

Allen, W.E. 1941a. Depth relationships of plankton diatoms in sea water. *J. Mar. Res.* 4:107–12.

Allen, W.E. 1941b. Twenty years' statistical studies of marine plankton dinoflagellates of southern California. *Am. Midl. Nat.* 26:603–35.

Allsopp, A. 1969. Phylogenetic relationships of the Procaryota and the origin of the eucaryotic cell. *New Phytol.* 68:591–612.

Balech, E. 1958. The changes in the phytoplankton population off the California coast. *Calif. Coop. Oceanic Fish. Invest. Repts.* 7:127–232.

Balech, E. 1976. Some Norwegian *Dinophysis* species (Dinoflagellata). *Sarsia* 61:75–94.

Biggley, W.H., E. Swift, R.J. Buchanan, and H.H. Seliger. 1969. Stimulable and spontaneous bioluminescence in the marine dinoflagellates, *Pyrodinium bahamense*, *Gonyaulax polyedra*, and *Pyrocystis lunula*. *J. Gen. Physiol.* 54:96–122.

Bouck, G.B., and B.M. Sweeney. 1966. The fine structure and ontogeny of trichocysts in marine dinoflagellates. *Protoplasma* 61:205–23.

Braarud, T. 1935. *The "Øst" Expedition to the Denmark Strait 1929. II. The Phytoplankton and its Conditions of Growth (Including Some Qualitative Data from the Arctic in 1930).* Hvalradets Skrifter, Akademi I Oslo Nr. 10, 173 p.

Braarud, T. 1945a. *A Phytoplankton Survey of the Polluted Waters of Inner Oslo Fjord.* Hvalradets Skrifter Nr. 28, 142 p.

Braarud, T. 1945b. *Morphological Observations on Marine Dinoflagellate Cultures.* Avhand. Utgitt Norsk Videnskaps-Akademi Oslo. I. Mat.-Naturv. Klasse 1944. No. 11, 18 p.

Braarud, T., and I. Pappas. 1951. *Experimental Studies on the Dinoflagellate* Peridinium triquetrum *(Ehrb.) Lebour.* Avhand. Utgitt Norsk Videnskaps-Akademi Oslo, I. Mat.-Naturv. Klasse 1951, No. 2, 23 p.

Buchanan, R.J. 1968. Studies at Oyster Bay in Jamaica, West Indies, IV. Observations on the morphology and asexual cycle of *Pyrodinium bahamense* Plate. *J. Phycol.* 4:272–77.

Bull, R.J., and B.H. Pringle. 1968. Saxitoxin as an example of biologically active marine substances. In: *Drugs from the Sea* (H.D. Freudenthal, ed.). Mar. Tech. Soc., Washington D.C., pp. 73–86.

Cao Vien, M. 1968. Sur la Germination du zygote et sur un mode particulier de multiplication végétative chez le Péridinien libre *Amphidinium carteri*. *C.R. Acad. Sci. Paris, Ser. D* 267:701–3.

Cox, E.R., and H.J. Arnott. 1971. The ultrastructure of the theca of the marine dinoflagellate, *Ensiculifera loeblichii* sp. nov. In: *Contributions in Phycology* (B.C. Parker and R.M. Brown, Jr., eds.). Allen Press, Lawrence, Kansas, pp. 121–36.

Dodge, J.D. 1963. The nucleus and nuclear division in the dinophyceae. *Arch. Protistenk.* 106:442–52.

Dodge, J.D. 1966. The Dinophyceae. In: *The Chromosomes of the Algae* (M. Godward, ed.). St. Martin's Press, New York, pp. 121–136, and Edward Arnold, London, pp. 96–115.

Dodge, J.D. 1968. The fine structure of chloroplasts and pyrenoids in some marine dinoflagellates. *J. Cell Sci.* 3:41–48.

Dodge, J.D. 1969. A review of the fine structure of algal eyespots. *Br. Phycol. J.* 4:199–210.

Dodge, J.D. 1971. Fine structure of the Pyrrophyta. *Bot. Rev.* 37:481–508.

Dodge, J.D. 1972. The ultrastructure of the dinoflagellate pusule: A unique osmo-regulatory organelle. *Protoplasma* 75:285–302.

Dodge, J.D. 1975. The Prorocentrales (Dinophyceae). II. Revision of the taxonomy within the genus *Prorocentrum*. *Bot. J. Linn. Soc.* 71:103–25.

Dodge, J.D., and R.M. Crawford. 1968. Fine structure of the dinoflagellate *Amphidinium carteri* Hulbert. *Protistologica* 4:231–41.

Dodge, J.D., and R.M. Crawford. 1970. A survey of thecal fine structure in the Dinophyceae. *Bot. J. Linn. Soc.* 63:53–67.

Dragovich, A., J.A. Kelly, Jr., and H.G. Goodell. 1968. Hydrological and biological characteristics of Florida's west coast tributaries. *Fish. Bull. U.S. Fish. Wildl. Serv.* 66:463–77.

Drebes, G. 1969. *Dissodinium pseudocalani* sp. nov., ein parasitischer Dinoflagellat auf Copepodeneiern. *Helgol. wiss. Meeres.* 19:58–67.

Eisenack, A. 1964. Erörterungen über einige Gattungen fossiler Dinoflagellaten und über die Einordnung der Gattungen in das System. *Neues Jahrb. Geol. Pal., Monatsh.* 6:321–36.

Eppley, R.W., O. Holm-Hansen, and J.D.H. Strickland. 1968. Some observations on the vertical migration of dinoflagellates. *J. Phycol.* 4:333–40.

Evitt, W.R. 1963. A discussion and proposals concerning fossil dinoflagellates, hystrichospheres, and acritarchs, I. *Proc. Nat. Acad. Sci.* 49:158–64, 298–302.

Evitt, W.R., and S.E. Davidson. 1964. Dinoflagellate studies. I. Dinoflagellate cysts and thecae. *Stanford University Publ. Geol. Sci.* 10(1):1–12.

Forward, R.B., Jr., and D. Davenport. 1970. The circadian rhythm of a behavioral photoresponse in the dinoflagellate *Gyrodinium dorsum*. *Planta* 92:259–66.

Gocht, H., and H. Netzel. 1974. Scanning electron microscope studies on the theca of *Peridinium* (Dinoflagellata). *Arch. Protistenk.* 116:381–410.

Gough, L.H. 1905. Report on the plankton of the English Channel in 1903. In: *Mar. Biol. Assoc. U.K. Intern. Fish. Invest., 1st Report Fish. Hydrographic Invest. in the North Sea and Adjacent waters (Southern Area).* 1902–03, pp. 325–77.

Graham, H.W. 1942. *Studies in the Morphology, Taxonomy, and Ecology of the Peridiniales.* Carnegie Inst. Publ. No. 542, 129 p.

Gran, H.H., and T. Braarud. 1935. A quantitative study of the phytoplankton in the Bay of Fundy and the Gulf of Maine (including observations on hydrography, chemistry and turbidity). *J. Biol. Bd. Canada (J. Fish. Res. Bd. Canada)* 1:279–467.

Greuet, C. 1968. Organisation ultrastructurale de l'ocelle de deux peridiniens warnowiidae, *Erythropsis pavillardi* Kofoid et Swezy et *Warnowia pulchra* Schiller. *Protistologica* 4:209–30.

Guillard, R.R.L. 1973. Methods for microflagellates and nanoplankton. In: *Handbook of Phycological Methods* (J.R. Stein, ed.). Cambridge University Press, New York and London, pp. 69–85.

Guillard, R.R.L., and J.H. Ryther. 1962. Studies of marine planktonic diatoms. I. *Cyclotella nana* Hustedt, and *Detonula confervacea* (Cleve) Gran. *Can. J. Microbiol.* 8:229–39.

Hart, T.J. 1966. Some observations on the relative abundance of marine phytoplankton populations in nature. In: *Some Contemporary Studies in Marine Science* (H. Barnes, ed.). Allen and Unwin, London, pp. 375–93.

Harvey, E.N. 1952. *Bioluminescence.* Academic Press, London and New York, 632 p.

Hasle, G.R. 1950. Phototactic vertical migrations in marine dinoflagellates. *Oikos* 2:162–75.

Herdman, E.C. 1920–24. Notes on dinoflagellates and other organisms causing discolouration of the sand at Port Erin. *Proc. Trans. Liverpool Biol. Soc.* I. 35:59–63; II. 36:15–30; III. 38:58–63; IV. 38:75–84.

Herdman, W.A. 1922. Spolia Runiana. V. Summary of results of continuous investigation of the plankton of the Irish Sea during fifteen years. *J. Linn. Soc. London, Botany*, 46:141–70.

Jahn, T.L., W.M. Harmon, and M. Landman. 1963. Mechanisms of locomotion in flagellates. I. *Ceratium*. *J. Protozool.* 10:358–63.

Jörgensen, E. 1911. Die Ceratien. Eine kurze Monographie der Gattung *Ceratium* shrank. *Int. Revue ges. Hydrobiol. Hydrogr.* 4 (Biol. Suppl. 1):1–124.

Kalley, J.P., and T. Bisalputra. 1970. *Peridinium trochoideum:* The fine structure of the theca as shown by freeze-etching. *J. Ultrastr. Res.* 31:95–108.

Kelly, M.G., and S. Katona. 1966. An endogenous diurnal rhythm of bioluminescence in the natural population of dinoflagellates. *Biol. Bull.* 131:115–26.

Kofoid, C.A. 1909. On *Peridinium steinii* Jörgensen, with a note on the nomenclature of the skeleton of the Peridineae. *Arch. Protistenk.* 16:25–61.

Kofoid, C.A., and O. Swezy. 1921. *The Free-living Unarmoured Dinoflagellata, Vol. 5.* Mem. Univ. California, Berkeley, 563 p.

Kofoid, C.A., and T. Skogsberg. 1928. The dinoflagellate: The dinophysidae. *Mem. Mus. Comp. Zool. Harv.* 51:1–766.

Kowallik, K. 1969. The crystal lattice of the pyrenoid matrix of *Prorocentrum micans*. *J. Cell. Sci.* 5:251–69.

Kubai, D.F., and H. Ris. 1969. Division in the dinoflagellate

Gyrodinium cohnii (Schiller), a new type of nuclear reproduction. *J. Cell. Biol.* 40:508–28.

Leadbeater, B.S.C., and J.D. Dodge. 1967a. Fine structure of the dinoflagellate transverse flagellum. *Nature* 213:421–22.

Leadbeater, B.S.C., and J.D. Dodge. 1967b. An electron microscope study of dinoflagellate flagella. *J. gen. Microbiol.* 46:305–14.

LeBlond, P.H., and F.J.R. Taylor. 1976. The propulsive mechanism of the dinoflagellate transverse flagellum reconsidered. *Biosystems* 8:33–39.

Lebour, M.V. 1925. *The Dinoflagellates of the Northern Seas.* Mar. Biol. Assoc. U.K., Plymouth, 250 p.

LoCicero, V.R. (ed.). 1975. *Proceedings of the First International Conference on Toxic Dinoflagellate Blooms.* Mass. Sci. Tech. Foundation, Wakefield, Mass., 541 p.

Loeblich, A.R., Jr., and A.R. Loeblich III. 1966. *Index to the Genera, Subgenera, and Sections of the Pyrrophyta.* Stud. Trop. Oceanogr., Miami, No. 3, 94 p.

Loeblich, A.R. III. 1976. Dinoflagellate evolution: Speculation and evidence. *J. Protozool.* 23:13–28.

Mandelli, E.F. 1968. Carotenoid pigments of the dinoflagellate *Glenodinium foliaceum* Stein. *J. Phycol.* 4:347–48.

Nordli, E. 1951. Resting spores in *Gonyaulax polyedra* Stein. *Nytt. Mag. Naturvidensk.* 88:207–12.

Norris, D.R. 1969. Thecal morphology of *Ornithocercus magnificus* (Dinoflagellata) with notes on related species. *Bull. Mar. Sci.* 19:175–93.

Norris, D.R., and L.D. Berner, Jr. 1970. Thecal morphology of selected species of *Dinophysis* (Dinoflagellata) from the Gulf of Mexico. *Contr. Mar. Sci.* 15:145–92.

Parke, M., and P.S. Dixon. 1968. Check-list of British marine algae—second revision. *J. mar. biol. Assoc. U.K.* 48:783–832.

Parke, M., and P.S. Dixon. 1976. Check-list of British Marine Algae—third revision. *J. mar. biol. Assoc. U.K.* 56:527–94.

Provasoli, L., and I.J. Pintner. 1953. Ecological implications of *in vitro* nutritional requirements of algal flagellates. *Ann. N.Y. Acad. Sci.* 56:839–51.

Provasoli, L., J.J.A. McLaughlin, and I.J. Pintner. 1954. Relative and limiting concentrations of major mineral constituents for the growth of algal flagellates. *Trans. N.Y. Acad. Sci. Ser. II* 16:412–17.

Provasoli, L., J.J.A. McLaughlin, and M.R. Droop. 1957. The development of artificial media for marine algae. *Arch. Mikrobiol.* 25:392–28.

Reid, P.C. 1974. Gonyaulacean dinoflagellate cysts from the British Isles. *Nova Hedwigia* 25:579–637.

Roberts, T.M., R.C. Tuttle, J.R. Allen, A.R. Loeblich III, and L.C. Klotz. 1974. New genetic and physicochemical data on structure of dinoflagellate chromosomes. *Nature* 248:446–49.

Ryther, J.H. 1954. The ecology of phytoplankton blooms in Moriches Bay and Great South Bay, Long Island, New York. *Biol. Bull.* 106:198–209.

Ryther, J.H. 1955. Ecology of autotrophic marine dinoflagellates with reference to red water conditions. In: *The Luminescence of Biological Systems* (F.H. Johnson, ed.). AAAS, Washington D.C., pp. 387–414.

Schiller, J. 1933, 1937. Dinoflagellatae (Peridineae). In: *Dr. L. Rabenhorst's Kryptogamen-Flora,* Vol. 10(3), Teil 1, Teil 2. Akad. Verlag, Leipzig.

Seliger, H.H., J.H. Carpenter, M. Loftus, and W.D. McElroy. 1970. Mechanisms for the accumulation of high concentrations of dinoflagellates in a bioluminescent bay. *Limnol. Oceanogr.* 15:234–45.

Shimizu, Y. 1978. Dinoflagellate toxins. In: *Marine Natural Products: New Perspectives* (P.J. Scheuer, ed.). Academic Press, New York and London, pp. 1–42.

Silva, P.C. 1962. Classification of algae. In: *Physiology and Biochemistry of Algae* (R.A. Lewin, ed.). Academic Press, New York and London, pp. 827–37.

Soyer, M-O., and O.K. Haapala. 1974. Division and function of dinoflagellate chromosomes. *J. Microscopie* 19(2):137–46.

Steidinger, K.A., and J. Williams. 1970. *Memoirs of the Hourglass Cruises. II. Dinoflagellates.* Mar. Res. Lab., Florida Dept. Nat. Res., St. Petersburg, Fla., 251 p.

Sweeney, B.M., and J.W. Hastings. 1957. Characteristics of the diurnal rhythm of luminescence in *Gonyaulax polyedra. J. Cell. Comp. Physiol.* 49:115–28.

Sweeney, B.M., and J.W. Hastings. 1958. Rhythmic cell division in populations of *Gonyaulax polyedra. J. Protozool.* 5:217–24.

Sweeney, B.M., and J.W. Hastings. 1962. Rhythms. In: *Physiology and Biochemistry of Algae* (R.A. Lewin, ed.). Academic Press, New York and London, pp. 687–700.

Swift, E. 1974. *Dissodinium pseudolunula* n. sp. *Phycologia* 12:90–91.

Swift, E., and C.C. Remsen. 1970. The cell wall of *Pyrocystis* spp. (Dinococcales). *J. Phycol.* 6:79–86.

Swift, E., and W.R. Taylor. 1967. Bioluminescence and chloroplast movement in the dinoflagellate *Pyrocystis lunula. J. Phycol.* 2:77–81.

Tai, L-S., and T. Skogsberg. 1934. Studies on the Dinophysoidae marine armored dinoflagellates of Monterey Bay, California. *Arch. Protistenk.* 82:380–482.

Taylor, F.J.R. 1971. Scanning electron microscopy of thecae of the dinoflagellate genus *Ornithocercus. J. Phycol.* 7:249–58.

Taylor, F.J.R. 1973a. Topography of cell division in the structurally complex dinoflagellate genus *Ornithocercus. J. Phycol.* 9:1–10.

Taylor, F.J.R. 1973b. General features of dinoflagellate material collected by the "Anton Bruun" during the International Indian Ocean Expedition. In: *Ecological Studies, Analysis and Synthesis, Vol. 3* (B. Zeitzschel, ed.). Springer-Verlag, Berlin, pp. 155–69.

Taylor, F.J.R. 1975. Non-helical transverse flagella in dinoflagellates. *Phycologia* 14:45–47.

Taylor, F.J.R. 1976. Dinoflagellates from the International Indian Ocean Expedition. *Bibliotheca Botanica* (H. Melchior, ed.), No. 132, 234 p.

Tomas, R.N., E.R. Cox, and K.A. Steidinger. 1973. *Peridinium balticum* (Levander) Lemmerman, an unusual dinoflagellate with a mesocaryotic and an eucaryotic nucleus. *J. Phycol.* 9:91–98.

von Stosch, H.A. 1965. Sexualität bei *Ceratium cornutum* (Dinophyta). *Naturwissenschaften* 52:112–13.

von Stosch, H.A. 1967. Vegetative Fortpflanzung, Parthogenese und Apogamie bei Algen. D. Dinophyta. In: *Handbuch der Pflanzenphysiologie* (H. Ruhland, ed.), Vol. 18. Springer-Verlag, Berlin, pp. 626–36.

Wall, D., and B. Dale. 1968. Modern dinoflagellate cysts and evolution of the Peridiniales. *Micropaleontology* 14:265–304.

Wall, D., and B. Dale. 1969. The "hystrichosphaerid" resting spore of the dinoflagellate *Pyrodinium bahamense* Plate 1906. *J. Phycol.* 5:140–49.

Wall, D., R.R.L. Guillard, and B. Dale. 1967. Marine dinoflagellates cultures from resting spores. *Phycologia* 6:83–86.

Wall, D., R.R.L. Guillard, B. Dale, E. Swift, and N. Watabe. 1970. Calcitic resting cysts in *Peridinium trochoideum* (Stein) Lemmermann, an autotrophic marine dinoflagellate. *Phycologia* 9:151–56.

Zahl, P.A. 1960. Sailing a sea of fire. *Natl. Geogr.* 118(1):120–29.

Zingmark, R.G. 1970. Ultrastructural studies on two kinds of mesocaryotic dinoflagellate nuclei. *Amer. J. Bot.* 57:586–92.

CHAPTER 12

Haptophytes and Prasinophytes

A. HAPTOPHYTES

These phytoflagellates are small (usually less than 20 μm) and have one or two golden-brown chromatophores and reserves of leucosin and lipid (see Plate 12-2). Most have a unique filiform appendage, the haptonema, which is thinner than a flagellum (Parke, Manton, and Clarke, 1955) and arises near the pair of smooth acronematic (slender tipped) flagella of equal or subequal length (Plates 12-1 and 12-2). The haptonema and/or the distinctively marked scales, which are organic or calcareous in nature (Boney, 1970), distinguish the members of the Haptophyceae from the Chrysophyceae (Chapter 13,A). The problems involving the inclusion of such microorganisms in the Chrysophyceae (Parke, 1961) were resolved by the establishment (Christensen, 1962, 1966) and adoption (Parke and Dixon, 1968) of the class Haptophyceae. In addition to the unicellular flagellated forms, palmelloid colonies (groups of single nonmotile cells in a gelatinous mass) or filaments are produced by some species. A notable example is *Phaeocystis poucheti*, which forms brief dense blooms of brownish globular colonies several millimeters in diameter, but which also has a zooid stage indistinguishable from the unicellular *Prymnesium.* This is analogous to the large cyst form of the Prasinophycean *Halosphaera* and its *Pyramimonas*-like zooid stage, discussed in Section B. The haptophytes are among the faster growing phytoplankters. In the planktonic form, *P. poucheti* has up to two divisions per day (Kayser, 1970), whereas the fastest growth recorded for the coccolithophorids is 1.85 divisions per day for *Emiliania huxleyi* (= *Coccolithus huxleyi,*) and 2.25 for *Cricosphaera elongata* (Paasche, 1968).

The motile phase of these microalgae is distinguished by a haptonema and organic body scales, as in the neritic *Chrysochromulina* shown in Plate 12-1. The haptonema can be short (*Prymnesium*), long (*Chrysochromulina*), or absent (*Coccolithus*). Used for temporary attachment, this organelle ranges in form from stiff to tightly coiled. It can be seen between the flagella in

PLATE 12-1. A shadowed transmission electron micrograph of the typical haptophyte cell of *Chrysochromulina kappa*, showing the organic body scales and the thin coiled haptonema, with its bulbous base situated between the flagella (bar = 1 μm). From M. Parke, I. Manton, and B. Clarke, 1955, *Journal Mar. Biol. Ass. UK* 34:579–609.

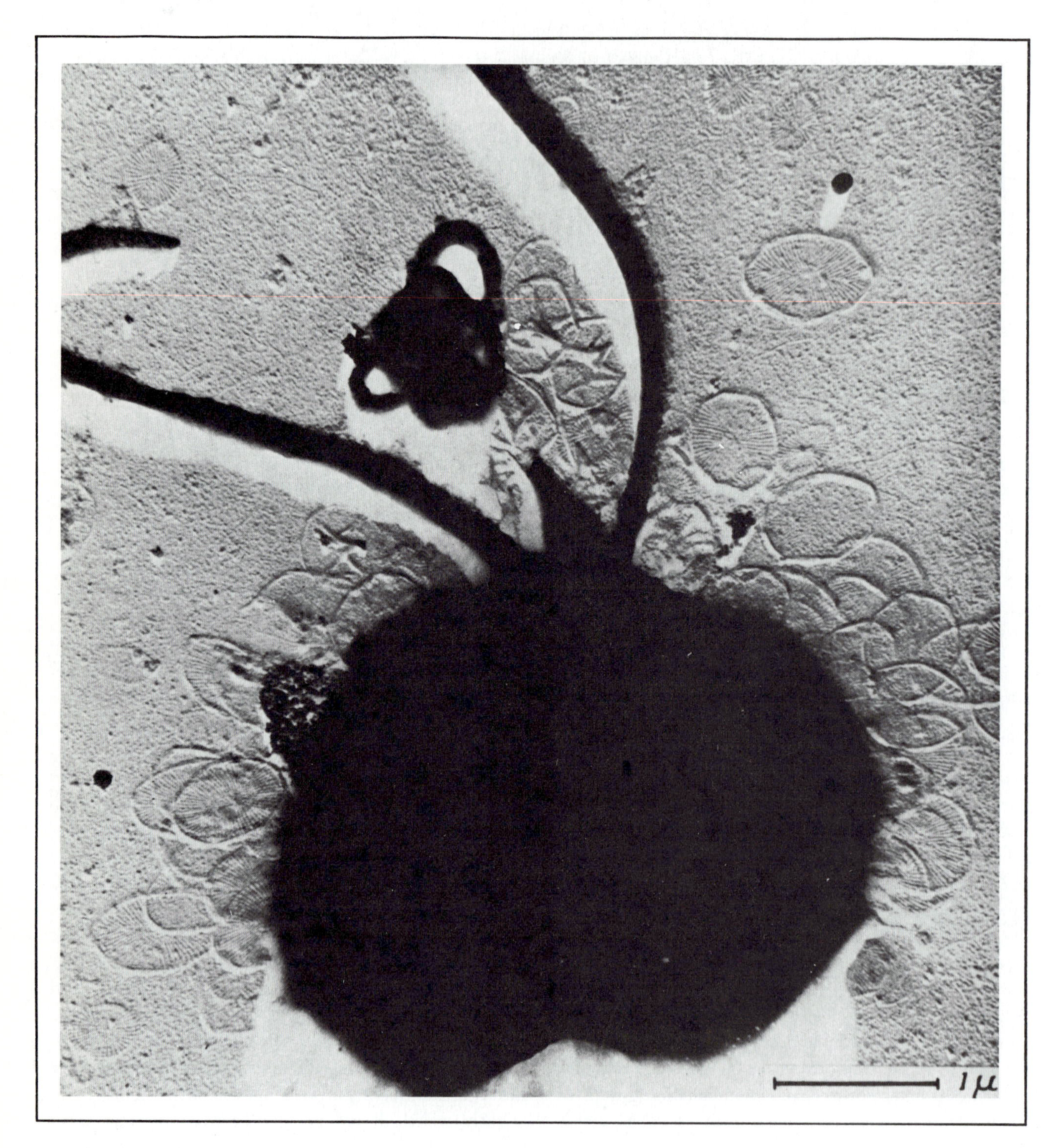

Chrysochromulina kappa, *Prymnesium parvum*, and *Phaeocystis poucheti*, as shown in Plates 12-1, 12-2, and 12-4, respectively. In *Prymnesium parvum*, the haptonema has a cavity, a simple tip, and a complex base, which increases from seven core fibers to nine at the extreme base; its movements are probably controlled by hydrostatic pressure (Manton, 1964a). In the coccolithophorids, there are six microtubules (fibers) in the external haptonema (Klaveness, 1973). The scales of *P. parvum* are apparently formed in "hairy" pits in the Golgi cisternae in a regular sequence from immature to mature scales, which have a regular orientation until they migrate, via a deep pit adjacent to the flagella, to the attached plasmalemma (Manton, 1966a). The scales of *Chrysochromulina chiton* have a fibrillar construction of radiating spokes and a rim, vary in morphology with nutrition, and are composed of a polysaccharide consisting mainly of ribose and galactose (Green and Jennings, 1967).

Haptophytes are present in the open sea and in coastal and neritic waters, where they can be a major component of the phytoplankton. The most numerous quantitatively, and the most ubiquitous, are the coccolithophorids, which are often less than 10 μm in diameter and have tiny 1 μm calcareous coccoliths (scales), with an individualistic micro-architecture by electron microscopy (Black, 1965; Gaarder and Hasle, 1971; Plate 12-3). Some genera, such as *Cricosphaera*, are limited to inshore waters (Gaarder, 1971), since they undergo an alternation of generations between a diploid planktonic phase as a coccolithophorid and a haploid benthic phase as a palmelloid colony (Rayns, 1962; Plate 12-4). The benthic phase of *Cricosphaera* is identical with that of the benthic genera in the family Chrysotilaceae (Parke, 1961; Boney and Burrows, 1966). These benthic forms (Parke and Dixon, 1968) have been described in the upper splash zone of British chalk cliffs (Anand, 1937a,b), to depths of several meters in rocky habitats (Pascher, 1925; Waern, 1952), and in certain estuaries as an epiphyte on grasses and seaweeds (Magne, 1957).

Most of the some 200 species of marine coccolithophorids are strictly oceanic, some being restricted to certain areas, whereas others, such as *Emiliania huxleyi*, are cosmopolitan, equally at home in the open sea (Okada and Honjo, 1973) or in Norwegian fjords where they can achieve population densities that discolor the water (Birkenes and Braarud, 1952). Identification at the species level is valid only when the coccolith morphology has been determined by electron microscopy (Gaarder, 1971). The data on global distribution have been reviewed by Gaarder (1971). The most studied area is the Mediterranean Sea, where Lohmann conducted the studies that led to the first, and basic, monograph on coccolithophorids (Lohmann, 1902). The Atlantic Ocean is considered to have the richest coccolithophorid flora. Norwegian workers have studied the seasonal variation of populations in the North Atlantic and have shown that, in inshore waters, there is a seasonal mass occurrence of a few species, whereas in offshore waters, the warm and more saline waters from the lower latitudes produce a diverse flora during the summer. In both inshore and offshore waters, coccolithophorids are very often second only to the diatoms in population and may even outnumber them at times.

Many oceanic areas, such as the Pacific, have been long ignored in regard to the coccolithophorids, but recent papers indicate that they are now receiving the attention they deserve (Okada and Honjo, 1973; Honjo and Okada, 1974). High standing crops of 10^5 coccolithophorids/liter in the high latitudes decreased to 10^3/liter in temperate and subtropical areas in the upper 200 m of the photic zone. In the aphotic zone, there were some coccolithophorids at most stations, with populations of 10^2/liter occurring down to 4000 m; at 5° N latitude, populations of 10^4/liter occurred at 600 m. In the photic layer, different community structures appear to occur with depth (Honjo and Okada, 1974). Although *Cyclococcolithus fragilis* has been reported to exist heterotrophically for at least one phase of its life cycle, (Bernard, 1963), this report has been challenged on the basis of particulate organic carbon determinations (Fournier, 1968).

The coccolithophorid cell, for at least one phase in its life cycle has an outer covering of plates (coccoliths) made of calcium carbonate (See Plate 12-3). Coccolith patterns may be formed by simple calcitic microcrystals, mostly as rhombohedrons grouped to form holococcoliths (Klaveness, 1973) or by diverse morphological elements with the property of single crystals, which form heterococcoliths (Klaveness, 1972a). Although it is stated that the coccoliths only occur in the form of calcite (Black, 1961–62), there is evidence that coccoliths occur in three forms: calcite, aragonite, and vaterite, the proportions depending upon the temperature of incubation and the strain studied (Wilbur and Watabe, 1963). Coccolith formation is a light-dependent process (Paas-

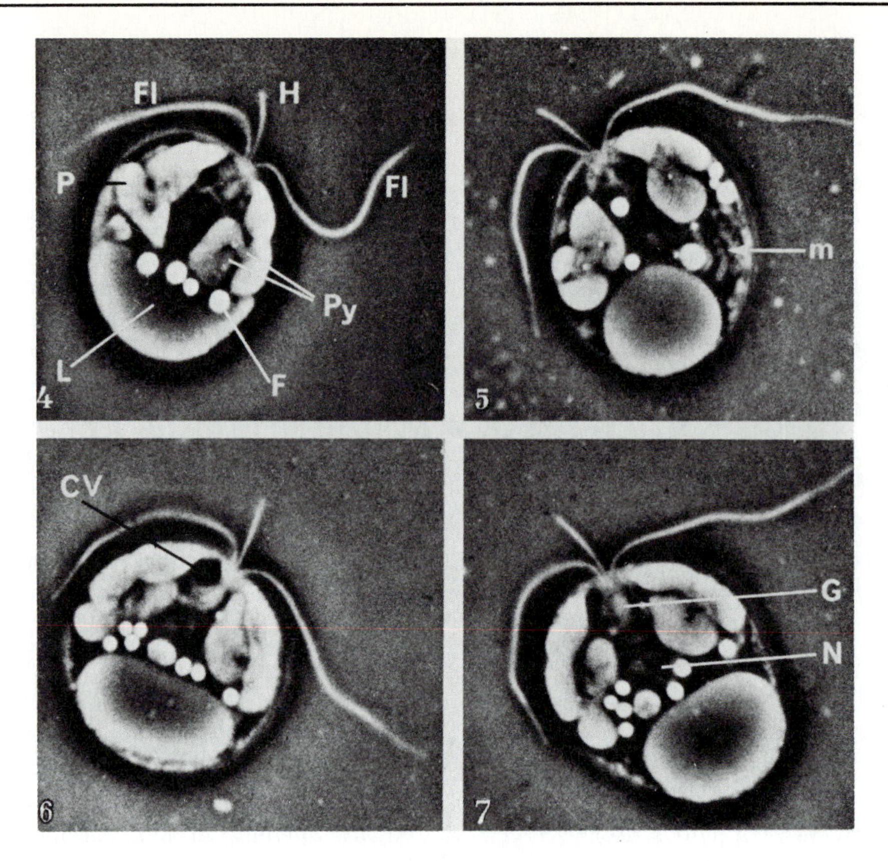

A

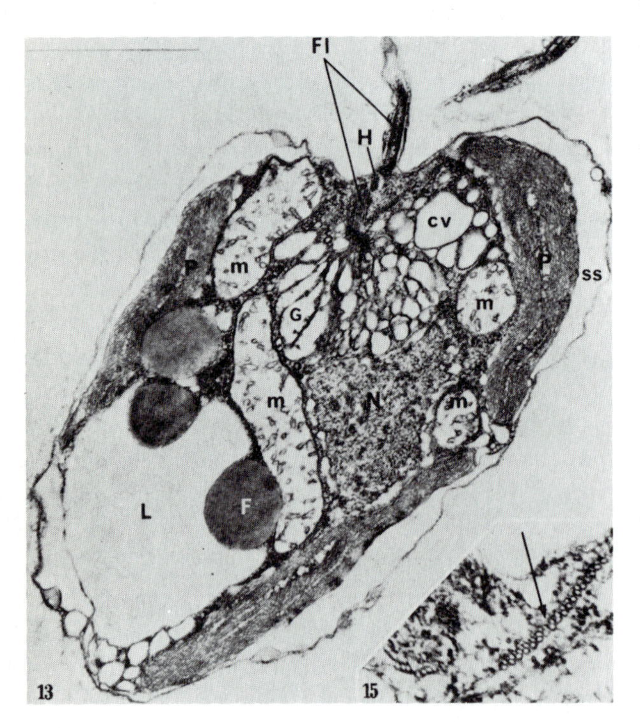

B

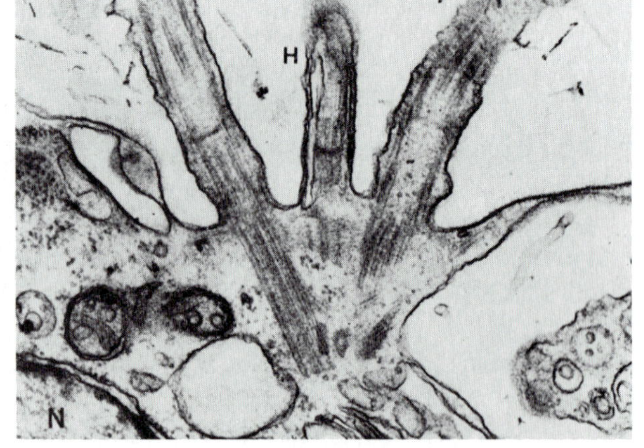

C

PLATE 12-2. *Prymnesium parvum*, showing the morphology of living cells and ultrastructure in thin sections. (*A*) Light microscopy of living cells flattened by coverslip pressure (Fl, flagellum; H, haptonema; P, plastid; Py, pyrenoid; L, leucosin vesicle; F, fat body; m, mitochondria; CV, contractile vacuole; G, Golgi body, and N, nucleus) (anoptral contrast, 2,370 X). (*B*) Thin section, showing the same organelles. (*C*) Longitudinal section through flagellar bases and haptonema (30,200 X). (*A,B*) from I. Manton and G. Leedale, 1963, *Arch. Mikrobiol.* 45:285–303; (*C*) from I. Manton, 1964a, *Arch. Mikrobiol.* 49:315–30.

che, 1964), taking place within the endomembrane system of the cell (Klaveness, 1976). The formation of coccoliths in *E. huxleyi* occurs by a process similar to organic scale formation (Klaveness, 1972a). The formation of body and flagellar scales and flagellar hairs in the prasinophytes, discussed in the next section, also occurs in Golgi bodies. Golgi bodies are large aggregates of membranes often seen in certain regions of cells. The involvement of the Golgi bodies (Golgi apparatus) in the internal synthesis of polysaccharide and the transferral of the polysaccharide to the cell wall occurs not only in algae but in such higher plants as wheat (Northcote and Pickett-Heaps, 1966).

Propagation in the coccolithophorids is by several modes. One is by fission of the mother cell, usually longitudinally, followed by regeneration of the coccoliths. Another is by repeated divisions within the coccolith-encrusted mother cell, with the subsequent release of motile or nonmotile daughter cells (Klaveness, 1972b). The coccoliths regenerated may be of the same or a different type. The well-known, nonmotile species *Coccolithus pelagicus*, which bears heterococcoliths, has been shown to have a motile stage bearing holococcoliths, which had been described as *Crystallolithus hyalinus* (Parke and Adams, 1960). The relationship of the motile stage of *C. pelagicus* to the holococcolithophorids has also been studied by Klaveness (1973). The life histories of coccolithophorids can be quite complex, and a possible life history of the cosmopolitan *E. huxleyi* is shown in Plate 12-3. Three types of cells can be formed: the regular nonmotile coccolithophorid cell (C), a nonmotile naked cell (N), or a flagellated coccolithophorid cell (S). Each of these forms can apparently reproduce by normal vegetative propagation or reproduce aberrantly by discarding the theca to become amoeboid swarmers before division occurs (Klaveness, 1972b).

In the original description of *Phaeocystis poucheti* (Plate 12-4), Pouchet recognized both a colonial form and a motile stage; he observed that *P. poucheti* was similar in some respects to the prasinophyte *Halosphaera viridis* common in the Mediterranean (Pouchet, 1892). There are two planktonic forms: (1) the palmelloid colony stage consisting of non-flagellated cells arranged in a regular manner on the inside of a membrane, the colony being mucoid and either spherical or lobed, and (2) the single cell stage released from the colony, which has a motile swarmer stage. The single cells tend to become sessile, attaching themselves to solid surfaces from which they liberate both free single cells and small colonies. In addition to this cycle (Kayser, 1970), microspores and macrospores have been described (Kornmann, 1956). The cell morphology of the swarmer stage is identical to the cell morphology of *Prymnesium*.

Phaeocystis poucheti is best known in the cold waters of the higher latitudes, where it can form dense spring blooms of the colonial form; these are gummy blobs up to 2 to 3 mm in diameter, which rapidly clog plankton nets. On both sides of the North Atlantic, dense blooms can occur for short periods from March to June depending upon the latitude (Bigelow, 1924; Jones and Haq, 1963), whereas in the Antarctic, *Phaeocystis* blooms are more persistent, often dominating the phytoplankton in protected waters (Burkholder and Sieburth, 1961; Kashkin, 1963). In the Thames estuary of England, fishermen call it "baccy juice" and associate poor trawling with its presence (Orton, 1923). The exclusion of herring from dense *Phaeocystis* blooms has been confirmed by Savage (1930). The alga is apparently restricted to inshore and coastal waters by a benthic resting phase, which presumably carries it over from year to year. A tropical form of *Phaeocystis* is also known (Guillard and Hellebust, 1971).

This writer was introduced to marine microbiology by this microorganism. As a bacteriologist studying the gastrointestinal microflora of birds aboard Argentine Naval vessels during the International Geophysical Years of 1957–1959 (Sieburth, 1958, 1965), I traced the cause of the apparent "bacterial sterility" in the gut of penguins (Sieburth, 1959a) through their diet of euphausids (planktonic crustaceans) to dense blooms of *P. poucheti*. The antibacterial activity (Sieburth, 1959b) was caused by the acrylic acid (Sieburth, 1960, 1961) released during hydrolysis of dimethyl β-propiothetin. The dimethyl

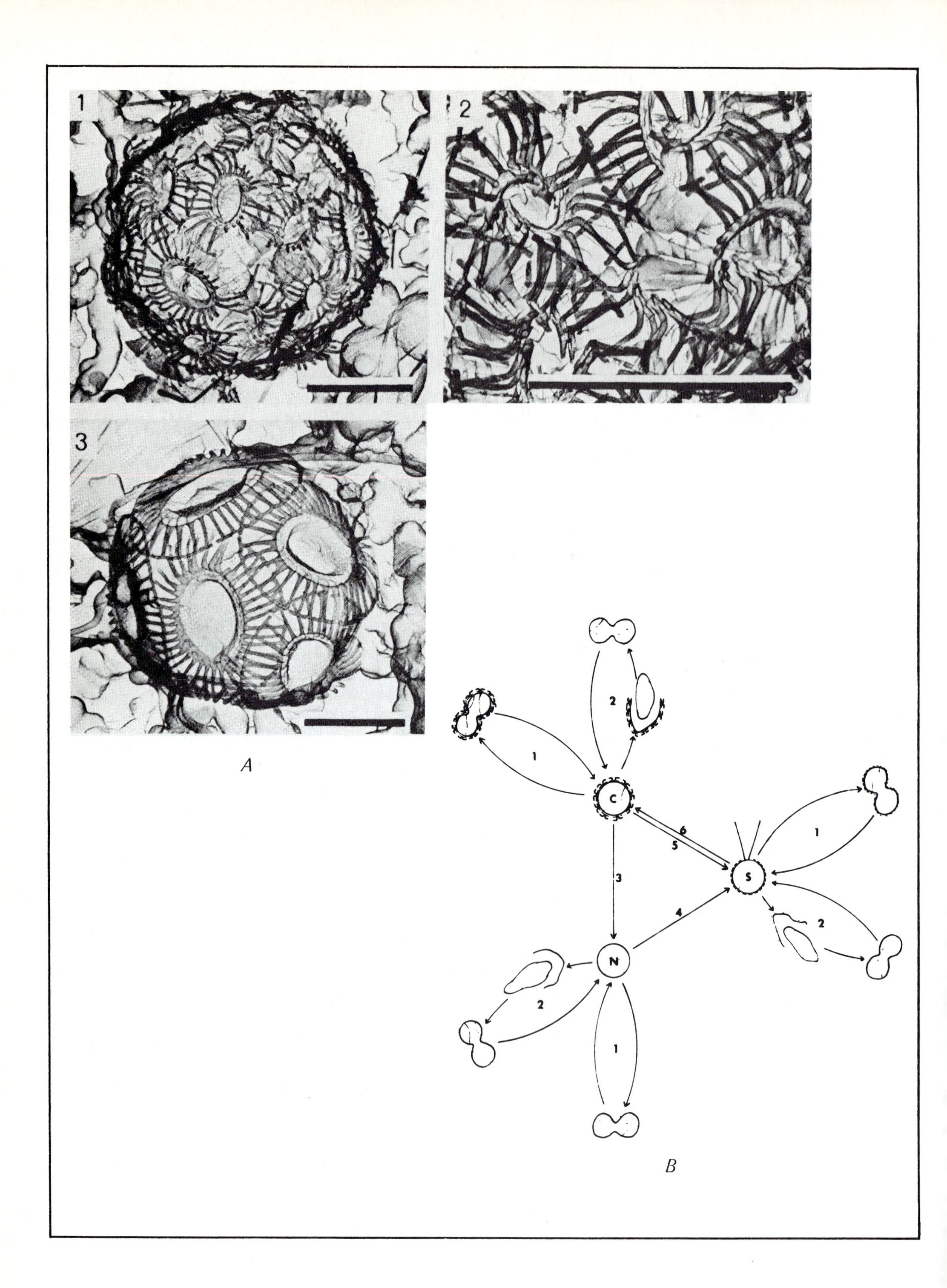
1
2
3
A
2
1
C
6
5
1
S
3
2
4
N
2
1
B

PLATE 12-3. The coccolithophorid *Coccolithus (Emiliania) huxleyi.* (*A*) Vegetative cells showing details of coccoliths (bars = 3 μm). (*B*) The possible life history of *Coccolithus huxleyi:* (1) normal vegetative propagation; (2) aberrant mode of vegetative propagation: the cell leaves the cover as an amoeboid swarmer before division; (3) loss of the ability to make calcified coccoliths, N-cells appear; (4) flagellated cells appear in pure cultures of N-cells; (5) flagellated cells appear in pure cultures of C-cells; (6) coccolith-covered cells appear in pure cultures of S-cells (C, coccolithophorid nonmotile phase; S, motile coccolithophorid phase; N, naked nonmotile phase). (*A*) from H. Okada and S. Honjo, 1973, *Deep-Sea Res.* 20:355–74; (*B*) from D. Klaveness, 1972b, *Br. Phycol. J.* 7:309–18.

sulfide also produced by this hydrolysis, which gives this and many other algae their characteristic odor, may be the cause of fish exclusion from dense blooms. The role of dimethyl sulfide in the marine environment has not been studied, but the compound is related to the alkyl sulfides of onions that not only give them their eye-watering properties but that have an antibacterial activity, as discovered by Pasteur. Dimethyl sulfide is also related to dimethyl sulfoxide, which is known to have an antiviral activity. From 16 to 64% of the photosynthate of the colonial form of *P. poucheti* is released into the sea. In terms of dissolved organic matter, this would yield concentrations of up to 300 μg of polysaccharide/liter daily and 7 μg of acrylic acid/liter daily. In contrast, the microorganism in its unicellular phase releases only 3% of its photosynthate (Guillard and Hellebust, 1971).

The microorganisms in the haptophytes are another excellent example of how important it is to bring microorganisms into culture to observe their life cycles, in order to explain what we see in the sea. Ultrastructural studies that could only have been conducted with cultures have permitted the solution of such paradoxes as a non-thecate, diploid benthic phase alternating generations with haploid coccolithophorid cells (Rayns, 1962; Boney and Burrows, 1966), the alternation between holococcolith and heterococcolith forms (Parke and Adams, 1960), and the relationship between swarming and colonial forms (Kayser, 1970). Only from similar studies on the oceanic species (Klaveness, 1972a,b, 1973) will information be obtained that will eventually determine the systematic and phylogenetic relationships among these fascinating microorganisms and permit a more substantial taxonomy than the one based on coccolith microstructure. The biological activities of the haptophytes have been studied in cultures, and studies on coccolithophorid physiology have been reviewed by Paasche (1968). Both *Prymnesium parvum* (Rahat and Spira, 1967) and the coccolithophorids (Blankley, 1971) are capable of heterotrophic growth in the dark in the presence of relatively high concentrations of glycerol. Apparently the only member of the Haptophyceae that is toxic to fish is *P. parvum*, a euryhaline and common neritic species that can be a problem in certain inshore waters and especially in brackish fish farms such as those in Israel (Shilo, 1967). These endemic areas of algal toxicity are ideal for study. The division cycle of *P. parvum* and its ultrastructural details have been described by Manton (1964b), and the organelles, which are clearly seen in both the phase-contrast micrographs and the thin section shown in Plate 12-2, have been fully described by Manton and Leedale (1963). Other physiological studies on the Haptophyceae include a study by Guillard and Hellebust (1971) on *Phaeocystis* and the many studies in which the tide-pool form *Isochrysis* has been used.

The classification of the non-calcified genera is based on cell morphology and life cycles. The coccolithophorids have been classified by coccolith morphology and the arrangement and shape of the cell. Some cells have characteristic appendages for flotation, but as we have seen, species in different genera have been found to be different phases in the life cycle of some other species. Further life cycle and ultrastructural studies will undoubtedly lead to revised taxonomic schemes.

B. PRASINOPHYTES

This homogeneous group of chlorophyll-*b*–containing algae (mainly green, but some yellowish green to brown) are characterized by motile zooid cells that have a double layer of scales on both the body and the thick flagella. The family Pyramimonadales contains species of free-living motile unicells, 10 to 30 μm in diameter, although some species can also form nonmotile dendroid colonies. The family Halosphaerales contains microorganisms conspicuous by their large nonmotile phycoma stage (cyst), 200 to 800 μm in diameter, which also have a motile zooid stage indistinguishable from forms in the genus *Pyramimonas.* Sexual reproduction in the class is unknown, so far.

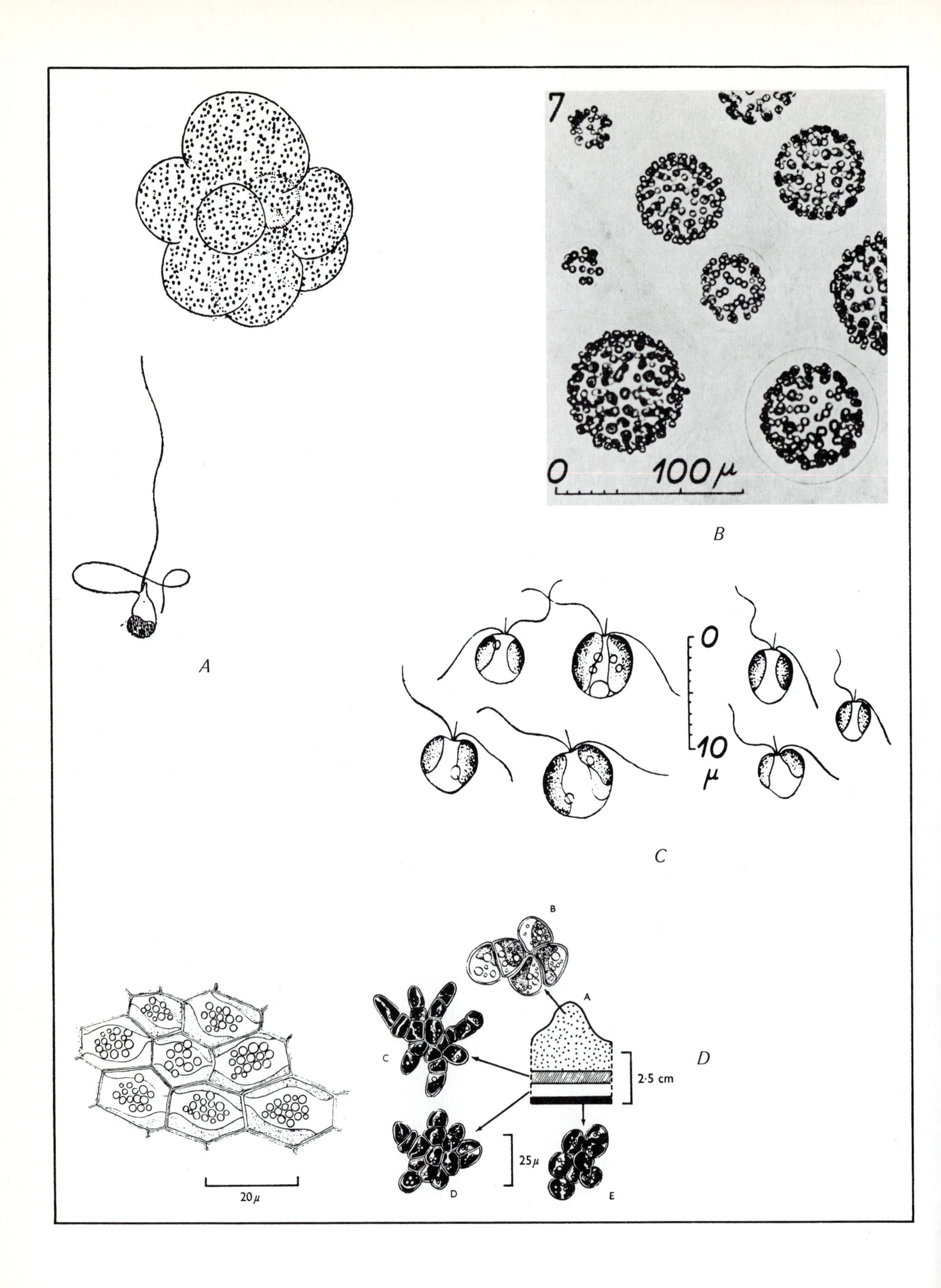
7
0
100 μ
B
A
0
10
μ
C
B
A
C
2·5 cm
D
25 μ
D
E
20 μ

PLATE 12-4. A comparison of two different stages of two very different haptophytes: the planktonic stages of *Phaeocystis poucheti* and the benthonic stages of the coccolithophorid *Cricosphaera*. (*A*) The colonial form and flagellated cell of *Phaeocystis poucheti*, as illustrated in the original description, note haptonema. (*B*) A micrograph of the colonial phase of *P. poucheti*, as seen in culture. (*C*) *P. poucheti* swarming cells (left) and microzoospores (right), note haptonema between flagella. (*D*) The benthic phase of a species of *Cricosphaera*, showing (left) the appearance of an 11-month-old culture from which coccolithophorids were released and (right) the zonation of a 6-month-old benthic culture. (*A*) from M.G. Pouchet, 1892, *C.R. Mem. Soc. Biol.* 44:34–36; (*B,C*) from P. Kornmann, 1956, *Helgoländer wiss. Meeres.* 5:218–33; (*D*) A.D. Boney and A. Burrows, 1966, *J. mar. biol. Ass. UK* 46:295–319.

Very little is known about the habitat and distribution of these microalgae, since recent studies on their life histories and classification show that species identification in the older literature is suspect. The phytoflagellates in the Pyramimonadales are known mainly from cultures, but they are presumably inshore forms. It must have been a surprise to discover that some *Pyramimonas*-like flagellates, isolated from the marine plankton near Plymouth, England and maintained for two decades in culture, were actually part of the life cycle of the nonmotile green alga *Halosphaera* (Manton, Oates, and Parke, 1963). Blooms of *Halosphaera* are so frequent in the Mediterranean that it has the common name *punti verdi* among Neapolitan fishermen (Steuer, 1910). *Halosphaera* species, especially *H. viridis*, have often been observed in plankton records, and *H. viridis* has been regarded as a tropical or subtropical form that is carried to temperate and polar waters (Yashnov, 1965). Although there are probably more, four species have been clearly identified (Parke and Hartog-Adams, 1965; Boalch and Mommaerts, 1969). All distribution records for *Halosphaera* that have not taken the new criteria for species identification into account (Travers and Travers, 1973), using living material, are suspect. It is believed that *H. minor* is the most common species in the North Atlantic and that the waters just inside and outside the Mediterranean Sea could be the nursery for the production of the cyst phase, which is then carried northward by ocean currents (Parke and Hartog-Adams, 1965).

Although *Halosphaera* has long been known (Schmitz, 1879), records on it and the *Halosphaera*-like genera *Pachysphaera* and *Pterosperma* are few and scanty (Boalch and Parke, 1971). Methods of plankton sedimentation for concentrating and examining plankton would miss the pelagic forms that float because of their large lipid content. The buoyant phycoma stage seen in the surface may represent senescent cells. The functional nonmotile stage may be a "shade-form" occurring at greater depths (Yashnov, 1965), and the cysts found at depths to 2000 m may be prey for deep-sea benthos (Wiebe, Remsen, and Vaccaro, 1974). The poroid walls of some species of *Pachysphaera* could be mistaken for centric diatoms, under low magnification (Parke, 1966).

The illustration of the life history of *H. viridis* from the original description is shown in Plate 12-5. Species in the genus *Halosphaera*, as well as the *Halosphaera*-like genera *Pachysphaera* and *Pterosperma*, have two distinct phases, the motile phase with four flagella, which persists in an independent form and reproduces by fission, and the larger, nonmotile uninucleate cyst phase, which develops directly from the motile phase. Once the motile cells round off, lose their flagella, and initiate the cyst phase, the cell gradually increases to a maximum size at maturity, which varies with the species, season, and geographical location (Parke and Hartog-Adams, 1965). As the cyst increases in size, the contents form distinct discs (rosettes); the outer double wall grows with the cell, being cast off as the rosettes undergo fission and are liberated as motile cells. The number and size of the rosettes vary with the species. The cysts of *Pterosperma* have alae (wings) on the outer surfaces, *Halosphaera* has the cell contents concentrated in the periphery, whereas the cell content of *Pachysphaera* fills the cell. The species in these three genera are distinguished by the fine structure of the outer wall of the phycoma phase in either fresh or preserved material. The cysts of the living genus *Pachysphaera* and the fossil genus *Tasmanites* (Boalch and Parke, 1971) are closely related, and further study may show that some species are identical, just as the resting cysts of a number of species of contemporary dinoflagellates have been found to be identical to fossil dinoflagellates and hystrichospheres (see Chapter 11).

The motile phase of all members of this class are similar. The flagella and body scales that occur in a double layer are shown in Plate 12-6. There are apparently slight variations in the size, pattern, and shape of the scales between species. The scales and flagellar hairs are formed by the Golgi body, and detailed thin sections of *Pyramimonas amylifera* show the production of separate types of body and flagellar scales within the

PLATE 12-5. *Halosphaera viridis*, as illustrated in the original description by Schmitz (1879). (*1*) Individual cell whose nucleus has formed a number of nuclei by repeated binary fission (122 X). (*2*) The plasma of the cell envelope, with the numerous nuclei now gathered into a smoothly hemispherical mass (122 X). (*3*) Complete maturation of the daughter cells, with ruptured outer membrane of the cell and daughter cells detached from the membrane (122 X). (*4*) Formation of daughter cells within a colony (366 X). (*6*) Daughter cells after fission of the mother cell (366 X) (*7–9*) Division of daughter cells to form zoospores (366 X). (*10–11*) Zoospores. (*12*) Division of daughter cell to form four zoospores. (*5*, *13–15*) Abnormal cell divisions and development. From F. Schmitz, 1879, *Mitt. Zool. Sta. Neapel* 1:67–92, Plate 1.

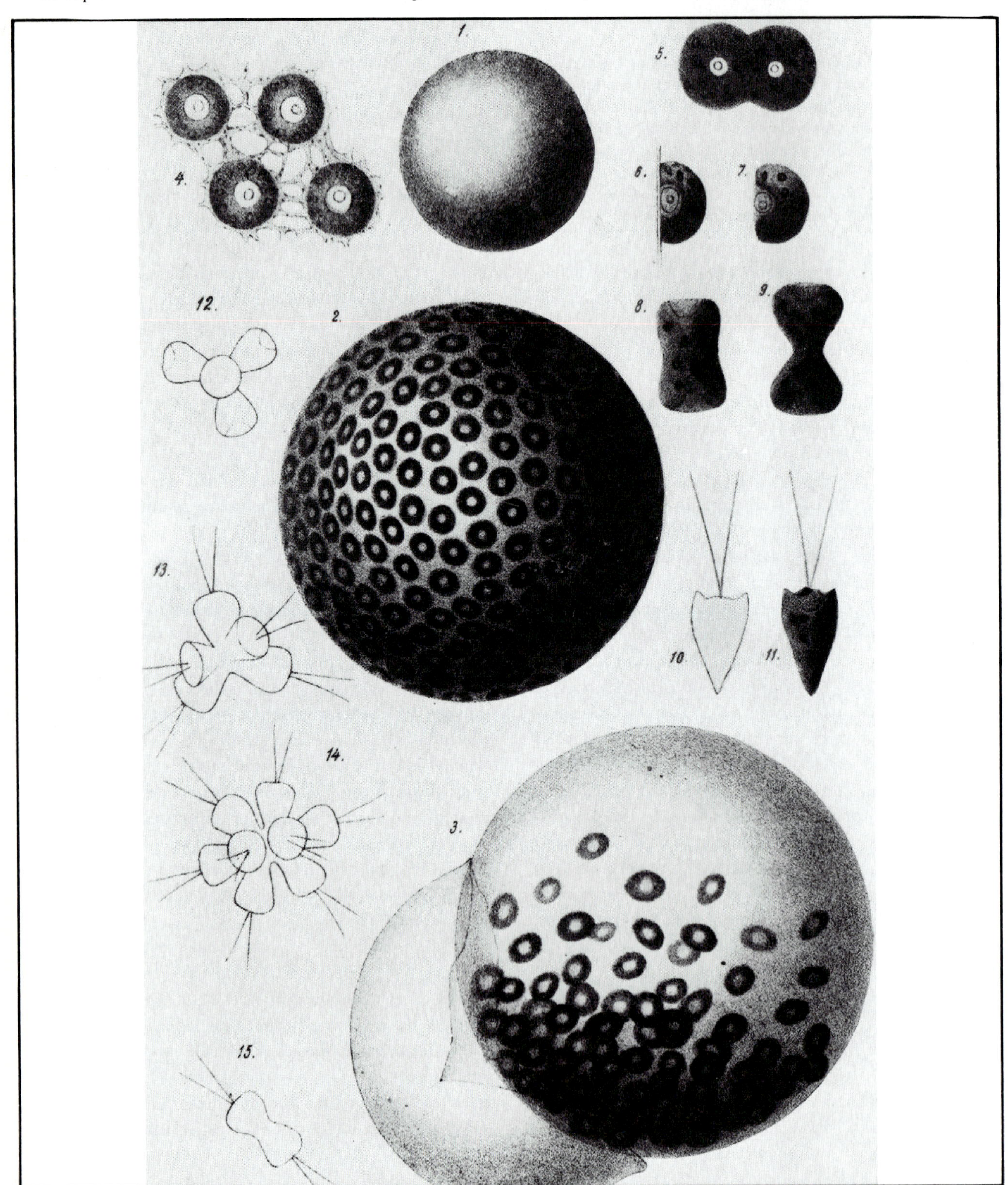

PLATE 12-6. The characteristic body and flagellar scales of the Prasinophyceae, as seen in *Pyramimonas amylifera*. (*A*) Bright-field micrograph of Lugol's stained cells of Gardiner's clone, showing multiple flagella. (*B*) Transmission electron micrograph of the entire cell, showing the eight flagella and some loose body scales. (*C*) Smaller scale (common to body and flagella) and larger horseshoe crab-like flagellar scales and one large basket- or crown-like body scale. (*D*) Thin section of flagella in longitudinal and cross section, showing the two types of scales. (*A-C*) Micrographs courtesy of Paul E. Hargraves; (*D*) electron micrograph courtesy of Paul W. Johnson.

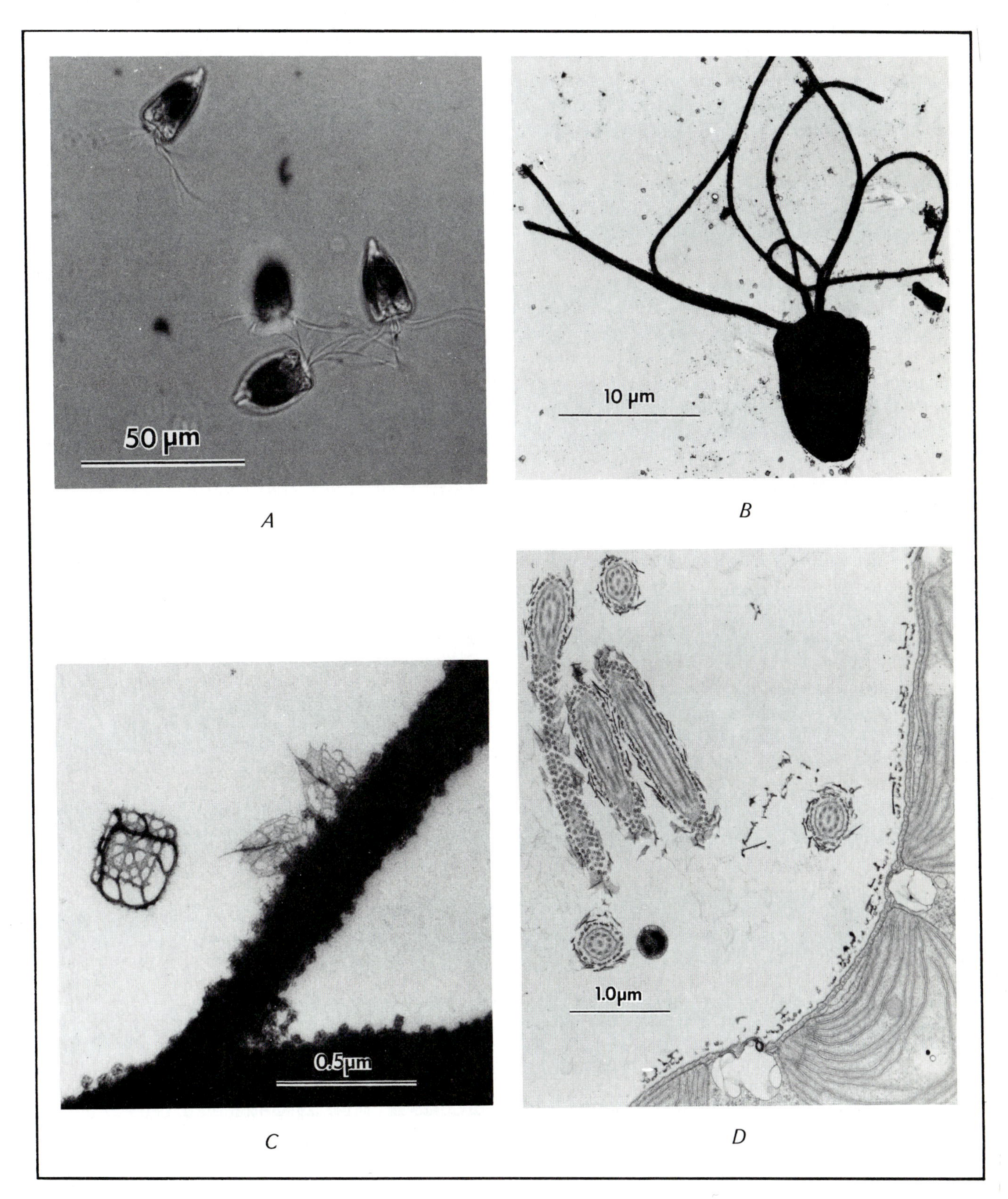

Golgi cisternae, their internal reservoirs, and their mode of escape (Manton, 1966b). This resembles scale production in the Haptophyceae. Spherical ejectile organelles, approximately 1 μm in diameter, are seen in the peripheral cytoplasm beneath the scale layers. The urn-shaped chloroplast has two or four pyrenoids surrounded by a starch sheath. Swimming is rapid and the body is quickly rotated by one, two, four, or eight flagella, which originate either anteriorly or posteriorly. The points of similarity and difference in the morphology and microanatomy of the different species in the family Pyramimonadales are discussed in papers on the genera *Heteromastix* (Manton et al., 1965), *Micromonas* (Manton and Parke, 1960), *Nephroselmis* (Parke and Rayns, 1964), *Pyramimonas* (Manton, Oates, and Parke, 1963), *Platymonas* (Manton and Parke, 1965; McLachlan and Parke, 1967), and *Prasinocladus* (Parke and Manton, 1965). Most of the taxonomic studies conducted at the Plymouth Laboratory and at Leeds University have been with unialgal cultures brought into culture and/or maintained in an *Erdschreiber* medium together with their associated bacteria.

The class Prasinophyceae, as suggested by Christensen (1962, 1966), contains this group of phytoflagellates, which were formerly assigned to a number of classes (see the discussions on taxonomy in the papers cited above for the different genera). Their similarity is indicated by the production of mannitol in all the species so far tested and the presence of dimethyl-β-propiothetin in all species except *Pyramimonas* (which might be expected, since its production and release is associated with the colonial form *Pyramimonas* lacks) (McLachlan and Craigie, 1967; Craigie et al., 1967). The "Check-list of British Marine Algae" (Parke and Dixon, 1976) serves as a species list for this class.

REFERENCES

Anand, P.L. 1937a. A taxonomic study of the algae of the British chalk-cliffs. *J. Bot. (London)* 75 (Suppl.):1–51.

Anand, P.L. 1937b. An ecological study of the algae of the British chalk-cliffs. Part II. *J. Ecology* 25:344–67.

Bernard, F. 1963. Density of flagellates and myxophyceae in the heterotrophic layers related to environment. In: *Marine Microbiology* (C.H. Oppenheimer, ed.). C. C. Thomas, Springfield, Ill., pp. 215–28.

Bigelow, H.B. 1924. Plankton of the offshore waters of the Gulf of Maine. *Bull. U.S. Bur. Fish.* 40(2):1–509.

Birkenes, E., and T. Braarud. 1952. *Phytoplankton in the Oslo Fjord during a "Coccolithus huxleyi-*summer." Avhand. Utgitt. Det. Norske Vidensk.-Akademi. I Oslo 1. Mat. Natur. Klasse 1952, No. 2, 23 p.

Black, M. 1961–62. The fine structure of the mineral parts of Coccolithophoridae. *Proc. Linn. Soc. London* 174:41–46.

Black, M. 1965. Coccoliths. *Endeavour* 24:131–37.

Blankley, W.F. 1971. Auxotrophic and heterotrophic growth and calcification in coccolithophorids. Ph.D. Thesis, University of California, San Diego, 186 p.

Boalch, G.T., and J.P. Mommaerts. 1969. A new species of *Halosphaera*. *J. mar. biol. Assoc. U.K.* 49:129–39.

Boalch, G.T., and M. Parke. 1971. The prasinophycean genera (*Chlorophyta*) possibly related to fossil genera, in particular the genus *Tasmanites*. In: *Proc. 2nd Plankt. Conf., Rome 1970* (A. Farinacci, ed.), pp. 99–105.

Boney, A.D. 1970. Scale-bearing phytoflagellates: An interim review. *Oceanogr. Mar. Biol. Ann. Rev.* 8:251–305.

Boney, A.D., and A. Burrows. 1966. Experimental studies on the benthic phases of Haptophyceae. I. Effects of some experimental conditions on the release of coccolithophorids. *J. mar. biol. Assoc. U.K.* 46:295–319.

Burkholder, P.R., and J.McN. Sieburth. 1961. Phytoplankton and chlorophyll in the Gerlache and Bransfield straits of Antarctica. *Limnol. Oceanogr.* 6:45–52.

Christensen, T. 1962, 1966. Alger. In: *Botanik (Systematisk Botanik)* (T.W. Bocher, M. Lange, and T. Sørensen, eds.), No. 2, Vdg. 2 (1st Ed., 178 p.) (2nd Ed., 180 p.) (in Danish), Munksgaard, Copenhagen.

Craigie, J.S., J. McLachlan, R.G. Ackman, and C.S. Tocher, 1967. Photosynthesis in algae. III. Distribution of soluble carbohydrates and dimethl-β-propiothetin in marine unicellular Chlorophyceae and Prasinophyceae. *Can. J. Bot.* 45: 1327–34.

Fournier, R.O. 1968. Observations of particulate organic carbon in the Mediterranean Sea and their relevance to the deep-living coccolithophorid *Cyclococcolithus fragilis*. *Limnol. Oceanogr.* 13:693–97.

Gaarder, K.R. 1971. Comments on the distribution of Coccolithophorids in the oceans. In: *The Micropalaeontology of Oceans* (B.M. Funnell and W.R. Riedel, eds.). Cambridge University Press, New York and London, pp. 97–103.

Gaarder, K.R., and G.R. Hasle. 1971. Coccolithophorids of the Gulf of Mexico. *Bull. Mar. Sci.* 21:519–44.

Green, J.C., and D.H. Jennings. 1967. A physical and chemical investigation of the scales produced by the Golgi apparatus within and found on the surface of the cells of *Chrysochromulina chiton* Parke et Manton. *J. Exp. Bot.* 18:359–70.

Guillard, R.R.L., and J.A. Hellebust. 1971. Growth and the production of extracellular substances by two strains of *Phaeocystis poucheti*. *J. Phycol.* 7:330–38.

Honjo, S., and H. Okada. 1974. Community structure of coccolithophores in the photic layer of the mid-Pacific. *Micropaleontology* 20:209–230.

Jones, P.G.W., and S.M. Haq. 1963. The distribution of *Phaeocystis* in the eastern Irish Sea. *J. du Conseil* 28:8–20.

Kashkin, K.I. 1963. Materials on the ecology of *Phaeocystis pouchetii* (Hariot) Lagerheim, 1893 (Chrysophyceae). II. Habitat and specifications of biogeographical characteristics. *Okeanologiya* 3:697–705.

Kayser, H. 1970. Experimental-ecological investigations on *Phaeocystis poucheti* (Haptophyceae): Cultivation and waste water test. *Helgoländer wiss. Meeresunters.* 20:195–212.

Klaveness, D. 1972a. *Coccolithus huxleyi* (Lohmann) Kamptner I. Morphological investigations on the vegetative cell and the process of coccolith formation. *Protistologica* 8:335–46.

Klaveness, D. 1972b. *Coccolithus huxleyi* (Lohm.) Kamptn. II. The flagellate cell, aberrant cell types, vegetative propagation and life cycles. *Br. phycol. J.* 7:309–18.

Klaveness, D. 1973. The microanatomy of *Calyptrosphaera sphaeroidea*, with some supplementary observations on the motile stage of *Coccolithus pelagicus*. *Norw. J. Bot.* 20:151–62.

Klaveness, D. 1976. *Emiliania huxleyi* (Lohmann) Hay & Mohler. III. Mineral deposition and the origin of the matrix during coccolith formation. *Protistologica* 12:217–24.

Kornmann, P. 1956. Beobachtungen an *Phaeocystis*-Kulturen. *Helgoländer wiss. Meeres.* 5:218–33.

Lohmann, H. 1902. Die Coccolithophoridae, eine Monographie der Coccolithen bildenden Flagellaten. *Arch. Protistenk.* 1:89–165.

Magne, F. 1957. Sur un biotype marin favorable aux Chrysophycées benthiques. *C.R. Acad. Sci. Paris* 245:983–85.

Manton, I. 1964a. Further observations on the fine structure of the haptonema in *Prymnesium parvum*. *Arch. Mikrobiol.* 49:315–30.

Manton, I. 1964b. Observations with the electron microscope on the division cycle in the flagellate *Prymnesium parvum* Carter. *J. Royal Microsc. Soc.* 83:317–25.

Manton, I. 1966a. Observations on scale production in *Prymnesium parvum*. *J. Cell Sci.* 1:375–80.

Manton, I. 1966b. Observations on scale production in *Pyramimonas amylifera* Conrad. *J. Cell Sci.* 1:429–38.

Manton, I., and G.F. Leedale. 1963. Observations on the fine structure of *Prymnesium parvum* Carter. *Arch. Mikrobiol.* 45:285–303.

Manton, I., and M. Parke. 1960. Further observations on small green flagellates with special reference to possible relatives of *Chromulina pusilla* Butcher. *J. mar. biol. Assoc. U.K.* 39:275–98.

Manton, I., and M. Parke. 1965. Observations on the fine structure of two species of *Platymonas* with special reference to flagellar scales and the mode of origin of the theca. *J. mar. biol. Assoc. U.K.* 45:743–54.

Manton, I., K. Oates, and M. Parke. 1963. Observations on the fine structure of the *Pyramimonas* stage of *Halosphaera* and preliminary observations on three species of *Pyramimonas*. *J. mar. biol. Assoc. U.K.* 43:225–38.

Manton, I., D.G. Rayns, H. Ettl, and M. Parke. 1965. Further observations on green flagellates with scaly flagella: The genus *Heteromastix* Korshikov. *J. mar. biol. Assoc. U.K.* 45:241–55.

McLachlan, J., and J.S. Craigie. 1967. Photosynthesis in algae containing chlorophyll *b*. *Br. phycol. Bull.* 3:408–9.

McLachlan, J., and M. Parke. 1967. *Platymonas impellucida* sp. nov. from Puerto Rico. *J. mar. biol. Assoc. U.K.* 47:723–33.

Northcote, D.H., and J.D. Pickett-Heaps. 1966. A function of the Golgi apparatus in polysaccharide synthesis and transport in the root-cap cells of wheat. *Biochem. J.* 98:159–67.

Okada, H., and S. Honjo. 1973. The distribution of oceanic coccolithophorids in the Pacific. *Deep-Sea Res.* 20:355–74.

Orton, J.H. 1923. The so-called "baccy-juice" in the waters of the Thames oyster-beds. *Nature* 111:773.

Paasche, E. 1964. A tracer study of the inorganic carbon uptake during coccolith formation and photosynthesis in the coccolithophorid *Coccolithus huxleyi*. *Physiol. Plant.* (Suppl. No. 3):1–82.

Paasche, E. 1968. Biology and physiology of coccolithophorids. *Ann. Rev. Microbiol.* 22:71–86.

Parke, M. 1961. Some remarks concerning the class Chrysophyceae. *Br. phycol. Bull.* 2:47–55.

Parke, M. 1966. The genus *Pachysphaera* (Prasinophyceae). In: *Some Contemporary Studies in Marine Science* (H. Barnes, ed.). G. Allen and Unwin, London, pp. 555–63.

Parke, M., and I. Adams. 1960. The motile (*Crystallolithus hyalinus* Gaarder & Markali) and non-motile phases in the life history of *Coccolithus pelagius* (Wallich) Schiller. *J. mar. biol. Assoc. U.K.* 39:263–74.

Parke, M., and I. den Hartog-Adams. 1965. Three species of *Halosphaera*. *J. mar. biol. Assoc. U.K.* 45:537–57.

Parke, M., and P.S. Dixon. 1968. Check-list of British marine algae—second revision. *J. mar. biol. Assoc. U.K.* 48:783–832.

Parke, M., and P.S. Dixon. 1976. Check-list of British marine algae—third revision. *J. mar. biol. Assoc. U.K.* 56:527–94.

Parke, M., and I. Manton. 1965. Preliminary observations on the fine structure of *Prasinocladus marinus*. *J. mar. biol. Assoc. U.K.* 45:525–36.

Parke, M., and D.G. Rayns. 1964. Studies on marine flagellates. VII. *Nephroselmis gilva* sp. nov. and some allied forms. *J. mar. biol. Assoc. U.K.* 44:209–17.

Parke, M., I. Manton, and B. Clarke. 1955. Studies on marine flagellates. II. Three new species of *Chrysochromulina*. *J. mar. biol. Assoc. U.K.* 34:579–609.

Pascher, A. 1925. Die braune Algenreihe der Chrysophyceen. *Arch. Protistenk.* 52:489–564.

Pouchet, M.G. 1892. Sur une algue pélagique nouvelle. *C.R. Mém. Soc. Biol.* 44:34–36.

Rahat, M., and Z. Spira. 1967. Specificity of glycerol for dark growth of *Prymnesium parvum*. *J. Protozool.* 14:45–48.

Rayns, D.G. 1962. Alternation of generations in a coccolithophorid, *Cricosphaera carterae* (Braarud & Fagerl.) Braarud. *J. mar. biol. Assoc. U.K.* 42:481–84.

Savage, R.E. 1930. The influence of *Phaeocystis* on the migrations of the herring. *Fish. Invest. Ser. 2*, 12(2):1–14.

Schmitz, Fr. 1879. *Halosphaera,* eine neue Gattung grüner Algen aus dem Mittelmer. *Mitt. Zool. Sta. Neapel.* 1:67–92.

Shilo, M. 1967. Formation and mode of action of algal toxins. *Bacteriol. Rev.* 31:180–93.

Sieburth, J.McN. 1958. Antarctic microbiology. A study of Antarctic birds conducted during the 1957–58 Argentine Antarctic Expedition. *AIBS Bull.* June, 10–12.

Sieburth, J.McN. 1959a. Gastrointestinal microflora of Antarctic birds. *J. Bacteriol.* 77:521–31.

Sieburth, J.McN. 1959b. Antibacterial activity of Antarctic marine phytoplankton. *Limnol. Oceanogr.* 4:419–24.

Sieburth, J.McN. 1960. Acrylic acid, an "antibiotic" principle in *Phaeocystis* blooms in Antarctic waters. *Science* 132:676–77.

Sieburth, J.McN. 1961. Antibiotic properties of acrylic acid, a factor in the gastrointestinal antibiosis of polar marine animals. *J. Bacteriol.* 82:72–79.

Sieburth, J.McN. 1965. Microbiology of Antarctica. In: *Biogeography and Ecology in Antarctica* (P. Van Oye and J. Van Mieghem, eds.), Vol. 15. Monographiae Biologicae, Dr. W. Junk, Publ., The Hague, pp. 267–95.

Steuer, A. 1910. *Planktonkunde,* Vol. 2. B.G. Teubner, Stuttgart, 723 p.

Travers, A., and M. Travers. 1973. Le genre *Halosphaera* Schmitz dans le golfe de Marseille. *Rapp. Proc.-Verb. Réun. Cons. Int. Explor. Mer* 21:425–28.

Waern, M. 1952. Rocky-shore algae in the Öregrund archipelago. *Acta Phytogeogr. Suecica* 30:85–93.

Wiebe, P.H., C.C. Remsen, and R. Vaccaro. 1974. *Halosphaera viridis* in the Mediterranean Sea: Size range, vertical distribution, and potential energy source for deep-sea benthos. *Deep-Sea Res.* 21:657–67.

Wilbur, K.M., and N. Watabe. 1963. Experimental studies on calcification in molluscs and the alga *Coccolithus huxleyi. Ann. N.Y. Acad. Sci.* 109:82–112.

Yashnov, V.A. 1965. Water masses and plankton. 3. *Halosphaera viridis* as an indicator of Mediterranean waters in the North Atlantic. *Oceanology* 5(5):94–98.

CHAPTER 13

Other Microalgae

A. CHRYSOPHYTES

The algae in this class are a heterogeneous grouping of golden-brown cells, which occur as unicells, colonies, and filaments. Some genera have cell walls with siliceous scales and endogenous siliceous cysts. The unicells have either a long single flagellum or, when paired flagella are present, they are heterokont, with a short smooth flagellum and a long flagellum with hairs. The marine algae of this class are underrepresented after the removal to the Haptophyceae of those genera having a haptonema (Christensen, 1962, 1966; Parke and Dixon, 1968, 1976). Three recently described species of marine chrysophytes with characteristic scales and flagella (Boney, 1970) are colorless phagotrophs and are therefore discussed in Chapter 22,D. The marine chrysophyte-like microorganisms of note still remaining in this class are the silicoflagellates and the taxonomic orphan, *Olisthodiscus luteus*.

The silicoflagellates, a small group within the Chrysophyceae, have a unique, radially arranged internal skeleton of tubular silica elements, a single flagellum, and golden-brown chromatophores. They are more plentiful in the fossil record (Loeblich et al., 1968) than in present-day populations. In the Mediterranean Sea off the French (Nival, 1965; Travers and Travers, 1968), Spanish (Herrara and Margalef, 1963), and Greek (Ignatiades, 1970) coasts, they appear in low numbers in the fall, increase to maximal numbers during the winter and spring, and migrate to the cool, deeper waters in the summer. During the winter and spring, concentrations of hundreds to a thousand silicoflagellates/liter are common, with a maxima of 5 to 10,000/liter at water temperatures of 12 to 15 °C (Nival, 1965; Travers and Travers, 1968). During the Great Barrier Reef Expedition (Australia), silicoflagellates occurred sporadically in numbers up to 30/liter (Marshall, 1934). Living specimens from the Clyde Sea have been beautifully described by Marshall (1934). In Narragansett Bay, Rhode Island, silicoflagellates were present all year except during periods of coldest water, with temperatures below 5 °C (Pratt, 1959).

Observations on these microorganisms were restricted to such field studies until a strain of *Dictyocha fibula* (a common species) from the Washington Pacific coast was maintained in culture (Van Valkenburg and Norris, 1970). This clone was quite specific in its requirements for 160-footcandle illumination, in 24 to 26 o/oo S seawater enriched with 7.5 to 8 ml of Provasoli's solution, and an incubation temperature of 10 °C, under which conditions it had a 49-hour generation time. In addition to its narrow range of acceptable growing conditions, *D. fibula* is difficult to handle and study due to such idiosyncracies as failure to grow when shaken or when bubbled with air, death at certain culture densities, uneven distribution in and adherence to counting chambers, and the similarity between its skeletal remains and its live cells, which disintegrate rapidly under a compound microscope. The cell with and without skeleton and flagellum is shown in Plate 13-1. The great variation in the 200 skeletons examined would put the progeny from this one clone into as many as seven taxa as defined by Gemeinhardt (1930), which suggests that most silicoflagellates belong to one genus, *Dictyocha*, with at least *Distephanus* and *Cannopilus* being identical (Van Valkenburg and Norris, 1970). The variability of the skeletal characteristics indicates that a new basis for classification of living silicoflagellates is needed.

Van Valkenburg (1971a,b) used the same *D. fibula* clone to describe the ultrastructure of the siliceous skeleton and the protoplast (Plate 13-1). The skeleton consists of a basal ring with a central arch, radiating spines, and supporting bars. Unbroken ends of the growing skeleton show delicate spinose endings. A diagrammatic representation of the typical sunburst form of *D. fibula* shows the large central nucleus, surrounded by vesicles and dictyosomes, which is connected, via fine cytoplasmic strands, to the perinuclear cytoplasm, which contains the chloroplasts, pyrenoids, and mitochondria. The flagellum lacks the usual microtubules and flagellar root structure observed in the dinoflagellates. The phototrophic silicoflagellates should not be confused with the phagotrophic ebridians discussed in Chapter 22,F, which they resemble superficially.

Olisthodiscus luteus is a pale yellow-pigmented, heterokont flagellate (with unequal length flagella), which is a taxonomic orphan, having been variously classified in the Xanthophyceae, Chrysophyceae, and Cryptophyceae, although it has been recommended, on the basis of ultrastructure, that it be temporarily assigned to the Chrysophyceae (Leadbeater, 1969). It is not easily recognized because its described dorsoventral flattening and subovoid shape varies, as do its dimensions (Plate 13-2). The longer anterior flagellum bears a bilateral array of fine 2 μm long hairs, whereas the shorter posterior flagellum is hairless. In swimming, the anteriorly diverted flagellum is thrown into a series of waves, and the shorter, more posteriorly directed, flagellum is held out rigidly so that this microorganism glides gracefully like a swimming flatfish but with a rigid body. This distinctive flagellate was first described from a population in a small brackish pool on the Isle of Wight, along with several other interesting microflagellates (Carter, 1937). Subsequently, it was found in brackish waters from the shore of Belgium (Conrad, 1939) and in the vicinity of Woods Hole, Massachusetts (Hulbert, 1965), where two new species of *Olisthodiscus* were observed and described. *Olisthodiscus luteus* has been found and cultured on both sides of the Atlantic and extends from temperate into subtropical waters. An *Olisthodiscus* species in culture was completely suppressed at 2.5 to 5 o/oo S, had maximal growth at 15 o/oo S, and did well at 20 to 35 o/oo S (McLachlan, 1961), growing better in natural than in artificial seawater (McLachlan, 1964). Tomas (1977) describes a 3 to 50 o/oo range with no distinct salinity growth optimum.

In a 5-year study of the phytoplankton in Narragansett Bay, Rhode Island, *O. luteus* was considered one

PLATE 13-1. The cell and siliceous skeleton of the silicoflagellate *Dictyocha fibula*. (*A*) Diagram of sunburst form, showing the system of fine protoplasmic strands that extend into the wall to connect the perikaryon with the peripheral cytoplasmic components (skeleton incomplete). Phase-contrast micrographs of living cells from cultures, showing (*B*) swimming cell, with skeleton (450 X); (*C*) swimming cell, with flagellum but no skeleton (520 X); and (*D*) single nucleated cell, with no skeleton or flagellum (420 X). (*E*) A carbon replica of a skeleton, showing a basal ring (BR), central arch (CA), radial spine (RS), supporting bar (SB), and spinose endings (S), and a diagram of the sequence of formation. (*F*) and (*G*) Phase-contrast micrographs, showing skeletal variations (520 X). (*H*) Scanning electron micrograph of silicoflagellate skeleton, with debris. (*A,E*) from S.D. Van Valkenburg, 1971a, *J. Phycol.* 7:113–18; (*B,C,D,F,G*) from S.D. Van Valkenburg and R.E. Norris, 1970, *J. Phycol.* 6:48–54; (*H*) from J.McN. Sieburth, 1975, *Microbial Seascapes*, University Park Press, Baltimore, Md.

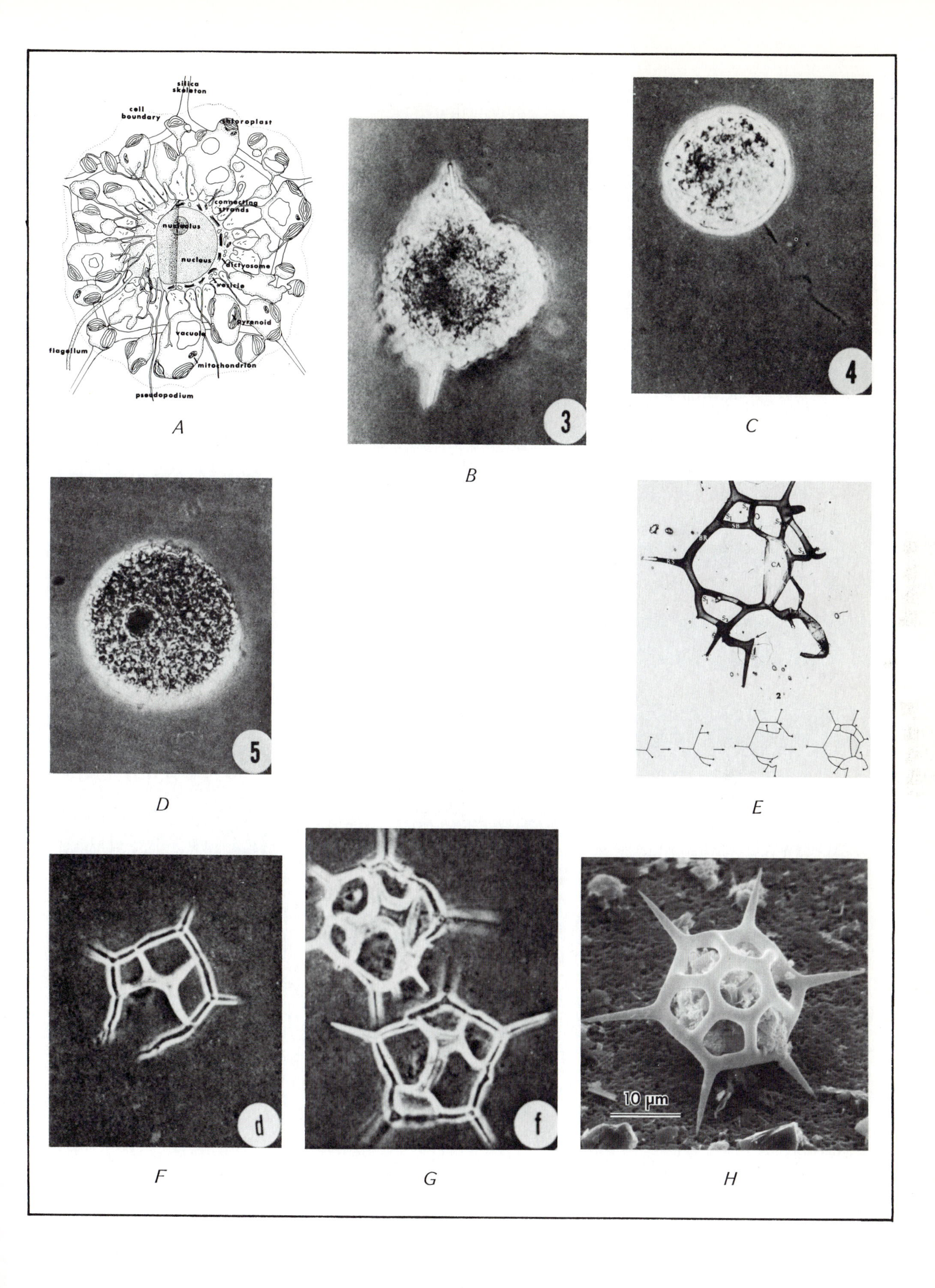
silica skeleton
cell boundary
chloroplast
connecting strands
nucleolus
nucleus
dictyosome
vesicle
pyrenoid
vacuole
flagellum
mitochondrion
pseudopodium
A
3
B
4
C
5
D
CA
2
E
d
F
f
G
10 µm
H

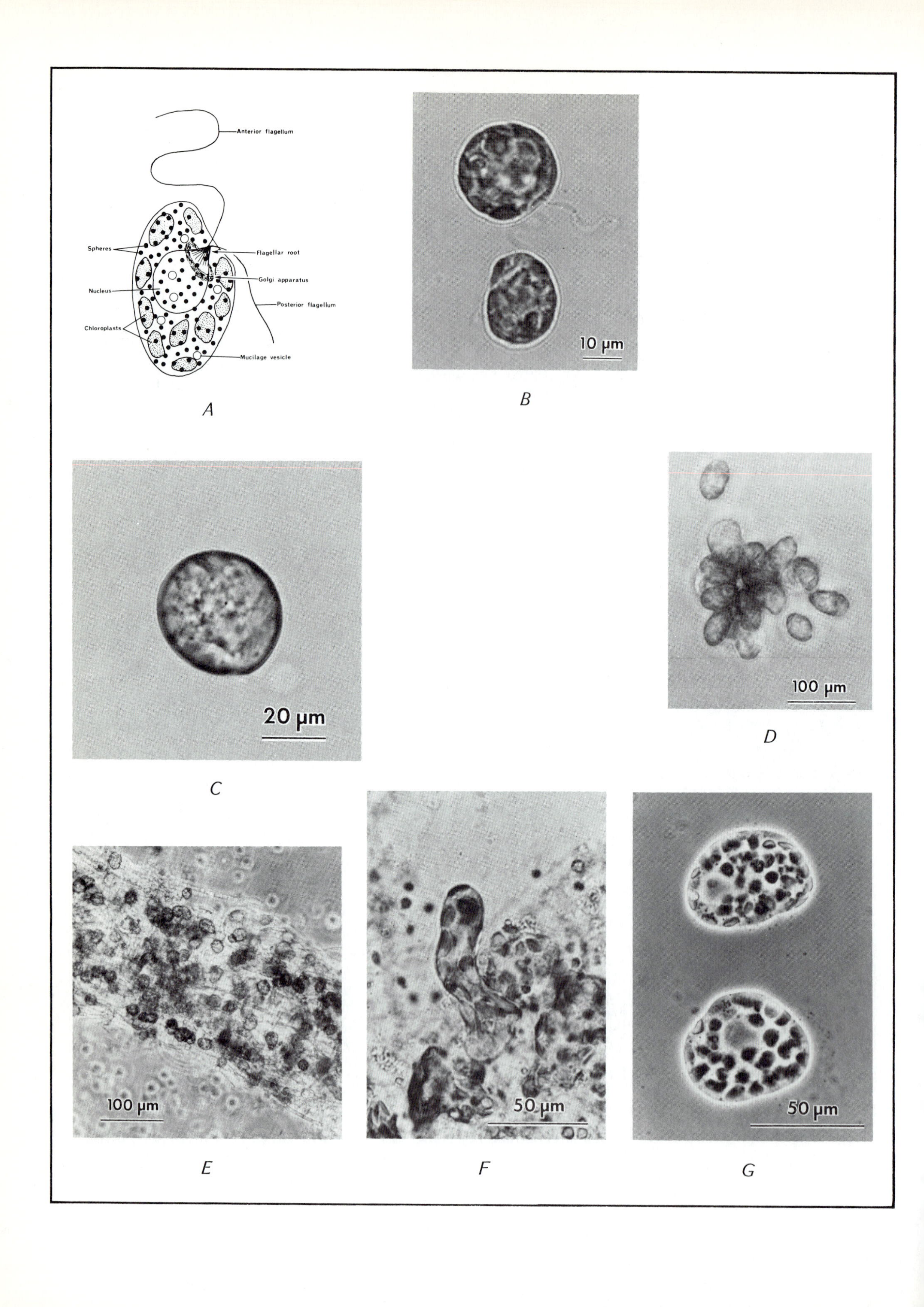
Anterior flagellum
Spheres
Flagellar root
Golgi apparatus
Nucleus
Posterior flagellum
Chloroplasts
Mucilage vesicle
A
10 μm
B
20 μm
C
100 μm
D
100 μm
E
50 μm
F
50 μm
G

PLATE 13-2. The cell and parts of the life cycle of *Olisthodiscus luteus*, showing both its planktonic and benthonic phases. (*A*) Diagram of the general arrangement of the organelles. Micrographs of (*B*) motile cells, (*C*) a nonmotile cell, (*D*) clustering of motile cells, (*E*) the benthonic phase in a loosely packed gelatinous strand, (*F*) a benthonic cell being released from the gelatinous matrix, and (*G*) a newly released, nonmotile benthonic cell. (*A*) from B.S.C. Leadbeater, 1969, *Br. Phycol. J.* 4:3–17; (*B–G*) courtesy of Carmelo Tomas, 1977.

of the important dominant species recurring annually as a temporary bloom at temperatures between 8 and 24 °C, usually in the late spring (Pratt, 1959). In a subsequent 7-year study, Pratt (1965) noted that after the winter–spring (December–May) diatom flowering dominated by *Skeletonema costatum*, for the summer–fall period May through October, when the water temperature was between 14 and 22 °C, there were alternate blooms of *S. costatum* and *O. luteus*, the two species reaching populations of 10^6 to 10^7 cells/liter; they were almost never abundant at the same time. Pratt (1966) has studied this competition in culture. In culture, the temperature optimum for growth is between 15 and 20 °C (Throndsen, 1976; Tomas, 1977).

Clones of *O. luteus* from both sides of the Atlantic have been found to be ultrastructurally similar (Leadbeater, 1969). A diagrammatic representation of the subovoid shape and general arrangement of the organelles is shown in Plate 13-2. The cell is bounded by a single plasma membrane without surface ornamentation. Beneath this membrane is a layer of regular 0.35 μm electron-opaque spheres, which can be observed by phase-contrast microscopy. They are produced in small membrane-bounded vesicles near the Golgi body. Mucilage vesicles, interspersed irregularly between the vesicles containing the spheres, discharge their contents when the cell is squashed, the mucilage being altered to form an interwoven fibrous network around the cell. The subspherical nucleus just anterior to the central part of the cell projects on one side toward the flagellar root and in this area is encircled by the Golgi body. Mitochondria are scattered in the cytoplasm adjacent to the nucleus. From seven to twenty-five individual chloroplasts are arranged peripherally around the cell with the pyrenoids on their inner surface. The two flagella that enter the cell anterio-laterally have a complicated system of striated fibers and microtubules anchored by the nucleus. Such observations have been used to amend the description of this species and to debate its taxonomy (Leadbeater, 1969). A detailed study of the autecology of *Olisthodiscus luteus* by Tomas (1977) has shown that a benthonic or palmelloid phase can also occur (Plate 13-2).

B. CRYPTOPHYTES

This small but discretely defined class (Butcher, 1967) consists of ovoid or bean-shaped microorganisms with a furrow, groove, or gullet. Dorso-ventral flattening gives them an asymmetrical outline as they turn on their vertical axis. They are also characterized by two subequal flagella, and although flagellated cells are the norm, certain species have a palmelloid state. The color of cryptophycean cells was formerly used to characterize the genera, e.g., the colorless *Chilomonas* (discussed in Chapter 22,F), the blue *Chroomonas*, the red *Rhodomonas*, and the yellow or brown *Cryptomonas*. Their small size (less than 20 μm and often only 3 to 6 μm), rapid darting motion, and fragility during collection by net tows make them rather difficult to study.

Although cryptomonads are seldom dominant, they are numerous and widely distributed in marine waters. In Narragansett Bay, the smaller than 15 μm microflagellates are a constant feature of the phytoplankton, slowly and slightly changing their population from a summer maximum to a winter minimum. This population contains species of cryptophycean algae the year round, including *Cryptomonas*, *Chroomonas*, and *Chilomonas* (Pratt, 1959). In brackish British waters, the genera *Chroomonas* and *Cryptomonas* are widespread in the pelagic zone, the former being found in low numbers in the open sea. Cryptomonads have also been observed in Norwegian coastal waters (Throndsen, 1969).

A diagrammatic representation of the main features of four genera of cryptophycean algae is shown in Plate 13-3,*A*. Ultrastructure has been studied in species of *Cryptomonas* (Butcher, 1967; Lucas, 1970a), *Chroomonas* (Butcher, 1967; Dodge, 1969; Lucas, 1970b; Gantt, 1971), and *Hemiselmis* (Butcher, 1967; Lucas, 1970b). The furrow, which is either vertical or horizontally oblique, ranges in complexity from a shallow depression to a tubular gullet. This furrow–groove–gullet system is usually lined with ejectosomes (trichocysts) in various patterns, which discharge to project the cells into their

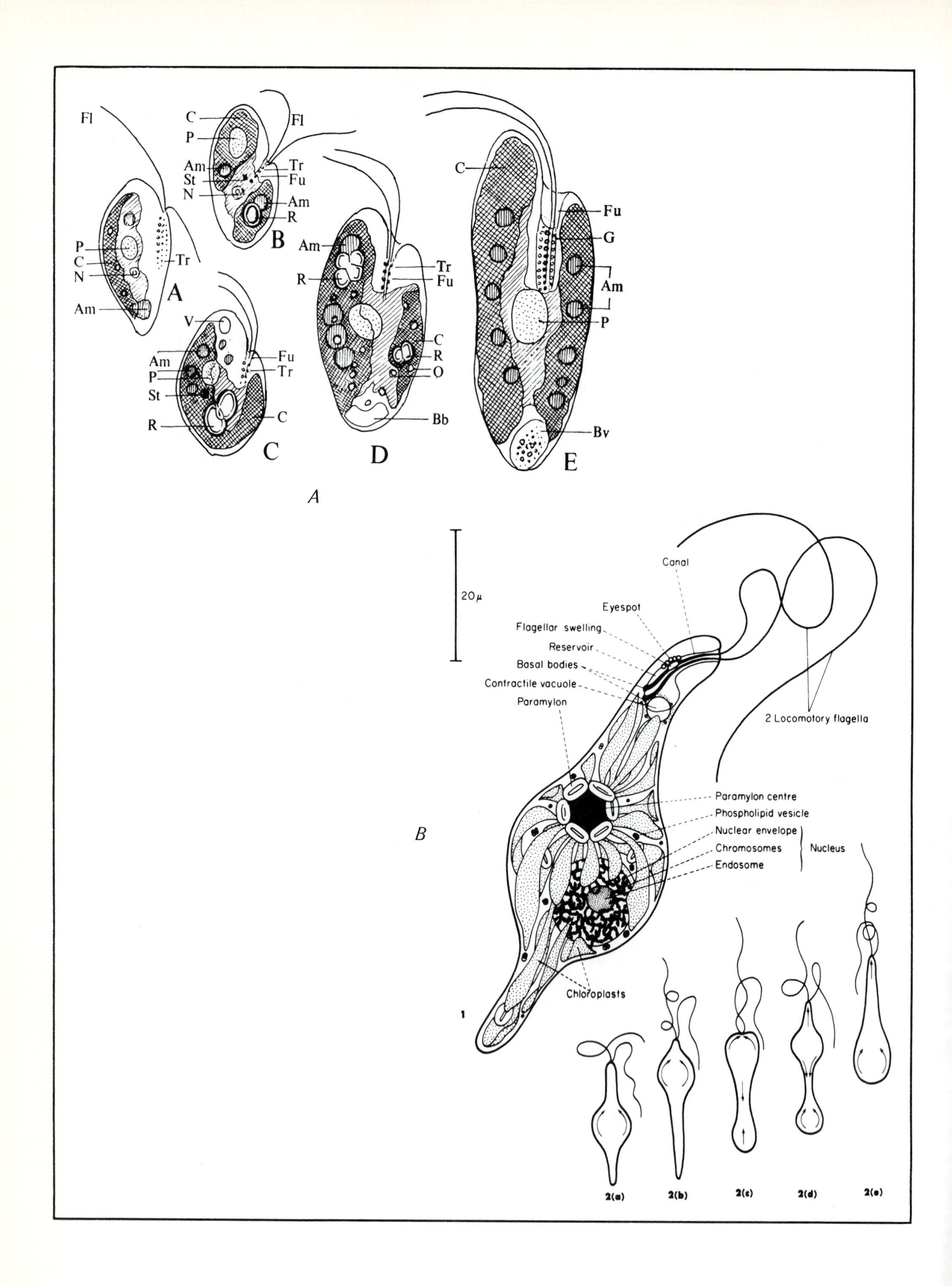

Fl
C
P
Fl
Am
St
N
Tr
Fu
Am
R
B
P
C
N
Tr
A
Am
V
Am
P
St
R
Fu
Tr
C
C
Am
R
Tr
Fu
C
R
O
Bb
D
C
Fu
G
Am
P
Bv
E
A
20μ
Canal
Eyespot
Flagellar swelling
Reservoir
Basal bodies
Contractile vacuole
Paramylon
2 Locomotory flagella
Paramylon centre
Phospholipid vesicle
Nuclear envelope
Chromosomes
Endosome
Nucleus
B
Chloroplasts
1
2(a)
2(b)
2(c)
2(d)
2(e)

PLATE 13-3. Diagrams of the distinctive features of cryptophyte and euglenophyte cells. (*A*) The cryptophyte cell is characterized by the furrow–gullet system lined with trichocysts, as shown in the genera *Plagioselmis* (A), *Hemiselmis* (B), *Chroomonas* (C,D), and *Cryptomonas* (E) (Am, stroma starch grains; Bb, a basal body of a leucosinic nature; Bv, basal vesicle, with vibrating granules; C, chromatophore; Fl, flagella; Fu, furrow; G, gullet; N, nucleus; O, small granules of a lipoid nature; P, pyrenoid; *R*, refractive bodies, *corp de Maupas;* St, stigma or masses of carotin; Tr = trichocysts) (magnifications not given). (*B*) Living cells of *Eutreptia pertyi,* showing the major organelles (1) and stages in euglenoid movement (2) (arrows, direction of cytoplasmic flow). (*A*) from R.W. Butcher, 1967, Fishery Investigations, Series IV, London. Reproduced with permission of Controller of Her Britannic Majesty's Stationery Office; (*B*) from G.F. Leedale, 1967, *Euglenoid Flagellates,* Prentice-Hall, Englewood Cliffs, N.J.

characteristic darting movements, leaving behind a trail of mucilage. These trichocysts differ from those in the dinoflagellates in both structure and method of ejection. The heterokont flagellation may be unusual in that the shorter flagellum is of the bilateral flimmer type, whereas the other has unilateral hairs (Hibberd, Greenwood, and Griffiths, 1971). The structure of the periplast may also be distinctive (Hibberd, Greenwood, and Griffiths, 1971).

The cryptomonads are relatively easy to culture, being euryhaline and having an optimum temperature between 18 and 20 °C. They are quite photosensitive. Some species can be grown on agar media, and *Chroomonas salina* can grow on glycerol in the dark, but it grows better photoheterotrophically in the light where a variety of organic substances can stimulate growth (Antia, Cheng, and Taylor, 1969). Another species can use urea, glycine, and some of the purines as a source of nitrogen (Antia and Chorney, 1968). In addition to chlorophyll *a* and *c*, they have unique xanthophylls including alloxanthin, which comprises some 75 to 90% of the total carotenoids. Their red and blue coloration is due to the phycobilin pigments phycoerythrin and phycocyanin (Butcher, 1967). The location of the phycobilins in the chloroplasts differs from the location in the red algae (Gantt, Edwards, and Provasoli, 1971). The reserve food is starch, which is deposited as free grains around the pyrenoid. Cryptophyte taxonomy is discussed in detail in the monograph by Butcher (1967). Coloration is so variable, especially with age, that it has been disregarded for classification purposes. Primary classification is based on the nature and structure of the depression–furrow–gullet system and the arrangement of ejectosomes, with ultrastructural studies substantiating observations based on light microscopy studies.

C. EUGLENOPHYTES

The euglenoid flagellates are a well-defined group, with very few marine species. They can be pigmented or colorless and either naked or encased within an elastic or a firm periplast. They are usually free swimming, with one, two, or occasionally more flagella that are always terminal, originating in a chamber or reservoir and passing to the exterior via a narrow canal. Features for classification include a unilateral row of fine hairs on the emergent portion of each locomotory flagellum and paramylon (a β-1:3-linked glucan) as reserve food storage, which is deposited as solid granules in a helical array. Nuclear division is by an odd form of mitosis, and the eyespot, when present, is independent of the chloroplasts (Leedale, 1967). This is shown in Plate 13-3, *B;* also shown is the cytoplasmic flow in nonswimming cells that is responsible for their characteristic euglenoid movement. A number of phototrophic species in the genera *Eutreptia*, *Eutreptiella*, and *Euglena*, among others, are common inhabitants of brackish and eutrophic marine waters. The heterotrophic species are discussed in Chapter 22,A. Various species occur throughout the year in temperate waters (Pratt, 1959; Throndsen, 1969), and yearly maxima in Norwegian waters range from 200 to 20,000 euglenids/liter (Throndsen, 1969). A species of *Eutreptiella* has been observed to form a bloom in the waters of the Seto Inland Sea (Okaichi, 1969). *Euglena proxima*, a phototrophic species, can reach considerable densities under certain high organic, low oxygen situations (Miller, 1972) under which anoxyphotobacteria (Chapter 8,A) also occur.

A diagrammatic representation of the main organelles visible in living cells of *Eutreptia pertyi* is shown in Plate 13-3,*B*. The typical euglenoid form and movement are representative. The ultrastructure of euglenid flagellates is discussed by Leedale (1967), and Throndsen (1973) has described the ultrastructure of *Eutreptiella gymnastica*. The phototrophic euglenids are cultured in the same manner as other phytoflagellates (Cook, 1968; Throndsen, 1969; Guillard, 1973). These flagellates, particularly the freshwater *Euglena gracilis*, must be among the most studied of phototrophic microorganisms. Like the algae included in the Chlorophyta, they contain both chlorophyll *a* and *b*. Classification is based upon the presence or absence and/or the number and form of the organelles and structures including flagella, chloroplasts, eyespot and flagellar swelling, ingestion apparatus, envelope, colonial habit, paramylon muciferous bodies, contractile vacuoles, nucleus and cell size, shape and rigidity, and position of canal and reservoir. Euglenoid classification has been discussed in general by Leedale (1967) and for the marine forms by Butcher (1961). New species have been described from Norwegian waters by Throndsen (1969).

D. CHLOROPHYTES

The green algae are characterized by the true green color of chlorophyll *a* and *b*, which is not masked by accessory pigments. The multicellular green algae include such genera as the filamentous *Cladophora*, the sheetlike *Ulva* and *Monostroma*, and the spongy *Codium fragile*. In addition to these macroalgae, there are a number of unicellular green algae, both nonmotile and flagellated. The green microalgae in the marine environment are mainly brackish water forms found in saline ditches, marshes, ponds, and especially polluted and slow flushing bays. An example is Great South Bay, Long Island where, during the 1950s before pollution abatement, an eutrophic growth of green algae was caused by wastes from the duck farming industry (Ryther, 1954). These algae are also found the year round in temperate inshore waters in detectable numbers (Pratt, 1959; Throndsen, 1969). Some of the common nonmotile pond forms are shown in Plate 13-4,*A* and include the marine species of *Chlorella*. The small nonmotile forms (2 to 4 μm), like *Nannochloris* and *Stichococcus*, are impossible to detect in the routine examination of natural populations. The motile forms are illustrated in Plate 13-4,*B* to *E*. All are green unicells with hairs on the terminal end of the flagella. The genus *Dunaliella* is without a visible cell wall and reproduces by direct longitudinal division of motile cells. *Dunaliella* species have two equal flagella, an elastic cell envelope, and a variety of cell shapes. They also reproduce by forming a zygote from two motile cells. The genus *Chlamydomonas* has two equal flagella, like *Dunaliella*, but the cell envelope is rigid with a medium to thick cell wall. Other motile marine chlorophytes are *Brachiomonas* and *Carteria*. Vegetative division occurs in a nonmotile stage within a mother cell to form two to sixty-four flagellated daughter cells. Sexual reproduction is by the fusion of two, equal size motile gametes, which divide to form four to sixteen daughter cells.

PLATE 13-4. Motile and nonmotile species of Chlorophyceae. (*A*) Nonmotile species, showing *Chlorella marinus* (6–10) *Chlorella salina* (11–14), *Nannochloris maculatus* (23–25), *Nannochloris atomus* (27–29), and *Stichococcus cylindricus* (26) (all 1500 X). Motile species, showing (*B*) *Carteria crucifera;* (*C*) schematic diagram of *Chlamydomonas*, indicating cell wall polar papilla (b), plasmalemma (m), chloroplast (ch), eyespot (s), pyrenoid, with starch granules (p), pulsating vacuole below the flagellar bases (pv), nucleus, with nucleolus (n), and cytoplasm (c); (*D*) *Brachiomonas westiana;* and (*E*) *Dunaliella salina*, (magnifications not given).
(*A*) from R.W. Butcher, 1952, *J. Mar. Biol. Ass. UK* 31:175–91; (*B–E*) from B. Fott, 1971, *Algenkunde*, Gustav Fischer Verlag, Jena.

Many of these small cells do not lend themselves to the usual pipette isolation and washing procedures. These and other small phytoflagellates are often obtained from replicate serial decimal dilution in culture media, which can result in unialgal cultures of the dominant forms (Butcher, 1952, 1959; Throndsen, 1969; Guillard, 1973). The small forms like *Nannochloris* and *Stichococcus* (see Plate 13-4,*A*) can be readily isolated on agar (Ryther, 1954). Classification is based upon flagellation (Ryther, 1954), reproduction, size, and shape, as well as cell structure. The works by Butcher (1952, 1959) remain the standard for classifying the green marine phytoflagellates, although a number of his green flagellates have had their names emended and have been placed in other classes, particularly the prasinophytes. The genus *Chlamydomonas* is the subject of an extensive review by Ettl (1965).

E. RHODOPHYTES

The red algae are characterized by the red phycobilin pigment phycoerythrin, which masks the green color of chlorophyll *a*. Other characteristics are their lack of flagella in any stage of the life cycle, and their reserve food material, floridean starch. Floridean starch is a highly branched polyglucan (α-1,4) with 1,6 side

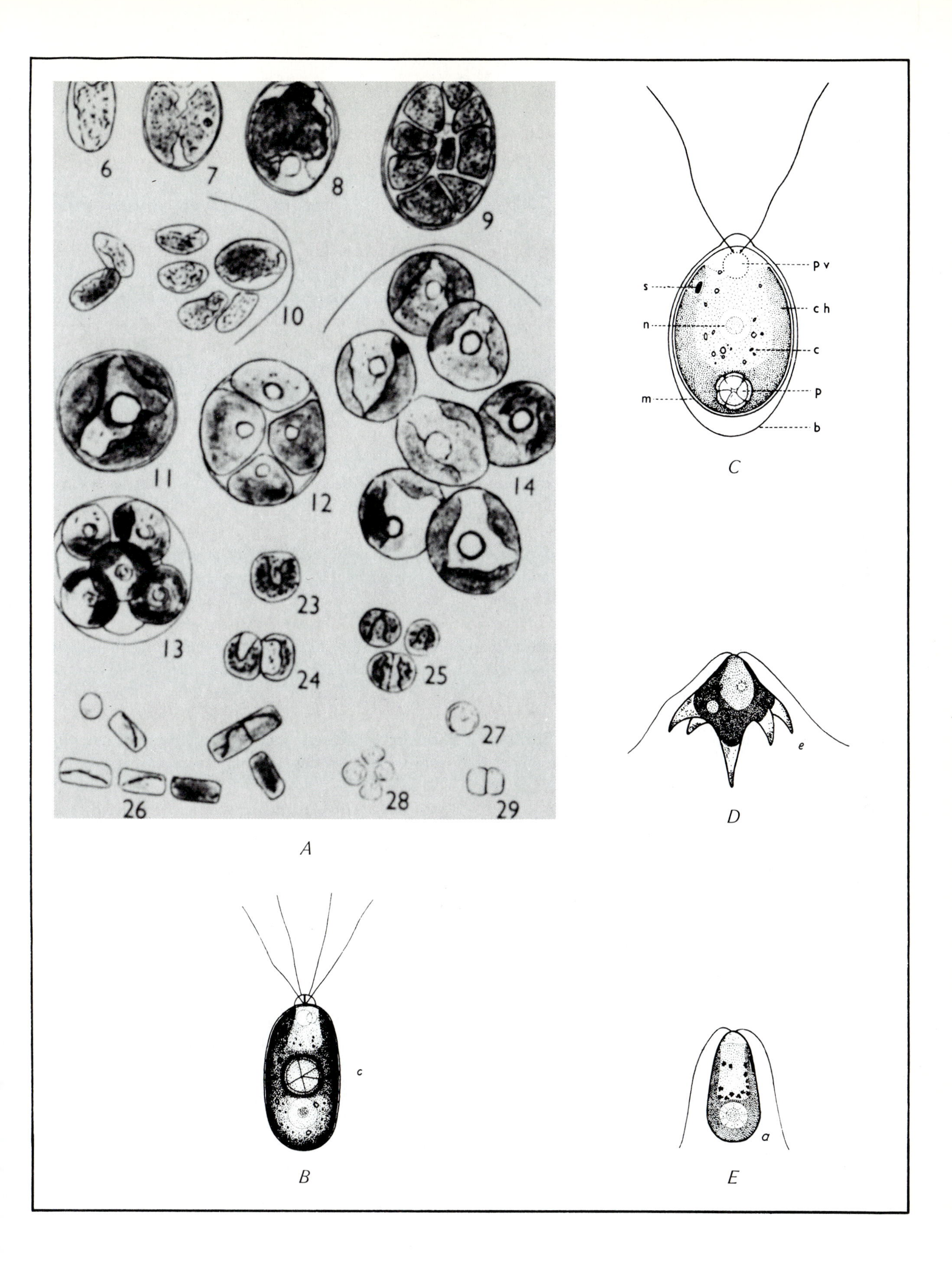
6
7
8
9
10
11
12
13
14
23
24
25
26
27
28
29
A
B
c
C
p v
s
c h
n
c
p
m
b
D
e
E
a

chains, which resembles amylopectin and cyanophycean starch.

The term Rhodophyceae, or red algae, usually brings to mind such larger red algae as the agarophytes *Gelidium*, *Ahnfeltia*, and *Euchema*, which are used to produce "agar" for bacteriological and biochemical uses, or *Chondrus crispus*, the source of another important phycocolloid, carageenin. In addition to these larger leaved multicellular red algae and the finer filamentous forms whose floating beauty disappears upon removal from the sea, there are several unicellular red algae.

One such, *Porphyridium cruentum*, occurs on rocks, soil, and walls in humid areas on land, and indistinguishable marine ecotypes have been isolated from a number of marine environments (Jones, Speer, and Kury, 1963). The cells average 6 μm in diameter, have no cell wall, grow well in artificial seawater from 4.5 to 35 o/oo S, have specific growth rates of 0.5 to 1.1, grow only during the light period of light–dark cycles, and do not grow in the dark on an organic carbon source (Jones, Speer, and Kury, 1963; Sommerfeld and Nichols, 1970). Reproduction is by (1) dividing of regular size unicells into two daughter cells, (2) division of large multinucleate cells (up to 60 μm in diameter) in more than one plane into four, eight, or sixteen cells, and (3) division of large multinucleate cells by constrictions into small buds (Sommerfeld and Nichols, 1970). The cytology, ultrastructure, and movement of cells of *P. cruentum* on agar surfaces have been reviewed by Sommerfeld and Nichols (1970) and the taxonomy of the genus was reviewed by Ott (1972).

Rhodella maculata, a similar red unicellular alga, has been separated from the genus *Porphyridium* because of differences in ultrastructure, especially the branched chloroplasts, position of the pyrenoid, and the presence of numerous electron-dense globules in the chloroplast lobes (Evans, 1970). This microorganism is also euryhaline; it requires an exogenous source of vitamin B_{12}, it is unable to utilize acetate or glucose in the dark, and it has a limited capacity to use organic nitrogen sources (Turner, 1970). The fine structure (Evans, 1970; Paasche and Throndsen, 1970) is shown in Plate 13-5. This alga is capable of a truly planktonic existence. It can be conveniently detected in fresh plankton by concentrating through centrifugation and streaking the concentrate on a 1% agar medium upon which it grows readily. Giant cells exceeding 30 μm are frequently formed (Paasche and Throndsen, 1970). *Rhodella maculata* may be identical to *Porphyridium violaceum* isolated from the North Sea coast of Germany by Kornmann (1965).

Another unicellular red is *Rhodosorus marinus*, which developed in cultures made at Las Palmas, Canary Islands by Geitler (1930). It is distinguished from the above species by its cup-shaped parietal plastid with (usually) four blunt appendages forming the walls of the cup. From the interior of this cup mushrooms a large, prominent, and unique pyrenoid structure. Using an isolate from Banyuls, France, Giraud (1958, 1962, 1963) studied the ultrastructure of this microorganism as well as its pigments, the latter being identical to those of *Porphyridium cruentum*. In stagnant cultures, the 4 to 7 μm cells are imbedded in a clear, homogeneous, fragile matrix. When grown in dim light, these colonies have a rich wine-red color, but when they are grown in bright light, they fade to a pink color, and the cells die (Giraud, 1958; Ott, 1967). Both Giraud and Ott have noted that upon strong aeration, homogeneous cell suspensions are formed. The strain studied by Ott was associated with *Digenea simplex* and had earlier been tentatively identified as the red-pigmented, chroococcoid cyanobacterium *Anacystis aeruginosa* (Drouet and Daily, 1948); and erroneusly bore this identification after unialgal clones were established. When these strains were used to study the effect of salinity upon growth and morphology, the cells became enlarged at the lower salinity, making cytological observations easier and revealing the characteristic structure of the plastids and pyrenoids. The fact that pigmentation was associated with a particular structure indicated that the microorganism was not a cyanobacterium and led to its proper identification. *Rhodosorus marinus* is usually found at salinities of 36.5 to 37.0 o/oo S in warm climates, does well up to 60 o/oo S, but dies off at concentrations below 15 o/oo S (Ott, 1967). This tolerance to hypersalinity that occurred in evaporating dishes during each of its four recorded isolations may have favored its growth over competing forms.

These relatively obscure unicellular rhodophytes are described in some detail because they have apparently gone unrecognized in the search for phycocolloids. The increasing demand for carageenin- and agarose-producing species and their dwindling supply suggests that such unicells, if they produce useful phycocolloids, might lend themselves to high density, large-scale continuous cultures.

PLATE 13-5. The cytology and fine structure of unicellular red algae. (*A*) *Rhodella maculata*, showing the nucleus (n) and pyrenoid (p), with a shell of floridean starch grains, as seen by phase-contrast microscopy (800 X). (*B*) Median section through the cell, showing the organelles. Within the mucilaginous sheath (speckled background) are a bounding plasma membrane, chloroplast lobes (c), nucleus (n), with a nucleolus and nuclear projection (arrow) penetrating the pyrenoid (p), floridean starch grains (s), part of a branched mitochondrion (m), a Golgi body (g), and electron-dense bodies in the peripheral part of chloroplasts (triangles) (10,000 X). (*C*) A colony of *Rhodosorus marinus*, showing the four-lobed, cup-shaped chromatophore and the large round pyrenoid mushrooming from inside the chromatophore (magnification not given). (*A,B*) from L.V. Evans, 1970, *Br. Phycol. J.* 5:1–13; (*C*) from L. Geitler, 1930, *Arch. Protistenkunde* 69:615–36.

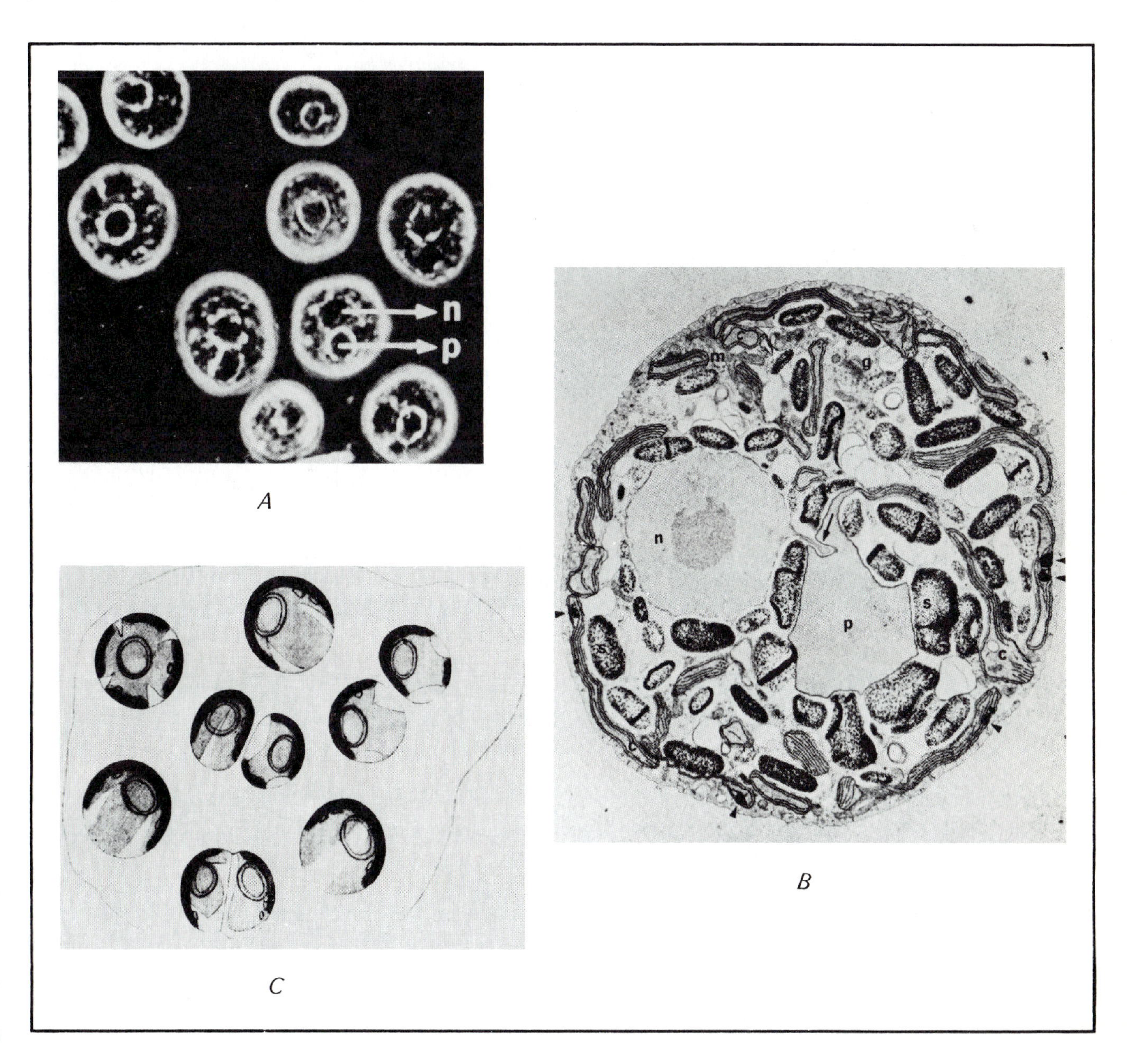

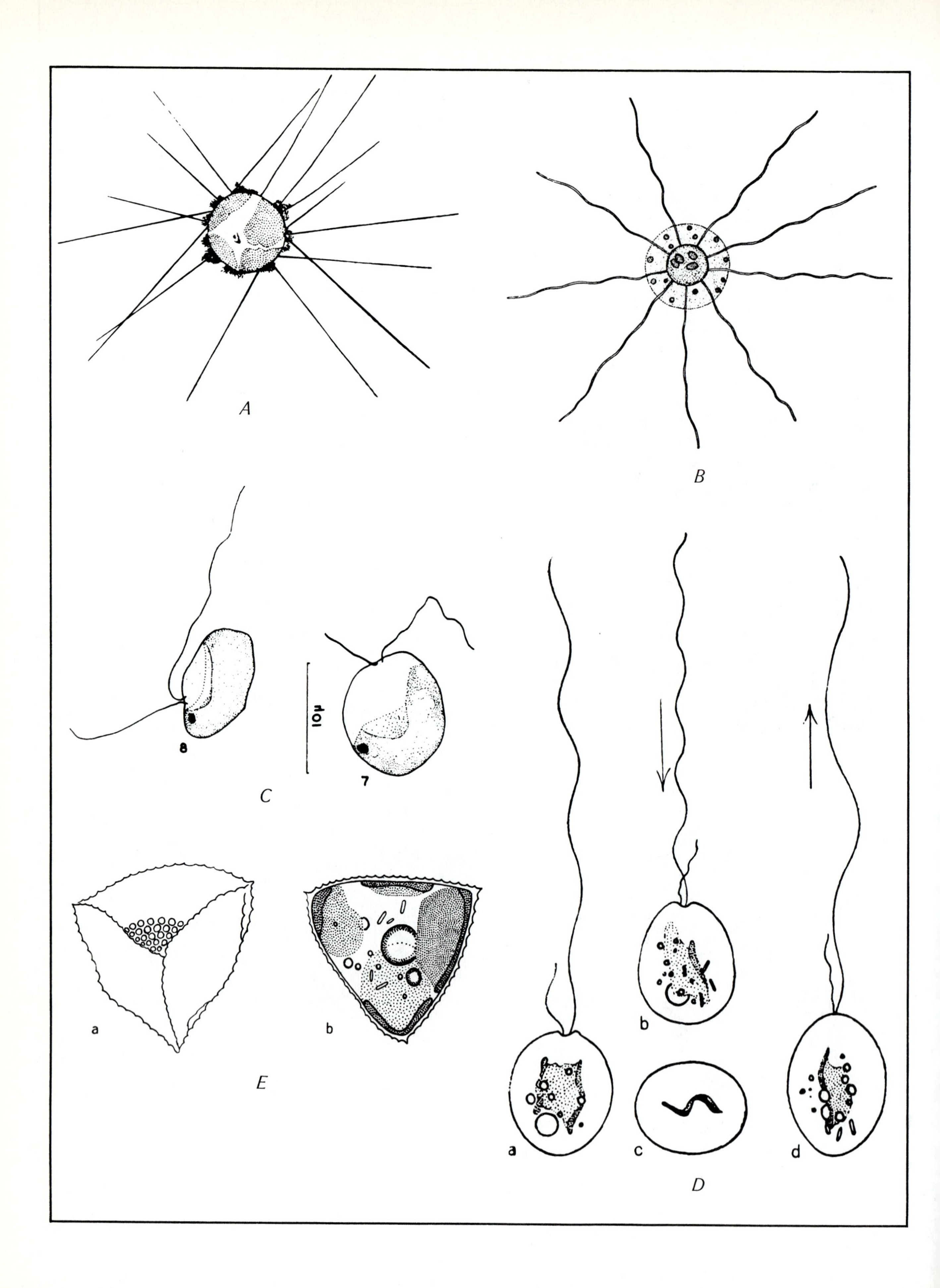
A
B
8
10 µ
7
C
a
b
E
a
b
c
d
D

PLATE 13-6. The diversity of marine xanthophytes. The spined siliceous cells of (*A*) *Meringosphaera tenerrima* (cell; 4 to 6 μm) and (*B*) *M. mediterranea* (cell: 5–9 μm). The heterokont flagellates (*C*) *Xanthomonas thalassoides* and (*D*) *Chloromeson agile* (cells; 8 to 13 μm long by 6 to 10 μm wide). (*E*) *Tetraedriella acuta*, showing its sculptured cell walls and peripheral chromatophore, as seen in optical section (cell; 12 × 16 μm). (*A,B*) from A. Pascher, 1932, *Arch. Protistenkunde* 77:195–218; (*C*) from R.F. Scagel and J.R. Stein, 1961, *Can. J. Bot.* 39:1205–13; (*D,E*) from A. Pascher, 1930, *Arch. Protistenkunde* 69:401–51.

F. XANTHOPHYTES

These heterokont algae (one short smooth flagellum, one longer flagellum with hairs) are also represented in the marine environment by only a few species. The genus *Meringosphaera* (Pascher, 1932, 1939) has spherical and somewhat irregular cells, with a rigid to thick cell wall of two hemispherical shells. These consist of cellulose and pectin with deposited silica and can be smooth or sculptured. Long, straight to undulating spines radiate from the cells and can be firm or fragile; they are distributed uniformly at one pole or in one plane. The presence of pyrenoids and the number of chromatophores vary with the species. Drawings of two species occurring in the Mediterranean Sea and in the coastal waters of the North Sea are shown in Plate 13-6,*A,B*. *Meringosphaera tenerrima*, described from the Adriatic Sea, can be the dominant phytoplankton in Kings Bay, Spitzbergen (Halldal and Halldal, 1973) and in the Norwegian fjords. The 4 to 6 μm diameter cells have a thin, smooth siliceous membrane and the spines, three times the cell diameter, are very fragile, a large portion of the cells being denuded. The cells contain four yellow–green chromatophores and reproduce with two to four auxospores. *Meringosphaera mediterranea* has been reported from brackish waters in the Mediterranean, Adriatic, Baltic, and North seas. The 5 to 9 μm diameter cells have a smooth but fragile siliceous outer covering with undulating spines that are four times the cell diameter. Other xanthophytes in the marine environment are species of *Chlorarachnion*, *Chloromeson* (Plate 13-6,*D*), *Helminthogloea*, *Heterochloris*, *Skiadosphaera*, and *Tetraedriella* (Plate 13-6,*E*), which are listed by Fott (1971).

Among the more recent descriptions of xanthophytes were those obtained in culture from Indian Arm, British Colombia. This inlet, like similar fjords in Norway, has a primary productivity in which nanoplankton plays a significant role. During a preliminary study of natural and cultured preparations from this inlet, Scagel and Stein (1961) encountered a great variety of microorganisms, many that posed difficult taxonomic problems. *Xanthomonas thalassoides* is a unicellular spherical cell, 10 μm in diameter, with a single, pale green, cup-shaped chromatophore that half fills the cell. An eyespot on the anterior of the chromatophore lacks pyrenoids, with oil being stored as droplets. It also has two unequal flagella, one 10 μm and the other 25 μm in length. *Xanthomonas thalassoides* bears a superficial resemblance to the *Thalassomonas* of Butcher (1959).

G. EUSTIGMATOPHYTES AND RHAPHIDOPHYTES

Eustigmatophyceae is a new algal class split from the Xanthophyceae (Hibberd and Leedale, 1970). The distinguishing characteristics of the motile cells are an eyespot at the anterior end, which is independent of the single chloroplast; one emergent anterior flagellum, with a bilateral array of stiff hairs; a second basal body in the region of the eyespot; motile cells with no Golgi bodies; and a unique form of polygonal pyrenoid in vegetative (nonmotile) cells. The microorganism known as Clone GSB STICHO from the WHOI collection (Guillard and Lorenzen, 1972) probably belongs in this class, and others may now be thus reclassified as further criteria, such as pigment, are developed.

The class Rhaphidophyceae (Chloromonadophyceae) includes one marine species notable for its biological activity. *Chattonella subsalsa* is a large, naked, flagellate, first described from brackish waters by Biecheler (1936). It was subsequently redescribed under the synonym *Hornellia marina* from the Malabar Coast by Subrahmanyan (1954), where it caused green discoloration and marine animal kills. The mucoid material released from decaying cells is reported to have a curare-like activity, killing both animals and humans (Bernard, 1967). Its ultrastructure has been studied by Mignot (1976). Recently Tomotoshi Okaichi (pers. comm.) has isolated normal (50 μm) and large (100 μm) strains of *Chattonella* from the Seto Inland Sea that were associated with massive fish kills. The flagellate attaches to the gills, apparently with its trichocysts, and the toxic factor has been identified as palmitic acid, which presumably prevents the passage of oxygen and suffocates the fish.

REFERENCES

Antia, N.J., and V. Chorney. 1968. Nature of nitrogen compounds supporting phototrophic growth of the marine cryptomonad *Hemiselmis virescens*. *J. Protozool.* 15:198–201.

Antia, N.J., J.Y. Cheng, and F.J.R. Taylor. 1969. The heterotrophic growth of a marine photosynthetic cryptomonad (*Chroomonas salina*). *Intl. Seaweed Symp. Proc.* 6:17–29.

Bernard, F. 1967. Research on phytoplankton and pelagic protozoa in the Mediterranean Sea from 1953 to 1966. *Oceanogr. Mar. Biol. Ann. Rev.* 5:205–29.

Biecheler, B. 1936. Sur une Chloromonadine nouvelle d'eau saumatre *Chattonella subsalsa* n. gen. n. sp. *Arch. Zool. exp. gén.* 78:79–83.

Boney, A.D. 1970. Scale-bearing phytoflagellates: An interim review. *Oceanogr. Mar. Biol. Ann. Rev.* 8:251–305.

Butcher, R.W. 1952. Contributions to our knowledge of the smaller marine algae. *J. mar. biol. Assoc. U.K.* 31:175–91.

Butcher, R.W. 1959. *An Introductory Account of the Smaller Algae of British Coastal Waters. I. Introduction and Chlorophyceae.* Fish. Invest. (London) Ser. IV, 74 p.

Butcher, R.W. 1961. *An Introductory Account of the Smaller Algae of British Coastal Waters. Part VIII. Euglenophyceae-Eugleninae.* Fish. Invest. (London) Ser. IV, 17 p.

Butcher, R.W. 1967. *An Introductory Account of the Smaller Algae of British Coastal Waters. Part IV. Cryptophyceae.* Fish. Invest. (London) Ser. IV, 54 p.

Carter, N. 1937. New or interesting algae from brackish water. *Arch. Protistenk.* 90:1–68.

Christensen, T. 1962, 1966. Alger. In: *Botanik (Systematisk Botanik)* (T.W. Böcher, M. Lange, and T. Sørensen, eds.) No. 2, Vdg. 2 (1st Ed., 178 p) (2nd Ed., 180 p.) (in Danish), Munksgaard, Copenhagen.

Conrad, W. 1939. Notes Protistologiques. 10, sur le schorre de Lilloo. *Bull. Mus. Hist. Nat. Belg.* 14:1–18.

Cook, J.R. 1968. The cultivation of Euglena. In: *The Biology of Euglena* (D.E. Buetow, ed.), Vol. I. Academic Press, London and New York, pp. 244–309.

Dodge, J.D. 1969. The ultrastructure of *Chroomonas mesostigmatica* Butcher (Cryptophyceae). *Arch. Mikrobiol.* 69:266–80.

Drouet, F., and W.A. Daily. 1948. Nomenclatural transfers among coccoid algae. *Lloydia* 11:77–79.

Ettl, H. 1965. Beitrag zur Kenntnis der Morphologie der Gattung *Chlamydomonas* Ehrenberg. *Arch. Protistenk.* 108:271–430.

Evans, L.V. 1970. Electron microscopical observations on a new red algal unicell, *Rhodella maculata* gen. nov., sp. nov. *Br. phycol. J.* 5:1–13.

Fott, B. 1971. *Algenkunde.* Gustav Fischer Verlag, Jena. 581 p.

Gantt, E. 1971. Micromorphology of the periplast of *Chroomonas* sp. (Cryptophyceae). *J. Phycol.* 7:177–84.

Gantt, E., M.R. Edwards, and L. Provasoli. 1971. Chloroplast structure of the Cryptophyceae. *J. Cell Biol.* 48:280–90.

Geitler, L. 1930. Ein grünes Filarplasmodium und andere neue Protisten. *Arch. Protistenk.* 69:615–36.

Gemeinhardt, K. 1930. Silicoflagellatae. In: *Dr. L. Rabenhorst's Kryptogamen-Flora*, Vol. 10. Akademie Verlag, Leipzig, pp. 1–87.

Giraud, G. 1958. Sur la Vitesse de croissance d'une Rhodophycée monocellulaire marine, le *Rhodosorus marinus* Geitler, cultivée en milieu synthétique. *C.R. Acad. Sci. Paris* 246:3501–4.

Giraud, G. 1962. Les Infrastructures de quelques algues et leur physiologie. *J. Microscopie* 1:251–54.

Giraud, G. 1963. La Structure, les pigments et les caractéristiques functionnelles et l'appariel photosynthétique des diverses algues. *Physiol. Vég.* 1:203–55.

Guillard, R.R.L. 1973. Methods for microflagellates and nanoplankton. In: *Handbook of Phycological Methods* (J.R. Stein, ed.). Cambridge University Press, London and New York, pp. 69–85.

Guillard, R.R.L., and C.J. Lorenzen. 1972. Yellow-green algae with chlorophyllide *c*. *J. Phycol.* 8:10–14.

Halldal, P., and K. Halldal. 1973. Phytoplankton, chlorophyll and submarine light conditions in Kings Bay, Spitzbergen, July 1971. *Norw. J. Bot.* 20:99–108.

Herrera, J., y R. Margalef. 1963. Hidrografía y fitoplancton de la costa comprendida entre Castellón y la desembocadura del Ebro, de Julio de 1960 a Junio de 1961. *Inv. Pesq.* 24:33–112.

Hibberd, D.J., and G.F. Leedale. 1970. Eustigmatophyceae—a new algal class with unique organization of the motile cell. *Nature* 225:758–60.

Hibberd, D.J., A.D. Greenwood, and H.B. Griffiths. 1971. Observations on the ultrastructure of the flagella and periplast in the Cryptophyceae. *Br. phycol. J.* 6:61–72.

Hulbert, E.M. 1965. Flagellates from brackish waters in the vicinity of Woods Hole, Massachusetts. *J. Phycol.* 1:87–94.

Ignatiades, L. 1970. The relationship of the seasonality of the silicoflagellates to certain environmental factors. *Bot. Mar.* 13:44–46.

Jones, R.F., H.L. Speer, and W. Kury. 1963. Studies on the growth of the red alga *Porphyridium cruentum*. *Physiol. Plant.* 16:636–43.

Kornmann, P. 1965. *Porphyridium violaceum* eine marine neue art. *Helgoländer wiss. Meeres.* 12:420–23.

Leadbeater, B.S.C. 1969. A fine structural study of *Olisthodiscus luteus* Carter. *Br. phycol. J.* 4:3–17.

Leedale, G.F. 1967. *Euglenoid Flagellates.* Prentice-Hall, Englewood Cliffs, N.J., 242 p.

Loeblich, A.R., III, L.A. Loeblich, H. Tappan, and A.R. Loeblich, Jr. 1968. *Annotated Index of Fossil and Recent Silicoflagellates and Ebridians with Descriptions and Illustrations of Validly Proposed Taxa.* Geol. Soc. Amer. Mem. 106, Boulder, Colo., 319 p.

Lucas, I.A.N. 1970a. Observations on the fine structure of the Cryptophyceae. I. The genus *Cryptomonas*. *J. Phycol.* 6:30–38.

Lucas, I.A.N. 1970b. Observations on the ultrastructure of

representatives of the genera *Hemiselmis* and *Chroomonas* (Cryptophyceae). *Br. phycol. J.* 5:29–37.

Marshall, S.M. 1934. The Silicoflagellata and Tintinnoinea. In: *Brit. Mus. Nat. Hist. Great Barrier Reef Exped. 1928–29.* Sci. rept. 4, pp. 623–64.

McLachlan, J. 1961. The effect of salinity on growth and chlorophyll content in representative classes of unicellular marine algae. *Can. J. Microbiol.* 7:399–406.

McLachlan, J. 1964. Some considerations of the growth of marine algae in artificial media. *Can. J. Microbiol.* 10:769–82.

Mignot, J-P. 1976. Compléments a l'étude des chloromonadines. Ultrastructure de *Chattonella subsalsa* Biecheler flagellé d'eau saumatre. *Protistologica* 12:279–93.

Miller, B.T. 1972. The phytoplankton and related hydrography in the south basin of the Pettaquamscutt River. M.S. Thesis, University of Rhode Island, 119 p.

Nival, P. 1965. Sur le Cycle de *Dictyocha fibula* Ehrenberg dans les eaux de surface de la rade de Villefranche-sur-Mer. *Cahiers Biol. Mar.* 6:67–82.

Okaichi, T. 1969. Water bloom due to marine *Eutreptiella* sp. in the Seto Inland Sea. *Bull. Plankton Soc. Japan* 16:115–20.

Ott, F.D. 1967. *Rhodosorus marinus* Geitler: A new addition to the marine algal flora of the Western hemisphere. *J. Phycol.* 3:158–59.

Ott, F.D. 1972. A review of the synonyms and the taxonomic positions of the algal genus *Porphyridium* Näegli 1849. *Nova Hedwigia* 23:237–89.

Paasche, E., and J. Throndsen. 1970. *Rhodella maculata* Evans (Rhodophyceae, Porphyridiales) isolated from the plankton of the Oslo Fjord. *Nytt. Mag. Bot. (Norw. J. Bot.)* 17:209–12.

Parke, M., and P.S. Dixon. 1968. Check-list of British marine algae—second revision. *J. mar. biol. Assoc. U.K.* 48:783–832.

Parke, M., and P.S. Dixon. 1976. Check-list of British marine algae—third revision. *J. mar. biol. Assoc. U.K.* 56:527–94.

Pascher, A. 1930. Zur Kenntnis der heterokonten Algen. *Arch. Protistenk.* 69:401–51.

Pascher, A. 1932. Zur Kenntnis mariner Planktonten. I. *Meringosphaera* und ihre Verwandten. *Arch. Protistenk.* 77:195–218.

Pascher, A. 1939. Heterokonten. In: *Dr. L. Rabenhorst's Kryptogamen-flora,* Vol. 11. Akademische Verlag, Leipzig, pp. 536–49.

Pratt, D.M. 1959. The phytoplankton of Narragansett Bay. *Limnol. Oceanogr.* 4:425–40.

Pratt, D.M. 1965. The winter-spring diatom flowering in Narragansett Bay. *Limnol. Oceanogr.* 10:173–84.

Pratt, D.M. 1966. Competition between *Skeletonema costatum* and *Olisthodiscus luteus* in Narragansett Bay and in culture. *Limnol. Oceanogr.* 11:447–55.

Ryther, J.H. 1954. The ecology of phytoplankton blooms in Moriches Bay and Great South Bay, Long Island, New York. *Biol. Bull.* 106:198–209.

Scagel, R.F., and J.R. Stein. 1961. Marine nanoplankton from a British Columbia fjord. *Can. J. Bot.* 39:1205–13.

Sieburth, J.McN. 1975. *Microbial Seascapes.* University Park Press, Baltimore, Md., 248 p.

Sommerfeld M.R., and H.W. Nichols. 1970. Comparative studies in the genus *Porphyridium* Naeg. *J. Phycol.* 6:67–78.

Subrahmanyan, R. 1954. On the life-history and ecology of *Hornellia marina* gen. et sp. nov. (Chloromonadineae), causing green discoloration of the sea and mortality among marine organisms of the Malabar Coast. *Indian J. Fish.* 1:182–203.

Throndsen, J. 1969. Flagellates of Norwegian Coastal waters. *Nytt. Mag. Bott. (Norw. J. Bot.)* 16:161–216.

Throndsen, J. 1973. Fine structure of *Eutreptiella gymnastica* (Euglenophyceae). *Norw. J. Bot.* 20:271–80.

Throndsen, J. 1976. Occurrence and productivity of small marine flagellates. *Norw. J. Bot.* 23:269–93.

Tomas, C.R. 1977. A study of the autecology of *Olisthodiscus luteus* Carter. Ph.D. Thesis, University of Rhode Island, 146 p.

Travers, A., and M. Travers. 1968. Les Silicoflagellés du Golfe de Marseille. *Mar. Biol.* 1:285–88.

Turner, M.F. 1970. A note on the nutrition of *Rhodella. Br. phycol. J.* 5:15–18.

Van Valkenburg, S.D. 1971a. Observations on the fine structure of *Dictyocha fibula* Ehrenberg. I. The skeleton. *J. Phycol.* 7:113–18.

Van Valkenburg, S.D. 1971b. Observations on the fine structure of *Dictyocha fibula* Ehrenberg. II. The protoplast. *J. Phycol.* 7:118–32.

Van Valkenburg, S.D., and R.E. Norris. 1970. The growth and morphology of the silicoflagellate *Dictyocha fibula* Ehrenberg in culture. *J. Phycol.* 6:48–54.

PART V

Osmotrophic Procaryotes

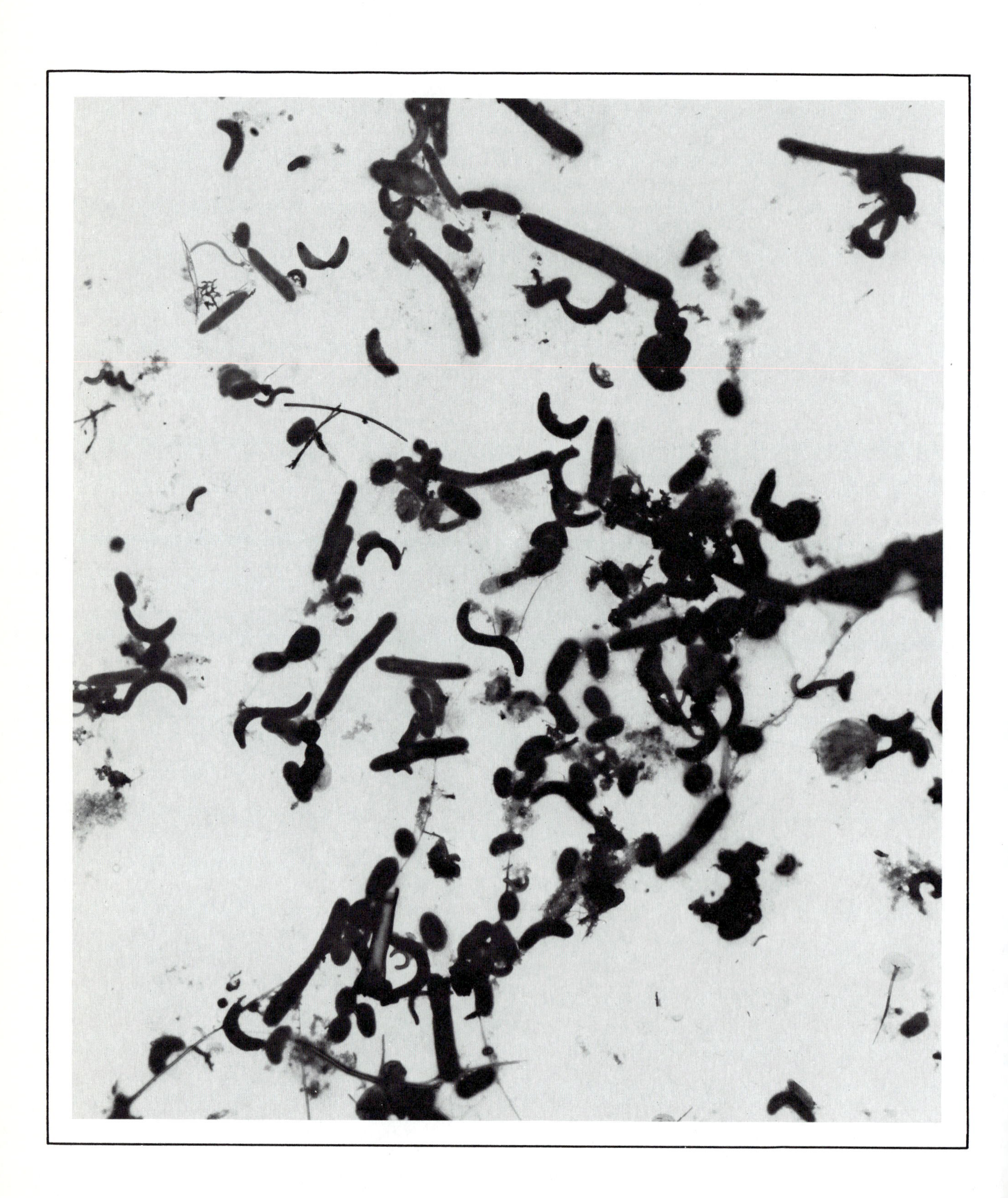

The recent development of techniques to visualize bacteria and differentiate them from organic debris with acridine-orange epifluorescence, to determine the activities of natural populations with autoradiography, to isolate planktonic bacteria from the protist plankton with sieve-like, perforated polycarbonate filters, to examine this filtered fraction carefully in negatively stained preparations by transmission electron microscopy, and to also determine the viable biomass of this fraction of the plankton with the assay of ATP, has opened up the field. From the recent literature and the work in the writer's laboratory, it is becoming apparent that there are three distinct populations of bacteria in the sea: a dominant population of smaller osmotrophic planktobacteria that have little affinity for nutrient-rich surfaces, including agar media; a population of larger osmotrophic epibacteria associated with seston and shallow waters that often have organelles for locomotion and attachment and are mostly amenable to culture on agar; and a population of phototrophic chroococcoid cyanobacteria that can become a significant biomass in oceanic waters.

It is impossible to reconcile the two distinct populations of osmotrophic marine bacteria with Winogradsky's autochthonous and zymogenous bacteria. The writer believes that both forms of marine bacteria favor the more labile substrates and utilize them while co-metabolizing more refractory substrates. The basic difference in nutrient requirements between the two populations of marine bacteria is probably not due to substrate lability (Winogradsky postulate) nor to substrate concentration (K_s chemostat postulate) but to a primary difference in cell structure and function for the utilization of soluble and solid food. Although the epibacteria that form colonies on agar media will grow in liquid media, most are motile, can form organelles for attachment, and tend to accumulate at nutrient-rich and solid interfaces. Therefore, the osmotrophic bacteria in the sea may be thought of as indigenous (autochthonous) bacteria, but with primary affinities for either soluble or solid substrates. The "autochthonous planktobacteria" would be the free-living (non-attaching) forms utilizing dissolved organic matter above threshold concentrations (below which they cannot survive) and having little affinity for solid surfaces. The "autochthonous epibacteria" would be the often larger, motile forms having a great affinity for surfaces and capable of secreting the exogenous enzymes needed to utilize solid organic substrates. Organic matter is either dissolved or particulate. It is logical that there will be two different forms of bacteria to better utilize one or the other form. This property of one group of microorganisms apparently being adapted to a planktonic existence and another group to an epibiotic existence, in which the microorganisms attach upon or glide over surfaces, would not be unique to the bacteria. We have seen in Chapter 10 that the diatoms can be separated naturally into centric diatoms, which are mainly planktonic, and pennate diatoms, which are mainly epibiotic. The protozoa can also be separated into either planktonic or epibiotic forms. Why not also the bacteria?

The evidence for the existence of the autochthonous planktobacteria is presented in Chapter 14, whereas the autochthonous epibacteria, which are usually associated with surfaces but stray into the plankton as starving transients, are discussed in Chapters 15 through 18. The first draft of my taxonomic groupings was written in 1973–1974, before the appearance of the long-awaited 8th edition of *Bergey's Manual of Determinative Bacteriology* (Buchanan and Gibbons, 1974). Therefore, the writer came to his own decisions without being influenced by the decisions of the editorial board of *Bergey's 8th.* We both did away with a phylogenetic approach and chose vernacular groupings of somewhat similar genera. [In general, the arrangement in *Bergey's 8th* is quite pleasing and effective. But some logical suggestions like redefining the genus *Beneckea* to include some of the vibrios unlike *Vibrio cholerae*, and the creation of the genus *Altermonas* for pseudomonads that fall outside a reasonable G + C range, have been ignored, whereas the ill-defined genus *Flavobacterium* has been retained despite good evidence that a significant proportion of the strains are *Cytophaga.* The editorial board of *Bergey's 8th* also used vernacular groupings for the higher taxa of bacteria and failed to support formal names for the subdivisions of the Kingdom Procaryotae. This has now been corrected with a formal proposal by Gibbons and Murray (1978) that is simple but elegant. Three equal divisions house the bacteria with gram-negative cell walls (*Gracilicutes* = thin-skinned), the bacteria with gram-positive cell walls (*Firmacutes* = thick-skinned), and the bacteria lacking cell walls (*Mollicutes* = tender-skinned). The division of gram-negative bacteria is further divided into two classes, those with (*Photobacteria* = light-requiring bacteria) and those without (*Scotobacteria* = dark bacteria) a phototrophic metabolism. The latter class includes the predominant gram-negative bacteria in the sea, both planktobacteria and epibacteria.]

In the chapters that follow, the genera recognized by *Bergey's 8th* will be used, but the author's preferences will be noted.

REFERENCES

Buchanan, R.E., and N.E. Gibbons (eds.). 1974. *Bergey's Manual of Determinative Bacteriology* (8th Ed.). Williams & Wilkins, Baltimore, Md., 1246 p.

Conn, J.H. 1948. The most abundant groups of bacteria in soil. *Bacteriol Rev.* 12:257–74.

Gibbons, N.E., and R.G.E. Murray, 1978. Proposals concerning the higher taxa of bacteria. *Int. J. Syst. Bacteriol.* 28:1–6.

Jannasch, H.W., and G.E. Jones. 1959. Bacterial populations in sea water as determined by different methods of enumeration. *Limnol. Oceanogr.* 4:128–39.

Morris, J.G. 1960. Studies on the metabolism of *Arthrobacter globiformis*. *J. gen. Microbiol.* 22:564–82.

Winogradsky, S. 1925. Études sur la microbiologie du sol. I. Sur la Méthode. *Ann. Inst. Pasteur* 39:299–354.

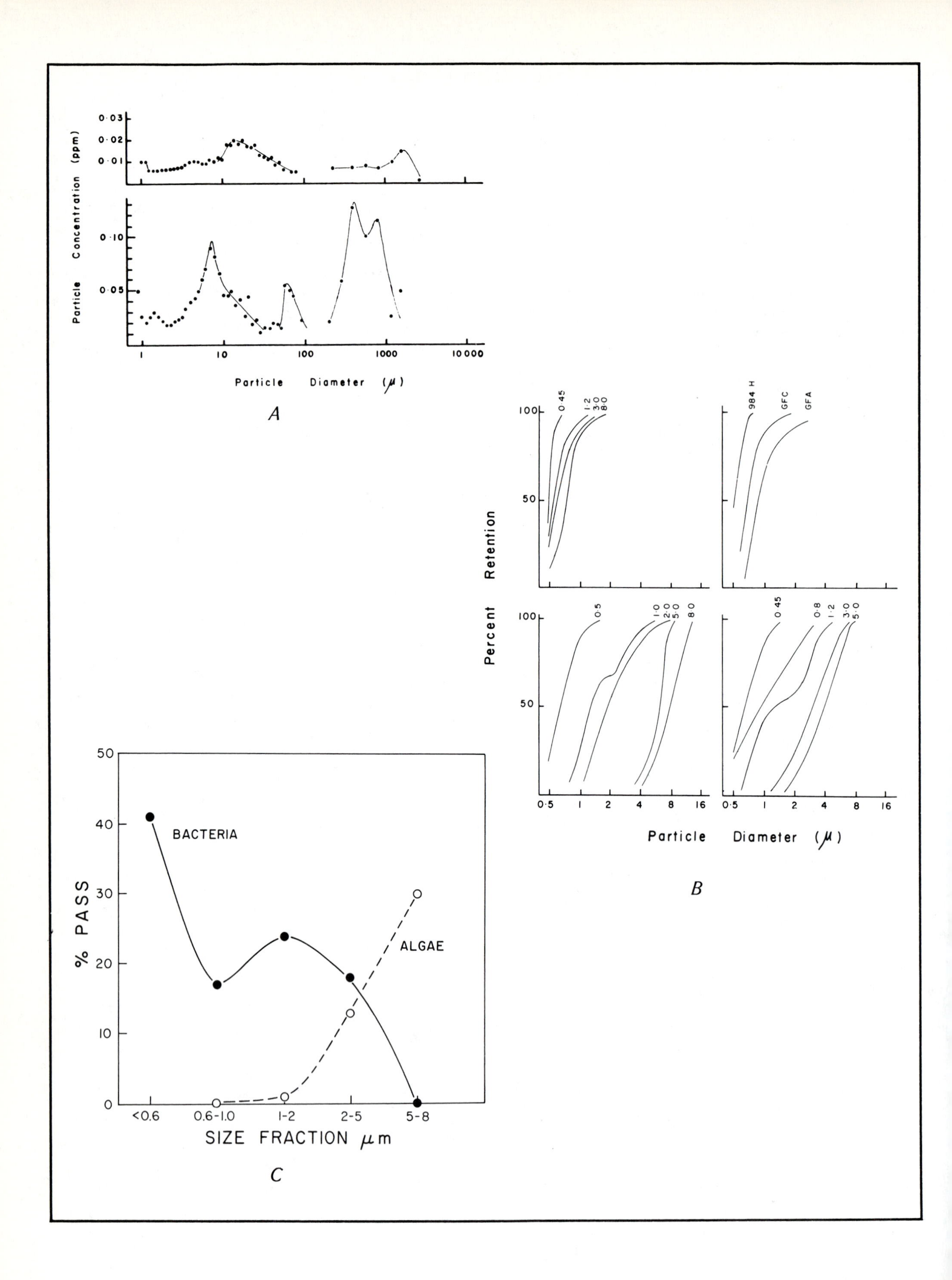
Particle Concentration (ppm)
0·03
0·02
0·01
0·10
0·05
1
10
100
1000
10 000
Particle Diameter (μ)
A
Percent Retention
100
50
0·45
1·2
3·0
8·0
984 H
GFC
GFA
0·5
1·0
2·0
5·0
0·8
0·5
1
2
4
8
16
Particle Diameter (μ)
B
% PASS
50
40
30
20
10
0
BACTERIA
ALGAE
<0.6
0.6-1.0
1-2
2-5
5-8
SIZE FRACTION μm
C

CHAPTER 14

Autochthonous Planktobacteria

Studies on the size distribution and formation of particles in seawater have indicated that particles under 4 μm in diameter can be considered to be bacteria (Sheldon and Parsons, 1967; Sheldon, Evelyn, and Parsons, 1967). In commenting on the bulk distribution of particles in the sea, Sheldon, Prakash, and Sutcliffe (1972) stated that in the size range 1 μm to 1 mm, one-third would be 1 to 10 μm, one-third 10 to 100 μm, and one-third 100 μm to 1 mm. This is illustrated for low and high standing stocks in Plate 14-1,*A*. Although particles under 1 μm in diameter are difficult to enumerate in the Coulter Counter, Sheldon, Prakash, and Sutcliffe (1972) suggested that the concentration of particles 0.5 to 1.0 μm is similar to the concentration of particles in the 1 to 100 μm size range. This would indicate that particles under 1 μm could account for a third of the particulate mass that is under 1 mm.

PLATE 14-1. The size distribution of particles in the sea and the effectiveness of perforated polycarbonate membranes (Nuclepore filters) in passing bacteria-sized particles while retaining algae-sized particles. (*A*) The size distribution of particles in the complete spectrum over the range 1 to 4000 μm for low (top; 10°09′S, 150°02′W) and high (bottom; 35°00′S, 73°00′W) standing stocks. (*B*) Retention curves of particles less than 16 μm for cellulose ester (Millipore, upper left), glass-fiber (upper right), perforated polycarbonate (Nuclepore, lower left), and metal (Selas Flotronics, lower right) membranes. (*C*) A comparison of the percentage of lake bacteria and algae passing through five size ranges of Nuclepore filters. (*A*) from R.W. Sheldon, A. Prakash, and W.H. Sutcliffe, Jr., 1972, *Limnol. Oceanogr.* 17:327–40. (*B*) from R.W. Sheldon, 1972, *Limnol. Oceanogr.* 17:494–98; (*C*) adapted from K. Salonen, 1974, *Ann. Bot. Fennici* 11:133–35.

Although some of these small particles are organic debris, most are bacteria, as seen in the preceding micrograph (Part V Plate). Sheldon (1972) showed that, unlike the conventional cellulose ester membrane (Millipore) and glass-fiber filters, which fail to pass many of the

PLATE 14-2. Epifluorescent photomicrographs of acridine-orange–stained bacteria from Narragansett Bay, R.I. (*A*) Free bacterioplankton concentrated after passing a 2 μm Nuclepore filter. Note the diversity of size and shape, approximately 5 × 10^5 cells/ml. (*B*), (*C*), and (*D*) Bacteria on organic debris (epibacteria) that pass through a 20 μm screen but are retained by a 2 μm membrane. Photomicrographs courtesy of Paul W. Johnson.

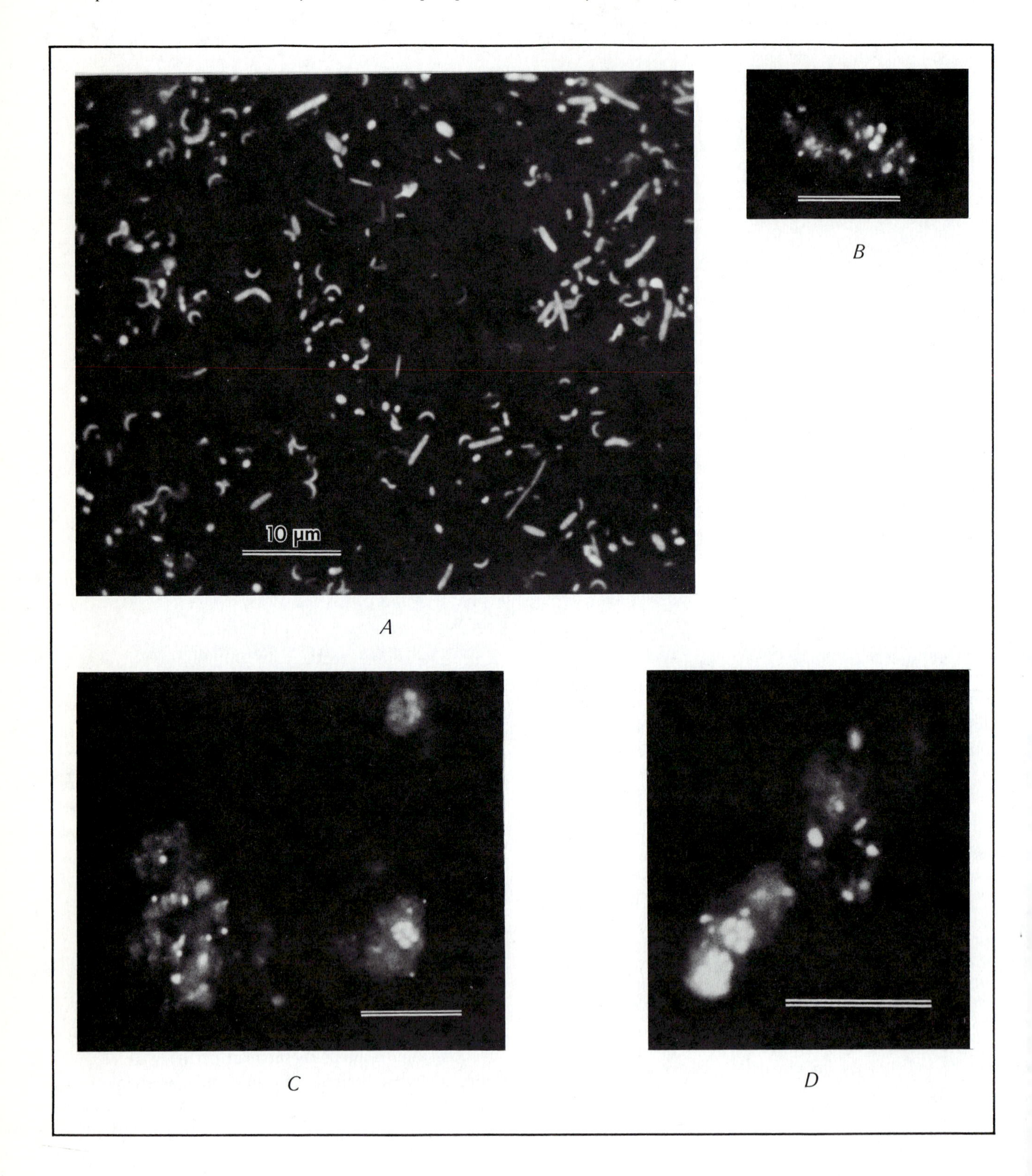

bacteria-sized particles that should pass their stated pore size, perforated polycarbonate (Nuclepore) and metal (Selas Flotronics) membranes act as screens, and that the stated pore size is a good indicator of the size particle that can pass (Plate 14-1,*B*). The studies on the relation of particle size to heterotrophic utilization of dissolved organic carbon by Williams (1970a,b) are unfortunately affected by his use of cellulose acetate filters for size separation, and the results must be interpreted in context with the retention values of Sheldon (1972). Salonen (1974) confirmed that cellulose ester membrane filters were unsatisfactory for separating bacteria from algae, whereas the perforated polycarbonate membranes (Nuclepore) in the size range 2 to 5 μm were satisfactory in allowing a significant proportion of the bacteria to pass through while retaining practically all the phytoplankton. This is illustrated in Plate 14-1,*C*.

Nuclepore membranes of pore size between 2 and 5 μm were recommended by Salonen (1974) for separating bacteria from the rest of the plankton. The data of Sheldon (1972) indicate that a 2 μm Nuclepore membrane will pass almost all the 1μm particles while retaining 90% of the 4 μm particles. Although preparations that pass a 2 μm porosity Nuclepore membrane appear satisfactory for such uses as microscopic examination (Part V Title Plate; Plates 14-2,*A*, 14-5), and may also be satisfactory for cultural purposes and ATP assays under many conditions, technically they are unsatisfactory, since I have found that some of the smaller microflagellates appear in them. A 1 μm porosity Nuclepore membrane will retain all the flagellates while allowing some 90% of the heterotrophic bacteria, as measured by activity, to pass through (Hodson, Azam, and Lee, 1977; Azam and Hodson, 1977). This is due to the size distribution of the bacterioplankton, as shown in Plate 14-3. Using tritium-labeled amino acids and autoradiography to detect active cells, which were compared with colony-forming units, Hoppe (1976) found that most of the active cells were in the 0.2 to 0.6 μm size fraction, whereas the colony-forming units peaked in the 0.8 to 1.0 μm size fraction. Although it would be useful to be able to separate the planktobacteria from the epibacteria on the basis of size, this is impossible, because colony-forming bacteria occur in all size fractions, but are more concentrated in the larger size fraction (0.8 to 1.0 μm). Bacteria active on low levels of soluble organic substrates appear to be mainly smaller cells free in the water (planktobacteria), whereas the larger colony-forming bacteria (epibacteria) seem to occur both as inactive free cells and attached to organic debris, where they are presumably active. This two-population distribution of bacteria was recognized by Ferguson and Rublee (1976).

Epifluorescence microscopy of acridine-orange–stained preparations is very effective for detecting and differentiating bacteria from organic debris (Zimmerman and Meyer-Reil, 1974; Daley and Hobbie, 1975). But some investigators who advocate radioisotope labeling for autoradiography (Peroni and Laverello, 1975; Hoppe, 1976) feel that all the cells present may not be active and that only the labeled cells are active. Hoppe (1974), however, has shown that the number of labeled bacteria varied markedly with the season, the water sample, and the specific substrate used. Substrates such as acetate and glucose only marked 6% and 30% of the bacteria, respectively, whereas a mixture of tritium-labeled amino acids marked 41% of the cells enumerated by acridine-orange epifluorescence (Hoppe, 1976). The writer feels that no mixture of radioisotope-labeled substrates will label 100% of the active cells, due to substrate specificity and to substrate inhibition (Smith, 1976). Acridine-orange epifluorescence counts are probably a good index for the enumeration of total viable bacteria. Although Ferguson and Rublee (1976) used acridine-orange–stained preparations to count cells according to size and frequency and to describe their morphology, the smallest cells were probably lost due to the filter used and only very gross cell morphology can be seen in such preparations. Scanning electron microscopy of sputter-coated, air dried preparations of picoplankton could also be used for enumeration and would be helpful for a preliminary characterization of cell morphology. This technique may yield information similar to the highly informative, but more laborious negative stains and replicas (Watson et al., 1977) of such preparations seen by transmission electron microscopy.

Peroni and Laverello (1975) used ^{32}P autoradiography to study microbial activities as a function of depth in the Ligurian Sea. Although they showed a three- to four-order of magnitude discrepancy between autoradiographic counts and colony-forming units on poured plates (see Chapter 7,C), for some reason the procedure failed to demonstrate the significant concentrations of bacteria-sized particles known to occur in deep waters (Sheldon, Prakash, and Sutcliffe, 1972). Autoradiogra-

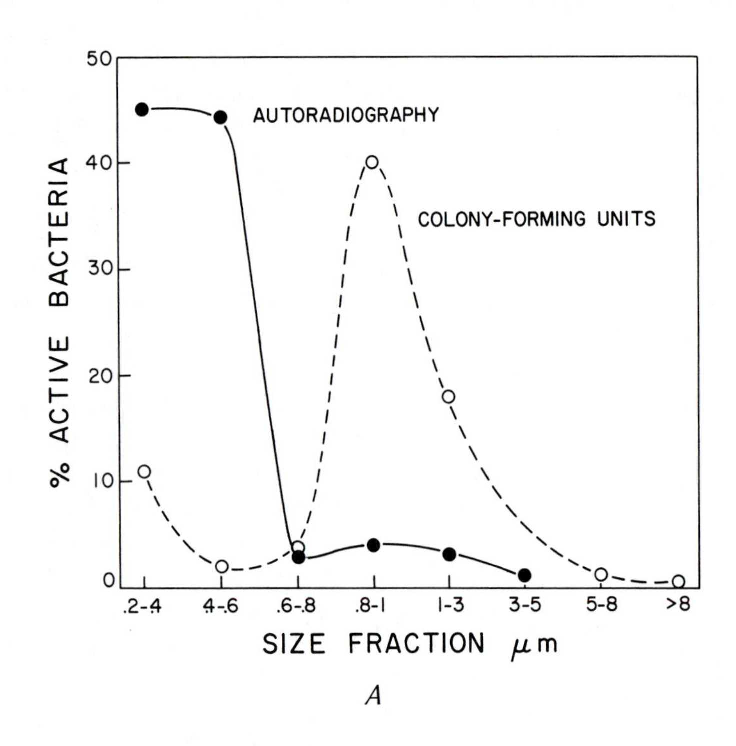
% ACTIVE BACTERIA
AUTORADIOGRAPHY
COLONY-FORMING UNITS
SIZE FRACTION μm
.2-.4
.4-.6
.6-.8
.8-1
1-3
3-5
5-8
>8

A

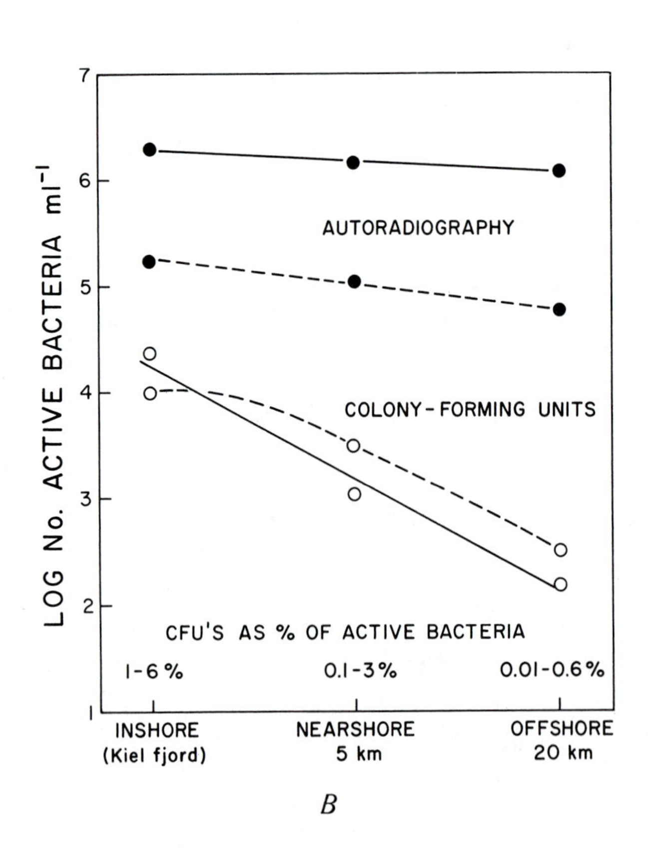
LOG No. ACTIVE BACTERIA ml⁻¹
AUTORADIOGRAPHY
COLONY-FORMING UNITS
CFU'S AS % OF ACTIVE BACTERIA
1-6 %
0.1-3%
0.01-0.6%
INSHORE
(Kiel fjord)
NEARSHORE
5 km
OFFSHORE
20 km

B

PLATE 14-3. A comparison of populations of cells actively utilizing labeled amino acids and those forming colonies on an agar medium. (*A*) The size distribution of active minicells and the larger colony-forming units from the middle of the Kiel Bight. (*B*) The persistence of active minicells with distance from shore; the number of colony-formings units, however, decrease rapidly. Adapted from H-G. Hoppe, 1974, *Mar. Biol.* 36:291–302.

phy of natural populations, as pioneered by Brock and Brock (1968), has been used to show that the heterotrophic uptake of dissolved organic matter is mainly by bacteria (Munro and Brock, 1968; Hoppe, 1976) and that its measurement can be a valuable ecological tool. Its limitations are the nutrient specificity of the populations studied (Hoppe, 1974, 1976), and in the case of ^{32}P labeling, a possible non-linear response that could lead to lower concentrations being underestimated (Peroni and Lavarello, 1975). Such studies, however, have indicated that the inactive cells that give rise to colony-forming bacteria are a small fraction of the active cells in nearshore waters and that they decrease markedly in number with distance from the shore in the Baltic Sea (Plate 14-3,*B*; Hoppe, 1976). The colony-forming bacteria inshore accounted for 1 to 6% of the population and decreased to 0.01 to 0.6% 20 km offshore depending upon the season.

Bacteria occurring as unattached cells (bacterioplankton) can be selectively obtained from the eucaryotic plankton on the basis of size by Nuclepore filtration, as discussed above. As we have seen, this population in the open sea is dominated by planktobacteria failing to form colonies on the usual agar media. Adenosine triphosphate (ATP) assays of these filtrates would give a direct indication of viable biomass. Although expensive to perform, ATP assays eliminate the laborious, time-consuming, and somewhat questionable autoradiographic method of quantification. This is illustrated in Plate 14-4,*A*, which shows a biomodal peak of ATP (below 3 μm) that reflects the biomodal peak differentiating the active planktobacteria and colony-forming bacteria in Plate 14-3,*A*. Preliminary studies show that the biomass of bacterioplankton refractory to cultivation in batch culture is significant throughout the water column, decreasing only one order of magnitude in the aphotic zone. The biomass of viable particles less than 3 μm varied from 3 to 80% of the total ATP of particles under 1 mm, with mean values of 30% for the photic zone and 40% for the aphotic zone (Burney et al., in press). Similar values have been obtained using 1 μm membranes to exclude the smallest microflagellates. Although it is possible to determine carbon and the carbon-to-ATP ratios for these natural populations of bacterioplankton, such studies have not been completed. When the ratio of 1:250, obtained by Hamilton and Holm-Hansen (1967), is used [but corrected for the cell volume of natural populations, (Ferguson and Rublee, 1976; Watson et al., 1977)] bacterial populations in the surface skin of up to 10^8/ml decrease to 10^5 to 10^6/ml in euphotic subsurface waters but remain at a constant 10^4/ml throughout the aphotic zone. These population figures held across the North Atlantic, and the expected areas of poorer populations were not observed. These observations on viable biomass agree with fluorescent direct counts on oceanic waters.

From the foregoing discussion, it can be seen that two distinct populations of heterotrophic bacteria appear to occur in water samples: cells capable of forming colonies on agar media, which will be discussed in the following four chapters, and a more numerous population and large biomass of presumably smaller cells incapable of forming colonies on the agar media in use at this time. This latter bacterial flora must be the indigenous or autochthonous population of planktobacteria occurring as free cells living a planktonic existence, with no requirement for solid surfaces. The minor population of epibacteria in the bacterioplankton must be considered to be stray or transient bacteria from the richer solid surfaces of such organic debris as fecal fragments and senescent phytoplankton.

The planktobacteria must be considered to be a truly planktonic and buoyant population dependent upon dissolved organic matter for nutrients. Their growth pattern must be sporadic and controlled by diel pulses of labile nutrients. Such substrates as polysaccharides, which are broken down by extracellular enzymes, occur at concentrations down to 0 μg of carbon/liter whereas the more labile monosaccharides, which are transported into the cell for breakdown by intracellular enzymes, apparently have a lower concentration of 25 to 50 μg of carbon/liter (Burney et al., in press). This "threshold concentration" appears to reflect a bacterial feedback control mechanism whereby

PLATE 14-4. The size range of selectively filtered bacterioplankton in Narragansett Bay and its distribution with depth compared to the protist plankton in the North Atlantic as indicated by adenosine triphosphate (ATP). (*A*) Percent ATP of particles concentrated on a series of Nuclepore membranes with a pore size between 0.2 and 8 μm. Note the minimum at the 1 to 3 μm size fraction. A series of depth profiles showing the relative abundance of procaryote ATP (<3 μm) (x) to protist ATP (>3 μm) (o) in the upper layers along the outer continental shelf (*B*,*C*,*D*) and in deep water (*E*,*F*). (*A*) R. Heffernan, unpubl.; (*B–E*) data from R/V *Trident* Cruise 170.

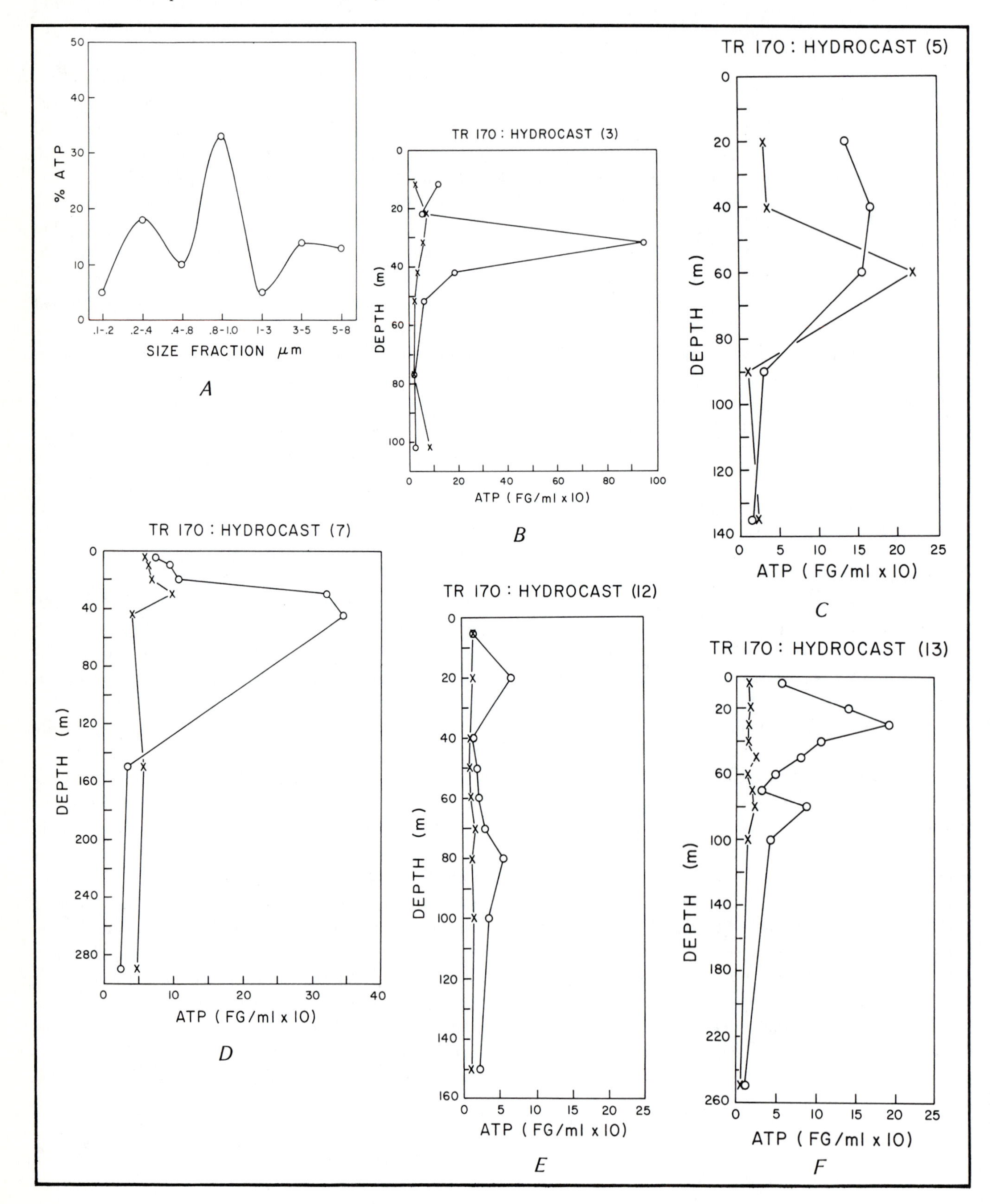

PLATE 14-5. The ultrastructure of bacterioplankton from Narragansett Bay, R.I. that has passed through a 2 μm Nuclepore membrane. (*A*) Transmission electron micrograph of a thin section showing the diversity of size and form of the bacteria. The smallest cells are marked by arrows. (*B*) and (*C*) Higher magnifications showing that the smallest cells (minibacteria) possess the typical gram-negative cell envelope; note the size of a protist flagellum in (*B*) for comparison. Electron micrographs courtesy of Paul W. Johnson.

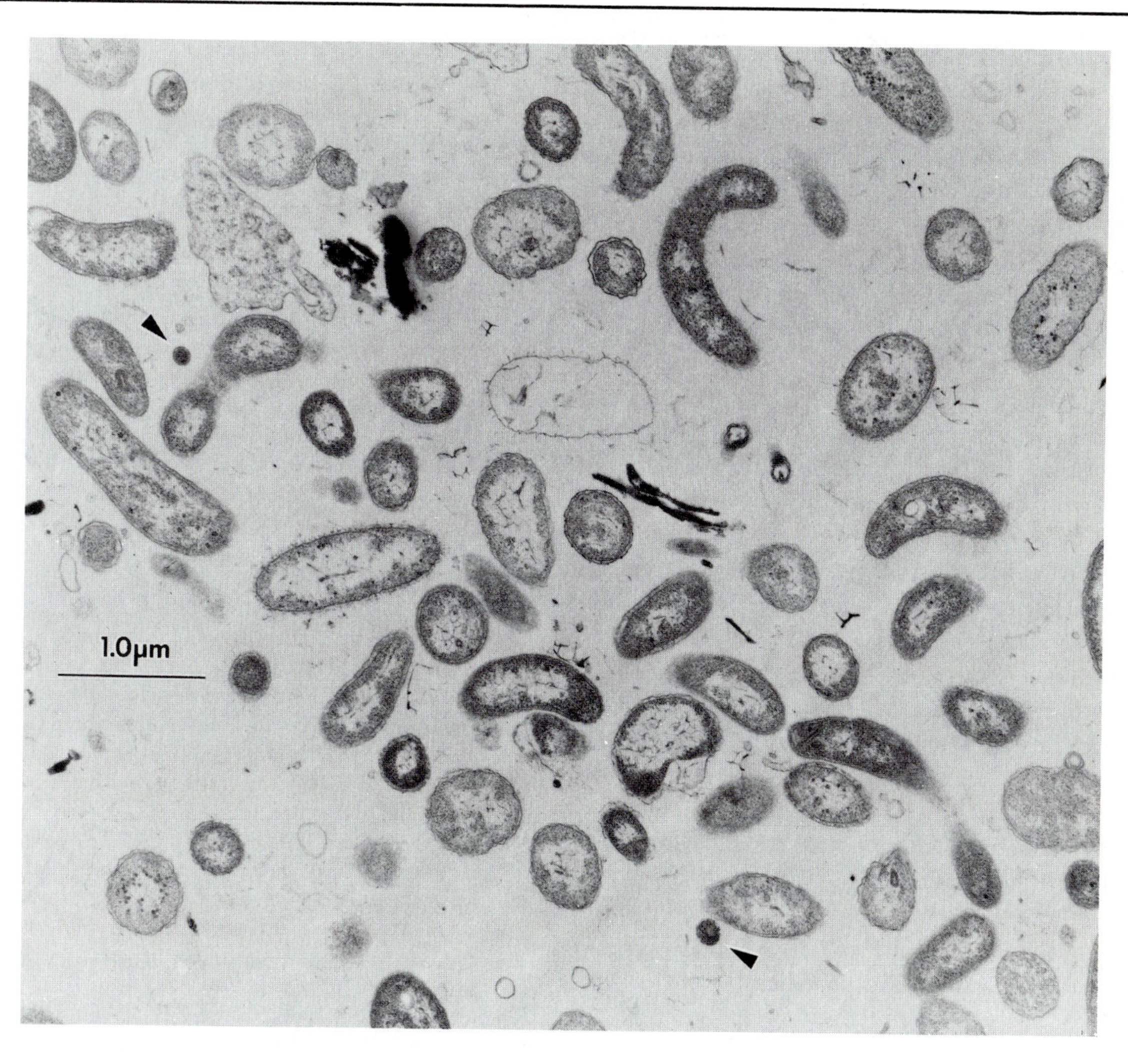

A

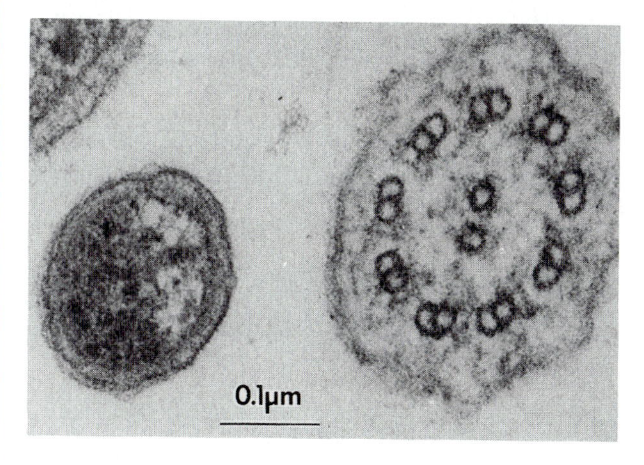

B

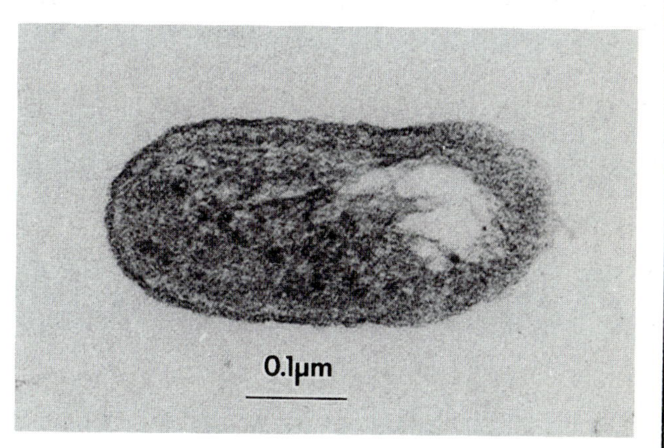

C

for each substrate there is a concentration below which each bacterium will shut down its active transport and dissimilatory enzymes to conserve energy. As we have seen in Chapter 7,C,3 (Plate 7-2), growth on *in-situ* water is not constant and coincides with the transient release of dissolved organic carbon that occurs during specific periods (diel periodicity) (Sieburth et al., 1977). These ubiquitous, numerous, and typical bacterial forms have gone unrecognized long enough. It is time to give them a name and to establish their identifying characteristics. The vernacular term planktobacteria seems appropriate now. When more is known about their ultrastructure and physiology, they can be placed in existing genera or new genera can be formed, such as "*Planktococcus*," "*Planktobacterium*," "*Planktovibrio*," and "*Planktocyclus*." The tiniest of the smaller forms shown in the Part V title plate are curved rods 0.2 μm or less in diameter, a cell size barely large enough to accommodate bacteriophage particles. These minibacteria are at the limit of resolution of the light microscope. In thin section, these minute bacterial cells can be seen to have typical gram-negative cell walls, resembling the other cells in the bacterioplankton (Plate 14-5).

Planktobacteria, therefore, may be defined as small cells of differing morphologies that are gram negative, have little affinity for attaching to surfaces, do not grow on conventional agar media, and are nurtured by dissolved organic matter above threshold values. Selectively filtered natural populations can grow on *in-situ* water by diffusion culture. Growth patterns depend upon the interactions of the phototrophic and phagotrophic populations and are regulated by diel periodicity.

REFERENCES

Azam, F., and R.E. Hodson. 1977. Size distribution and activity of marine microheterotrophs. *Limnol. Oceanogr.* 22:492–501.

Brock, M.L., and T.D. Brock. 1968. The application of micro-autoradiographic techniques to ecological studies. *Mitt. int. Verein, theor. angew. Limnol.* 15:1–29.

Burney, C.M., K.M. Johnson, D.M. Lavoie, and J.McN. Sieburth. Dissolved carbohydrates in the North Atlantic: concentrations and microbial cycling. *Deep-Sea Res.* in press.

Daley, R.J., and J.E. Hobbie. 1975. Direct counts of aquatic bacteria by a modified epifluorescence technique. *Limnol. Oceanogr.* 20:875–82.

Ferguson, R.L., and P. Rublee, 1976. Contribution of bacteria to standing crop of coastal plankton. *Limnol. Oceanogr.* 21:141–145.

Hamilton, R.D., and O. Holm-Hansen. 1967. Adenosine triphosphate content of marine bacteria. *Limnol. Oceanogr.* 12:319–24.

Hodson, R.E., F. Azam, and R.F. Lee. 1977. Effects of four oils on marine bacterial populations: Controlled ecosystem population experiment. *Bull. Mar. Sci.* 27:119–26.

Hoppe, H-G. 1974. Untersuchungen zur Analyze mariner Bakterien-populationen mit einer autoradiographischen Methode. *Kieler Meeresforsch.* 30:107–16.

Hoppe, H-G. 1976. Determination and properties of actively metabolizing heterotrophic bacteria in the sea, investigated by means of micro-autoradiography. *Mar. Biol.* 36:291–302.

Jannasch, H.W., and G.E. Jones. 1959. Bacterial populations in sea water as determined by different methods of enumeration. *Limnol. Oceanogr.* 4:128–39.

Munro, A.L.S., and T.D. Brock. 1968. Distinction between bacterial and algal utilization of soluble substances in the sea. *J. gen Microbiol.* 51:35–42.

Peroni, C., and O. Lavarello. 1975. Microbial activities as a function of water depth in the Ligurian Sea: An autoradiographic study. *Mar. Biol.* 30:37–50.

Salonen, K. 1974. Effectiveness of cellulose ester and perforated polycarbonate membrane filters in separating bacteria and phytoplankton. *Ann. Bot. Fennici* 11:133–35.

Sheldon, R.W. 1972. Size separation of marine seston by membrane and glass-fiber filters. *Limnol. Oceanogr.* 17:494–98.

Sheldon, R.W., and T.R. Parsons. 1967. A continuous size spectrum for particulate matter in the sea. *J. Fish. Res. Bd. Canada* 24:909–15.

Sheldon, R.W., T.P.T. Evelyn, and T.R. Parsons. 1967. On the occurrence and formation of small particles in seawater. *Limnol. Oceanogr.* 12:367–75.

Sheldon, R.W., A. Prakash, and W.H. Sutcliffe, Jr. 1972. The size distribution of particles in the ocean. *Limnol. Oceanogr.* 17:327–40.

Sieburth, J.McN., K.M. Johnson, C.M. Burney, and D.M. Lavoie. 1977. Estimation of *in-situ* rates of heterotrophy using diurnal changes in dissolved organic matter and growth rates of picoplankton in diffusion culture. *Helgoländer wiss. Meeresunters.* 30:565–74.

Smith, W.O. 1976. The regulation and ecological significance of phytoplankton excretion. Ph.D. Diss., Duke University.

Watson, S.W., T.J. Novitsky, H.L. Quinby, and F.W. Valois. 1977. Determination of bacterial number and biomass in the marine environment. *Appl. Environ. Microbiol.* 33:940–54.

Williams, P.J. LeB. 1970a. Heterotrophic utilization of dissolved organic compounds in the sea. I. Size distribution of populations and relationship between respiration and

incorporation of growth substrates. *J. mar. biol. Assoc. U.K.* 50:859–70.
Williams, P.J. Le B. 1970b. Heterotrophic utilization of dissolved organic compounds in the sea. II. Observations on the responses of heterotrophic marine populations to abrupt increase in amino acid concentration. *J. mar. biol. Assoc. U.K.* 50:871–81.
Zimmerman, R., and L-A. Meyer-Reil. 1974. A new method for fluorescence staining of bacterial populations on membrane filters. *Kieler Meeresforsch.* 30:24–27.

CHAPTER 15

Gram-Negative Non-distinctive Epibacteria

The non-distinctive bacteria, usually seen on particulate material, occur as transients in the water column, especially in shallow waters, grow readily on a variety of culture media, and have an undistinguished morphology. Without distinctive cellular or colonial morphologies or biochemical activities, they can be thought of as the "silent majority" of colony-forming bacteria that must be largely responsible for the degradation of particulate organic matter in the sea. They are gram negative and coccoid to rod shaped.

The majority of bacteria found in the sea are, in fact, gram negative. The gram-negative cell is well adapted for life in the low nutrient, aqueous environment of the sea because its degradative enzymes are retained in a highly protective association with the complex cell wall (Costerton, Ingram, and Cheng, 1974). These cell wall-associated enzymes enable the gram-negative cell to digest a wide variety of complex substrates as "food" in a zone immediately surrounding the cell, with the products of digestion being readily available for active transport into the cell without being lost to the sea. Gram-positive bacteria, on the other hand, are usually found in such protected habitats as sediments (Chapter 3) containing high concentrations of organic matter and in the organic matter-rich surface microlayer. Their distribution may be due to the release of large amounts of extracellular enzymes (Rogers, 1970), which would be wasted in more dilute habitats, such as the water column. The gram-positive bacteria only become a significant part of the colony-forming units on agar plates under selective cultural conditions that inhibit the dominant gram-negative bacteria, such as plating on media made with distilled water, with high concentrations of salt, or incubating at temperatures of 37 °C or above.

The procaryotic bacterial cell, compared to the eucaryote, has a rather simple organization: cytoplasm surrounded by a cell envelope (Pollard, 1967; Henneberg, 1969). Examination by transmission electron microscopy shows that the cytoplasm of bacteria contains nuclear material (DNA) in a fibrillar pattern and electron-dense particles of ribosomal material (RNA and protein) responsible for protein synthesis (Plate 15-1,*A*,*B*). Some bacterial types also have an internal membrane system or mesosomes. The ultrastructure of the cell envelopes of gram-negative and gram-positive bacteria is found to be distinctly different by transmission electron microscopy. The gram-positive cell envelope consists of an inner cytoplasmic membrane surrounded by a thick cell wall or peptidoglycan layer (Plate 15-1,*B*). The gram-negative cell envelope, however, is more complex, with a number of layers. The cytoplasmic membrane, which is similar to that of gram-positive cells, is surrounded by a cell wall consisting of a thin peptidoglycan layer and a unique outer membrane layer (Plate 15-1,*A*). A number of gram-negative bacteria also have an extra protective layer external to the outer membrane.

The peptidoglycan or murein component of the cell wall of both gram-negative and gram-positive bacteria is very similar, being composed of polysaccharide (glycan) chains cross-linked by a peptide bridge. The peptidoglycan of the bacterial cell wall confers rigidity and shape, prevents osmotic lysis, and binds certain enzymes in gram-positive bacteria. It comprises some 50 to 90% of the cell wall in gram-positive bacteria, but only 5 to 20% of the cell wall in gram-negative bacteria. The cytoplasmic membrane of both gram-positive and gram-negative bacteria is similar in chemical composition to other biological membranes. Certain enzymes are present on the surface of and within the cytoplasmic membrane. These include cell-wall synthesizing enzymes, the permeases of active transport, and enzymes with a cytoplasmic function, as well as structural and binding proteins. The cytoplasmic membrane also acts as a barrier to toxic molecules from the external environment.

The unique outer membrane layer is found only in the cell envelope of gram-negative bacteria. It is made up of phospholipids, proteins, and lipopolysaccharide and bounds a region known as the "periplasmic space" outside the cytoplasmic membrane. The degradative enzymes of gram-negative bacteria are localized in this periplasmic space. The outer membrane layer is not a site of active transport, but acts more as a "molecular sieve" to control the passage of molecules into and out of the periplasmic space, which is an additional barrier to exclude a wide variety of molecules. The complex, multilayered cell envelope, with each layer possessing different structural and physical–chemical characteristics, enables the gram-negative bacteria to live in a wide range of environments, including the sea.

The success of the ubiquitous, gram-negative non-distinctive epibacteria is also attributed to the wide variety of particulate substrates they can colonize and utilize. Although most of the water mass is oxygenated and these bacteria are aerobes, some are also facultative anaerobes and grow well in such oxygen-depleted environments as anaerobic sediments, the animal gut, and anoxic water masses. Population estimates depend upon cultural methods (Jannasch and Jones, 1959) and the conditions of culture (Sieburth, 1967a, 1971). The procedures that are used to estimate colony-forming units, rather than biomass or individual cells, therefore only indicate the relative abundance of these bacteria. Larger populations usually occur in shallow nearshore waters, with their higher concentrations of suspended particulates and resuspended bottom sediments (Sieburth, 1967a). Oceanic populations are usually a few colony-forming units/1 to 10 ml (Jannasch and Jones, 1959), except in the thermocline and surface film, in which samples can occasionally contain populations approaching 10^3 to 10^5 colony-forming units/ml (Sieburth, 1971; Sieburth et al., 1976). Even in open ocean waters, with their dominant populations of autochthonous planktobacteria, there always appears to be a minimal background population sufficient to populate any readily degradable solid organic substrate. Most studies on distribution were limited to total count estimates (Jannasch and Jones, 1959), until James Shewan and his colleagues at the Torry Research Station in Aberdeen, Scotland devised a system (Shewan, Hobbs, and Hodgkiss, 1960) that would sort into generic groups the more common forms encountered in fishery products, on the basis of a few simple tests applicable to large numbers of cultures. This system has been improved, with the accumulation of knowledge and the availability of better tests (Hendrie, Hodgkiss, and Shewan, 1964; Hendrie and Shewan, 1966; Hendrie, Mitchell, and Shewan, 1968; Bain and Shewan, 1968). This system was used in a few studies to characterize the seasonal variation in these groups of bacteria in inshore waters (Simidu and Aiso, 1962; Altschuler and Riley, 1967; Sieburth, 1967a, 1968b) and in the open sea (Sieburth, 1971). The use of arbitrary groupings, as shown in the scheme in Table 15-1, indicates the differences between the larger populations that occur sporadically and the low background microflora in offshore waters (Sieburth, 1971),

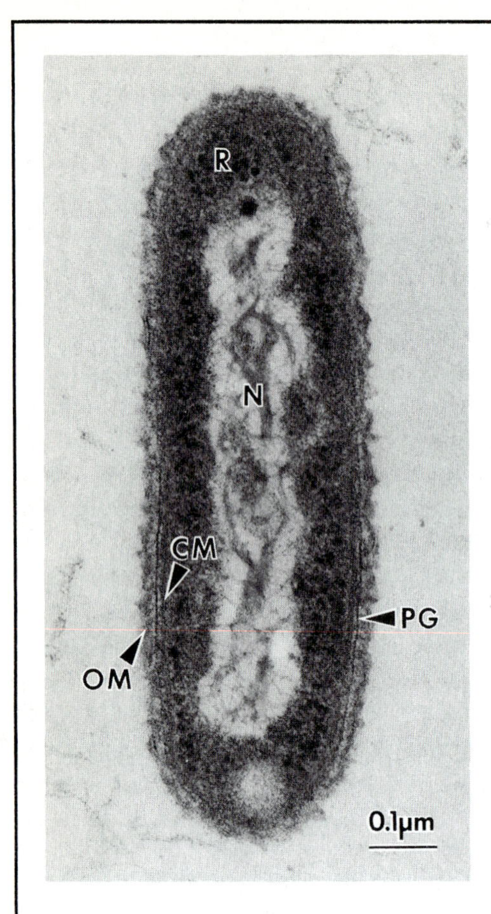
R
N
CM
PG
OM
0.1µm

A

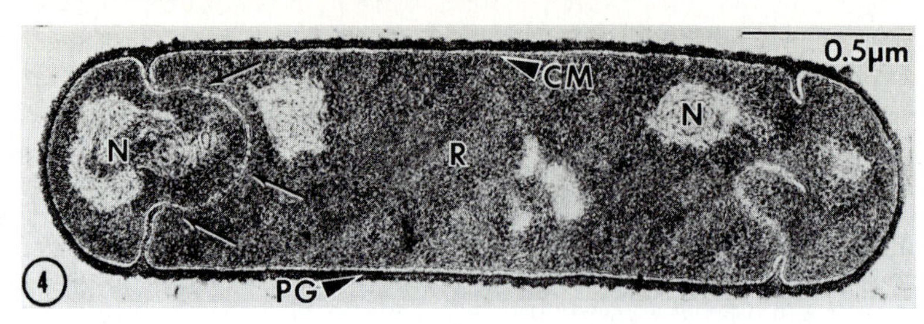
0.5µm
CM
N
R
N
4
PG

B

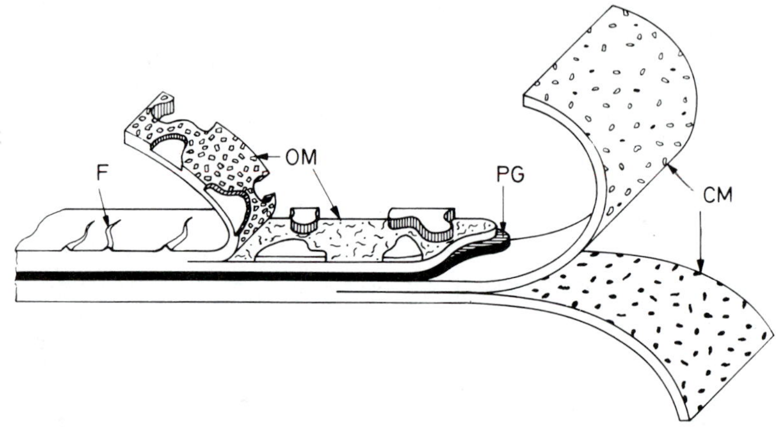
OM
F
PG
CM

C

% of Dry Weight of Whole Cell.
Loosely Bound Layer---4·7%
Outer Double-track Layer---7·9%
Underlying Layer ---6·1 %
Peptidoglycan Layer---
Cytoplasmic Membrane------

D

PLATE 15-1. Comparison of the ultrastructure of cell envelopes of gram-negative and gram-positive bacteria. (*A*) Transmission electron micrograph of a thin secion of a typical gram-negative marine bacterium. The multilayered cell envelope consists of an outer membrane layer (OM), a thin peptidoglycan layer (PG), and the cytoplasmic membrane (CM). The cytoplasmic constituents visible are ribosomal (R) and nuclear (N) material. (*B*) Transmission electron micrograph of a thin section of a gram-positive bacterium, *Bacillus subtilus*, undergoing spore formation (arrows). The cell envelope of this bacterium consists of a thick peptidoglycan layer (PG) and the cytoplasmic membrane (CM). The ribosomal (R) and nuclear (N) material are also visible. (*C*) Reconstruction of a freeze-fractured *Escherichia coli* (gram-negative) cell envelope showing the relative positions of fracture faces in the cytoplasmic membrane (CM), the peptidoglycan layer (PG), fracture faces in the outer membrane (OM), and flagella (F). (*D*) Thin section and model of the cell envelope of the marine pseudomonad B-16 (gram negative), showing the contribution of the outer three layers to the cell dry weight. (*A*) electron micrograph courtesy of Paul. W. Johnson; (*B*) from L. Santo, T.J. Leighton, and R.H. Doi, 1972, *J. Bacteriol.* 111:248–53; (*C*) from A.P. Van Gool and N. Nanninga, 1971, *J. Bacteriol.* 108:474–81; (*D*) from C.W. Forsberg, J.W. Costerton, and R.A. MacLeod, 1970, *J. Bacteriol.* 104:1354–68.

differences in inshore waters (Simidu and Aiso, 1962; Altschuler and Riley, 1967; Sieburth, 1967a), differences in oceanic waters (Sieburth, 1971), and the abrupt seasonal changes in generic composition of the very constant populations of an estuarine bacterial flora (Sieburth, 1967a, 1968b). The more precise, but much more laborious method of numerical taxonomy, which compares small collections of several hundred isolates with each other and authentic cultures, has eclipsed the simpler approach. But although numerical taxonomy is very useful and has its application, it could never be used on the thousands of isolates required for ecological studies (Sieburth, 1967a, 1971). Numerical taxonomy, however, can show the species clusters (Hendrie, Hodgkiss and Shewan, 1970; Reichelt and Baumann, 1973), and the thousands of isolates of ecological interest can be classified by their dominant characteristics (Ruby and Nealson, 1978).

A. FERMENTATIVE RODS AND COCCOIDS

The fermentative bacteria can be readily distinguished from other bacteria with non-distinctive morphologies by culture on the oxidative–fermentative (O/F) medium of Hugh and Leifson (1953), modified for marine bacteria by Leifson (1963). When these bacteria ferment carbohydrates in this semisolid medium, they produce acid or acid and gas from the bottom of the tube or throughout the tube, thus indicating that they are facultative aerobes and have an anaerobic pathway for the dissimilation of carbohydrates. The fermentative rods from the marine environment studied by Shewan, Hobbs, and Hodgkiss (1960) were classified as vibrios. These microorganisms are associated with zooplankters, the gut of animals feeding on zooplankton, and sediments that are probably enriched by fecal debris. Zooplankton and the gut of fishes are a rich source of chitin-decomposing bacteria (Lear, 1963), and most chitin-decomposing bacteria studied by Brisou et al., (1964) were also classified as vibrios. Although vibrios accounted for some 35 to 48 % of the total isolates from the intestines of newly caught fish (Shewan, 1966) and up to 100% of the gut flora of flatfish (Liston, 1957), these fermentative bacteria with mixed flagellation are common in the intestines of other marine animals as well (Leifson et al., 1964). Vibrios are common in the gut of teredos (wood boring molluscs), where they help decompose cellulose (Kadota, 1959). Vibrios also represent the main, aerobic cellulose-decomposing microflora in the water and sediments of Hiroshima Bay (Kadota, 1959). The incidence of vibrios in the waters of Narragansett Bay ranges from 0 to 38%, with a mean of 14% (Sieburth, 1967a); peaks occur during the senescent periods of diatom blooms (Sieburth, 1968b), which apparently coincide with periods of zooplankton abundance (Martin, 1965). A study of motile marine bacteria showed that those producing acid from glucose anaerobically accounted for some 10% of the population from water and from 36 to 85% of the population from animals (Leifson et al., 1964). This group was not detected in the open ocean microflora (Sieburth, 1971). The marine vibrios, including the human pathogen *Vibrio parahaemolyticus*, which is also associated with mud and zooplankton (Kaneko and Colwell, 1973) and the fish pathogen *Vibrio anguillarum* (Levin, Wolke, and Cabelli, 1972), are genetically (Colwell, 1970) and therefore generically distinct from *Vibrio cholerae*. Tubiash, Colwell, and Sakazaki (1970) considered but discarded the possibility of forming a new genus for the marine vibrios, whereas Johnson, Katarski and Weisrock (1968) and Ruger (1972) suggested the genus *Marinovibrio*. Baumann, Baumann, and Mandel (1971), however, had earlier proposed the removal of the species distinct from *V. cholerae*, with peritrichous flagellation, which would be placed in a redefined genus *Beneckea*. Fermentative chitin-decomposing bacteria from mud, described as a species of *Achromobacter* by Campbell and Williams (1951), were subsequently placed in a new genus *Beneckea* in *Bergey's 7th* (Breed, Murray, and Smith, 1957). Such bacteria were first isolated by Benecke (1905), a pioneer in marine bacteriology (Benecke, 1933). The genus *Beneckea*, as originally described, contained only microorganisms with peritrichous flagella. But many strains of *Beneckea* that have a single, sheathed polar flagellum in liquid media also produce unsheathed peritrichous flagella, the "mixed flagellation" of Leifson (1960), when grown on a solid medium (Allen and Baumann, 1971; Plate 15-2). The emendation of the genus *Beneckea* to include both the polar flagellated fermenting bacteria (Baumann, Baumann, and Mandel, 1971) and the pigmented forms (Baumann et al., 1971) or the formation of the genus *Marinovibrio* (Johnson, Katarski, and Weisrock, 1968; Rogers 1970) appear to be sensible solutions to a difficult taxonomic problem, and either genus would encompass a rather homogeneous group of microorganisms with a more or less specific habitat, the marine enterobacteria (Baumann and Baumann, 1977). Unfortunately, Shewan and Veron in *Bergey's 8th* fail to do this.

All vibrio strains other than *V. cholerae* can grow in artificial seawater with glycerol as the sole source of carbon and energy and ammonia as the sole source of nitrogen. Most strains reduce nitrate to nitrite, but none are able to denitrify or to fix molecular nitrogen. Most strains produce extracellular amylase, gelatinase lipase, and chitinase and hemolyze red blood cells. Many strains can utilize a variety of organic compounds for both carbon and energy; these include pentoses, hexoses, disaccharides, sugar alcohols, and C_2-C_{10} monocarboxylic fatty acids. Among the more common substrates they can utilize are L-valine, L-lysine, aromatic amino acids, purines, pyrimidines, formate, C_6-C_{10} dicarboxylic acids, agarose, and cellulose. The remarkable

TABLE 15-1. The division of the dominant, non-distinctive gram-negative rods into preliminary and arbitrary generic groupings.

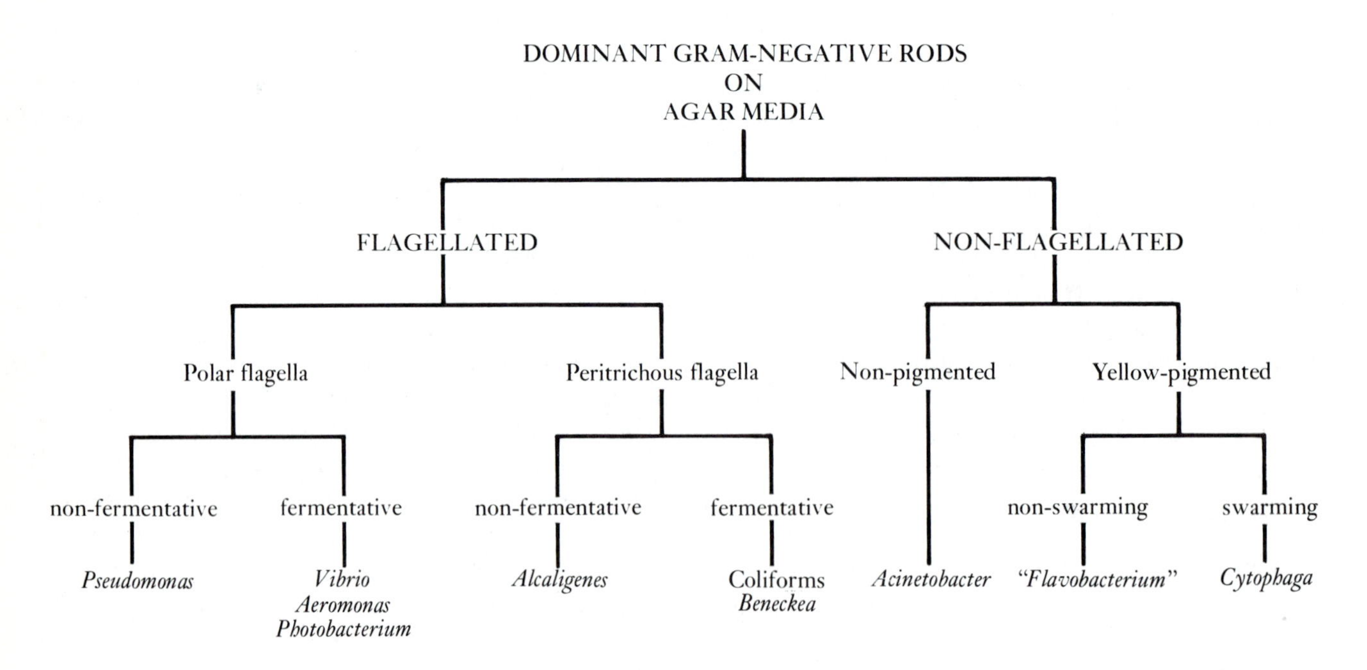

properties of *Beneckea natriegens*, with a 10-minute generation time and a marked susceptibility to bacteriophage, make it a useful teaching tool (*Anon.*, 1976).

A group of vibrio-like bacteria, which exhibited a pronounced pleomorphism in complex media, were termed "gut-group vibrios" by Liston (1957) and Colwell (1962). The pleomorphic rods described by Tsiklinsky from the gut of fish and from other marine materials brought back by Jean Charcot from the 1903–1905 French Antarctic Expedition (Tsiklinsky, 1908) may have belonged to this group, as did some 8% of the isolates from the waters of Narragansett Bay (Sieburth, 1967a). Among them were some obligate psychrophiles, which, in culture, were gram-negative pleomorphic rods, with some large lobate forms at temperatures below 15 °C that were transformed into gram-positive rods at elevated temperatures and fragmented to form gram-positive cocci suggestive of the arthrobacters (Chapter 18,C). At the time, this possible life cycle seemed sufficient grounds to provisionally designate them as arthrobacters (Sieburth, 1964, 1967a,b, 1968b). An obligate psychrophile MP-1 isolated in the Pacific (Morita and Haight, 1964) and named *Vibrio marinus* (Colwell and Morita, 1964) also has exhibited a pleomorphism (Felter, Colwell, and Chapman, 1969) similar to that of the marine arthrobacters (Plate 15-3), but its gram reaction in older cultures and at elevated temperatures has not been reported. *Vibrio marinus* (strain MP-1) (Citarella and Colwell, 1970) and the B-2 group of *Beneckea* (Baumann et al., 1971) are distinct from the other vibrios, having a 41 mole % G + C content. On this basis alone, these organisms obviously don't belong in the genus *Arthrobacter*, which has a 60 to 64 mole % G + C content (Veldkamp, 1970; Bousfield, 1972). They may belong to the genus *Acinetobacter*, which contains pleomorphic rods with a 40 to 47 mole % G + C content. A recent study of the coryneforms demonstrated that the final transformation into cocci, which is the most important feature in the "life cycle" of "typical" arthrobacters, is common in other genera (Bousfield, 1972). This group of pleomorphic vibrio-like psychrophiles associated with the gut of fish requires more study before it can be placed in a legitimate taxon.

The genus *Aeromonas* contains bacteria similar to those in the genus *Vibrio*, which may have delayed gas production. They are distinguished from the vibrios by their insensitivity to the "vibriostat" pteridine 0/129 (Bain and Shewan, 1968) and by a 57 to 63 mole % G + C DNA content in contrast to that of 45 to 50 for the genus *Vibrio*. Species of *Aeromonas* are associated with furuncles in salmon and other infections.

Bacteria exhibiting luminescence are grouped in or with the vibrios. The history of luminescence and bioluminescence from the earliest times has been compiled by Harvey (1952, 1957). Aristotle knew that dead fish emitted light and Robert Boyle demonstrated that air was required for the process. But it was not until 1875 that Pfluger showed that bacteria were responsible for such luminescence. The first in-depth studies on luminescent marine bacteria were conducted by Bernard Fischer, a medical officer on various German ships and on the Humboldt Plankton Expedition, who became an early worker in marine bacteriology (Fischer, 1887, 1888, 1894). Luminescent bacteria usually occur in low numbers on agar media inoculated with seawater, but they are not rare. Reliable enrichments can be prepared by holding moist (partially immersed and loosely covered) fish or squid in a 15 °C incubator overnight or in a refrigerator for several days until bright colonies are apparent in the dark. The viewer's eyes must be dark-adapted before the luminescent colonies can be seen. A simple device for the enumeration and isolation of luminescent colonies has been developed (Cosenza and Buck, 1966) in which a 10-W red light bulb in a Quebec bacterial colony counter was dimmed by a powerstat so that the non-luminescent colonies were slightly visible but those luminescing assumed a bluish tinge accented by the red light. In a study on the bacterial flora of flatfish, luminescent bacteria were found to belong to the gut-group vibrios, just discussed, which often approached 100% of the gut flora and were present in greater number in the gills than on the skin (Liston, 1957).

The ATP-dependent luciferin–luciferase bioluminescence system in these bacteria has been studied extensively (Harvey, 1952; McElroy, 1961; Nealson, Platt, and Hastings, 1970; Nealson and Hastings, 1972; Makemson, 1973). Recent interest in this group is rectifying a chaotic taxonomy. The genus *Photobacterium*, proposed and used by Beijerinck (1899a,b), is an ecological group based solely on light emission, although he (1916) conceded that some luminous species were closely related to the cholera vibrio. The first review of taxonomy by Spencer (1955), concluded that they could be regarded as species of *Vibrio* and *Aeromonas*. In a more

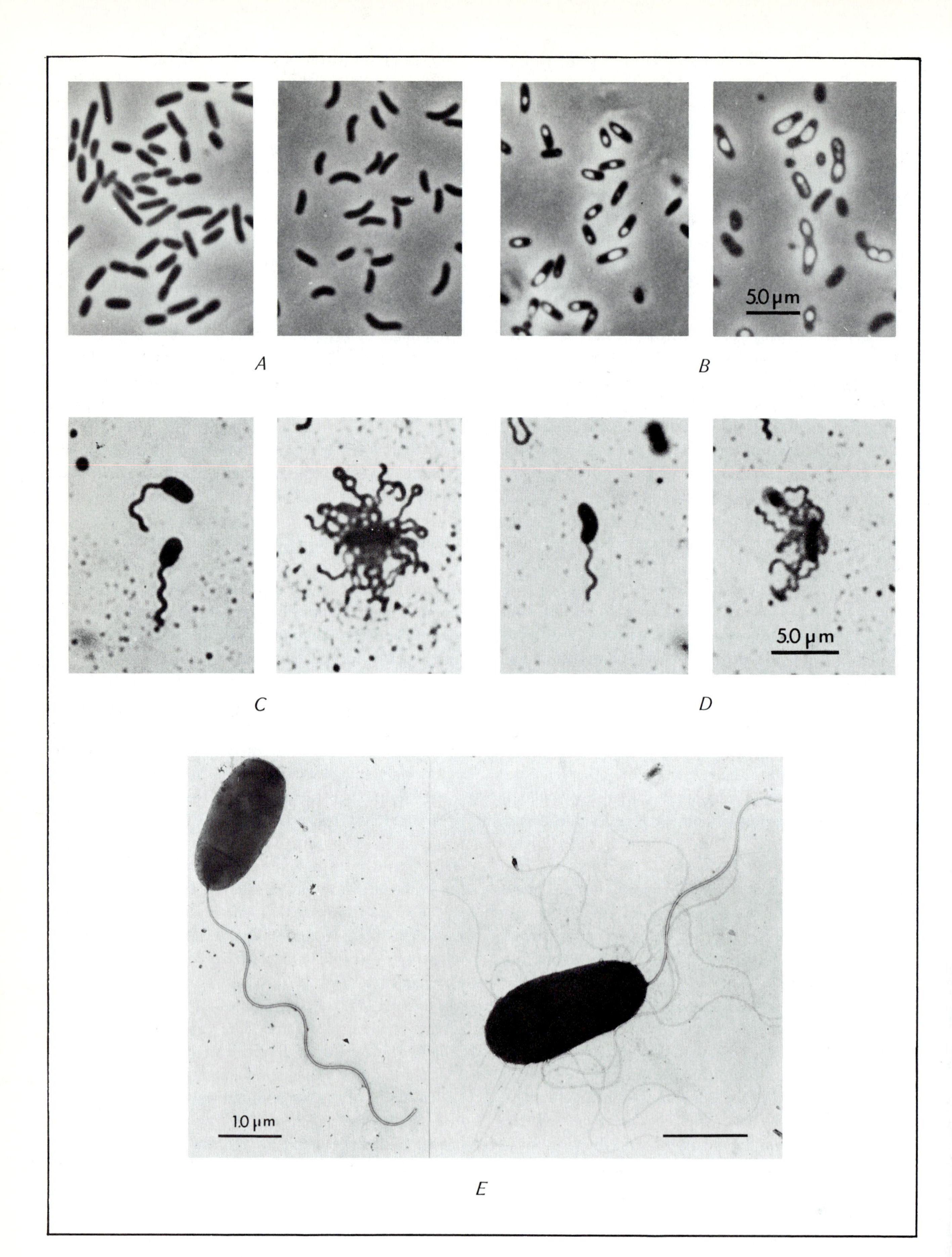

A
B
5.0 µm
C
D
5.0 µm
1.0 µm
E

PLATE 15-2. Examples of the cells and unique flagellation of *Beneckea* strains, which have a polar flagellation in broth culture but peritrichous flagellation on a solid agar medium. Phase-contrast photomicrographs of the straight to bent rods during exponential growth (*A*), which change to older cells with storage granules of polyhydroxy butyrate (*B*). Leifson flagella-stained preparations, showing the single polar flagellum in liquid culture (left) and peritrichous flagella (right) for the same strain grown on a solid medium (*C,D*). (*E*) Transmission electron micrographs of negatively stained *Vibrio (Beneckea) parahaemolyticus*, showing the sheathed polar flagellum in broth culture (left) and the unsheathed peritrichous flagella (right) on a solid medium. (*A–D*) from P. Baumann, L. Baumann, and M. Mandel, 1971, *J. Bacteriol.* 107:268–94; (*E*) from R.D. Allen and P. Baumann, 1971. *J. Bacteriol.* 107:295–302.

recent study, with the advantages of more strains and much improved criteria, 51 strains could be placed in three major groups (Hendrie, Hodgkiss, and Shewan, 1970). Twenty-eight strains were put in an emended genus *Photobacterium*, with *P. phosphoreum* and *P. mandapamensis* as the two species (aerogenic and oxidase negative, with 39.5 to 42.0 mole % G + C content). Ten strains were considered to be *Vibrio*, with the type species *V. (Photobacterium) fischeri.* Since they are closely related to *V. marinus*, they may be of the gut-group vibrio type, which is known to be luminescent. Thirteen strains with petitrichous flagella and a 45.5 to 46.5 mole % G + C content were placed in a new genus *Lucibacterium*, with *L. (Photobacterium) harveyi* the type species.

Reichelt and Baumann (1973) used more characteristics to conduct an in-depth study with 173 strains of marine luminous bacteria isolated from seawater, from the surface and intestines of fish, and from the luminous light organs of fish and squid. They came to the same conclusion as Hendrie, Hodgkiss, and Shewan (1970)—that there are four distinct species—but for good reasons they placed two of them in different genera. There is no trouble with the taxonomy of either *Photobacterium phosphoreum* or *P. mandapamensis*, except that *P. mandapamensis* should be assigned the earlier species name *leiognathi* (Reichelt and Baumann, 1975). Since the marine vibrios are still a problem, taxonomically, *Vibrio fischeri* was retained as *Photobacterium fischeri* and *P. harveyi* was put in the previously described genus *Beneckea* for peritrichous bacteria rather than in the genus *Lucibacterium*, which was created especially for this one species. There is also an enzymatic argument for putting the luminous bacteria in two rather than three genera. Hastings and Mitchell (1971) found that *B. harveyi* contains a luciferase with "slow" turnover kinetics, whereas the luciferase in several strains of *P. fischeri* and *P. leiognathi* had "fast" enzyme kinetics. In their study of 1800 isolates, Ruby and Nealson (1978) found "slow" enzyme kinetics for all isolates of *B. harveyi* and "fast" enzyme kinetics for all *Photobacterium* isolates. They stated that luciferase kinetics can be used in taxon placement.

The characteristics listed by Reichelt and Baumann (1973) really seem to work in assigning wild isolates to clusters and have permitted some tentative conclusions on the distribution and ecology of the luminous marine bacteria to be made. The coastal waters entering Mission Bay in Southern California had a constant population of 1 to 5 cells/liter of *P. fischeri*, whereas *B. harveyi* predominated in the warmer summer waters (Ruby and Nealson, 1978). *Beneckea harveyi* was also the sole or predominant luminous bacterium in the inshore waters of Oahu, Hawaii, but in deeper waters (50 to 500 m) some 8 km from the coast, *B. harveyi*, *P. phosphoreum*, and *P. leiognathi* were found in approximately equal numbers, with few *P. fischeri* (Reichelt and Baumann, 1973). *Beneckea harveyi* can be isolated directly from the surfaces of fish, squid, and octopus, whereas enrichment cultures of these materials yield *P. phosphoreum.*

In addition to the luminescent bacteria from seawater, the fish gut, and decaying fish, which are easily cultivated, there are luminous bacteria that apparently grow symbiotically within the light organs of some species of squid (Pierantoni, 1918) and certain fish (Harvey, 1921; Haneda and Tsuji, 1971). The light from these organs matches the background light from above, obscuring the silhouette of the host (Hastings, 1971); this light can be turned on and repressed by the host animal (Morin et al., 1975). Luminous bacteria from these light organs were reported to be fastidious in their growth requirements and difficult to culture. They were considered to be analogous to the symbiotic bacterioids of insects (Haneda and Tsuji, 1971), which have not been cultured. Although Pierantoni (1918) claimed to have cultured the luminescent bacteria from the light organ of certain squid, attempts to culture the bacteria in the light organs of the marine fishes *Photoplepharon* and *Anomalops* from the Banda Islands in the

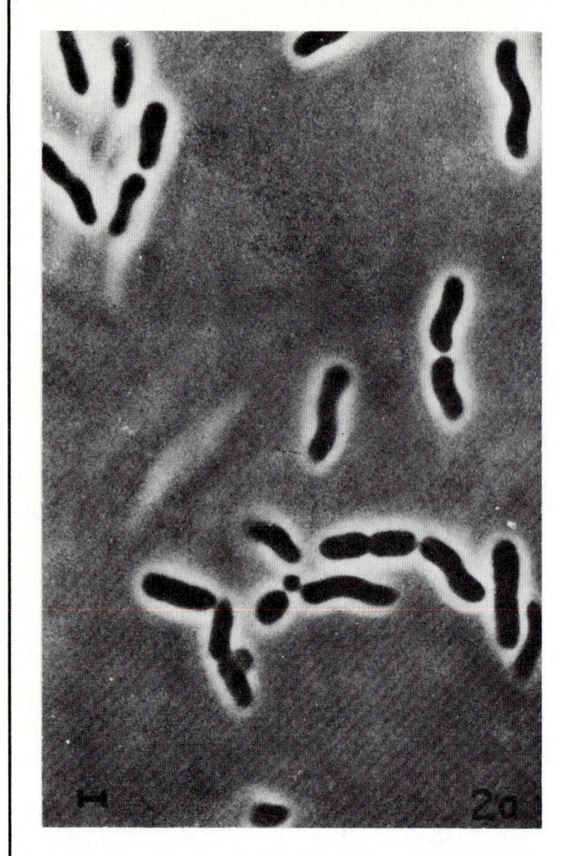

A

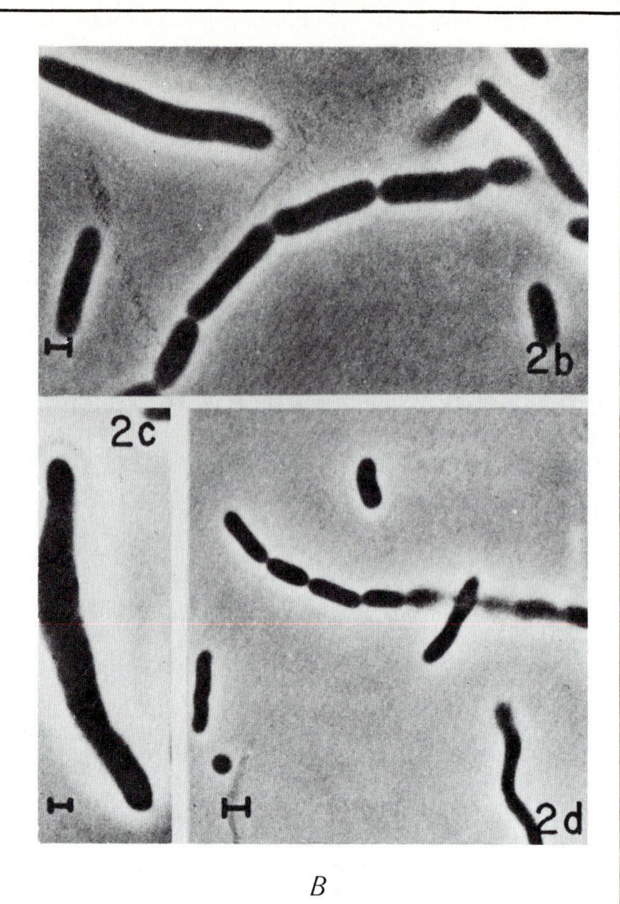

B

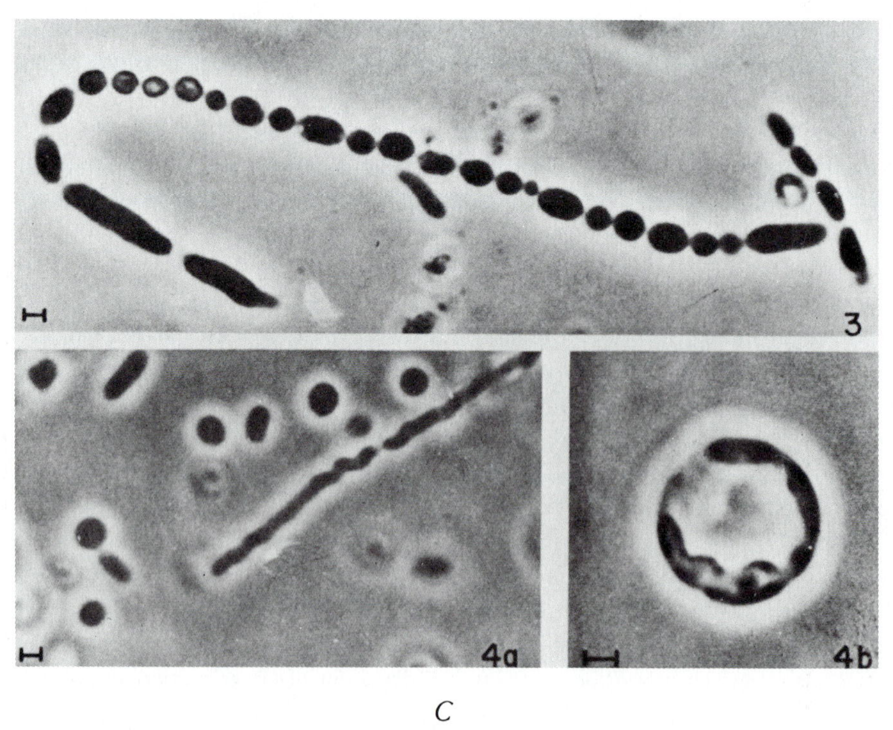

C

PLATE 15-3. The pleomorphism of *Vibrio marinus* (strain MP-1). (*A*) Growth after 2 hours at 15 °C. (*B*) Pleomorphism after 24 hours. (*C*) Pleomorphism after 48 hours, showing the tendency for myceloid formation as well as coccoid cells (marker bars = 1.0 μm). From R.A. Felter, R.R. Colwell, and G.B. Chapman, 1969, *J. Bacteriol.* 99:326–35.

eastern end of the Indonesian Archipelago failed (Harvey, 1921; Haneda and Tsuji, 1971). More recently, a number of well-characterized isolates have been obtained from luminescent organs. Six strains of *P. phosphoreum* and twelve of *P. leiognathi* were obtained from the light organs of fish and squid by Reichelt and Baumann (1973). *Photobacterium fischeri* has been isolated from the interior of the light organ of the Japanese pinecone fish *Monocentris japonica* (Ruby and Nealson, 1976) and the Australian pinecone fish *Cleidopus gloriamaris* (Fitzgerald, 1977), whereas *P. leiognathi* has been isolated from several leiognathid fish (Boisvert, Chatelain, and Bassot, 1967; Reichelt, Nealson, and Hastings, 1977). Ruby and Nealson (1976) developed a speculative model for the *Photobacterium fischeri–Monocentris japonica* symbiosis based on bacterial physiology, which showed that *P. fischeri* is an ideal symbiont, whereas *B. harveyi* could not serve as a symbiont under the conditions imposed by the host. Bacterial bioluminescene has been reviewed by Hastings and Nealson (1977).

B. POLAR-FLAGELLATED OXIDATIVE RODS

The polar-flagellated oxidative rods known as pseudomonads are a diverse and heterogeneous group, with many taxonomic problems. They are often the dominant bacteria on agar media inoculated with fish material (Colwell, 1962) and water from inshore (Simidu and Aiso, 1962; Sieburth, 1967a) and offshore (Sieburth, 1971) areas. Biochemically, they are very versatile, degrading a number of monomeric, aliphatic, and aromatic compounds as well as polymeric compounds.

The aerobic pseudomonads, as defined by Stanier, Palleroni, and Doudoroff (1966) in their classic taxonomic study of this group, are unicellular gram-negative rods, which are bent or straight on the long axis (never helical), and motile by one or more polar flagella; they are chemoorganotrophs that obtain their energy through respiratory metabolism. Molecular oxygen is used as a terminal oxidant, even in the facultatively anaerobic denitrifiers that have an anaerobic respiratory mechanism and utilize the oxygen in nitrate. Some are facultative chemolithotrophs using hydrogen or carbon monoxide as energy sources. They do not obtain their energy through fermentative or photosynthetic reactions, nor do they form spores, sheaths, or stalks. This definition of aerobic pseudomonads covers many bacteria formerly classified in the genera *Acetomonas*, *Alginomonas*, *Cellumonas*, *Cellvibrio*, *Comamonas*, *Gluconobacter*, *Hydrogenomonas*, *Pseudomonas*, and *Xanthomonas*. These authors consider the splitting of the aerobic pseudomonads into so many genera as excessive and state that they all could be put provisionally into the genus *Pseudomonas* with the same definition, which has now been done in *Bergey's 8th*. Pseudomonad DNA has a 58 to 70 mole % G + C content. Baumann et al. (1972) conducted a similar study of some 200 nonfermentative flagellated marine bacteria lacking cell-associated yellow pigments. The polar flagellated strains with the physiological characteristics of the genus *Pseudomonas* fell into three ranges: 30.5, 43.2 to 48, and 52 to 64.7 mole % G + C. Only those with a 57.8 to 64.7 mole % G + C content could be considered to belong to the genus *Pseudomonas* without widening the range of values. Those with a 52 to 56.4 value, as well as those with a 30.5 value, were unassigned. The strains forming a distinct group with a 43 to 48 mole % G + C content were put into a new genus, *Altermonas*, which is not recognized in *Bergey's 8th*.

The taxonomy of the pseudomonads at the species level is very complex. To get around this problem, some studies have been limited to such groups as the fluorescent pseudomonads, which produce a chemically undefined, water-soluble yellow-green fluorescent pigment, to such species as *Pseudomonas aeruginosa*, which also produce a type-specific phenazine pigment, to the *Pseudomonas fluorescens* group, or to such denitrifiers as *Pseudomonas stutzeri*. The original grouping of marine pseudomonads into four groups (Shewan, Hobbs, and Hodgkiss, 1960) has now apparently been supplanted by sorting by species identification (Hendrie and Shewan, 1966). The current trend is of sorting into groups on the basis of common but independent phenotypic traits through numerical taxonomy and of defining certain taxons by the presence or absence of constant characteristics (Stanier, Palleroni, and Doudoroff, 1966; Baumann et al., 1972). Such an approach has led to a long species list in *Bergey's 8th*, but one on which the species are grouped so that they can be reconciled with the older taxonomy as well as with morphological, physiological, and ecological characteristics and with pigmentation.

C. PERITRICHOUS OXIDATIVE RODS

Nonpigmented marine bacteria with no apparent flagella or with peritrichous flagella have customarily been placed into an *Achromobacter–Alcaligenes* grouping (Hendrie, Hodgkiss, and Shewan, 1964). The original confusion over the microorganisms in this group was due to imprecise definitions because of the lack of authentic cultures and the negative results obtained with these bacteria in the usual biochemical tests used for classification. More careful study and redefinition of the nondistinctive gram-negative rods are simplifying the situation. All nonfermentative bacteria with peritrichous flagella are now considered to be in the genus *Alcaligenes* (Davis et al., 1969) and to account for 10% of the nonfermentative flagellated rods, which are divided into four main species on the basis of their mole % G + C content (Baumann et al., 1972). There are few data on the occurrence of these bacteria with simple nutrient requirements but no distinctive degrading properties, since most studies have reported them as *Achromobacter* or *Achromobacter–Alcaligenes* without specifying their motility. An exception is the study by Simidu and Aiso (1962), which shows that 12 % of its *Achromobacter* group or some 2.5% of its total isolates were motile and could be considered to be *Alcaligenes* by the current definition. The *Achromobacter–Alcaligenes* group accounted for 6.5% of the isolates from Narragansett Bay and the surface film of the open Pacific Ocean, but only 0.6% of the open sea isolates from the Atlantic (Sieburth, 1967a, 1971). The genus *Alcaligenes* is grouped with the pseudomonads in *Bergey's 8th* but as a genus with uncertain affiliations.

D. NONMOTILE RODS AND COCCOIDS

Now that the nonpigmented nonfermentative bacteria with peritrichous flagella have been removed from the defunct *Achromobacter–Alcaligenes* group and put in the redefined genus *Alcaligenes*, the nonflagellated members of the group are unclassified. A new genus, *Acinetobacter*, proposed for nonmotile bacteria previously classified as *Achromobacter* (Brisou and Prevot, 1954), was a possible genus for such bacteria, but as originally described, the genus included obviously unrelated species, no type species was designated within the genus, and the genus was apparently illegitimate. *Achromobacter anitratum* has now been designated as the type species (Brisou, 1957), a type strain proposed, and the genus redescribed (Steel and Cowan, 1964). These bacteria form distinct clusters in numerical analysis studies (Thornley, 1967, 1968). Baumann, Doudoroff, and Stanier (1968) conducted a similar study, which used terrestrial strains but also included many strains enriched from soil and water, and came to a similar conclusion; they modified the description of the genus and proposed *A. calcoaceticus* as the type species (regarding *A. anitratum* as a synonym or variety). These modifications are incorporated into *Bergey's 8th*, in which the genus is grouped with the *Neisseria* and other nonmotile coccobacilli. This group, which grows in a mineral medium with a single carbon source (Warskow and Juni, 1972), has been shown genetically to be a ubiquitous genus (Juni, 1972). *Acinetobacter* from terrestrial sources have a 37 to 47 mole % G + C content (Thornley, 1967; Baumann, Doudoroff, and Stanier, 1968). The marine forms, some 89% of the "*Achromobacter*" from Kamogawa Bay (Simidu and Aiso, 1962), have not been studied as a group in any detail.

Yellow pigmented rods are widely distributed, occurring in some 12 bacterial genera (Ciegler, 1965). In addition to the motile strains falling into *Vibrio* (*Beneckea*) and *Pseudomonas*, many are apparently nonflagellated. The genus *Flavobacterium* in *Bergey's 7th* (Breed, Murray, and Smith, 1957) was inadequately characterized, and the alleged re-isolate of the type species has been proven to be a *Cytophaga* (Hendrie, Mitchell, and Shewan, 1968). The original scheme of Shewan, Hobbs, and Hodgkiss (1960) grouped these bacteria into a *Flavobacterium–Cytophaga* group. Although these microorganisms have not been detected in the open sea (Sieburth, 1971), they can dominate such eutrophic inshore waters as Long Island Sound and Narragansett Bay (Altschuler and Riley, 1967; Sieburth, 1967a), where they appear to be directly associated with phytoplankton blooms (Sieburth, 1968a). A marked seasonal variation of yellow and orange pigmented strains correlates with seasonal changes in solar radiation (Sieburth, 1968b).

The first really good taxonomic look at these bacteria used 61 marine strains classified as *Flavobacterium*. These studies indicated that eight could be ascribed to *Pseudomonas*, "*Vibrio*," or *Corynebacterium*; twenty-one to *Cytophaga*; which left thirty-two designated as *Flavobacterium*, divided into two well-defined groups, A and G (Hayes, 1963; Floodgate and Hayes, 1963). A vexing problem was detecting the gliding bacteria in the genus *Cytophaga*, in which swarming was difficult to induce. The 20 strains in Group G have a 35 to 40 mole % G + C content and are now also considered *Cytophaga* (De Ley and Van Muylem, 1963; Hendrie, Mitchell, and Shewan, 1968). The recommended approach to identifying yellow pigmented rods is to look vigorously for possible relationships to well-defined genera. Provisionally, the *Flavobacterium* may be considered to be microorganisms not ascribable to any other genus, especially *Cytophaga*. These would be the bacteria in the Group A of Floodgate and Hayes (1963), which have a 63 to 64 mole % G + C (De Ley and Van Muylem, 1963). Of the sixty-two isolates examined as *Flavobacterium*, forty-one are now considered to be *Cytophaga* and eight belong to other genera, whereas only sixteen are provisionally considered to belong to the ill-defined genus *Flavobacterium* (Hendrie, Mitchell, and Shewan, 1968). (The persistence of the genus *Flavobacterium* in *Bergey's 8th* according to the old groupings with two distinct G + C ranges is illogical in view of the work cited above. Grouping these bacteria with a respiratory metabolism in with the fermentative enteric bacteria is curious.)

REFERENCES

Allen, R.D., and P. Baumann. 1971. Structure and arrangement of flagella in species of the genus *Beneckea* and *Photobacterium fischeri*. *J. Bacteriol.* 107:295–302.

Altschuler, S.J., and G.A. Riley. 1967. Microbiological studies in Long Island Sound. *Bull. Bingham Oceanogr. Coll.* 19, Art. 2, pp. 81–88.

Anon. 1976. *Beneckea natriegens*, an educational bonanza. *Amer. Soc. Microbiol. News* 42:480–81.

Bain, N., and J.M. Shewan. 1968. Identification of *Aeromonas*, *Vibrio* and related organisms. In: *Identification Methods for Microbiologists* (B.M. Gibbs and D.A. Shapton, eds.), Part B. Academic Press, London and New York, pp. 79–84.

Baumann, L., P. Baumann, M. Mandel, and R.D. Allen. 1972. Taxonomy of aerobic marine eubacteria. *J. Bacteriol.* 110:402–29.

Baumann, P., and L. Baumann. 1977. Biology of the marine enterobacteria: Genera *Beneckea* and *Photobacterium*. *Ann. Rev. Microbiol.* 31:39–61.

Baumann, P., M. Doudoroff, and R.Y. Stanier. 1968. A study of the *Moraxella* group. II. Oxidative-negative species (genus *Acinetobacter*). *J. Bacteriol.* 95:1520–41.

Baumann, P., L. Baumann, and M. Mandel. 1971. Taxonomy of marine bacteria: The genus *Beneckea*. *J. Bacteriol.* 107:268–94.

Baumann, P., L. Baumann, M. Mandel, and R.D. Allen. 1971. Taxonomy of marine bacteria: *Beneckea nigrapulchrituda* sp. n. *J. Bacteriol.* 108:1380–83.

Beijerinck, M.W. 1889a. Le *Photobacterium luminosum*, bactérie lumineuse de la Mer du Nord. *Arch. Néerland Sci. Exact. Natur.* 23:401–15.

Beijerinck, M.W. 1889b. Les Bactéries lumineuses dans leurs rapports avec l'oxygène. *Arch. Néerland Sci. Exact. Natur.* 23:416–27.

Beijerinck, M.W. 1916. Die Leuchtbaktérien der Nordsee im August und Septembre. *Folia Microbiol.* (*Delft*) 4:15–40.

Benecke, W. 1905. Uber *Bacillus chitinovorus*, einen Chitin zersetzenden Spaltpilz. *Bot. Zeits.* 63:227–42.

Benecke, W. 1933. Baktériologie des Meeres. *Abderhalden's Handb. biol. Arbeitsmethoden* IX Abt., T. 5:717–854.

Boisvert, H., R. Chatelain, and J.M. Bassot. 1967. Étude d'un *Photobacterium* isolé de l'organe lumineux de poissons Leiognathidae. *Ann. Inst. Pasteur* 112:520–24.

Bousfield, I.J. 1972. A taxonomic study of some coryneform bacteria. *J. gen. Microbiol.* 71:441–55.

Breed, R.S., E.G.D. Murray, and N.R. Smith (eds.). 1957. *Bergey's Manual of Determinative Bacteriology* (7th Ed.). Williams & Wilkins., Baltimore, Md., 1094 p.

Brisou, J. 1957. Comparison a l'étude de la systématique des Pseudomonadaceae. *Ann. Inst. Pasteur* 93:397–404.

Brisou, J., and A-R. Prévot. 1954. Études de systématique bactérienne. X. Révision des espèces réunies dans le genre *Achromobacter*. *Ann. Inst. Pasteur* 86:722–28.

Brisou, J., C. Tysset, Y. de Rautlin de La Roy, R. Curcier, and R. Moreau. 1964. Étude sur la chitinolyse en milieu marin. *Ann. Inst. Pasteur* 106:469–78.

Campbell, L.L., and O.B. Williams. 1951. A study of chitin-decomposing micro-organisms of marine origin. *J. gen Microbiol.* 5:894–905.

Ciegler, A. 1965. Microbial carotogenesis. *Adv. Appl. Microbiol.* 7:1–34.

Citarella, R.V., and R.R. Colwell. 1970. Polyphasic taxonomy of the genus *Vibrio*: Polynucleotide sequence relationships among selected *Vibrio* species. *J. Bacteriol.* 104:434–42.

Colwell, R.R. 1962. The bacterial flora of Puget Sound fish. *J. Appl. Bacteriol.* 25:147–58.

Colwell, R.R. 1970. Polyphasic taxonomy of the genus *Vibrio*: Numerical taxonomy of *Vibrio cholerae*, *Vibrio parahaemolyticus* and related *Vibrio* species. *J. Bacteriol.* 104:410–33.

Colwell, R.R., and R.Y. Morita. 1964. Reisolation and emendation of description of *Vibrio marinus* (Russell) Ford. *J. Bacteriol.* 88:831–37.

Cosenza, B.J., and J.D. Buck. 1966. Simple device for enumeration and isolation of luminescent bacterial colonies. *Appl. Microbiol.* 14:692.

Costerton, J.W., J.W. Ingram, and K.J. Cheng. 1974. Structure and function of the cell envelope of gram-negative bacteria. *Bacteriol Rev.* 38:87–110.

Davis, D.H., M. Doudoroff, R.Y. Stanier, and M. Mandel. 1969. Proposal to reject the genus *Hydrogenomonas*: Taxonomic implications. *Int. J. Syst. Bacteriol.* 19:375–90.

De Ley, J., and J. Van Muylem. 1963. Some applications of deoxyribonucleic acid base composition in bacterial taxonomy. *Antonie van Leeuwenhoek* 29:344–58.

Felter, R.A., R.R. Colwell and G.B. Chapman. 1969. Morphology and round body formation in *Vibrio marinus*. *J. Bacteriol.* 99:326–35.

Fischer, B. 1887. Bacteriologische Untersuchungen auf einer Reise nach Westindien. II. Ueber einen lichtentwickelunden im Meerwasser gefundenen Spaltpilz. *Zeits. Hyg.* 2:54–95.

Fischer, B. 1888. Ueber einen neuen lichtentwickelunden *Bacillus*. *Centralbl. Bakteriol. Parasit.* 3:105–8, 137–41.

Fischer, B. 1894. Die Bakterien des Meeres nach den Untersuchungen der Plankton-Expedition unter gleichzeitiger Berücksichtigung einiger älterer und neuerer Untersuchungen. *Ergeb. Plankton-Expedition Humboldt-Stiftung.* 4:1–83.

Fitzgerald, J.M. 1977. Classification of luminous bacteria from the light organ of the Australian pinecone fish, *Cleidopus gloriamaris*. *Arch. Microbiol.* 112:153–56.

Floodgate, G.D., and P.R. Hayes. 1963. The Adansonian taxonomy of some yellow pigmented marine bacteria. *J. gen. Microbiol.* 30:237–44.

Forsberg, C.W., J.W. Costerton, and R.A. MacLeod. 1970. Quantitation, chemical characteristics and ultrastructure of the three outer cell wall layers of a gram-negative bacterium. *J. Bacteriol.* 104:1354–68.

Haneda, Y., and F.I. Tsuji. 1971. Light production in the luminous fishes *Photoblepharon* and *Anomalops* from the Banda Islands. *Science* 173:143–45.

Harvey, E.N. 1921. A fish with a luminous organ, designed for the growth of luminous bacteria. *Science* 53:314–15.

Harvey, E.N. 1952. *Bioluminescence*. Academic Press, London and New York, 649 p.

Harvey, E.N. 1957. *A History of Luminescence.* Mem. 44, Amer. Philosoph. Soc., Philadelphia, 692 p.

Hastings, J.W. 1971. Light to hide by: Ventral luminescence to camouflage the silhouette. *Science* 173:1016–17.

Hastings, J.W., and G. Mitchell. 1971. Endosymbiotic bioluminescent bacteria from the light organ of pony fish. *Biol. Bull.* 141:261–68.

Hastings, J.W., and K.H. Nealson. 1977. Bacterial bioluminescence. *Ann. Rev. Microbiol.* 31:549–95.

Hayes, P.R. 1963. Studies on marine flavobacteria. *J. gen. Microbiol.* 30:1–19.

Hendrie, M.S., and J.M. Shewan. 1966. The identification of certain *Pseudomonas* species. In: *Identification Methods for Microbiologists* (B.M. Gibbs and F.A. Skinner, eds.), Part A. Academic Press, London and New York, pp. 1–7.

Hendrie, M.S., W. Hodgkiss, and J.M. Shewan. 1964. Considerations on organisms of the *Achromobacter–Alcaligenes* group. *Ann. Inst. Pasteur de Lille* 15:43–59.

Hendrie, M.S., T.G. Mitchell, and J.M. Shewan. 1968. The identification of yellow-pigmented rods. In: *Identification Methods for Microbiologists* (B.M. Gibbs and D.A. Shapton, eds.), Part B. Academic Press, London and New York, pp. 67–78.

Hendrie, M.S., W. Hodgkiss, and J.M. Shewan. 1970. The identification, taxonomy and classification of luminous bacteria. *J. gen. Microbiol.* 64:151–69.

Henneberg, G. 1969. *Pictorial Atlas of Pathogenic Microorganisms*, Vol. 3. Gustav Fischer Verlag, Stuttgart.

Hugh, R., and E. Leifson. 1953. The taxonomic significance of fermentative versus oxidative metabolism of carbohydrates by various gram-negative bacteria. *J. Bacteriol.* 66:24–26.

Jannasch, H.W., and G.E. Jones. 1959. Bacterial populations in sea water as determined by different methods of enumeration. *Limnol. Oceanogr.* 4:128–39.

Johnson, R.M., M.E. Katarski, and W.P. Weisrock. 1968. Correlation of taxonomic criteria for a collection of marine bacteria. *Appl. Microbiol.* 16:708–13.

Juni, E. 1972. Interspecies transformation of *Acinetobacter*: Genetic evidence for a ubiquitous genus. *J. Bacteriol.* 112:917–31.

Kadota, H. 1959. Cellulose decomposing bacteria in the sea. In: *Marine Boring and Fouling Organisms* (D.L. Ray, ed.). University of Washington Press, Seattle, pp. 332–40.

Kaneko, T., and R.R. Colwell. 1973. Ecology of *Vibrio parahaemolyticus* in Chesapeake Bay. *J. Bacteriol.* 113:24–32.

Lear, D.W., Jr. 1963. Occurrence and significance of chitinoclastic bacteria in pelagic waters and zooplankton. In: *Marine Microbiology* (C.H. Oppenheimer, ed.). C.C. Thomas, Springfield, Ill., pp. 594–610.

Leifson, E. 1960. *Atlas of Bacterial Flagellation.* Academic Press, London and New York.

Leifson, E. 1963. Determination of carbohydrate metabolism of marine bacteria. *J. Bacteriol.* 85:1183–84.

Leifson, E., B.J. Cosenza, R. Murchelano, and R.C. Cleverdon. 1964. Motile marine bacteria. I. Techniques, ecology, and general characteristics. *J. Bacteriol.* 87:652–66.

Levin, M.A., R.E. Wolke, and V.J. Cabelli. 1972. *Vibrio anguillarum* as a cause of disease in winter flounder (*Pseudopleuronectes americanus*). *Can. J. Microbiol.* 18:1585–92.

Liston, J. 1957. The occurrence and distribution of bacterial types on flatfish. *J. gen. Microbiol.* 16:205–16.

Makemson, J.C. 1973. Control of *in vivo* luminescence in psychrophilic marine photobacterium. *Arch. Mikrobiol.* 93:347–58.

Martin, J.H. 1965. Phytoplankton-zooplankton relationships in Narragansett Bay. *Limnol. Oceanogr.* 10:185–91.

McElroy, W.D. 1961. Bacterial luminescence. In: *The Bacteria, Vol. 2 Metabolism* (I.C. Gunsalus and R.Y. Stanier, eds.). Academic Press, London and New York, pp. 479–508.

Morin, J.G., A. Harrington, K. Nealson, N. Krieger, T.O. Baldwin, and J.W. Hastings. 1975. A light for all reasons: Versatility in the behavioral repertoire of the flashlight fish. *Science* 190:74–76.

Morita, R.Y., and R.D. Haight. 1964. Temperature effects on the growth of an obligate psychrophilic marine bacterium. *Limnol. Oceanogr.* 9:103–6.

Nealson, K.H., and J.W. Hastings. 1972. The inhibition of bacterial luciferase by mixed function oxidase inhibitors. *J. Biol. Chem.* 247:888–94.

Nealson, K.H., T. Platt, and J.W. Hastings. 1970. Cellular control of the synthesis and activity of the bacterial luminescent system. *J. Bacteriol.* 104:313–22.

Pierantoni, U. 1918. I microorganismi fisiologici e la luminescenza degli animali. *Scientia* (*Bologna*) 23:102–10.

Pollard, E.C. 1967. The degree of organization in the bacterial cell. In: *Formation and Fate of Cell Organelles* (K.B. Warren, ed.), Vol. 6, Symp. Int. Soc. Cell Biol., New York, pp. 291–303.

Reichelt, J.L., and P. Baumann. 1973. Taxonomy of the marine luminous bacteria. *Arch. Mikrobiol.* 94:283–330.

Reichelt, J.L., and P. Baumann. 1975. *Photobacterium mandapamensis* Hendrie et al., a later synonym of *Photobacterium leiognathi* Boisvert et al. *J. Syst. Bacteriol.* 25:208–9.

Reichelt, J.L., K.H. Nealson, and J.W. Hastings. 1977. The specificity of symbiosis: Pony fish and luminescent bacteria. *Arch. Mikrobiol.* 112:157–61.

Rogers, H.J. 1970. Bacterial growth and the cell envelope. *Bacteriol. Rev.* 34:194–204.

Ruby, E.G., and K.H. Nealson. 1976. Symbiotic association of *Photobacterium fischeri* with the marine luminous fish *Monocentris japonica*: A model of symbiosis based on bacterial studies. *Biol. Bull.* 151:574–86.

Ruby, E.G., and K.H. Nealson. 1978. Seasonal changes in the species composition of luminous bacteria in nearshore seawater. *Limnol. Oceanogr.* 23:530–33.

Rüger, H-J. 1972. Taxonomic studies of marine bacteria from the North Sea sediments: Genus *Vibrio*. Proposal for combining marine vibrio-like bacteria into a new genus *Marinovibrio*. *Veröff. Inst. Meeresforsch. Bremerh.* 13:239–354.

Santo, L., T.J. Leighton, and R.H. Doi. 1972. Ultrastructural analysis of sporulation in a conditional serine protease mutant of *Bacillus subtilis*. *J. Bacteriol.* 111:248–53.

Shewan, J.M. 1966. I. Some factors affecting the bacterial flora of marine fish. *Med. Norske Vet.*, Suppl. No. 11, pp. 1–12.

Shewan, J.M., G. Hobbs, and W. Hodgkiss. 1960. A determinative scheme for the identification of certain genera of Gram-negative bacteria, with special reference to the Pseudomonadaceae. *J. Appl. Bacteriol.* 23:379–90.

Sieburth, J.McN. 1964. Polymorphism of a marine bacterium (*Arthrobacter*) as a function of multiple temperature optima and nutrition. *In: Proc. Symp. Exp. Mar. Ecol.*, Occas. Publ. No. 2, Grad. Sch. Oceanogr., Univ. R.I., Kingston, R.I., pp. 11–16.

Sieburth, J.McN. 1967a. Seasonal selection of estuarine bacteria by water temperature. *J. exp. mar. Biol. Ecol.* 1:98–121.

Sieburth, J.McN. 1967b. Inhibition and agglutination of arthrobacters by pseudomonads. *J. Bacteriol.* 93:1911–16.

Sieburth, J.McN. 1968a. The influence of algal antibiosis on the ecology of marine microorganisms. In: *Advances in Microbiology of the Sea* (M.R. Droop and E.J.F. Wood, eds.). Academic Press, London and New York, pp. 63–94.

Sieburth, J.McN. 1968b. Observations on bacteria planktonic in Narragansett Bay, Rhode Island; a résumé. *Bull. Misaki Biol. Inst., Kyoto Univ.* 12:49–64.

Sieburth, J.McN. 1971. Distribution and activity of oceanic bacteria. *Deep-Sea Res.* 18:1111–21.

Sieburth, J.McN., P-J. Willis, K.M. Johnson, C.M. Burney, D.M. Lavoie, K.R. Hinga, D.A. Caron, F.W. French III, P.W. Johnson, and P.G. Davis. 1976. Dissolved organic matter and heterotrophic microneuston in the surface microlayers of the North Atlantic. *Science* 194:1415–18.

Simidu, U., and K. Aiso. 1962. Occurrence and distribution of heterotrophic bacteria in sea water from Kamogawa Bay. *Bull. Jap. Soc. Sci. Fish.* 28:1133–41.

Spencer, R. 1955. The taxonomy of certain luminous bacteria. *J. gen. Microbiol.* 13:111–18.

Stanier, R.Y., N.J. Palleroni, and M. Doudoroff. 1966. The aerobic pseudomonads: A taxonomic study. *J. gen. Microbiol.* 43:159–271.

Steel, K.J., and S.T. Cowan. 1964. Le rattachement de *Bacterium anitratum*, *Moraxella lwoffi*, *Bacillus mallei* et *Haemophilus parapertussis* au genre *Acinetobacter* Bisou et Prévot. *Ann. Inst. Pasteur* (*Paris*) 106:479–83.

Thornley, M.J. 1967. A taxonomic study of *Acinetobacter* and related genera. *J. gen. Microbiol.* 49:211–57.

Thornley, M.J. 1968. Properties of *Acinetobacter* and related genera. In: *Identification Methods for Microbiologists* (B.M. Gibbs and D.A. Shapton, eds.), Part B. Academic Press, London and New York, pp. 29–50.

Tsiklinsky. 1908. *La flore microbienne dans les régions du Pôle sud. Expédition Antarctique Française (1903–1905).* Masson et Cie., Paris, 34 p.

Tubiash, H.S., R.R. Colwell, and R. Sakazaki. 1970. Marine vibrios associated with bacillary necrosis, a disease of larval and juvenile bivalve molluscs. *J. Bacteriol.* 103:272–73.

Van Gool, A.P., and N. Nanninga. 1971. Fracture faces in the cell envelope of *Escherichia coli. J. Bacteriol.* 108:474–81.

Veldkamp, H. 1970. Saprophytic coryneform bacteria. *Ann. Rev. Microbiol.* 24:209–40.

Warskow, A.L., and E. Juni. 1972. Nutritional requirements of *Acinetobacter* strains isolated from soil, water and sewage. *J. Bacteriol.* 112:1014–16.

CHAPTER 16

Gram-Negative Distinctive Epibacteria

These morphologically distinct bacteria have a marked affinity for surfaces upon which some glide or oscillate back and forth with a corkscrew or wiggling motion, whereas others attach themselves and form stalks, filaments, or an extensive hyphal network between cells. Although these are some of the unusual bacteria, they are not rarities and they are important in the degradation of organic debris wherever it accumulates.

A. GLIDING BACTERIA

All the gliding (and flexing) bacteria are now grouped together in the order Cytophagales. These bacteria are either unicellular filaments or pluricellular filaments, which are slender, flexing, and aflagellate. Morphologically, they range from the thin unicellular cells of *Cytophaga* and *Flexibacter* that are differentiated by their ability to utilize polysaccharides to the pluricellular filaments containing sulfur inclusions, such as the individual trichomes of *Beggiatoa* or the bundles of similar filaments of *Thioploca*, which are visible to the naked eye. Their gliding motility, at speeds up to 2 to 10 μm/minute, is similar to that of the cyanobacteria (Chapter 8,B) to which some of them may be related (Pringsheim, 1949). The filaments are usually 0.5 to 2.0 μm in diameter, with most exceeding 10 μm in length; some are 50 to 100 μm or longer. Although gliding bacteria are morphologically distinct, they are often too few in number to detect during the direct microscopic examination of the natural materials from which they can be isolated. The flexing forms are conspicuous in the nearshore bacterioplankton. Members of the cytophagales are often inhibited in the rich organic media used to cultivate the dominant gram-negative bacteria (Lewin, 1969), but they can be obtained by streaking fresh material on a low nutrient medium, such as a 1% seawater agar containing 0.02% tryptone for nutrients and 0.01% actidione to inhibit the growth and spreading of myxomycetes, amoebae and diatoms. The

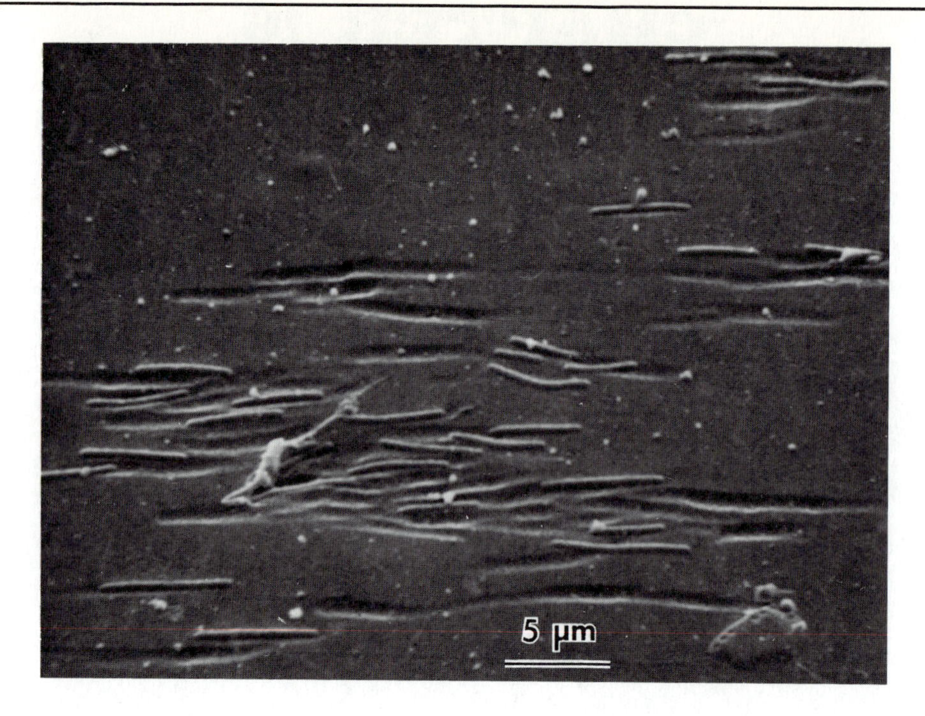

A

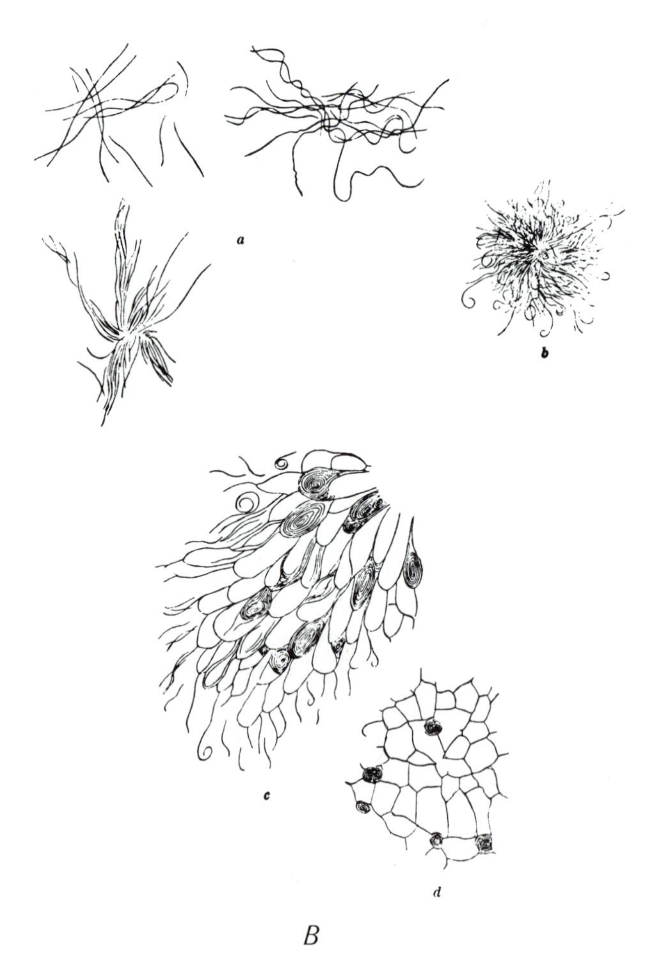

B

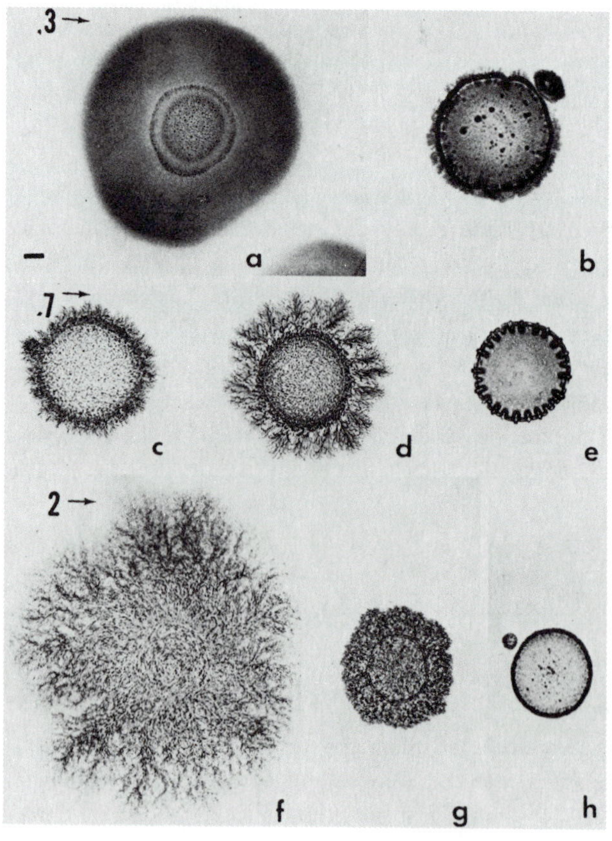

C

PLATE 16-1. Colonial growth characteristics for members of the Cytophagaceae. (*A*) Cytophaga-like cells digesting cellulose-ester dialysis sac material after 48 hours in 19 °C running seawater. (*B*) *Flexibacter (Microscilla) marina* showing (a) young colonies on agar (175 X); (b) older colony (100 X); (c) edge of old colony (50 X), and (d) sparse growth on seawater agar with Benecke solution but no organic additions (100 X). (*C*) Influence of agar concentration on the colonial appearance of mutants of *Cytophaga columnaris*. Appearance on 0.3, 0.7, and 2.0% agars is shown for mutant 1-R-43 (a, c, and f), NG-1 (a, d, and g), and SP-10 (b, e, and h) (bar = 1.3 mm). (*A*) from Dennis M. Lavoie, 1975, Master's Thesis, University of Rhode Island; (*B*) from E. Pringsheim, 1951, *J. gen. Microbiol.* 5:124–49. (*C*) from J. Glaser and J.L. Pate, 1973, *Arch. Mikrobiol.* 93:295–309.

agar plates are observed at 2- to 4-day intervals for 2 to 3 weeks for the characteristic colonies with a rhizoid or fimbriate margin (Lewin and Lounsbery, 1969), as shown in Plate 16-1, and for the whorls that some species tend to form. Different genera appear to occupy different habitats, and the microorganisms obtained will depend in some measure upon the habitat and the materials sampled. Only the dominant aerobic gliding bacteria will be discussed here. Anaerobic *Cytophaga*-like bacteria, which can form sizable populations in anaerobic sediments, have been cultivated in the laboratories of Eimhjellen and of the writer but have not been described in the literature. A convenient way to discuss the gliding bacteria is to take a middle ground between a splitter (Lewin, 1969) and a grouper (Colwell, 1969) and to discuss the unicells in the three genera *Cytophaga*, *Flexibacter*, and *Saprospira* and the pluricells in the five genera *Leucothrix*, *Herpetosiphon*, *Beggiatoa*, *Thioploca*, and *Flexithrix*.

1. Unicellular Gliding Bacteria

The genus *Cytophaga* contains bacteria, usually under 30 μm in length, that have marked polysaccharolytic properties, digesting cellulose, agar, and alginate in addition to other carbohydrates. Small delicate pigmented colonies within a concave depression of an agar plate are characteristic of these bacteria. The biology of terrestrial cytophagas, especially in relation to cellulose decomposition, is covered in a masterly monograph by Stanier (1942). It is difficult to demonstrate cellulose activity using filter paper, but dialysis tubing is easily colonized and degraded and makes a suitable surface for scanning electron microscopy of cellulose digesters (Plate 16-1). Cytophagas accounted for 0 to 33% of the total aerobic cellulose-digesting bacteria in Maizuru and Hiroshima bays, while vibrios dominated this microflora (Kadota, 1959). The abundance of *Cytophaga* appeared to correlate with the amount of suspended matter. The population of the *Cytophaga–Flavobacterium* group in Narragansett Bay appeared to correlate with the phytoplankton (Sieburth, 1968). Although Kadota (1959) found that the cytophagas were less abundant than the other bacteria in the water, they were enriched on cotton nets, even when there were no detectable cytophagas in the water, and constituted some 22 to 40% of the net microflora, which Kadota attributed to their tendency for "tenacious attachment." Kadota reported populations of 1 to 100 cells/ml of seawater and 10 to 10,000 cells/g of mud in inshore environments. The genus *Sporocytophaga*, which is indistinguishable from *Cytophaga* except for the occurrence of a resting stage, the microcyst, should probably be classified with the cytophagas, which also produce aberrant microcyst-like cells under certain cultural conditions (Lewin, 1969; Mitchell, Hendrie, and Shewan, 1969). The taxonomy of the genus *Cytophaga* is discussed by Lewin (1969) and by Mitchell, Hendrie, and Shewan (1969), as well as in *Bergey's 8th*.

The genus *Flexibacter* is distinguished by being somewhat carbohydrate independent, and it uses organic nitrogen sources; the genus *Cytophaga* is carbohydrate (polysaccharide) dependent and mainly uses inorganic nitrogen sources (Lewin and Lounsbery, 1969; Soriano, 1973). The genus *Flexibacter* was proposed by Soriano (1945) for terrestrial isolates, the genus *Microscilla* by Pringsheim (1951) for marine isolates (see Plate 16-1). The genus *Flexibacter* in *Bergey's 8th* includes both *Flexibacter* and *Microscilla* species, although Lewin (1969) and Lewin and Lounsbery (1969) considered the former mainly a freshwater genus and the latter marine. Colwell (1969) concluded that, except for a few distinct species, most strains would fit equally well into the genera *Cytophaga*, *Flexibacter*, or *Microscilla*.

The genus *Saprospira* includes helical apochlorotic gliding cells encountered on decaying algae and littoral mud and may be the colorless equivalent of the helical cyanobacterium *Spirulina*. Although they may appear to be spirochetes at first glance, their gliding speed of up to 2 μm/second contrasts markedly to the wriggling swimming speed of up to 50 μm/second achieved by

PLATE 16-2. The gliding bacterium *Saprospira grandis*. Phase-contrast photomicrographs showing (*A*) its spirilla-like helical filaments when first isolated and (*B*) appearance after 2 years of culture (bar = 10 μm). Transmission electron micrographs of shadowed preparations of particles associated with *S. grandis*, showing (*C*) the ever-present rhapidosomes (with hollow and wicked forms), which are not lysogenic (bar = 100 nm), and (*D*) lysogenic phage (same magnification as *C*). (*A,B*) from R.A. Lewin, 1962, *Can. J. Microbiol.* 8:555–63; (*C*) from D.L. Correll and R.A. Lewin, 1964, *Can. J. Microbiol.* 10:63–74; (*D*) from R.A. Lewin, D.M. Crothers, D.L. Correll, and B.E. Reimann, 1964, *Can. J. Microbiol.* 10:75–85, reproduced by permission of the National Research Council of Canada.

spirochetes (Lewin, 1962). The cells are usually 50 to 500 μm in length, but the cross-walls that occur at 1 to 2.5 μm intervals are only seen by phase-contrast microscopy in dying cultures. Most strains liquefy gelatin but not agar and do not attack carbohydrates. The marine species are *Saprospira grandis* and *Saprospira toviformis* (Lewin and Mandel, 1970). The characteristic helical morphology that distinguishes *Saprospira* from the *Flexibacter–Microscilla* group can be lost on culture (Lewin, 1962) (see Plate 16-2). A striking feature of all 19 strains of *Saprospira* studied so far (Lewin and Mandel, 1970) is that after vigorous growth for several days they spontaneously lyse to liberate rhapidosomes, which are non-

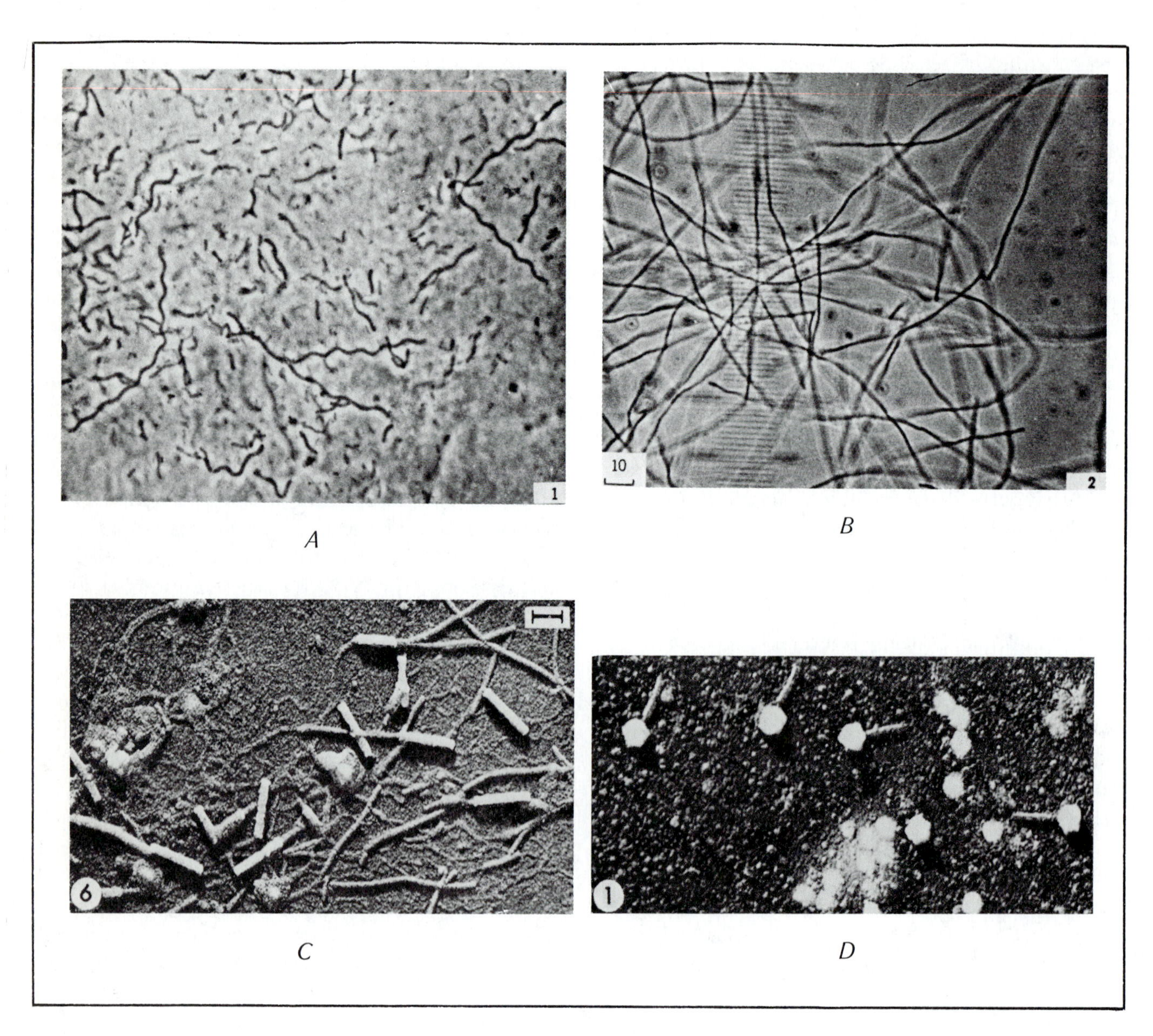

infective rods that resemble some plant viruses (Correll and Lewin, 1964). One strain of *S. grandis* is also susceptible to premature lysis by a phage (Lewin et al., 1964). These two particles are compared in Plate 16-2. The taxonomic position of this group is discussed by Soriano (1973) and Lewin (1969).

2. Pluricellular Gliding Bacteria

Leucothrix mucor, a distinctive filamentous microorganism discovered in Danish tide pools by Oersted in 1844, was subsequently described as a new species by other investigators until it was rediscovered in enrichments of the decomposing thallus of the green alga *Ulva lactuca* on the California coast by Harold and Stanier (1955). They described how the tapering filaments, which are attached to a substrate by a holdfast near the air–water interface, terminate in ovoid cells called gonidia. The gonidia separate, settle on a surface, and then creep together to form aggregates, with characteristic rosettes, from which new tufts of filaments are formed. Harold and Stanier also noted the characteristic thumbprint appearance of young microcolonies observed at 50 to 100 X (see Plate 16-3). Subsequently, Pringsheim (1957) reported the isolation of *Leucothrix* from similar infusions of the brown alga *Fucus vesiculosus* from the Baltic Sea. Lewin (1959) readily obtained isolates at Woods Hole, Massachusetts by streaking freshly collected species of the red alga *Callithamnion* on a seawater agar medium.

When either stagnant algal enrichments or fresh algae are used as an inoculum for obtaining *Leucothrix* isolates on complex agar media, the *Leucothrix* colonies are rapidly overgrown by the non-distinctive epibacteria just discussed. The author has solved this problem by removing the polysaccharide impurities of agarose in commercial agars (see chapter 7, 8): The sheets of 1.5% agar, up to 2 cm thick, are immersed in a 10 X volume of cold filtered seawater and refrigerated for several days; the seawater is changed several times. Then 2 g of mannitol and 1 g of ammonium chloride/liter of melted seawater agar are added. In this medium, the colonial development of the non-distinctive epibacteria becomes as slow as that of *Leucothrix*, permitting the easy detection and picking of isolated thumbprint colonies of *Leucothrix*.

Brock (1966) reported that the habitat of *L. mucor* was red and filamentous green algae upon which it is commonly epiphytic in cold and temperate waters, but not in warmer seawater (Kelly and Brock, 1969). Reports on the infestation and killing of fish eggs by filamentous microorganisms (Dannevig, 1919; Oppenheimer, 1955) prompted a study that showed that *L. mucor* readily infests fish eggs and the eggs and larvae of benthic crustaceans, which can lead to high mortalities in intensive aquaculture, and that *L. mucor* is widespread in tropical waters (Johnson et al., 1971). Using Nomarski interference optics and thin wet mount preparations to examine algal materials, Bland and Brock (1973) observed that *L. mucor* was most prevalent in the intertidal zone, especially on the red algae *Bangia* and *Porphyra*. When polypropylene strips with a laboratory-induced *L. mucor* flora were exposed in the marine environment, the bacteria apparently obtained little nourishment from the dissolved organic matter in seawater, indicating a requirement for a living host. Scanning electron microscopy has permitted the examination of the thick and large algal species to show that *L. mucor* is as abundant on the larger brown algae as it is on the thinner and filamentous species of red and green algae (Tootle, 1974; Sieburth et al., 1974). Scanning electron microscopy of such surfaces as the exoskeleton of lobster larvae has also shown that, contrary to Brock, gonidia do aggregate to form rosettes under natural conditions (Plate 16-3; Sieburth, 1975). Autoradiographic studies have shown that preferential growth does not occur at the holdfast or apical portions of the filament, but rather the whole filament in culture, whereas *in-situ* samples from some areas had filaments with regions where the cells were inexplicably dormant (Brock, 1967). The flexible filaments of *Leucothrix* can form knots (Brock, 1964). Although knots have been seen in natural materials, they are quite rare; knotting is apparently an idiosyncracy of certain strains under certain cultural conditions that may lead to aberrant forms (Snellen and Raj, 1970). Metabolic studies have shown that *L. mucor* degrades simple carbohydrates, intermediary metabolites, and amino acids by the Entner–Doudoroff pathway, in conjunction with the tricarboxylic acid cycle (Raj, 1967; Biggins and Dietrich, 1968). The DNA base compositions of 49 $\pm$1.1 mole % G + C indicated that strains from a number of different geographical areas were homogeneous (Brock and Mandel, 1966). The literature on *L. mucor* has been summarized in the illustrated review by Raj (1977a).

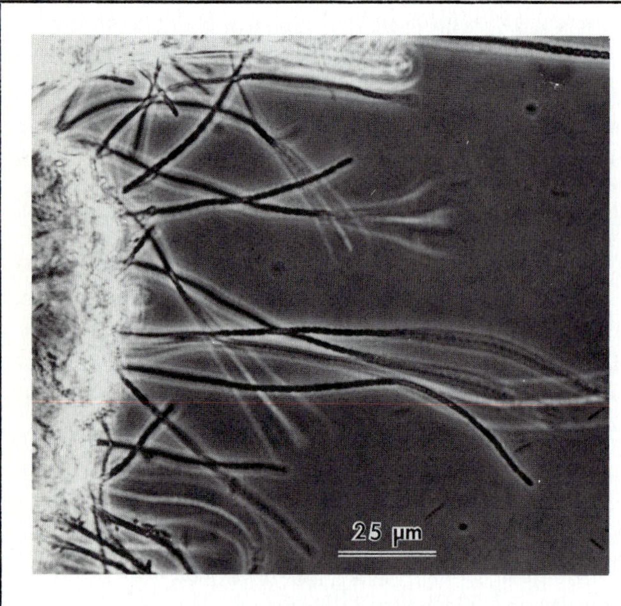

A

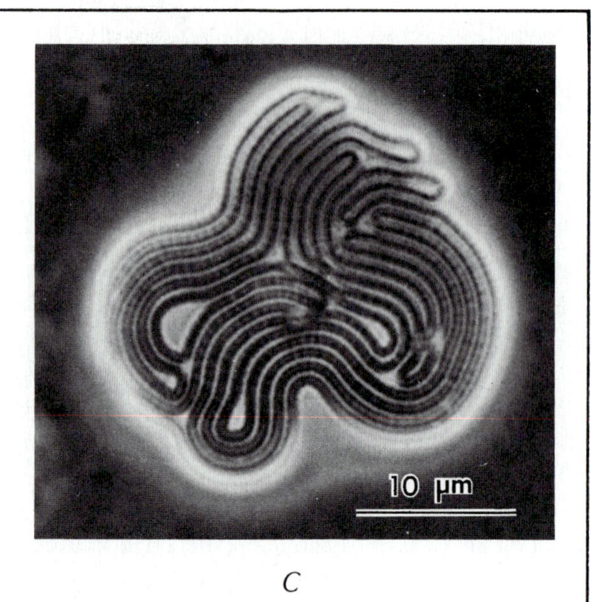

C

B

PLATE 16-3. The filamentous bacterium *Leucothrix mucor*, showing its distinctive morphology in nature (as observed on the surface of the larvae of the lobster *Homarus americanus*) and in culture. (*A*) Phase-contrast photomicrograph of flexible *Leucothrix* filaments with terminal gonidia, which will detach to colonize a new surface. (*B*) Scanning electron micrograph showing recently settled gonidia that aggregate by gliding motility into rosettes, form holdfasts, and divide to form tufts of filaments. (*C*) Typical thumbprint appearance of a young (24-hour) colony on nutrient agar. (*A,C*) photomicrographs courtesy of Paul W. Johnson; (*B*) from J.McN. Sieburth, 1975, *Microbial Seascapes*, University Park Press, Baltimore, Md.

A *Leucothrix* look-alike is *Thiothrix* (Harold and Stanier, 1955; Pringsheim, 1957), which, in such marine sulfide-containing habitats as rotting accumulations of seaweeds, produces cells with internal sulfur granules; these filaments resemble nonmotile *Beggiatoa* filaments. *Leucothrix* and *Thiothrix* are often found together in hydrogen sulfide-deficient habitats where *Thiothrix* is reported to lose its sulfur inclusions and thus to resemble *Leucothrix*. *Thiothrix*, described by Winogradsky, is possibly an obligate autotroph belonging with the colorless sulfur bacteria (Chapter 9,B) and not with the gliding bacteria (this chapter). Its freshwater habitat is the flowing waters of sulfur springs and sewage treatment plants.

Beggiatoa are unattached gliding trichomes (pluricellular filaments) that contain refractile sulfur granules. These filaments aggregate to form visible white mats on muds and surfaces, where a dissolved oxygen content of less than 0.16 mg/liter permits the accumulation of hydrogen sulfide (Lackey, 1961a). *Beggiatoa* is rather uncommon in freshwater, occurring mainly in sulfur springs and one type of sewage treatment plant and flooded soils like rice paddies and marshes. It is widespread though sparse in marine environments, with aggregations being associated with decaying vegetation, in which populations exceeding 20,000 filaments/ml of sediment have been found (Lackey and Clendenning, 1965). Generally *Beggiatoa* will grow on any organic substrate that is undergoing decomposition with the liberation of sulfides, as on compacted solid wastes in the sea and in aquaculture facilities where overfeeding occurs. This organism was originally believed to be a sulfide-dependent chemolithotroph and was used by Winogradsky to develop his concept of chemolithotrophy. It has been isolated from enrichment cultures in the absence of sulfides (Faust and Wolfe, 1961) and found to possess hydrogen sulfide-insensitive enzymes capable of utilizing acetate (Burton, Morita, and Miller, 1966). Pringsheim (1967) has also found that, although some strains of *Beggiatoa* can grow autotrophically, they do grow better in the presence of low levels of acetate. *Beggiatoa* may be neither an obligate chemolithotroph nor a heterotroph but a mixotroph as defined by Rittenberg (1969). The addition of the enzyme catalase to agar media, which destroys peroxides, increased the yield of *Beggiatoa* cells and extends viability from 1 week to 2 months (Burton and Morita, 1964). The hydrogen sulfide requirement may be as much for the inhibition of peroxide formation as for sulfur oxidation.

James B. Lackey, who has examined marine environments in California, Florida, Rhode Island, Massachusetts, and Plymouth, England, as well as a spectrum of inland aquatic habitats for *Beggiatoa*, recognizes the six species described in *Bergey's 7th* (Lackey, 1961a,b; Lackey and Lackey, 1963; Lackey, Lackey, and Morgan, 1965). He has confirmed *Beggiatoa gigantea* and *Beggiatoa mirabilis* as marine species, whereas the remaining four *Beggiatoa* species are both freshwater and marine species. These species, however, are distinguished solely on the basis of filament width (see Plate 16-4), and environmental conditions can markedly affect such morphology. The validity of the six species found in the marine environment, and continued in *Bergey's 8th*, will not be proven until they have been established in pure culture.

Filamentous gliding bacteria with sulfur inclusions similar to those in *Beggiatoa*, but which occur as many parallel or braided filaments in the form of bundles or fascicles within a common sheath (Plate 16-4), were first described from the sediments of the Rhine by Lauterborn (1907), who created the genus *Thioploca*. Wislouch (1912) described a second species from brackish as well as from freshwater as *Thioploca ingrica*, and on the basis of the greenish blue color of the trichomes (filaments of cylindrical cells), he postulated a relationship to such cyanobacteria as *Microcoleus* of the Oscillatoriaceae. Accumulations of bacteria dominated by *Thioploca* have been extracted and shown to have a spectrum typical of chlorophyll *a* which further suggests a relationship to the cyanobacteria (Gallardo, 1977) although the pigments could be due to other microorganisms in these natural assemblages. A comparison of the ultrastructure of *T. ingrica* with that of strains of *Beggiatoa* showed a relationship between the two, but distinct differences in

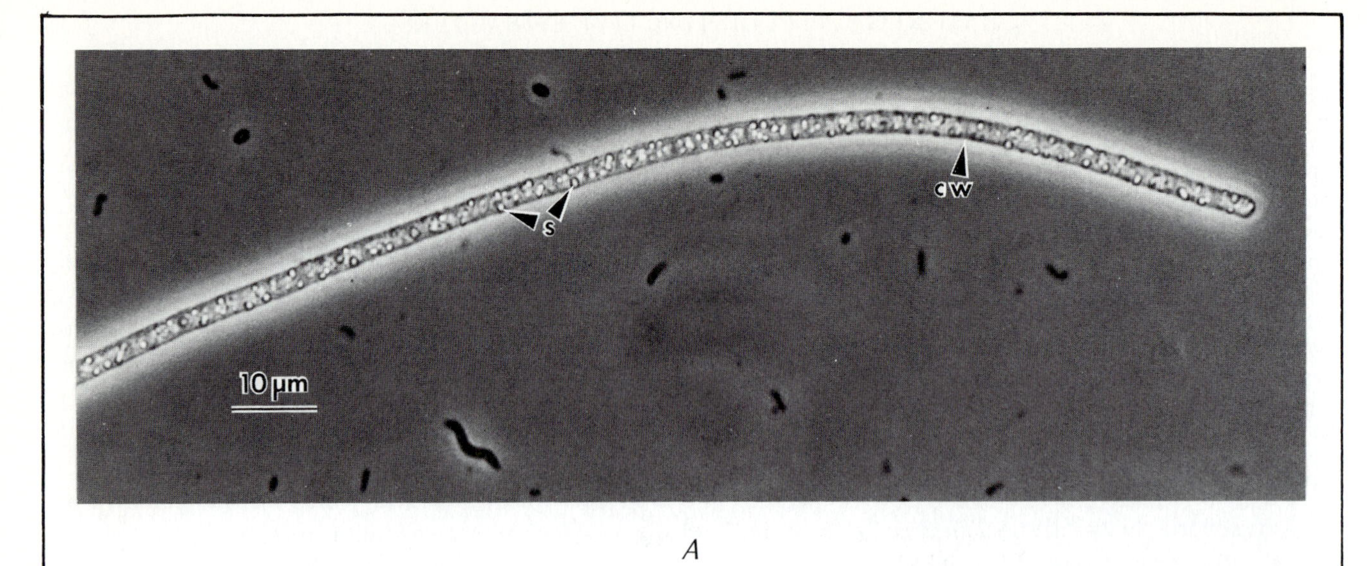

A

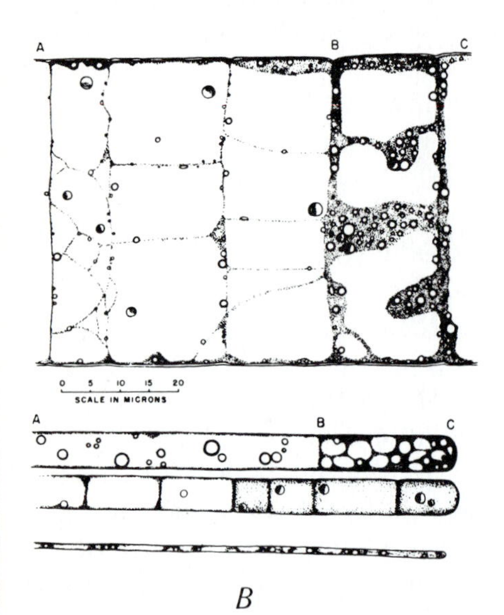

B

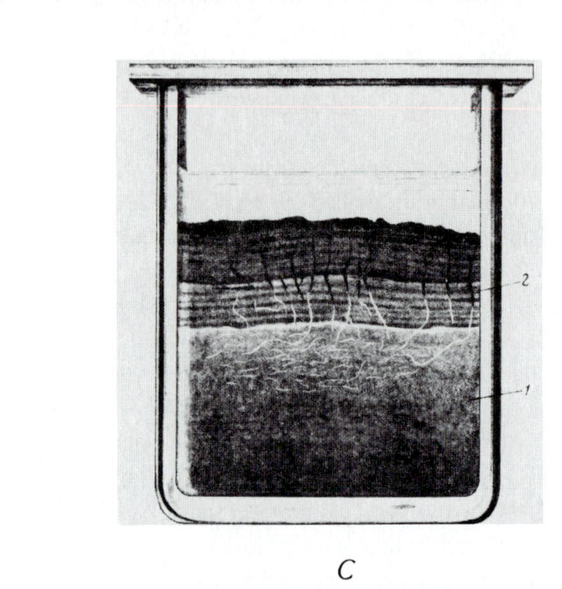

C

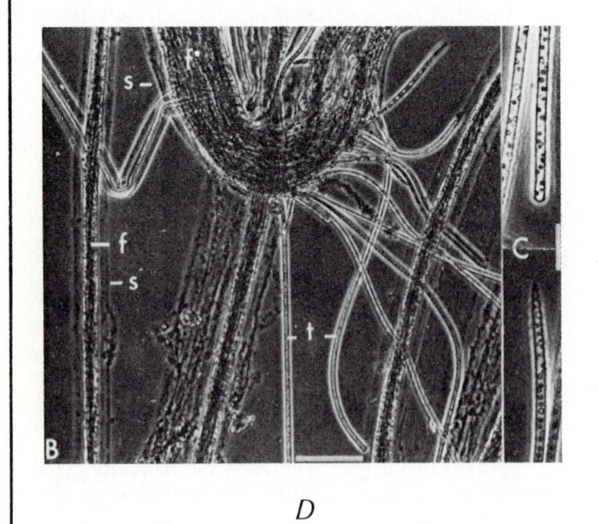

D

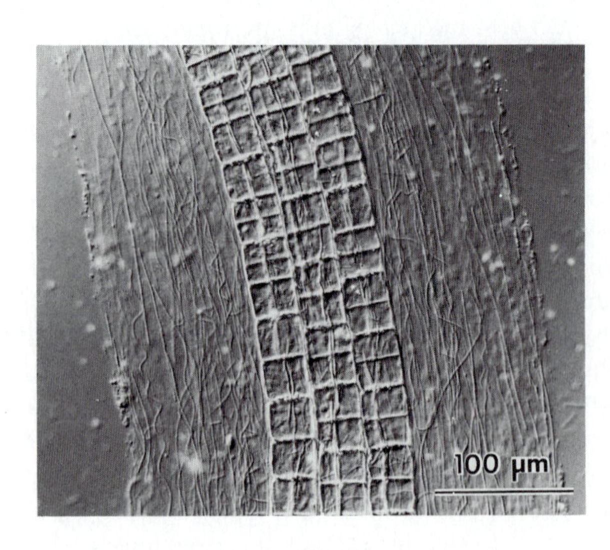

E

PLATE 16-4. Gliding bacteria with sulfur inclusions, as they occur in sulfide muds. (*A*) Phase-contrast photomicrograph of a *Beggiatoa* filament, showing intracellular sulfur granules (s) and cross-walls (c-w) from a naturally enriched microcosm. (*B*) *Beggiatoa* species in a diagrammatic representation, showing their great size variation in the single-occurring filaments: *B. gigantea* (top) to *B. leptomitiformis* (bottom) in transverse (A-B) and surface (B-C) views. (*C*) Fascicles of *Thioploca* filaments with their vertical orientation in the upper oxidized horizon (2) and a horizontal orientation in the lower, sulfide-containing reducing horizon (1) of lake mud. (*D*) Fascicle of *Thioploca* filament (f) enclosed in a sheath (s), free filament (t) emerging from a break in sheath. Lake Erie material, phase contrast (bar in B = 50 μm) (*E*) Three *Thioploca*-like filaments in sheath from bottom sediments collected by Gallardo (1977) off the coast of Chile. (Nomarski differential interference contrast.) (*A,E*) photomicrographs courtesy of Paul W. Johnson; (*B*) from J.B. Lackey, 1961, *Water & Sewage Works* 108:29–31; (*C*) from B.V. Perfil'ev, 1965, in *Applied Capillary Microscopy*, Plenum New York, pp. 1–8; (*D*) from S. Maier, 1974, in *Bergey's Manual of Determinative Bacteriology* (8th Ed.), Williams & Wilkins, Baltimore, Md.

both cell wall and cytoplasm were noted (Maier and Murray, 1965). Koppe (1924) described an additional two species, and *Bergey's 8th* distinguishes all four species by filament size, since they have never been cultured.

The individual cylindrical cells 1 to 8 μm in length occur in filaments 1 to 9 μm in width, whereas the sheath-bound fascicles, which are 50 to 160 μm wide, grow to several centimeters in length. Perfil'ev (1965) used capillary microscopy to observe the distribution of *Thioploca* between the oxidized and reduced layers of mud from Lake Kinchezero. Within the sheath, some filaments move to one end while others move to the opposite end, a movement related to the intake of nutrients. Perfil'ev observed that the fascicles were always oriented with the upper ends vertical and extended into the oxidizing horizon, whereas the lower portions of the fascicles penetrated into the lower reducing horizon where they bend and lie almost horizontally. The hydrogen sulfide of the lower reducing horizon is obtained by the filaments of *Thioploca* that then glide into the upper part of the fascicle in the oxidized layer where it is oxidized to provide free energy. This migration is similar to that occurring in *Thiovulum majus* (Chapter 9,B). The sheath of *Thioploca*, which is colorless in the reduced layer, becomes darker in the oxygenated horizon due to the deposition of iron and manganese oxides.

Our only knowledge of the distribution of *Thioploca* fascicles in the open sea has been provided by Gallardo (1977). A bacterial flora dominated by fascicles of *Thioploca* is often widely distributed in sulfide-containing bottom sediments at depths of 50 to 280 m, where they are covered by the low-oxygen waters of the Peru–Chile subsurface countercurrent. They are so thick that the fishermen who bring them up in their nets call them *Estopa* (oakum—loose hemp or jute fibers used for caulking wooden ships). Associated with the fascicles of *Thioploca* are the filaments of oscillatoria-like cyanobacteria and of other gliding bacteria. A high concentration of *Spirilla* associated with the filaments were observed within the sheath of *Thioploca* samples provided by Gallardo and examined in the author's laboratory. This is reminiscent of the spirochetes and spirilla associated with *Beggiatoa* filaments. (The ecology of this microorganism and its impact on the economy of this particular environment would be an intriguing study.)

Other pluricellular gliding bacteria with hyaline sheaths revealed by phase-contrast microscopy have been placed in two genera: the unbranched forms in the genus *Herpetosiphon* (Lewin, 1970b) and the forms with false branches in the genus *Flexithrix* (Lewin, 1970a). They are classified with the family Cytophagaceae and the family Beggiatoaceae, respectively. These bacteria appear to be apochlorotic forms of the filamentous cyanobacterium *Lyngbya*. They are also isolated from littoral sand and mud and have a colonial morphology on agar media similar to that of other members of the Cytophagaceae.

The Cytophagales have had a varied and long taxonomic history, the unicells being classified as non-fruiting members of the myxobacteria in *Bergey's 7th*, having been placed in the ill-defined and possibly nonexistent genus *Flavobacterium* (Hayes, 1963; Floodgate and Hayes, 1963; Mitchell, Hendrie, and Shewan, 1969) and even in the Spirochaetales (Lewin, 1962), whereas the pluricellular forms were placed in the trichome-forming orders Beggiatoales and Chlamydobacteriales. Soriano (1945, 1947) proposed that these forms with nonrigid flexing cells scattered among different orders might logically be placed in a single order, Flexibacteriales. This proposal was rejected by Pringsheim (1949, 1951) and nothing further was done until Lewin studied these microorganisms (Lewin, 1962), updated the order Flexibacteriales with Soriano (Soriano and Lewin, 1965), and with his colleagues, conducted an intensive and productive program of isolation and characterization of many members of this group (Lewin and Lounsbery, 1969; Mandel and Lewin, 1969; Fager, 1969; Lewin, 1969). Lewin and his colleagues have been more concerned with characterization at the genus and species levels, but Lewin (1969) did offer a provisional taxonomic scheme for the flexibacteria with the marine families Cytophagaceae (Flexibacteriaceae), Leucothrichaceae, and Beggiatoacheae. His work has had a major impact on the revision of the taxonomy of the gliding bacteria in *Bergey's 8th*. A simplified key to the Cytophagales is given in Table 16-1.

B. APPENDAGED BACTERIA

Appendaged bacteria, also called prosthecate bacteria, are distinguished from other bacteria by their distinctive shapes and complex life cycles. Prosthecae are semi-rigid appendages, such as stalks and hyphae, that are extensions of the cell, being bound by a membrane or a membrane and the cell wall. They differ structurally from pili or fimbrae (fine hairlike protuberances associated with sexuality and used for interface attachment by gram-negative bacteria), flagella, and the acellular appendages produced by the genera *Gallionella* and *Nevskia* (Henrici and Johnson, 1935; Staley, 1973b). The known appendaged bacteria in the sea are usually attached to both living and nonliving surfaces (see Plate 16-5) and even to the sea–air interface. Since most natural marine surfaces are opaque, or too thick to be studied by light microscopy, these bacteria have been detected only occasionally, and they have not received the attention they deserve even many decades after their accidental discovery by Henrici (1933). The recent discovery of new forms, by transmission electron microscopy (Nikitin and Kuznetsov, 1967; Staley, 1968) has refocused attention on this group (Staley, 1973a). The

Table 16-1. A simplified key to the genera of gliding bacteria in the order Cytophagales.

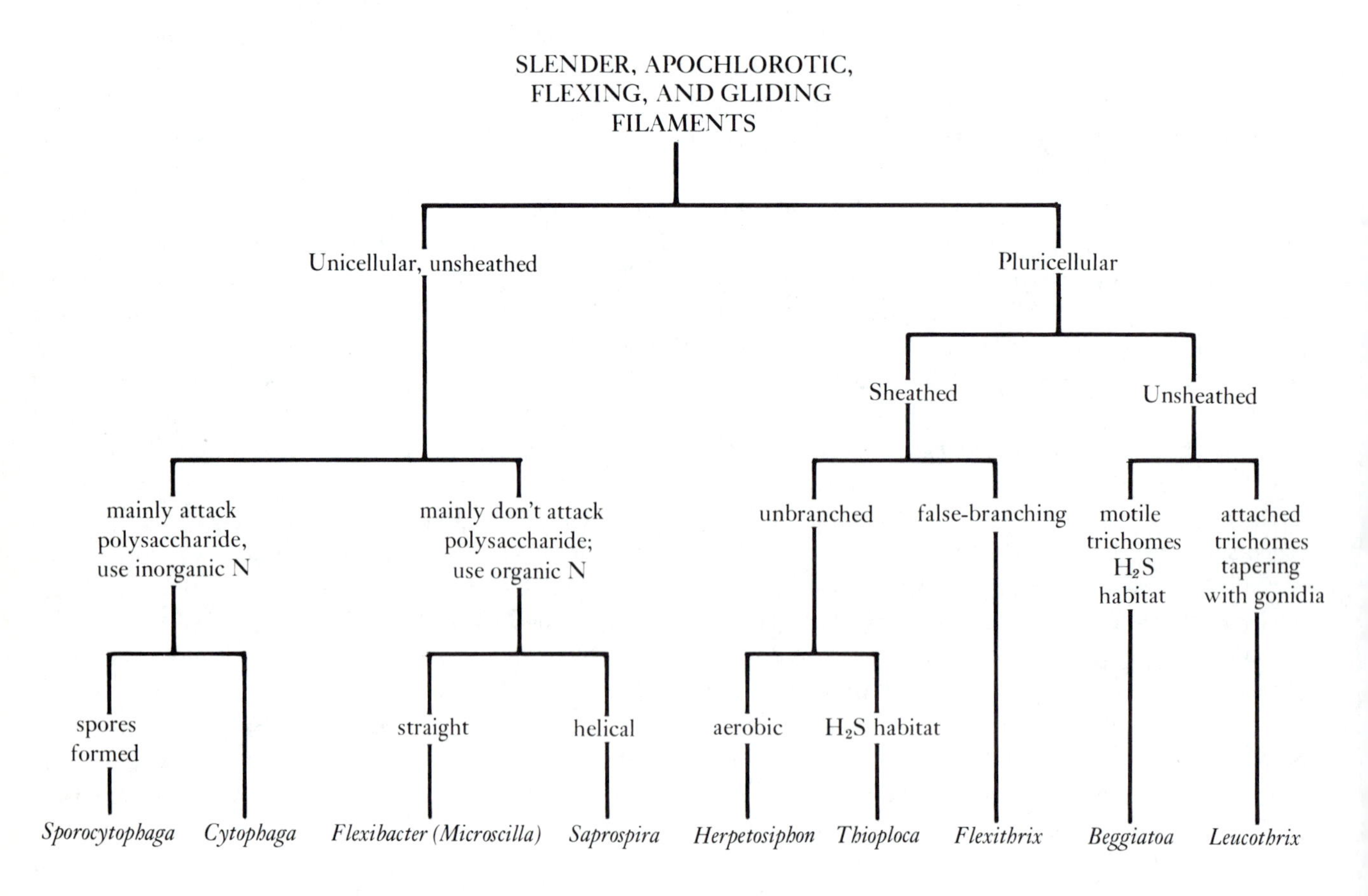

scanning electron microscope is also an ideal tool for studying these forms in the epimicrobiota that grow on immersed marine surfaces (Sieburth, 1975).

Henrici (1933) discovered the genus *Caulobacter* because he was curious about the algae growing on the walls of an aquarium in his laboratory. He placed some microscope slides in the water in the expectation that he could make permanent preparations showing the algae *in situ*. Examination of the slides after a week's immersion revealed, in addition to the algae, a variety of bacteria including some with an unusual morphology. This experiment was repeated with similar results in other aquaria, a lily pond in the university greenhouse, and eventually in the waters of Lake Alexander, Minnesota (Henrici, 1933; *Anon.*, 1960). In further study, Henrici compared the stalked bacteria with those previously described in the literature and created the genus *Caulobacter* for the stalked forms that could occur in characteristic rosettes (Henrici and Johnson, 1935). ZoBell and Allen (1935) studied the bacteria growing on submerged slides from the pier at the Scripps Institution of Oceanography, but they were reluctant to describe or name the microorganisms solely on a morphological basis. It was not until the 1950s, when *Caulobacter* turned up in electron-microscope preparations as contaminants from distilled water, after having been interpreted as invertebrate chromosomes (Houwink, 1952; Kandler, Zehender and Huber, 1954), that they began to receive some attention (Poindexter, 1964). The first unambiguous evidence for the occurrence of caulobacters in the sea was in the electron micrographs prepared from offshore water samples by Jannasch and Jones (1960). The dearth of information on this interesting group was rectified when Jeanne Poindexter used the enrichment and culture techniques of Houwink (1951) to obtain a culture collection from a number of sources, including seawater, which she used in her superb monograph on the characteristics and classification of *Caulobacter* (Poindexter, 1964). Other marine isolates have also been obtained (Leifson et al., 1964). Caulobacters are enriched by the addition of 0.01% casein hydrolysate, peptone, or yeast extract to the stored water samples until the population is sufficient to permit the streaking of plates of 1% seawater agar containing 0.05% peptone. *Caulobacter* is receiving attention as a model system for studying the molecular basis of cell differentiation (Shapiro et al., 1971). No ecological studies on the distribution of *Caulobacter in situ* have been conducted in the marine environment, but the continuing examination of immersed surfaces by scanning electron microscopy in the author's laboratory indicates that they are not rare and are sometimes present in goodly number (Plate 16-5). *Caulobacter*-like cells are also seen in negatively stained preparations of selectively filtered bacterioplankton observed by transmission electron microscopy.

Poindexter (1964) concluded that the caulobacters are a somewhat fastidious group of aerobic chemoheterotrophs, with a relatively slow rate of growth. Many strains require growth factors, and the number of compounds that can be used as a principal carbon source is usually quite limited. She believed that the reason they are widespread and can compete with bacteria, such as the pseudomonads, despite these limitations, is that they have an attaching organ, the holdfast. By attaching to surfaces exposed to a stream of nutrients or to other organisms, including microorganisms, they become ectocommensals and are ensured a continuing supply of nutrients. The classification of *Caulobacter* species is based primarily on the morphology of the cell and secondarily on requirements for growth. All the marine species characterized so far are bacteroid (rod-shaped) forms. The genus *Caulobacter* has a 64 to 67 mole % G + C content, which overlaps with that of the genus *Pseudomonas* with which they share many characteristics.

The appendages (hyphae) of *Hyphomicrobium* were first observed in soil samples enriched with ammonium salts, where they were thought to be a new form of autotrophic nitrifiers (Rullman, 1897). The microorganism was named *Hyphomicrobium vulgare* (Stutzer and Hartleb, 1898). The cells on the hyphae were thought to be spores, and the reproduction of these microorganisms through budding was not recognized until much later (Kingma-Boltjes, 1936). Their heterotrophic utilization of single-carbon compounds was investigated by Large, Peel, and Quayle (1961). The first report of a marine *Hyphomicrobium* repeated the story of its original discovery when it was found as a contaminant on another autotroph (Guillard and Watson, 1962). This time it was a contaminant in a culture of diatoms from the Sargasso Sea, which was believed to be bacteria-free, since the bacterium did not grow in the high nutrient media used to detect bacterial contaminants. Other *Hyphomicrobium* strains have subsequently been isolated from seawater (Hirsch and Conti, 1964; Leifson, 1964; Hirsch, 1968). These microorganisms appear to be quite widespread on marine surfaces; they require mineral

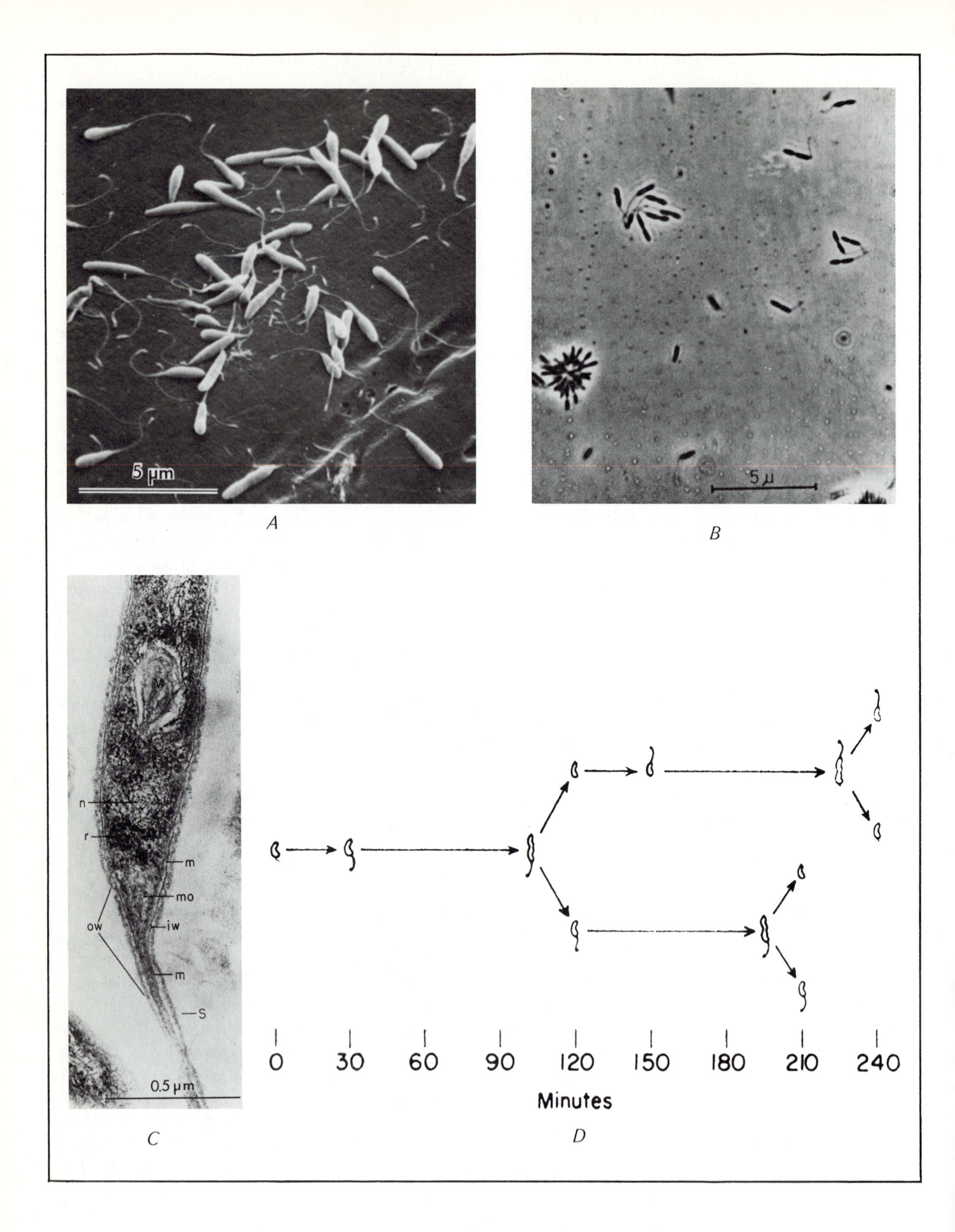
5 μm
A
5 μ
B
M
n
r
m
mo
ow
iw
m
s
0.5 μm
C
0
30
60
90
120
150
180
210
240
Minutes
D

PLATE 16-5. The stalked bacterium *Caulobacter*. (*A*) Scanning electron micrograph of bacteroid form, associated with a culture of the centric diatom *Skeletonema costatum*, growing on the inner surface of dialysis sac placed in running seawater. (*B*) Phase-contrast photomicrograph of a bacteroid strain. (*C*) Transmission electron micrograph of a thin section of *Caulobacter bacteroides* showing that the wavy outer layer of cell wall (OW) is continuous with the smooth outer envelope of the stalk (S) but that the inner wall (iw) is not clearly visible in the stalk. Cytoplasm contains a large mesosome (M), nucleoplasm (n), and ribosomes (r). A membranous organelle (mo) completely fills the stalked end and continues into the core of the stalk. (*D*) Diagram of behavior within a clone of *Caulobacter*, showing the dimorphism and time sequence of stalked and flagellated cells. (*A*) Electron micrograph courtesy of Paul W. Johnson; (*B*) from J.S. Poindexter, 1964, *Bacteriol. Revs.* 28:231–95; (*C*) from J.S. Poindexter and G. Cohen-Bazire, 1964, *J. Cell Biol.* 23:587–607; (*D*) from J.L. Stove and R.Y. Stanier, 1962, *Nature* 196:1189–92.

nutrients for growth and actively deposit iron epicellularly (Hirsch, 1968). They are conspicuous on surfaces examined by scanning electron microscopy (Sieburth, 1975; Plate 16-6). Hirsch and Rheinheimer (1968) also concluded that hyphomicrobia are widespread in marine environments, although they are not usually detected until after prolonged incubation of enrichment cultures.

Hirsch and Conti (1964) used protein content and dry weight as an index of growth and found that methanol, urea, and methylamine could serve as carbon sources in the presence of calcium ions, trace elements, and either nitrite or ammonium salts. In addition, hyphomicrobia can grow "oligocarbophilically" without added carbon sources, presumably obtaining carbon from volatile organic compounds in the atmosphere. This ability accounts for the reports of its apparent utilization of such substrates as methane and oxalate. An analysis of the DNA base composition of 70 hyphomicrobia strains showed a range of 59.2 to 66.8 mole % G + C; it also indicated that they could be divided into three groups, with *H. neptunium* occurring in Group 1. The DNA base-sequence homologies of a number of prosthecate and budding bacteria show that there is no homology (genetic relatedness) between the genera and that, within the hyphomicrobia isolated from seawater, freshwater, and soil, there is very little or no homology (Moore and Hirsch, 1972). The iron-depositing (epicellular) strains of hyphomicrobia from the marine environment form multiple hyphae in the rod-shaped swarmer cells after settlement (Hirsch, 1968) and would thus belong to the genus *Pedomicrobium* (Aristovskaya, 1961) and not *Hyphomicrobium*, which has only one hypha. But hyphomicrobia from manganese deposits in freshwater pipelines (Tyler and Marshall, 1967a,b) showed an astonishing degree of pleomorphism that, under varying cultural conditions, span both genera (Tyler and Marshall, 1967c). For this reason, Tyler and Marshall concluded that *Pedomicrobium* is not a valid genus and that both forms should be considered *Hyphomicrobium*.

The relationship of the appendaged cells of *Caulobacter* and *Hyphomicrobium* is shown in Plates 16-5 and 16-6. The cytoplasmic membrane and cell wall are continuous between the cell and stalk or hyphae in both types of cell. The morphological differences between these microorganisms in nature are apparent at once, but in culture, the two appear to be the same at first glance. Cells prepared in the same manner are shown in Plates 16-5 and 16-6 to indicate the basic differences in cell form and reproductive cycle. The life cycle of *Caulobacter* (Plate 16-5,*D*) shows that the mature stalked cell elongates to form a daughter cell at the pole opposite the stalk. The daughter cell forms a flagellum at its free end, and after binary fission and separation, this swarmer cell settles on a surface where the flagellum then enlarges to form a stalk. The process, which can take only 2 hours in culture (Stove and Stanier, 1962), is then repeated. A diagrammatic representation of the *Hyphomicrobium* life cycle in a marine strain, *H. neptunium*, is shown in Plate 16-6,*B*. The mature mother cell may also form branching hyphae before additional bud formation occurs, to produce the structures seen by scanning electron microscopy.

A number of other prosthecate and similar budding forms occur in freshwater. These include such genera as *Blastobacter*, *Pasteuria*, *Planctomyces*, *Blastocaulis*, *Metallogenium*, *Asticcaulis* (an eccentric caulobacter), *Prosthecomicrobium* and *Ancalomicrobium* (Staley, 1973a,b; *Bergey's 8th*). The examination of water samples by transmission electron microscopy reveals the existence of star-shaped knobby and spiny forms, which were apparently missed by light microscopy (Nikitin and Kuznetsov, 1967; Staley, 1968). An example of these forms is shown in Plate 16-7. Some of the marine forms with projections appear to be prosthecate bacteria, such as *Prosthecomicrobium*. *Planctomyces*, a common inhabitant of lake water, which has never been isolated (Staley, 1973a), has also been reported to occur in seawater (Hirsch and Rheinheimer, 1968). A study on the occurrence of prosthecate forms in freshwater, using both direct and viable count-

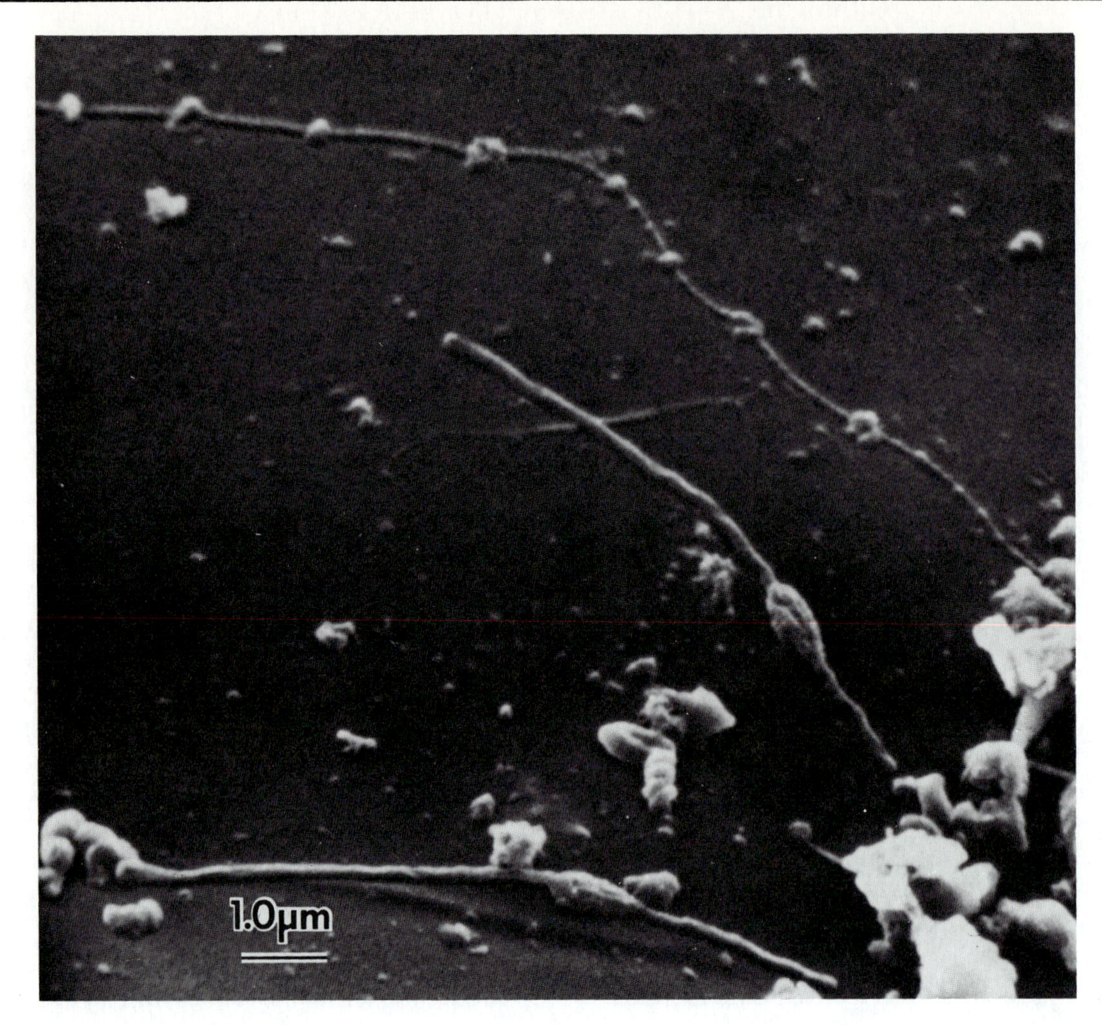
1.0µm

A

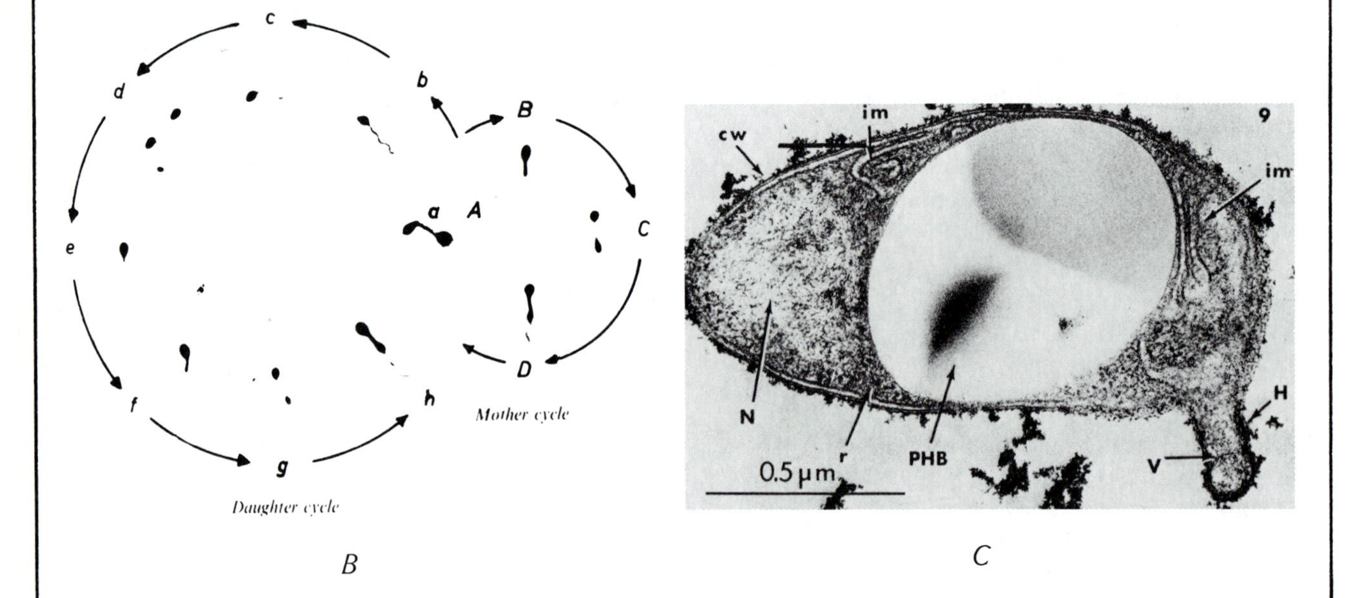
c
b
B
d
a
A
C
e
D
f
h
g
Mother cycle
Daughter cycle
im
9
cw
im
N
r
PHB
H
V
0.5 µm

B

C

PLATE 16-6. Budding bacteria of the genus *Hyphomicrobium*. (*A*) Scanning electron micrograph of budding hyphae of *Hyphomicrobium*-like microorganisms on a gel coat of reinforced fiber glass immersed in running seawater. (*B*) Flagella-stained preparations arranged in sequence to show the stages in the mother (A-D) and daughter (a-h) cycles of *Hyphomicrobium* (*Hyphomonas*) *neptunium*. (*C*) Transmission electron micrograph of a thin section of *Hyphomicrobium vulgare*, showing that the cytoplasmic membrane of the rod portion of the cell and the hypha (H) are continuous. The main structural elements are the cell wall (cw), the internal membrane (im), the nucleoplasm (N), ribosomes (r), a poly beta hydroxybutyrate granule (PHB), and a membrane-bounded vesicle within the hypha (V). (*A*) electron micrograph courtesy of David A. Caron; (*B*) from E. Liefson, 1964, *Antonie van Leeuwenhoek* 30:249–56; (*C*) from S.F. Conti and P. Hirsch, 1965, *J. Bacteriol.* 89:503–12.

ing techniques, indicated that these microorganisms accounted for 0.62 to 1.1% of the total microflora of water samples. Caulobacters were dominant. Future studies should also include populations on representative surfaces, the usual habitat for many of these microorganisms. Non-prosthecate tubular and spinelike appendages have been reported on marine pseudomonads by Moll and Ahrens (1970) and McGregor-Shaw et al. (1973) (Plate 16-7). Such appendages occurring outside the cell wall are present only on nonflagellated cells, and their nature and function are unknown. These forms have been seen in electron micrographs of natural populations of bacterioplankton from Narragansett Bay.

C. CURVED BACTERIA

Two forms of curved bacteria with uncertain affiliations are *Bdellovibrio*, which parasitizes and lyses bacteria, and *Flectobacillus*, which is an odd horseshoe- to doughnut-shaped bacterium with easily recognizable, distinctive forms. The only reason for grouping them together here is that they are both curved forms, with no direct relationship to other curved forms (vibrios and spirillas).

The lysis of bacteria by bacteriophage, the host-specific bacterial viruses, is well known. The lysis of bacteria from a number of habitats has erroneously been ascribed to bacteriophage, until Stolp and Petzold (1962) detected the existence of obligately parasitic bacteria with lytic activity. This group of bacteria, which attack, invade, and lyse other bacteria, with some degree of host specificity, was named *Bdellovibrio* by Stolp and Starr (1963). The bdellovibrios isolated from soil, rivers, fishponds, and sewage lagoons have been studied in regard to the developmental process, physiology, and host dependence and independence; they have been reviewed by Shilo (1969), Starr and Seidler (1971), and Stolp (1973). Although the presence of bdellovibrio in the sea is reported in the literature (Shilo, 1966; Mitchell, Yanofsky and Jannasch, 1967; Mitchell and Morris, 1969), there was no indication of specificity for these marine bacteria until the extensive study by Taylor et al. (1974) on bdellovibrios in the coastal waters of Oahu, Hawaii. A similar study on isolates from the Israeli littoral of the Mediterranean Sea was reported by Marbach, Varon, and Shilo (1976). These two studies showed that, except for salt requirements and a somewhat lower minimal temperature for growth, the marine isolates have the same appearance and activities as the terrestrial isolates, but with distinctive host specificities and a lower mole % G + C range. All the marine isolates are vibroid gram-negative cells, 0.22 to 0.45 μm in width and 0.8 to 1.4 μm in length (Plate 16-8,*A*). By phase-contrast microscopy, they show the jerky motion typical of bdellovibrios. For cultivation, an inoculum concentrated by filtration/centrifugation is plated by the double layer technique on a lawn of suitable host strain. Pure cultures are obtained by a series of single plaque transfers. In the two-membered culture systems, the typical developmental cycle of attachment, penetration, and intraperiplasmic growth takes place (Plate 16-8,*B*), with the host cell taking on the rounded bdelloplast form while, within the cell, the bdellovibrio undergoes elongation and multiple fission to form several daughter cells that swarm from the remnants of the lysed host cell. As with freshwater bdellovibrios, marine isolates die off very rapidly after the host cell is lysed. Host-independent or axenic cultures that live on the usual complex organic agar media are obtained from mutants that develop on opaque layers of autoclaved cells; they are detected as colonies surrounded by a clear zone. These facultative strains derived from host-independent mutants are chemoorganotrophs with a respiratory, rather than a fermentative metabolism, unlike the genus *Vibrio*. But two of the three recognized species in *Bergey's 8th* are sensitive to vibriostat 0/129. The sensitivity of marine isolates to 0/129 has not been reported. The host-dependent bdellovibrios have been estimated to occur as

PLATE 16-7. Appendages of nonstalked and nonbudding bacteria. (*A*) Flagella-stained marine bacteria, showing spine-like projections (2,700 X). Transmission electron micrographs of a negatively stained freshwater species of *Prosthecomicrobium*, showing prosthecate projections and electron-transparent gas vacuoles (*B*); of a shadowed preparation of pseudomonad isolate from decomposing seaweeds, showing randomly arranged hollow, tubular, spinelike nonprosthecate appendages (*C*); and a negatively stained agrobacterium-like isolate from the Elbe estuary, showing identical projections. (*D*). (*A*) from E. Leifson, B.J. Cosenza, R. Murchelano, and R.C. Cleverdon, 1964, *J. Bacteriol.* 87:652–66; (*B*) from J.T. Staley, 1968, *J. Bacteriol.* 95:1921–42; (*C*) from J.B. McGregor-Shaw, K.B. Easterbrook, and R.P. McBride, 1973, *Int. J. Syst. Bacteriol.* 23:267–70; (*D*) from G. Moll and R. Ahrens, 1970, *Arch. Mikrobiol.* 70:361–68.

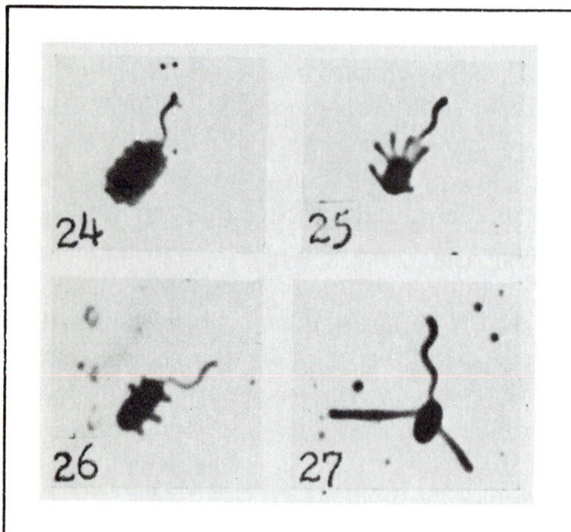

A

1.0 µm

B

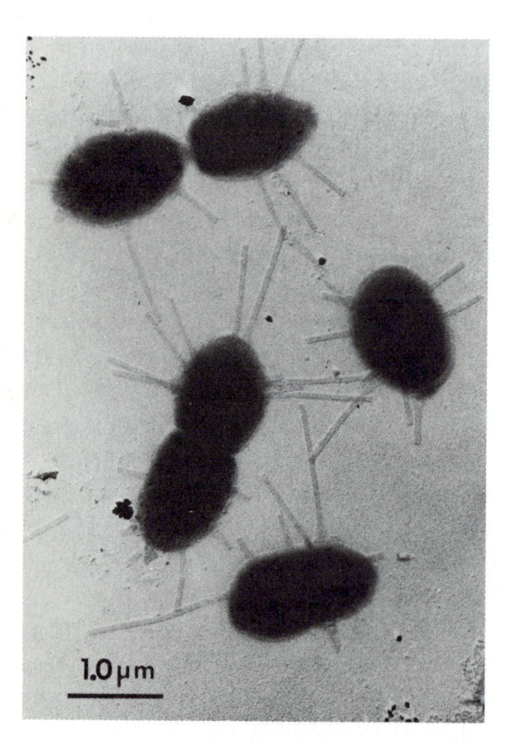

C

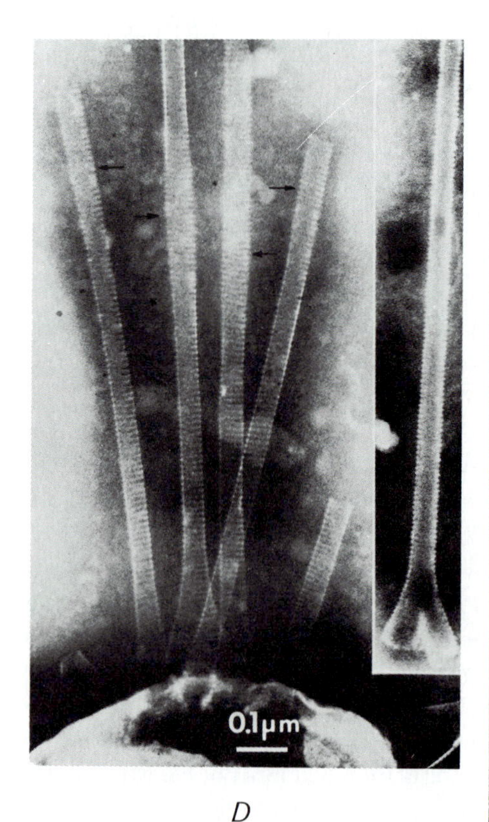

D

populations of 100 to 200/liter in the waters off Oahu, Hawaii (Taylor et al., 1974). The presence of populations of host-independent forms in natural samples would be nearly impossible to determine (Marbach et al., 1976). In selectively filtered natural populations of bacterioplankton, however, bdellovibrio-like cells are readily observed. If these cells are bdellovibrios, then either the host-dependent cells may be more numerous than plaque assays indicate or there is a fairly large bdellovibrio flora in which the host-dependent forms are a minority.

Taylor et al. (1974) and Marbach, Varon, and Shilo (1976) have shown that the marine isolates also exhibit a host-range specificity. Taylor et al. (1974) found that bacteria of marine origin were better hosts than terrestrial strains, but that *Vibrio cholerae*, *Aeromonas formicans*, and *Salmonella typhimurium* were lysed by some strains of *Bdellovibrio* at the same rate and intensity as the marine bacterial hosts are lysed. In contrast, the gram-positive bacilli and a terrestrial pseudomonad were not attacked by the ten types of bdellovibrios isolated by Marbach, Varon, and Shilo (1976). Taylor et al. (1974) found that sodium chloride (optimal concentration, 125 to 140 m*M*) was required for plaque formation and could not be replaced with potassium chloride. Marbach, Varon, and Shilo (1976) came to the same conclusion and found that optimal plaque formation was only obtained when calcium, potassium, and magnesium salts were also present, in addition to sodium at an optimal concentration.

Terrestrial bdellovibrios have two ranges of G + C content. Most cultures of the type species *Bdellovibrio bacteriovorus* have a 50.4 ±0.9 mole % G + C content, whereas a second group of facultatively parasitic strains have a 42 to 43 mole % G + C content. Of the four host-independent strains tested by Taylor et al. (1974), one had a 43.5 and the others a 38.6 mole % G + C content. All ten types of marine bdellovibrios characterized by Marbach, Varon, and Shilo (1976) had a 33 to 38 mole % G + C content. The marine bdellovibrios are apparently a distinct group within the genus *Bdellovibrio* and must have had a different origin than their terrestrial counterparts.

Ring-forming bacteria (Plate 16-9) have been sporadically reported since their discovery by Weibel (1887), but a genus was not created until Ørskov (1928) described *Microcyclus aquaticus*. *Bergey's 8th* lists three species of respiratory, strictly aerobic, nonmotile, curved rods in the genus *Microcyclus*. These dissimilar microorganisms, however, have a wide range in DNA base composition, being 66.3 to 68.4 mole % G + C for *M. aquaticus*, 51.0 to 51.5 mole % G + C for *M. flavus*, and 39.5 mole % G + C for *M. major*. The only marine isolate of a similar nature was obtained from a sand dollar (*Dendraster*) off Newport Beach, California, and named *Microcyclus marinus* (Raj, 1970, 1976). Larkin, Williams, and Taylor (1977) have made the much needed realignment of the members of this genus, placing each of the species in *Bergey's 8th* in an emended or a new genus. The genus *Microcyclus* Ørskov 1928 was redefined and contains only *M. aquaticus*, the genus *Spirosoma* Migula 1894 was reintroduced and emended and *M. flavus* became *S. linguale*, and a new genus, *Flectobacillus*, was created to house *M. major*. The marine species *M. marinus* is more closely related to *Flectobacillus* than to *Microcyclus*, and Borrall and Larkin (1978) have emended the description of *Flectobacillus* to accommodate *M. marinus* as *F. marinus*.

Little is known about the distribution of marine ring-forming bacteria with the morphology of *Flectobacillus*. Similar forms are present on the surfaces of the red alga *Rhodymenia palmata* from Rhode Island waters and on the seagrasses *Thallassia testudinum* and *Syringodium filiforme* and decaying mangrove leaves from the waters of the Florida Keys (Sieburth, 1975). When natural populations of bacterioplankton are selectively filtered, negatively stained, and examined by transmission electron microscopy, a fairly high number of curved rods are seen. Horseshoes, rings, and spirals are ubiquitous in such preparations and appear as dominant forms in open ocean samples (Plate 16-9,B). There is a marked difference in size between the forms in the plankton (planktobacteria) and those on surfaces (epibacteria), and they may be two very distinct types of bacteria.

When *Flectobacillus marinus* is grown on complex agar media it produces pinkish, circular, convex, smooth, mucoid, opaque colonies, less than 2 mm in diameter, which have a colorless peripheral zone. The cells are consistently nonmotile, nonflagellated, and gram negative. They occur singly and have the appearance of either closed rings with a 0.9 to 1.5 μm outer diameter and a hollow center 0.3 to 0.6 μm in diameter, or of horseshoes or curved rods of similar size, 0.3 to 0.7 μm wide, with rounded ends. Transmission electron microscopy shows that the ends of the ring- or doughnut-shaped cells are not completely joined. Spiral

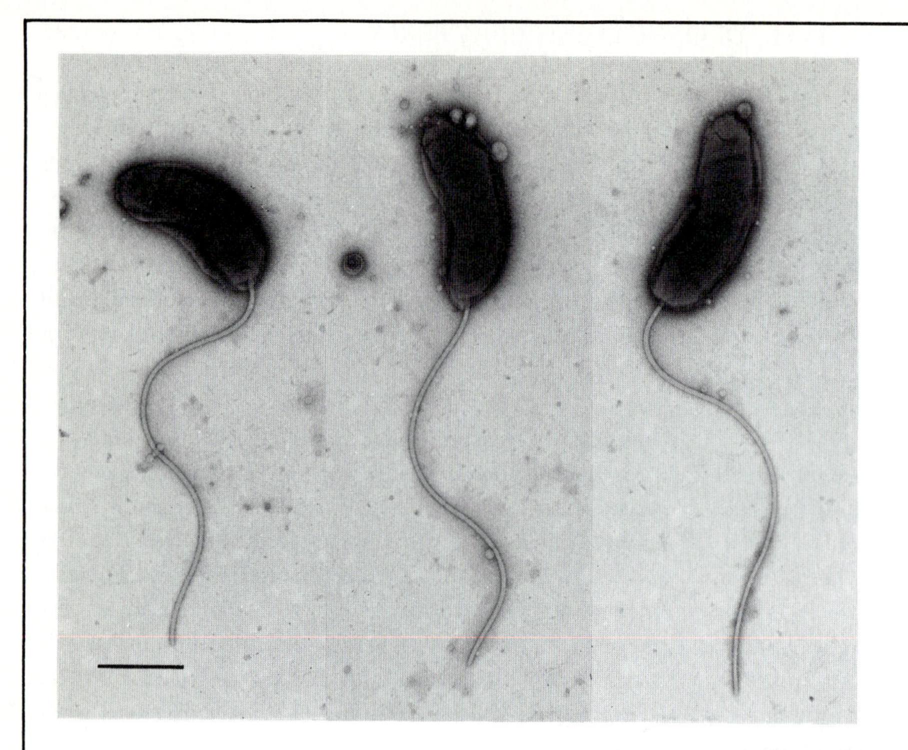

A

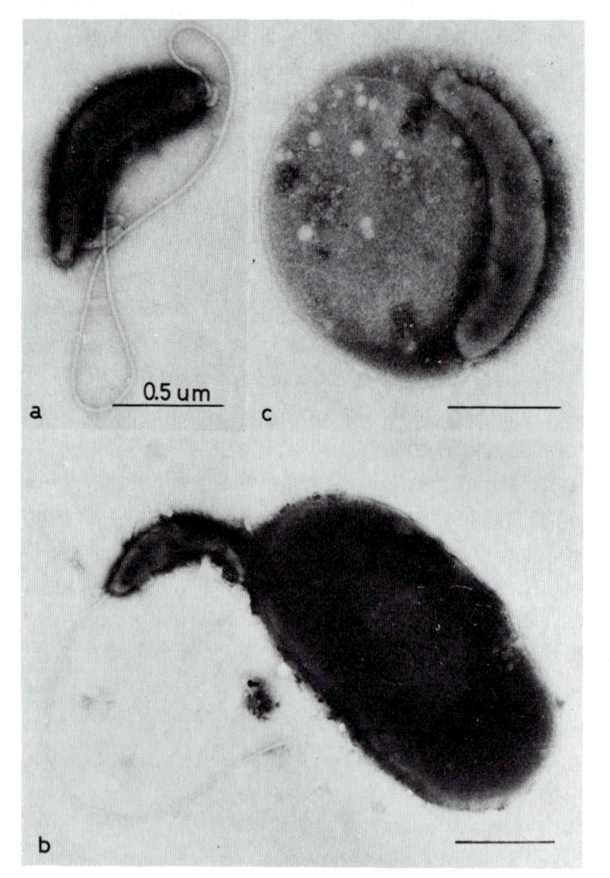

B

PLATE 16-8. Transmission electron micrographs of predatory, parasitic, and bacteriolytic bacteria of the genus *Bdellovibrio*. (*A*) Negatively stained isolates from Hawaiian waters with a single polar flagellum 24 to 27 nm wide composed of a central core surrounded by a sheath (bar = 0.5 μm). (*B*) The life cycle of bdellovibrios as illustrated by negative stains of an isolate from the Mediterranean Sea: (a) solitary flagellated cell; (b) bdellovibrio attached to the host cell; (c) bdellovibrio inside the periplasmic space of the host cell, where it elongates, undergoes multiple fission, and forms daughter cells, which swarm away from the host cell (bdelloplast). (*A*) from V.I. Taylor, P. Baumann, J.L. Reichelt, and R.D. Allen, 1974, *Arch. Mikrobiol.* 98:101–14; (*B*) from A. Marbach, M. Varon, and M. Shilo, 1976, *Micro. Ecol.* 2:284–95.

forms and coils of ten cells or more are common in culture (Plate 16-9,C). *Flectobacillus marinus* utilizes a variety of organic acids, and carbohydrates are respired, not fermented. The ring-forming bacteria have been reviewed by Raj (1977b).

D. SPIRILLA

Spirilla are helical-shaped, rigid-walled, motile bacteria usually with tufts of bipolar flagella (Plate 16-10). They are not to be confused with the spirochetes in Section E, which also have helical cells but are flexible bacteria without flagella. The spirilla were first described by Moeller in 1786, but the genus *Spirillum* was first used to identify these bacteria by Ehrenberg (1838) in his marvelously illustrated monograph on microorganisms in nature. Marine spirilla are apparently associated with estuarine mud and the digestive system of shellfish, where they can be obtained by direct culture (Leifson et al., 1964) and enrichment culture (Watanabe, 1959). They occur frequently in marine materials, but difficulties are often encountered in their culture (Dimitroff, 1926a; Williams and Rittenberg, 1957). Williams and Rittenberg (1957) used the green alga *Ulva* to obtain enrichments of spirilla from seawater. Spirilla can be isolated from enrichment cultures or samples rich in spirilla by the capillary method of Jannasch (1965). Here, a glass tube is heated in the middle and nearly flattened with a pair of forceps before drawing it out into an oval capillary. One end is then broken off, the culture drawn up into the capillary, and the end is sealed. The capillary is then viewed with high dry phase-contrast microscopy, and, as the very motile spirilla come in view, the end of the capillary with the mixed culture is broken off and the drops containing the spirilla are cultured.

A continuing taxonomic study of most of the extant strains of spirilla, including those from the marine environment, is being conducted by Krieg and his colleagues at Virginia Polytechnical Institute and State University, using uniform current methods to classify them by nutritional and morphological characteristics and their ecology, as well as DNA base composition (Hylemon et al., 1973). The diversity of these strains was too great to place them in a single genus. The freshwater strains, which are usually from stagnant pools, were placed either into an emended genus *Spirillum* with *S. volutans* the type species (which includes the very large obligately microaerophilic spirilla with a 38 mole % G + C content) or into a new genus *Aquaspirillum*, with *A. serpens* the type species (which includes the aerobic spirilla with a 50 to 65 mole % G + C content).

All the marine isolates studied by Hylemon et al. (1973), except one, were put in the new genus *Oceanospirillum*, which includes spirilla that cannot attack carbohydrates, cannot grow anaerobically with nitrate, and cannot use urea or nitrate as a sole source of nitrogen; they have a 42 to 48 mole % G + C content. The *Oceanospirillum* species are *O. maris*, which includes three previously undescribed strains isolated from seawater by Jannasch; *O. atlanticum* and *O. beijerinckii*, isolated from seawater by Williams and Rittenberg (1957); and *O. japonicum* and *O. minultulum*, isolated from shellfish (Watanabe, 1959); the latter is the type species. A single marine strain that attacks a variety of carbohydrates, grows anaerobically with nitrate, uses urea or nitrate as a sole source of nitrogen, and which has a 63 mole % G + C content, was put in a fourth genus *Marinospirillum*, with *M. (spirillum) lunatum* the type species (Williams and Rittenberg, 1956). This large species was used to observe both the formation of microcysts and their germination (see Plate 16-10). Microcyst formation may be common in spirilla (Dimitroff, 1926a; Williams and Rittenberg, 1956).

In *Bergey's 8th*, Krieg departed from the position of Hylemon et al. (1973) and kept all the spirilla in the genus *Spirillum*. He does this while pointing out that the great range of G + C content values and the differences in size, nutrition, physiology, and serology suggest that the genus could be divided into two genera or more. In

PLATE 16-9. *Flectobacillus* and *Flectobacillus*-like bacteria in the shape of rings and horseshoes. (*A*) Scanning electron micrograph of the surface of the red alga *Rhodymenia palmata*, showing ring forms. (*B*) Transmission electron micrograph of negatively stained bacterioplankton from the Western Sargasso Sea, showing horseshoe and ring forms. (*C*) Scanning electron micrographs of *Flectobacillus marinus* showing rings, horseshoes, and spirals in lab media. (*D*) Photomicrograph of a stained preparation of *M. marinus* from broth culture. (*A*) from J.McN. Sieburth, 1975, *Microbial Seascapes*, University Park Press, Baltimore, Md.; (*B,D*) micrographs courtesy of Paul W. Johnson; (*C*) from H.D. Raj, 1976, *Int. J. Syst. Bacteriol.* 26:528–44.

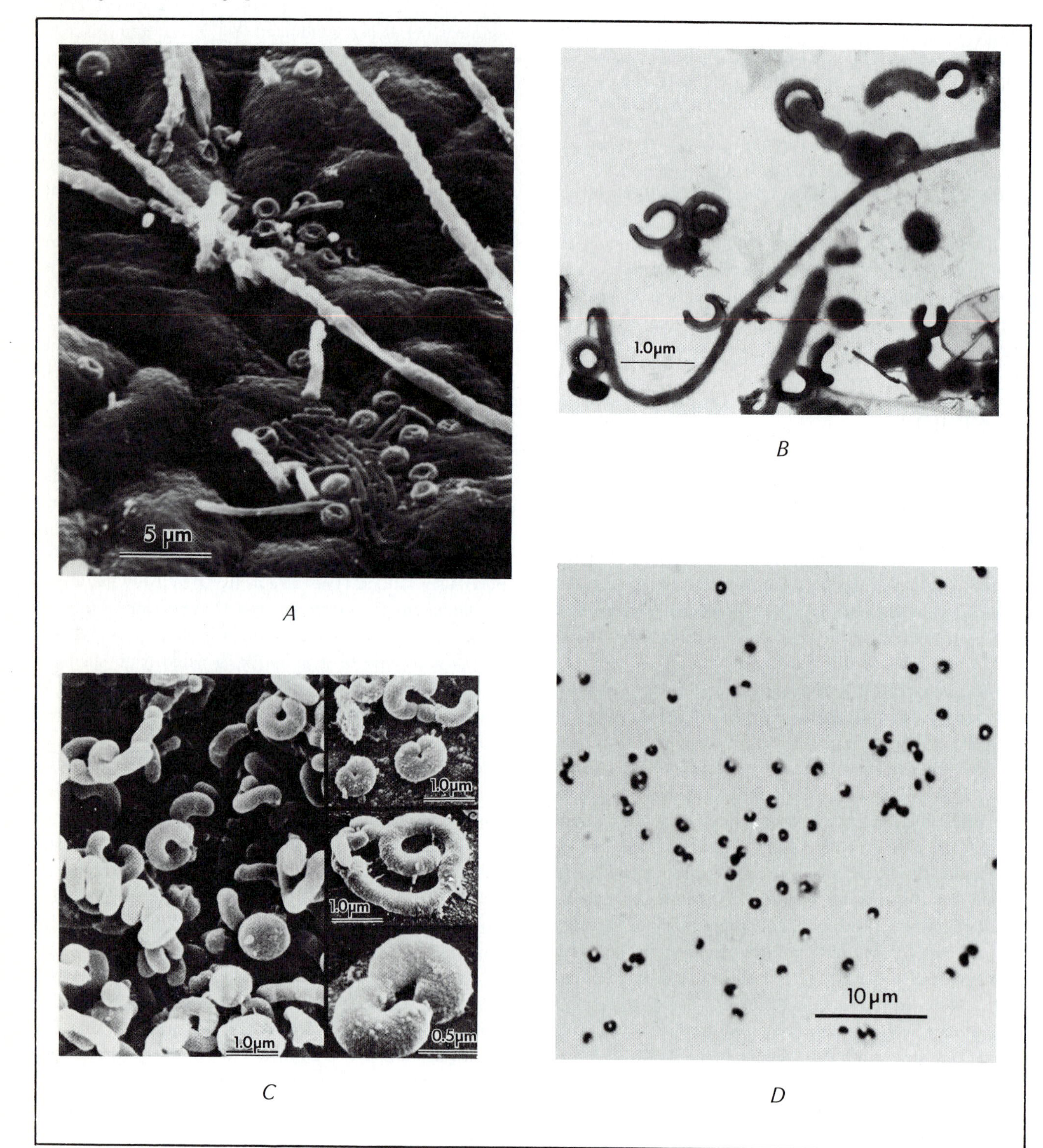

Bergey's 8th, five major groups of spirilla are distinguished, with the marine species occurring in two groups, the genera *Marinospirillum* and *Oceanospirillum* of Hylemon et al. (1973). One problem with spirilla taxonomy is the small number of isolates, often one per species. This is due to a poor understanding of the conditions needed to bring these microorganisms into culture. *Spirillum ostreae*, which occurs with *Cristispira* in the crystalline style of the eastern oyster (Dimitroff, 1926b) has never been cultured. Spirilla are found in stagnant water and associated with such sulfur bacteria as the white patches of *Beggiatoa* and the sheathed bundles of *Thioploca*. Some of these forms may be anaerobic or oxygen sensitive.

E. SPIROCHETES

The large spirochetes were among the first bacteria to be described from tooth scrapings by Antonie van Leeuwenhoek (Hoole, 1800; Dobell, 1932) and in rotting vegetation from ponds by Ehrenberg (1834, 1838). The spirochetes are slender, helically coiled, spiral flexuous cells, 30 to 500 μm in length. The protoplasmic cylinder is intertwined with bundles of axial fibrils that originate at the ends of the cylinder; it is enclosed by a thin outer membrane. Only recently have some of these microorganisms been brought into culture and characterized. The marine spirochetes fall into two genera, free-living forms associated with decomposing vegetative matter and black sulfide muds in the genus *Spirochaeta* and the commensals associated with the crystalline style (enzymatic pestle) of bivalve molluscs (lamellibranchs) in the genus *Cristispira*. *Saprospira*, which occurs in the same habitats as the *Spirochaeta* and *Cristispira*, was erroneously included in the family Spirochaetaceae (*Bergey's 7th*) until its relationship with the cytophagales was recognized (Lewin, 1962). The spirochetes could, and perhaps should, be included with the anaerobic bacteria in Chapter 17, but they are discussed with the spirilla and other distinctive bacteria for convenience.

A working definition of the genus *Spirochaeta* (Canale-Parola, Udris, and Mandel, 1968), as modified by subsequent observations (Hespell and Canale-Parola, 1970; *Bergey's 8th*), would be a unicellular or pluricellular helical bacterium 0.20 to 0.75 μm wide and 5 to 500 μm long, with or without cross-walls, but with no cross-striations visible under the light microscope. A spinning motility, with lashing, bending, undulating, curling, and serpentine movements, characterizes the spirochetes; these movements probably originate in the bundle of axial fibrils. Multiplication is by transverse fission. These facultative and strict anaerobes are found free-living in sulfide muds with *Beggiatoa* in freshwater and marine environments. Their DNA base composition range is 50 to 67 mole % G + C.

The type species *Spirochaeta plicatilis* was first described from freshwater by Ehrenberg in 1834 and later included in his magnificent 1838 monograph. This species and a slightly thinner and shorter spirochete, *S. marina*, from marine muds, were described and illustrated with photomicrographs by Zuelzer (1912). Dobell (1912), working independently with different methods, failed to observe these two species in marine muds. The first, free-living obligately anaerobic spirochete to be obtained in pure culture [*Spirochaeta (Treponema) zuelzerae*] was detected as a contaminant in an enrichment culture for anoxyphotobacteria from a freshwater pond (Veldkamp, 1960). The successful isolation encouraged Canale-Parola, Holt, and Udris (1967) to attempt the isolation of free-living anaerobic spirochetes from enrichments of three marine muds. Using black sulfide-rich muds from a seawater marsh, Blakemore and Canale-Parola (1973) found that individual samples rich in *Beggiatoa* contained either *S. plicatilis* or another form, the type B spirochete (Plate 16-11). These large spirochetes have not been obtained in pure culture, possibly because they may require the presence of *Beggiatoa* (Hespell and Canale-Parola, 1970). The difference in size of these spirochetes, as well as the characteristic primary and secondary coiling of the large forms, is shown in Plate 16-11. It is ironic that until the ultrastructure of natural populations of *S. plicatilis* was studied (Blakemore and Canale-Parola, 1973), it was not known whether the type species actually had the structural features that characterize the genus *Spirochaeta*. The fine structure of the large forms *S. plicatilis* and the type B spirochete (Blakemore and Canale-Parola, 1973) shows that they really belong to the genus *Spirochaeta*. Because of their cross-walls, large coils of *S. plicatilis* fragment easily upon mechanical agitation. The metabolic activity of the spirochetes studied so far indicates that they are saccharolytic and do not utilize organic acids, fatty acids, amino acids, or alcohols as energy sources (Hespell and Canale-Parola, 1973). The freshwater species *Spirochaeta aurantia* is the

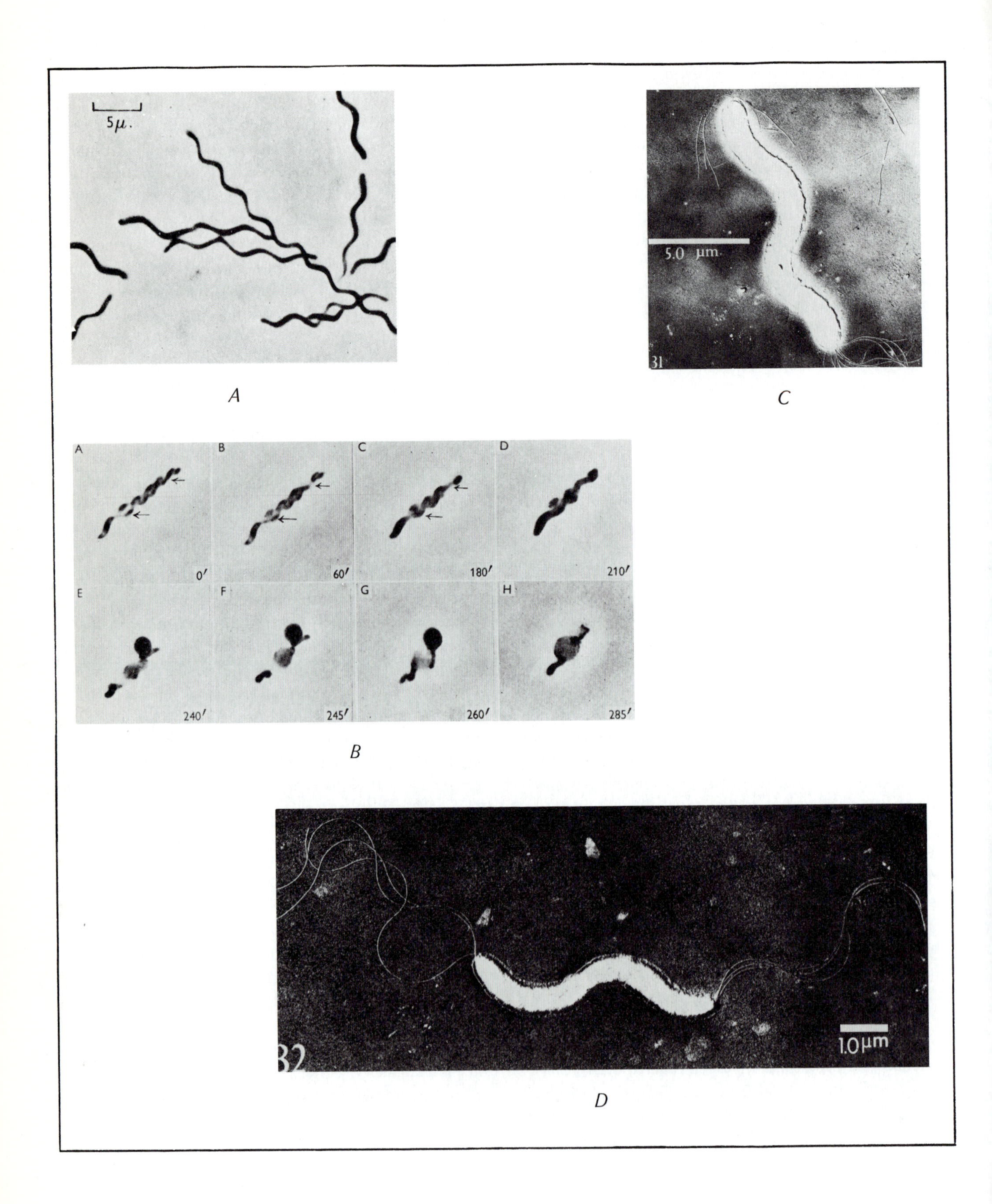
5μ.
A
5.0 μm
31
C
A
B
C
D
0′
60′
180′
210′
E
F
G
H
240′
245′
260′
285′
B
32
1.0μm
D

PLATE 16-10. Morphology and life cycle of spiral-shaped bacteria of the genus *Spirillum*. (*A*) Vegetative cells of *Spirillum lunatum* in a 26-hour culture (phase contrast). (*B*) Microcyst formation following fusion of two spirilla in a 24-hour culture. Transfer of old microcyst cultures to new media results in unipolar and bipolar germination in 10 to 12 hours, which reverses the sequence of events. (*C*) and (*D*) Transmission electron micrographs (Tungsten-oxide shadowed) of spirilla, showing cell morphology and bipolar flagellation. (*A*,*B*) from M.A. Williams and S.C. Rittenberg, 1956, *J. gen. Microbiol.* 15:205–9; (*C*,*D*) from P.H. Hylemon, J.S. Wells, Jr., N.R. Krieg, and H.W. Jannasch, 1973, *Int. J. Syst. Bacteriol.* 23:340–80.

only facultative anaerobe isolated so far (Breznak and Canale-Parola, 1972), although *S. plicatilis* is apparently an anaerobe that can tolerate low oxygen tension or a microaerophile able to metabolize or even grow in the absence of oxygen (Blakemore and Canale-Parola, 1973). As would be expected in an anaerobe, *Spirochaeta litoralis* catabolizes glucose via the Embden–Meyerhoff pathway, with carbon dioxide, hydrogen, acetate, and ethanol the major metabolic by-products (Hespell and Canale-Parola, 1973). It also possesses a clostridial-type, pyruvate clastic system, although catalytic amounts of coenzyme A are required, which indicates impaired coenzyme A biosynthesis. The non-heme iron proteins ferrodoxin and rubredoxin, which participate in the metabolism of many anaerobic bacteria, have been characterized from two spirochetes (Johnson and Canale-Parola, 1973); the red protein of *S. litoralis* is a similar rubredoxin (Hespell and Canale-Parola, 1973).

Cristispira are spirochetes with flexing 0.5 to 3.0 μm diameter cells 30 to 150 μm in length that form coarse spirals in two to ten complete turns (Plate 16-12). They are distinguishable from the *Spirochaeta* by a typical forward and backward locomotion, sometimes by ovoid occlusions, which appear as cross-striations, and by the ready detachment of the axial filaments within the cell envelope to form crests or crista. These intriguing bacteria have a highly specialized niche, the crystalline style, a labile structure in the gut of bivalve molluscs (lamellibranchs). Certes (1882) reported that amid the gut contents of European oysters are numerous highly motile microorganisms that he considered to be parasitic or commensal protozoans. One microorganism found at the entrance to the stomach was described as *Trypanosoma balbianii* (*Cristispira balbianii*) (see Plate 16-12,*A*). Möbius (1883) reported that he had seen this microorganism lodged in the crystalline style of oysters from Schleswig-Holstein as early as August 1869.

An elongated sac adjacent to the posterior wall of the stomach of oysters is attached along its length to the mid gut. In actively feeding oysters, the bulging head of the crystalline style, which is a gelatinous rod, protrudes into the stomach and its pointed tail extends into the distal part of the sac (Plate 16-12,*D*). The ciliary activity of the epithelium in the sac rotates the style to carry food particles down the gut or to tangle them in the substance of the style, which is carried back to the stomach. The style rotates and rubs against the gastric shield, mixing and triturating food particles, and in the process, slowly dissolves at the head end to yield digestive enzymes (Galtsoff, 1964). Similar styles are found in all bivalves and are apparently very labile, dissolving soon after the animal is removed from the sea. Regeneration of the style takes hours (Kuobomura, 1965) to days (Berkeley, 1959).

The early literature on *Cristispira* is summarized by Dimitroff (1926a, 1926b). Perrin (1906) and Fantham (1908) noted how quickly the crystalline style dissolves after the shellfish is removed from the sea and emphasized the need to use very fresh material only. The bacterial nature of these style-inhabiting spirochetes was recognized by Gross (1910) who placed them in a new genus, *Cristispira*. Taxonomic studies by Kuhn (1970) indicate that Gross's *Cristispira pectinis* should be the type species. The species *incertae sedis* of *Cristispira*, their host, and their geographical area are listed in *Bergey's 8th*.

Although these spirochetes are found in many species of bivalves (Berkeley, 1959), other species have been reported to be spirochete-free (Noguchi, 1921). The common clam of British Columbian waters, *Saxidomus giganteus*, contains large populations of *Cristispira*, and this clam has been used to follow the colonization of newly generated styles (Berkeley, 1959). The absence of *Cristispira* in the functional anterior end has been ascribed to the toxicity of glucosone, an oxidative metabolite of glucose in the food (Berkeley, 1962). The common clam of Japan, *Venerupis phillipinarum*, was used to determine that the crystalline style has a 2% fructose content, that the *Cristispira* from this clam have a positive chemotaxis to fructose, and that they can first multiply and then survive for a week in a fructose medium (Kuobomura, 1965). The commensal or parasitic nature of these bacteria, as well as their culture, has not been

PLATE 16-11. Spirochetes from marine mud. (*A*) Phase-contrast photomicrographs of wet mount preparations comparing (1) *Spirochaeta plicatilis*, (2) type B spirochete, and (3) *S. litoralis* (bar = 100 μm). (*B*) Type B spirochetes associated with the trichomes of *Beggiatoa* (bar = 100 μm). (*C*) Transmission electron micrographs of negatively stained spirochetes: (1) one end of a type B spirochete with its bundle of axial fibrils (AF) wound around the protoplasmic cylinder (PC), both being enclosed within an outer sheath of transversely arranged fibrils (S) (bar = 1 μm); (2) another type B spirochete, showing the bundles of axial fibrils (arrows) wound around the cell, with secondary coils superimposed upon the primary coils (bar = 2 μm); (3) the end of a *S. plicatilis* cell, with septa evenly spaced along the cell (arrows) (bar = 2 μm). From R.P. Blakemore and E. Canale-Parola, 1973, *Arch. Mikrobiol.* 89:273–89.

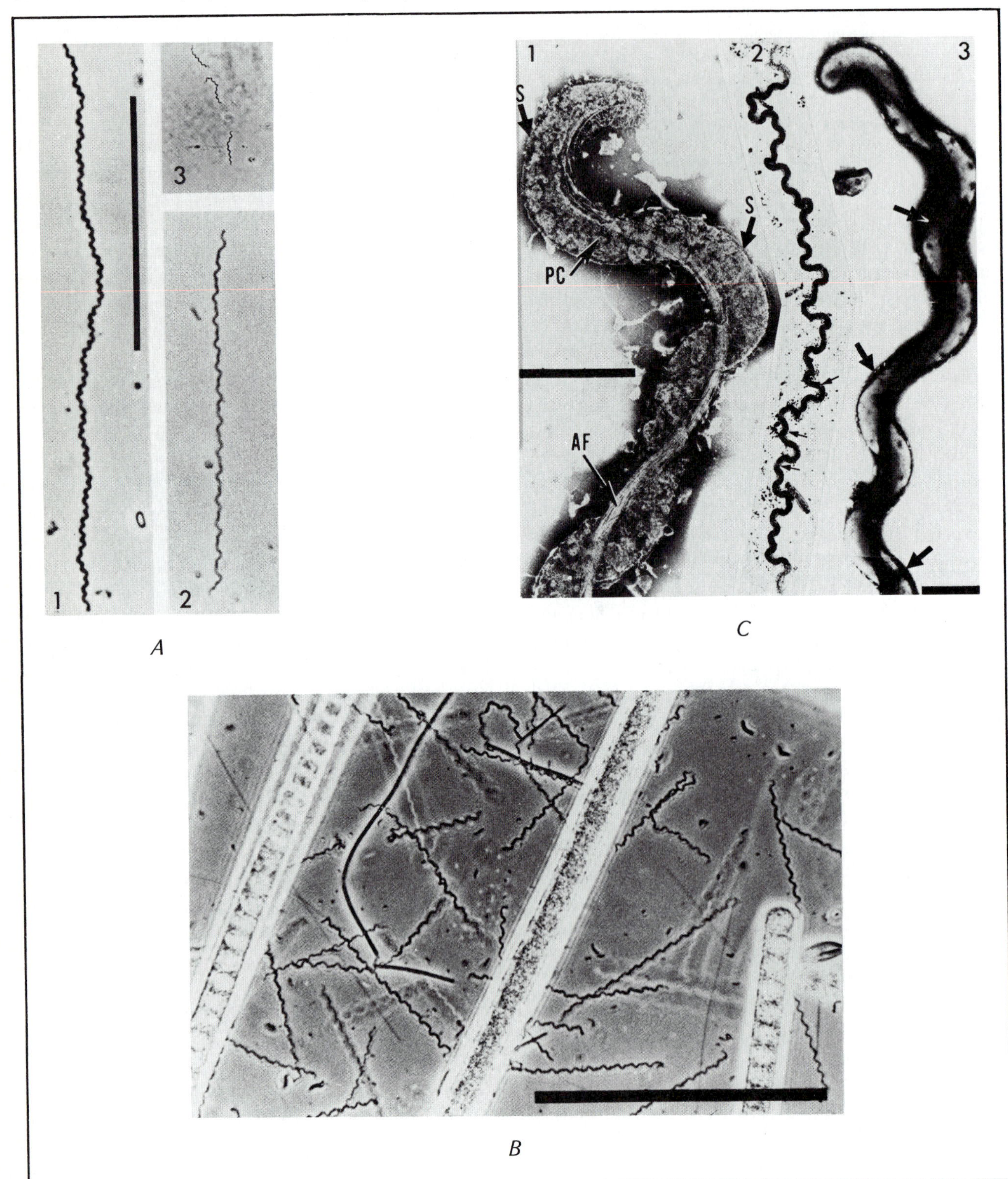

PLATE 16-12. *Cristispira*, the spirochete indigenous to the enzyme-secreting crystalline style of oysters. (*A*) *Cristispira (Trypanosoma) balbianii* as shown in the original description by Certes (1882) for the European oyster *Ostrea edulis* (340 X). (*B*) Phase-contrast photomicrograph of *Cristispira* within the crystalline style of the American oyster *Ostrea virginiana*. (*C*) Transmission electron micrograph (thin section) of *Cristispira* within the oyster crystalline style (CS), showing the flexed protoplasmic cylinder (PC) and axial fibrils (AF), both of which are enclosed by the outer sheath (OS). The relaxing of the axial fibrils in moribund cells forms typical cristi. (*D*) Dissected American oyster, showing the position of the crystalline style (CS) against the gastric shield and within its sac attached to the posterior wall of the stomach.

(*A*) from A. Certes, 1882, *Bull. Soc. Zool. France* 7:347–53; (*B,C*) micrographs courtesy of Paul W. Johnson; (*D*) from P.S. Galtsoff, 1964, *U.S. Fish. Wildlife Serv. Bull.* 64:1–480.

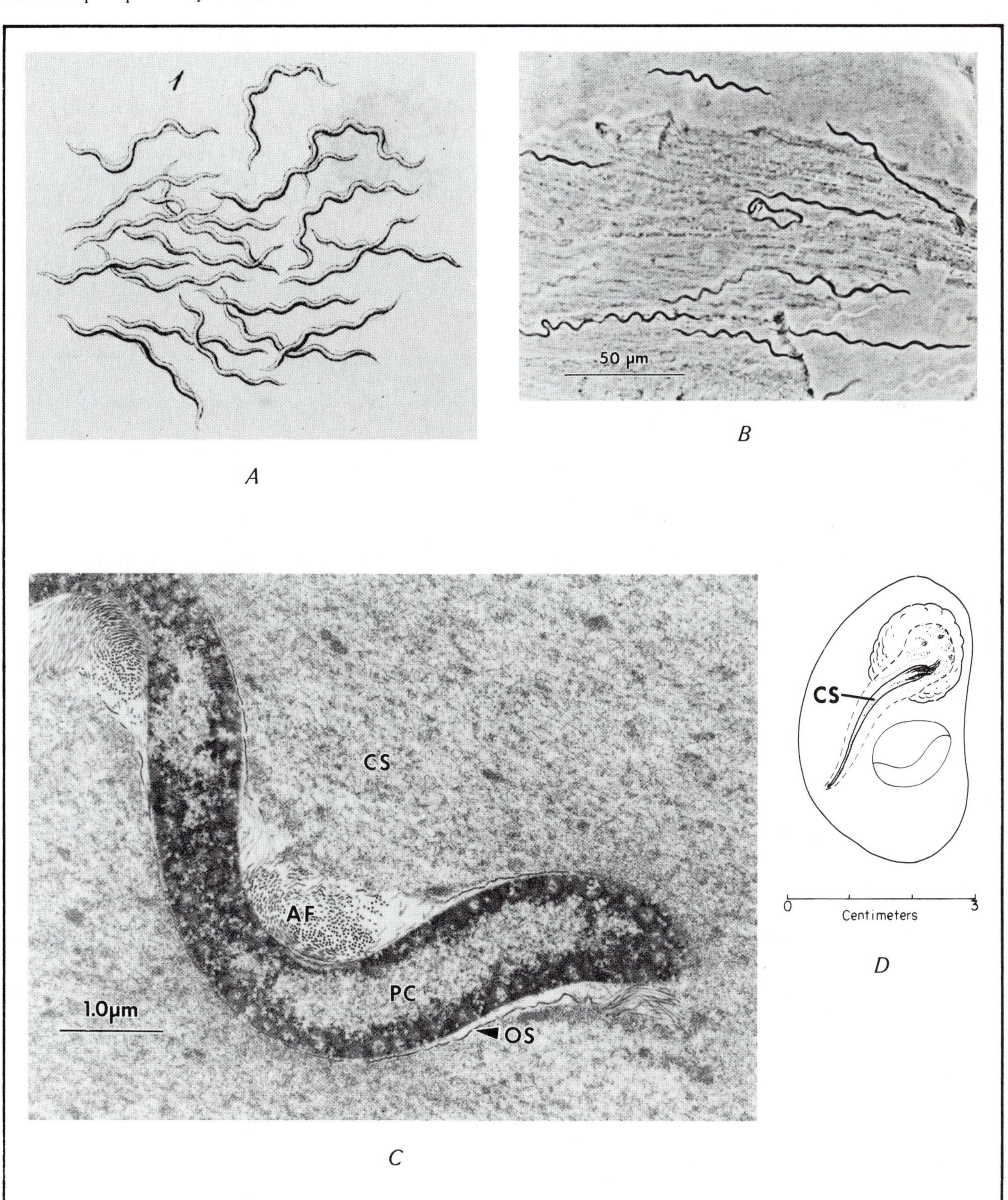

resolved. Observations made in our laboratory on the microflora of the crystalline style of the eastern oyster *Ostrea virginiana* indicate that the style and its microflora are present at 8 but not at 4 °C. In addition to *Cristispira*, a spirillum with a unique morphology [possibly the *Spirilla ostreae* of Dimitroff (1926b)], was its only, but numerous, companion. Both helical bacteria are oriented parallel to the axis of the style and have a similar reciprocal motion within the gel of the style. *Cristispira* can be maintained in a seawater–serum broth for several weeks (the surface-to-volume ratio is critical), but growth has not yet been achieved in our laboratory.

REFERENCES

Anon. 1960. *Henrici. Recollections by Some Close Friends and Associates.* Burgess Publ. Co., Minneapolis, 32 p.

Aristovskaya, T.V. 1961. Accumulation of iron in breakdown of organomineral humus complexes by microorganisms. *Doklody Akad. Nauk. SSSR* (Trans.) 136:111–14.

Berkeley, C. 1959. Some observations on *Cristispira* in the crystalline style of *Saxidomus giganteus* Deshayes and in that of some other Lamellibranchiata. *Can. J. Zool.* 37:53–58.

Berkeley, C. 1962. Toxicity of plankton to *Cristispira* inhabiting the crystalline style of a mollusk. *Science* 135:664–65.

Biggins, J., and W.E. Dietrich, Jr. 1968. Respiratory mechanisms in the Flexibacteriaceae. I. Studies on the terminal oxidase system of *Leucothrix mucor. Arch. Biochem. Biophys.* 128:40–50.

Blakemore, R.P., and E. Canale-Parola. 1973. Morphological and ecological characteristics of *Spirochaeta plicatilis. Arch. Mikrobiol.* 89:273–89.

Bland, J.A., and T.D. Brock. 1973. The marine bacterium *Leucothrix mucor* as an algal epiphyte. *Mar. Biol.* 23:283–92.

Borrall, R., and J.M. Larkin, 1978. *Flectobacillus marinus* (Raj) comb. nov., a marine bacterium previously assigned to *Microcyclus. Int. J. Syst. Bacteriol.* 28:341–43.

Breznak, J.A., and E. Canale-Parola. 1972. Metabolism of *Spirochaeta aurantia. Arch. Mikrobiol.* 83:261–77.

Brock, T.D. 1964. Knots in *Leucothrix mucor. Science* 144:870–72.

Brock, T.D. 1966. The habitat of *Leucothrix mucor*, a widespread marine microorganism. *Limnol. Oceanogr.* 11:303–7.

Brock, T.D. 1967. Mode of filamentous growth of *Leucothrix mucor* in pure culture and in nature, as studied by tritiated thymidine autoradiography. *J. Bacteriol.* 93:985–90.

Brock, T.D., and M. Mandel. 1966. Deoxyribonucleic acid base composition of geographically diverse strains of *Leucothrix mucor. J. Bacteriol.* 91:1659–60.

Burton, S.D., and R.Y. Morita. 1964. Effect of catalase and cultural conditions on growth of *Beggiatoa. J. Bacteriol.* 88:1755–61.

Burton, S.D., R.Y. Morita, and W. Miller. 1966. Utilization of acetate by *Beggiatoa. J. Bacteriol.* 91:1192–1200.

Canale-Parola, E., S.C. Holt, and Z. Udris. 1967. Isolation of free-living, anaerobic spirochetes. *Arch. Mikrobiol.* 59:41–48.

Canale-Parola, E., Z. Udris, and M. Mandel. 1968. The classification of free-living spirochetes. *Arch. Mikrobiol.* 63:385–97.

Certes, A. 1882. Les Parasites et les commensaux de l'huître. *Bull. Soc. Zool. France* 7:347–53.

Colwell, R.R. 1969. Numerical taxonomy of the flexibacteria. *J. gen. Microbiol.* 58:207–15.

Conti, S.F., and P. Hirsch. 1965. Biology of budding bacteria. III. Fine structure of *Rhodomicrobium* and *Hyphomicrobium* spp. *J. Bacteriol.* 89:503–12.

Correll, D.L., and R.A. Lewin. 1964. Rod-shaped ribonucleoprotein particles from *Saprospira. Can. J. Microbiol.* 10:63–74.

Dannevig, A. 1919. Canadian fish eggs and larvae. In: *Canadian Fisheries Expedition 1914–15.* Dept. Naval Serv., Ottawa, p. 48.

Dimitroff, V.T. 1926a. *Spirillum virginianum* nov. spec. *J. Bacteriol* 12:19–45.

Dimitroff, V.T. 1926b. Spirochaetes in Baltimore market oysters. *J. Bacteriol.* 12:135–65.

Dobell, C. 1912. Researches on spirochaetes and related organisms. *Arch. Protistenk.* 26:117–240.

Dobell, C. 1932. *Anthony van Leeuwenhoek and his "Little Animals."* Harcourt Brace, New York, 435 p.

Ehrenberg, C.G. 1834. *Organisation in der Richtung des kleinster Raumes.* Druckerei Konigl. Akad. Wissenschaft, Berlin, 169 p., 92 seit.

Ehrenberg, C.G. 1838. *Die Infusionsthierchen als vollkommene Organismen.* Verlag Leopold Voss, Leipzig, 547 p., Atlas.

Fager, E.W. 1969. Recurrent group analysis in the classification of flexibacteria. *J. gen. Microbiol.* 58:179–87.

Fantham, H.B. 1908. *Spirochaeta (Trypanosoma) balbianii* (Certes) and *Spirochaeta anodontae* (Keysselitz): Their movements, structures, and affinities. *Quart. J. Micros. Sci.* 52:1–73.

Faust, L., and R.S. Wolfe. 1961. Enrichment and cultivation of *Beggiatoa alba. J. Bacteriol.* 81:99–106.

Floodgate, G.D., and P.R. Hayes. 1963. The Adansonian taxonomy of some yellow pigmented marine bacteria. *J. gen. Microbiol.* 30:237–44.

Gallardo, V.A. 1977. Large benthic microbial communities in sulfide biota under the Peru–Chile subsurface countercurrent. *Nature* 268:331–32.

Galtsoff, P.S. 1964. The American oyster *Crassostrea virginica* Gmelin. *Fish. Bull. (Fish & Wildlife Serv.)* 64:1–480.

Glaser, J., and J.L. Pate. 1973. Isolation and characterization of gliding motility mutants of *Cytophaga columnaris. Arch. Mikrobiol.* 93:295–309.

Gross, J. 1910. *Cristispira* nov. gen. Ein Beitrag zur Spirochätenfrage. *Mitt. Zoo. Stat. Neapel.* 20:41–93.

Guillard, R.R.L., and S.W. Watson. 1962. A new marine bacterium. *Oceanus* 8(3):22–23.

Harold, R., and R.Y. Stanier. 1955. The genera *Leucothrix* and *Thiothrix. Bacteriol. Rev.* 19:49–58.

Hayes, P.R. 1963. Studies on marine flavobacteria. *J. gen. Microbiol.* 30:1–19.

Henrici, A.T. 1933. Studies of freshwater bacteria. I. A direct microscopic technique. *J. Bacteriol.* 25:277–287.

Henrici, A.T., and D.E. Johnson. 1935. Studies on freshwater bacteria. II. Stalked bacteria, a new order of Schizomycetes. *J. Bacteriol.* 30:61–93.

Hespell, R.B., and E. Canale-Parola. 1970. *Spirochaeta litoralis* sp. n., a strictly anaerobic marine spirochete. *Arch. Mikrobiol.* 74:1–18.

Hespell, R.B., and E. Canale-Parola. 1973. Glucose and pyruvate metabolism of *Spirochaeta litoralis*, an anaerobic marine spirochete. *J. Bacteriol* 116:931–37.

Hirsch, P. 1968. Biology of budding bacteria. IV. Epicellular deposition of iron by aquatic budding bacteria. *Arch. Mikrobiol.* 60:201–16.

Hirsch, P., and S.F. Conti. 1964. Biology of budding bacteria. I. Enrichment, isolation and morphology of *Hyphomicrobium* spp. *Arch. Mikrobiol.* 48:339–57.

Hirsch, P., and G. Rheinheimer. 1968. Biology of budding bacteria. V. Budding bacteria in aquatic habitats: Occurrence, enrichment, and isolation. *Arch. Mikrobiol.* 62:289–306.

Hoole, S. 1800. *The Select Works of Anthony van Leeuwenhoek Containing His Microscopical Discoveries in Many of the Works of Nature* (Transl. from Dutch and Latin editions by the author). G. Sidney, London, 344 p.

Houwink, A.L. 1951. *Caulobacter* versus *Bacillus* spec. div. *Nature* 168:654–655.

Houwink, A.L. 1952. Contamination of electron microscope preparations. *Experientia* 8:385.

Hylemon, P.H., J.S. Wells, Jr., N.R. Krieg, and H.W. Jannasch. 1973. The genus *Spirillum*: A taxonomic study. *Int. J. Syst. Bacteriol.* 23:340–80.

Jannasch, H.W. 1965. Die isolierung Heterotrophen Aquatischer Spirillen. In: *Anreicherungskultur und Mutantenauslese* (H.G. Schlegel, ed.), Gustav Fischer, Stuttgart, Zentralbl. Bakt. Parasit. Infekt. Hyg., 1 Abt. Suppl. 1, pp. 198–203.

Jannasch, H.W., and G.E. Jones. 1960. *Caulobacter* sp. in sea water. *Limnol. Oceanogr.* 5:432–33.

Johnson, P.W., and E. Canale-Parola. 1973. Properties of rubredoxin and ferredoxin isolated from spirochetes. *Arch. Mikrobiol.* 89:341–53.

Johnson, P.W., J.McN. Sieburth, A Sastry, C.R. Arnold, and M.S. Doty. 1971. *Leucothrix mucor* infestation of benthic crustacea, fish eggs and tropical algae. *Limnol. Oceanogr.* 16:962–69.

Kadota, H. 1959. Cellulose decomposing bacteria in the sea. In: *Marine Boring and Fouling Organisms* (D.L. Ray, ed.). University of Washington Press, Seattle, p. 332–40.

Kandler, O., C. Zehender, and O. Huber. 1954. Über das Vorkommen von *Caulobacter* spec. in destilliertem Wasser. *Arch. Mikrobiol.* 21:57–59.

Kelly, M.T., and T.D. Brock. 1969. Warm-water strain of *Leucothrix mucor. J. Bacteriol.* 98:1402–03.

Kingma Boltjes, T.Y. 1936. Über *Hyphomicrobium vulgare* Stutzer et Hartleb. *Arch. Mikrobiol.* 7:188–205.

Koppe, F. 1924. Die Schlammflora der Ostholsteinschen Seen und des Bodensees. *Arch. Hydrobiol.* 14:619–72.

Kuhn, D.A. 1970. Proposal of *Cristispira pectinis* Gross 1910, 44 as the type species of the genus *Cristispira. Int. J. Syst. Bacteriol.* 20:301–3.

Kuobomura, K. 1965. Fructose medium for the cultivation of *Cristispira* sp., a flagellate living in the crystalline style of bivalves. *Science Rept. Saitama Univ. Ser. B* 5(1):1–5.

Lackey, J.B. 1961a. Occurrence of *Beggiatoa* species relative to pollution. *Water & Sewage Works* 108(1):29–31.

Lackey, J.B. 1961b. Bottom sampling and environmental niches. *Limnol. Oceanogr.* 6:271–79.

Lackey, J.B., and K.A. Clendenning. 1965. Ecology of the microbiota of San Diego Bay, California. *Trans. San Diego Soc. Nat. Hist.* 14:9–40.

Lackey, J.B., and E.W. Lackey. 1963. Microscopic algae and protozoa in the waters near Plymouth in August 1962. *J. mar. biol. Assoc. U.K.* 43:797–805.

Lackey, J.B., E.W. Lackey, and G.B. Morgan. 1965. Taxonomy and ecology of the sulfur bacteria. *Engineering Progr. Univ. Florida*, Ser. No. 119, 23 p.

Large, P.J., D. Peel, and J.R. Quayle. 1961. Microbial growth on C_1 compounds. 2. Synthesis of cell constituents by methanol- and methanol-grown *Pseudomonas* A.M. 1, and methanol-grown *Hyphomicrobium vulgare. Biochem.* 81:470–80.

Larkin, J.M., P.M. Williams, and R. Taylor. 1977. Taxonomy of the genus *Microcyclus* Orskov 1928: reintroduction and emendation of the genus *Spirosoma* Migula 1894 and proposal of a new genus, *Flectobacillus. Int. J. Syst. Bacteriol.* 27:147-56.

Lauterborn, R. 1907. Eine neue Gattung der Schwefelbakterien (*Thioploca schmidlei* nov. gen. nov. spec.). *Berl. Deutsch. Bot. Ges.* 25:238–42.

Lavoie, D.M. 1975. Application of diffusion culture to ecological observations on marine microorganisms. M.S. Thesis, University of Rhode Island, 91 p.

Leifson, E. 1964. *Hyphomicrobium neptunium* sp. n. *Antonie van Leeuwenhoek* 30:249–56.

Leifson, E., B.J. Cosenza, R. Murchelano, and R.C. Cleverdon. 1964. Motile marine bacteria. I. Techniques, ecology, and general characteristics. *J. Bacteriol.* 87:652–66.

Lewin, R.A. 1959. *Leucothrix mucor. Biol. Bull.* 117:418.

Lewin, R.A. 1962. *Saprospira grandis* Gross; and suggestions for reclassifying helical, apochlorotic, gliding organisms. *Can. J. Microbiol.* 8:555–63 (4 Pl.).

Lewin, R.A. 1969. A classification of flexibacteria. *J. gen. Microbiol.* 58:189–206.

Lewin, R.A. 1970a. *Flexithrix dorotheae* gen. et sp. nov. (Flexibacteriales); and suggestions for reclassifying sheathed bacteria. *Can. J. Microbiol.* 16:511–15.

Lewin, R.A. 1970b. New *Herpetosiphon* species (Flexibacteriales). *Can. J. Microbiol.* 16:517–20.

Lewin, R.A., and D.M. Lounsbery. 1969. Isolation, cultivation and characterization of flexibacteria. *J. gen. Microbiol.* 58:145–170.

Lewin, R.A., and M. Mandel. 1970. *Saprospira toviformis* nov. spec. (Flexibacteriales) from a New Zealand seashore. *Can. J. Microbiol.* 16:507–10.

Lewin, R.A., D.M. Crothers, D.L. Corell, and B.E. Reimann. 1964. A phage infecting *Saprospira grandis. Can. J. Microbiol.* 10:75–85.

Maier, S., and R.G.E. Murray. 1965. The fine structure of *Thioploca ingrica* and a comparison with *Beggiatoa. Can. J. Microbiol.* 11:645–56.

Mandel, M., and R.A. Lewin. 1969. Deoxyribonucleic acid base composition of flexibacteria. *J. gen. Microbiol.* 58:171–78.

Marbach, A., M. Varon, and M. Shilo. 1976. Properties of marine bdellovibrios. *Microbial Ecology* 2:284–95.

McGregor-Shaw, J.B., K.B. Easterbrook, and R.P. McBride. 1973. A bacterium with echinuliform (nonprosthecate) appendages. *Int. J. Syst. Bacteriol.* 23:267–70.

Mitchell, R., and J.C. Morris. 1969. The fate of intestinal bacteria in the sea. In: *Advances in Water Pollution Research* (S.H. Jenkins, ed.). Pergamon, Oxford, pp. 811–17.

Mitchell, R., S. Yanofsky, and H.W. Jannasch. 1967. Lysis of *Escherichia coli* by marine micro-organisms. *Nature* 215:891–92.

Mitchell, T.G., M.S. Hendrie, and J.M. Shewan. 1969. The taxonomy, differentiation and identification of *Cytophaga* species. *J. Appl. Bacteriol.* 32:40–50.

Möbius, K. 1883. *Trypanosoma balbianii* im Krystallstiel schleswigholsteinischer Austern. *Zool. Anzeiger* 6:148.

Moll, G., and R. Ahrens. 1970. Ein neuer Fimbrientyp. *Arch. Mikrobiol.* 70:361–68.

Moore, R.L., and P. Hirsch. 1972. Deoxyribonucleic acid base sequence homologies of some budding and prosthecate bacteria. *J. Bacteriol.* 110:256–61.

Nikitin, D.I., and S.I. Kuznetsov. 1967. Electron microscope study of the microflora of water. *Microbiology* (Transl.) 36:789–94.

Noguchi, H. 1921. *Cristispira* in North American shellfish. A note on a spirillum found in oysters. *J. Exp. Med.* 34:295–315.

Oppenheimer, C.H. 1955. The effect of marine bacteria on the development and hatching of pelagic fish eggs, and the control of such bacteria by antibiotics. *Copeia* 1955(1):43–49.

Ørskov, J. 1928. Beschreibung eines neuen Mikroben, *Microcyclus aquaticus* mit eigentumlicher Morphologie. *Zentralbl. Bakteriol. Parasitenk. Infektionskr.* Abt. I, 107–80.

Perfil'ev, B.V. 1965. The capillary microbial-landscape method in geomicrobiology. In: *Applied Capillary Microscopy. The Role of Microorganisms in the Formation of Iron-Manganese Deposits* by B.V. Perfil'ev, D.R. Gabe, A.M. Gal'perina, V.A. Rabinovich, A.A. Sapotnitskii, É.É. Sherman, and É.P. Troshanov (Transl. by F.L. Sinclair). Plenum Press, New York, pp. 1–8.

Perrin, W.S. 1906. Researches upon the life-history of *Trypanosoma balbianii* (Certes). *Arch. Protistenk.* 7:131–56.

Poindexter, J.S. 1964. Biological properties and classification of the *Caulobacter* group. *Bacteriol. Rev.* 28:231–95.

Poindexter, J.L.S., and G. Cohen-Bazire. 1964. The fine structure of stalked bacteria belonging to the family Caulobacteriaceae. *J. Cell Biol.* 23:587–607.

Pringsheim, E.G. 1949. The relationship between bacteria and myxophyceae. *Bacteriol. Rev.* 13:47–98.

Pringsheim, E.G. 1951. The Vitreoscillaceae: A family of colorless, gliding, filamentous organisms. *J. gen. Microbiol.* 5:124–49.

Pringsheim, E.G. 1957. Observations on *Leucothrix mucor* and *Leucothrix cohaerens* nov. sp. *Bacteriol. Rev.* 21:69–76.

Pringsheim, E.G. 1967. Die mixotrophie von *Beggiatoa. Arch. Mikrobiol.* 59:247–54.

Raj, D.H. 1967. Radiorespirometric studies of *Leucothrix mucor. J. Bacteriol.* 94:615–23.

Raj, H. D. 1970. A new species—*Microcyclus flavus. Int. J. Syst. Bacteriol.* 20:61–81.

Raj, H.D. 1976. A new species: *Microcyclus marinus. Int. J. Syst. Bacteriol.* 26:528–44.

Raj, H.D. 1977a. *Leucothrix*. In: *Critical Reviews in Microbiology*, Vol. 5. CRC Press, Cleveland, Ohio, pp. 271–304.

Raj, H.D. 1977b. *Microcyclus* and related ring-forming bacteria. In: *Critical Reviews in Microbiology*, Vol. 5. CRC Press, Cleveland, Ohio, pp. 243–69.

Rittenberg, S.C. 1969. The roles of exogenous organic matter in the physiology of chemolithotrophic bacteria. *Adv. Microbial Physiol.* 3:159–96.

Rullmann, W. 1897. Ueber ein Nitrosobakterium mit neuen Wuchsformen. *Central. Bakteriol. Parasit. Infekt.* 2 Abt 3:228–31.

Shapiro, L., N. Agabian-Keshishian, and I. Bendis. 1971. Bacterial differentiation. *Science* 173:884–92.

Shilo, M. 1966. Predatory bacteria. *Science Journal* 2(9):33–37.

Shilo, M. 1969. Morphological and physiological aspects of the interaction of *Bdellovibrio* with host bacteria. *Curr. Topics Microbiol. Immunol.* 50:174–204.

Sieburth, J.McN. 1968. The influence of algal antibiosis on the ecology of marine microorganisms. In: *Advances in Microbiology of the Sea.* (M.R. Droop and E.J.F. Wood, eds.). Academic Press, London and New York, pp. 63–94.

Sieburth, J.McN. 1975. *Microbial Seascapes.* University Park Press, Baltimore, Md., 248 p.

Sieburth, J.McN., R.D. Brooks, R.V. Gessner, C.D. Thomas, and J.L. Tootle. 1974. Microbial colonization of marine plant surfaces as observed by scanning electron microscopy. In: *Effect of the Ocean Environment on Microbial Activities* (R.R. Colwell and R.Y. Morita, eds.). University Park Press, Baltimore, Md., pp. 418–32.

Snellen, J.E., and H.D. Raj. 1970. Morphogenesis and fine structure of *Leucothrix mucor* and effects of calcium deficiency. *J. Bacteriol* 101:240–49.

Soriano, S. 1945. El nuevo orden Flexibacteriales y la classificacíon de los órdenes de las bacterias. *Revta. argent. Agron.* 12:120–40.

Soriano, S. 1947. The Flexibacteriales and their systematic position. *Antonie van Leeuwenhoek* 12:215–22.

Soriano, S. 1973. Flexibacteria. *Ann. Rev. Microbiol.* 27:155–70.

Soriano, S., and R.A. Lewin. 1965. Gliding microbes: Some taxonomic reconsiderations. *Antonie van Leeuwenhoek* 31:66–80.

Staley, J.T. 1968. *Prosthecomicrobium* and *Ancalomicrobium*: New prosthecate freshwater bacteria. *J. Bacteriol.* 95:1921–42.

Staley, J.T. 1973a. Budding and prosthecate bacteria. In: *Handbook of Microbiology, Vol. I. Organismic Microbiology* (A.I. Laskin and H.A. Lechevalier, eds.). CRC Press, Cleveland, Ohio, pp. 29–49.

Staley, J.T. 1973b. Bacteria with acellular appendages. In: *Handbook of Microbiology, Vol. I. Organismic Microbiology* (A.I. Laskin and H.A. Lechevalier, eds.). CRC Press, Cleveland, Ohio, pp. 51–55.

Stanier, R.Y. 1942. The cytophaga group: A contribution to the biology of myxobacteria. *Bacteriol. Rev.* 6:143–96.

Starr, M.P., and R.J. Seidler. 1971. The bdellovibrios. *Ann. Rev. Microbiol.* 25:649–78.

Stolp, H. 1973. The Bdellovibrios: Bacterial parasites of bacteria. *Ann. Rev. Phytopath.* 11:53–76.

Stolp, H., and H. Petzold. 1962. Untersuchungen über einen obligat parasitischen Mikroorganismus mit lytischer Aktivität für *Pseudomonas*-Bakterien. *Phytopathol. Z.* 45:364–90.

Stolp, H., and M.P. Starr. 1963. *Bdellovibrio bacteriovorus* gen. et sp. n., a predatory, ectoparasitic and bacteriolytic microorganism. *Antonie van Leeuwenhoek* 29:217–48.

Stove, J.L., and R.Y. Stanier. 1962. Cellular differentiation in stalked bacteria. *Nature* 196:1189–92.

Stutzer, A., and R. Hartleb. 1898. Untersuchungen über die bei der Bildung von Saltpeter beobacteten Mikroorganismen. I. Abhandlungen. *Mitt. Landw. d. Konigl. Univ. Breslau* 1:75–100.

Taylor, V.I., P. Baumann, J.L. Reichelt, and R.D. Allen. 1974. Isolation, enumeration, and host range of marine bdellovibrios. *Arch. Mikrobiol.* 98:101–14.

Tootle, J.L. 1974. The fouling microflora of intertidal seaweeds: Seasonal, species and cuticular regulation. M.S. Thesis, University of Rhode Island, 65 p.

Tyler, P.A., and K.C. Marshall. 1967a. Microbial oxidation of manganese in hydro-electric pipelines. *Antonie van Leeuwenhoek* 33:171–83.

Tyler, P.A., and K.C. Marshall. 1967b. Form and function in manganese-oxidizing bacteria. *Arch. Mikrobiol.* 56:344–53.

Tyler, P.A., and K.C. Marshall. 1967c. Pleomorphy in stalked budding bacteria. *J. Bacteriol.* 93:1132–36.

Veldkamp, H. 1960. Isolation and characteristics of *Treponema zuelzerae* nov. spec., an anaerobic free-living spirochete. *Antonie van Leeuwenhoek* 26:103–25.

Watanabe, N. 1959. On four new halophile species of *Spirillum*. *Bot. Mag. (Tokyo)* 72:77–86.

Weibel, E. 1887. Untersuchungen über Vibrionen. *Zentralbl. Bakteriol.* II Abt. 16, p. 465.

Williams, M.A., and S.C. Rittenberg. 1956. Microcyst formation and germination in *Spirillum lunatum*. *J. gen. Microbiol.* 15:205–9.

Williams, M.A., and S.C. Rittenberg. 1957. A taxonomic study of the genus *Spirillum* Ehrenberg. *Int. Bull. Bacteriol. Nomencl. Tax.* (*Int. J. Syst. Bacteriol.*) 7:49–111.

Wislouch, S.M. 1912. *Thioploca ingrica* nov. sp. *Ber. Deut. Botan. Ges.* 30:470–74.

ZoBell, C.E., and E.C. Allen. 1935. The significance of marine bacteria in the fouling of submerged surfaces. *J. Bacteriol.* 29:239–51.

Zuelzer, M. 1912. Uber *Spirochaeta plicatilis* Ehrbg. und deren Verwandtschaftsbeziehungen. *Arch. Protistenk.* 24:1–59.

CHAPTER 17

Obligately Anaerobic Epibacteria

In the freely circulating waters of the oceans, the photosynthetic production of oxygen and the solution of oxygen from air more than meet the oxygen demand of the respiring organisms in the sea (Redfield, 1958). Most of the heterotrophic bacteria discussed in Chapters 15, 16, and 18 grow aerobically and do well in this oxygenated water column. Exceptions to this general rule of oxygen sufficiency occur under conditions that permit organic matter to accumulate (nutrient traps), decay, and utilize oxygen at rates faster than it can be replenished. There are four types of nutrient traps in the sea. One occurs at certain depths in the water column in geographical areas where an oxygen minimum layer is formed (Wyrtki, 1962; Kester, Crocker, and Miller, 1973), presumably when the oxygen requirements of the bacteria, protist-plankton, and animals is not met by the oxygen supply. Another major nutrient trap is the sediment in productive areas, in which organic debris undergoes marked microbial colonization and decay to consume the available oxygen in the upper few millimeters or centimeters of sediment. In these two oxygen-depleted areas, many of the bacteria discussed in Chapters 15, 16, and 18 that are not strict aerobes are able to survive and grow. Some species are microaerophilic and adapted to reduced oxygen tensions. Others are facultative anaerobes, and in addition to oxidative dissimilatory pathways, also utilize anaerobic pathways in the absence of free oxygen. Many of the bacteria growing in the sea are of this type. Such groups as the spirilla and spirochetes, which were just discussed, are associated with decomposing organic matter and include species that prefer or require the absence of air and thus must be anaerobes.

A third type of nutrient trap is the semi-enclosed basin, such as the fjord with its sill near the entrance or a deep-sea trench where the circulatory replenishment of oxygenated water is limited and the biochemical oxygen demand of entrapped organic matter is unusually high. Examples of these environments are the Black Sea, the fjords of British Columbia and Norway, and the Cariaco Trench in the Caribbean Sea off Venezuela. A fourth type of nutrient trap is the gut of animals that retain ingesta for prolonged periods of time, for example, marine mammals and birds. In sediments, anoxic waters, and the gut of certain animals, the persistent anaerobic conditions are conducive to the development of obligate anaerobes, which are sensitive to oxygen.

The fragments of terrestrial plant material carried into the sea, the detritus from seagrasses, marsh grasses, and seaweeds found at the edge of the sea, and the phytoplanktic debris of shallow eutrophic waters all contribute to the coastal sediments (Perkins, 1974). The structural carbohydrates are the main energy source left in such autolyzing and leaching material. A simplified scheme for the complete anaerobic fermentation of these carbohydrates, and some of the bacterial groups involved, is shown in Table 17-1. Anaerobic processes are basically the same, irrespective of habitat, and the same types of microorganisms are involved. There must be much in common between the processes that occur in the rumen of the cow (Bryant, 1964; Hungate, 1966) and those in salt marshes and marine sediments. But information on these groups and their activities in the marine environment is meager. The amount of information presented here reflects the almost virginal state of anaerobic work in the marine environment.

TABLE 17-1. A simplified scheme for the complete anaerobic fermentation of carbohydrates and some of the bacterial groups involved (adapted from an unpublished scheme by M.P. Bryant).

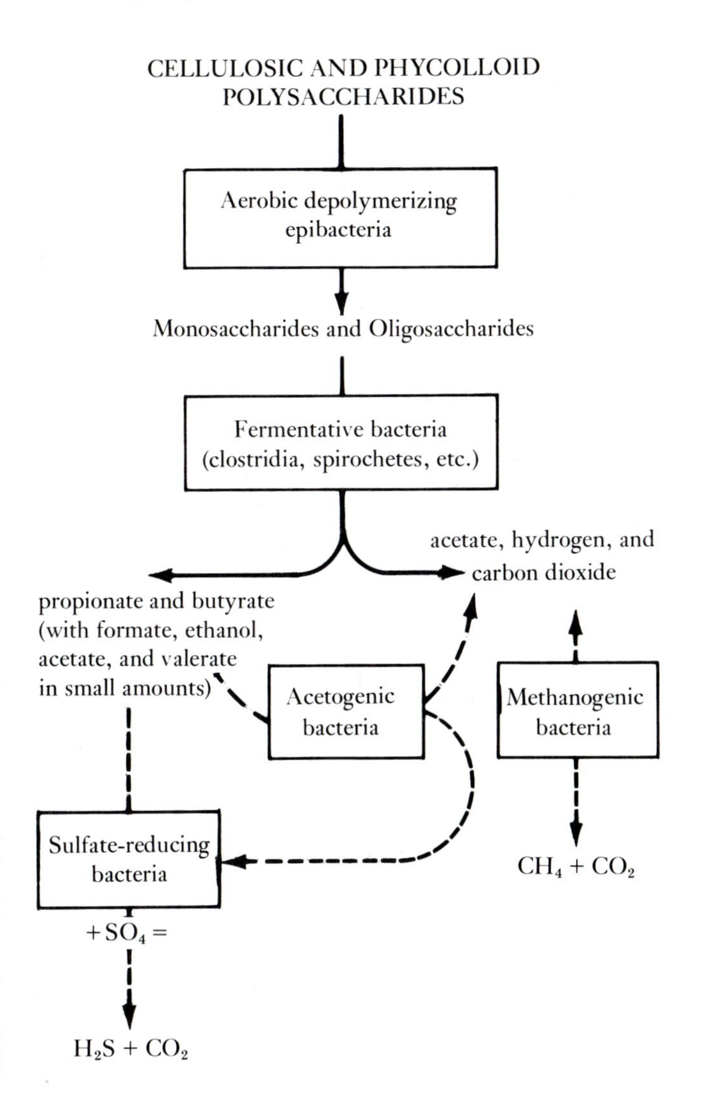

A. ANAEROBIC CULTURE

Some species and genera of anaerobic bacteria can survive temporary exposure to oxygen but require an oxygen-free environment for growth. Examples are *Desulfovibrio* and some of the species of *Clostridium*. For these microorganisms, freshly prepared and pre-reduced media can be inoculated in the presence of air and incubated in anaerobe jars. Air is replaced with gas or gas mixtures from which oxygen has been scrubbed and the remaining oxygen is adsorbed or catalyzed to form water (Holdeman and Moore, 1972). An alternate method, which eliminates the need for tanks of gas and oxygen scrubbers, is the Gaspak system (Martin, 1971; Collee et al., 1972), but such procedures cannot be used to cultivate the strict anaerobes, which are extremely sensitive to oxygen (McMinn and Crawford, 1970; Vervaeke and Van Nevel, 1972). For these microorganisms, Hungate (1950) developed a culture technique in which media are prepared, sterilized, inoculated, and incubated in the total absence of oxygen. There have been a number of minor modifications and refinements in the succeeding 25 years (Bryant and Robinson, 1971; Fulghum, 1971; Wolfe, 1971; Holdeman and Moore, 1972), including a commercial version of Moore's "VPI Anaerobic Culture System" available from Belco Glass, Inc. (Vineland, N.J.). But the basic technique has remained the same; a simple version is shown in Plate 17-1. The heart of the technique is the gassing cannula or probe, which keeps air out of the medium and culture vessels with a stream of sterile, oxygen-free carbon dioxide or other appropriate gas or gas mixture. The object is to prepare and cap tubes of media from which all oxygen has been excluded and to autoclave these tubes while the black rubber stoppers are retained in place with a clamp or a press. The fluid, hot agar medium is rotated while it cools to form a thin film of gelled medium on the walls of the test tubes. These roll or spin tubes are conveniently inoculated, during gassing while rotating in a streaking device. Details on techniques and sources of supply are given by Holdeman and Moore (1972). Although the Hungate technique is effective, the necessity of using tubes rather than Petri dishes and of creating and maintaining strict anaerobic conditions separately in each tube makes it a difficult and laborious procedure, especially for the novice. It was only a matter of time and technology until the glove box was modified as an anaerobic cabinet to maintain oxygen levels below 10 μl/liter (Aranki and Freter, 1972). For the strictest anaerobes, such as the methanogenic bacteria, a separate, ultra-low oxygen chamber is housed in the anaerobic cabinet to attain and maintain the ultra-low oxygen concentrations in the media required for anaerobic growth (Edwards and McBride, 1975). The advantages of an anaerobic cabinet with an ultra-low oxygen chamber are that a medium can be prepared in the oxygenated atmosphere and be ready for use 24 hours later. The plates, as well as the original samples, can be observed with microscopes within the chamber, and standard methods can be used to prepare, streak, and pick the plates. The Hungate method has been used to examine marine sediments for obligate anaerobes (Iannotti, Elgart, and Curtis, 1975). (The development of anaerobic cabinets to permit the handling and examination of anaerobes from core to culture should encourage marine microbiologists to adopt these techniques for studying the microorganisms of highly reduced sediments and anoxic water basins. Then perhaps more of the anaerobic spirilla, gliding bacteria, spirochetes, methane-producing bacteria, clostridia, and other fermenters from anaerobic marine habitats will be obtained in culture and studied.)

PLATE 17-1. Techniques for the anaerobic culture and isolation of strict anaerobes. (*A*) Medium is prepared in a round-bottomed flask in which the boiling medium is sparged with an oxygen-scrubbed gas mixture until reduction, as indicated by resazurin, has taken place; it is then transferred by pipette to a gassed tube. The filled tube is carefully fitted with a solid black rubber stopper (bung), the probe is removed, and a coat hanger wire clamp used to hold the stopper in place during autoclaving. Liquid agar medium is solidified in a thin film on the side of the tubes by rotating under a cold-water faucet or on a spinner, as shown. (*B*) The roll, or spin tube, is inoculated by removing the stopper and inserting a gassing probe, placing the tube in a rotating streaker, which turns the tube at 60 to 85 rpm while a loop of inoculum is applied to the film of medium from bottom to top. (*C*) The distribution of colonies in such a streaked tube. (*D*) In the anaerobic chamber (F), conventional pour plates are stored in an ultra-low oxygen cabinet (E) and flushed with the gas mixture until use. Such plates can be manipulated in the usual way. (*E*) Colonies of methanogenic bacteria on agar plates are fluorescent under long-wave ultraviolet light. (*A,B,C*) from R.S. Wolfe, 1971, *Adv. Microbial Physiol.* 6:107–46; (*D,E*) from T.L. Edwards and B.C. McBride, 1975, *Appl. Microbiol.* 29:540–45.

B. SPOREFORMERS

The rod-shaped clostridia are distinguished by the presence of ovoid to spherical endospores that usually distend these microorganisms. They are generally motile by means of peritrichous flagella and usually gram positive. Most are strict anaerobes, but some strains may grow in the presence of air just as certain strains in the aerobic sporeforming genus *Bacillus* are able to grow in the reduced oxygen tension of some anaerobe jars. Clostridia do not reduce sulfate by dissimilation or respiration. A few species can fix nitrogen and do not require other nitrogen sources. Some species are saccharolytic, some are proteolytic, some are both, and some are neither. The analyses of the acids produced during fermentation are a great aid in species identification (Holdeman and Moore, 1972; *Bergey's 8th*). The G + C DNA base ratio is 23 to 43 mole %.

Over one-quarter of the some 61 recognized species have been recorded from marine sediments, and five species were first isolated and described from marine sediments. Among the distinguishing properties of these isolates are the formation of orotic acid by *Clostridium oroticum* (Wachsman and Barker, 1954), the fermentation of amino acids and coupled oxidation-reduction reactions between certain amino acid pairs (Strickland reaction) by *Clostridium sticklandii* (Stadtman and McClung, 1957), the production of propionic acid by *Clostridium propionicum* (Cardon and Barker, 1946a,b), the production of caproic acid from ethanol by *Clostridium kluyveri* (Barker and Taha, 1942), and the marked proteolytic activity in *Clostridium oceanicum*, which often has

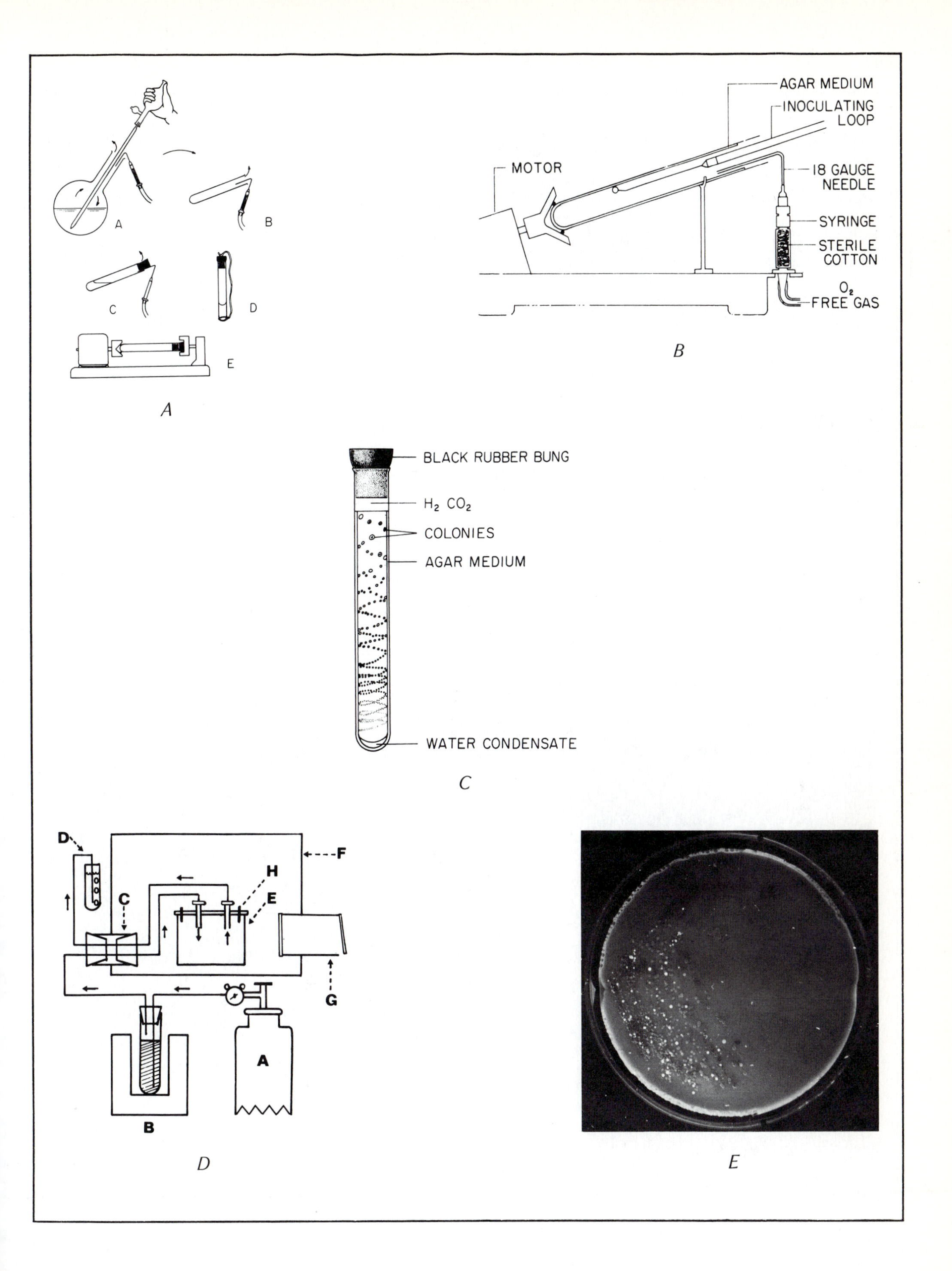
A
B
C
D
E
A
AGAR MEDIUM
INOCULATING LOOP
MOTOR
18 GAUGE NEEDLE
SYRINGE
STERILE COTTON
O_2 FREE GAS
B
BLACK RUBBER BUNG
H_2 CO_2
COLONIES
AGAR MEDIUM
WATER CONDENSATE
C
D
F
H
C
E
G
A
B
D
E

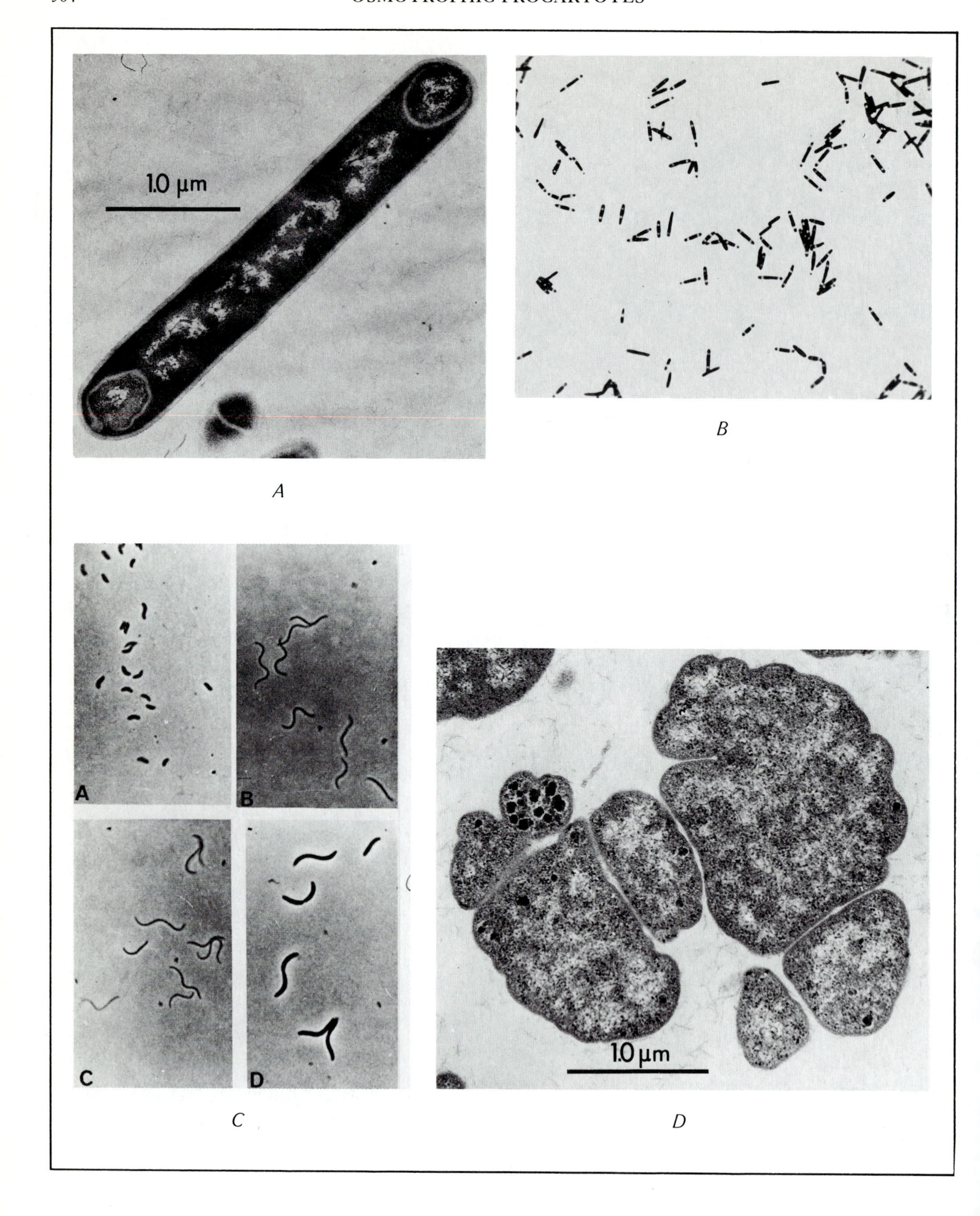
1.0 µm
B
A
A
B
C
D
C
1.0 µm
D

PLATE 17-2. Some marine sediment anaerobes. (*A*) Transmission electron micrograph of a sporulating cell of *Clostridium oceanicum*, showing double terminal spores. (*B*) Light micrograph of *C. oceanicum*, also showing double terminal spores. (*C*) A comparison of the morphology of three species of *Desulfovibrio* as seen in phase-contrast micrographs in a 2% agar film: A, *D. vulgaris*; B and C, *D. africans*; D, *D. gigas* (all 1,030 X). (*D*) Transmission electron micrograph of a thin section of *Methanococcus vannielii*. (*A*) from L. DS. Smith, 1970, *J. Bacteriol.* 103:811–13; (*B*) courtesy of Dr. L. DS. Smith; (*C*) from L.L. Campbell, M.A. Kasprzycki, and J.R. Postgate, 1966, *J. Bacteriol.* 92:1122–27; (*D*) courtesy of Dr. T.C. Stadtman.

two terminal spores (Smith, 1970; see Plate 17-2). The other species of *Clostridium* that occur in the sea are listed by Davies (1967), Smith (1968), and *Bergey's 8th*. All the anaerobic bacteria from marine mud capable of chitinolytic activity that were studied by Timmis, Hobbs, and Berkeley (1974) were species of *Clostridium* with an inducible chitinase that broke chitin down to *N*-acetylglucosamine.

For convenience, this genus has been divided into five groups on the basis of spore position, gelatin liquefaction, and special growth requirements. There are marine occurring clostridia in all five groups. The main interest in clostridia in the marine environment has been in such nitrogen-fixing species as *Clostridium pasteurianum* (Waksman, Hotchkiss, and Carey, 1933) and the toxigenic strains of *Clostridium botulinum* (Ward et al., 1967; Miller, Clark, and Kunkle, 1972). True psychrophilic strains, which can grow well in the cold of the deep bottom sediments, appear to be quite rare (Roberts and Hobbs, 1968). *Clostridium perfringens* appears to be associated with polluted waters only and apparently makes a fine tracer for the incursion of pollution into the sea (Bonde, 1968), where it can be incorporated into the human food system through fish caught in polluted waters (Matches, Liston, and Curran, 1974). In North Sea sediments Davies (1967) found that, among the clostridia, *C. perfringens*, *C. bifermentans*, and *C. sordelli* were the dominant species. Smith (1968) found *Clostridium perfringens* and *Clostridium novyi* type A in the sediments of the Peruvian current but not in the Atlantic–Caribbean samples, whereas the clostridia *C. sporogenes*, *C. bifermentans*, *C. stricklandii*, and *C. oceanicum* were often present in both types of sediments. There appears to be a tendency toward greater diversity in eutrophic areas, which contain both saccharolytic and proteolytic clostridial species, whereas the sediments of oligotrophic waters contain mainly proteolytic species. [Studies on the ecology of clostridia and other obligate anaerobes in rich inshore muds underlying eutrophic waters are long overdue. Marine sediments would lend themselves to observations similar to those conducted on ruminants (Hungate, 1966) in regard to carbon balance, types of products formed, and types of fermenting microorganisms present.]

C. SULFATE-REDUCERS

Sulfate-reducing bacteria in the genus *Desulfovibrio* are obligate anaerobes that tolerate oxygen but only grow anaerobically at low oxidation–reduction potentials (Eh) (0 to − 350 MV) on such substrates as lactate and pyruvate, which are formed during the bacterial decay (fermentation) of organic matter in sediments. In the process, sulfate, used as a hydrogen acceptor, is reduced to hydrogen sulfide. The genus *Desulfovibrio* contains gram-negative organisms with forms ranging from straight to sigmoid to spirilloid (see Plate 17-2,*C*), that are motile by polar flagella. They contain two pigments: desulfoviridin, which dissociates in alkaline solution to yield a photosensitive chromatophore with a strong red fluorescence in the ultraviolet region at 365 nm (Postgate, 1959), and cytochrome C_3, which contains heme with "pyridine hemochromogen" of the hematoheme class, as indicated by absorption at 552 to 554 nm (Postgate and Campbell, 1966). These pigments distinguish all but one strain (Miller and Saleh, 1964) of this genus from the mesophilic sporeforming sulfate-reducer *Desulfotomaculum*, which is terrestrial.

Sulfate-reducing bacteria (*Desulfovibrio*) are abundant and widely distributed in the benthic marine environment (ZoBell and Rittenberg, 1948). A medium containing lactate, yeast extract, and sulfate and poised to an appropriate Eh with glycolate and ascorbate, apparently gives valid colony counts of sulfate-reducing bacteria from both cultures and natural samples (Postgate, 1963). The Furusaka method of estimating sulfate-reducing activity by measuring the increase in hydrogenase activity upon the addition of sulfite (samples are prewashed to remove sulfate) has been applied to marine muds (Asakawa, 1972). A survey of the bottom muds of the Seto Inland Sea indicated that sulfate-reducing activity was absent from the upper 2 cm and maximal at 10 cm, being higher in fall than in summer, but absent in winter.

Sulfate ion accounts for 7.68% of the total salts of the sea and is the third major constituent after chloride and sodium (Sverdrup, Johnson, and Fleming, 1942). This sulfate is used by the majority of bacteria, fungi, yeasts, algae, and plants that assimilate sulfate ions as the sole source of cellular sulfur (Roy and Trudinger, 1970) for the synthesis of sulfur-containing amino acids and such other substances as sulfonium compounds and the sulfated polysaccharides produced by certain algae in relatively high concentration. Terrestrial erosion puts some 368 megatons of sulfate annually in river water, which finds its way into the sea (Berner, 1972). A constant proportion of salts in seawater (Sverdrup, Johnson and Fleming, 1942) is maintained because this contribution of sulfate ion from land is removed either by precipitation with calcium ions to form calcium sulfate or by reduction by *Desulfovibrio* to form hydrogen sulfide. This hydrogen sulfide is either used by the anoxyphotobacteria (Chapter 8,A) and the colorless sulfur bacteria (Chapter 9,B) or it is precipitated with iron to yield pyrite (FeS_2), which discolors the sediments to a bluish-black. This annual uptake of dissolved sulfate is estimated at 0.01 mg of sulfur/cm^2, or an annual removal of 6 megatons of sulfur (Berner, 1972). The pioneering oceanographer Murray (Murray and Irvine, 1892–1893) gave an eloquent and accurate account of the cause and effect of "blue mud" and its relationship to anoxic waters, such as those of the Black Sea. A modern account is the conceptual model of a tidal flat system of Bella, Ramm, and Peterson (1972).

Desulfovibrio desulfuricans, the type species of the genus, was described as *Spirillum desulfuricans* by Beijerinck (1895) from a freshwater ditch at Delft, Holland and obtained in pure culture as *Microspira aesturaii* from mud and seawater from the North Sea by van Delden (1903). A different type of sulfate-reducing bacterium, which forms spores and is now included in the genus *Desulfotomaculum*, was isolated from a Dutch ditch and named *Vibrio thermodesulfuricans* by Elion (1924). At this point, the taxonomy becomes muddled. Baars (1930), in his Ph.D. thesis, believed that he could interconvert freshwater and saltwater spore-formers and thermophilic isolates by acclimatization in the laboratory and, therefore, considered all three microorganisms to be strains of *Vibrio desulfuricans*. His concept was substantiated by Starkey (1938), who proposed that the genus name be changed to *Sporovibrio*. In his Ph.D. thesis, Rittenberg (ZoBell and Rittenberg, 1948) was unsuccessful in obtaining sporeformers from the marine environment and in *Bergey's 6th*, ZoBell retained the name *Desulfovibrio* but put all marine isolates in *Desulfovibrio aestuarii*. Campbell and Postgate (1965), noting the consistent differences in the sporeforming sulfate-reducers, created the new genus *Desulfotomaculum* (41.7 to 45.5 mole % G + C) and emended the earlier descriptions. There are no described marine species. Then Postgate and Campbell (1966) classified the non-sporeforming sulfate-reducing bacteria as *Desulfovibrio* (46.1 to 61.2 mole % G +C), put the strains that exist in recognized collections into five species, and listed their synonyms. All five species recognized in *Bergey's 8th* have marine representatives.

Sulfate reduction occurs by two pathways (Peck, 1961; Roy and Trudinger, 1970). In microorganisms that are "assimilatory sulfate-reducers," that is, microorganisms that reduce only enough sulfur to satisfy their nutrient requirement, the substrate is PAPS (3^1-phosphoadenosine-5^1-phosphosulfate) and the enzyme is PAPS-reductase. In microorganisms that are "dissimilatory sulfate-reducers," that is, microorganisms that utilize sulfate as their terminal electron-acceptor in anaerobic respiration, the substrate is APS (adenosine-5^1-phosphosulfate) and the enzyme is APS-reductase. The pathway of dissimilatory sulfate reduction in *Desulfovibrio* is shown in Table 17-2. But dissimilatory sulfate reduction in *Desulfovibrio* is not an obligatory function. *Desulfovibrio* grows well in sulfate-free media containing ethanol or lactate as an energy source if a hydrogen-utilizing methanogenic bacterium is also present (Bryant, 1969).

The enzyme APS-reductase also occurs in the genus *Thiobacillus* and catalyzes the oxidation of reduced sulfur compounds to sulfate. Thiobacilli-like pseudomonads of the *Thiobacillus trautweinii* type isolated from seawater (Tuttle and Jannasch, 1972) reduce inorganic sulfur compounds other than sulfate (Tuttle and Jannasch, 1973). An apparatus devised for the continuous culture of sulfate-reducing bacteria (Hallberg, 1970) has been tested for the removal of the toxic metals mercury and copper from polluted waters by precipitation as harmless metal sulfides (Vosjan and Van Der Hoek, 1972). Although the desulfovibrios may prove useful in this application, they can also cause economic loss by corrosion of concrete and iron structures buried in marine sediments (Roy and Trudinger, 1970).

D. METHANE-PRODUCING BACTERIA

The effective oxidation of major organic materials, such as cellulose, proteinaceous matter, organic acids, and alcohols, to acetate and carbon dioxide, by fermenting non-methanogenic anaerobes requires that the hydrogen produced be kept at low, non-inhibitory concentrations. In the absence of such electron acceptors as nitrate and sulfate, this is accomplished by interacting bacterial species, such as the methanogenic bacteria, which are a highly specialized group. They are not only essential in the production of the terminal products of fermentation (methane and carbon dioxide), but they also maintain the overall fermentation process by utilizing excess hydrogen by methane formation. They do this by cleaving acetate to carbon dioxide, which is reduced with hydrogen to form methane. Some 60 to 70% of the methane comes from the methyl group of acetate. Methane is thus produced by methanogenic bacteria in such anaerobic environments as marshes and the gut of animals where fermentative bacteria are vigorously fermenting organic matter.

Methane is present in trace amounts in pelagic waters and usually decreases in concentration with depth, indicating that the sea may be a sink for methane (Swinnerton and Linnenbom, 1967; Swinnerton, Linnenbom, and Cheek, 1969). At certain stations in the Gulf of Mexico, however, five- to fourteen-fold increases were found at depths of from 30 to 50 m (see Plate 17-3,*A*), suggesting that methane may be released in the euphotic zone from the anaerobic gut of animals. A more usual environment for methane is the anoxic waters of the Black Sea, fjords, and deep trenches, which have been characterized by Atkinson and Richards (1967) (see Plate 17-3,*B* to *D*). Here there is a tendency for a gradual increase with depth, to values of 12 to 14 μmoles of methane/liter in the Cariaco Trench and Black Sea, and to 30 to 70 μmoles of methane in Lake Nitinat, an anoxic fjord in British Columbia. The distribution of methane in Chesapeake sediments was observed by Reeburgh (1969) (Plate 17-3,*E*). Concentrations in the interstitial or pore water of sediments became appreciable at a depth of about 10 cm and increased to 40 to 75 μmoles/liter at depths of 80 to 100 cm. This has been confirmed by Martens and Berner (1974), who found that about 90% of the seawater sulfate must be reduced in the pore water by sulfate-reducing bacteria before appreciable amounts of methane can be formed. The ability of *Desulfovibrio* to grow with methanogenic bacteria in the absence of sulfate (Bryant, 1969) fits in with this observation. Despite the well-known occurrence of methane production in anaerobic marine environments, only a few methanogenic species have been described from the marine environment.

Wolfe's excellent review article (1971) on this small but common group of bacteria includes discussions on their isolation, characteristics, mass culture, and biochemistry. Methane bacteria do not utilize the usual carbohydrate and proteinaceous substrates or other organic compounds (except those listed below) as a source of energy; they obtain their energy by the formation of methane from the reduction of carbon dioxide. Carbon dioxide is reduced by utilizing the electrons generated by the oxidation of hydrogen and formate or the fermentation of acetate and methanol in which both methane and carbon dioxide are products. Although the classification of these highly specialized bacteria is in need of drastic revision, all the validly described species have now been pulled together into one physiological family, the Methanobacteriaceae, in *Bergey's 8th*. The other known species and genera require further taxonomic study and description, but are only encountered by

TABLE 17-2. Pathway of dissimilatory sulfate reduction in *Desulfovibrio* (Roy and Trudinger, 1970).

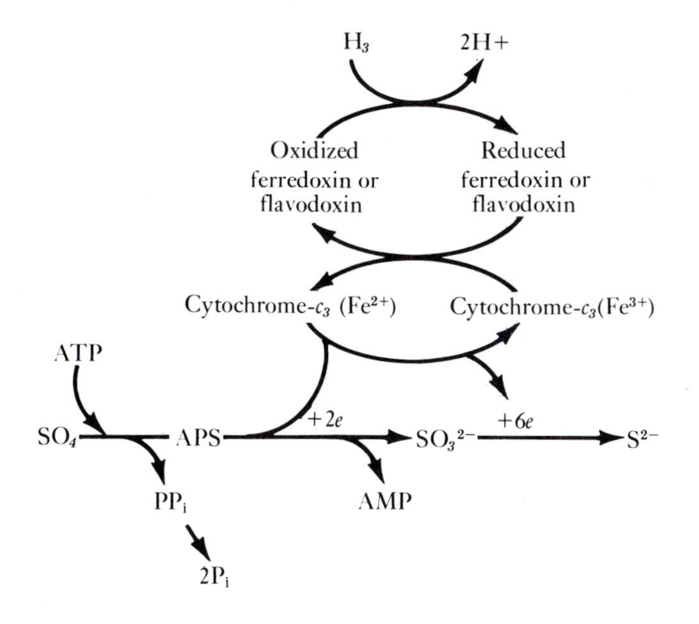

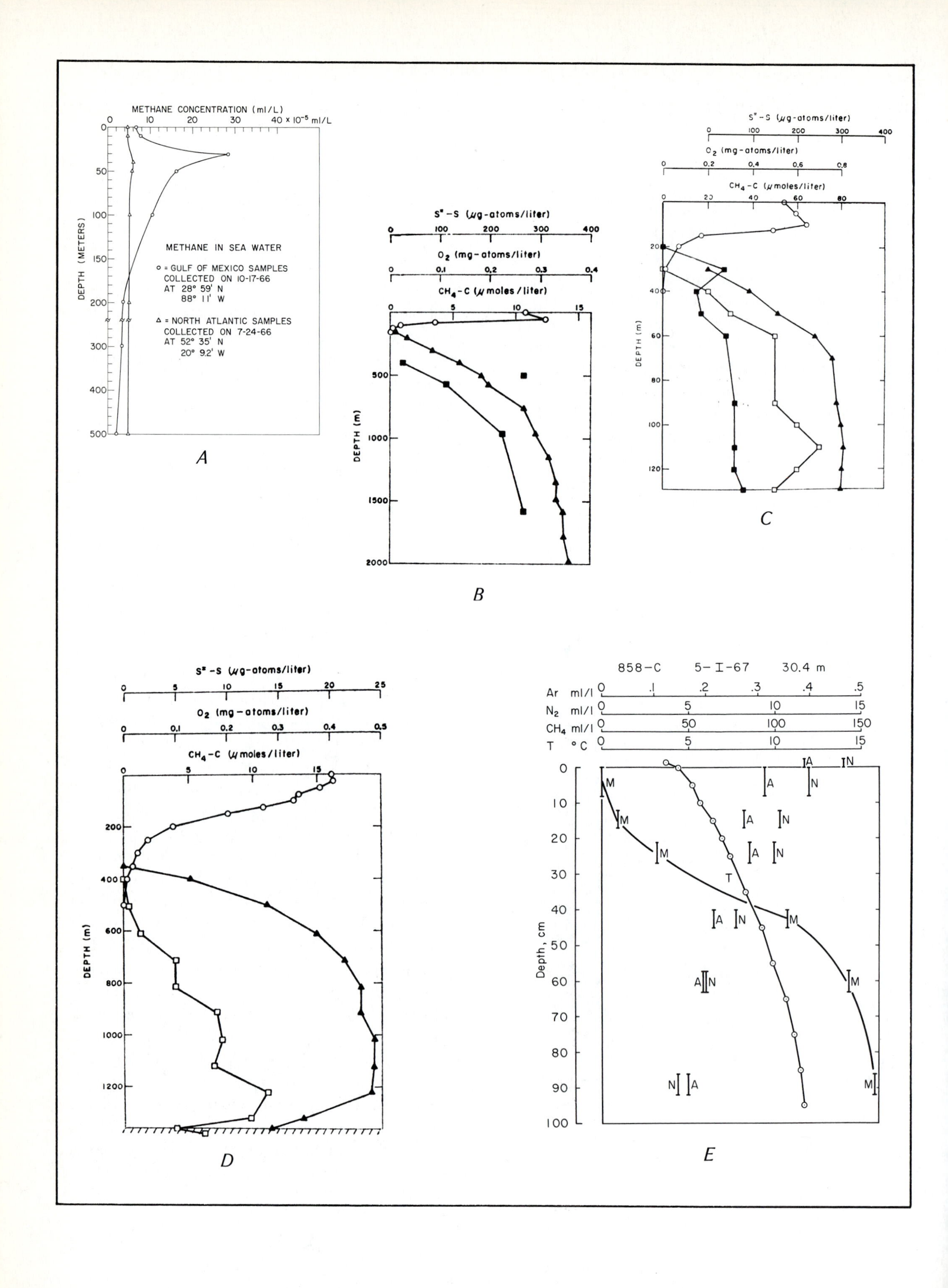

METHANE CONCENTRATION (ml/L)
0 10 20 30 40 x 10^-5 ml/L
DEPTH (METERS)
METHANE IN SEA WATER
○ = GULF OF MEXICO SAMPLES COLLECTED ON 10-17-66 AT 28° 59' N 88° 11' W
△ = NORTH ATLANTIC SAMPLES COLLECTED ON 7-24-66 AT 52° 35' N 20° 9.2' W
A
S=-S (μg-atoms/liter)
O2 (mg-atoms/liter)
CH4-C (μ moles/liter)
DEPTH (m)
B
C
D
858-C 5-I-67 30.4 m
Ar ml/l
N2 ml/l
CH4 ml/l
T °C
Depth, cm
E

PLATE 17-3. The occurrence of methane in the marine environment. (*A*) Concentration depth profile for methane ($x10^{-5}$ ml/liter) in oxygenated water from the Atlantic Ocean and the Gulf of Mexico. Vertical distribution of oxygen (O), hydrogen sulfide (▲), and methane (□, stripper; ϕ, extractor), in (*B*) the Black Sea; (*C*) the fjord, Lake Nitinat, British Columbia; and (*D*) the Cariaco Trench. (*E*) The distribution of argon (A), nitrogen (N), and methane (M) in a Chesapeake Bay sediment in January; temperature range (T). (*A*) from J.W. Swinnerton and V.J. Linnenbom, 1967, *Science* 156:1119–20, copyright 1967 by the American Association for the Advancement of Science; (*B,C,D*) from L. Atkinson and F.A. Richards, 1967, *Deep-Sea Res.* 14:673–84; (*E*) from W.S. Reeburgh, 1969, *Limnol. Oceanogr.* 14:368–75.

workers searching for them, since they are very strict anaerobes and thus highly susceptible to oxygen. Their sources of energy are limited and they are extremely difficult to obtain in pure culture.

A study on methane fermentation by microorganisms enriched from the marine environment was conducted by Stadtman and Barker (1951a). Primary cultures enriched from a black mud from San Francisco Bay in a formate–mineral, salts–carbonate medium required 2 weeks' incubation at 30 °C for gas production, but successive transfers using a 10% inoculum gave a surprisingly vigorous methane fermentation in 16 to 36 hours. The most conspicuous microorganism was a motile coccus. Attempts to isolate it were hampered by greater populations of a formate-fermenting rod, which produced carbon dioxide and hydrogen but not methane. The elevation of the culture medium pH from 7.0 to 8.0 facilitated the isolation of this rapidly growing coccus, which can only ferment formate to methane, carbon dioxide, and hydrogen (Stadtman and Barker, 1951b). This marine representative of the methane-producing bacteria was named *Methanococcus vannielii* in honor of C.B. van Niel, who developed the carbon-dioxide reduction theory of methane formation that stimulated research on this process (Barker, 1956). A second example of the problems these difficult bacteria present is the culture of *Methanobacterium omelianskii*, also isolated from marine muds by Barker (1936, 1940). It had been maintained and studied for over 30 years in an ethanol–carbonate medium as an example of a pure culture utilizing ethanol and certain aliphatic alcohols. Bryant et al. (1967), however, found that this culture really involved a pair of symbiotic bacteria and that neither bacterium could grow alone in this medium. A non-methanogenic species, designated "S" organism, produces hydrogen by oxidizing ethanol to acetate, but unless this hydrogen is removed by the methanogenic strain *Methanobacterium* M-O-H to reduce carbon dioxide and produce methane, there is hydrogen inhibition of the "S" organism.

There has been a recent interest in methane production in seagrass and marsh communities. Oremland (1975) observed that dark incubation of sediments overestimated methane production in beds of the turtle grass *Thalassia testudinum* and that methane production is depressed by photosynthetic oxygen production. More extensive observations on methanogenesis, extending from the marsh into estuarine sediments, have been made by Jones and Paynter (1975). These authors also developed a simple and reliable method to estimate populations of methanogenic bacteria. A negative pressure in a dilution culture that contains formate and a gas mixture of 80% hydrogen and 20% carbon dioxide indicates methanogenesis. A more elegant culture method, which also yields isolated colonies, is the use of Petri dishes in an ultra-low oxygen chamber and the detection of colonies of methanogens by a long-wave ultraviolet light (Edwards and McBride, 1975; Plate 17-1).

The methane bacteria in anaerobic sediments have a generally beneficial effect in the final step of the anaerobic decomposition of organic matter. But their presence may be disastrous under certain circumstances: the enzymes of methanogenic bacteria that actively convert the methyl group of methylated vitamin B_{12} (methyl cobalamin) to methane can apparently also catalyze the methylation of arsenate (McBride and Wolfe, 1969) and mercury (Wood, Kennedy, and Rosen, 1968) to toxic arsines and methyl mercury compounds, respectively. The outbreak in Japan of Minamata disease (Ui, 1969) is an example of what can happen when these substances enter the food chain at appreciable concentrations. An acetaldehyde factory using mercury as a catalyst began discharging a mercury-containing effluent into Minamata Bay in 1950. The mercury apparently became methylated by the methane bacteria in the anaerobic waste sludge and the bottom sediments and passed up the food chain to fish and shellfish and then to man. By 1953, the intake of the contaminated seafood had caused crippling, insanity, deafness, blindness, and death in at least 106 victims. Not until 1963 did a panel of investigators finally trace the "disease" to the mercury effluents.

REFERENCES

Aranki, A., and R. Freter. 1972. Use of anaerobic glove boxes for the cultivation of strictly anaerobic bacteria. *Am. J. Clin. Nutr.* 25:1329–34.

Asakawa, S. 1972. An experiment on the sulfate reducing activity of sea bottom mud. *J. Fac. Fish. Anim. Husb. Hiroshima Univ.* 11:59–64.

Atkinson, L.P., and F.A. Richards. 1967. The occurrence and distribution of methane in the marine environment. *Deep-Sea Res.* 14:673–84.

Baars, J.K. 1930. *Over sulfaatreductie door Bakterien.* W.D. Meinema, N.V., Delft, Holland, 164 p.

Barker, H.A. 1936. Studies upon the methane-producing bacteria. *Arch. Mikrobiol.* 7;420–38.

Barker, H.A. 1940. Studies upon the methane fermentation. IV. The isolation and culture of *Methanobacterium omelianskii. Antonie van Leeuwenhoek* 6:201–20.

Barker, H.A. 1956. *Bacterial Fermentations.* Wiley, New York, 95 p.

Barker, H.A., and S.M. Taha, 1942. *Clostridium kluyverii*, an organism concerned in the formation of caproic acid from ethyl alcohol. *J. Bacteriol.* 43:347–63.

Beijerinck, W.M. 1895. Ueber *Spirillum desulfuricans* als Ursache von Sulfatreduction. *Centralb. Bakteriol. Parasit.*, 2 Abt. 1:1–9.

Bella, D.A., A.E. Ramm, and P.E. Peterson. 1972. Effects of tidal flats on estuarine water quality. *Water Poll. Cont. Fed. J.* 44(4):541–56.

Berner, R.A. 1972. Sulfate reduction, pyrite formation, and the oceanic sulfur budget. In: *The Changing Chemistry of the Oceans* (D. Dryssen and D. Jagner, eds.). Wiley, New York, pp. 347–61.

Bonde, G.J. 1968. Studies on the dispersion and disappearance phenomena of enteric bacteria in the marine environment. *Rev. Intern. Oceanogr. Med.* 9:17–44.

Bryant, M.P. 1964. Some aspects of the bacteriology of the rumen. In: *Principles and Applications in Aquatic Microbiology* (H. Heukelekian and N.C. Dondero, eds.). Wiley, New York, pp. 366–93.

Bryant, M.P. 1969. Symbiotic associations of certain ethanol and lactate fermenting bacteria with methanogenic bacteria. Abstr. 158, *Natl. Mtg. Amer. Chem. Soc.* MICR-18.

Bryant, M.P., and I.M. Robinson. 1961. An improved nonselective culture medium for rumen bacteria and its use in determining diurnal variation in numbers of bacteria in the rumen. *J. Dairy Sci.* 44:1446–56.

Bryant, M.P., E.A. Wolin, M.J. Wolin, and R.S. Wolfe. 1967. *Methanobacillus omelianskii*, a symbiotic association of two species of bacteria. *Arch. Mikrobiol.* 59:20–31.

Campbell, L.L., and J.R. Postgate 1965. Classification of the spore-forming sulfate-reducing bacteria. *Bacteriol. Rev.* 29:359–63.

Campbell, L.L., M.A. Kasprzycki, and J.R. Postgate. 1966. *Desulfovibrio africanus* sp. n., a new dissimilatory sulfate-reducing bacterium. *J. Bacteriol.* 92:1122–27.

Cardon, B.P., and H.A. Barker. 1946a. Two new amino acid fermenting bacteria, *Clostridium propionicum* and *Diplococcus glycinophilus. J. Bacteriol.* 52:629–34.

Cardon, B.P., and H.A. Barker. 1946b. Amino fermentations by *Clostridium propionicum* and *Diplococcus glycinophilus. Arch. Biochem. Biophys.* 12:165–80.

Collee, J.G., B. Watt, E.B. Fowler, and R. Brown. 1972. An evaluation of the Gaspak system in the culture of anaerobic bacteria. *J. Appl. Bacteriol.* 35:71–82.

Davies, J.A. 1967. Clostridia from North Sea sediments. Ph.D. Thesis, University of Aberdeen.

Edwards, T., and B.C. McBride. 1975. New methods for the isolation and identification of methanogenic bacteria. *Appl. Microbiol.* 29:540–45.

Elion, L. 1924. A thermophilic sulphate-reducing bacterium. *Central. Bakteriol. Parasit. Infekt.* 2 Abt. 63:58–67.

Fulghum, R.S. 1971. Mobile anaerobe laboratory. *Appl. Microbiol.* 21:769–70.

Hallberg, R.O. 1970. An apparatus for the continuous cultivation of sulfate-reducing bacteria and its application to geomicrobiological purposes. *Antonie van Leeuwenhoek* 36:241–54.

Holdeman, L.V., and W.E.C. Moore (eds.). 1972. *Anaerobe Laboratory Manual.* Virginia Polytechnic. Inst., Blacksburg, Va., 132 p.

Hungate, R.E. 1950. The anaerobic mesophilic cellulolytic bacteria. *Bacteriol. Rev.* 14:1–49.

Hungate, R.E. 1966. *The Rumen and its Microbes.* Academic Press, New York and London, 533 p.

Iannotti, E.L., R. Elgart, and L.L. Curtis. 1975. Strict anaerobes and fecal indicators from marine sediments of the New York Bight. Abstr. Q33 Annual Meeting of the American Society of Microbiology, New York, p. 210.

Jones, W.J., and M.J.B. Paynter. 1975. Populations of methanogenic organisms in salt marsh and creek sediments. Abstr. I127, Annual Meeting of the American Society of Microbiology, New York, p. 138.

Kester, D.R., K.T. Crocker, and G.R. Miller, Jr. 1973. Small-scale oxygen variations in the thermocline. *Deep-Sea Res.* 20:409–12.

Martens, C.S., and R.A. Berner. 1974. Methane production in the interstitial waters of sulfate-depleted marine sediments. *Science* 185:1167–69.

Martin, W.J. 1971. Practical method for isolation of anaerobic bacteria in the clinical laboratory. *Appl. Microbiol.* 22:1168–71.

Matches, J.R., J. Liston, and D. Curran. 1974. *Clostridium perfringens* in the environment. *Appl. Microbiol.* 28:655–60.

McBride, B.C., and R.S. Wolfe. 1969. Biosynthesis of alkylated arsenic from CH_3-B_{12} in cell extracts of methane bacteria. *Bacteriol. Proc.* P85, p. 130 (Abstr.).

McMinn, M.T., and J.J. Crawford. 1970. Recovery of anaerobic microorganisms from clinical specimens in prereduced media versus recovery by routine clinical laboratory methods. *Appl Microbiol.* 19:207–13.

Miller, J.D.A., and A.M. Saleh. 1964. A sulphate-reducing bacterium containing cytochrome C_3 but lacking desulfiviridin. *J. gen. Microbiol.* 37:419–23.

Miller, L.G., P.S. Clark, and G.A. Kunkle. 1972. Possible origin of *Clostridium botulinum* contamination of Eskimo foods in Northwestern Alaska. *Appl Microbiol.* 23:427–28.

Murray, J., and R. Irvine. 1892–93. On the chemical changes which take place in the composition of the sea-water associated with blue muds on the floor of the ocean. *Trans. Royal Soc. Edinburgh* 37:481–507.

Oremland, R.S. 1975. Methane production in shallow-water, tropical marine sediments. *Appl Microbiol.* 30:602–8.

Peck, H.D. 1961. Enzymatic basis for assimilatory and dissimilatory sulfate reduction. *J. Bacteriol.* 82:933–39.

Perkins, E.J. 1974. The Marine Environment. In: *The Biology of Plant Litter Decomposition* (C.H. Dickinson and G.J.F. Pugh, eds.). Academic Press, London and New York, pp. 683–721.

Postgate, J.R. 1959. A diagnostic reaction of *Desulphovibrio desulphuricans. Nature* 183:481–82.

Postgate, J.R. 1963. Versatile medium for the enumeration of sulfate-reducing bacteria. *Appl. Microbiol.* 11:265–67.

Postgate, J.R., and L.L. Campbell. 1966. Classification of *Desulfovibrio* species, the non-sporulating sulfate-reducing bacteria. *Bacteriol. Rev.* 30:732–38.

Redfield, A.C. 1958. The biochemical control of chemical factors in the environment. *Amer. Scient.* 46:205–21.

Reeburgh, W.S. 1969. Observations of gases in Chesapeake Bay sediments. *Limnol. Oceanogr.* 14:368–75.

Roberts, T.A., and G. Hobbs. 1968. Low temperature growth characteristics of Clostridia. *J. Appl Bacteriol.* 31:75–88.

Roy, A.B., and P.A. Trudinger. 1970. *The Biochemistry of Inorganic Compounds of Sulphur*. Cambridge University Press, pp. 275–88.

Smith, L. DS. 1968. The clostridial flora of marine sediments from a productive and from a non-productive area. *Can. J. Microbiol.* 14:1301–04.

Smith, L. DS. 1970. *Clostridium oceanicum*, sp. n., a spore-forming anaerobe isolated from marine sediments. *J. Bacteriol.* 103:811–13.

Stadtman, T.C., and H.A. Barker. 1951a. Studies on the methane fermentation. VIII. Tracer experiments on fatty acid oxidation by methane bacteria. *J. Bacteriol.* 61:67–80.

Stadtman, T.C., and H.A. Barker. 1951b. Studies on the methane fermentation. X. A new formate-decomposing bacterium, *Methanoccocus vannielii. J. Bacteriol.* 62:269–80.

Stadtman, T.C., and L.S. McClung. 1957. *Clostridium stricklandii* nov. spec. *J. Bacteriol.* 73:218–19.

Starkey, R.L. 1938. A study of spore-formation and other morphological characteristics of *Vibrio desulfuricans. Arch. Mikrobiol.* 9:268–304.

Sverdrup, H.A., M.W. Johnson, and R.H. Fleming. 1942. *The Oceans: Their Physics, Chemistry and General Biology*. Prentice-Hall, Englewood Cliffs, N.J., 1087 p.

Swinnerton, J.W., and V.J. Linnenbom. 1967. Gaseous hydrocarbons in seawater: Determination. *Science* 156:1119–20.

Swinnerton, J.W., V.J. Linnenbom, and C.H. Cheek. 1969. Distribution of methane and carbon monoxide between the atmosphere and natural waters. *Environm. Sci. Technol.* 3:836–38.

Timmis, K., G. Hobbs, and R.C.W. Berkeley. 1974. Chitinolytic clostridia isolated from marine mud. *Can. J. Microbiol.* 20:1284–85.

Tuttle, J.H., and H.W. Jannasch. 1972. Occurrence and types of *Thiobacillus*-like bacteria in the sea. *Limnol. Oceanogr.* 17:532–43.

Tuttle, J.H., and H.W. Jannasch. 1973. Dissimilatory reduction of inorganic sulfur by facultatively anaerobic marine bacteria. *J. Bacteriol.* 115:732–37.

Ui, J. 1969. Minamata disease and water pollution by industrial waste. *Rev. Intern. Océanogr. Méd* 13/14:37–44.

van Delden, A. 1903. Beitrag zur Kenntnis der Sulfatreduction durch Bakterien. *Central Bakteriol. Parasit. Infekt.* 2 Abt. 11:81–94.

Vervaeke, I.J., and C.J. Van Nevel. 1972. Comparison of three techniques for the total count of anaerobes from intestinal contents of pigs. *Appl. Microbiol.* 24:513–15.

Vosjan, J.H., and G.J. Van Der Hoek. 1972. A continuous culture of desulfovibrio on a medium containing mercury and copper ions. *Netherlands J. Sea Res.* 5:440–44.

Wachsman, J.T., and H.A. Barker. 1954. Characterization of an orotic acid fermenting bacterium, *Zymobacterium oroticum*, nov. gen., nov. spec. *J. Bacteriol.* 68:400–404.

Waksman, S.A., M. Hotchkiss, and C.L. Carey. 1933. Marine bacteria and their role in the cycle of life in the sea. II. Bacteria concerned in the cycle of nitrogen in the sea. *Biol. Bull.* 65:137–67.

Ward, B.Q., B.J. Carroll. E.S. Garrett, and G.B. Reese. 1967. Survey of the U.S. Gulf Coast for the presence of *Clostridium botulinum. Appl. Micribiol.* 15:629–36.

Wolfe, R.S. 1971. Microbial formation of methane. *Adv. Microbial Physiol.* 6:107–46.

Wood, J.M., F.S. Kennedy, and C.G. Rosen. 1968. Synthesis of methyl-mercury compounds by extracts of a methanogenic bacterium. *Nature* 220:173–74.

Wyrtki, K. 1962. The oxygen minima in relation to ocean circulation. *Deep-Sea Res.* 9:11–23.

ZoBell, C.E., and S.C. Rittenberg. 1948. Sulfate-reducing bacteria in marine sediments. *J. Mar. Res.* 7:602–17.

CHAPTER 18

Aerobic Gram-positive Epibacteria

Water samples, the most studied of all marine material, usually contain few gram-positive bacteria. Most investigators ignore them, or at best throw them into convenient catchall classifications, such as micrococci, bacilli, and corynebacteria. For these reasons, our information on the gram-positive bacteria in the sea is meager. Of the 475 isolates examined from Kamogawa Bay water, only 6% were gram-positive (Simidu and Aiso, 1962). In studying water from mid-Narragansett Bay, Sieburth (1967a) examined an average of 53 isolates from each of 21 semi-monthly samples for a total of 1119 isolates. Of the twenty-one samples, ten contained gram-positive forms, the highest occurrence being 26%, the average occurrence 2.6%. In such shallow waters, the source of the gram-positive bacteria is probably resuspended sediments rich in organic matter. Gram-positive forms are always present in nearshore samples, but they are not usually detected, since they are crowded out by the dominant gram-negative forms (Chapter 15) that also dominate the detritus on the surface of bottom sediments (Sieburth, 1967a). This does not mean that there are not sizable populations of gram-positive microorganisms; it means rather that when the gram-negative population is 10^8/g, the gram-positive population will be approximately 10^6/g. If culture plates are incubated at 36 °C instead of the usual 15 to 18 °C, many gram-negative forms are inhibited, and the gram-positive forms increase from 2 to 26% (Sieburth, 1967a). The incidence of gram-positive forms on fish from Puget Sound was mouth, 9%; gut, 11.5%; gills, 12.6% and skin 15.6% (Colwell, 1962). The high incidence of gram-positive bacteria on fish, especially the bacilli and micrococci, appears to indicate contact with bottom sediments. The occurrence of these microorganisms was 5 to 10% for purse-seine and offshore otter trawl and handline-caught fish, but 43% for those from an inshore otter trawl, in which the fish are tumbled and dragged through rich bottom sediments (Colwell, 1962). Spencer (1959, 1960) thought that the gram-positive bacteria on fish were derived from both the net and the bottom mud. With the exception of a survey on North Sea sediments (Boeyé, Wayenbergh, and Aerts, 1975) and a taxonomic analysis of the resulting isolates of *Bacillus* (Boeyé and Aerts, 1976), the occurrence of different groups of gram-positive bacteria in marine sediments has apparently not been recorded in recent years. Their evident association with inshore sediments is probably due to the continual incorporation of organic debris into these sediments, which meets their high nutrient requirement. (The gram-positive microorganisms in sediments are long overdue for an intensive study of their occurrence and *in-situ* activity.)

The recent observation that the surface microlayer could contain dissolved organic matter at a concentration equal to that of laboratory media (Sieburth et al., 1976) prompted further observations on this microhabitat as a possible niche where enteric microorganisms could persist and proliferate. On a transect from Narragansett Bay to the western Sargasso Sea, consistently high background levels of gram-positive cocci were obtained on a selective medium for enterococci inoculated with glass-plate collected material but not with subsurface water from a depth of 5 m. This indicates that the sea surface as well as the sea floor may be a reservoir for marine gram-positive bacteria (Sieburth and Dufour, unpublished data).

A. MICROCOCCI

Only one field study has been concerned with the nature of marine micrococci. Three samples of water from the shallow and well-mixed North Sea, which were plated on a relatively rich organic medium, yielded 205 isolates of micrococci from some 700 randomly picked colonies, an incidence of approximately 29% (Anderson, 1962). These catalase-positive cocci were irregular in size, ranging from 0.5 to 3.0 μm in diameter, and usually occurring in irregular clumps or occasionally in pairs. In a modified glucose oxidation/fermentation medium, 2.5% were inactive, 7.5% produced acid oxidatively, whereas 90% were fermentative. The fermentative micrococci were the most irregular in size. Of the fermenters, 53% of the isolates were most active physiologically, being characterized by the clotting of litmus milk and the production of acid from galactose but not mannitol. Anderson was reluctant to put these micrococci into taxonomic groups. But Evans, Bradford, and Niven (1955) had already shown that the catalase-positive cocci could be divided into the genus *Micrococcus*, which only uses glucose aerobically, and the genus *Staphylococcus*, which also uses glucose anaerobically. This was later confirmed by Baird-Parker (1963). This physiological division has been verified by the determination of DNA base composition values. The species of *Staphylococcus* have a 30.8 to 36.5 mole % G + C content (Silvestri and Hill, 1965); the species of *Micrococcus* lie within the range of 65 to 75 mole % (Boháček, Kocur, and Martinec, 1967). A comparison of 15 strains of marine cocci from culture collections indicated that, in addition to these two genera, a third group of cocci were distinguishable by being flagellated and having a 48 to 52 mole % G + C content (Boháček, Kocur, and Martinec, 1968). They are also serologically distinct from staphylococci and micrococci (Oeding, 1971). These strains have been put in an emended genus, *Planococcus* (Plate 18-1,*C*), and include the *Micrococcus aquivivus* of ZoBell and Upham (1944), the flagellated cocci from a clam intestine (Leifson et al., 1964) previously described as *M. eucinetus* (Leifson, 1964), as well as three species of micrococci isolated from fish by investigators at the Torry Research Station (Kocur et al., 1970). Although it is only speculative, one might deduce from Anderson's data that 7.5% of his isolates were either *Micrococcus* or *Planococcus*, whereas some 90% were strains of *Staphylococcus*. The taxonomy of the micrococci, as clearly laid out in *Bergey's 8th*, is such that in future studies there should be little trouble in characterizing these marine isolates. It will be interesting to see in which group(s) the cocci of the sea–air interface fall.

B. SPOREFORMING BACILLI

The genus *Bacillus* contains aerobic, gram-positive, sporeforming catalase-positive rods. Like the micrococci just discussed, they can be aerial contaminants and are usually not taken too seriously when only an occasional colony is detected on plates inoculated with seawater. In cultures from marine sediments, the characteristic colonies of *Bacillus* are quite common (ZoBell, 1946; Wood, 1953). But just like their anaerobic counterparts in the genus *Clostridium*, their presence may only indicate the existence of resting spores, which have accumulated to detectable numbers over the years. An indica-

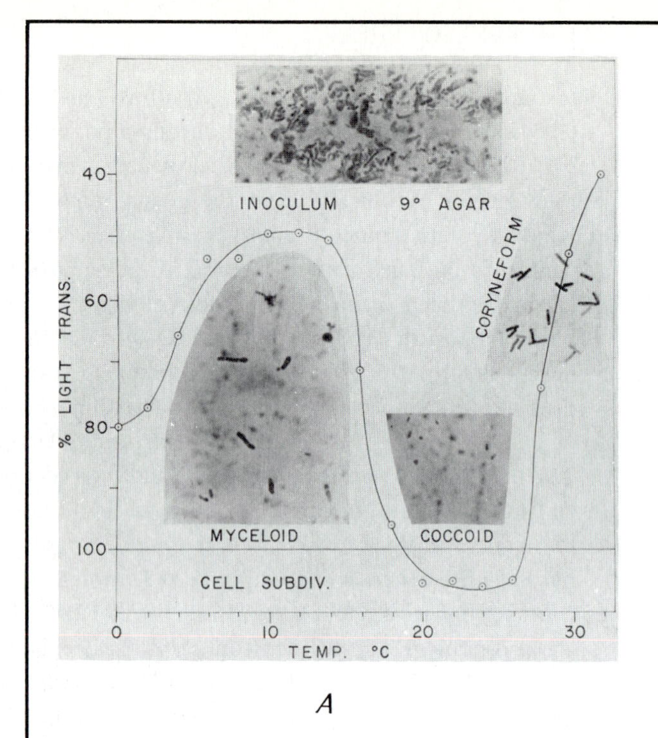
INOCULUM
9° AGAR
CORYNEFORM
MYCELOID
COCCOID
CELL SUBDIV.
% LIGHT TRANS.
TEMP. °C
40
60
80
100
0
10
20
30

A

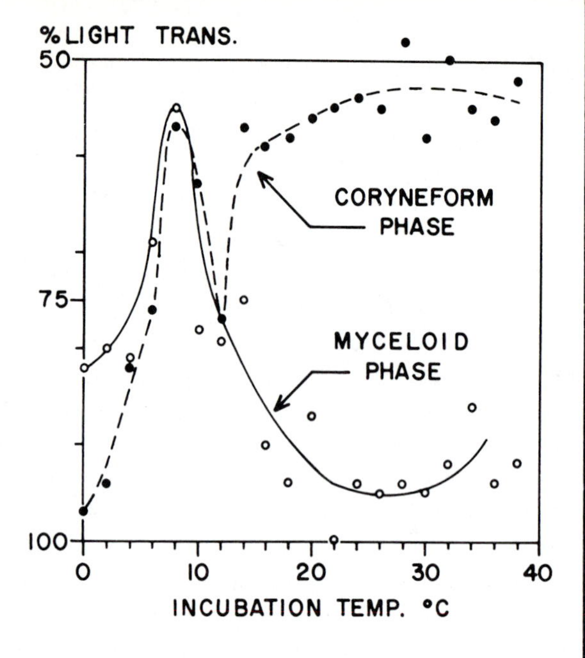
%LIGHT TRANS.
50
75
100
CORYNEFORM
PHASE
MYCELOID
PHASE
0
10
20
30
40
INCUBATION TEMP. °C

B

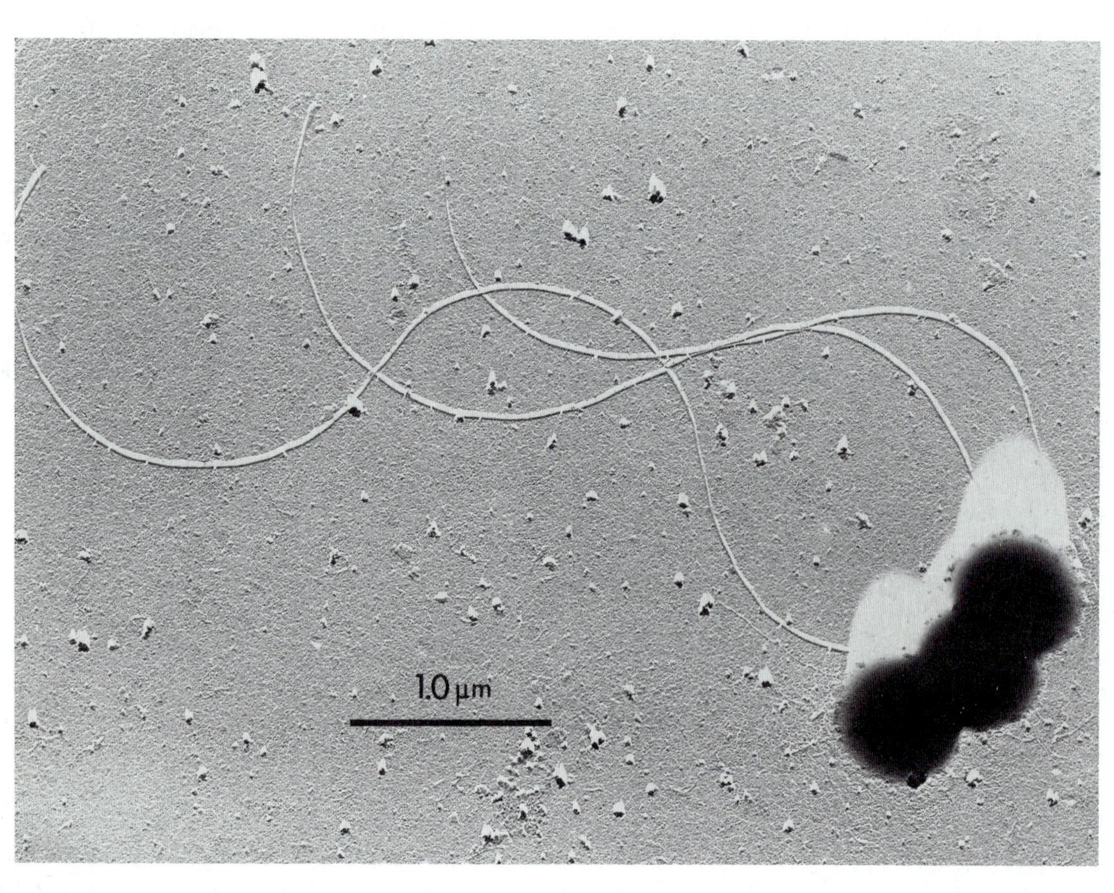
1.0 µm

C

PLATE 18-1. Unusual gram-positive bacteria, the gram-variable *Arthrobacter*-like forms, and a motile coccus. (*A*) A pleomorphic gram-negative bacterium isolated from seawater develops an *Arthrobacter*-like life cycle as a function of temperature. Below 20 °C the gram-negative myceloid stage persists; a gram-positive coryneform stage grows above 26 °C and apparently divides to form a gram-positive coccoid resting stage (arthrospore) between 20 and 26 °C. (*B*) Growth pattern in broth of an *Arthrobacter*-like isolate in which the myceloid form maintained on agar at 9 °C and the coryneform form maintained at 10 °C are grown across a temperature spectrum. (*C*) A group of three cocci of *Planococcus aquivivus*, isolated from clams and fish, in a shadowed preparation, showing flagella with a regular sine curve. (*A,B*) J.McN. Sieburth, 1964, In: *Proc. Symp. Exp. Mar. Ecol.*, Occasional Publication No. 2, Graduate School of Oceanography, University of Rhode Island, pp. 11–16; (*C*) from M. Kocur, Z. Páčová, W. Hodgkiss, and T. Martinec, 1970, *Int. J. Syst. Bacteriol.* 20:241–48.

tion of this is the presence of obligately thermophilic sporeforming bacilli identical or similar to *Bacillus stearothermophilus*, which can only grow at temperatures from 55 to 70 °C, in populations of tens to hundreds of bacilli per gram of sediment on the continental shelf where the temperature is usually a constant 4 °C (Bartholomew and Paik, 1966). There are three choices as to the origin of such bacteria in cores from depths exceeding 150 cm. They could be living fossils many thousands of years old. They could arise from spores originating from the dungheaps and the other usual sources of thermophiles that are carried with dust into the atmosphere and then scrubbed by rain to fall into the sea, slowly settle to the seafloor, and there permeate the porous sediment by diffusion. A third source could be past or current hot spots on the seafloor associated with such fracture zones as the Galápagos rift that has recently been discovered as a productive abyssal area (Corliss and Ballard, 1977; Lonsdale, 1977). The question whether sporeforming bacteria actually grow in the marine environment is still unanswered. The genus *Bacillus* is a large group that contains many species, that can grow in brines up to saturation, but only in the presence of adequate nutrients. The current taxonomy of this group has been discussed by Gordon (1973) and in *Bergey's 8th*.

In a study that precluded the isolation of psychrophilic and thermophilic bacteria, Boeyé, Wayenberg, and Aerts (1975) observed the population of mesophilic aerobes in cool (6 and 14 °C) North Sea sediments that survived frozen storage and subsequent plating in a somewhat overrich 2216E medium (discussed by Sieburth, 1967a) at 44 °C (see Chapter 9). They were impressed with the predictable occurrence of 16 sporulating gram-positive rods out of 47 random isolates. This led to an in-depth study of 138 *Bacillus* strains by Boeyé and Aerts (1976). As one might expect from similar studies on other groups, conventional taxonomy failed at the species level, and groupings were formed by computer analysis. The isolates fell in a relatively small part of the *Bacillus* spectrum, forming two groups in the *B. subtilus* spectrum; one group was Voges-Proskauer positive and included the known strain *B. licheniformis*, and the other was V-P negative and included the known strains of *B. firmus*. Boeyé and Aerts (1976) have compared their taxonomic results and groupings with those of Bonde (1975), who also included isolates from coastal muds in his study of *Bacillus*.

C. ARTHROBACTERS

Gram-positive rods that stain irregularly and show some degree of pleomorphism fall into a number of genera. Gram-variable pleomorphic bacteria from soil that have rod-shaped, coccoid, and myceloid (mycelium-like) phases were considered by Jensen (1934) to be in the genera *Mycobacterium* and *Corynebacterium*, which include pathogenic bacteria. Conn (1948), who recognized these indigenous soil forms on the basis of both colonial and cell morphology, objected to placing saprophytic microorganisms in genera including pathogens and proposed an emendation of the genus *Arthrobacter* to include these forms (Conn and Dimmick, 1947). Wood (1953) classified about 15% of his some 3000 marine isolates as belonging to this group, but considered them to be *Corynebacterium*, 50% being identified as *C. globiforme*, since he objected to the genus *Arthrobacter* being emended on an ecological basis. Whether or not strains showing a pronounced tendency to have a "life cycle" with coccoid forms are put in the genus *Arthrobacter* depends upon the describer's orientation. A bacterium described as *Arthrobacter marinus* by Cobet, Wirsen, and Jones (1970) was considered a pseudomonad by Baumann et al. (1972). A further study by Rake and Jones (1977) failed to show the gram-positive stage and the characteristic morphological cycle of *Arthrobacter*, and this species has been reclassified as *Alcaligenes marinus*.

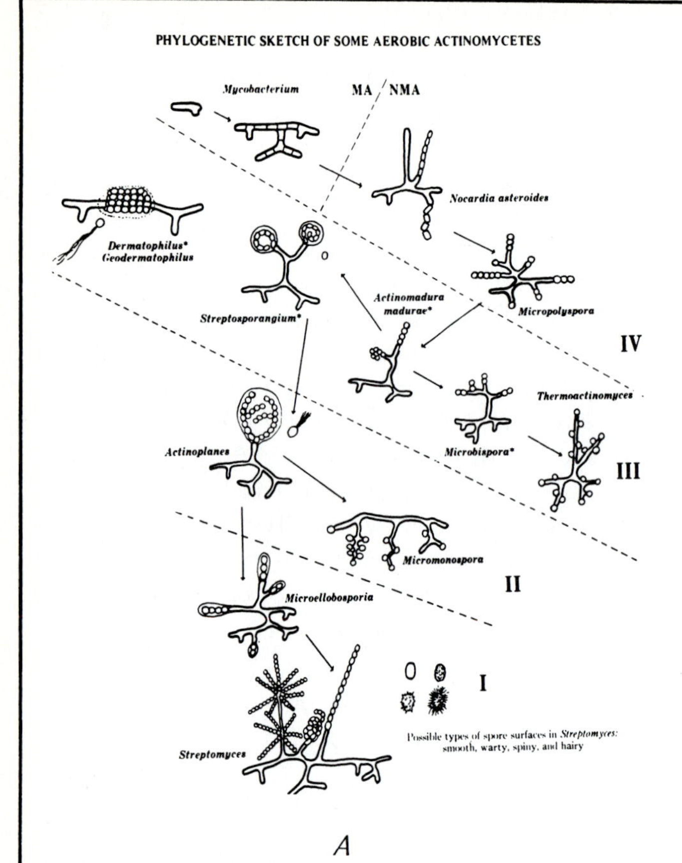
PHYLOGENETIC SKETCH OF SOME AEROBIC ACTINOMYCETES
Mycobacterium
MA
NMA
Nocardia asteroides
Dermatophilus*
Geodermatophilus
Streptosporangium*
Actinomadura
madurae*
Micropolyspora
IV
Thermoactinomyces
Microbispora*
III
Actinoplanes
Micromonospora
II
Microellobosporia
I
Streptomyces
Possible types of spore surfaces in Streptomyces:
smooth, warty, spiny, and hairy

A

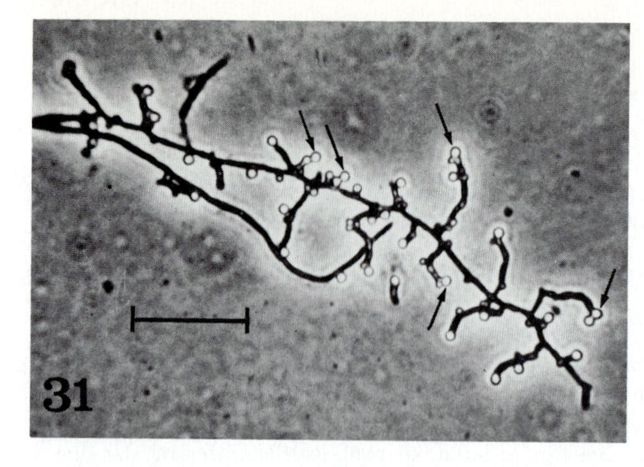
31

B

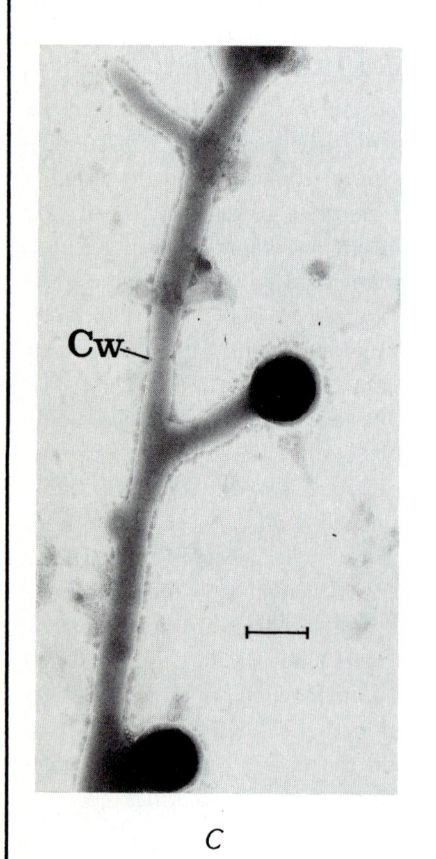
Cw

C

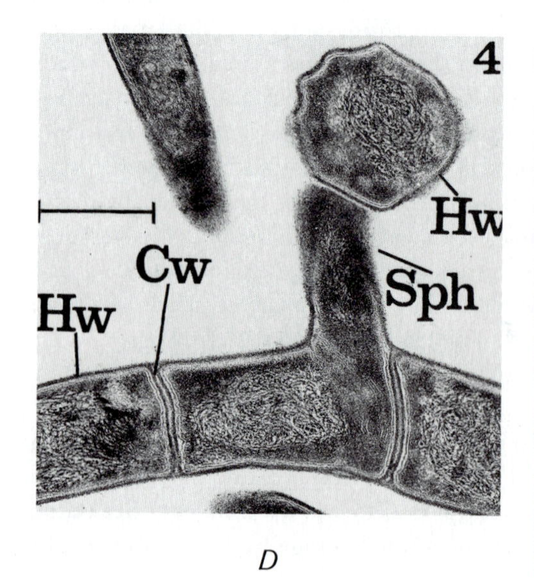
4
Hw
Cw
Sph
Hw

D

PLATE 18-2. The aerobic actinomycetes. (*A*) A phylogenic sketch of some of the dominant genera, many of which occur in marine niches. The morphology and ultrastructure of *Micromonospora chalcea*, the type species: (*B*) Phase-contrast micrograph of sporulating mycelium, showing spore pairs (arrows), (bar = 20 μm). (*C*) Negative stain of a whole mount, showing a single spore on a lateral side branch (bar = 1 μm). (*D*) Longitudinal section of hypha, lateral sporophore, and very young (top) and more mature (bottom) spores (Bars = 1 μm) (Hw, hyphal cell wall; Cw, cross-wall; Sph, sporophore; C, cortex; Osc, outer spore coat). (*A*) from H. Lechevalier, 1967, *Laval Medical* 38:740; (*B,C,D*) from G.M. Luedemann and C.J. Casmer, 1973, *Int. J. Syst. Bacteriol.* 23:243–55.

Pleomorphic gram-negative bacteria with a characteristic gram-positive *Arthrobacter* life cycle induced by temperature and cultural conditions (Sieburth, 1964, 1967a,b, 1968) (see Plate 18-1) may be arthrobacters or they may belong to the gut-group vibrios, of dubious taxonomic status, discussed in Chapter 15,*A*. From studies on fish and marine muds (Wood, 1953; Colwell, 1962; Shewan, 1966), pleomorphic gram-variable bacteria apparently make up a substantial fraction of these microfloras. Veldkamp (1970) came to the conclusion that our state of knowledge is still too limited to provide a basis for a satisfactory classification of the bacteria falling into this group. Bousfield (1972) included a number of isolates from fish and other marine niches in a taxonomic study of these bacteria. Many of the marine isolates fell into the same computer grouping as the named species of *Arthrobacter*, showing a typical "life cycle." Although this was the most nearly homogeneous group, Bousfield concluded that the life cycle criterion was not sufficient to characterize the subgroup and that additional characteristics, such as nutrition and cell wall composition, would be more useful. But so far, taxonomic analysis has failed to indicate a solution for the unsuccessful taxonomy of this group. In *Bergey's 8th*, the genus *Arthrobacter* is grouped with the actinomycetes.

D. ACTINOMYCETES

This is a highly diverse group of gram-positive bacteria, which sometimes have acid-fast branching filaments, with colonies that range from typical bacterial colonies to colonies having a well-defined coherent mycelium, with specialized structures and spores (see Plate 18-2). The saprophytic actinomycetes have an aerobic metabolism and do not accumulate acids from carbohydrate substrates, whereas the parasitic forms are usually micro-aerophilic and convert about 50% of their substrate carbon to acids. *Mycobacterium marinum* has been isolated from spontaneous tubercular lesions from fish dying in seawater aquaria (Aronson, 1926). The incidence of skin sensitivity to *Mycobacterium intracellulare*, which is concentrated along the U.S. southeastern coast from Virginia to Florida and along the Gulf of Mexico, and which extends for some distance inland, is thought to be due to the ability of this bacterium to survive in seawater, to be concentrated into surface films, and to be projected into aerosols and carried by prevailing winds (Gruft, Katz, and Blanchard, 1975). Saprophytic actinomycetes in the genera *Mycobacterium*, *Actinomyces*, *Nocardia*, and *Micromonospora*, as well as pseudomonads in marine sediments, have been reported to oxidize the hydrocarbons in petroleum (ZoBell, Grant, and Haas, 1943). Two species of *Actinomyces* associated with kelp and marine muds were described by ZoBell and Upham (1944). Two species of *Proactinomyces*, and one of *Actinomyces* that digests agar, were described from intertidal seaweed and marine sediments by Humm and Shepard (1946). Strains of *Nocardia* and *Streptomyces* were obtained from cordage and fishnets by Freitas and Bhat (1954). Air-dried seaweed yielded one strain of *Nocardia* and six of *Streptomyces* (Siebert and Schwartz, 1956). Seventeen strains of alginate- and laminarin-decomposing actinomycetes, including *Streptomyces*, were obtained from decomposing seaweeds and sediments by Chesters, Apinis, and Turner (1953–54).

The first survey for actinomycetes in marine sediments was conducted by Grein and Meyers (1958). Ninety-six samples yielded some 262 isolates, and about 50% of the 200 isolates tested showed antibacterial activity. Approximately 80% were strains of *Streptomyces*, with *S. griseus* being quite common; the remaining isolates belonged to *Nocardia* and *Micromonospora*. In addition to the sediment isolates, some were also obtained from water samples. Such sterilized baits as basswood panels, cellulose disks, and sections of banana stalk supported the growth of *Streptomyces*. Maximum development of the isolates occurred in a medium containing 25 to 50% seawater. *Streptomyces griseus* showed more salinity tolerance than strains of *Streptomyces coelicolor*. Grein and Meyers were of the opinion that actinomycetes in the sea may be terrestrial forms that have adapted to the saline sedi-

PLATE 18-3. The growth cycle of *Blastococcus aggregatus*. Transmission electron micrographs of (*A*) a flagellated swarmer cell (negative stain), (*B*) coccal aggregates with buds (negative stain) and (*C*) in thin section. (*D*) Schematic representation, showing how *B. aggregatus* reproduces by budding to form flagellated swarmer cells and by multiple fission to form cell aggregates. (*A–D*) from R. Ahrens and G. Moll, 1970, *Arch. Mikrobiol.* 70:243–65.

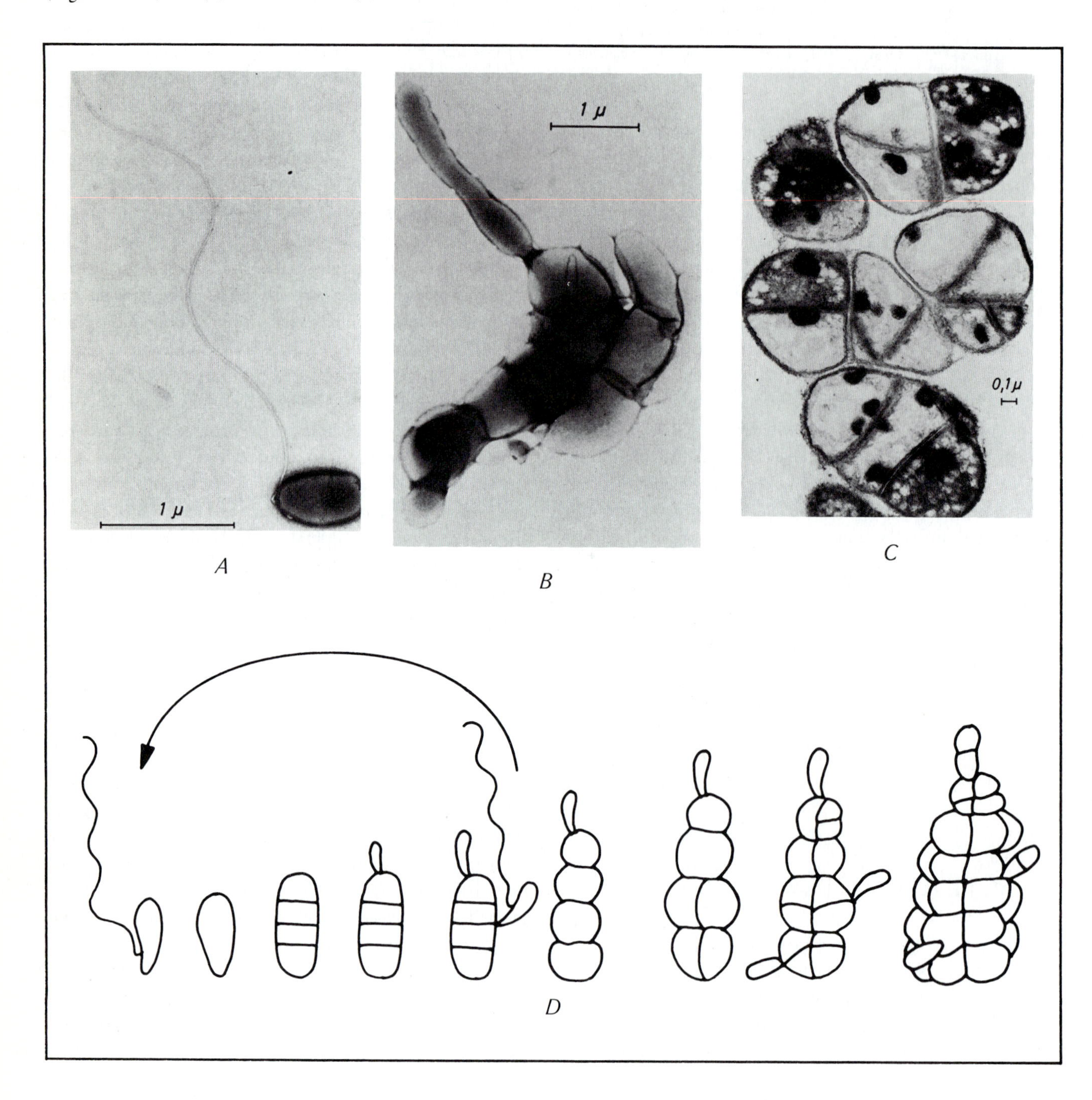

ments and seawater. The first study to enumerate actinomycetes in fresh sediment samples and to examine offshore samples was conducted in the North Sea and in the open Atlantic Ocean by Weyland (1969). In the North Sea, 102 of 107 stations yielded colonies of actinomycetes, with counts ranging from 39 to 2909/cm^3. In the Atlantic, from depths of 25 to 3362 m, nine out of twelve samples contained actinomycetes, the number ranging from 136/cm^3 at a 299-m depth, 40 km from shore, to 23/cm^3 at a 3662-m depth, 300 km from shore. The 1348 isolates included strains of *Nocardia*, *Micromonospora*, *Microbispora*, and *Streptomyces*. About one-half of the strains developed aerial mycelium. The typical morphology and ultrastructure of *Micromonospora* is shown in Plate 18-2. In a survey of Sagami Bay, 136 strains of actinomycetes were obtained from 37 samples, of which 27% had antimicrobial activity, and 17% inhibited a sarcoma cell. The saline tolerance of representative isolates was also tested (Okazaki and Okami, 1972).

It would appear from the above that actinomycetes belonging to four or five genera are associated with vegetative matter and sediments in the sea, and apparently take an active part in the benthic microflora. In addition to antibiotic production, some actinomycetes are used industrially for the chemical transformation of such substrates as steroids (Waksman, 1967).

E. BLASTOCOCCUS

An unusual estuarine microorganism has been described from the Baltic Sea (Ahrens and Moll, 1970). It has the appropriate name *Blastococcus aggregatus*—budding coccus that forms aggregates. Its growth cycle is shown in Plate 18-3. The flagellated swarmer cells lose their flagella and undergo multiple division to form a fat rod. The rods then swell to form cocci, and new swarmer cells are produced by budding. Thin sections show the invagination of the cytoplasmic membrane to form lamellar structures. This species has a characteristic, rose-colored pigment and a salinity preference of 15 o/oo. A similar rose-colored, budding, and aggregating coccus (Haskins Strain No. M_{23}) has been isolated from seaweeds on the Connecticut coast by Irma Pintner. This strain restores normal morphology to algae that are aberrant in axenic culture (Provasoli and Pintner, 1972). The prevalence, habitat, and physiology of these bacteria have yet to be determined. They may be related to the actinomycete-like microorganisms, which have a similar life cycle, in the family Dermatophilaceae (see Plate 18-2 and *Bergey's 8th*).

REFERENCES

Ahrens, R., and G. Moll. 1970. Ein neues knospendes Bakterium aus der Ostsee. *Arch. Mikrobiol.* 70:243–65.

Anderson, J.I.W. 1962. Studies on micrococci isolated from the North Sea. *J. appl. Bacteriol.* 25:362–68.

Aronson, J.D. 1926. Spontaneous tuberculosis in salt water fish. *J. Infect. Dis.* 39:315–20.

Baird-Parker, A.C. 1963. A classification of micrococci and staphylococci based on physiological and biochemical tests. *J. gen. Microbiol.* 30:409–27.

Bartholomew, J.W., and G. Paik. 1966. Isolation and identification of obligate thermophilic sporeforming bacilli from ocean basin cores. *J. Bacteriol.* 92:635–38.

Baumann, P., L. Baumann, M. Mandel, and R.D. Allen. 1972. Taxonomy of aerobic marine eubacteria. *J. Bacteriol* 110:402–29.

Boeyé, A., and M. Aerts. 1976. Numerical taxonomy of *Bacillus* isolates from North Sea sediments. *Int. J. Syst. Bacteriol.* 26:427–41.

Boeyé, A., M. Wayenbergh, and M. Aerts. 1975. Density and composition of heterotrophic bacterial populations in North Sea sediments. *Mar. Biol.* 32:263–70.

Bohaćček, J., M. Kocur, and T. Martinec. 1967. DNA base composition and taxonomy of some micrococci. *J. gen. Microbiol.* 46:369–76.

Bohaćcek, J., M. Kocur, and T. Martinec. 1968. Deoxyribonucleic acid base composition of some marine and halophilic micrococci. *J. Appl. Bacteriol.* 31:215–19.

Bonde, G.J. 1975. The genus *Bacillus*. *Dan. Med. Bull.* 22:41–61.

Bousfield, I.J. 1972. A taxonomic study of some coryneform bacteria. *J. gen. Microbiol.* 71:441–55.

Chesters, C.G.C., A. Apinis, and M. Turner. 1953–54. Studies on the decomposition of seaweeds and seaweed products by micro-organisms. *Proc. Linn. Soc. London* 166:87–97.

Cobet, A.B., C. Wirsen, Jr., and G.E. Jones. 1970. The effect of nickel on a marine bacterium *Arthrobacter marinus* sp. nov. *J. gen. Microbiol.* 62:159–69.

Colwell, R.R. 1962. The bacterial flora of Puget Sound fish. *J. Appl. Bacteriol.* 25:147–58.

Conn, H.J. 1948. The most abundant groups of bacteria in soil. *Bacteriol. Rev.* 12:257–73.

Conn, H.G., and I. Dimmick. 1947. Soil bacteria similar in morphology to *Mycobacterium* and *Corynebacterium*. *J. Bacteriol.* 54:291–303.

Corliss, J.B., and R.D. Ballard. 1977. Oases of life in the cold abyss. Natl. Geogr. 152:440–453.

Evans, J.B., W.L. Bradford, and C.F. Niven, Jr. 1955. Comments concerning the taxonomy of the genera *Micrococcus* and *Staphylococcus*. *Int. Bull. Bacteriol. Nomencl. Tax.* (*Int. J. Syst. Bacteriol.*) 5:61–66.

Freitas, Y.M., and J.V. Bhat. 1954. Microorganisms associated with the deterioration of fishnets and cordage. *J. Univ. Bombay* 23:53–59.

Gordon, R.E. 1973. The genus *Bacillus*. In: *Handbook of Microbiology, Vol. I, Organismic Microbiology* (A.I. Laskin and H.A. Lechevalier, eds.). CRC Press, Cleveland, Ohio, pp. 71–88.

Grein, A., and S.P. Meyers. 1958. Growth characteristics and antibiotic production of actinomycetes isolated from littoral sediments and materials suspended in seawater. *J. Bacteriol.* 76:457–63.

Gruft, H., J. Katz, and D.C. Blanchard. 1975. Postulated source of *Mycobacterium intracellulare* (Battey) infection. *Amer. J. Epidemiol.* 102:311–18.

Humm, H.J., and K.S. Shepard. 1946. Three new agar-digesting actinomycetes. *Duke Univ. Mar. Station Bull.* 3:76–80.

Jensen, H.L. 1934. Studies on saprophytic mycobacteria and corynebacteria. *Proc. Linn. Soc. New So. Wales* 59:19–61.

Kocur, M., Z. Páčová, W. Hodgkiss, and T. Martinec. 1970. The taxonomic status of the genus *Planoccocus* Migula 1894. *Int. J. Syst. Bacteriol.* 20:241–48.

Lechevalier, H. 1967. Les actinomycetes, grands producteurs d'antibiotiques. *Laval Médical, Université Laval, Québec,* 38:740.

Leifson, E. 1964. *Micrococcus ecinetus* n. sp. *Int. Bull. Bacteriol. Nomencl. Taxon.* (*Int. J. Syst. Bacteriol.*) 14:41–44.

Leifson, E., B.J. Cosenza, R. Murchelano, and R.C. Cleverdon. 1964. Motile marine bacteria. I. Techniques, ecology, and general characteristics. *J. Bacteriol.* 87:652–66.

Lonsdale, P. 1977. Clustering of suspension-feeding macrobenthos near abyssal hydrothermal vents of oceanic spreading centers. *Deep-Sea Res.* 24:857–863.

Luedemann, G.M., and C.J. Casmer. 1973. Electron microscope study of whole mounts and thin sections of *Micromonospora chalcea* ATCC 12457. *Int. J. Syst. Bacteriol.* 23:243–55.

Oeding, P. 1971. Serological investigations of *Planococcus* strains. *Int. J. Syst. Bacteriol.* 21:323–25.

Okazaki, T., and Y. Okami. 1972. Studies on marine microorganisms. II. Actinomycetes in Sagami Bay and their antibiotic substances. *J. Antibiotics* 25:461–66.

Provasoli, L., and I.J. Pintner. 1972. Effects of bacteria on seaweed morphology. *Abstr., J. Phycol.* 8 (Suppl.):10.

Rake, J.B., and G.E. Jones. 1977. Reclassification of *Arthrobacter marinus* (Cobet et al., 1970) as an anomolous species of Allcaligenes: *A. marinus*. Abstract of Annual Meeting 1977, New Orleans, LA, American Society for Microbiology I 33, p. 160.

Shewan, J.M. 1966. Some factors affecting the bacterial flora of marine fish. *Med. Den Norske Vet.*, (Suppl.) 11:1–12.

Siebert, G., and W. Schwartz. 1956. Untersuchungen über das Vorkommen von Mikroorganismen in entstehenden Sedimenten. *Arch. Hydrobiol.* 52:321–66.

Sieburth, J.McN. 1964. Polymorphism of a marine bacterium (*Arthrobacter*) as a function of multiple temperature optima and nutrition. *Proc. Symp. Exp. Mar. Ecol.* Occas. Publ. No. 2, GSO, University of Rhode Island, pp. 11–16.

Sieburth, J.McN. 1967a. Seasonal selection of estuarine bacteria by water temperature. *J. exp. mar. Biol. Ecol.* 1:98–121.

Sieburth, J.McN. 1967b. Inhibition and agglutination of arthrobacters by pseudomonads. *J. Bacteriol.* 93:1911–16.

Sieburth, J.McN. 1968. The influences of algal antibiosis on the ecology of marine microorganisms. In: *Advances in Microbiology of the Sea* (M.R. Droop and E.J.F. Wood, eds.). Academic Press, London and New York, pp. 63–94.

Sieburth, J.McN., P-J. Willis, K.M. Johnson, C.M. Burney, D.M. Lavoie, K.R. Hinga, D.A. Caron, F.W. French III, P.W. Johnson, and P.G. Davis. 1976. Dissolved organic matter and heterotrophic microneuston in the surface microlayers of the North Atlantic. *Science* 194:1415–18.

Silvestri, L.G., and L.R. Hill. 1965. Agreement between deoxyribonucleic acid base composition and taxometric classification of Gram-positive cocci. *J. Bacteriol.* 90:136–40.

Simidu, U., and K. Aiso. 1962. Occurrence and distribution of heterotrophic bacteria in sea water from Kamogawa Bay. *Bull. Jap. Soc. Sci. Fish.* 28:1133–41.

Spencer, R. 1959. The sanitation of fish boxes. I. The quantitative and qualitative bacteriology of commercial wooden fish boxes. *J. Appl. Bacteriol.* 22:73–84.

Spencer, R. 1960. The sanitation of fish boxes. II. The efficiency of various sanitizers in the cleaning of commercial wooden fish boxes. *J. Appl. Bacteriol.* 23:10–17.

Veldkamp, H. 1970. Saprophytic coryneform bacteria. *Ann. Rev. Microbiol.* 24:209–40.

Waksman, S.A. 1967. *The Actinomycetes.* Ronald Press, New York.

Weyland, H. 1969. Actinomycetes in North Sea and Atlantic Ocean sediments. *Nature* 223:858.

Wood, E.J.F. 1953. Heterotrophic bacteria in marine environments of Eastern Australia. *Austr. J. Mar. Freshw. Res.* 4:160–200.

ZoBell, C.E. 1946. *Marine Microbiology.* Chronica Botanica Co., Waltham, Mass., 240 p.

ZoBell, C.E., and H.C. Upham. 1944. A list of marine bacteria including descriptions of sixty new species. *Bull. Scripps Inst. Oceanogr.* 5:239–92.

ZoBell, C.E., C.W. Grant, and H.F. Haas. 1943. Marine microorganisms which oxidize petroleum hydrocarbons. *Bull. Amer. Assoc. Petrol. Geol.* 27:1175–93.

PART VI

Osmotrophic Eucaryotes

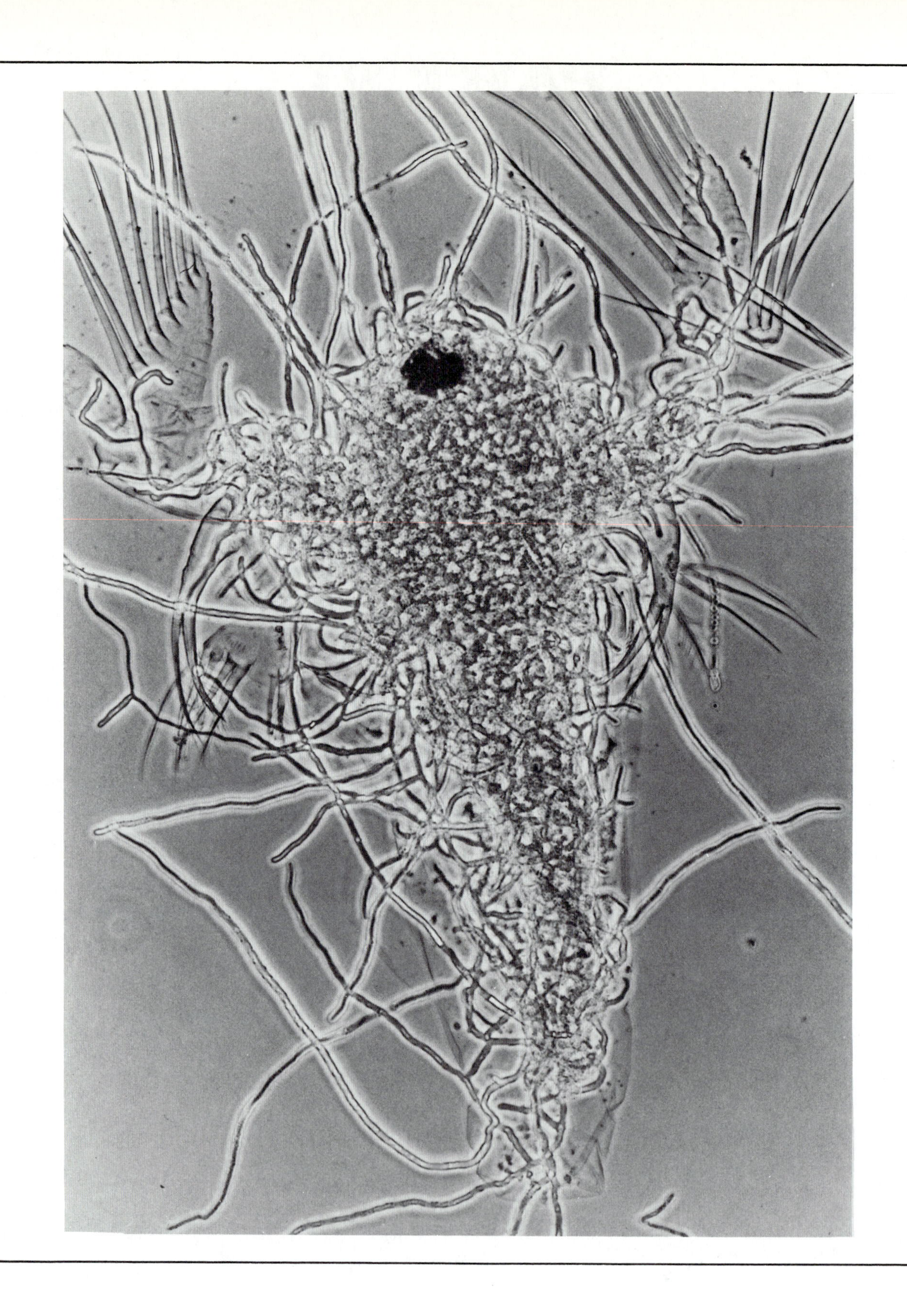

The mycelium of the phycomycete *Lagenidium callinectes* infesting the brine shrimp *Artemia salina* (anterior end with eyespot, top) 5 days post inoculation (250X). Phase-contrast micrograph courtesy of Paul W. Johnson and James Fontaine.

The fungi can occupy the same niche as the osmotrophic epibacteria (discussed in Chapters 15 to 18) and live on the rich organic surfaces of organisms as well as their organic debris. Both groups not only utilize soluble substrates but can break down solid substrates into soluble compounds for active transport into their cells. Pennate diatoms, with a mixotrophic to obligately osmotrophic nutrition, can also occupy the same niche (see Chapter 10). The first time a bacteriology student encounters these nonbacterial forms on rich organic media intended for bacterial culture, it may come as a shock that some microorganisms don't respect disciplinary boundaries. But filamentous fungi and yeasts as culture contaminants are quite common.

The fungi considered here are the Ascomycetes, Basidiomycetes, Deuteromycetes, and Phycomycetes. The filamentous forms in the first three classes, considered in Chapter 19, are found mainly on plant and woody material as epiphytes, saprophytes, and mild pathogens. The yeasts, which are very distinct from the filamentous fungi but are placed in the same fungal classes, are discussed in Chapter 20. Yeasts are also associated with plant material as epiphytes and saprophytes. The zoosporic Phycomycetes, reviewed in Chapter 21, are a heterogeneous group; some form extensive mycelia, others are thalli that convert entirely to a single sporangium, and still others are amoeboid in nature and produce slime tracks. What they share in common is a flagellated reproductive stage with either uniflagellate or biflagellate zoospores. They may be mainly epiphytes or saprophytes, although a number of these species are implicated in diseases of invertebrates, molluscs, seagrasses, and algae (Chapter 3, C).

A group of filamentous fungi not included here are the Zygomycetes. They have been reported from inshore marine sediments by Harder and Uebelmesser (1955) and Siepmann (1959a,b), who believe them to be associated with plant and animal debris. Since they are not detected visually in marine material, as the other higher filamentous fungi are, they may be terrestrial fungi washed into the marine environment. They are distinctive enough to be detected if they occurred to any extent. The Zygomycetes are distinguished by their sexual spore or zygote formed by the fusion of gametangia. In addition to an extensively branched mycelium, which usually penetrates the substrate, there is asexual reproduction by chlamydospores formed in the mycelium and by nonmotile sporangiospores borne in sporangia at the end of sporangiophores.

REFERENCES

Harder, R., and E. Uebelmesser. 1955. Über marine saprophytische Chytridiales und einige andre Pilze vom Meeresboden und Meerestrand. *Arch. Mikrobiol.* 22:87–114.

Siepmann, R. 1959a. Ein Beitrag zur saprophytischen Pilzflora des Wattes der Wesermündung. I. Systematischer Teil. *Veröff. Inst. Meeresforsch Bremerh.* 6:213–81.

Siepmann, R. 1959b. Ein Beitrag zur saprophytischen Pilzflora des Wattes der Wesermüngung. Zweiter Teil. *Veröff. Inst. Meeresforsch. Bremerh.* 6:283–301.

CHAPTER 19

Filamentous Fungi

The filamentous fungi are unmistakably eucaryotic cells. The rigid cell walls, which are responsible for the stability of their vegetative and reproductive structures, are multi-laminate, with the microfibrils in each layer having a different orientation. These microfibrils are usually chitin, sometimes cellulose, and occasionally a mixture of chitin and cellulose. The microfibrils are embedded in an amorphous matrix that includes glucans, mannans, and polyuronides, among other substances (Bartnicki-Garcia, 1968). Between the cell wall and the cell membrane (plasmalemma) are characteristic fungal structures, the lomasomes, which are membrane-bound tubules or particles that occur singly and in groups; they may be elaborations of the plasma membrane. Although their function is not known, it has been suggested that they increase the surface area at the periphery of the cell and may be involved in excretion. The nuclei of fungi are very small, and their nuclear envelopes and nuclear division are distinctive. The endoplasmic reticulum forms bubble-like vesicles in young cells and is generally sparse in mature cells. The cytoplasmic inclusions, i.e., ribosomes, mitochondria, and storage vacuoles, are similar to those in other cells. Golgi bodies occur infrequently in fungi. The ultrastructure of the fungi has been reviewed by Ainsworth and Sussman (1965) and by Bracher (1967).

Unlike the yeasts, in which the same cells accomplish both the assimilative (vegetative) and reproductive phases of growth, most filamentous fungi have a division of labor, with assimilation carried out by an extensive mycelium that penetrates the substrate and reproduction by specific structures for conjugation and/or dissemination of spores. Fungi reproduce both asexually and sexually. The formation of large numbers of asexual spores takes place under conditions that permit a rapid buildup of biomass. Spores resulting from sexual reproduction, however, are often thick-walled, resistant structures. Sexual reproduction occurs when haploid nuclei are united by the simple union of hyphae, the union of a female gametangium (organ) with motile, passively

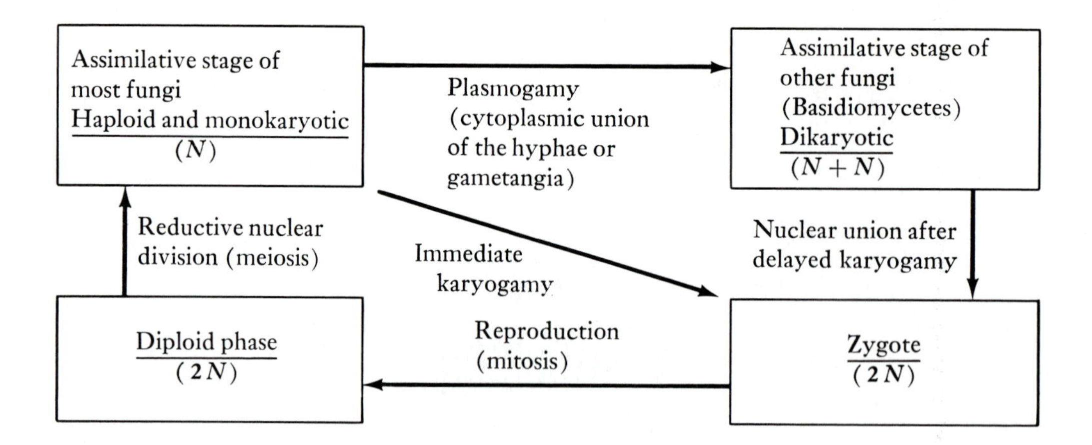

motile or nonmotile gametes, the conjugation of differentiated multinucleated female and male gametangia, or the fusion of motile gametes. A diagrammatic sequence of events in sexual reproduction of the fungi is shown below.

The largest group of filamentous fungi are the Ascomycetes, which produce sexual spores within an ascus. The filamentous Basidiomycetes have sexual spores borne externally on a basidium, with only three species described from the marine environment (Kohlmeyer and Kohlmeyer, 1971a). The filamentous Deuteromycetes are imperfect fungi, and most of them are presumably asexual forms of the Ascomycetes.

The filamentous fungi have been recovered from a variety of materials in the sea. They appear to be associated mainly with decomposing algal and plant tissues including the intertidal and benthic algae, seagrasses, and mangroves as well as the wide variety of cellulosic materials that flow from land, such as driftwood, pinecones, and leaves (Jones and Eltringham, 1971; Jones, 1976). Certain aquatic fungi from freshwater streams have processes on their spores that enable them to be preferentially concentrated in foam (Ingold, 1973). This trapping of spores by air bubbles is nicely illustrated with photomicrographs by Iqbal and Webster (1973). Similar ecological observations on the spores of marine species from sandy habitats (arenicolous fungi) that occur in sea-foam have been made by Kohlmeyer (1966). Calcium carbonate deposits are often attacked and are apparently actively reworked by the filaments of the boring fungi (Kohlmeyer, 1969), which are more widespread and abundant than the endolithic algae (see Chapter 8,B; Perkins and Halsey, 1971; Rooney and Perkins, 1972).

Filamentous fungi, especially those developing on plant materials, attract animal predators and become important in the food chain. The fungal mycelium of Ascomycetes and Deuteromycetes growing on cellulosic debris is able to support the requirements for growth and reproduction of one species of nematode (roundworm) (Meyers, Feder, and Tsue, 1964), and to attract the gravid females of another species of nematode (Meyers and Hopper, 1966).

The role of fungi and boring organisms in the destruction of submerged wood has been considered in two multi-authored volumes, one edited by Ray (1959) and the other by Jones and Eltringham (1971). The extensive study on the role of fungi in the nutrition of the boring crustacean (gribble) *Limnoria* by Kohlmeyer, Becker, and Kampf (1959) has been largely overlooked. The larvae of the boring mollusc, *Toredo pedicellata*, however, are not attracted chemotactically by the filamentous fungi, but when given the choice between fresh wood and microbially colonized and "predigested" wood, they settle on the latter (Kampf, Becker, and Kohlmeyer, 1959). The toredo may be repulsed by such secondary substances of fresh wood as terpenoid hydrocarbons and phenolic substances, which are known to affect invertebrate larvae (Sieburth, 1968) and which would be largely leached out or metabolized in the "predigested" wood.

The synoptic plates of Kohlmeyer and Kohlmeyer (1971a) are invaluable in identifying the filamentous fungi from marine materials. Jones (1972), in reviewing the Kohlmeyers' concise booklet, states that the keys do work and that the characteristic spores of marine fungi enable the marine mycologist to accurately identify 98% of the fungi encountered. But the omission of the hierarchy to which they belong left the novice, including this writer, adrift until a hierarchy was supplied in the exhaustive taxonomic treatise edited by Ainsworth, Sparrow, and Sussman (1973a,b) and that by Kohlmeyer (1974a). As Kohlmeyer (1974a) points out, marine fungi are not a taxonomic group, and they must be related to terrestrial taxa to make classification meaningful. The ultrastructural studies that are just getting under way should do much to solve at least some of the remaining taxonomic problems.

Now that the taxonomy of this group is better understood, there appears to be a trend away from taxonomy and toward ecological study. Marine mycology is, one hopes, past the stage of drawing a line between the land and the sea (Apinis and Chesters, 1964) and beyond the question of what marine species are (Jones and Jennings, 1964) and of being content to examine the colonization of wood baits for the preparation of floristic records (Poole and Price, 1972). Scanning electron microscopy has been used to indicate the nature of fungal colonization and the decomposition of *Spartina* and wood debris (Gessner, Goos, and Sieburth, 1972; Brooks, Goos, and Sieburth, 1972) and has opened up a whole new method of investigation (Sieburth, 1975). The extensive mangrove forests that invade the sea are now receiving attention (Kohlmeyer and Kohlmeyer, 1971b; Lee and Baker, 1972a,b, 1973; Newell, 1972, 1976), as are animal substrates (Kohlmeyer, 1969, 1972). Calcareous materials in addition to corals are also receiving attention (Alderman and Jones, 1971). The recognition of the nutrient enrichment of cellulosic substrates by mycelial development has only opened the door to food web studies. The fungi are dealt with in great detail in the multi-author, multi-volume treatise edited by Ainsworth and Sussman (1965, 1966, 1968) and Ainsworth, Sparrow, and Sussman (1973a,b), whereas the glossaries of Snell and Dick (1971) and Ainsworth, James, and Hawksworth (1971) are indispensable aids to deciphering the jargon in this field. The monograph *Fungi in Oceans and Estuaries* by Johnson and Sparrow (1961), considered the cornerstone work for the marine mycology student, summarizes the early literature. It has been updated in an extensive review by Hughes (1975), which emphasizes the ecology of the fungi. The multi-authored volume on aquatic mycology edited by Jones (1976) includes freshwater forms and gives equal attention to the zoosporic fungi.

A. ASCOMYCETES

The Ascomycetes are the largest group of fungi, with over 2000 genera. Their possible evolutionary relationship to the red algae has been discussed by Kohlmeyer (1975). They produce sexual spores (ascospores) following karyogamy and meiosis within a specialized structure called the ascus. The ascus, and the ultrastructure of the spores, varies with the order of Ascomycetes; the cytochemistry of the latter has been reviewed by Kirk (1976). Ingold (1968) has commented on the differences between the rich submerged ascomycete microfloras that develop in marine water and those that develop in freshwater. Sea Ascomycetes are mainly Loculoascomycetes or Pyrenomycetes, whereas Discomycetes (Ascomycetes with open cup-shaped ascocarps) only occur in freshwater. Though most freshwater Ascomycetes release their spores, like terrestrial species, by bursting their ascus, the mechanism of spore release varies in some marine occurring Ascomycetes. Some species in the Pyrenomycetes have ascus walls that deliquesce, freeing the ascospores into the interior of the perithecium (fruiting body) from which they subsequently ooze out; and over one-third of the Ascomycetes belong to the Pseudosphaeriales, which have thick-walled, non-deliquescing asci, mostly with a "jack-in-the-box" mechanism of spore release. Of the 199 species of higher fungi from the marine environment listed by Kohlmeyer (1974a), 144 are Ascomycetes. All but nine of the species fall into just two of the six Ascomycete classes (Ainsworth, Sparrow, and Sussman, 1973a), the Pyrenomycetes and the Loculoascomycetes. The main order in each of these classes, the Sphaeriales and Pseudosphaeriales, include 83% of the species of marine Ascomycetes (Kohlmeyer, 1974a).

Pyrenomycetes in the marine environment grow preferentially on such cellulosic debris as wood, leaves, and pinecones, but they can also develop on algae and seagrasses. The fruiting bodies, known as perithecia or ascocarps, are spherical, hemispherical, or flask-shaped structures (Plate 19-1), with walls that range from bright and fleshy to dark and membranous. They may occur as solitary, aggregated, or connected forms or rest on a dense felt of hyphae. These fruiting bodies consist of an enveloping coat or peridial wall containing spherical, clavate, fusiform, or cylindrical cells called asci, and a pore (ostiole) for the release of ascospores. Each ascus usually contains eight one-celled or septate ascospores, which can be hyaline or opaque and colored or colorless. The ascus wall in the Pyrenomycetes is single layered or unitunicate, and the asci usually occur in a palisade-like arrangement (Plate 19-1,*E*). The apex of the ascus wall in some species is thickened to form the ascus apical apparatus, a ring-like structure that is distinctive in each species so far examined (Griffiths, 1973). The ascospores themselves (Plate 19-1,*B*,*E*) are

PLATE 19-1. The perithecia (fruiting body) and ascospores of two marine Pyrenomycetes. *Corollospora intermedia*, showing (*A*) a perithecium with mycelium and (*B*) mycelium with released ascospores; *Biconiosporella corniculata*, showing (*C*) a perithecium, (*D*) a transverse section through a perithecium, and (*E*) an ascus bundle with unitunicate asci and ascospores. (*A,B*) from K. Schaumann, 1972a, *Veroff. Inst. Meeresforsch. Bremerhaven* 14:13–22; (*C–E*) from K. Schaumann, 1972b, *Veroff. Inst. Meeresforsch. Bremerhaven* 14:23–44.

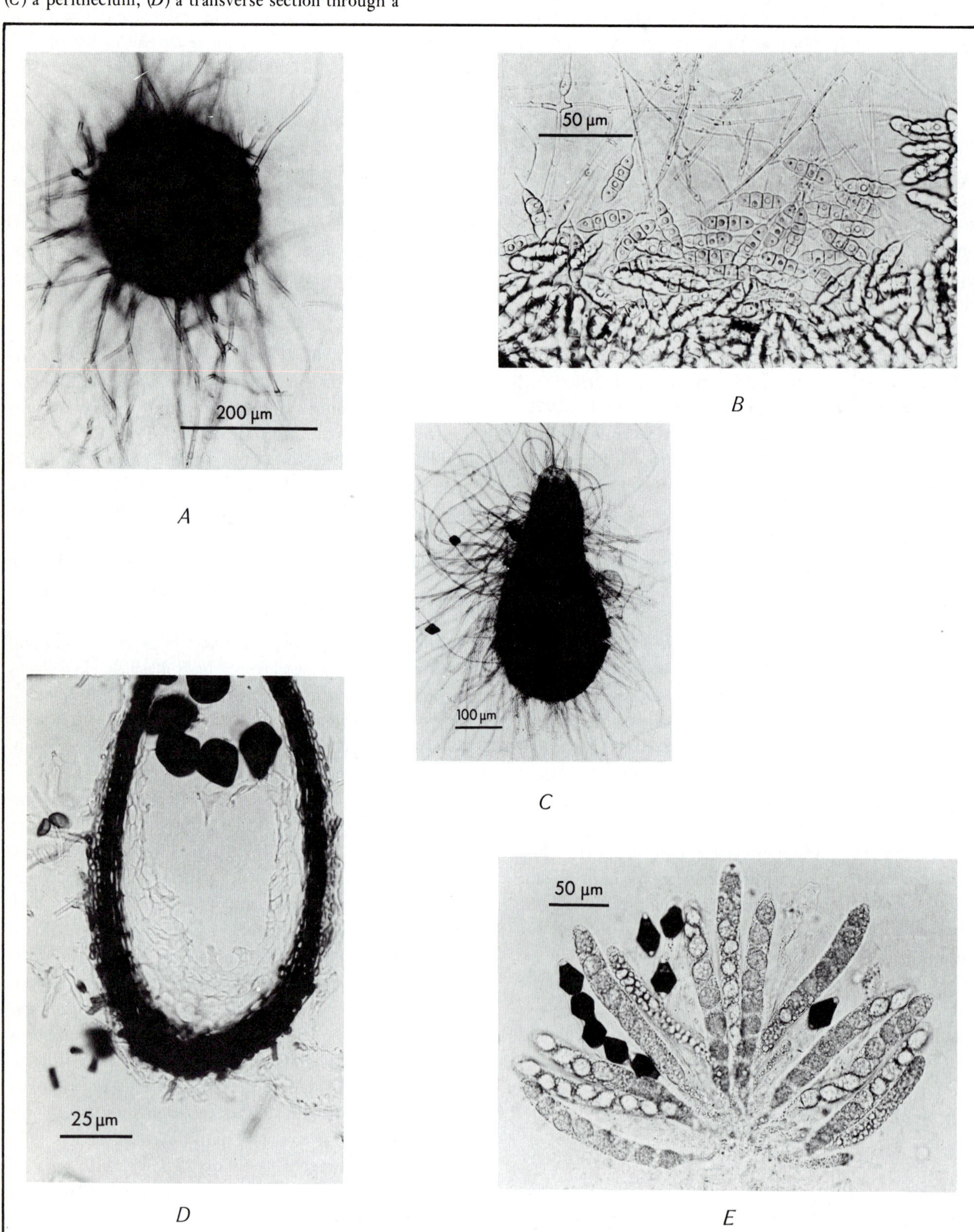

PLATE 19-2. A comparison of the ascospores from different species of *Corollospora*, showing those of (a) *C. comata*, (b) *C. cristata*, (c) *C. pulchella*, (d) *C. lacera*, (e) *C. intermedia*, (f) *C. maritima*, (g) *C. tubulata*, and (h) *C. trifurcata*. From K. Schaumann, 1972a, *Veroff. Inst. Meeresforsch. Bremerhaven* 14:13–22.

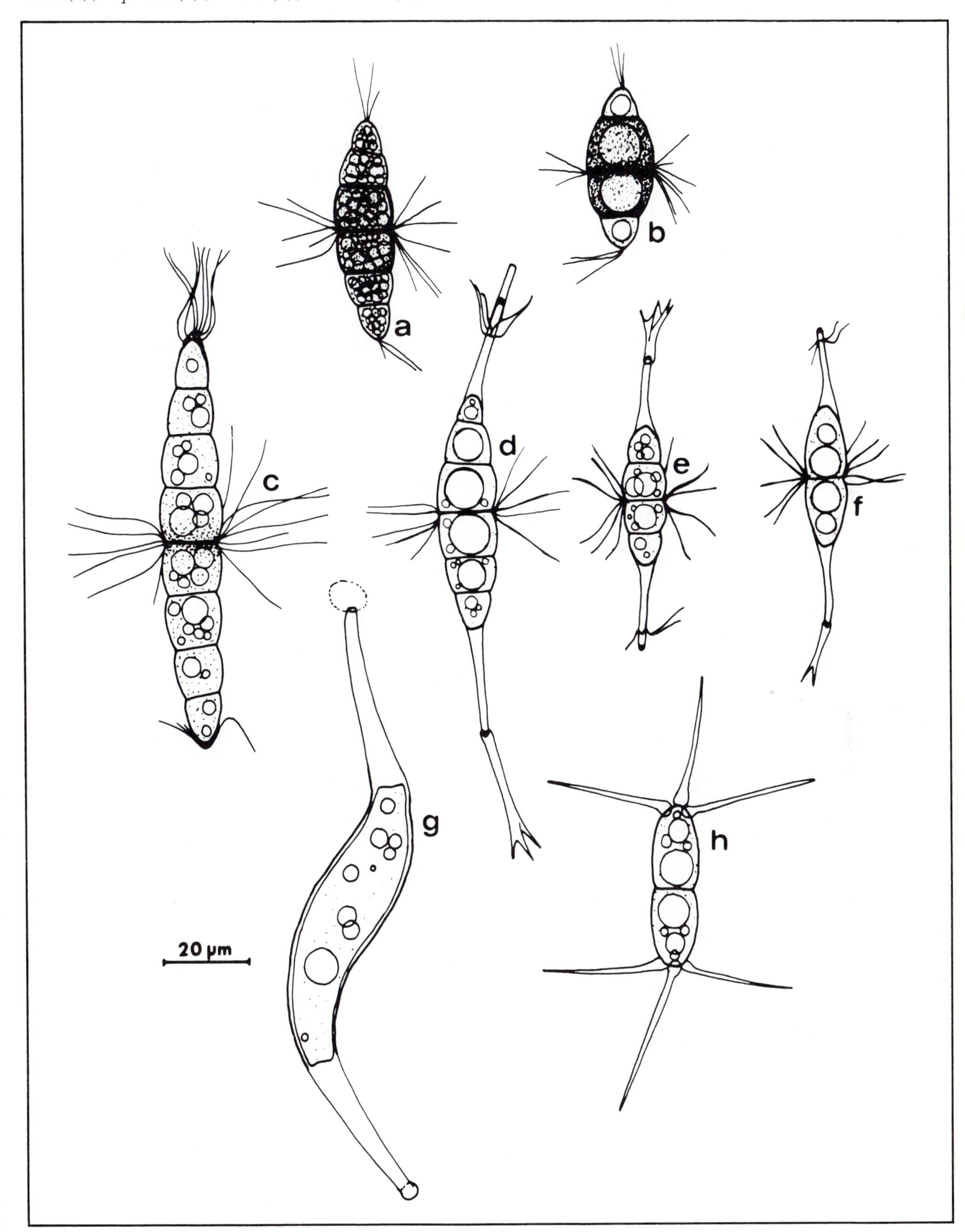

characteristic for each species. This is shown further in Plate 19-2, in which ascospores from a number of marine species of *Corollospora* are compared. Their distinctive structure is the basis of the keys in Ainsworth, Sparrow, and Sussman (1973a,b) for fungi in general and in Kohlmeyer and Kohlmeyer (1971a) for marine fungi. The development and fine structure of the ascospores of marine Pyrenomycetes in the Halosphaeriaceae have been observed by Lutley and Wilson (1972a,b).

PLATE 19-3. A comparison between a Pyrenomycete (*Chaetosphaeria chaetosa*) (*A*) with its unitunicate ascus, and a Loculoascomycete (*Leptosphaeria contecta*) (*B*) with its bitunicate ascus. The ascocarps (peritheca) are shown in external view (16 and 22) and in cross section (17 and 23), the asci with their ascospores are shown within the ascocarp and in greater detail outside the ascocarp. From J. Kohlmeyer, 1963, *Nova Hedwigia* 6:297–329.

The Loculoascomycetes, the other large class of Ascomycetes, usually have dark ascocarps, but in some genera, they are brightly colored. The genera in this class are distinguished by bitunicate asci. The bitunicate ascus consists of two layers: a thin inextensible outer layer (ectotunica) and a thick extensible inner layer (endotunica). The two layers differ only in the direction of their microfibrils, and during ascospore expulsion, there is a rapid reorientation of the microfibrils in the endotunica (Reynolds, 1971). The immature ascus is thin-walled, but during most of its development, the endoascus is conspicuously thickened especially toward the apex, where the ascospores will later be ejected through a pore. During ascospore discharge, the endoascus and the ectoascus separate. A Pyrenomycete with a unitunicate ascus and a Loculoascomycete with a bitunicate ascus, drawn by the same investigator, are compared in Plate 19-3. Other bitunicate asci are shown

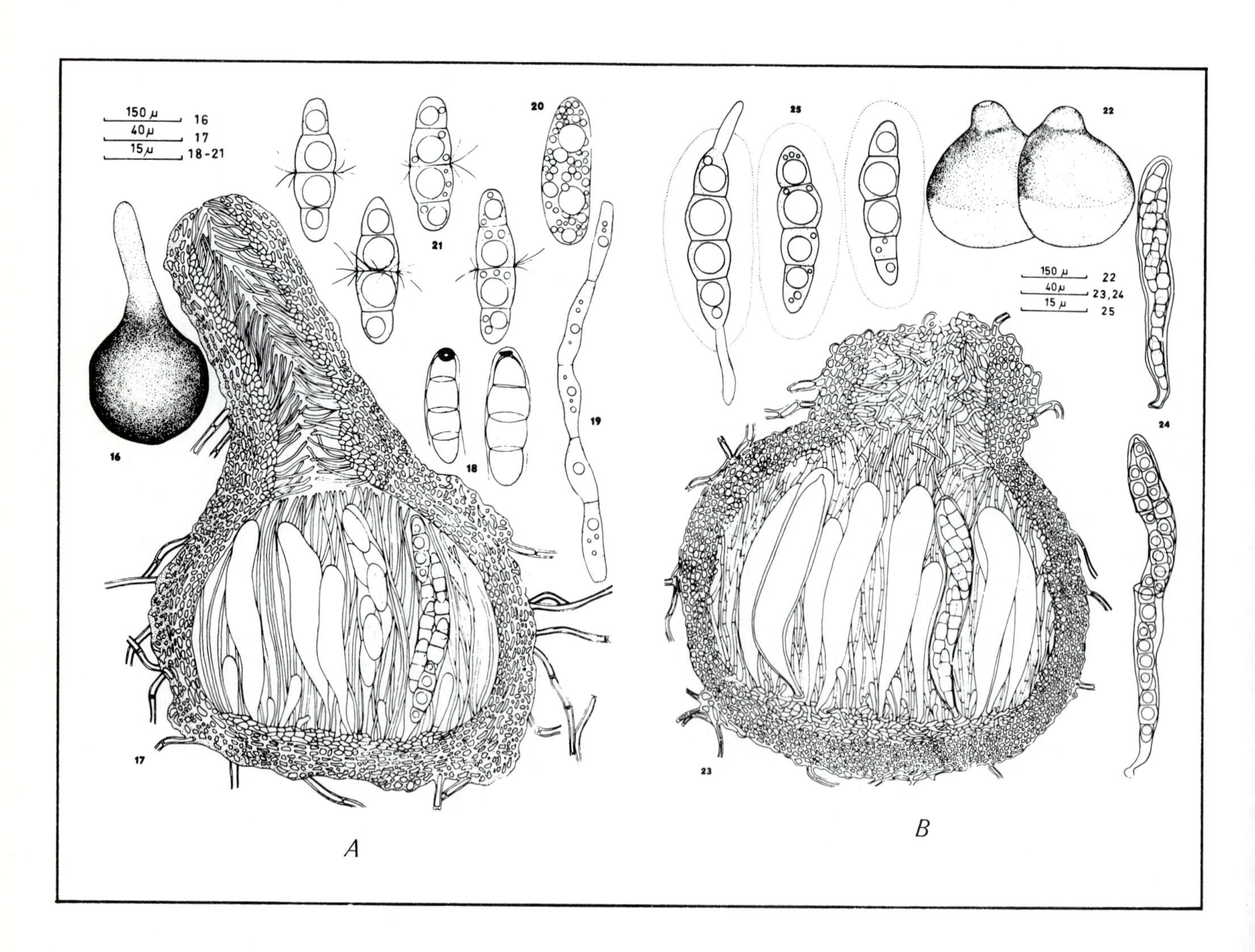

in the classic work on the fungi of submerged wood by Barghoorn and Linder (1944) and in the paper by Johnson (1956).

The Pyrenomycete and Loculoascomycete genera recovered from the sea so far are about equal in number. There is apparently no strong generic preference among the Loculoascomycetes for algae, higher plants, or cellulosic debris as substrates. Exceptions are the genera in the Dothideaceae, which have species intimately associated with intertidal algae and, to a lesser degree, seagrasses. Living cordgrass, *Spartina alterniflora*, is colonized by the weak parasite *Buergenerula spartinae* (Gessner, 1976), which persists on the knocked down and submerged thallus in the fall as it decomposes (Gessner, Goos, and Sieburth, 1972; Gessner, 1977). Another example is *Mycosphaerella.* Terrestrial species cause leaf and stem spot disease in such angiosperms as the chrysanthemum. Two species, which are probably synonymous, are constant symbionts of the fucoid brown algae *Pelvetia canaliculata* and *Fucus vesiculosus.* Long overdue and comprehensive studies on the structure, life history, and biology of *M. ascophylli* have been conducted by Webber (1967) and Kohlmeyer and Kohlmeyer (1972), who again ask the question whether this algal–fungal relationship should be considered a lichen (Sutherland, 1915); this relationship, along with other examples of symbiosis, is discussed further by Kohlmeyer (1974b). The occurrence of mycelia between the cells and the development of perithecia are shown in Plate 19-4. When chemical or biochemical studies on marine plants are conducted (Madgwick, Haug, and Larsen, 1973), the presence of such fungal symbionts as well as epibionts (Sieburth, 1975) must be considered (Sieburth, 1968).

Other frequently occurring species of Ascomycetes listed by Kohlmeyer and Kohlmeyer (1971a) are the Plectomycetes, which are not as clearly differentiated from the Pyrenomycetes as the Loculoascomycetes are. A combination of characteristics rather than one is used for differentiation. Marine species are in the order Eurotiales (Fennell, 1973). These fungi have small ascocarps, which are usually spherical and non-ostiolate (having no pore for spore release) and are produced on a more or less well developed septate mycelium. Ascogenous hyphae ramify throughout the centrum and bear asci at all levels, either singly or in chains. The ascocarps and spore formation are shown for *Amylocarpus encephaloides* and *Eiona tunicata* in Plate 19-5.

B. BASIDIOMYCETES

The Basidiomycetes, which are the most highly evolved class of fungi, are distinguished by the formation of external basidiospores, on structures called basidia. On land, they are represented by the rusts, smuts, jelly fungi, mushrooms, puffballs, shelf fungi, coral fungi, stinkhorns, earthstars, and birds-nest fungi. In the marine environment, there are three recognized filamentous species (Kohlmeyer and Kohlmeyer, 1971a) in addition to a few yeast-like forms with heterobasidiomycetous life cycles discussed in Chapter 20. *Melanotaenium ruppiae*, a smut on the marine grass *Ruppia maritima* attacks the rhizomes and leaf bases (Feldmann, 1959). A saprophytic species developing on submerged wood, *Digitatispora marina*, was described by Doguet (1962) and is diagrammatically shown in Plate 19-6,*A*. The fruiting bodies appear to be quite sensitive to the stresses caused by emersion, since this winter-occurring species in Point Judith Pond, Rhode Island is common on constantly immersed wood and rare on intermittently exposed substrates (Brooks, 1972). The third species is *Nia vibrissa*, which has been observed on wood and bark as well as on the mangrove *Rhizophora* and the cordgrass *Spartina.* Although originally described as a Deuteromycete (Moore and Meyers, 1959), Doguet (1967) recognized that the "conidia" were basidiospores and described the various stages of its development (see Plate 19-6,*B*). This marine heterothallic Homobasidiomycete is unique in that it belongs to the Gasteromycetes; thus, its fruiting bodies are tiny puffballs less than 2 mm in diameter. Like *D. marina*, *N. vibrissa* also appears to require an (almost continuously) immersed substratum for its development. Both these species have a special structure, the dolipore septum (Brooks, 1975), in the cross-walls between the cells, which is another characteristic of the Basidiomycetes.

C. DEUTEROMYCETES

The class Deuteromycetes (Fungi imperfecti) is an artificial, or a "wastebasket," subdivision for fungi producing only conidial states. These fungi include the imperfect, asexual, or conidial states of the Ascomycetes, Basidiomycetes, and even certain Zygomycetes, as well as the hypothetical asexual fungi, which have no perfect

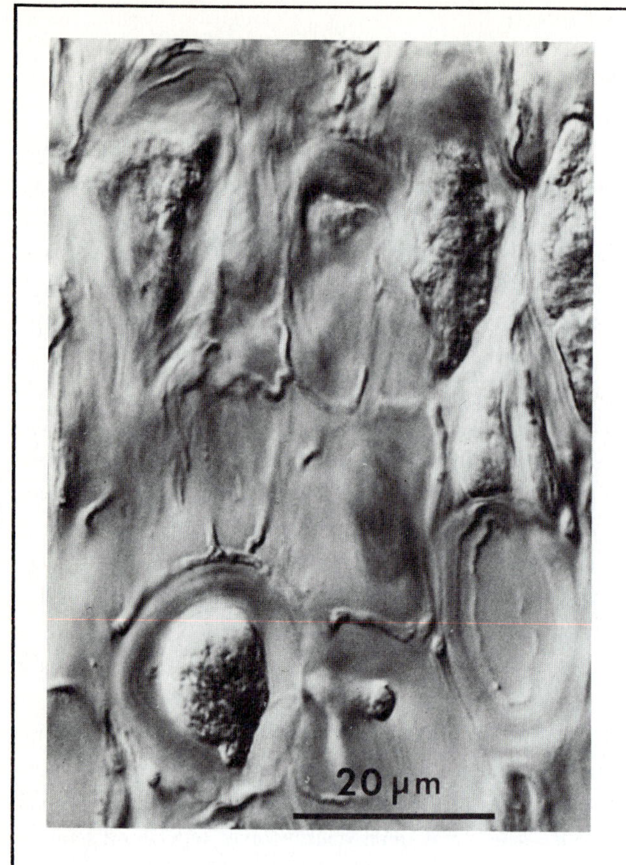

A

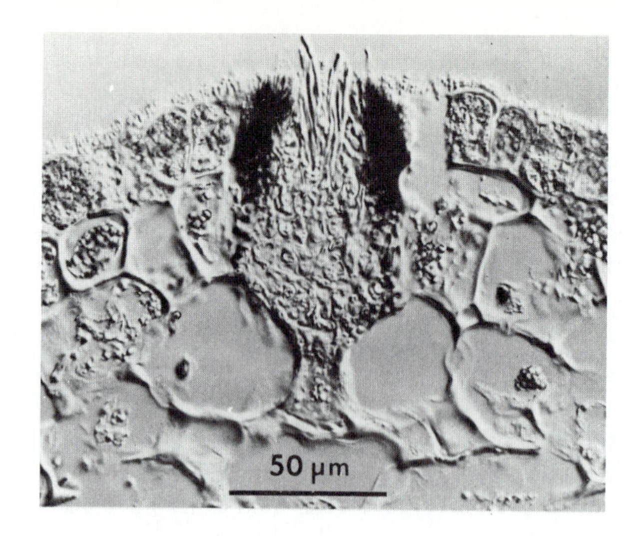

B

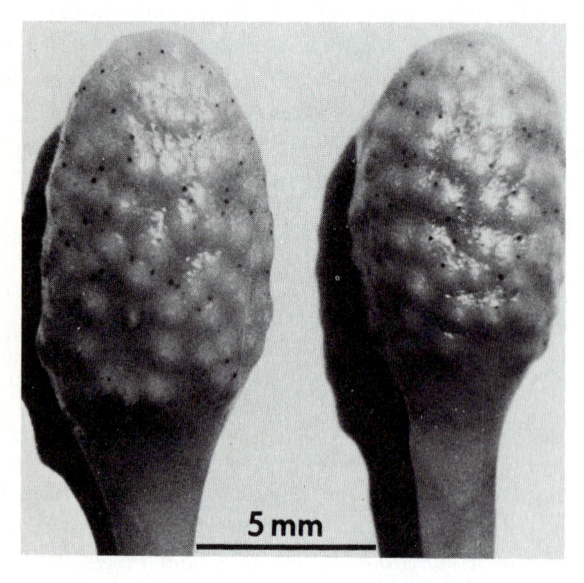

C

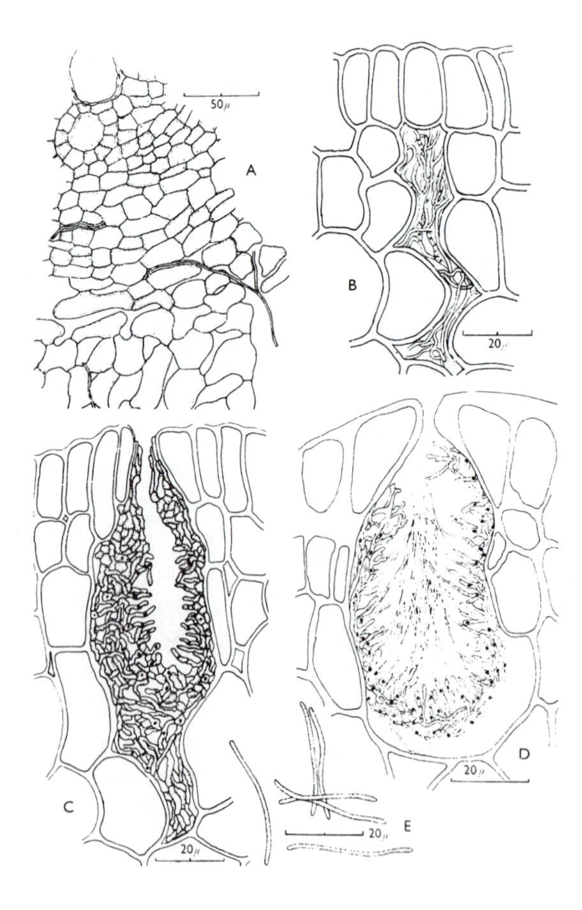

D

PLATE 19-4. The symbiont *Mycosphaerella ascophylli* in its brown algal host *Ascophyllum nodosum*. Nomarski interference-contrast micrographs, showing (*A*) hyphae between host cells and (*B*) immature ascocarp. Photograph of algal receptacles with black ascocarps (*C*); and diagrammatic representation of the sequence of pycnidium (spermogonium) development (*D*). (*A,B,C*) from J. Kohlmeyer and E. Kohlmeyer, 1972, *Botanica Marina* 15:109–12; (*D*) from F.C. Webber, 1967, *Trans. Br. Mycol. Soc.* 50:583–601.

states. When water, sand, mud, and other marine material are cultured on rich organic media for fungi, the Deuteromycetes usually dominate (Sparrow, 1937; Steele, 1967). The presence of the filamentous fungi is usually detected by using baits or incubating the substrates in moist chambers and allowing the fruiting bodies or conidia to develop. It has been stated by mycologists that perhaps one-third of the Deuteromycetes have named perfect states (which take precedence), one-third have perfect states not yet named, and one-third have no such states. Since the perfect-state taxa they represent are not known until the sexual phase is dis-

PLATE 19-5. Two species of Plectomycetes showing the asci being borne on ascogenous septate hyphae among the sterile hyphae in the fruiting body. (*A*) *Eiona tunicata*, as seen in cross section of the fruiting body (11), showing the hyaline outer layer (A) and the brown inner layer (B), ascogenous hyphae and immature ascospores released during sectioning (12), ascogenous hyphae with young ascus (13), mature ascus (14), and ascospores (15–20). (*B*) *Amylocarpus encephaloides*, showing the fruiting bodies just emerging (45), partially dessicated (46), and with dissolving wall (47), scale a. The cross section of a fruiting body (48), showing the outer layer above and the inner layer below, with loosely interwoven sterile hyphae between which occur broader ascogenous hyphae that form asci, scale b. The nearly mature ascus is shown in (49), the prickly ascospores in (50); note the one at right is germinating, scale c. (*A*) from J. Kohlmeyer, 1968, *Ber. Deutsche Bot. Ges.* 81:53–61; (*B*) from J. Kohlmeyer, 1960, *Nova Hedwigia* 2:293–343.

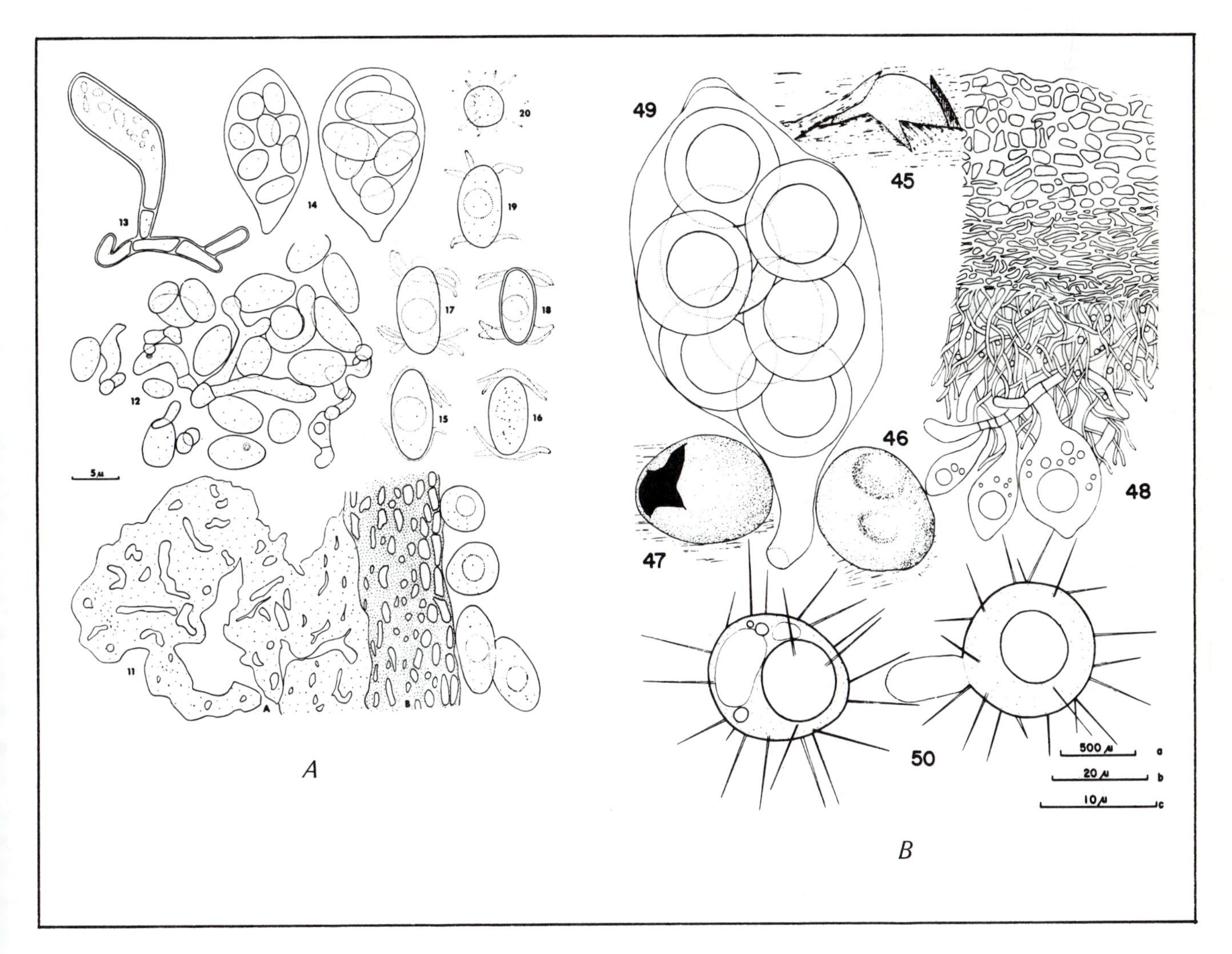

PLATE 19-6. Diagrammatic representation of two marine Basidiomycetes. (*A*) *Digitatispora marina*, showing (1) mature thallus on wood substrate in cross section (r.m., medullary rays of the wood; c, cushion-like hymenophore; h, hymenium; sp, spores); (2) vegetative hyphae of the hymenophore; (3) basidia formed at the tip of a hypha; (4–8) sequence of development of a basidium and basidiophores; (9) basidium after spore detachment. (*B*) *Nia vibrissa*, showing (1) mature fruiting bodies (puffballs) on wood substrate; (2) tips of hairs from the basidiocarp surface; (3) six basidia in different stages of development; (4–7) sequence of development of basidiospores; (8) basidiospores; (9) a germinating spore; (10) primary mycelium with simple septa; (11) mycelium with clamp connections; (12) hyphae with incomplete clamp connections. (*A*) from G. Doguet, 1962, *C.R. Acad. Sci. Paris* 254:4336–38; (*B*) from G. Doguet, 1967, *C.R. Acad. Sci. Paris* 265:1780–83.

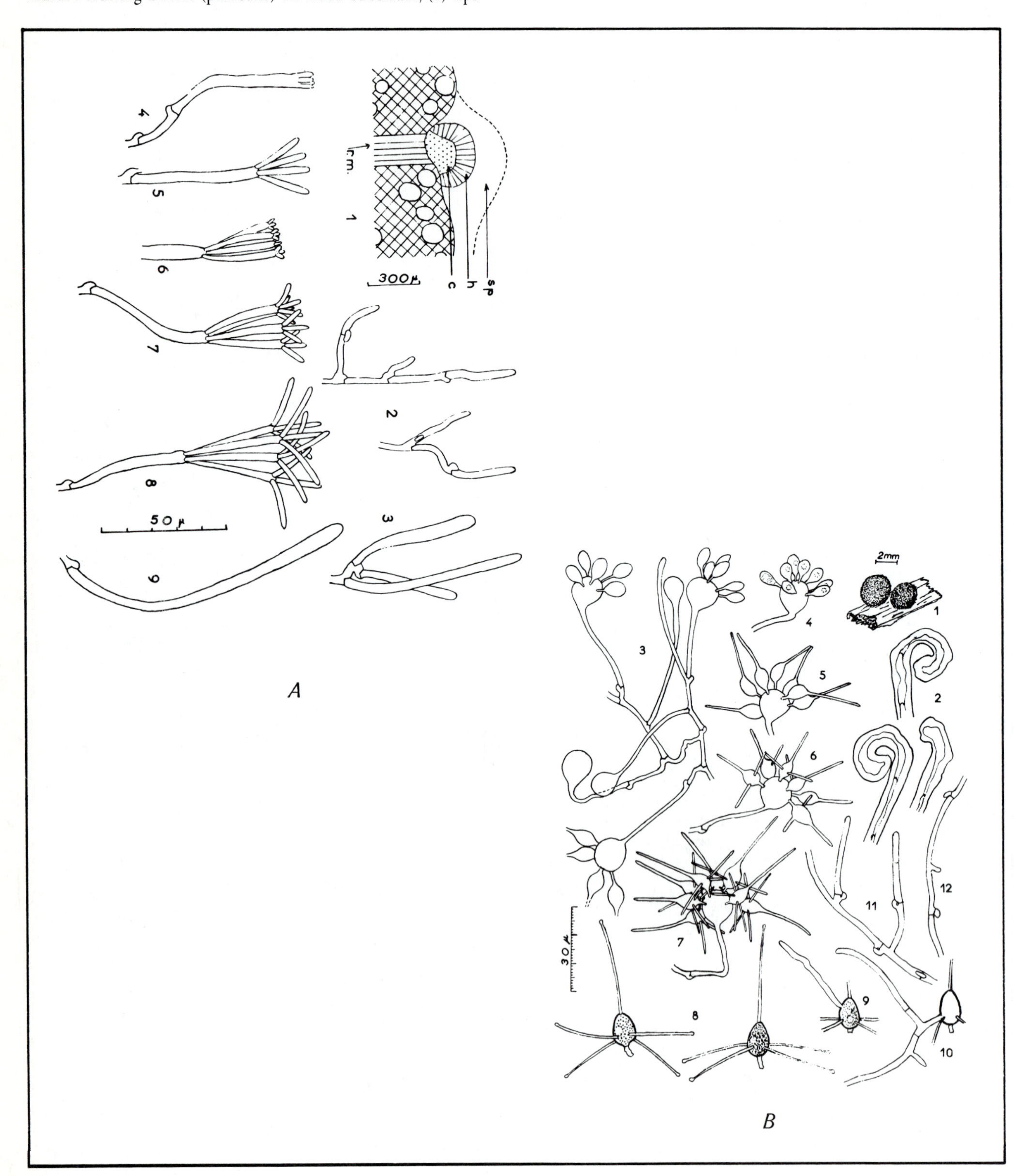

PLATE 19-7. Marine examples of the form-genera (based on conidial form and structure) in the Deuteromycete class Hyphomycetes. (*A*) The one-septate (amerosporous) species *Humicola alopallonella;* (*B*) the multi-cross septate (phragmosporous) species *Sporidesmium salinum;* (*C*) the cross and longitudinally septate (dictysporous) species *Alternaria maritima* (bar = 10 μm/unit); (*D*) the star-shaped (staurosporous) species *Orbimyces spectabilis* (bar B = 10 μm/unit); (*E*) the spirally coiled (helicosporous) species *Cirrenalia macrocephala*. (*A*) from S.P. Meyers and R.T. Moore, 1960, *Amer. J. Bot.* 47:345–49; (*B*) from E.B.G. Jones, 1963, *Trans. Br. Mycol. Soc.* 46:135–44, by permission, Cambridge University Press; (*C,D*) from E.S. Barghoorn and D.H. Linder, 1944, *Farlowia* 1:395–467; (*E*) from J. Kohlmeyer, 1958, *Ber. Deutschen Bot. Ges.* 71:98–116.

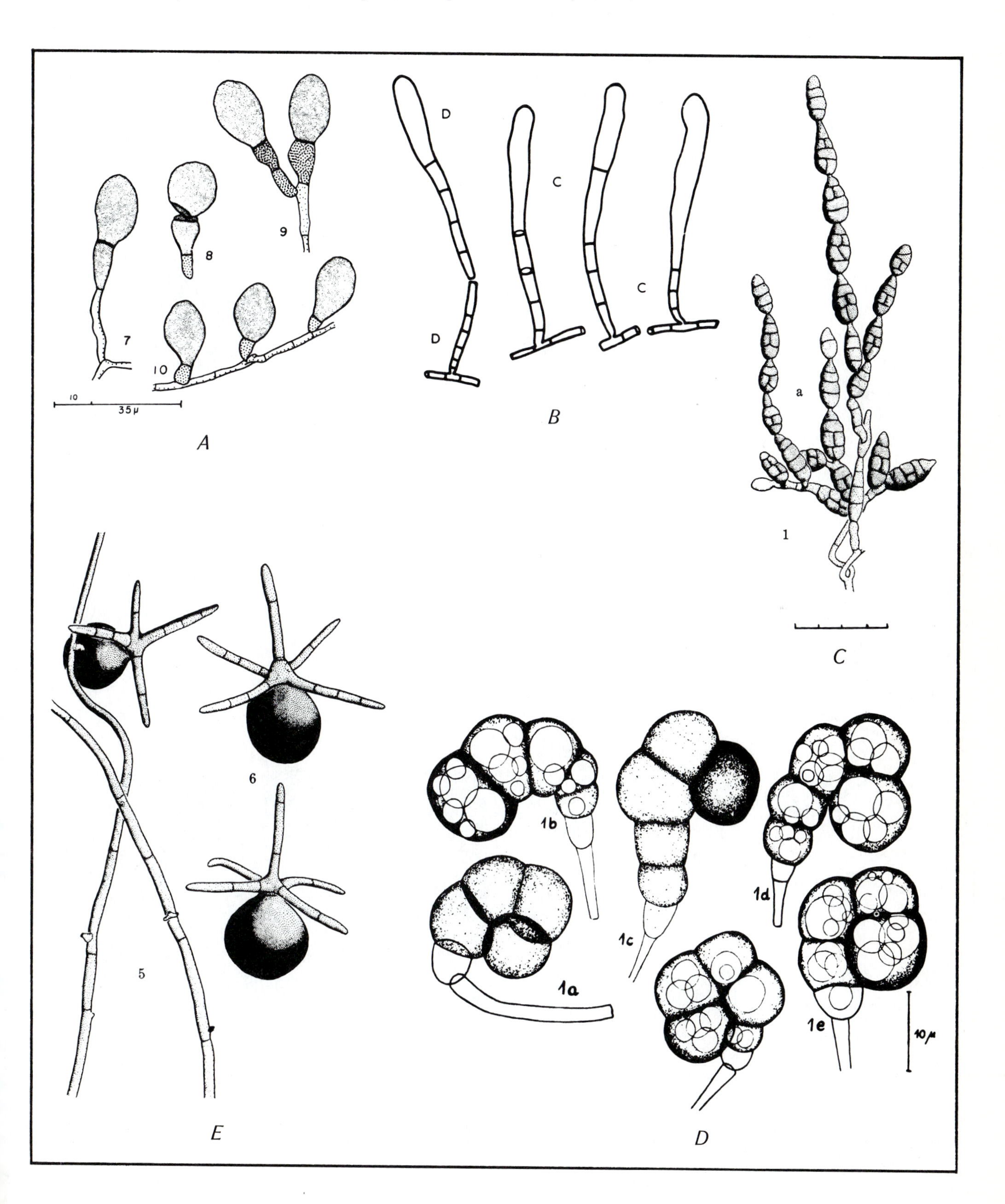

Plate 19-8. Several form-species of Coelomycetes showing how the conidiospores are borne within a cavity (pycnidium) in the substrate. (*A*) *Dinemasporium marinum*, on driftwood; (*B*) *Phialophorophoma littoralis*, on submerged wood; (*C*) *Diplodina laminariana*, from the fronds of the brown alga *Laminaria;* (*D*) *Rhabdospora aricenniae*, from the mangrove pneumatophore (longitudinal section, interference-contrast micrograph) (magnifications not given). (*A*) from S. Nilsson, 1957, *Bot. Notiser* 110:321–24; (*B*) from E.S. Barghoorn and D.H. Linder, 1944, *Farlowia* 1:395–467; (*C*) from G.K. Sutherland, 1916, *New Phytol.* 15:35–48; (*D*) from J. Kohlmeyer and E. Kohlmeyer, 1971b, *Mycologia* 63:831–61.

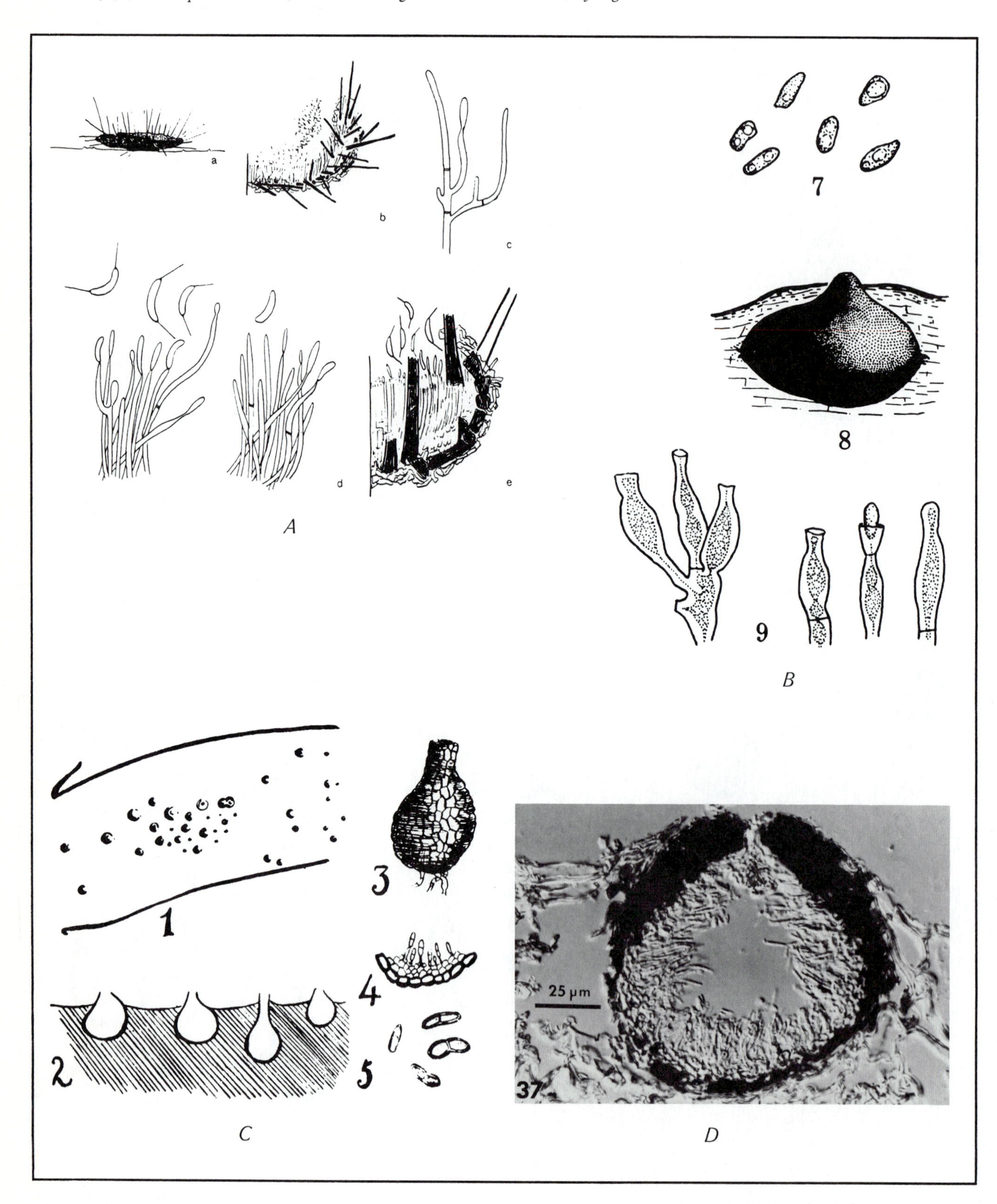

covered, the genera and species are termed "form-genera" and "form-species," based on the similarity of the conidiospores, and are not related to the main taxonomic classification of the fungi with sexual stages. The marine Deuteromycetes are divided into two classes on the basis of whether their conidia are borne on hyphae or are inside fruiting bodies.

1. Hyphomycetes

The class Hyphomycetes are the imperfect fungi that bear their conidia on conidiophores (or directly on hyphae) and not inside a fruiting body. The genera are based on the seven spore groups of Saccardo as defined by Ainsworth, James, and Hawksworth (1971). The taxonomy of this group is discussed by Kendrick (1971) and Kendrick and Carmichael (1973). Examples of the conidia of the form-genera in which marine species occur are shown in Plate 19-7. The Hyphomycetes appear to prefer the woody substrates of higher plants but are also associated with algae.

2. Coelomycetes

The class Coelomycetes are imperfect fungi that form their conidia within a pycnidium (acervulus). This more or less enclosed fruiting body is formed from fungal tissues or host tissues or both. There are perhaps some 900 non-redundant genera. Perfect states correlated with the Coelomycetes belong to the Ascomycetes. The criteria and classification of the Coelomycetes has been discussed by Sutton (1973). Boerema and Dorenbosch (1968), in discussing the taxonomy of species of *Phoma* from marine sediments, emphasized that, for form-species of Coelomycetes, the host (substrate) is the primary criterion for identification, whereas other criteria should only depend upon a clear, stable, and generally recognized morphology. Examples of superficial and immersed pycnidia of Coelomycetes on a variety of substrates are shown in Plate 19-8. This group has a decided preference for higher marine plants and submerged wood as substrate.

REFERENCES

Ainsworth, G.C., and A.S. Sussman. (eds.) 1965. *The Fungi. Vol. 1. The Fungal Cell.* Academic Press, New York and London, 748 p.

Ainsworth, G.C., and A.S. Sussman. (eds.) 1966. *The Fungi. Vol. 2. The Fungal Organism.* Academic Press, New York and London, 805 p.

Ainsworth, G.C., and A.S. Sussman. (eds.) 1968. *The Fungi. Vol. 3. The Fungal Population.* Academic Press, New York and London, 738 p.

Ainsworth, G.C., P.W. James, and D.L. Hawksworth. 1971. *Ainsworth and Bisby's Dictionary of the Fungi, including the Lichens* (6th Ed.). Commonwealth Mycological Institute, Kew, Surrey, England, 663 p.

Ainsworth, G.C., F.K. Sparrow, and A.S. Sussman (eds.). 1973a. *The Fungi. Vol. 4A. A Taxonomic Review with Keys: Ascomycetes and Fungi imperfecti.* Academic Press, New York and London, 621 p.

Ainsworth, G.C., F.K. Sparrow, and A.S. Sussman (eds.). 1973b. *The Fungi. Vol. 4B. A Taxonomic Review with Keys: Basidiomycetes and Lower Fungi.* Academic Press, New York and London, 504 p.

Alderman, D.J., and E.B.G. Jones. 1971. Shell disease of oysters. *Fish. Invest. (Great Britain)* Ser. II, vol. 26, No. 8, 19 p.

Apinis, A.E., and C.G.C. Chesters. 1964. Ascomycetes of some salt marshes and sand dunes. *Trans. Br. mycol. Soc.* 47:419–35.

Barghoorn, E.S., and D.H. Linder. 1944. Marine fungi, their taxonomy and biology. *Farlowia* 1:395–467.

Bartnicki-Garcia, S. 1968. Cell wall chemistry, morphogenesis and taxonomy of fungi. *Ann. Rev. Microbiol.* 22:87–108.

Boerema, G.H., and M.M.J. Dorenbosch. 1968. Some *Phoma* species recently described from marine soils in India. *Trans. Br. mycol. Soc.* 51:145–46.

Bracher, C.E. 1967. Ultrastructure of fungi. *Ann. Rev. Phytopath.* 5:343–74.

Brooks, R.D. 1972. The occurrence and distribution of wood-inhabiting marine fungi in Point Judith Pond. M.S. Thesis, University of Rhode Island, 91 p.

Brooks, R.D. 1975. The presence of dolipora septa in *Nia vibrissa* and *Digitatispora marina*. *Mycologia* 47:172–74.

Brooks, R.D., R.D. Goos, and J.McN. Sieburth. 1972. Fungal infestation of the surface and interior vessels of freshly collected driftwood. *Mar. Biol.* 16:274–78.

Doguet, G. 1962. *Digitatispora marina*, n.g., n. sp., Basidiomycete marin. *C.R. Acad. Sci., Paris* 254:4336–38.

Doguet, G. 1967. *Nia vibrissa* Moore et Meyers, remarquable Basidiomycète marin. *C.R. Acad. Sci., Paris* 265:1780–83.

Feldmann, G. 1959. Une ustilaginale marine, parasite du *Ruppia maritima* L. *Rev. Gén. Bot.* 66:35–39.

Fennell, D.I. 1973. Plectomycetes: Eurotiales. In: *The Fungi, Vol. 4A* (G.C. Ainsworth, F.K. Sparrow, and A.S. Sussman, eds.). Academic Press, New York and London, pp. 45–68.

Gessner, R.V. 1976. *In vitro* growth and nutrition of *Buergenerula spartinae*, a fungus associated with *Spartina alterniflora*. *Mycologia* 68:583–99.

Gessner, R.V. 1977. Seasonal occurrence and distribution of fungi associated with *Spartina alterniflora* from a Rhode Island estuary. *Mycologia* 69:477–91.

Gessner, R.V., R.D. Goos, and J.McN. Sieburth. 1972. The

fungal microcosm of the internodes of *Spartina alterniflora*. *Mar. Biol.* 16:269–73.

Griffiths, H.B. 1973. Fine structure of seven unitunicate pyrenomycete asci. *Trans. Br. mycol. Soc.* 60:261–71.

Hughes, G.C. 1975. Studies of fungi in oceans and estuaries since 1961. I. Lignicolous, caulicolous and folicolous species. *Oceanogr. Mar. Biol. Ann. Rev.* 13:69–180.

Ingold, C.T. 1968. Spore liberation in *Loramyces*. *Trans. Br. mycol. Soc.* 51:323–25.

Ingold, C.T. 1973. Aquatic hyphomycete spores from West Scotland. *Trans. Br. mycol. Soc.* 61:251–55.

Iqbal, S.H., and J. Webster. 1973. The trapping of aquatic hyphomycete spores by air bubbles. *Trans. Br. mycol. Soc.* 60:37–48.

Johnson, T.W., Jr. 1956. Marine fungi. I. *Leptosphaeria* and *Pleospora*. *Mycologia* 48:495–505.

Johnson, T.W., Jr., and F.K. Sparrow, Jr. 1961. *Fungi in Oceans and Estuaries*. J. Cramer, Weinheim, and Hafner Publ., New York, 668 p.

Jones, E.B.G. 1963. Marine fungi. II. Ascomycetes and Deuteromycetes from submerged wood and drift *Spartina*. *Trans. Br. mycol. Soc.* 46:135–44.

Jones, E.B.G. 1972. Book review of *Synoptic Plates of Higher Marine Fungi* by Kohlmeyer and Kohlmeyer (1971a-*q.v.*). *Trans. Br. mycol. Soc.* 58:533–34.

Jones, E.B.G. (ed.). 1976. *Recent Advances in Aquatic Mycology*. Wiley, New York, 749 p.

Jones, E.B.G., and D.H. Jennings. 1964. The effect of salinity on the growth of marine fungi in comparison with non-marine species. *Trans. Br. mycol. Soc.* 47:619–25.

Jones, E.B.G., and S.K. Eltringham (eds.). 1971. *Marine borers, fungi and fouling organisms of wood*. OECD, Paris. 367 p.

Kampf, W-D., G. Becker, and J. Kohlmeyer. 1959. Versuche über das Auffinden und den Befall von Holz durch Larven der Bohrmuschel *Teredo pedicellata* Qutrf. *Zeitschr. angew. Zool.* 46:257–83.

Kendrick, W.B. (ed.). 1971. Taxonomy of Fungi Imperfecti. *Proc. 1st Intl. Spec. Workshop Conf. at Kananaskis, Alberta, Canada*. University of Toronto Press, Toronto, 306 p.

Kendrick, W.B., and J.W. Carmichael. 1973. Hyphomycetes. In: *The Fungi, Vol. 4A* (G.C. Ainsworth, F.K. Sparrow, and A.S. Sussman, eds.). Academic Press, New York and London, pp. 323–509.

Kirk, P.W., Jr. 1976. Cytochemistry of marine fungal spores. In: *Recent Advances in Aquatic Mycology* (E.B.G. Jones, ed.). Wiley, New York, pp. 177–92.

Kohlmeyer, J. 1958. Beobachtungen über Mediterrane Meerespilze sowie das Vorkommen von marinen Moderfäule-Erregern in Aquariumszuchten holzzerstörender Meerestiere. *Ber. Deutsch. Bot. Ges.* 71:98–116.

Kohlmeyer, J. 1960. Wood-inhabiting marine fungi from the Pacific Northwest and California. *Nova Hedwigia* 2:293–343.

Kohlmeyer, J. 1963. Fungi marini novi vel critici. *Nova Hedwigia* 6:297–329.

Kohlmeyer, J. 1966. Ecological observations on arenicolous marine fungi. *Zeits. Allg. Mikrobiol.* 6:95–106.

Kohlmeyer, J. 1968. Dänische Meerespilze (Ascomycetes). *Ber. Deutsch. Bot. Ges.* 81:53–61.

Kohlmeyer, J. 1969. The role of marine fungi in the penetration of calcareous substances. *Amer. Zool.* 9:741–46.

Kohlmeyer, J. 1972. Marine fungi deteriorating chitin of hydrozoa and keratin-like annelid tubes. *Mar. Biol.* 12:277–84.

Kohlmeyer, J. 1974a. On the definition and taxonomy of higher marine fungi. *Veröff. Inst. Meeresforsch. Bremerh.* (Suppl.) 5:263–86.

Kohlmeyer, J. 1974b. Higher fungi as parasites and symbionts of algae. *Veröff. Inst. Meeresforsch. Bremerh.* (Suppl.) 5:339–56.

Kohlmeyer, J. 1975. New clues to the possible origin of Ascomycetes. *BioScience* 25(2):86–93.

Kohlmeyer, J., and E. Kohlmeyer. 1971a. *Synoptic Plates of Higher Marine Fungi* (3rd Ed.). J. Cramer, Lehre, 87 p.

Kohlmeyer, J., and E. Kohlmeyer. 1971b. Marine fungi from tropical America and Africa. *Mycologia* 63:831–61.

Kohlmeyer, J., and E. Kohlmeyer. 1972. Is *Ascophyllum nodosum* lichenized? *Botanica Mar.* 15:109–12.

Kohlmeyer, J., G. Becker, and W-D. Kampf. 1959. Versuche zur Kenntnis der Ernährung der Holzbohrassel *Limnoria tripunctata* und ihre Beziehung zu holzzerstörenden Pilzen. *Zeitschr. angew. Zool.* 46:457–89.

Lee, B.K.H., and G.E. Baker. 1972a. An ecological study of the soil in a Hawaiian mangrove swamp. *Pac. Sci.* 26:1–10.

Lee, B.K.H., and G.E. Baker. 1972b. Environment and the distribution of microfungi in a Hawaiian mangrove swamp. *Pac. Sci.* 26:11–19.

Lee, B.K.H., and G.E. Baker. 1973. Fungi associated with the roots of red mangrove, *Rhizophora mangle*. *Mycologia* 65:894–906.

Lutley, M., and I.M. Wilson. 1972a. Development and fine structure of ascospores in the marine fungus *Ceriosporopsis halima*. *Trans. Br. mycol. Soc.* 58:393–402.

Lutley, M., and I.M. Wilson. 1972b. Observations on the fine structure of ascospores of marine fungi: *Halosphaeria appendiculata*, *Torpedospora radiata* and *Corollospora maritima*. *Trans. Br. mycol. Soc.* 59:219–27.

Madgwick, J., A. Haug, and B. Larsen. 1973. Alginate lyase in the brown alga *Laminaria digitata* (Hus.) Lamour. *Acta Chem. Scand.* 27:711–12.

Meyers, S.P., and B.E. Hopper. 1966. Attraction of the marine nematode, *Metoncholaimus* sp., to fungal substrates. *Bull. Mar. Sci.* 16:142–50.

Meyers, S.P., and R.L. Moore. 1960. Thalassiomycetes. II. New genera and species of Deuteromycetes. *Amer. J. Bot.* 47:345–49.

Meyers, S.P., W.A. Feder, and K.M. Tsue. 1964. Studies of relationships among nematodes and filamentous fungi in the marine environment. *Dev. Industr. Microbiol.* 5:354–64.

Moore, R.L., and S.P. Meyers. 1959. Thalassiomycetes. I. Principles of delimitation of the marine mycota with the description of a new aquatically adapted deuteromycete genus. *Mycologia* 51:871–76.

Newell, S.Y. 1972. Succession and role of fungi in the degradation of red mangrove seedlings. In: *Estuarine Microbial Ecology* (L.H. Stevenson and R.R. Colwell, eds.), Vol. I, Belle Baruch Library in Marine Sciences. University of South Carolina Press, pp. 467–80.

Newell, S.Y. 1976. Mangrove fungi: The succession in the mycoflora of red mangrove (*Rhizophora mangle* L.) seedlings. In: *Recent Advances in Aquatic Mycology* (E.B.G. Jones, ed.). Wiley, New York, pp. 51–91.

Nilsson, S. 1957. A new Danish fungus *Dinemasporium marinum. Bot. Not.* 110:321–24.

Perkins, R.D., and S.D. Halsey. 1971. Geologic significance of microboring fungi and algae in Carolina shelf sediments. *J. Sed. Petrol.* 41:843–53.

Poole, N.J., and P.C. Price. 1972. Fungi colonizing wood submerged in the Medway estuary. *Trans. Br. mycol. Soc.* 59:333–35.

Ray, D.L. (ed.). 1959. *Marine Boring and Fouling organisms.* University of Washington Press, Seattle, 536 p.

Reynolds, D.R. 1971. Wall structure of a bitunicate ascus. *Planta* 98:244–57.

Rooney, W.S., Jr., and R.D. Perkins. 1972. Distribution and geologic significance of microboring organisms within sediments of the Arlington Reef Complex, Australia. *Geo. Soc. Amer. Bull.* 83:1139–50.

Schaumann, K. 1972a. *Corollospora intermedia* (Ascomycetes, Halosphaeriaceae) vom Sandstrand der Insel Helgoland (Deutsche Bucht). *Veröff. Inst. Meeresforsch. Bremerh.* 14:13–22.

Schaumann, K. 1972b. *Biconiosporella corniculata* nov. gen. et nov. spec., ein holzbesiedelnder Ascomycet des marinen Litorals. *Veröff Inst. Meeresforsch Bremerh.* 14:23–44.

Sieburth, J.McN. 1968. The influence of algal antibiosis on the ecology of marine microorganisms. In: *Advances in Microbiology of the Sea* (M.R. Droop and E.J. Ferguson Wood, eds.). Academic Press, New York and London, pp. 63–94.

Sieburth, J.McN. 1975. *Microbial Seascapes.* University Park Press, Baltimore, Md., 246 p.

Snell, W.H., and E.A. Dick. 1971. *A Glossary of Mycology.* Harvard University Press, Cambridge, Mass., 181 p.

Sparrow, F.K., Jr. 1937. The occurrence of saprophytic fungi in marine muds. *Biol. Bull.* 73:242–48.

Steele, C.W. 1967. Fungus populations in marine waters and coastal sands of the Hawaiian, Line and Phoenix Islands. *Pac. Sci.* 21:317–31.

Sutherland, G.K. 1915. New marine fungi on *Pelvetia. New Phytol.* 14:33–42.

Sutherland, G.K. 1916. Marine fungi imperfecti. *New Phytol.* 15:35–48.

Sutton, B.C. 1973. Coelomycetes. In: *The Fungi. Vol. 4A* (G.C. Ainsworth, F.K. Sparrow, and A.S. Sussman, eds.). Academic Press, New York and London, pp. 513–82.

Webber, F.C. 1967. Observations on the structure, life history and biology of *Mycosphaerella ascophylli. Trans. Br. mycol. Soc.* 50:583–601.

CHAPTER 20

Yeasts

Yeasts are those fungi that, in a conspicuous stage of their life cycle, are in a unicellular phase that reproduces by budding or fission. They are recognized on agar media by a typical colonial morphology and are distinguished from the bacteria by their relatively larger cell size. Yeast populations decrease in numbers with distance from land (van Uden and Fell, 1968), falling off more rapidly than the bacteria (Hoppe, 1972a,b). Yeasts, however, are still the dominant fungi in the open ocean (Fell, 1968; Bahnweg and Sparrow, 1971). The filamentous fungi are mainly restricted to nearshore waters, due to the distribution of their substrates—the remnants of land plants, the seagrasses, and the seaweeds. Yeasts in inshore waters belong to a variety of species in all three divisions of the "higher" fungi, the Ascomycetes, the Basidiomycetes, and the Deuteromycetes. As one goes into the open sea, where the load of suspended organic debris is less, the variety of species decreases, and the Ascomycetes become rarer except for the ubiquitous species *Debaryomyces hansenii* (Fell, 1976).

The yeasts are not a natural taxonomic entity, but there is a great uniformity in their cell morphology and colonial appearance on nutrient agars. Yeasts are cultured by the same techniques used for bacteria, except that sugars are required for optimal growth. The ratio of yeasts to bacteria that form colonies on agar media in the western Baltic Sea has been reported as 1:124 (Hoppe, 1972a). Many yeasts grow well on the media used for the cultivation of the filamentous fungi. A medium commonly used by Fell and colleagues contains 2.3% nutrient agar, 2% glucose, and 0.1% yeast extract adjusted to pH 4.5 with hydrochloric acid to inhibit bacterial growth (Fell et al., 1969; Fell, Hunter, and Tallman, 1973). But marine yeasts grow better at a neutral pH or at the alkaline pH of seawater, and some strains are even acid sensitive (Seshadri and Sieburth, 1971, 1975; Seshadri, 1972). An alternate approach to culture is to use a neutral medium with antibiotics to inhibit bacteria and triphenyl tetrazolium chloride, which not only pigments the yeast colonies for easier detection on membrane filters but is also bacteriostatic (Seshadri and Sieburth, 1971) and decreases the amount of antibiotic required. Yeasts free in the water and associated with microparticulates suspended in the water column are concentrated from sample volumes up to 1 liter in oceanic waters. The yeasts are cultured directly on the membrane filters on a nutrient agar surface (Fell, Hunter, and Tallman, 1973) except when the inoculum contains inhibitory materials, such as the polyphenols from fucoid algae, which adsorb on the filter (Seshadri and Sieburth, 1971). Even though the yeasts are considered higher fungi, they have been avoided by the mold specialists (Kohlmeyer and Kohlmeyer, 1971) because they cannot be characterized by morphology alone, requiring biochemical tests similar to those used for bacteria. A standard set of tests and substrates are used to test for fermentation and assimilation, characteristics that are used for generic and/or species differentiation (van der Walt, 1970). The basis for the taxonomy and the systematics of the yeasts have been discussed by Kreger-van Rij (1969, 1973). The ultimate source book on the taxonomy of the 39 genera of yeasts is the costly 1385-page tome edited by Lodder (1970). The three volumes edited by Rose and Harrison (1969, 1970, 1971) contain useful discussions on yeast growth, cytology, life cycles, and genetics.

The history of the investigation of marine yeasts as well as their distribution in water and their occurrence on different substrates have been reviewed by Kriss et al. (1967), van Uden (1967), van Uden and Fell (1968), Morris (1968), and Fell (1974, 1976). The earliest observations were made by Bernard Fischer during his pioneering work on marine bacteria, which he started in the 1880s. The first comprehensive oceanic studies were those conducted by Kriss and his colleagues, which are reviewed by Sieburth (1960) and Kriss (1963). Their sampling methods are open to criticism, as discussed by Sieburth (1971). Virtually all the yeasts reported up to the 1960s were of the asporogenous type belonging to the Cryptococcaceae, such as species of *Rhodotorula*, *Candida*, and *Cryptococcus*, and more rarely the ascomycetous yeasts belonging to the Saccharomycetaceae. Since then, the genus *Metschnikowia*, with its distinctive ascospores, has been rediscovered in the sea by van Uden and Castelo-Branco (1961), and Fell and his associates have found the perfect stages of *Rhodotorula* (Newell and Fell, 1970) and *Candida* (Fell et al., 1969), which belong to the same basidiomycete family as the smuts.

The low dissolved organic carbon levels of inshore seawater must be inadequate for the active growth of yeast cells. A *Rhodotorula* strain grown in continuous culture in Baltic Sea water had a generation time of 178 hours at 25 °C (Hoppe, 1972b). Yeasts in highly polluted areas or yeasts associated with very dense algal blooms (Meyers et al., 1967) may be getting enough dissolved organic matter to support growth, but most yeasts appear to be associated with rich organic substrates, such as animal surfaces, the gut contents of fish, and vegetative debris (Roth et al., 1962; Meyers et al., 1970). The low number of yeasts associated with the living thallus of seaweeds (Roth et al., 1962; Seshadri and Sieburth, 1975) is only partially due to the inhibitory phenols in the fucoid brown algae (Seshadri and Sieburth, 1971). The low numbers may also be due to a labile proteinaceous cuticle (Hanic and Craigie, 1969), present in some algal species, which is sloughed off periodically, thereby removing the current crop of epibiotic microorganisms (Tootle, 1974).

A. ASCOMYCETOUS YEASTS

The ascomycetous yeasts are primitive fungi capable of forming ascospores in asci. They belong to the class Hemiascomycetes, which lack ascocarps and ascogenous hyphae, and the order Endomycetales, in which the zygotes or single cells are transformed directly into asci. The genera with marine representatives fall into two of the four families of Endomycetales, the Spermophthoraceae and the Saccharomycetaceae.

The species in the family Spermophthoraceae are distinguished by needle- or spindle-like ascospores, which are often larger than the vegetative yeast cells from which the ascus is formed. The marine isolates belong to the genus *Metschnikowia* and have budding vegetative cells, rudimentary pseudomycelium, and elongated asci with the characteristic needle-like spores, as shown in Plate 20-1. The checkered taxonomy of the five recognized species is discussed in detail by Miller and van Uden (1970). The type species *M. bicuspidata*, with its exceptional morphology of asci and needle-like spores pointed at both ends, was observed in the body cavity of the common freshwater crustacean *Daphnia magna* by the pioneer Russian microbiologist Elias Metschnikoff (1884). Metschnikoff illustrated the events

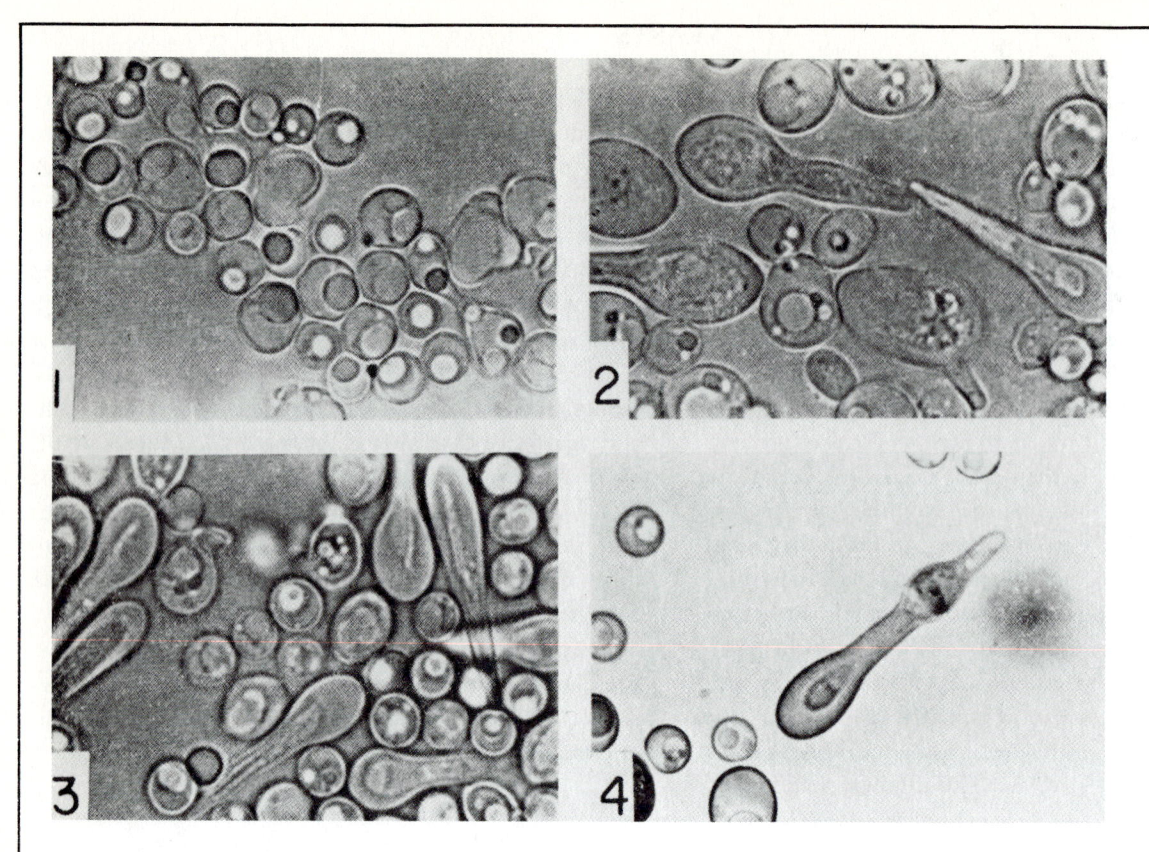

A

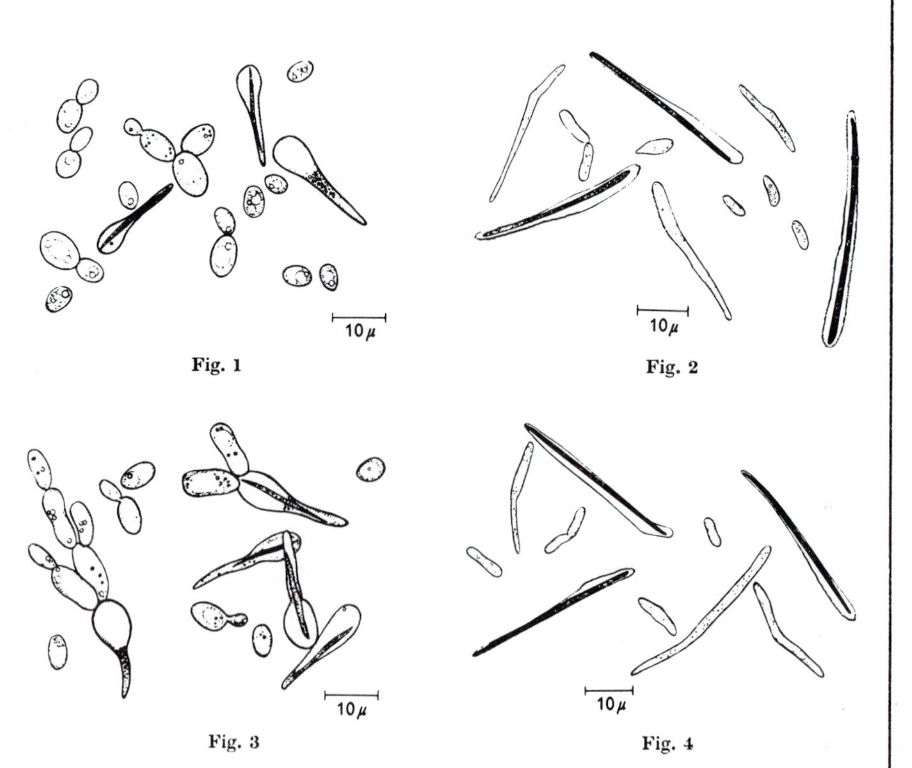

B

PLATE 20-1. Ascomycetous yeasts of the genus *Metschnikowia* capable of producing fatal infections in freshwater cladocerans and brackish water calanoid copepods. (*A*) *M. bicuspidata* var. *australis*, showing (1) conjugation, (2) concentration of cytoplasmic material, (3) ascospores, and (4) terminal swelling of ascus possibly due to spore abortion (all 1,500 X). (*B*) The vegetative cells and asci of *M. zobellii* on V-8 medium (1) and in *Daphnia magna* (2) and *M. krissii* on V-8 medium (3) and in *D. magna* (4). (*A*) from J.W. Fell and I.L. Hunter, 1968, *Antonie van Leeuwenhoek* 34:365–76; (*B*) from N. van Uden and R. Castelo Branco, 1961, *J. Gen. Microbiol.* 12:141–48.

of the budding yeast cells, the formation of the ascus and ascospores, and the infestation of susceptible young *Daphnia* that ingest detritus from diseased individuals. The ingested asci set free the spores, which, in turn, perforate the gut and penetrate the body cavity before becoming elongated and forming asci with ascospores that then perpetuate the cycle. Metschnikoff made his original observations on phagocytosis on this infested material. A synonymous species is associated with the brine shrimp *Artemia salina* (Spencer, Phaff, and Gardener, 1964). The first *Metschnikowia* yeasts obtained in pure culture and the first marine species, described as *M. zobellii* and *M. krissii*, were obtained off Southern California by van Uden and Castelo Branco (1961). *Metschnikowia krissii* was only isolated from seawater in which it occurred in minimal numbers of 1 to 57/100 ml; the population estimates for *M. zobellii* were 2 to 58/100 ml of seawater, 25 to 5730/ml of fish gut contents, and 520 to 39,200/g of the giant kelp *Macrocystis pyrifera* (van Uden and Castelo Branco, 1963). A number of haploid and heterothallic isolates of a marine variety of Metschnikoff's microorganism, *Metschnikowia bicuspidata* var. *australis*, were obtained near the South Shetland Islands in Antarctica by Fell and Hunter (1968).

Species of *Metschnikowia* appear to infest two species of copepods and to play a significant role in controlling their populations. Seki and Fulton (1969) described an infestation of *Calanus plumchrus*, a predominant copepod species in the Straits of Georgia, which was observed in February and March 1968. Yeasts identified as *M. krissii* were isolated and their physiology studied. Vegetative cells and asci with their single spores were found both as epibionts on the surface and as endobionts between the body wall and the wall of the gastrointestinal tract. Yeasts dominated by species of *Metschnikowia* were present in populations of 20 to 100/100 ml in the brackish waters of the Nanaimo River and its estuary; in the Straits of Georgia, yeasts were present in only the upper 100 m, with equally high populations at the sea surface where species of *Metschnikowia* accounted for some 40% of the yeast isolates. Fize, Manier, and Maurand (1970) described a similar infection of the calanoid copepod *Eurytemora velox* in the brackish waters east of the village of Saintes-Maries-de-la-Mer (the Golfe du Lion of the Mediterranean Sea) during April 1969. Although this species of *Metschnikowia* has a single spore per ascus, like *M. krissii* and *M. zobellii*, it is unique in having two slender ends. The yeast cells fill the hosts' bodies, kill them, and decimate the population, thereby influencing the food chain. There have been no similar observations to date on oceanic copepods. Seki and Fulton (1969) suggested that the presence of empty copepod exoskeletons in the deep aphotic zone of the sea (Wheeler, 1967) reflects kills by *Metschnikowia*, but normal molting and the efficient stripping of flesh from dead and dying copepods by carnivorous macrophagous ciliates (Chapter 24) could also account for their presence.

Most of the ascogenous yeasts are in the family Saccharomycetaceae, which reproduce vegetatively mainly by the budding of single cells, although fission occurs in two genera. The ascospores occur in a variety of shapes other than needle and are the main distinguishing feature of the genera. They are detected on a sporulating medium, such as the one made with V-8 (vegetable) juice. Species of *Debaryomyces*, which have round ascospores in an oval ascus, are quite common in some environments. *Debaryomyces globosus* has been isolated by Kriss and his colleagues in the Pacific Ocean and in the Greenland and Norwegian seas (Kriss, 1963), and *D. hansenii* was reported as common in Biscayne Bay (Roth et al., 1962), off the coast of Bombay (Bhat and Kachwalla, 1955), and in the waters around Helgoland (Meyers et al., 1967), being a ubiquitous species (Fell, 1976). Species of *Pichia* have vegetative cells of various shapes, reproduce by multiple budding, and form spherical Saturn- or hat-shaped spores usually within an internal oil drop. *Pichia guilliermondii* has been isolated from the marine environment, but it is not common. The genus *Saccharomyces*, which is vigorously fermentative, a feature put to good use by the alcoholic beverage and baking industries, has a few representatives in the sea. Other genera recovered from the sea include *Hansenula*, *Hanseniaspora*, *Endomycopsis*, and *Kluyveromyces*.

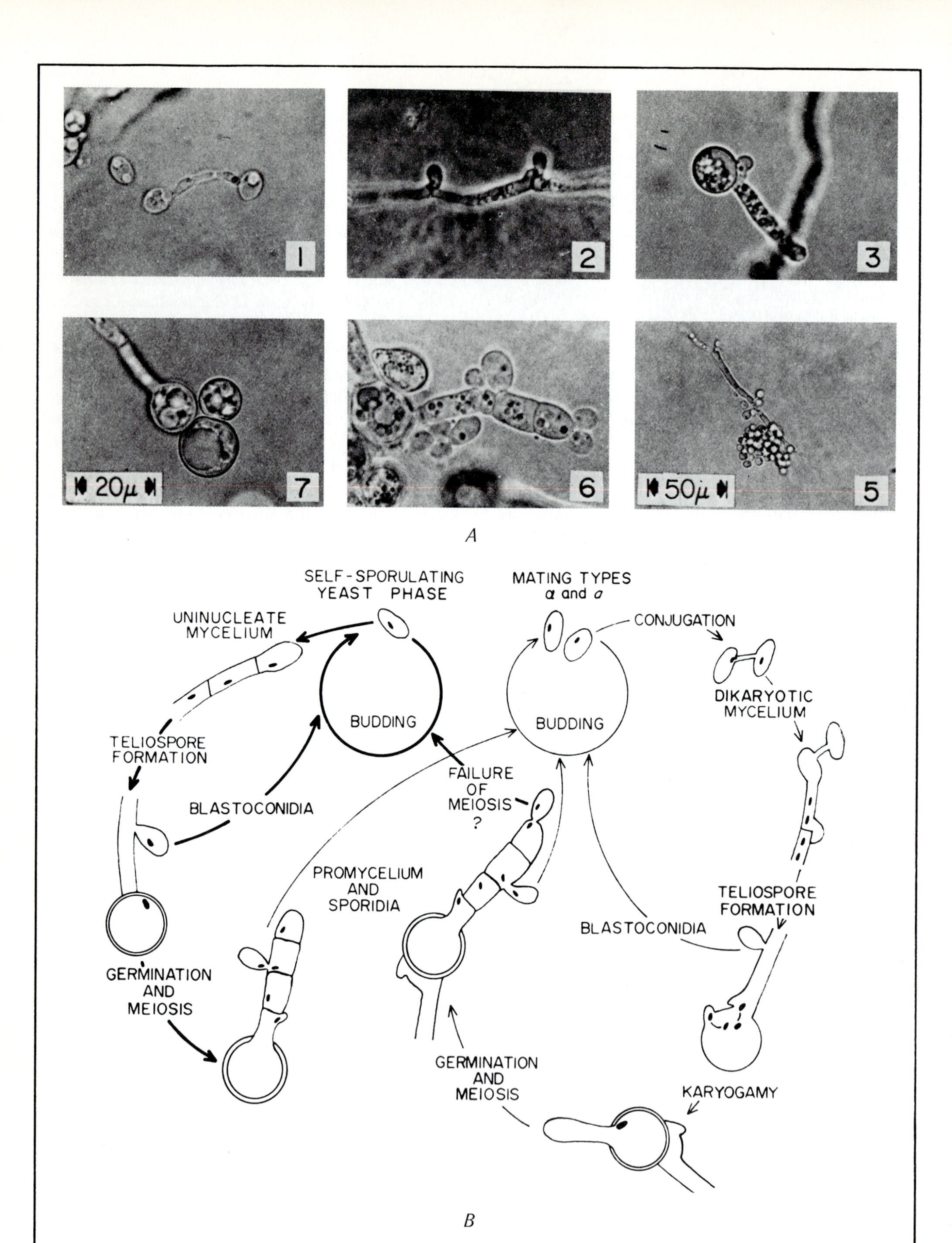

1
2
3
20μ
7
6
50μ
5
A
SELF-SPORULATING YEAST PHASE
MATING TYPES α and a
UNINUCLEATE MYCELIUM
CONJUGATION
BUDDING
BUDDING
DIKARYOTIC MYCELIUM
TELIOSPORE FORMATION
FAILURE OF MEIOSIS ?
BLASTOCONIDIA
PROMYCELIUM AND SPORIDIA
TELIOSPORE FORMATION
BLASTOCONIDIA
GERMINATION AND MEIOSIS
GERMINATION AND MEIOSIS
KARYOGAMY
B

PLATE 20-2. The life cycle of the heterobasidiomycetous yeast *Rhodosporidium sphaerocarpum*, shown in a diagrammatic representation and in micrographs. The micrographs show conjugation (1); hypha with clamp connections (2); teliospore with clamp at base (3); blastoconidia (5); teliospore germination with formation of promycelium and sporidia (6); and teliospores of a self-sporulating strain (7). [Scale as in (7), except (5).] From S. Newell and J. W. Fell, 1970, *Mycologia* 62:272–81.

B. BASIDIOMYCETOUS YEASTS

The existence of the basidiomycetous yeasts was only suspected until Nyland (1949) presented convincing evidence for a heterobasidiomycete-like sexual cycle for *Sporidiobolus*. More recently, Banno (1963, 1967) has shown that certain strains of *Rhodotorula* have a sexual life cycle, and these strains were placed in the genus *Rhodosporidium*. Some five genera have now been ascribed to the Ustilaginales, which are the smuts parasitic on grasses and sedges (Fell, 1970; Kreger-van Rij, 1973). The marine forms are the perfect stages of *Rhodotorula* and *Candida;* they are placed in *Rhodosporidium* and *Leucosporidium*, respectively. They have not been shown to be parasitic.

The genus *Rhodosporidium* was formed by Banno (1967) to include the perfect form of the red-pigmented asporogenous yeast *Rhodotorula glutinus*, which he named *Rhodosporidium toruloides*. Some of the 37 strains of *Rhodotorula glutinus* obtained by Newell and Fell (1970) from Antarctic and Caribbean waters had large numbers of unusually thick-walled spores, which led to the discovery of a sexual phase of the life cycle similar to that of *Rhodosporidium toruloides* but different enough to create the species *Rhodosporidium sphaerocarpum*. The photomicrographs and diagram in Plate 20-2 show the life cycle, in which the uninucleate yeast cells conjugate to form a binucleate mycelium with connecting "clamps." The clamps, a recurving outgrowth of a cell, act as a bridge at cell division, allowing the passage of one of the products of nuclear division into the penultimate cell, and thereby perpetuate the dikaryotic condition. The cells of the mycelium form teliospores, which are thick-walled spores in which nuclear fusion (karyogamy) occurs. The teliospores then germinate to form a transversely septate, four-celled promycelium, upon which sporidia bud at the tip of the ultimate cell as well as at the septa. These sporidia are in the uninucleate phase. The uninucleate phase is also produced from the mycelium as chains of blastoconidia of dimensions similar to those of yeast cells. In addition to this sexual cycle, there is a self-sporulating or homothallus yeast phase, in which certain strains form (with no apparent conjugation) uninucleate teliospore-bearing mycelia without clamp connections, which germinate just as those from the conjugating strains do.

The genus *Rhodosporidium* now contains eight species, of which seven were isolated from the sea. Studies on the mating systems of the genus have clarified the taxonomic status of the species and have strengthened the evidence of their relationship to the Basidiomycetes by the finding of both tetrapolar and multiple allelic bipolar systems in several species (Fell, Hunter, and Tallman, 1973). Two species are unique in that they not only form endospores but also produce both primary and secondary sporidia upon the elongated basidium.

The genus *Leucosporidium* was created by Fell et al. (1969) to include the *Candida*-like heterobasidiomycetous yeasts. The essential difference between *Rhodotorula* and *Candida* is that the former are colored orange to red by carotenoid pigments, whereas the latter are unpigmented, being white to cream. Pigmentation is also the distinguishing feature between *Rhodosporidium* and *Leucosporidium*. The life cycles of the two genera are essentially identical. The species of *Leucosporidium* have a distinct budding yeast phase and produce a pseudomycelium. Some species are heterothallic and homothallic, having both a mating phase and a self-sporulating phase. The *Leucosporidium scottii* and *Leucosporidium antarcticum*, which have both phases, are differentiated by their inability to intermate and by their different physiology. The four strictly self-sporulating species are differentiated on the basis of fermentation and assimilation of carbon compounds. The names of several species of the genus (*L. scottii*, *L. nivalis*, *L. frigidum*, *L. antarcticum*, and *L. gelidum*) indicate that they were isolated in cold Antarctic waters.

C. DEUTEROMYCETOUS YEASTS

The deuteromycetous yeasts also occur in the sea and are placed either in the family Sporobolomycetaceae, with characteristic ballistospores (which, as the name implies, are actively projected at maturity), or in

PLATE 20-3. Scanning electron micrographs of cryptococcoid yeast colonies on seaweeds. (*A*) Typical budding yeasts, on the red alga *Polysiphonia lanosa;* (*B*) pseudomycelia and blastospore-like structure of *Candida*, on the brown alga *Ascophyllum nodosum*. Micrographs by author.

A

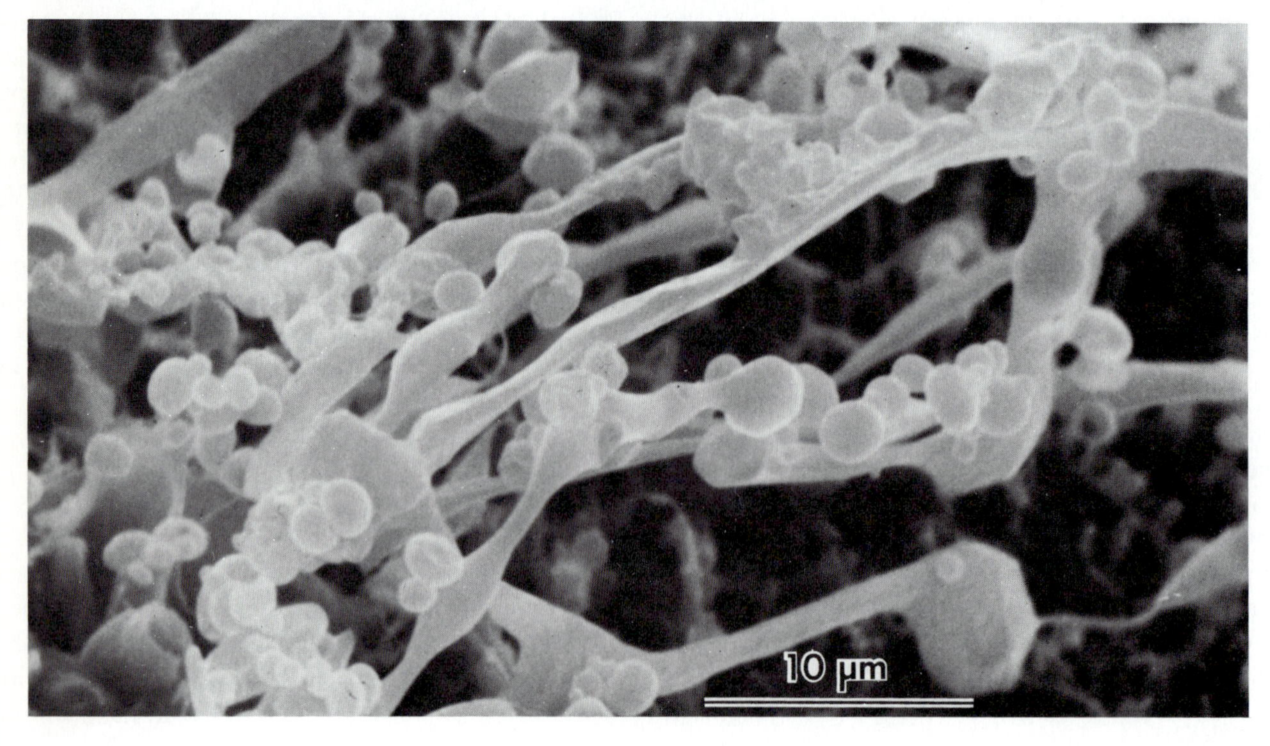

B

PLATE 20-4. Camera lucida drawings of unusual cryptococcoid marine yeasts reproducing by conidia. *Sympodiomyces parvus*, showing (*A*) mycelial formation on malt agar after 1 week at 12 °C; (*B*) sympodial formation of conidia in the yeast phase in liquid malt extract over a 4-week period at 12 °C; (*C*) *Sterigmatomyces halophilus*, showing the formation of conidia on sterigma-like structures. (*A*,*B*) from J.W. Fell and A.C. Statzell, 1971, *Antonie van Leeuwenhoek* 37:359–67; (*C*) from J.W. Fell, 1966, *Antonie van Leeuwenhoek* 32:99–104.

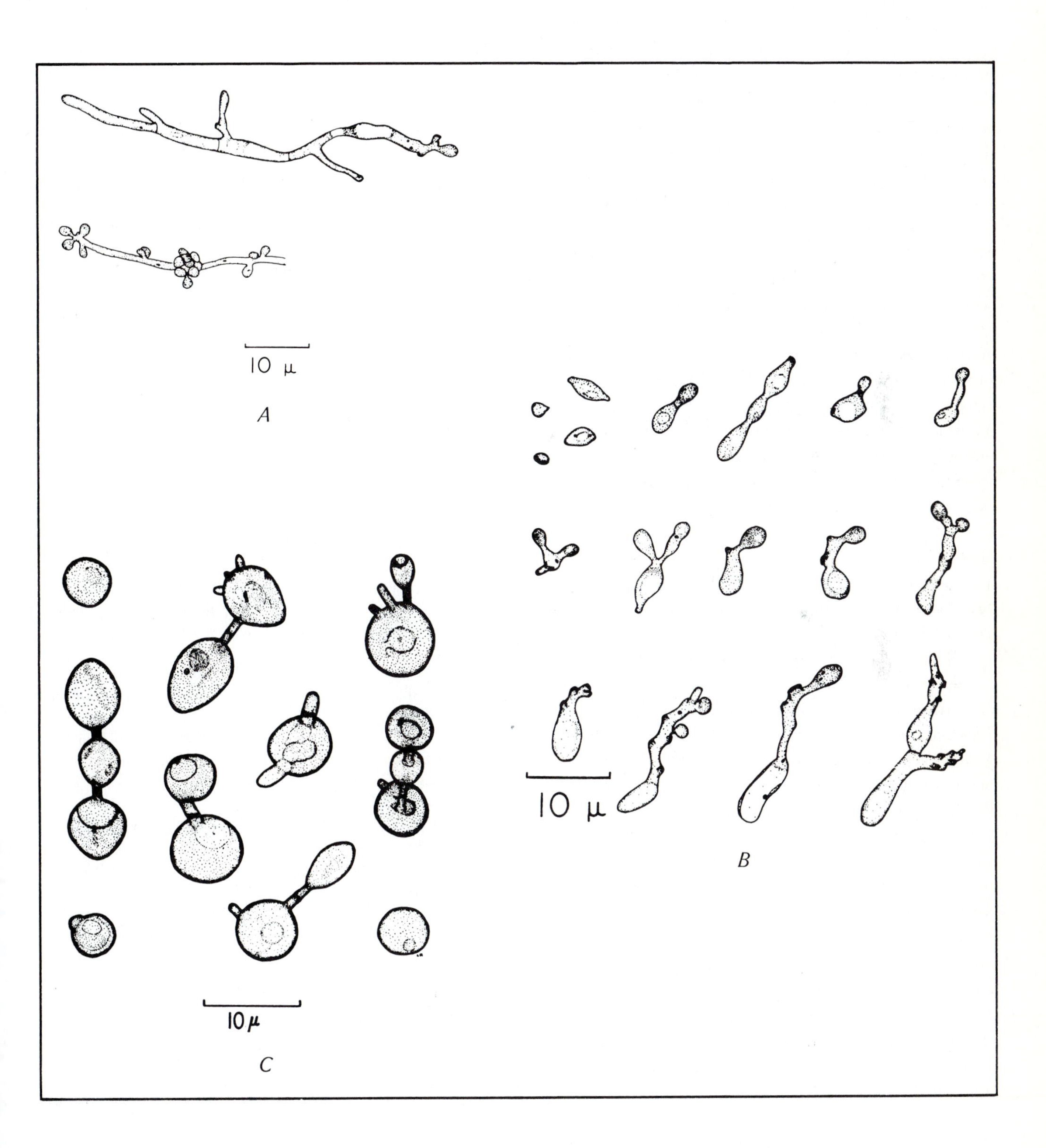

the family Cryptococcaceae, which lacks ballistospores. The Cryptococcaceae include cream, yellow, orange, and red yeasts that bud, with some forming pseudomycelium (by budding), and some forming true mycelium (by fission) or arthrospores (by hyphal fragmentation). The major yeast population of many nearshore materials is often these cryptococcoid yeasts. The proportion depends upon the nature of the inoculum. Some 95% of the yeast isolates from living seaweeds in Rhode Island were species of *Candida* (Seshadri and Sieburth, 1975). The genus *Candida*, which undergoes multipolar budding, is distinguished by the formation of pseudomycelium, often with pseudohyphae and blastospores (which arise by budding) (see Plate 20-3). Pseudomycelium and true mycelium are detected in slide cultures made with cornmeal or potato–dextrose agar (van der Walt, 1970). The some 80 recognized species of *Candida* (van Uden and Buckley, 1970), although including most of the possible permutations, appear too fragmented for the practical identification of natural populations. Out of 344 isolates from seaweeds, 300 keyed into 19 species, but 44 isolates were not identified. The numerical taxonomy of all these marine isolates and the eighty described species showed that there were seven discrete clusters, one containing only the described species, three containing both described and marine strains, and three almost exclusively containing marine strains (Seshadri and Sieburth, 1975). (Apparently much work on the taxonomy of the "common garden variety" of marine yeasts still lies ahead.)

The genera in the Cryptococcaceae (discussed in Lodder, 1970) are distinguished mainly by morphological rather than physiological differences. The genus *Torulopsis* is differentiated from *Candida* only by the poor development or lack of a pseudomycelium. The genus *Cryptococcus* not only lacks pseudomycelium, but it is also non-fermentative and assimilates inositol. The "pink" yeasts are immediately recognized as *Rhodotorula* by their visible carotenoid pigments and lack of fermentation. Species of *Trichosporon* are distinguished by the formation of true mycelium and arthrospores. Two genera of unusual yeasts reproduce by conidia. Fell (1966) obtained non-filamentous, yeast-like fungi from Indian Ocean as well as terrestrial sources; these reproduce by the development of conidia on a fine, sterigma-like structure. They have been put in the genus *Sterigmatomyces* (Plate 20-4,*C*). Another unusual genus, *Sympodiomyces* (Fell and Statzell, 1971), was obtained in the open waters of Antarctica. It is distinguished by the production of conidia on a conidiophore that develops sympodially (arising successively as branches, one from the other); it is shown in Plate 20-4,*A*,*B*.

REFERENCES

Bahnweg, G., and F.K. Sparrow, Jr. 1971. Marine fungi: Occurrence in the southern Indian Ocean. *Antarctic J. U.S.* 6:155.

Banno, I. 1963. Preliminary report on cell conjugation and mycelial stage in *Rhodotorula* yeasts. *J. Gen. Appl. Microbiol.* 9:249–51.

Banno, I. 1967. Studies on the sexuality of *Rhodotorula. J. Gen. Appl. Microbiol.* 13:167–96.

Bhat, J.V., and N. Kachwalla. 1955. Marine yeasts off the Indian coast. *Proc. Ind. Acad. Sci.* 41 (Sect. B):9–15.

Fell, J.W. 1966. *Sterigmatomyces*, a new fungal genus from marine areas. *Antonie van Leeuwenhoek* 32:99–104.

Fell, J.W. 1968. Distribution of Antarctic marine fungi. *Antarctic J. U.S. 3:157.*

Fell, J.W. 1970. Yeasts with heterobasidiomycetous life cycles. In: *Recent Trends in Yeast Research* (D.G. Ahearn, ed.), Vol. 1. Georgia State University, Atlanta, Ga., pp. 49–66.

Fell, J.W. 1974. Distributions of yeasts in the water masses of the southern Oceans. In: *Effect of the Ocean Environment on Microbial Activities* (R.R. Colwell and R.Y. Morita, eds.). University Park Press, Baltimore, Md., pp. 510–23.

Fell, J.W. 1976. Yeasts in oceanic regions. In: *Recent Advances in Aquatic Mycology* (E.B.G. Jones, ed.). Wiley, New York, pp. 93–124.

Fell, J.W., and I.L. Hunter. 1968. Isolation of heterothallic yeast strains of *Metschnikowia* Kamienski and their mating reactions with *Chlamydozyma* Wickerham spp. *Antonie van Leeuwenhoek* 34:365–76.

Fell, J.W., and A.C. Statzell. 1971. *Sympodiomyces* gen. n., a yeast-like organism from southern marine waters. *Antonie van Leeuwenhoek* 37:359–67.

Fell, J.W., A.C. Statzell, I.L. Hunter, and H.J. Phaff. 1969. *Leucosporidium* gen. n., the heterobasidiomycetous stage of several yeasts of the genus *Candida. Antonie van Leeuwenhoek* 35:433–62.

Fell, J.W., I.L. Hunter, and A.S. Tallman. 1973. Marine basidiomycetous yeasts (*Rhodosporidium* spp. n.) with tetrapolar and multiple allelic bipolar mating systems. *Can. J. Microbiol.* 19:643–57.

Fize, A., J.F. Manier, and J. Maurand. 1970. Sur un Cas d'infestation du copepode *Eurytemora velox* (Lillj) par une levure du genre *Metschnikowia* (Kamienski). *Ann. Parasit. hum. comp.* 45:357–63.

Hanic, L.A., and J.S. Craigie. 1969. Studies on the algal cuticle. *J. Phycol.* 5:89–102.

Hoppe, H-G. 1972a. Untersuchungen zur Ökologie der Hefen im Bereich der westlichen Ostsee. *Kiel. Meeresforsch.* 28:54–77.

Hoppe, H-G. 1972b. Taxonomische Untersuchungen an Hefen aus der westlichen Ostsee. *Kiel. Meeresforsch.* 28:219–26.

Kohlmeyer, J., and E. Kohlmeyer. 1971. *Synoptic Plates of*

Higher Marine Fungi (3rd ed.). Verlag Cramer, Lehre, Germany. 87 p.

Kreger-van Rij, N.J.W. 1969. Taxonomy and systematics of yeasts. In: *The Yeasts, Vol. 1. Biology of Yeasts* (A.H. Rose and J.S. Harrison, eds.). Academic Press, London and New York, pp. 5–78.

Kreger-van Rij, N.J.W. 1973. Endomycetales, Basidiomycetous yeasts, and related fungi. In:*The Fungi* (G.C. Ainsworth, F.K. Sparrow, and A.S. Sussman, eds.), Vol. 4A. Academic Press, London and New York, pp. 11–32.

Kriss, A.E. 1963. *Marine Microbiology (Deep-Sea).* (J.M. Shewan and Z. Kabata, transl.). Oliver and Boyd, Edinburgh. 536 p.

Kriss, A.E., I.E. Mishustina, I.N. Mitskevich, and E.V. Zemtsova. 1967. *Microbial Populations of Oceans and Seas,* (G.E. Fogg, ed.). St. Martin's Press, New York. 287 p.

Lodder, J. (ed.). 1970. *The Yeasts.* North-Holland Publ. Co., Amsterdam and London. 1385 p.

Metschnikoff, E. 1884. Ueber eine Sprosspilzkrankheit der Daphnien. Beitrag zur Lehre über den Kampf der Phagocyten gegen Krankheitserreger. *Virchows Arch. Path. Anat. Physiol. Klin. Med.* 96:177–95.

Meyers, S.P., D.G. Ahearn, W. Gunkel, and F.J. Roth, Jr. 1967. Yeasts from the North Sea. *Mar. Biol.* 1:118–23.

Meyers, S.P., M.L. Nicholson, J. Rhee, P. Miles, and D.G. Ahearn. 1970. Mycological studies in Barataria Bay, Louisiana, and biodegradation of oyster grass, *Spartina alterniflora.* Spec. Sea Grant Issue, *Louisiana State Univ. Coastal Stud. Bull.* 5:111–79.

Miller, M.W., and N. van Uden. 1970. *Metschnikowia* Kamienski. In: *The Yeasts* (J. Lodder, ed.). North-Holland Publ. Co., Amsterdam and London, pp. 408–29.

Morris, E.O. 1968. Yeasts of marine origin. *Oceanogr. Mar. Biol. Ann. Rev.* 6:201–30.

Newell, S.Y., and J.W. Fell. 1970. The perfect form of a marine occurring yeast of the genus *Rhodotorula. Mycologia* 62:272–81.

Nyland, G. 1949. Studies on some unusual heterobasidiomycetes from Washington State. *Mycologia* 41:686–701.

Rose, A.H., and J.S. Harrison (eds.). 1969, 1970, 1971. *The Yeasts. Vol. 1 Biology of Yeasts,* 508 p.; *Vol. 2 The Physiology and Biochemistry of Yeasts,* 571 p.; *Vol. 3 Yeast Technology,* 590 p. Academic Press, London and New York.

Roth, F.J., Jr., D.G. Ahearn, J.W. Fell, S.P. Meyers, and S.A. Meyer. 1962. Ecology and taxonomy of yeasts isolated from various marine substrates. *Limnol. Oceanogr.* 7:178–85.

Seki, H., and J. Fulton. 1969. Infection of marine copepods by *Metschnikowia* sp. *Mycopath. Mycol. Appl.* 38:61–70.

Seshadri, R. 1972. Seaweeds as a habitat for yeasts. Ph.D. Thesis, University of Rhode Island, 127 p.

Seshadri, R., and J.McN. Sieburth. 1971. Cultural estimation of yeasts on seaweeds. *Appl. Microbiol.* 22:507–12.

Seshadri, R., and J.McN. Sieburth. 1975. Seaweeds as a reservoir of *Candida* yeasts in inshore waters. *Mar. Biol.* 30:105–17.

Sieburth, J.McN. 1960. Soviet aquatic bacteriology: A review of the past decade. *Quart. Rev. Biol.* 35:179–205.

Sieburth, J.McN. 1971. Distribution and activity of oceanic bacteria. *Deep-Sea Res.* 18:1111–21.

Spencer, J.F.T., H.J. Phaff, and N.R. Gardener. 1964. *Metschnikowia kamienskii* sp. n., a yeast associated with brine shrimp. *J. Bacteriol.* 88:758–62.

Tootle, J.L. 1974. The fouling microflora of intertidal seaweeds: Seasonal, species and cuticular regulation. M.S. Thesis, University of Rhode Island, 65 p.

van Uden, N. 1967. Occurrence and origin of yeasts in estuaries. In: *Estuaries* (G.H. Lauff, ed.). Amer. Assoc. Adv. Sci. Publ. 83, pp. 306–310.

van Uden, N., and H. Buckley. 1970. *Candida* Berkhout. In: *The Yeasts* (J. Lodder, ed.). North-Holland Publ. Co., Amsterdam and London, pp. 893–1087.

van Uden, N., and R. Castelo Branco. 1961. *Metschnikowiella zobellii* sp. nov. and *M. krissii* sp. nov., two yeasts from the Pacific Ocean pathogenic for *Daphnia magna. J. gen. Microbiol.* 26:141–48.

van Uden, N., and R. Castelo Branco. 1963. Distribution and population densities of yeast species in Pacific water, air, animals, and kelp off Southern California. *Limnol. Oceanogr.* 8:323–29.

van Uden, N., and J.W. Fell. 1968. Marine Yeasts. In: *Advances in Microbiology of the Sea* (M.R. Droop and E.J. Ferguson Wood, eds.). Academic Press, London and New York, pp. 167–201.

van der Walt, J.P. 1970. Criteria and methods used in classification. In: *The Yeasts* (J. Lodder, ed.), North-Holland Publ. Co., Amsterdam and London, pp. 34–113.

Wheeler, E.H., Jr. 1967. Copepod detritus in the deep-sea. *Limnol. Oceanogr.* 12:697–702.

CHAPTER 21

Zoosporic Fungi

These osmotrophic fungi develop as mycelium or sporangia on the surface or within the tissues of plants and animals. They are distinguished by the production of motile cells, which may be either asexual spores or gametes. Motility is by one or two flagella. The labyrinthulas also have a gliding motility along slime tracks or tubes in a colonial arrangement called a net or plasmodium. The zoosporic Phycomycetes usually have been regarded as parasites. But Sparrow (1936), in his classic study on the occurrence of these forms in Eel Pond at Woods Hole, Massachusetts, points out that after they kill their hosts they persist as saprophytes, along with the bacteria and the phagotrophic protists, and aid in the decomposition of dead tissue. Such microbially enriched vegetative debris provides nutrients for benthic crustaceans and deposit-feeders. Vishniac (1956), working in the same habitat, demonstrated that some of the chytrid-like forms can be cultured using bacteriological techniques and suggested that they are more "associated with" than "parasitic upon" algae and that they compete with bacteria for the same niche. Fuller, Fowles, and McLaughlin (1964) modified Vishniac's methods and obtained some 200 pure cultures of zoosporic Phycomycetes (which they considered to be saprophytes) from filamentous red, green, and brown algae in Rhode Island and Maine waters. In addition to the ubiquitous fungi of the genus *Thraustochytrium*, frequent isolates from these algae were species of *Lagenidium*, *Haliphthoros*, and *Atkinsiella*, which are usually thought to be parasitic on invertebrates (Alderman, 1976). A survey in the South Indian Ocean, in which cultural techniques were used, indicated that these Phycomycetes were primarily recovered from the waters and benthic materials in the vicinity of islands, with only a few obtained from open seawater (Bahnweg and Sparrow, 1971). All were eucarpic forms, in which a portion of the thallus converts to a reproductive center, while a portion remains vegetative and usually rhizoidal. Five new species described from these isolations (Bahnweg and Sparrow, 1972, 1974) belong to the genera *Thraustochytrium* and *Aplanochytrium*.

The Phycomycetes also occur on calcareous materials (Golubic, Perkins, and Lukas, 1975) in association with the microboring green algae and cyanobacteria (Chapter 8,B). In the carbonate components along the continental margin of the Carolinas, Perkins and Halsey (1971) found that these fungi were the most widespread and abundant of the endolithic forms. Lukas (1973) reported the incidence of fungi in corals from the Pacific and Atlantic. Some were parasitic on the endolithic algae. A shell disease of the European oyster *Ostrea edulis* has been studied by Alderman and Jones (1971), and a fungus with a septate hyphae and ovoid swellings, identified as *Ostracoblabe implexa*, was implicated, although *Althornia crouchii*, a biflagellate with saprolegnian affinities, was also isolated (Jones and Alderman, 1971).

Increasing interest in the Phycomycetes is due in large part to the development of cultural procedures that permit their detection and study. Vishniac (1956) pointed out that some of the chytrid-like forms that grew on a nutrient agar were associated with detrital particles, including a stray pollen grain. The colonization of pine pollen grains has proved useful for the detection, enumeration, and study of at least some Phycomycetes (Clokie and Dickinson, 1972; Goldstein, 1973), whereas brine shrimp colonization is being used to supplement nutrient agar plates and pollen grain colonization in the detection and description of other marine Phycomycetes (Bahnweg and Sparrow, 1972, 1974; Ulken, 1972a,b; Artemtchuk, 1972). Some species of the amoeboid phycomycetes, the labyrinthulas, require at least a seawater agar medium containing serum; other labyrinthula species require particulate food (Amon and Perkins, 1968). The direct culture procedures of Fuller, Fowles, and McLaughlin (1964), as well as the concentration of detritus particles and zoospores from large water samples (Fuller and Poyton, 1964), appear to be well suited for studying the ecology of these overlooked and apparently numerous and active osmotrophs. Such studies on the populations and distribution of the lower marine fungi have been summarized by Bremer (1976). The population figures cited by Bremer, however, may be one or two orders of magnitude too high, since populations were estimated by the end point of the dilution series rather than by the number of positive tubes and negative tubes in conjugation with most probable number (MPN) tables.

The marine Phycomycetes may be placed in three groups: the chytridian forms, which have zoospores with a single flagellum; the assemblage of biflagellates with an affinity to the saprolegnian forms, which may have definite mycelia; and the assemblage of biflagellates with an affinity for the peronosporacean forms, which include the downy mildews and pythiaceous fungi (Sparrow, 1976). Further division is based upon life cycle as well as the type, position, and length of the flagella. The flagella have the usual nine peripheral plus two axial fibrils. Like the flagellated phototrophs described in Part IV, the marine Phycomycetes flagella are of two types, a smooth flagellum with a thin terminal portion called a whiplash flagellum and a flimmer-type with two lateral lines of fine hairs (mastigonemes) called a tinsel flagellum (see Plate 21-4,*C*,3).

A. UNIFLAGELLATES

The chytridian assemblage is divided into two groups, the Hyphochytriales and the Chytridiales. The marine uniflagellates in the order Hyphochytriales have an anterior, tinsel-type flagellum on their zoospores. A genus of hyphochytrids from the lower littoral of the White Sea was recently described by Artemtchuk (1972). Two species are diagrammatically shown developing on pine pollen, Plate 21-1,*A*. The species in the Chytridiales are similar to those in the Hyphochytriales, except that their zoospores are posteriorly flagellated, with a whiplash-type flagellum. Johnson and Sparrow (1961) list five families containing species in eight genera that are parasitic on red and green algae and associated with marine muds and their debris. A species of *Phlyctochytridium* isolated from mud in the Weser Estuary is shown in Plate 21-1,*B*, as it develops on an agar medium.

Chytridian species, as well as biflagellated species, can infest marine phytoplankton. Johnson (1965, 1966) has reviewed the early literature and described and illustrated the phycomycete species occurring on a few species of pennate and centric diatoms as well as a dinoflagellate in the brackish and coastal waters of North Carolina. Some of these are illustrated in Plate 21-2 to show the various stages of development and the similarity of the sporangia in the uniflagellate and biflagellate forms. Studies on the chytridian fungi in the coastal soils of British Columbia have been reported by Booth (1969, 1971a,b). Enrichments were at salinities of 0 and

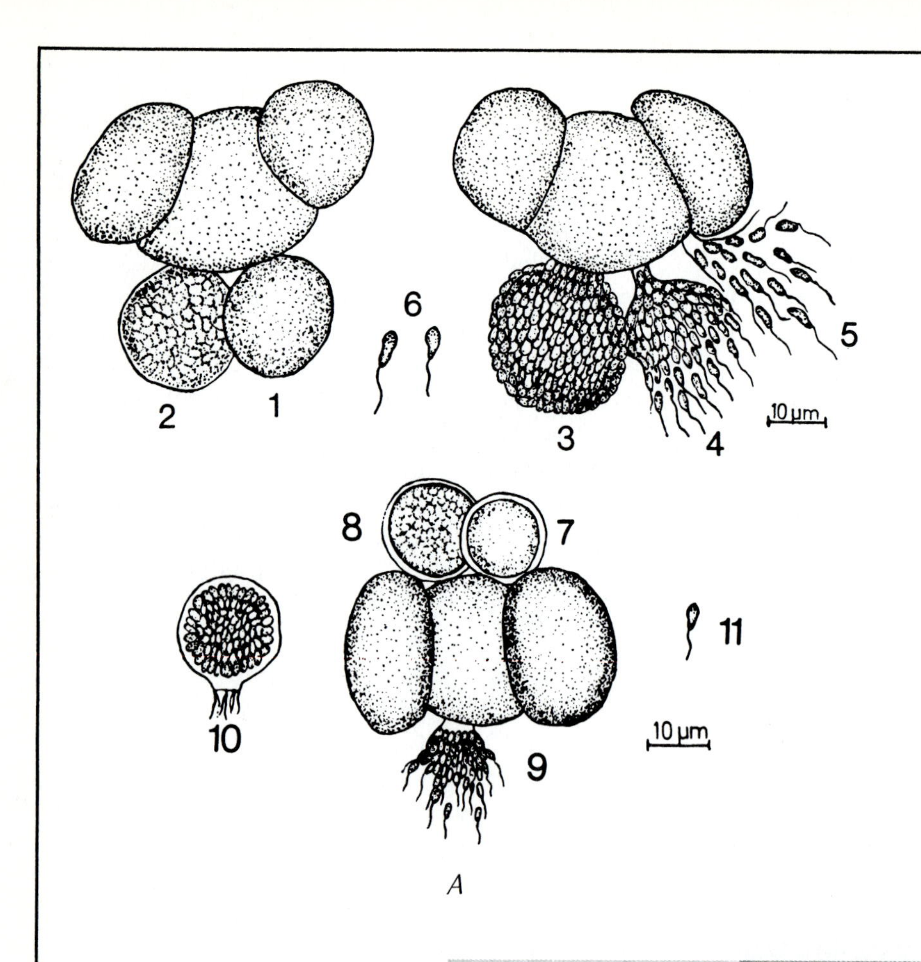

A

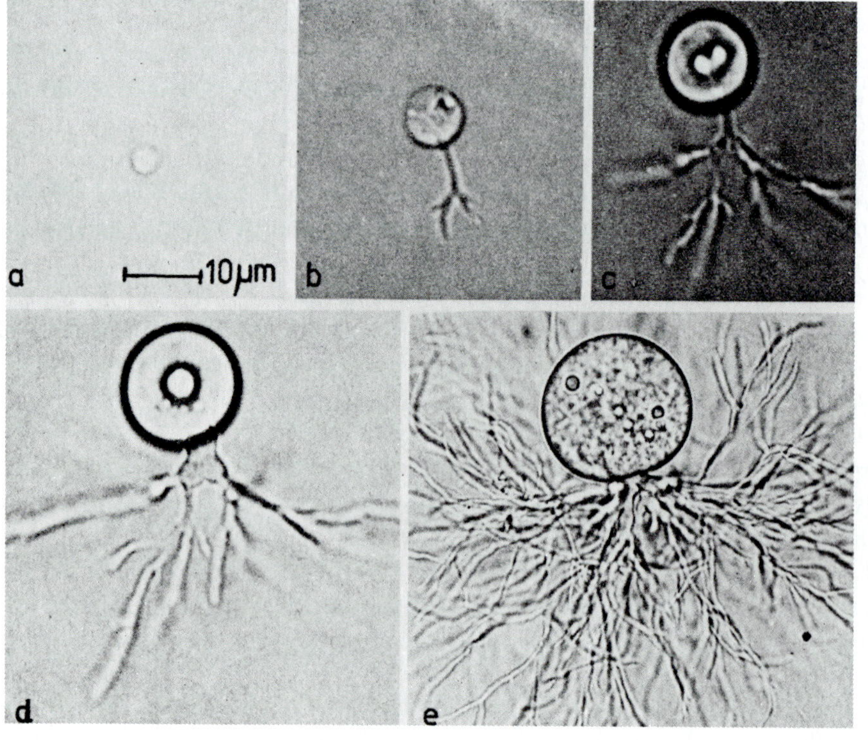

B

PLATE 21-1. Cultural characteristics of the uniflagellated chytridian Phycomycetes on pollen grains and agar culture. (*A*) The hyphochytrid *Elina*, which has zoospores with an anterior tinsel-type flagellum, shown in various stages of maturity growing on pine pollen grains. (*B*) The chytrid *Phlyctochytrium pleurigibbosum*, which has zoospores with a posterior whiplash flagellum, as shown in agar culture: (a) motile zoospore; (b) young germinating spore; (c) apophysis starting to form; (d) completed apophysis; (e) young thallus.

(*A*) from N.J. Artemtchuk, 1972, *Veröff. Inst. Meeresforsch. Bremerhaven* 13:231–37; (*B*) from A. Ulken, 1972, *Veröff. Inst. Meeresforsch. Bremerhaven* 13:205–16.

7 o/oo, since few recoveries were made at 28 o/oo S. Baits used for enrichment were pine pollen, dialysis membrane, snakeskin, shrimp exoskeletons, and blond baby hair.

B. SAPROLEGNIAN BIFLAGELLATES

The order Saprolegniales includes species in the Saprolegniaceae, Ectrogellaceae, and Atkinsiellaceae, among others. Some are saprophytes whereas other species infest diatoms (Sparrow, 1936), red and green algae (Johnson and Sparrow, 1961), the ova and embryos of molluscs and crustaceans (Atkins, 1954a,b; Vishniac, 1958) and fish. The most common form found by Sparrow (1936) in the waters of Woods Hole was a species of *Ectrogella* that quickly spread throughout the stands of the sessile pennate diatoms of the genera *Licmophora* and *Striatella*. The most common Phycomycete (the "S" forms), cultured by Vishniac (1956) on nutrient agar plates inoculated with either water or algal samples, also probably belong to the family Ectrogellaceae (Vishniac, 1961). *Ectrogella* was also present in the diatom *Synedra* in brackish water (Johnson, 1966).

Members of this group can also infest crustacean eggs and small crustaceans. Atkins (1954b), working at the Plymouth Laboratory in England, described a saprolegniaceous fungus parasitic on the eggs of several crab species as well as crustacea in other genera. Vishniac (1958), in describing a similar biflagellate from the egg cases of the oyster drill (*Urosalpinx*), named Atkins's fungus *Atkinsiella dubia*. Fuller, Fowles, and McLaughlin (1964) used solid culture media to isolate *Atkinsiella dubia* from marine algae in Narragansett Bay, Rhode Island and concluded that these and the other Phycomycetes were saprophytic. Aronson and Fuller (1969) isolated the cell walls from *A. dubia* and determined the carbohydrate constitution of the polysaccharide that made up 80% of the wall material. Sparrow (1973) observed *A. dubia* in the eggs of several crab species in the vicinity of Friday Harbor, Washington, but never in epidemic number. He concluded that *A. dubia* is doubtfully parasitic and carefully recorded its morphology and development in pure culture. Atkins (1954a) described another fungus of this group that was destructively parasitic in the pea-crab as *Leptolegnia marina* (see Plate 3-9,*B*).

C. PERONOSPORACEAN BIFLAGELLATES

The remaining orders to be discussed have been grouped by Sparrow (1976) as the peronosporacean biflagellates. More research on this group is being done, which seems to indicate the importance of these forms in the marine environment. The Thraustochytriales consist solely of marine forms that live a ubiquitous epibiotic existence. The largest genus, of over ten species, is *Thraustochytrium* (Booth and Miller, 1968; Jones, 1974). The first species, *T. proliferum*, was described by Sparrow (1936), who found it only on very disintegrated plants of the green algal genus *Bryopsis*. Species of *Thraustochytrium*, the second most prevalent Phycomycete developing on the agar plates of Vishniac (1956), were present on most plates streaked with algae by Fuller, Fowles, and McLaughlin (1964) and were among the most prevalent Phycomycetes associated with the nearshore waters and sediments cultured by Bahnweg and Sparrow (1971, 1974). Plate 21-3,*A*,*B*, are diagrammatic representations of the life cycles of two species on pollen grain and brine shrimp enrichments. The pollen grain culture method has been used to study the influence of environmental factors (Clokie and Dickinson, 1972) as well as to estimate populations (Bremer, 1976). Many of the characters used for speciation are substrate dependent, and the problems in the taxonomy of this genus have been reviewed by Booth and Miller (1968), Schneider (1969b), Gaertner (1972), and Goldstein (1973). Species of *Thraustochytrium* have been used to assay for cobalamin and thiamin in seawater (Vishniac, 1961; Vishniac and Riley, 1961). This sporozoic

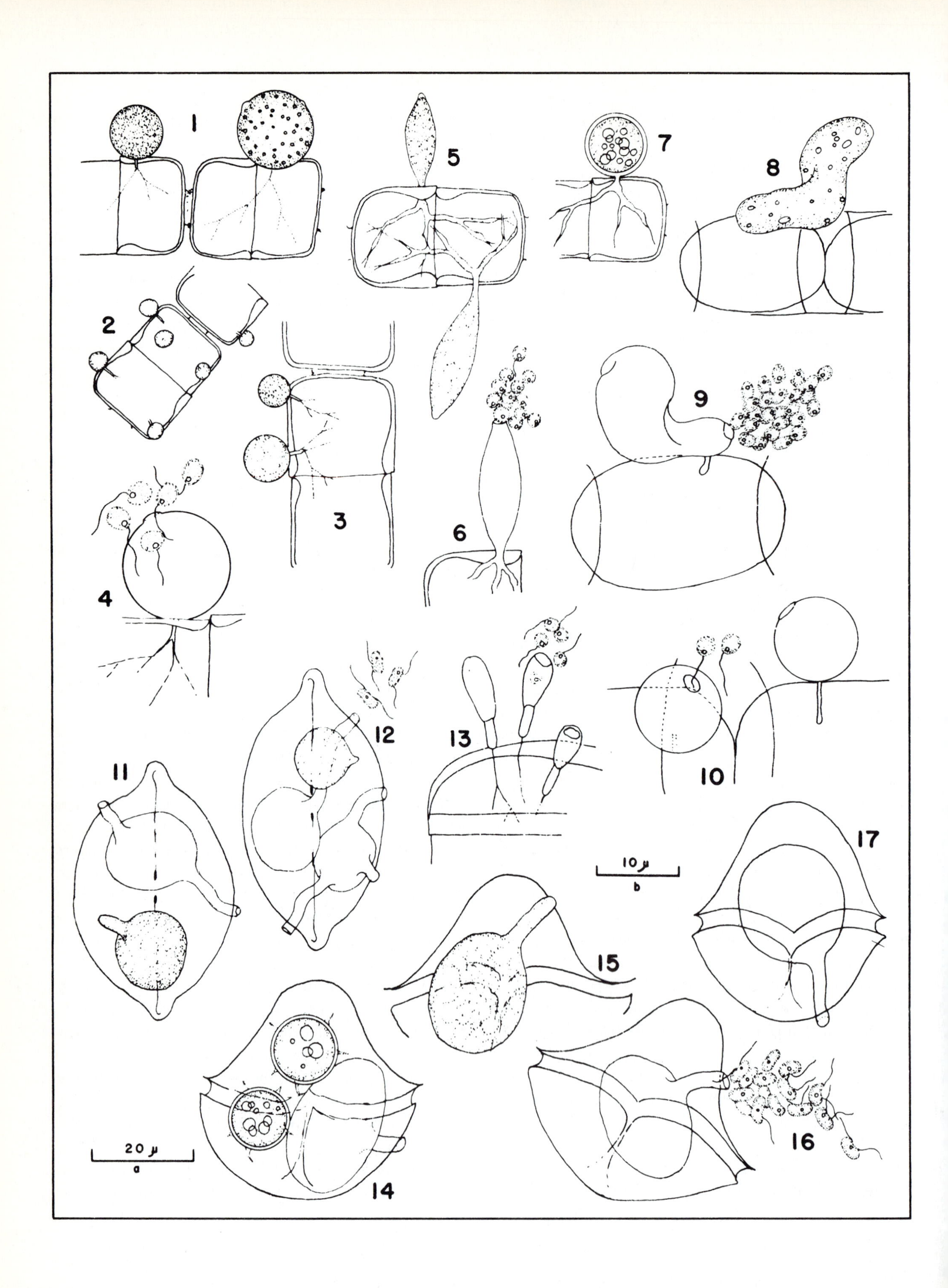

1
2
3
4
5
6
7
8
9
10
11
12
13
14
15
16
17
10 μ
b
20 μ
a

PLATE 21-2. Species of uniflagellate (1–10, 13) and biflagellate (11, 12, 14–17) Phycomycetes that infect marine and brackish water phytoplankton: The centric diatom *Melosira* infected with species of *Rhizophydium* (1–7) and *Phlyctidium* (8–10); the pennate diatom *Navicula* infected with a species of *Olpidiopsis* (11, 12) and *Podochytrium clavatum* (13); the dinoflagellate *Glenodinium* infected with *Olpidiopsis glenodinianum* (14–17). (Note: Although the uniflagellates shown have epibiotic sporangia and the biflagellates have endobiotic sporangia, the reverse can occur in both forms.) (Scale a:1,2,7,8,11–14; others, scale b.) From T.W. Johnson, Jr., 1965, *Nova Hedwigia* 10:581–88.

Phycomycete has been found to be infested by the net plasmodium of a species of *Labyrinthula* (Schneider, 1969a). Jones (1974) has included four other genera in the order Thraustochytriales. Some of these biflagellates, such as *Schizochytrium aggregatum* (Goldstein and Belsky, 1964) and *Aplanochytrium kerguelensis* (Bahnweg and Sparrow, 1972), have developmental sequences similar to those Labyrinthulales biflagellates that form slime tracks and plasmodia, called *Labyrinthulomyxa* (*Dermocystidium*) (Goldstein and Moriber, 1966). But some strains of *Labyrinthulomyxa* form only nonmotile aplanospores, and with their well-developed rhizoids and ability to use pollen as a substrate, they look very much like individual *Thraustochytrium*. The similarities between forms in the Thraustochytriales and those in the Labyrinthulales, as pointed out by Goldstein (1973), indicate that some major taxonomic revisions for these forms are due when there are sufficient comparative studies on development and ultrastructure. The few species of Phycomycetes in which fine structure has been studied are discussed by Perkins (1976). Sparrow (1976) has presented his views on the current taxonomic status of the biflagellated forms.

The Labyrinthulales contain the well-known genus *Labyrinthula*. These osmotrophic–phagotrophic microorganisms have been incriminated as the cause of wasting disease in eelgrass, *Zostera marina* (Renn, 1936; Young, 1943), but they are also regarded as commensals (Cox and Mackin, 1974). Characterized by uninucleate oval or spindle-shaped cells 1.5 to 8 μm wide by 8 to 30 μm long, these cells are motile and move along a network of hyaline slime filaments or tubes to form a "net-plasmodium" that is really a colony of pseudoplasmodia. Labyrinthulas have been cultured on seawater agar supplemented with serum, as well as on partially and completely synthetic media (Watson and Ordal, 1957). Sporulation, however, apparently only occurs when a medium, such as that of Fuller, Fowles, and McLaughlin (1964) containing 0.6 to 0.8% agar and *Rhodotorula* yeast cells as a food, is used. This medium has been used to study sporulation in a *Labyrinthula* isolated from *Spartina alterniflora* (Amon and Perkins, 1968). The spindle cells (Plate 21-4,*C*, 1) aggregate to form a coarse, yellow–orange tubular network and then round up and enlarge to form presporangia. The sporangia are compartmentalized within a large tube (Plate 21-4,*C*, 2) before eight zoospores are released from each sporangium. These zoospores (Plate 21-4,*C*, 3) have an anterior tinsel flagellum with two rows of mastigonemes and a posterior whiplash flagellum. After actively swimming for some 24 hours with a spiraling and undulating motion, the spore settles, rounds up, loses its flagella, and differentiates into a motile spindle cell, which again moves along its slime track (Perkins, 1972) to form a net-plasmodium. Details of the ultrastructural changes that occur during zoosporulation of this strain of *Labyrinthula* have been described by Perkins and Amon (1969). The presence of lomasomes indicates that the members of this group are indeed fungi. Ten marine species are recognized. These are associated with seagrasses; green, brown, and red algae; diatoms; water; and mud. Pokorny (1967) has reviewed the structure, physiology, ecology, and taxonomy of this genus. Two new species of *Labyrinthula* have recently been described. *Labyrinthula saliens* is apparently parasitic on the seagrass *Halophila englemanii*, since it is not found on other hosts and is not phagotrophic (Quick, 1974); its life cycle is shown in Plate 21-4,*B*. *Labyrinthula thais* is an ectocommensal on the marine gastropod *Thais haemostoma floridana* (Cox and Mackin, 1974); its life cycle is shown in Plate 21-4,*A*.

Another widespread group of microorganisms that might belong to the order Labyrinthulales are the shellfish pathogens and saprophytes belonging to *Labyrinthulomyxa* (*Dermocystidium*) (Bahnweg and Sparrow, 1971; Goldstein, 1973). Pérez (1913) created the genus *Dermocystidium* to describe *D. pusula*, which causes cutaneous lesions in the marbled newt in France. The presence of *Dermocystidium* in the gills of salmonid fish was reviewed by Davis (1947), and the first marine species known to cause mortality in stressed oysters was described as *D. marinum* by Mackin, Owen, and Collier (1950). This microorganism can be cultured easily in a thioglycollate seawater medium supplemented with antibiotics (Ray, 1952). The presence of the "parasite" in the cultured tissue is detected by staining with Lugol's iodine. This method has been used to show the widespread distribution of *D.*

PLATE 21-3. Developmental stages in the genus *Thraustochytrium*, non-thallus–forming, biflagellated "oomycetes," which are saprophytic and epibiotic. (*A*) *Thraustochytrium arudinmentale* on a pine pollen grain: (1) a mature sporangium; (2) a semi-empty sporangium and rudimentary zoospores at an apical pore; (3) a zoospore that becomes biflagellated after release. (*B*) *Thraustochytrium rossii*, shown developing on pine pollen grains and brine shrimp larvae: (A) a young sporangium, with one basal rudiment; (B) more rudiments are successively cleaved out of the cytoplasm as the sporangium enlarges; (C) the mature sporangium, containing a large number of rudiments, with the remaining cytoplasm cleaved into spores; (D) non-flagellated spores emerging through the deliquescing apical portion of the sporangium wall; (E) flagellated zoospores, dispersing after the complete disintegration of the sporangial wall; (F) an enlarging basal rudiment, with the cytoplasm cleaved entirely into spores; (F*) an enlarging basal rudiment, with new rudiments cleaved from its cytoplasm; (G) a mass of immobile spores set free by the disintegration of the rudiment wall. (*A*) from N.J. Artemtchuk, 1972, *Veröff. Inst. Meeresforsch. Bremerh.* 13:231–37; (*B*) from G. Bahnweg and F.K. Sparrow, 1974, *Amer. J. Bot.* 61:754–66.

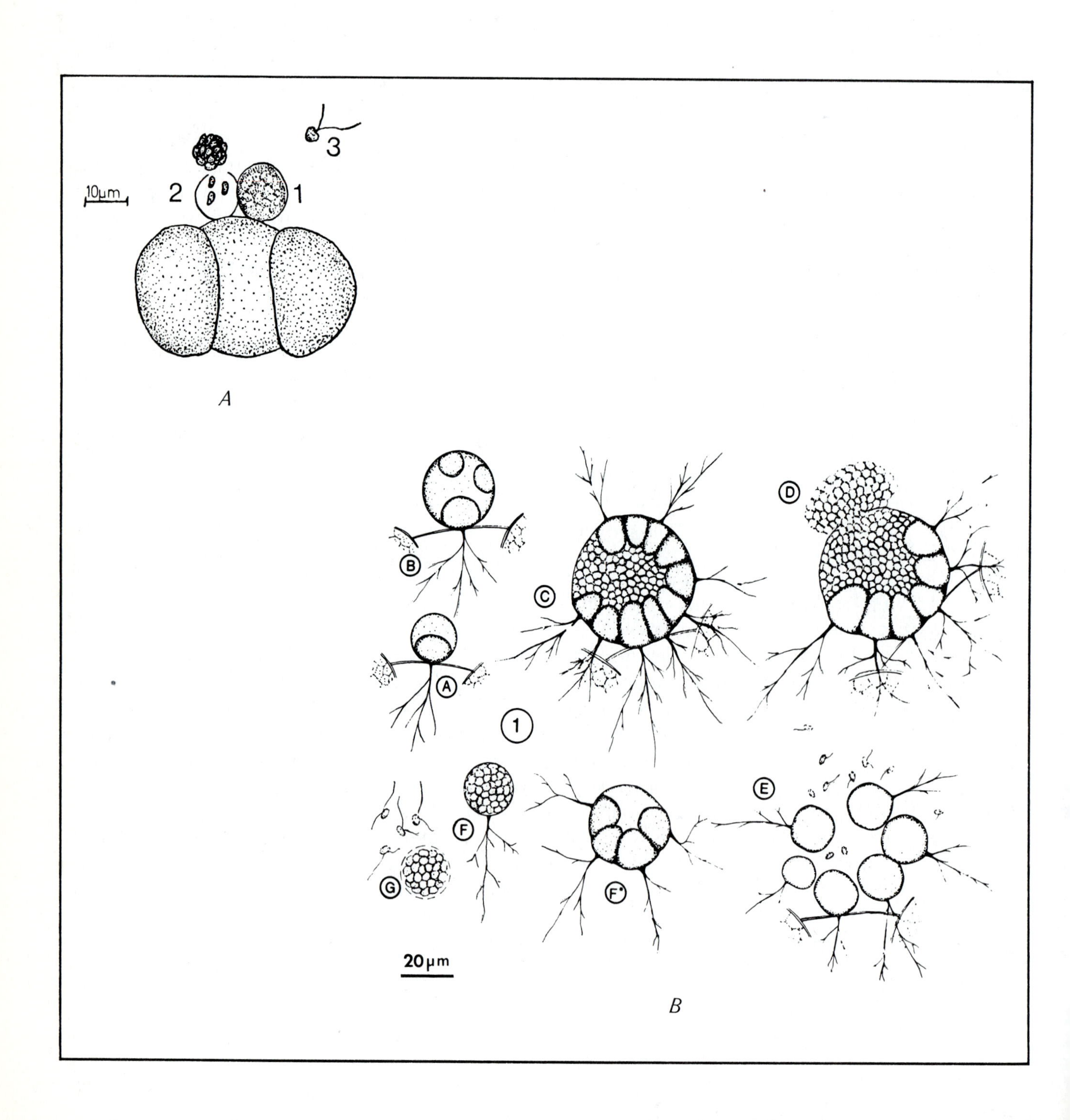

PLATE 21-4. Life cycles of species of *Labyrinthula*. (*A*) *L. thais*, an ectocommensal on a marine gastropod, showing its biphasic life cycle. (*B*) *L. saliens*, a nonphagotrophic species, with an obligate seagrass host, *Halophila englemanii;* note its unusual locomotion, shown as it develops in culture. The motile spindle cell (a) may become temporarily stationary and reproduce (b) or become an aplanospore (c) that enlarges to form either a cyst (d) or a presporangium (e). (*C*) Phase-contrast micrograph (1) of a *Labyrinthula* sp., showing the spindle cell with nucleus (N) in its mucoid tube (T) (bar $= 10\ \mu$m); light micrograph (2) of enlarged mucoid tubes, with zoosporangia and differentiated zoospores (circle) and stigmata (S) (bar $= 10\ \mu$m); transmission electron micrograph (3) of shadowed zoospore, showing an anterior flagellum, with mastigonemes (tinsel) and a posterior whiplash flagellum (bar $= 1\ \mu$m). (*A*) from B.A. Cox and J.G. Mackin, 1974, *Trans. Amer. Micros. Soc.* 93:62–70; (*B*) from J.A. Quick, Jr., 1974 *Trans. Amer. Micros. Soc.* 93:52–61; (*C*) from J.P. Amon and F.O. Perkins, 1968, *J. Protozool.* 15:543–46.

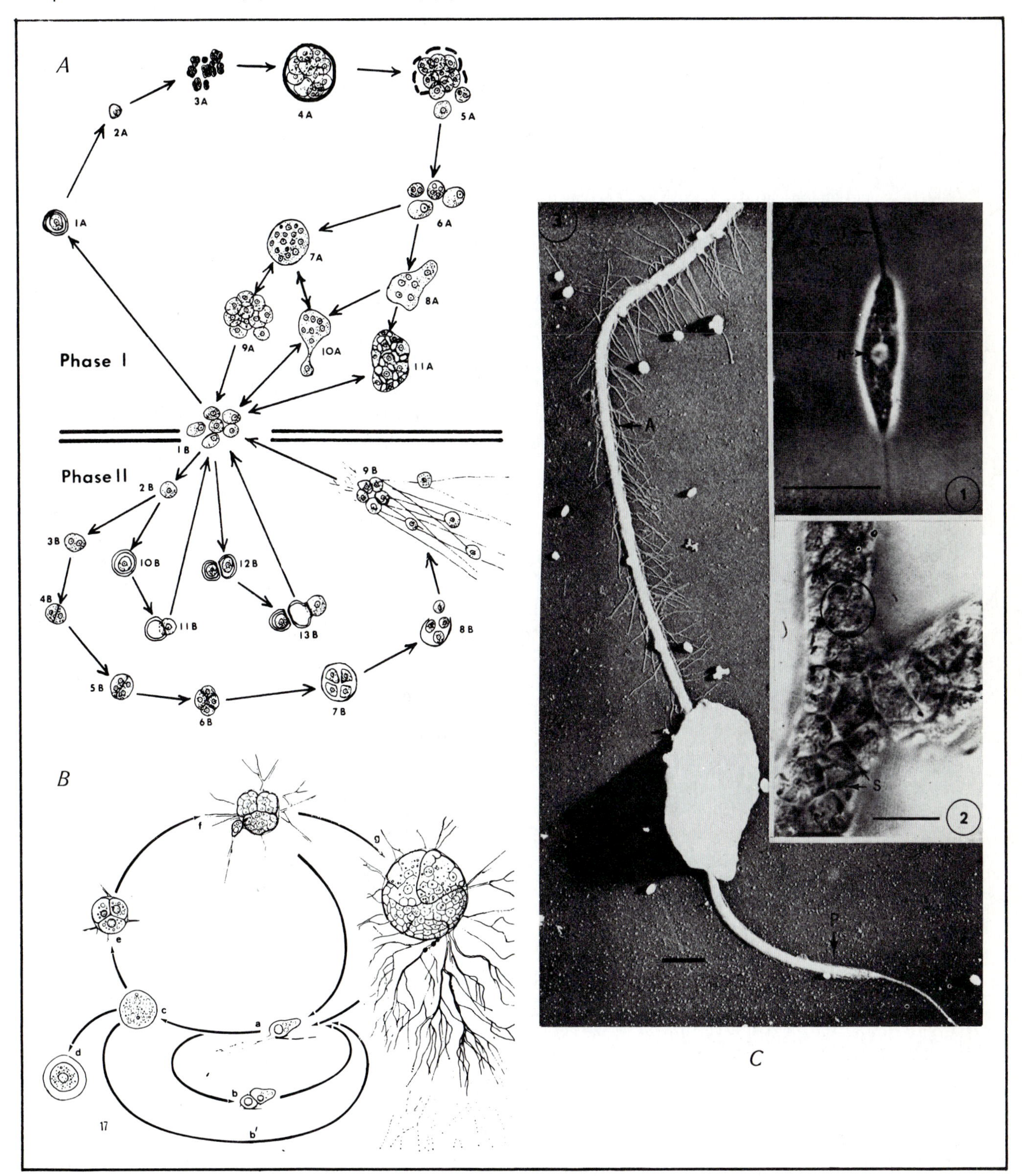

PLATE 21-5. The probable asexual developmental cycles of the spores of *Dermocystidium* (*Labyrinthulomyxa*) *marina* compared with those of *Hyalochlorella marina.* (*A*) In *L. marina* isolated from oysters the released spore (1) undergoes growth (2,3) to form a mature thallus that either undergoes bipartition to form two cells (6) or undergoes repeated karyokinesis followed by synchronous cytokinesis to yield eight- (5), sixteen- (7), or thirty-two- (8) celled sporangia. (*B*) In *H. marina* isolated from seaweeds, the released spore (A) undergoes growth and bipartition to form two cells (D), a four nucleate stage (B) to yield four spores, or greater nuclear division, growth, and cytokinesis to yield nuclear numbers that are usually an integral of 2^n (F, G, and C), except for a variation in which the four nucleate stage undergoes a degeneration of one nucleus before continuing with normal growth and division (H, I, J, and K). (*A*) from F.O. Perkins, 1969, *J. Invert. Pathol.* 13:199–212; (*B*) from R.O. Poyton, 1970, *J. gen. Microbiol.* 62:171–88.

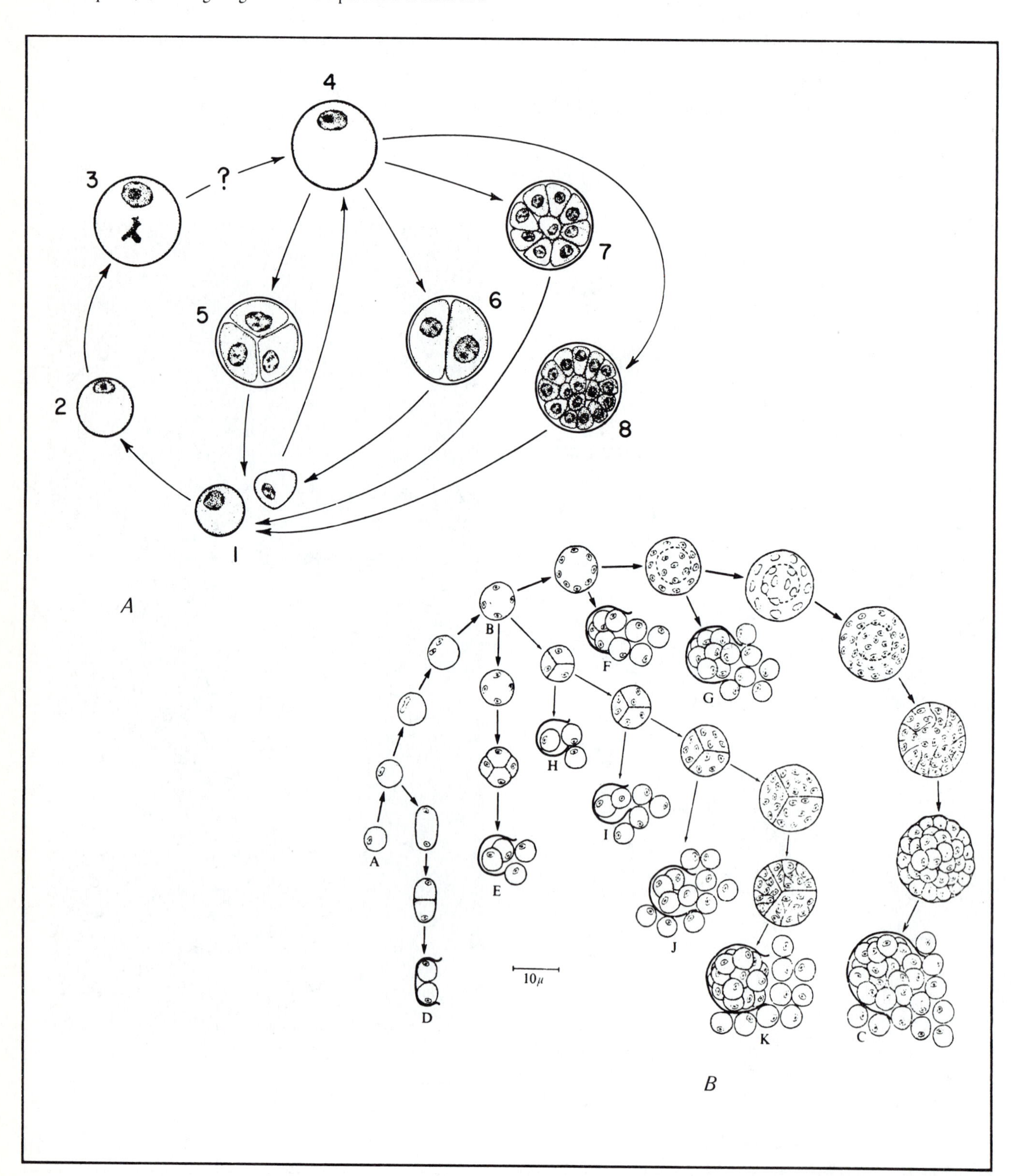

PLATE 21-6. The filamentous Phycomycete *Lagenidium callinectes* infesting the eggs of the blue crab *Callinectes sapida*. (*A*) Spore formation and discharge from a sporangium. (*B*) and (*C*) Electron micrographs, showing a partially retracted tinsel flagellum with mastigonemes and a partially retracted whiplash flagellum, respectively, of the zoospore. (*D*) An encysted spore, on the surface of an egg, penetrating the egg membrane with the germ tube and empty cyst left behind. (*E*) Hyphae, growing in and around the infested egg. From C.E. Bland and H.V. Amerson, 1973, *Mycologia* 65:310–20.

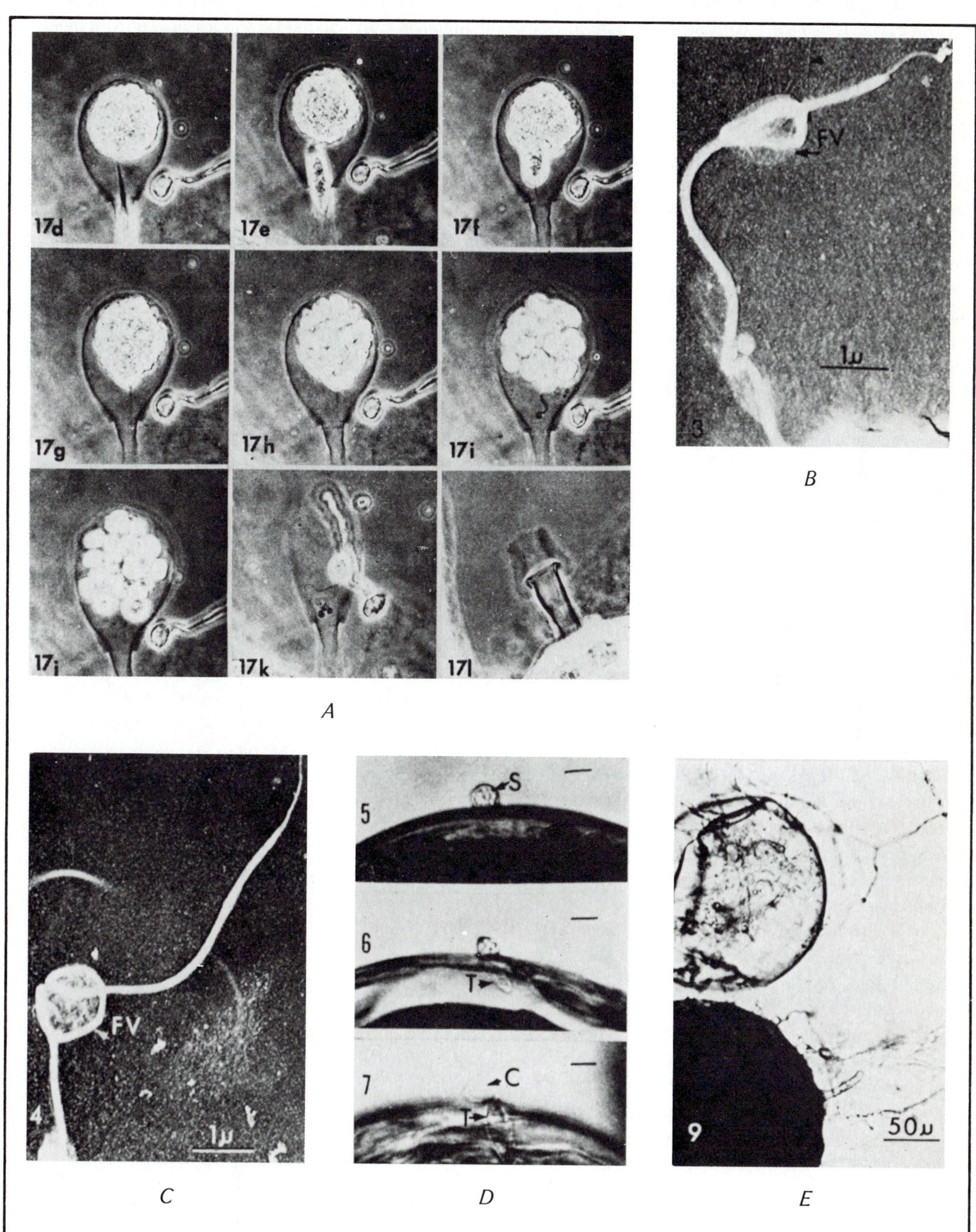

marinum as a weak parasite in a number of species from which it is easily eliminated during late winter and early spring (Ray and Chandler, 1955). A saprophytic species of *Dermocystidium* has been cultured axenically to study its life cycle (Goldstein and Moriber, 1966) and its nutrient requirements (Goldstein, Belsky, and Chosak, 1969). Reproduction in this strain is vegetative, involving cell enlargement and the shedding of an outer integument (ecdysis). This strain and those obtained by Bahnweg and Sparrow (1971) apparently lack the motile zoospores found in the oyster parasite. Mackin and Ray (1966) cultured the oyster parasite on serum–agar plates and, on the basis of the similar plasmodia, slime tracks, and zoospores, reclassified *Dermocystidium marinum* as *Labyrinthulomyxa marina*. The fine structure of the zoospores (Perkins, 1968) and the ultrastructure of the vegetative stages (Perkins, 1969) of *L. marina* have been studied, and the developmental sequence of *L. marina* is diagrammed in Plate 21-5,*A*. A number of isolates of a microorganism with a similar developmental cycle (Plate 21-5,*B*), isolated from marine sediment and detritus, have been described as *Hyalochlorella marina* by Poyton (1970a,b), who considers them apochlorotic forms of *Chlorella*. Jones (1974), however, considers *Hyalochlorella marina* a *Dermocystidium*-like form.

Species occurring in six genera in the order Lagenidiales have been reported as parasites of red, green, and brown algae and of the ova of clams, oysters, barnacles, and blue crabs (Johnson and Sparrow, 1961). An example is *Lagenidium callinectes*, which infests the ova of the blue crab. This species was brought into agar culture and its growth and sporogenesis have been studied by Bland and Amerson (1973). The vegetative thallus consists of both intramatrical hyphae, which can completely fill the infected ovum, and extramatrical hyphae, which may grow from egg to egg. The process of spore formation, the flagellation of zoospores, and the penetration of an egg are shown in Plate 21-6. The widespread occurrence of *L. callinectes* on filamentous red, green, and brown algae indicates that at least some stages of this microorganism are saprophytic (Fuller, Fowles, and McLaughlin, 1964). *Atkinsiella dubia*, which occurs on crab eggs (Sparrow, 1973), also occurs as a saprophyte on algae with other species of Phycomycetes (Fuller, Fowles, and McLaughlin, 1964).

In the family Pythiaceae, of the order Peronosporales, are species with slender well-branched hyphae, which often show irregular swellings. Members of the genus *Pythium* are prevalent in the marine environment as saprophytes or parasites on algae and larvae, and Höhnk (1957) has recorded some ten species and illustrated forms obtained from the muds of brackish waters. The taxonomy of the marine species of *Pythium* has been discussed by Sparrow (1960) and Johnson and Sparrow (1961). A species associated with leaf decay in the red mangrove was described by Fell and Master (1975). Although previously reported from the marine environment, members of the genus *Phytophora* have only recently been recognized as active inhabitants. Anastasiou and Churchland (1969) reported many *Phytophora vesicula* on decaying arbutus and laurel leaves submerged in brackish waters in the Vancouver, British Columbia area. Fell and Master (1975) found that this species, as well as four previously undescribed species, were an important component of red mangrove litter degradation systems throughout the tropics; these five species were also present on the litter from other species of mangrove and some of the other plant materials observed.

(This segment of the free-living osmotrophic population has been neglected by microbial ecologists for too long a time. Serious taxonomic problems persist, but this is no excuse for further neglect. This population can be concentrated and cultured with ease. Further studies should resolve the taxonomic problems and put the dual roles of free-living saprophyte and cell-dependent parasite into proper perspective. Studies comparing these biflagellates with the bacteria would indicate the relative importance of these two osmotrophic populations both in the plankton and on surfaces.)

REFERENCES

Alderman, D.J. 1976. Fungal diseases of marine animals. In: *Recent Advances in Aquatic Mycology* (E.B.G. Jones, ed.). Wiley, New York, pp. 223–60.

Alderman, D.J., and E.B.G. Jones. 1971. Shell disease of oysters. *Fish. Invest. (Great Britain) Ser. II*, 26(8):1–19.

Amon, J.P., and F.O. Perkins. 1968. Structure of *Labyrinthula* sp. zoospores. *J. Protozool.* 15:543–46.

Anastasiou, C.J., and L.M. Churchland. 1969. Fungi decaying leaves in marine habitats. *Can. J. Bot.* 47:251–57.

Aronson, J.M., and M.S. Fuller. 1969. Cell wall structure of the marine fungus, *Atkinsiella dubia*. *Arch. Mikrobiol.* 68:295–305.

Artemtchuk, N.J. 1972. The fungi of the White Sea. III. New phycomycetes, discovered in the Great Salma Strait of the Kandalakshial Bay. *Veröff. Inst. Meeresforsch. Bremerh.* 13:231–37.

Atkins, D. 1954a. Further notes on a marine member of the

Saprolegniaceae, *Leptolegnia marina* n. sp., infecting certain invertebrates. *J. mar. biol. Assoc. U.K.* 33:613–25.

Atkins, D. 1954b. A marine fungus *Plectospira dubia* n. sp. (Saprolegniaceae), infecting crustacean eggs and small crustacea. *J. mar. biol. Assoc. U.K.* 33:721–32.

Bahnweg, G., and F.K. Sparrow, Jr. 1971. Marine fungi: Occurrence in the southern Indian Ocean. *Antarctic J. U.S.* 6:155.

Bahnweg, G., and F.K. Sparrow, Jr. 1972. *Aplanochytrium kerguelensis* gen. nov., spec. nov., a new phycomycete from subantarctic waters. *Arch. Mikrobiol.* 81:45–49.

Bahnweg, G., and F.K. Sparrow, Jr. 1974. Four new species of *Thraustochytrium* from Antarctic regions, with notes on the distribution of zoosporic fungi in the Antarctic marine ecosystems. *Amer. J. Bot.* 61:754–66.

Bland, C.E., and H.V. Amerson. 1973. Observations on *Lagenidium callinectes:* Isolation and sporangial development. *Mycologia* 65:310–20.

Booth, T. 1969. Marine fungi from British Columbia: Monocentric chytrids and chytridiaceous species from coastal and interior soils. *Syesis* 2:141–61.

Booth, T. 1971a. Occurrence and distribution of some zoosporic fungi from soils of Hibben and Moresby Islands, Queen Charlotte Islands. *Can. J. Bot.* 49:951–56.

Booth, T. 1971b. Occurrence and distribution of zoosporic fungi and some Actinomycetales in coastal soils of southwestern British Columbia and the San Juan Islands. *Syesis* 4:197–208.

Booth, T., and C.E. Miller. 1968. Comparative morphologic and taxonomic studies in the genus *Thraustochytrium. Mycologia* 60:480–95.

Bremer, G.B. 1976. The ecology of marine lower fungi. In: *Recent Advances in Aquatic Mycology* (E.B.G. Jones, ed.). Wiley, New York, pp. 313–33.

Clokie, J.P., and C.H. Dickinson. 1972. The use of pollen colonization for growth studies of Thraustochytrids. *Veröff. Inst. Meeresforsch. Bremerh.* 13:195–204.

Cox, B.A., and J.G. Mackin. 1974. Studies on a new species of *Labyrinthula* (Labyrinthulales) isolated from the marine gastropod *Thais haemostoma floridana. Trans. Amer. Micros. Soc.* 93:62–70.

Davis, H.S. 1947. Studies on the protozoan parasites of freshwater fishes. *Fish. Bull. (U.S. Fish Wildl. Serv.)* 51(41):1–29.

Fell, J.W., and I.M. Master. 1975. Phycomycetes (*Phytophora* spp. nov. and *Pythium* sp. nov.) associated with degrading mangrove (*Rhizophora mangle*) leaves. *Can. J. Bot.* 53:2908–22.

Fuller, M.S., B.E. Fowles, and D.J. McLaughlin. 1964. Isolation and pure culture study of marine phycomycetes. *Mycologia* 56:745–56.

Fuller, M.S., and R.O. Poyton. 1964. A new technique for the isolation of aquatic fungi. *Bioscience* 14(9):45–46.

Gaertner, A. 1972. Characters used in the classification of thraustochytriaceous fungi. *Veröff. Inst. Meeresforsch. Bremerh.* 13:183–94.

Goldstein, S. 1973. Zoosporic marine fungi (Thraustochytriaceae and Dermocystidiaceae). *Ann. Rev. Microbiol.* 27:13–26.

Goldstein, S., and M. Belsky. 1964. Axenic culture studies of a new marine phycomycete possessing an unusual type of asexual reproduction. *Am. J. Botany* 51:72–78.

Goldstein, S., and L. Moriber. 1966. Biology of a problematic marine fungus, *Dermocystidium* sp. I. Development and cytology. *Arch. Mikrobiol.* 53:1–11.

Goldstein, S., M.M. Belsky, and R. Chosak. 1969. Biology of a problematic marine fungus, *Dermocystidium* sp. II. Nutrition and respiration. *Mycologia* 61:468–72.

Golubic, S., R.D. Perkins, and K.J. Lukas. 1975. Boring microorganisms and microborings in carbonate substrates. In: *The Study of Trace Fossils; a Synthesis of Principles, Problems and Procedures in Ichnology* (R.W. Frey, ed.). Springer-Verlag, New York, pp. 229–59.

Höhnk, W. 1957. Fortschritte der marinen Mykologie in jüngster. *Zeit. Natürwis. Rundschau*, 2 Abt, pp. 34–44.

Johnson, T.W., Jr. 1965. Chytridiomycetes and Oomycetes in marine phytoplankton. *Nova Hedwigia* 10:581–88.

Johnson, T.W., Jr. 1966. Fungi in planktonic *Synedra* from brackish waters. *Mycologia* 58:373–82.

Johnson, T.W., Jr., and F.K. Sparrow, Jr. 1961. *Fungi in Oceans and Estuaries.* J. Cramer, Weinheim, and Hafner Publ., New York. 668 p.

Jones, E.B.G. 1974. Aquatic fungi: Freshwater and marine. In: *Biology of Plant Litter Decomposition* (C.H. Dickinson and G.J.F. Pugh, eds.). Academic Press, London and New York, pp. 337–83.

Jones, E.B.G., and D.J. Alderman. 1971. *Althornia crouchii*, gen. et sp. nov., a marine biflagellate fungus. *Nova Hedwigia* 21:381–99.

Lukas, K.J. 1973. Taxonomy and ecology of the endolithic microflora of reef corals with a review of the literature on endolithic microphytes. Ph.D. Thesis, University of Rhode Island. 159 p.

Mackin, J.G., H.M. Owen, and A. Collier. 1950. Preliminary note on the occurrence of a new protistan parasite, *Dermocystidium marinum* n. sp. in *Crassostrea virginica* (Gmelin). *Science* 111:328–29.

Mackin, J.G., and S.M. Ray. 1966. The taxonomic relationships of *Dermocystidium marinum* Mackin Owen and Collier. *J. Invertebr. Path.* 8:544–45.

Pérez, C. 1913. *Dermocystidium pusula,* parasite de la peau des tritons. *Arch. Zool. Exp. Gén.* 52:343–57.

Perkins, F.O. 1968. Fine structure of zoospores from *Labyrinthulomyxa* sp. parasitizing the clam *Macoma balthica. Chesapeake Sci.* 9:198–202.

Perkins, F.O. 1969. Ultrastructure of vegetative stages in *Labyrinthulomyxa marina* (= *Dermocystidium marinum*), a commercially significant oyster pathogen. *J. Invertebr. Path.* 13:199–212.

Perkins, F.O. 1972. The ultrastructure of holdfasts, "rhizoids," and "slime tracks" in Thraustochytriaceous fungi. *Arch. Mikrobiol.* 84:95–118.

Perkins, F.O. 1976. Fine structure of lower marine and estuarine fungi. In: *Recent Advances in Aquatic Mycology* (E.B.G. Jones, ed.). Wiley, New York, pp. 279–312.

Perkins, F.O., and J.P. Amon. 1969. Zoosporulation in *Laby-*

rinthula sp.: An electron microscope study. *J. Protozool.* 16:235–57.

Perkins, R.D., and S.D. Halsey. 1971. Geologic significance of microboring fungi and algae in Carolina shelf sediments. *J. Sed. Petrol.* 41:843–53.

Pokorny, K.S. 1967. *Labyrinthula. J. Protozool.* 14:697–708.

Poyton, R.O. 1970a. The characterization of *Hyalochlorella marina* gen et sp. nov. a new colourless counterpart of *Chlorella. J. Gen. Microbiol.* 62:171–88.

Poyton, R.O. 1970b. The isolation and occurrence of *Hyalochlorella marina. J. Gen. Microbiol.* 62:189–94.

Quick, J.A., Jr. 1974. A new *Labyrinthula* with unusual locomotion. *Trans. Amer. Micros. Soc.* 93:52–61.

Ray, S.M. 1952. A culture technique for the diagnosis of infections with *Dermocystidium marinum* Mackin, Owen and Collier in oysters. *Science* 116:360–61.

Ray, S.M., and A.C. Chandler. 1955. *Dermocystidium marinum*, a parasite of oysters. *Exp. Parasit.* 4:172–200.

Renn, C.E. 1936. The wasting disease of *Zostera marina*. I. A phytological investigation of the diseased plant. *Biol. Bull.* 70:148–58.

Schneider, J. 1969a. Labyrinthula-befall an niederen Pilzen (*Thraustochytrium* spec.) aus der Flensburger Förde. *Kieler Meeresforsch.* 25:314–15.

Schneider, J. 1969b. Zur Taxonomie, Verbreitung und Ökologie einiger mariner Phycomyceten. *Kieler Meeresforsch.* 25:316–27.

Sparrow, F.K., Jr. 1936. Biological observations on the marine fungi of Woods Hole waters. *Biol. Bull.* 70:236–63.

Sparrow, F.K., Jr. 1960. *Aquatic Phycomycetes.* University of Michigan Press, Ann Arbor, Michigan. 1187 p.

Sparrow, F.K., Jr. 1973. The peculiar marine phycomycete *Atkinsiella dubia* from crab eggs. *Arch. Mikrobiol.* 93:137–44.

Sparrow, F.K. 1976. The present status of classification in biflagellate fungi. In: *Recent Advances in Aquatic Mycology* (E.B.G. Jones, ed.). Wiley, New York, pp. 213–22.

Ulken, A. 1972a. Über zwei Phycomyceten aus der Wesermündung und deren Entwicklung in der Kultur. *Veröff. Inst. Meeresforsch. Bremerh.* 13:205–16.

Ulken, A. 1972b. Physiological studies on a phycomycete from a mangrove swamp at Cananéia, São Paolo, Brazil. *Veröff. Inst. Meeresforsch. Bremerh.* 13:217–30.

Vishniac, H.S. 1956. On the ecology of the lower marine fungi. *Biol. Bull.* 111:410–14.

Vishniac, H.S. 1958. A new marine phycomycete. *Mycologia* 50:66–79.

Vishniac, H.S. 1960. Salt requirements of marine phycomycetes. *Limnol. Oceanogr.* 5:362–65.

Vishniac, H.S. 1961. A biological assay for thiamine in sea water. *Limnol. Oceanogr.* 6:31–35.

Vishniac, H.S., and G.A. Riley. 1961. Cobalamin and thiamine in Long Island Sound: Patterns of distribution and ecological significance. *Limnol. Oceanogr.* 6:36–41.

Watson, S.W., and E.J. Ordal. 1957. Techniques for the isolation of *Labyrinthula* and *Thraustochytrium* in pure culture. *J. Bacteriol.* 73:589–90.

Young, E.L., III. 1943. Studies on *Labyrinthula*, the etiologic agent of the wasting disease of eel-grass. *Amer. J. Bot.* 30:586–93.

PART VII

Phagotrophic Eucaryotes

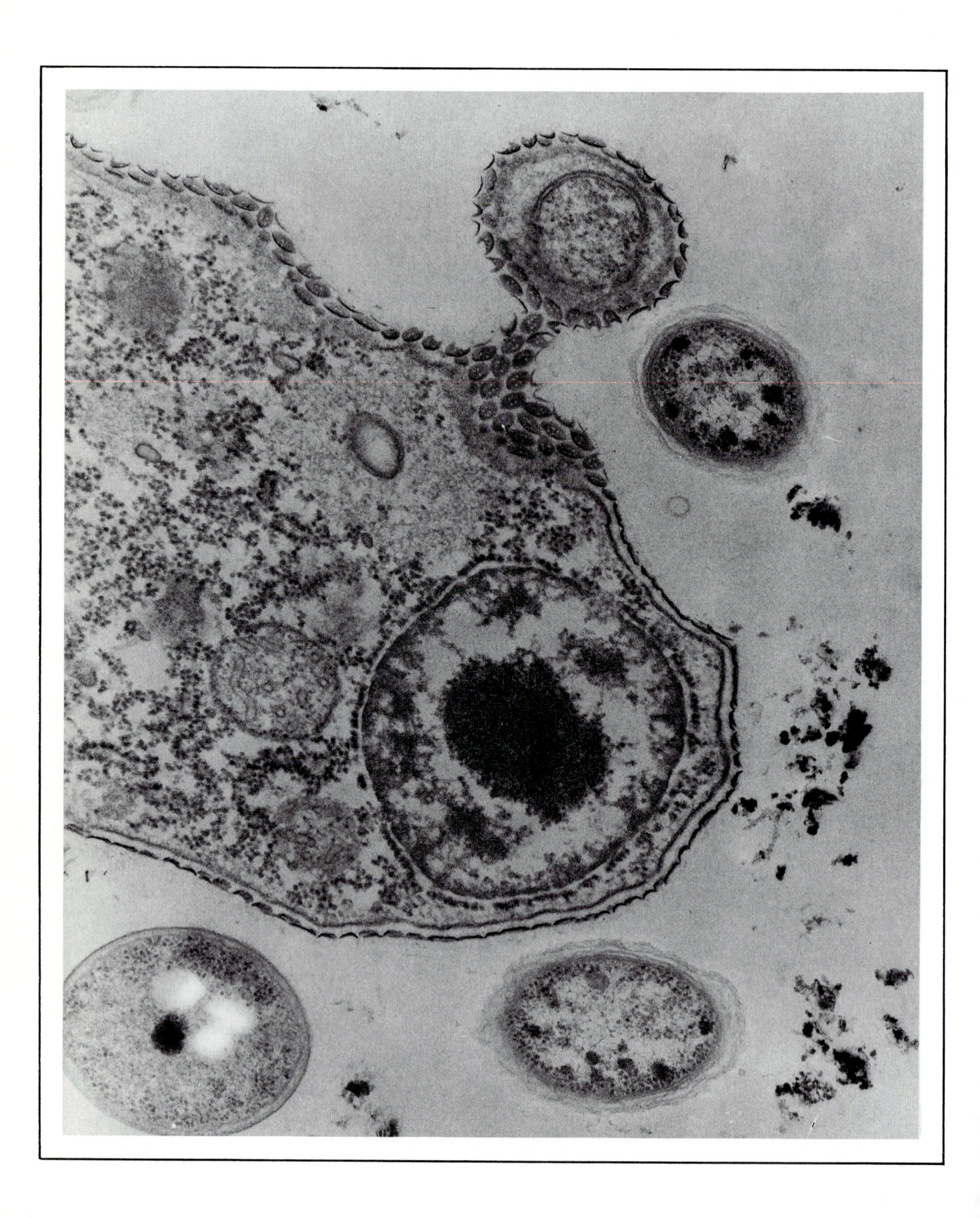

A scaled amoeba, with its eucaryotic cell structure, which has just engulfed a bacterial cell (upper right) in a film from the biological filter of a high-density, closed-system salmon culture system (transmission electron micrograph, thin section) (54,400 X). Photograph courtesy of Paul W. Johnson.

For our purposes, protozoa are defined as eucaryotes that obtain their nutrition heterotrophically, mainly from ingesting other microorganisms. They are motile, with flagellated, amoeboid, and ciliated stages. Those forms with prominent flagellated stages are discussed in Chapter 22. They exclude the flagellated fungi (Chapter 21), but include the apochlorotic forms of the phytoflagellates (Chapters 11 and 13), which can live osmotrophically as well as phagotrophically. Many are voracious predators (Pringsheim, 1959). Due to mixotrophy, it is impossible to distinguish osmotrophic and phagotrophic forms or even to guess to what degree osmotrophy is being practiced in those forms that are obviously phagotrophic.

The amoeboid forms discussed in Chapter 23 are noticably phagotrophic, with plastic protoplasm that actively snares or engulfs food particles. These amoeboid forms could just as well have been grouped with the flagellated forms, since they have flagellated stages. The presence of flagellated and amoeboflagellated stages in these forms indicates that there is no real boundary between these two apparently disparate groups. The presence of forms with scales in both groups is a further indication of relatedness. There are forms that range from naked amoebae to amoebae with chitinous, calcareous, siliceous, and strontium sulfate tests or houses.

The ciliates, Chapter 24, are a distinct and closely knit group; most of them are unmistakably phagotrophic, as shown by their oral cavities and oral ciliature. The phagotrophic, flagellated, amoeboid, and ciliated forms play a major role in microbial communities. No microbial microcosm is complete without them, and their appearance marks the maturity of the community. By preying on populations of bacteria and the smaller protists, they maintain those members of the community in a healthy state while providing the metazoans with a suitably sized prey. Sexual reproduction varies between and within groups and is discussed in detail by Grell (1967, 1973).

The distribution of the protozoa is obviously correlated with the abundance of their food supply. Noland (1925) studied the relationship of physical and chemical factors to protozoa in a freshwater stream. When conditions of high oxygen tension, low free carbon dioxide, and high pH indicated algal activity, then ciliates feeding on diatoms and other algae were present, but when conditions of low oxygen tension, high free carbon dioxide, and low pH indicated bacterial activities, bacterial-feeding ciliates were present. Noland concluded that these physical and chemical factors do not directly influence the protozoa, but that the sum total of these factors is indicative of either algal or bacterial growth, that maintains either algal-feeding or bacterial-feeding ciliates. Picken (1937) studied these algal and bacterial associations and showed that, in addition to these feeding types, the protozoan community also contains detritus-feeders and successive hierarchies of carnivorous forms, which create a pyramidal structure wherein a succession of carnivorous forms increase in size and diminish in numbers until one species feeds directly or indirectly on all the lesser elements of the population in a manner similar to that in the animal kingdom. The occurrence and distribution of protozoan forms in the inshore waters of Woods Hole also were related to their food supply (Lackey, 1936). The flagellates were the most prevalent swimming and suspended forms in the plankton whereas the tintinnids (motile loricate ciliates) varied greatly in number but were always present. The non-loricate ciliates formed the largest and most varied group on the bottom, but only a few were seen as swimming and floating forms. The colorless and often large euglenids were abundant in samples containing bottom sediments and debris. The amoeboid forms were the least obvious group. In these eutrophic and turbid waters, the rate of sedimentation into protected containers allowed the accumulation of algal food and debris over a 24-hour period, which resulted in a remarkable ciliate population. Lackey stressed the importance of this population in both the mineralization of organic matter and the provision of food for such larger forms as the copepods. The very numerous small flagellates that have an osmotrophic existence were believed to resemble the bacteria in "condensing" food for the larger ciliated forms. Lighthart (1969) has enumerated the bacteria and phagotrophic flagellates in inshore marine sediments, as well as in steady-state simulated seabed microcosms, and has observed a regression of 580 bacteria for every bacteria-eating flagellate.

Fauré-Fremiet (1950) has succinctly discussed the ecology of the protists in the littoral zone of the sea. He points out that their size limits them to a very small area, with further limitations being not only their doubling rates but their food source. In discussing the interrelationships and reciprocal dependency of the microorganisms, he points out the importance of spatial distribution and the specificity of food requirements that control the food chains, such as the partial one shown below. These rest upon a pyramid of numbers, each level depending

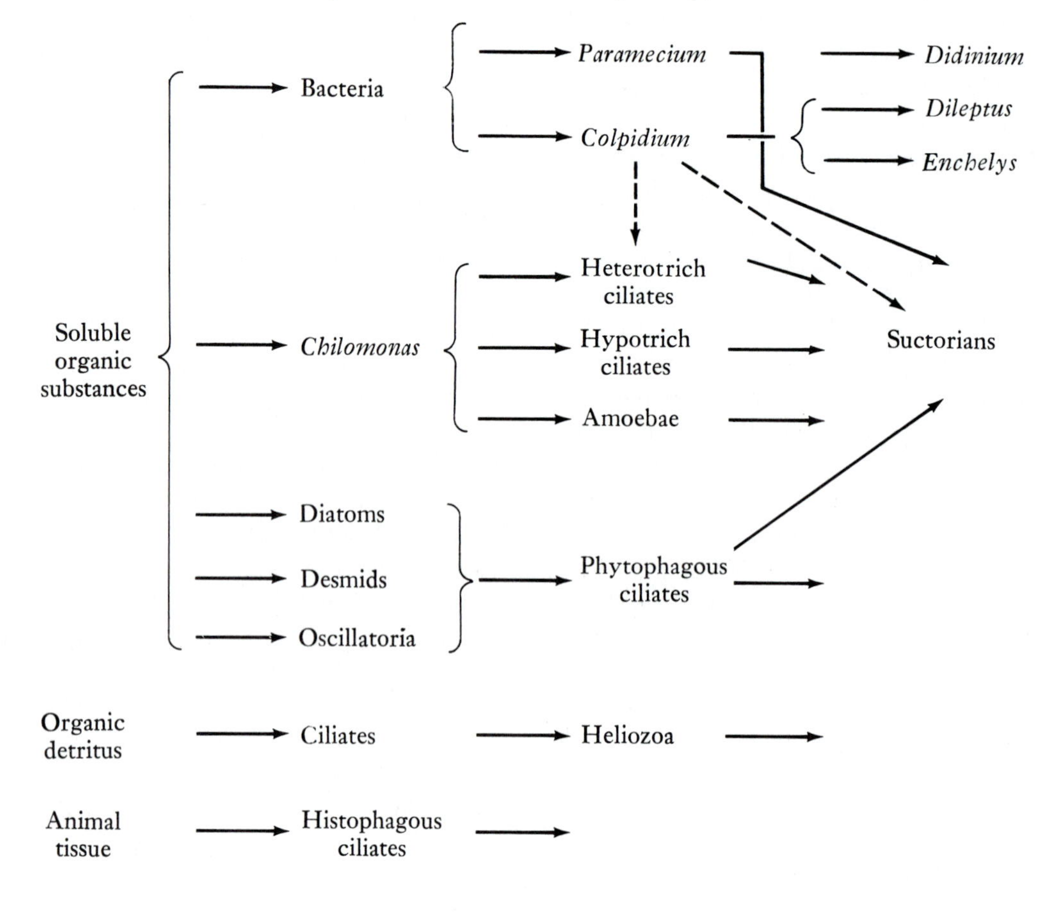

upon the reproductive rate of its food source. In addition to the localized microbiocenose, and the social and structural factors that hold it together, Fauré-Fremiet also discusses more diffuse microbiocenoses, such as those in the planktonic environment.

In scanning the literature of oceanic biology, we find that, in our obsession with phytoplankton and zooplankton, we often overlook the ecologically important protozooplankton. In addition to the well-studied dinoflagellates, tintinnids, foraminifera, and radiolaria, the protozooplankton also includes the microflagellates, amoebae, and non-loricate ciliates that graze upon the hordes of bacteria with which they work to decompose organic debris. The amoebae, foraminifera, testaceans, and surface-grazing ciliates are the janitors of fouling surfaces and organic debris. By removing the excess populations of the smaller heterotrophs, they help keep them in an actively growing, healthy state. The dynamics of any population depend upon the rate at which it is consumed as well as upon the rate at which it is produced. For this reason alone, it is surprising that ecologists studying the phototrophs (Parts III and IV) and the osmotrophs (Parts V and VI) have paid so little attention to the phagotrophs and their rate of utilization of the algal, bacterial, and fungal biomass. When the protozooplankton has been examined in relation with the other members of the plankton, it has been found that the protozooplankton (microzooplankton) plays a vital role in phototroph consumption (Margalef, 1967; Beers and Stewart, 1971; Eriksson, Sellei, and Wallström, 1977) and approaches or exceeds the copepods in importance as a grazing population in the euphotic zone. The protozooplankton may also play a vital role in the release of labile dissolved organic matter required by the extensive populations of bacterioplankton (Sieburth et al., 1976, 1977).

Protozoa are also important in epibiotic, neustonic, benthic, and endobiotic habitats, as discussed in Part I. The protozoa on surfaces (including the sea surface and subsurface films, Chapters 2,B and D) include choanoflagellates, amoebae, foraminifera, heliozoa, and ciliates, among others. The interstitial forms living in the benthos (Chapter 3,B) between sand grains also include foraminifera and amoebae, in addition to a great variety of highly specialized ciliates (Chapter 24). The parasitic forms are only mentioned here and in Chapter 3,C, since they are beyond the scope of this book.

REFERENCES

Beers, J.R., and G.L. Stewart. 1971. Micro-zooplankters in the plankton communities of the upper waters of the eastern tropical Pacific. *Deep-Sea Res.* 18:861–83.

Eriksson, S., C. Sellei, and K. Wallström. 1977. The structure of the plankton community of the Öregrundsgrepen (southwest Bothnian Sea). *Helgoländer wiss. Meeresunters.* 30:582–97.

Fauré-Fremiet, E. 1950. Ecologie des protistes littoraux. *Ann. Biol.* 27:437–47.

Grell, K.G. 1967. Sexual reproduction in protozoa. In: *Research in Protozoa* (T-T. Chen, ed.), Vol. 2. Pergamon, Oxford, pp. 147–213.

Grell, K.G. 1973. *Protozoology*. Springer-Verlag, Heidelberg and New York, 554 p.

Lackey, J.B. 1936. Occurrence and distribution of the marine protozoan species in the Woods Hole area. *Biol. Bull.* 70:263–78.

Lighthart, B. 1969. Planktonic and benthic bacteriovorous protozoa at eleven stations in Puget Sound and adjacent Pacific Ocean. *J. Fish. Res. Bd. Canada* 26:299–304.

Margalef, R. 1967. El ecosistema. In: *Ecologia Marina*, Vol. 14. *Fundatión la salle de ciencias naturales*. Caracas, pp. 377–453.

Noland, L.E. 1925. Factors affecting the distribution of fresh water ciliates. *Ecology* 6:437–52.

Picken, L.E.R. 1937. The structure of some protozoan communities. *J. Ecology* 25:368–84.

Pringsheim, E.G. 1959. Phagotrophie. In: *Handbuch der Pflanzenphysiologie* (W. Ruhland, ed.), Vol. II. Springer-Verlag, Berlin, pp. 179–97.

Sieburth, J.McN., P-J. Willis, K.M. Johnson, C.M. Burney, D.M. Lavoie, K.R. Hinga, D.A. Caron, F.W. French III, P.W. Johnson, and P.G. Davis. 1976. Dissolved organic matter and heterotrophic microneuston in the surface microlayers of the North Atlantic. *Science* 194:1415–18.

Sieburth, J.McN., K.M. Johnson, C.M. Burney, and D.M. Lavoie. 1977. Estimation of *in situ* rates of heterotrophy using diurnal changes in dissolved organic matter and growth rates of picoplankton in diffusion culture. *Helogoländer wiss. Meeresunters.* 30:565–74.

CHAPTER 22

Flagellates

Marine flagellates with an osmotrophic–phagotrophic nutritional mode include two kinds of microorganisms. The first group are the apochlorotic forms of the phototrophic flagellates discussed in Part IV, and include the obviously phagotrophic members of the euglenids, dinoflagellates, chrysophytes, and cryptomonads. The second group consists of miscellaneous flagellates. The most abundant of these are the choanoflagellates, with their distinctive collars, which live singly and in colonies both free in the water and attached to particles and surfaces. Other sporadically abundant forms are *Chilomonas*, *Bodo*, the bicoecids, and the ebridians.

A. EUGLENIDS

The euglenoid flagellates have been most intensively studied in the laboratory. Their structure and systematics are considered in the detailed but concise monograph by Leedale (1967), their biochemistry in the book by Wolken (1967), and their ultrastructure in the lengthy article by Mignot (1966). The observations made by James Lackey during summers spent at the Woods Hole Marine Biological Laboratory (Lackey, 1936), the Narragansett Marine Laboratory (Lackey, 1961), Great South Bay, Long Island (Lackey, 1963), Plymouth, England (Lackey and Lackey, 1963), and San Diego Bay (Lackey and Clendenning, 1965) give a very nice picture of the colorless euglenids that accumulate in detritus and sediments (Lackey, 1967). *Eutreptia* and *Eutreptiella* were the only common green euglenids, most of the forms encountered being colorless. Some colorless forms were sand-dwellers and apparently restricted to the upper 5 to 10 cm of the wet sand between tidemarks. Others were ubiquitous, colonizing the sand and the mud–water interface. Although the colorless euglenids did not penetrate beyond the first few millimeters of fine sediment, they were hydrogen-sulfide tolerant and apparently congregated in patches of sulfur bacteria.

The green euglenids can become apochlorotic in the dark and function as osmotrophs, but apparently no green members of the Euglenales can ingest particulate food. The only phagotrophs in this order are the species of the genus *Euglenopsis,* which lack a permanent cytostome and organelles for the capture and handling of food (see Plate 22-1,*A*). These euglenids are found associated with the plankton and debris, but they do not inhabit bottom sediments. Except for this one genus in the order Euglenales, all the other phagotrophic euglenids belong to just two orders of colorless euglenids. In the order Sphenomonadales which also lacks permanent cytostomes and organelles of ingestion, the genera with species commonly associated with detritus and bottom sediments are *Sphenomonas*, *Petalomonas*, *Tropidoscyphus*, *Notosolenus*, *Anisonema*, and possibly *Calkinsia* and *Pentamonas.* A second order, Heteronematales, includes the colorless euglenids, which have a permanent cytostome and ingestion rods to aid in the capture and handling of prey and of particulate food. Genera in this order with species commonly associated with detritus and bottom sediments are the large and common *Peranema* (Plate 22-1,*B*,*C*), *Heteronema*, *Dinema*, *Menoidium*, and *Entosiphon*. The text by Kudo (1966) and the monograph by Leedale (1967) are helpful for working to the genus level but for species descriptions, such monographs as the remarkable and classic ones for freshwater by Skuja (1939, 1948) are needed. Unfortunately, the taxonomic monograph on coastal forms by Butcher (1961) only covers the green euglenids. Some of the problems in the taxonomy of this group are discussed in a description of three new species occurring in Rhode Island marine habitats (Lackey, 1962, 1967). These are included in Plate 22-2 and show the diversity of form of this common, but little studied consumer population. Colorless euglenids from Hokkaido, Japan have been described by Ioriya (1976) and include arenacious species. Leedale (1967) stresses the phagotrophic nature of the colorless euglenids. Lackey (1967) has also pointed out that at least eight species grow abundantly if bacteria are present and that none has been grown in axenic culture. But the nutrient requirements of at least one euglenid can be met by "soluble" nutrients. The voracious *Peranema* has been grown in axenic culture, and it remained vigorous for over a year in a milk–lecithin medium and for 3 months in a nearly defined medium (Allen et al., 1966). This autoclavable biphasic medium contained linoleic acid, histidine, cholesterol, a mixture of long-chain saturated and unsaturated acids, crude lecithin, and a fat-soluble antioxidant. Voracity in axenic culture was retained as indicated by engorgement on latex spheres. The lipid requirements for this and other protozoa, such as *Oxyrrhis marina*, are of great interest, but the retained voracity raises the question of whether a biphasic medium with its viscosity increased by methyl cellulose might appear to be solid (on a micro-scale) to such voracious feeders.

PLATE 22-1. Diagrammatic representation of two types of colorless, phagotrophic euglenids. (*A*) A species of *Euglenopsis* (cells 6 to 9 by 30 to 35 μm) representative of those forms with no distinct cytostome or ingestion rods; note the paramylon grains at the anterior end and the larger ingested algal cells. (*B*) Three views of *Peranema trichophorum*, one of the large, common phagotrophic forms that ingest through a permanent cytostome (a), with the aid of a pair of ingestion rods used to capture and prod food particles and living prey. Note the stout anterior flagellum and the thin trailing flagellum. (*C*) Successive stages in the ingestion of *Euglena viridis* by *Peranema*. (*A*) from H.R. Christen, 1959, *J. Protozool.* 6:292–303; (*B*,*C*) from Y.T. Chen, 1950, *Quart. J. Micros. Sci.*, 3rd Series, 91:279–308.

B. DINOFLAGELLATES

Phagotrophic dinoflagellates are quite diverse and are found in every marine habitat. Even the chromatophore-bearing species *Gyrodinium pavillardi*, about 40 × 50 μm in size, that occurs on sandy bottoms and among the algae of brackish ponds has been observed to ingest other algae, bacteria, and protozoa. Although bearing numerous chromatophores, this rare dinoflagellate definitely appeared to be phagotrophic from the detailed observations of Berthe Biecheler (1952), which were published posthumously. *Gyrodinium pavillardi* attacked and ingested the oligotrichous ciliate *Strombidium* (25 × 40 μm), which itself attacks species of the dinoflagellate *Peridinium* as well as other protists (Biecheler, 1952). Both predator and prey have trichocysts. If the prey makes contact with the anterior end of the predator, nothing happens and after a while, both swim away. When the predator's posterior end makes contact, the ciliate is stunned before it can use its own trichocysts, and it is held in the region of the sulcus as the lips of the groove gape during ingestion, which takes some 10 min-

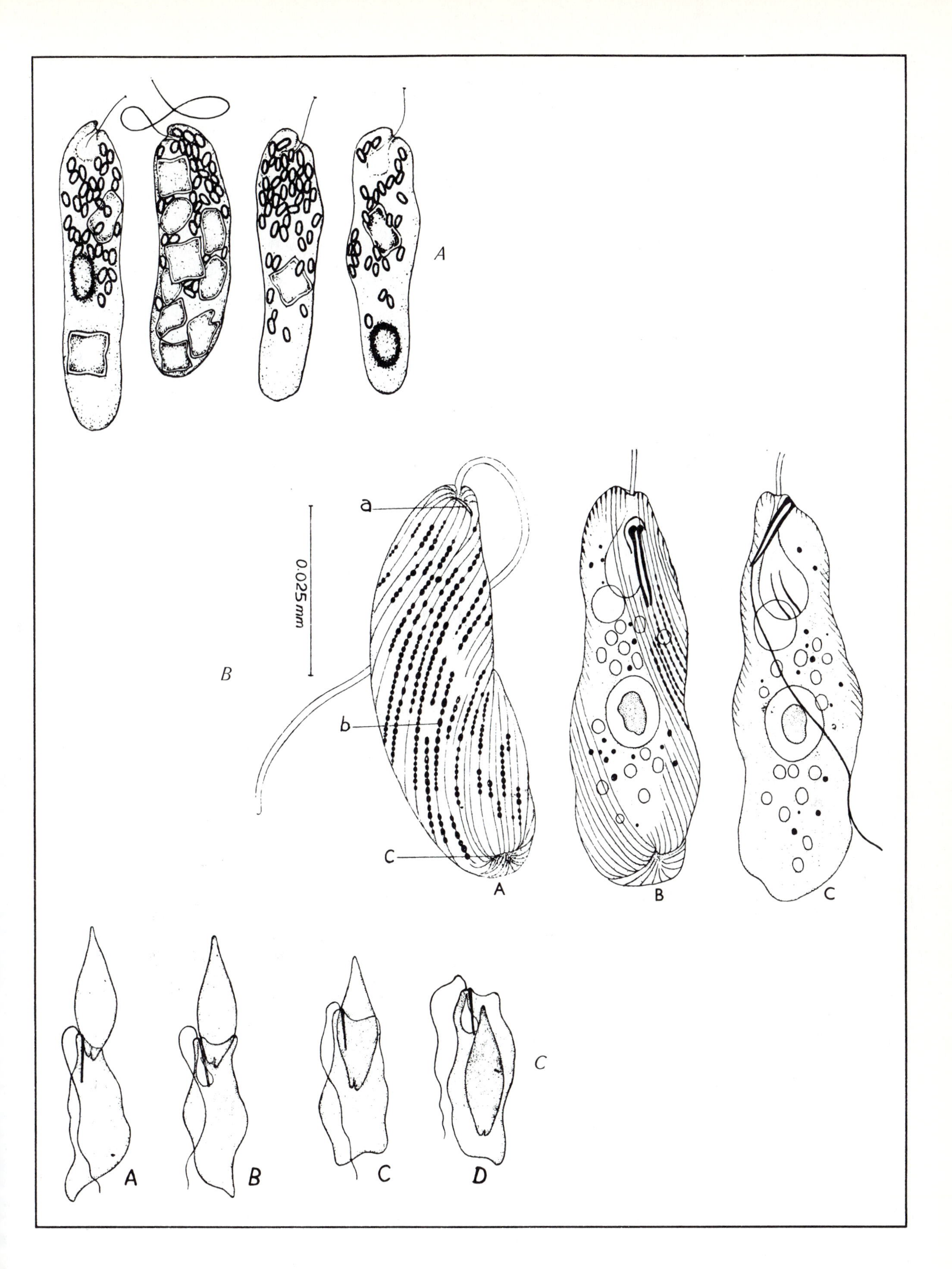
A
a
b
c
0.025 mm
B
A
B
C
A
B
C
D
C

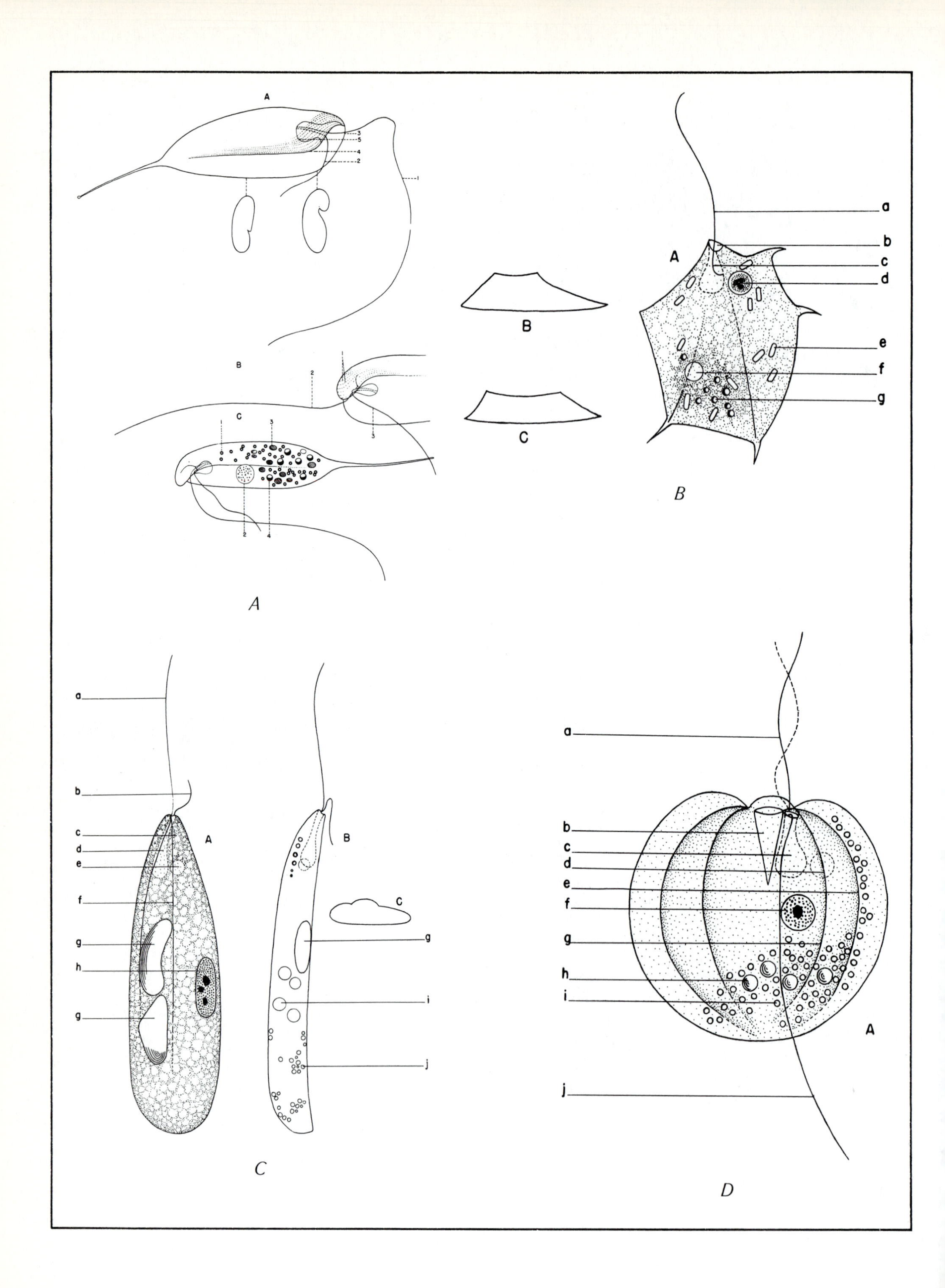
A
3
5
4
2
1
B
2
1
3
C
1
3
2
4
B
C
a
b
A
c
d
e
f
g
B
A
a
b
c
d
e
f
g
h
g
A
B
C
g
i
j
C
a
b
c
d
e
f
g
h
i
A
j
D

PLATE 22-2. The diversity of apochlorotic euglenids from inshore mud and sand with apparent osmotrophic nutrition. (*A*) The golden colored *Calkinsia aureus* from the interface of anaerobic mud (body proper, 40–50 μm long). (*B*) *Pentamonas spinifera*, occurring in sand, has characteristic spinous projections of the pellicle (440 X). (*C*) *Sphenomonas elongata* occurs in the intertidal sand and at the mud-water interface; it is a weak swimmer (300 X). (*D*) *Entosiphon cuneatiformis* is a distinctive circular form occurring in the sand and surf (250 X). (*A*) from J.B. Lackey, 1960, *Trans. Amer. Micros. Soc.* 79:105–7; (*B,C,D*) from J.B. Lackey, 1962, *Arch. Mikrobiol.* 42:190–95.

utes. This process, shown in Plate 22-3, is followed by a 5- to 6-hour digestion period. *Gyrodinium pavillardi* feeds while the dinoflagellate is motile, usually in the morning and in the early hours of the afternoon.

Other chloroplast-containing dinoflagellates are also phagotrophic (Biecheler, 1952). Norris (1969) cites the earlier reports and provides an illustration of *Ceratium lunula* with an engulfed *Peridinium* species. Dodge and Crawford (1970) showed that the heavily armored and common freshwater dinoflagellate *Ceratium hirundinella* is well adapted for particulate feeding, having a sulcal aperture bound by a membrane instead of plates. Food vacuoles contained the remains of heterotrophic bacteria, cyanobacteria, and diatoms. Cachon and Cachon (1974) studied the ultrastructure of the phagotrophic system in a noctilucoid dinoflagellate. The transverse flagellum of *Kofoidinium* sets up streams, which carry diatoms and other dinoflagellates to a groove at the bottom of which is a cytostome that opens to a cytopharynx with lamellar walls.

Oxyrrhis marina, a colorless dinoflagellate in coastal tide pools, salt marshes, and the colored scums and films of seawater aquaria, is easy to isolate in culture. It is an omnivore, ingesting bacteria, diatoms, and small flagellates, and is often distended into bizarre shapes by the rigid frustules of the ingested diatoms. Barker (1935) maintained an *Oxyrrhis* culture with the diatom *Nitzschia* as a food source for over a year without transfer. He describes how *O. marina* captured another dinoflagellate, a *Peridinium:* after making random contact with the *Peridinium*, which apparently adhered to one of the predator's flagella, *Oxyrrhis* swam around and around the prey as if winding it in a cocoon and then dragged it along as the prey was slowly drawn into a depression in the sulcal area, which was pushed in as *Oxyrrhis* gradually engulfed the *Peridinium*. Droop (1966) found that fourteen species of algae representing seven classes could all support growth in culture. Dodge and Crawford (1974) used electron microscopy to observe the natural food of *Oxyrrhis* in its natural habitat of high level tide pools and found that virtually any organism slightly smaller than itself, including smaller *Oxyrrhis*, would serve as food. When yeast cells were fed to starved *Oxyrrhis*, the yeast cells were ingested within 5 minutes: they distended *Oxyrrhis* within 30 minutes and were partly digested within 1 hour. When diatoms longer than *Oxyrrhis* were fed, they protruded from the posterior end of the cell in the vicinity of the flagellar bases, where the cell is covered by only a single membrane instead of thecal vesicles that cover the rest of the cell. Microbodies presumably containing enzymes aggregated evenly around the membrane bounding the food vacuoles.

Bacteria-free cultures of *Oxyrrhis* were obtained by Droop (1953), but there appeared to be a requirement for such living food as a yeast or an alga. Although McLaughlin at Haskins Laboratory succeeded at axenic culture (unpublished), his procedures failed in the hands of Michael Droop, who obtained his first successful axenic cultures when a micronutrient requirement was satisfied with neutralized lemon juice (Droop, 1959). The active factor is a colorless lipid that can be extracted, without the water-soluble components, from lemon rind. It is the only organic micronutrient required in addition to the three water-soluble vitamins thiamin, vitamin B_{12}, and biotin. Grass also contains the "lemon factor," but oils, sterols, and lecithins from a number of other sources did not support growth. Acetate or ethanol (but not carbohydrates or amino acids) serve as carbon sources; valine, alanine, and proline (but not nitrate, ammonia urea, or the other amino acids) serve as nitrogen sources. Except for the lemon factor, which turned out to be an absolute requirement for ubiquinone or plastiquinone (Droop and Doyle, 1966) and, to a lesser degree, sterol (Droop, 1970), *Oxyrrhis* is a typical acetate flagellate. Phagotrophy in *Oxyrrhis* may seem out of place, but such a relatively large microorganism has a limited cell surface and would therefore find phagotrophy the most efficient method of feeding. This much-studied colorless flagellate is therefore a very versatile microorganism, being both an osmotroph (acetate flagellate) and a phagotroph.

[The acetate flagellates are a minor group of phototrophic and phagotrophic eucaryotes that can use acetate preferably over other sources of dissolved organic matter and that belong to a number of different algal

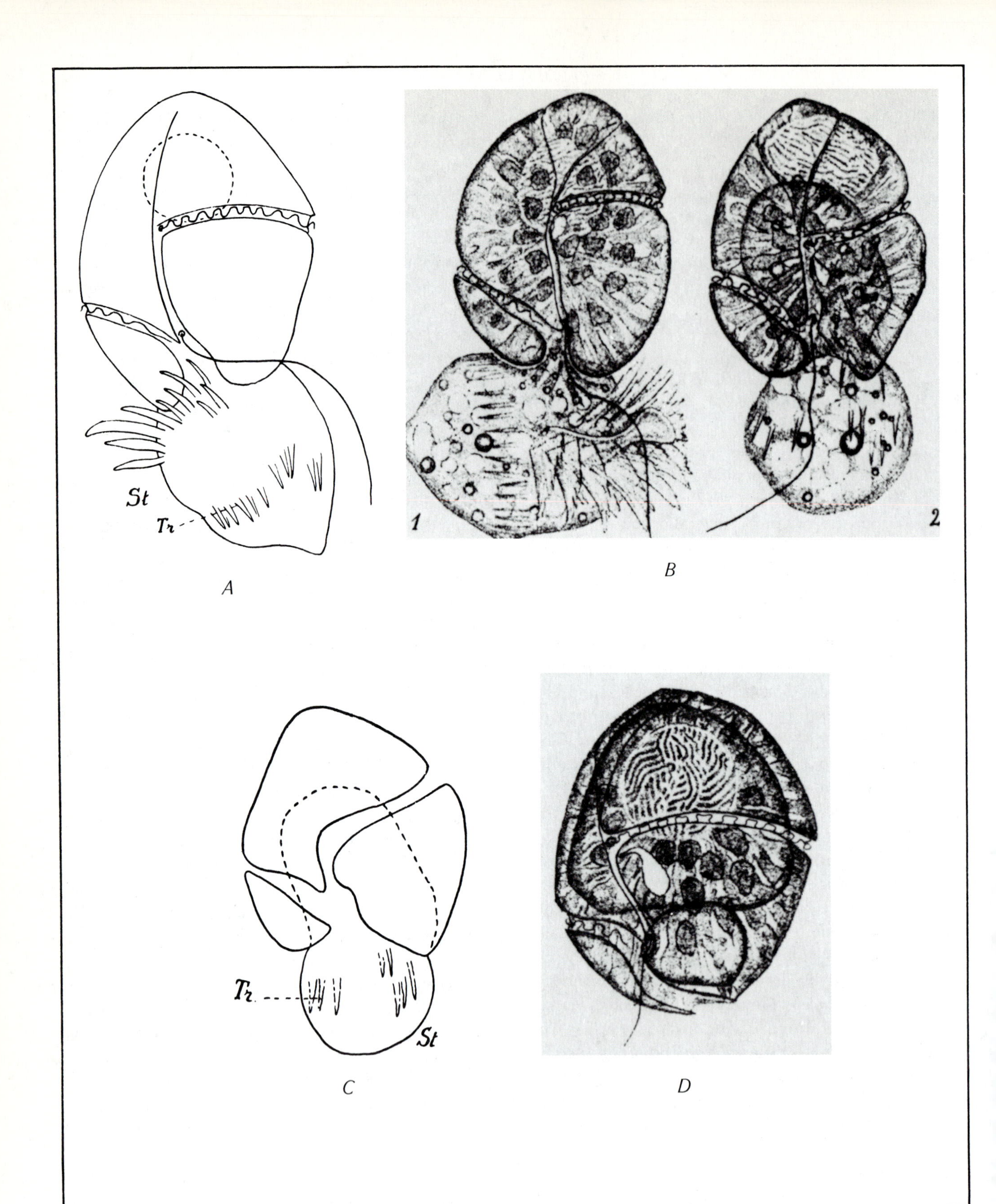
St
Tr
A
1
2
B
Tr
St
C
D

PLATE 22-3. The attack and ingestion of the ciliate *Strombidium* by the chromatophore-bearing dinoflagellate *Gyrodinium pavillardi.* (*A*), (*B*), and (*C*) Grasping and ingestion of prey, showing the position of the prey's trichocysts; (*D*) the digestion of a large and a small prey (magnifications not given). All from B. Biecheler, 1952, *Bull. Biol. France Belgique*, Suppl. 36.

and protozoan classes (Droop, 1974). Examples of acetate flagellates are *Brachiomonas submarina*, *Haematococcus pluvialis*, and *Euglena gracilis.*]

The general ultrastructure of *Oxyrrhis marina* was described in detail by Dodge and Crawford (1971a) and is shown diagrammatically in Plate 22-4,*C*. The cell is surrounded by an amphiesma, which consists of flattened vesicles and microtubules covered by a continuous outer membrane. Numerous typical dinoflagellate trichocysts are present. There are no plastids of any type. At the posterior end, two flagella emerge from the base of a projecting tentacle. The transverse flagellum is tightly coiled and bears complex mastigonemes that are composed of bundles of fibrils of three lengths, whereas the longitudinal flagellum is constructed in the normal manner without mastigonemes. Both flagella are covered (except at their proximal ends) with unusual ellipsoidal plate scales that form a tight spiral (Clarke and Pennick, 1972). Scales have also been seen on the body surface. The flagellar system has been studied in detail by Dodge and Crawford (1971c); the flagellar root system is more complicated than it is in other dinoflagellates.

Noctiluca scintillans (*N. miliaris*), an ubiquitous, colorless, luminescent dinoflagellate of nearshore waters, attains a diameter of 1.5 mm or more and is easily seen with the naked eye. *Noctiluca* is often the cause of the flashing luminescence seen in disturbed water at night, and it can fascinate an observer seeing his first night-haul of plankton. It is an aberrant form, which was once classified with the coelenterates as well as other invertebrate groups. This peach-shaped microorganism has a ventral side marked by a groove (Plate 22-4,*A*) and a thick, cross-striated tentacle, which originates at the mouth end and beats from four to eight times per minute. This tentacle, which rotates the whole microorganism, is the active feeding organ and has a sticky surface to which food adheres before it is brought into the mouth. Periodically, numerous orange droplets may form in the cytoplasm of the cell. In addition to asexual reproduction by binary fission, this dinoflagellate reproduces sexually by forming a cap of swarmers at the polar mass. These swarmers eventually break away and become flagellated isogametes when the parent is dying (Plate 22-4,*B*). Enomoto (1956) studied the occurrence of *Noctiluca*, and the type of food it was ingesting, on the west coast of Kyushu, Japan. Here, outbursts of *Noctiluca* followed the winter–spring diatom flowering to cause a marked decrease in phytoplankton populations. Species of diatoms made up some 90% of its food, the rest being mainly juvenile copepods. The eggs of only one species of fish, *Englauris japonica*, were eaten. Although the rate of consumption of the eggs of this fish species was from 7 to 67% during the spring bloom of *Noctiluca*, the consumption of eggs by *Noctiluca* for the year-long spawning season was under 1%. Prasad (1958) studied the occurrence and feeding habits of *Noctiluca* on the coast of India and also noted that outbursts were usually preceded by diatom blooms, which were rapidly grazed down. He also noted that a predominance of *Noctiluca* appeared to exclude other phytoplankton-eating members of the plankton, particularly the copepods. This was explained, according to the Hardy mutual-exclusion theory, as being due to the presence of an "ectocrine" released by the dinoflagellates similar to the toxic substance produced by "red tide" species. Prasad concluded that the presence of *Noctiluca* adversely affected important fisheries, such as sardine, anchovy, and mackerel. The occurrence and cause of *Noctiluca* "red tides" were reviewed by Le Fèvre and Grall (1970), who studied a swarm of *Noctiluca* off the western coast of Brittany. Although confirming the feeding observations of Prasad and Enomoto, these authors vigorously disagreed with Prasad on copepod exclusion, and explained the absence of copepods solely as competition due to the more rapid growth and higher grazing rates of *Noctiluca*. The large amount of ammonium contained in the vacuole has recently been identified as the cause of *Noctiluca* toxicity (Okaichi and Nishio, 1976).

In the family Noctiluciadae are a number of *Noctiluca* lookalikes that must cause much taxonomic confusion; they are put in the subfamily Kofoidininae by Cachon et al. (1967). The adult forms (sporonts) are characterized by a velum (hyposom flattened laterally). The morphological transformations that occur in the life cycle of a single species have affinities to the genera *Gymnodinium*, *Noctiluca*, and *Amphidinium*, have been

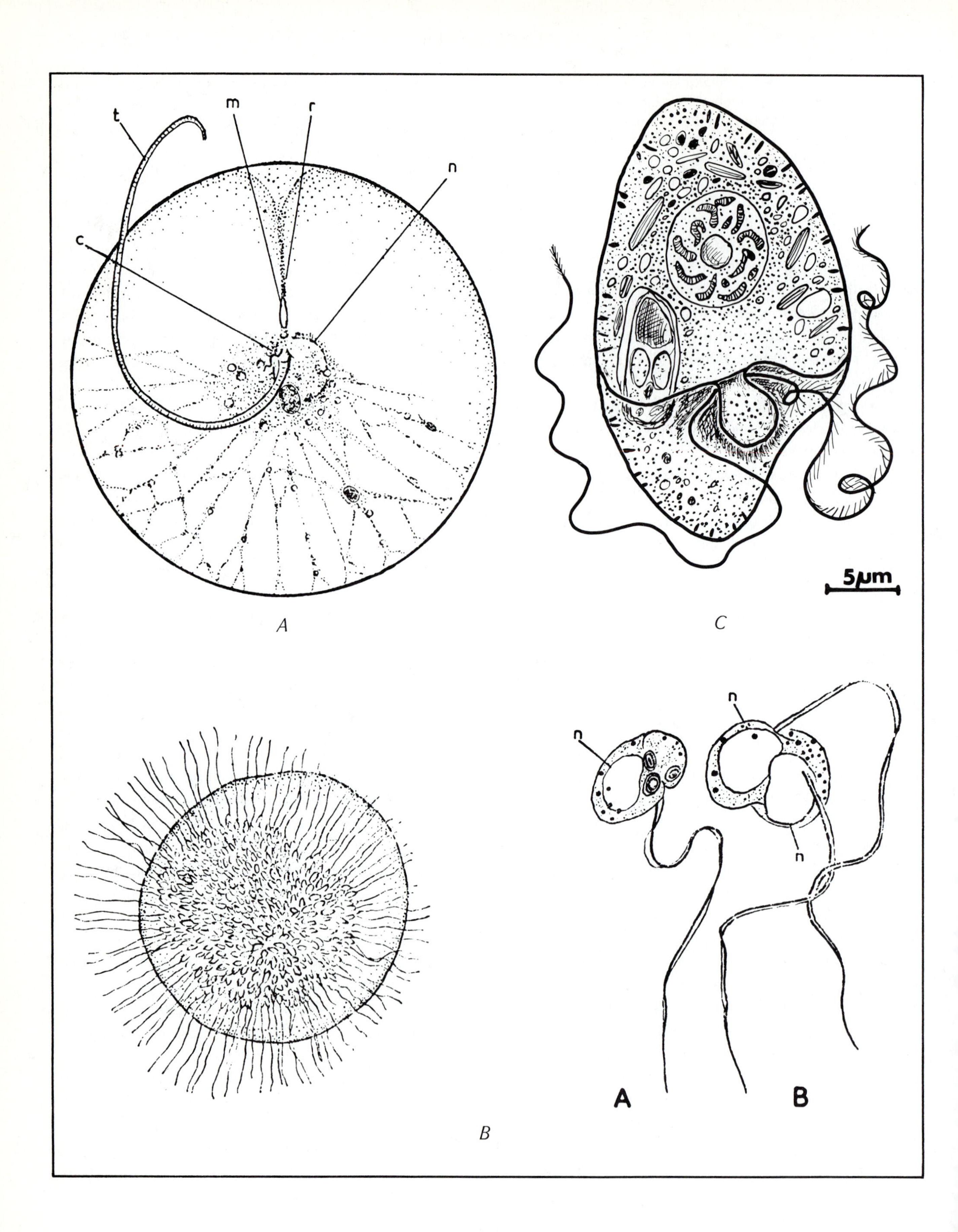
t
m
r
n
c
5µm
A
C
n
n
n
A
B
B

PLATE 22-4. Colorless dinoflagellates that compete with the copepods for diatoms. (*A*) *Noctiluca scintillans*, as viewed from the oral pole, showing the superficial view in the upper one-half and an optical section in the lower one-half (c, cilium; l, lip; m, mouth; n, nucleus; r, rod; t, tentacle) (745 X). (*B*) *Noctiluca* during gametogenesis, the relative size of the swarmers to the cell exaggerated 75 times; to the right, a single gamete (A) and syngamy (B) (940 X). (*C*) *Oxyrrhis marina*, in ventral view, showing engulfed diatom, coiled transverse flagellum with complex mastigonemes and longitudinal flagellum originating at tentacle. (*A,B*) from D.L. Mackinnon and P.S.J. Hawes, 1961, *An Introduction to the Study of Protozoa*, Clarendon Press, Oxford; (*B*) after A. Pratje, 1921, *Arch. Protistenk.* 42:1 (left) and F. Gross, 1934, *Arch. Protistenk.* 83:178; (*C*) from J.D. Dodge and R.M. Crawford, 1971, *Protistologica* 7:295–304.

previously described as individual species, and probably represent stages in their evolution. New species described by Cachon et al. (1967) include *Kofoidinium pavillardi*, *Kofoidinium velelloides*, and *Spatulodinium pseudonoctiluca*.

C. CHOANOFLAGELLATES

The choanoflagellates are a distinctive group of colorless flagellates. All have a distinct collar, which is formed by tentacles that aggregate to form a ring around an apical flagellum. There are three basic types of choanoflagellates, which are divided into three families (Boucaud-Camou, 1966; Plate 22-5). The family Codonosigidae includes species with naked cells as well as cells embedded in a mucilaginous matrix. They occur as solitary and colonial forms, either attached to substrates or free in the water. The species belonging to the Salpingoecidae are solitary cells encased in a lorica of resistant material ("chitin" or "cellulose"), with no visible striations or perforations. This large family is more homogeneous than the Codonosigidae. The family Acanthoecidae includes species in which the cells are housed in a large lorica of separate silica fibers that are bound together in a basket-like, network structure. These fibers are very distinctive under transmission and scanning electron microscopy (see Plate 22-6,*C*). Reproduction in the choanoflagellates is discussed by Ellis (1929) and Leadbeater (1975) and is shown diagrammatically in Plate 22-6,*F*.

Choanoflagellates are small cells (5 to 10 μm), which can easily be overlooked. It takes an experienced eye to detect the presence of stalked species attached to planktonic diatoms and debris, since they are barely detectable at the magnifications used to examine net plankton. Norris (1965), in studying the choanoflagellates of the neuston of tide pools in the upper littoral found that these species were attached to the air–water interface, as his illustrations with the cells in an inverted position indicate. Boucaud-Camou (1966) obtained her specimens from the surface of algae, hydroids, and bryozoans, but noted that they occurred in very small numbers. Throndsen (1969) reported the presence of several hundred choanoflagellates per milliliter in serial dilution cultures of water samples, from Norwegian coastal waters, although his objective was to enumerate the populations of phytoflagellates; therefore, his estimates of choanoflagellate numbers would presumably be low. Leadbeater (1972a) determined the number of free-living choanoflagellates in the nanoplankton by removing the coarse particulates by filtration before concentrating the small flagellates by centrifugation. The choanoflagellates observed by Lackey (1967) were mainly planktonic. Chretiennot (1974) used the Utermöhl sedimentation method to observe and enumerate the different classes of nanoplankton in the tide pools of the Marseilles region of France. She observed blooms of choanoflagellates ranging from 2000 to 29,000 cells/ml between May and November.

The choanoflagellates have been known for a long time, having been discovered by James-Clark (1867a, 1867b) and described by Stein (1878) and Kent (1880–82), whose accurate drawings are still useful for identification purposes. Ellis (1929) repeatedly points out that choanoflagellates feed on bacteria, and ingested bacteria are clearly shown in the micrographs of Leadbeater and Manton (1974). Lackey (1967) describes the choanoflagellates and other colorless flagellates as voracious bacteriovores and points out that, although they do not form single species blooms, their cumulative effect is important. The choanoflagellates are being included in surveys on nanoplankton in which transmission electron microscopy, which is required to resolve details not visible by light microscopy, is being used (Throndsen, 1970a,b; Leadbeater, 1972c, 1973, 1974). A description of particulate feeding by a marine choanoflagellate was obtained by studying numerous electron micrographs of thin sections showing the collar and prey (Laval, 1971; Plate 22-6,*D,E*). Currents created by the flagellum inside the perforated collar force the prey of large bacteria, including chroococcoid cyanobacteria, against the

PLATE 22-5. Choanoflagellates, which are active in bacteria-feeding, are represented diagrammatically to show the differences between the three choanoflagellate families. (*A*) The Codonosigidae contain species without a lorica (single and colonial, free and fixed), such as *Dicraspedella stokesi* (2) and *Desmarella moniloformis* (4). (*B*) The Salpingoecidae contain solitary cells (free and fixed) encased in a chitinous or cellulosic lorica, such as *Salpingoeca tuba* (1), *Salpingoeca inguillata* (5), and *Salpingoeca teres* (6). (*C*) The Acanthoecidae contain cells (single and colonial, free and fixed) encased in a large, basket-like lorica of silica strands, such as *Stephanoeca ampulla* (1), *Stephanoeca kenti* (4), *Stephanoeca campanula* (5), and *Diaphanoeca grandis* (6) (bars = 10 μm). All from E. Boucaud-Camou, 1966, *Bull. Soc. Linnéenne Normandie*, Ser. 10, 7:191–209.

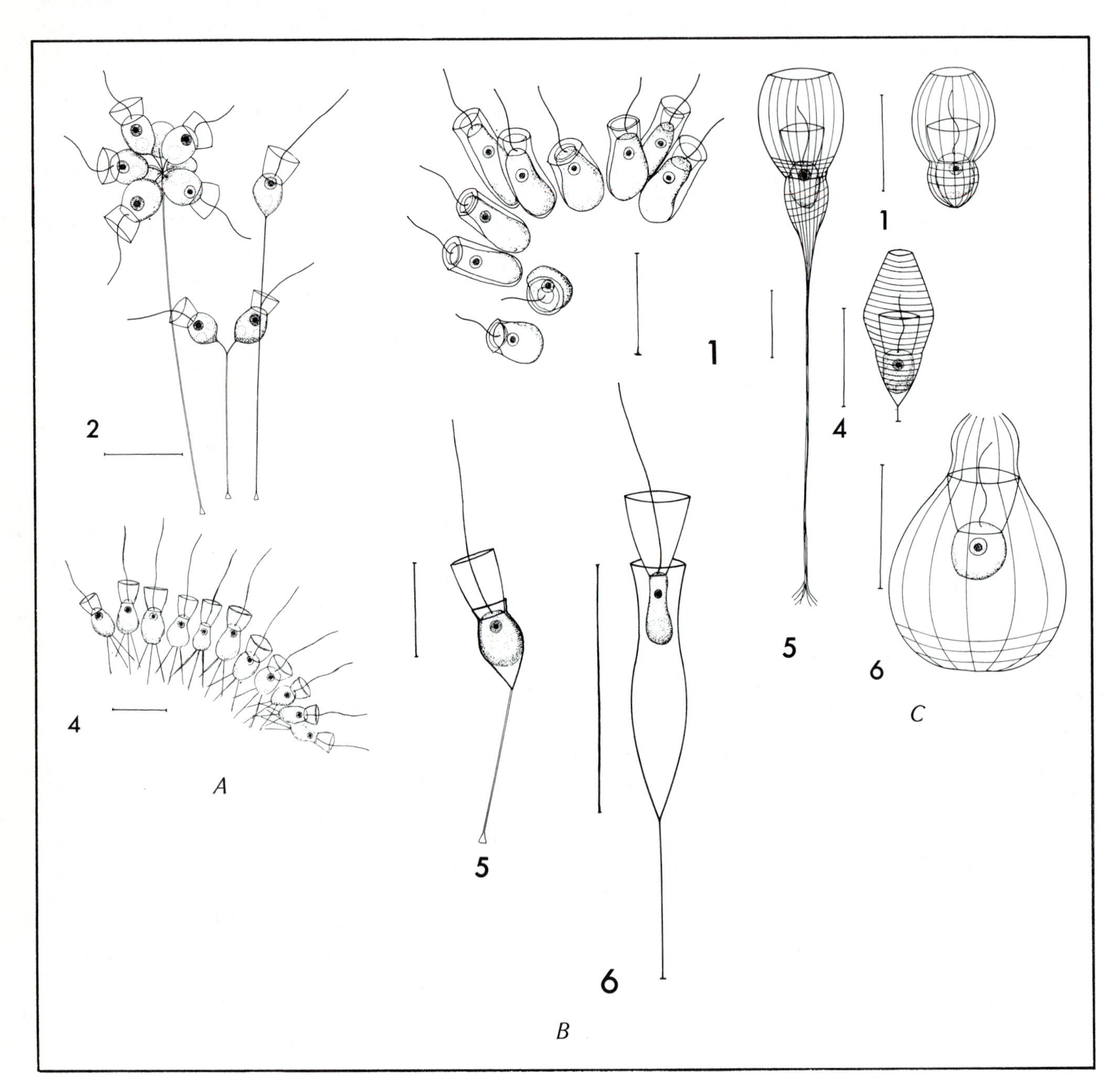

exterior surface of the collar where they are engulfed by pseudopodia arising from the base of the collar. The same currents that force prey against this area of the collar also create conditions within the collar favorable for the development of a microcosm of smaller bacteria. Such a microcosm of bacteria within the collar may have been mistaken for food by other investigators. Laval claims to be able to distinguish these associated bacteria from the prey bacteria by differences in size. The consumption of chroococcoid cyanobacteria by choanoflagellates observed by Laval (1971) is of interest, since these oxyphotobacteria are ubiquitous, can attain populations of 10^4 to 10^5 per milliliter in oceanic waters, and are apparently not digested by such zooplankters as calanoid copepods (see Chapter 8).

A semi-defined basal medium formulated for the axenic cultivation of a species of *Acanthoecopsis* gives high yields of choanoflagellates in culture (Gold, Pfister, and Liguori, 1970). Such cultures should be useful in studying osmotrophic and phagotrophic activities as well as determining whether the choanoflagellates could be used as food for experimentally raised ciliates (Gold, 1970). Ultrastructural studies in which the collars of choanoflagellates were compared with those of sponge choanocytes (Fjerdingstad, 1961a,b) showed points of similarity and dissimilarity. The ultrastructure of a whole choanoflagellate that occurs as an epibiont on the planktonic marine ciliate *Zoothamnium pelagicum* was illustrated by Laval (1971), and the ultrastructure of eight genera of choanoflagellates from a Danish salt marsh and seawater lagoon and from Norwegian fjords was described by Leadbeater (1972c,d). More recently, natural populations have been used to show that the costal strips forming the lorica are silica, that they originate in membrane-bound vesicles, and that the inside of the lorica is lined with an imperforate fibrillar membrane (Leadbeater and Manton, 1974; Leadbeater, 1975).

The taxonomic position of this group has been much debated, but agreement is slowly being reached. The phycologists point to a photosynthetic species, *Stylochromonas minuta* (Lackey, 1940), as an indication of an affinity to the photosynthetic Chrysophyceae, to place the choanoflagellates in the algal class Craspedomonadales (Parke and Dixon, 1968). Norris (1965) went even further and added a photosynthetic family, the Pedinellaceae, which contains naked cells with rhizopodia in a ring around the flagella, but lacks the definite collar of the choanoflagellates. The zoologists and protozoologists, on the other hand, point to the striking similarity of the choanoflagellate collar to the choanocytes (collar cells) of sponges (the order is named after these sponge cells) and place them in Choanoflagellida of the Zoomastigophorea (Honigberg et al., 1964; Boucaud-Camou, 1966). Throndsen (1969) takes a middle ground, putting them in the algal class Craspedophyceae of Christensen (1966); he uses the botanical rules of nomenclature while keeping the zoological subdivisions and keys of Boucaud-Camou (1966). The observation that the mitochondrial cristae of choanoflagellates are flat rather than tubular has led Leadbeater and Manton (1974) to conclude that the choanoflagellates are of animal rather than algal ancestry. As a consequence, Parke and Dixon (1976) have deleted the Craspedomonadales from their third Check-list of British algae.

D. CHRYSOMONADS

This once large class of protists was seriously depleted upon the removal of the haptophytes (Chapter 12). Current ultrastructural studies (Hibberd, 1977a,b) indicate that more and more protozoa (the choanoflagellates and bicoecids, for example), which were formerly grouped with the chrysomonads, should also be moved to other orders. The remaining group of colorless chrysomonads are in the genus *Paraphysomonas*. A colorless chrysomonad occurring as a stalked form in freshwater was described as *Physomonas vestita* in a paper by Stokes (1885), and in his exhaustive monograph on freshwater infusoria (Stokes, 1888). A more detailed and complete description of both the stalked and free-living forms occurring in the Kharkov Botanical Garden was obtained by Korshikov (1929); subsequently, the phagotrophs exhausted their supply of food, the purple sulfur bacterium *Chromatium okenii*, and either moved or died off. With the exception of one paper emending the name to *Paraphysomonas vestita*, and several other papers with electron micrographs of the distinctive body scales from preparations of natural samples of water, this apparently ubiquitous but furtive beast was overlooked until it discovered Manton and Leedale (1961). It was seen briefly as a culture contaminant before it destroyed itself by killing its host, a pigmented flagellate. These authors also reported its association with *Chromatium* in

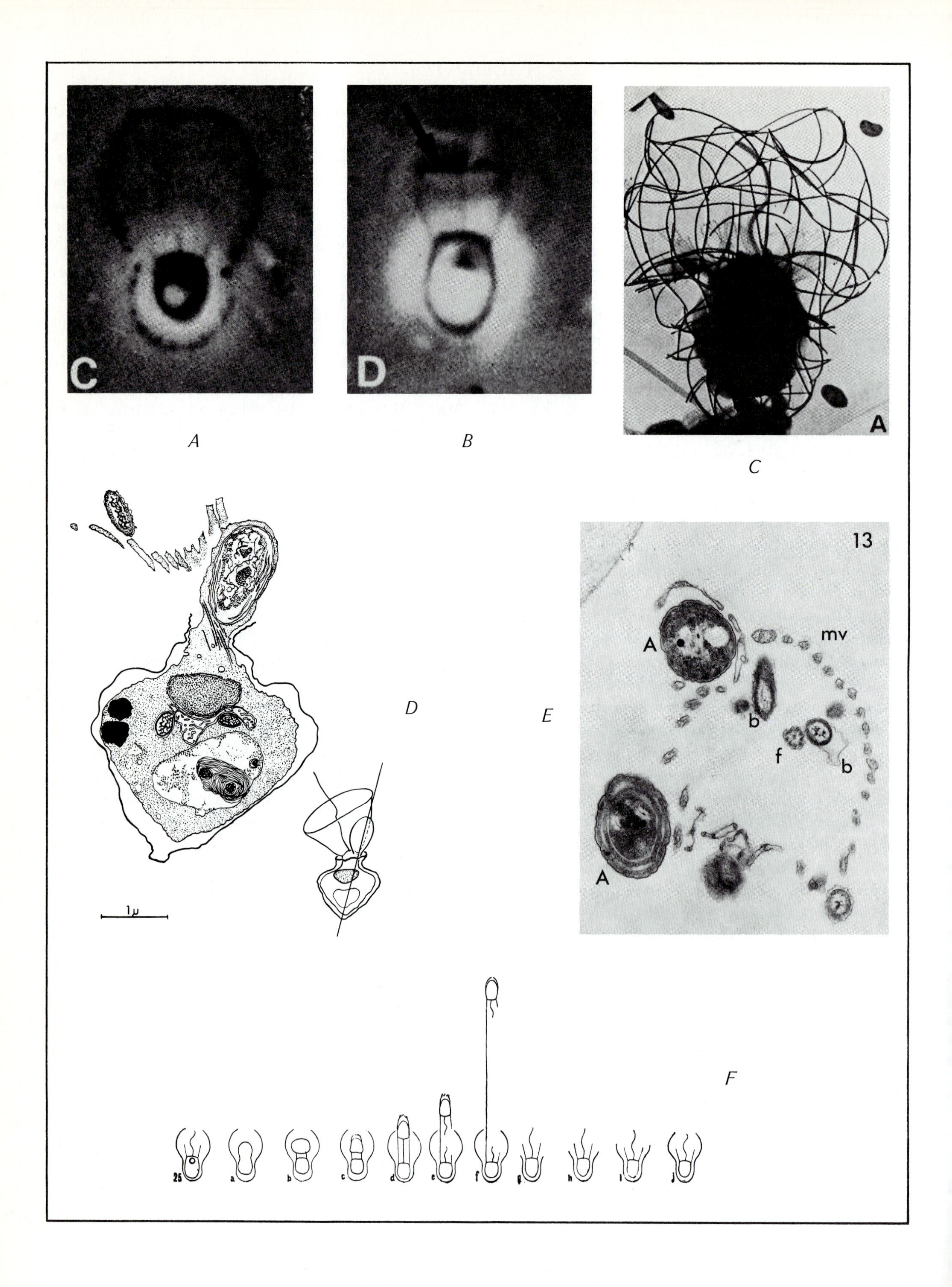
C
D
A
A
B
C
D
E
1μ
13
A
mv
b
f
b
A
F
25
a
b
c
d
e
f
g
h
i
j

PLATE 22-6. The appearance, feeding, and reproduction of choanoflagellates. Micrographs of *Stephanoeca diplocostata*, showing (*A*) a surface view of the living protoplast in phase contrast with the flagellum and the collar in focus and the lorica out of focus (4,150 X); (*B*) an optical section by phase-contrast microscopy of a living protoplast in position in the lorica (4,150 X); (*C*) an electron micrograph, shadow-cast whole mount of a complete cell, in which the arrangement of costae to form the lorica, the protoplast, and the flagellum and the tentacles that form the collar are clearly seen (5,600 X). (*D*) The feeding of *Salpingoeca pelagica* as observed by Laval (1971). A thin section, showing a free-living heterotrophic bacterium within the collar and a phototrophic cyanobacterium trapped on the exterior of the collar by the water currents generated by the flagellum and a diagram of the position of the thin section. (*E*) A transverse section through the collar of *S. pelagica*, showing heterotrophic bacteria living inside the collar and phototrophic cyanobacteria trapped on the outside, prior to ingestion (17,000 X). (*F*) Illustration of the stages in the reproduction of *Stephanoeca ampulla*, showing protoplast division (*a–f*) and the subsequent formation of the lorica by the daughter cell (*g–j*). (*A,B,C*) from B.S.C. Leadbeater and I. Manton, 1974, *J. Mar. biol. Ass. UK* 54:269–76; (*D,E*) from M. Laval, 1971, *Protistologica* 7:325–36; (*F*) from W.N. Ellis, 1929, *Ann. Soc. Roy. Zool. Belgique* 60:49–88.

a pond and described its ultrastructure, especially that of its body scales. Organisms of the genus *Paraphysomonas*, although colorless, are definitely chrysomonads because of their cell organization, e.g., heterokont flagella with the longer flagellum having flimmers (see Plate 22-7), and their reserves of lipid and leucosin. Since 1967, the type species (a freshwater form) as well as six additional species have been reported from marine waters. Four of these species originated as culture contaminants. Some of the marine isolates have rudimentary stalks, but others can attach without an organ at the posterior end. The genus *Monas* (Reynolds, 1934), referred to in the marine literature (Lackey, 1936; Lighthart, 1969), is apparently synonymous with the genera *Physomonas* and *Paraphysomonas*.

Virtually no information is available on the distribution and ecology of the paraphysomonads in marine waters. Their widespread occurrence from Norway to Italy and the simultaneous isolation of four species in culture (Leadbeater, 1972a) indicate that these small flagellates may be ubiquitous phagotrophs. Their ability to reproduce rapidly in the presence of photosynthetic bacteria and flagellates, which they decimate, suggests that they are voracious feeders. The reports on marine species are restricted to taxonomic observations (Lucas, 1967, 1968; Pennick and Clarke, 1972, 1973; Leadbeater, 1972a). The *Paraphysomonas* species fall into two distinct groups on the basis of their body scales (see Plate 22-7). This group certainly appears to be ecologically important.

E. BICOECIDS

Another group of colorless flagellates usually included with the chrysomonads because of their lorica and their heterokont flagellation (Bourrelly, 1968; Kristiansen, 1972) are the species placed in the genus *Bicoeca*. As a result of the study by Hibberd (1977b), they are considered separately here until a new order paralleling that for the choanoflagellates is created (Hibberd, 1977a). Bicoecids share characteristics with the choanoflagellates (Section C) and with *Bodo* (Section F). The genus *Bicoeca* was created by James-Clark (1867a,b) for a group of colorless, lorica-dwelling flagellates, which are kept in the lorica by a slender ligament believed to be contractile but subsequently shown to be a second flagellum that bends but does not contract. Of the forty described species of *Bicoeca* listed by Moestrup and Thomsen (1976), four have been described from the marine environment (*B. gracilipes*, *B. maris*, *B. mediterranea*, and *B. pontica*); some of these are shown in Plate 22-8,*B* to *F*. Lackey (1963) also lists *Bicoeca lacustris* from the marine environment and mentions (Lackey, 1967) that several species are found in brackish waters, thus contributing to the significant population of colorless flagellates. The most recent monographic study in which many of the freshwater forms are discussed is Hilliard's (1971) in which he described the flora of a small lake on the Kenai Peninsula of Alaska. The nature and diversity of species in the genus *Bicoeca* is shown in Plate 22-8,*A*.

The species of *Bicoeca* are distinguished mainly by the shape and size of their lorica, whose transparency lends itself to shadow-cast electron microscopy preparation. In order to observe the details of these hyaline flagellates, phase-contrast or electron microscopy is required. A good description of the marine species *B. maris*, which has been described under three different names, is provided by Moestrup and Thomsen (1976) (see Plate 22-8,*D* to *F*). Although these forms naturally occur on algal surfaces (Sieburth, 1975, Plate 6-7) and

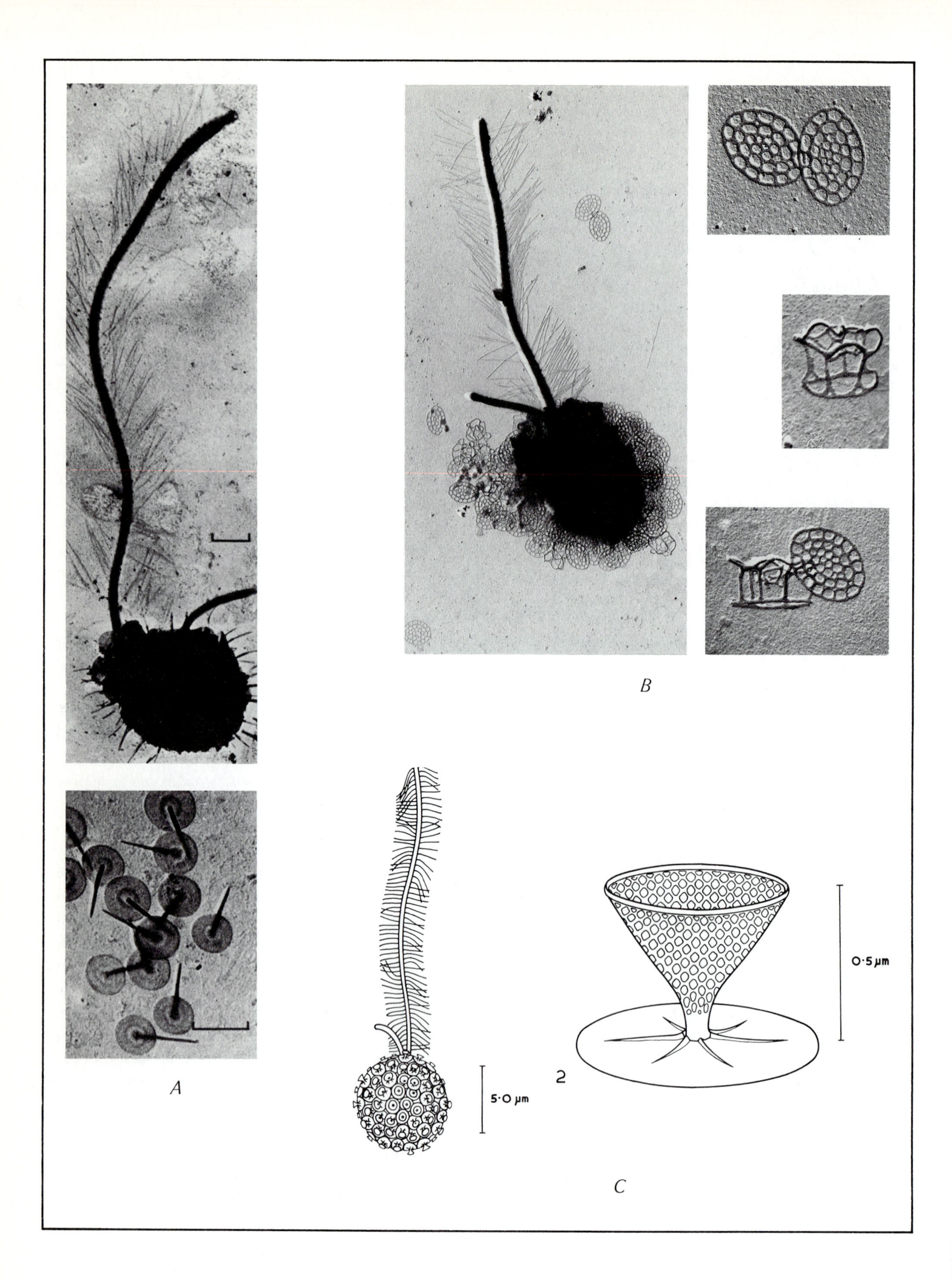
A
B
C
2
5·0 μm
0·5 μm

PLATE 22-7. Some species in the genus *Paraphysomonas*, with their characteristic heterokont flagellation of one short flagellum and one long flagellum with mastigonemes and species specific scales, which are colorless chrysomonads active in the ingestion of purple sulfur bacteria and phytoflagellates. (*A*) Transmission electron micrographs of shadowed *P. imperforata*, showing spiked scales (bar = 1 μm); (*B*) transmission electron micrographs of shadowed *P. butcheri* (left, (8,500 X), showing both the typical flat body scale and crown scale (right, 25,000 X); (*C*) a diagrammatic reconstruction of *P. cylicophora* from electron micrographs, showing the goblet-shaped body scales. (*A*) from I.A.N. Lucas, 1967, *J. mar. biol. Ass. UK* 47:329–34; (*B*) from N.C. Pennick and K.J. Clarke, 1972, *Br. Phycol. J.* 7:45–48; (*C*) from B.S.C. Leadbeater, 1972b, *Norweg. J. Bot.* 19:179–85.

on shells, in culture they attach to the sides of the culture tubes. Each cell is located in the middle of a cylindrical lorica 5×10 μm, whose rounded end is often attached eccentrically to the substrate. The rounded, but asymmetrical protoplast, 3 to 5 μm wide by 4 to 6 μm long, has two flagella inserted in an anterior depression. The longer flagellum (12 to 25 μm), with flimmer hairs, is directed out of the lorica, but the shorter flagellum (average length, 8 μm) is smooth. In its most expanded state, the short flagellum becomes almost straight to push the cell near the lorica aperture, but it apparently bends to retract the cell into the bottom of the lorica. The long flagellum is almost straight, with only the tip moving in the expanded state, but it undergoes a characteristic coiling when the cell is retracted. This extraordinarily complex flagellar apparatus is described by Moestrup and Thomsen (1976), who also cite other studies on *Bicoeca* ultrastructure.

Many of the microflagellates that are numerous in the nanoplankton of coastal and oceanic waters must be bicoecids, but since they are present as free-swimming forms (Plate 22-8, *D*) rather than as the loricate form, they are not usually identified.

Population estimates of about one thousand per milliliter have been obtained for Narragansett Bay, Rhode Island by inoculating bacterial suspensions with appropriate dilutions of raw seawater and examining the well cultures on an inverted microscope with phase-contrast optics. According to Moestrup and Thomsen (1976), *B. maris* can feed on bacteria that are recognized in thin sections of food vacuoles, but the more usual prey are larger eucaryotes.

F. OTHER FLAGELLATES

An in-depth study on the heterotrophic marine flagellates was conducted by Griessmann (1914), who observed the forms occurring in coastal waters and algal enrichments from the biological stations at Helgoland, Roscoff, Villefranche, Naples, Rovigno, Sevastopol, and Bergen. In addition to describing euglenids, bicoecids, and choanoflagellates, Griessmann also described a number of other miscellaneous flagellates, including the suborder Bodonina, which contains a few free-living marine forms. These forms are heterokont flagellates with one anterior flagellum and one posterior flagellum and a creeping motility. Griessmann also described two species of *Bodo* from seaweed enrichments at Roscoff and at Bergen. A species described as *Bodo marina* has been observed in low numbers in the Arctic waters of Greenland, along the Norwegian coast, on the east coast of North America, and year round near Trømso, Norway; it is also a regular and numerous inhabitant of the inner Oslo Fjord (Braarud, 1945). The occurrence of such flagellates in the nearshore waters of San Diego, California and Plymouth, England are listed by Lackey and Clendenning (1965) and by Lackey and Lackey (1963), respectively. In addition to a number of species of *Bodo*, *Rynchomonas nausta* also appears to be a commonly occurring flagellate. A few of the more common flagellate species are illustrated in Ruinen (1938); *Bodo* and *Rynchomonas* are shown in Plate 22-9, *B*.

The colorless cryptophycean *Chilomonas paramecium*, described by Ehrenberg from freshwater, has been widely studied (Kudo, 1966). This species (Butcher, 1967) and possibly others occur in brackish waters and are year-round residents of Narragansett Bay (Pratt, 1959). *Chilomonas paramecium* is shown in Plate 22-9, *A*. Its trichocysts or ejectosomes and other cytological features were studied by Anderson (1962).

Ebridians are rare flagellates with siliceous skeletons that superficially resemble those of silicoflagellates (Chapter 13, A). Like the silicoflagellates, they are more abundant in the fossil record than as modern forms. The ebridians are distinguished from the silicoflagellates by having a solid, rather than a hollow skeleton; by having two unequal flagella, instead of a single flagellum; and by having a dinokaryon nucleus, in which the chromosomes are condensed during interphase. The ebridians also lack the chloroplasts of the phototrophic silicoflagellates, and their nutrition is

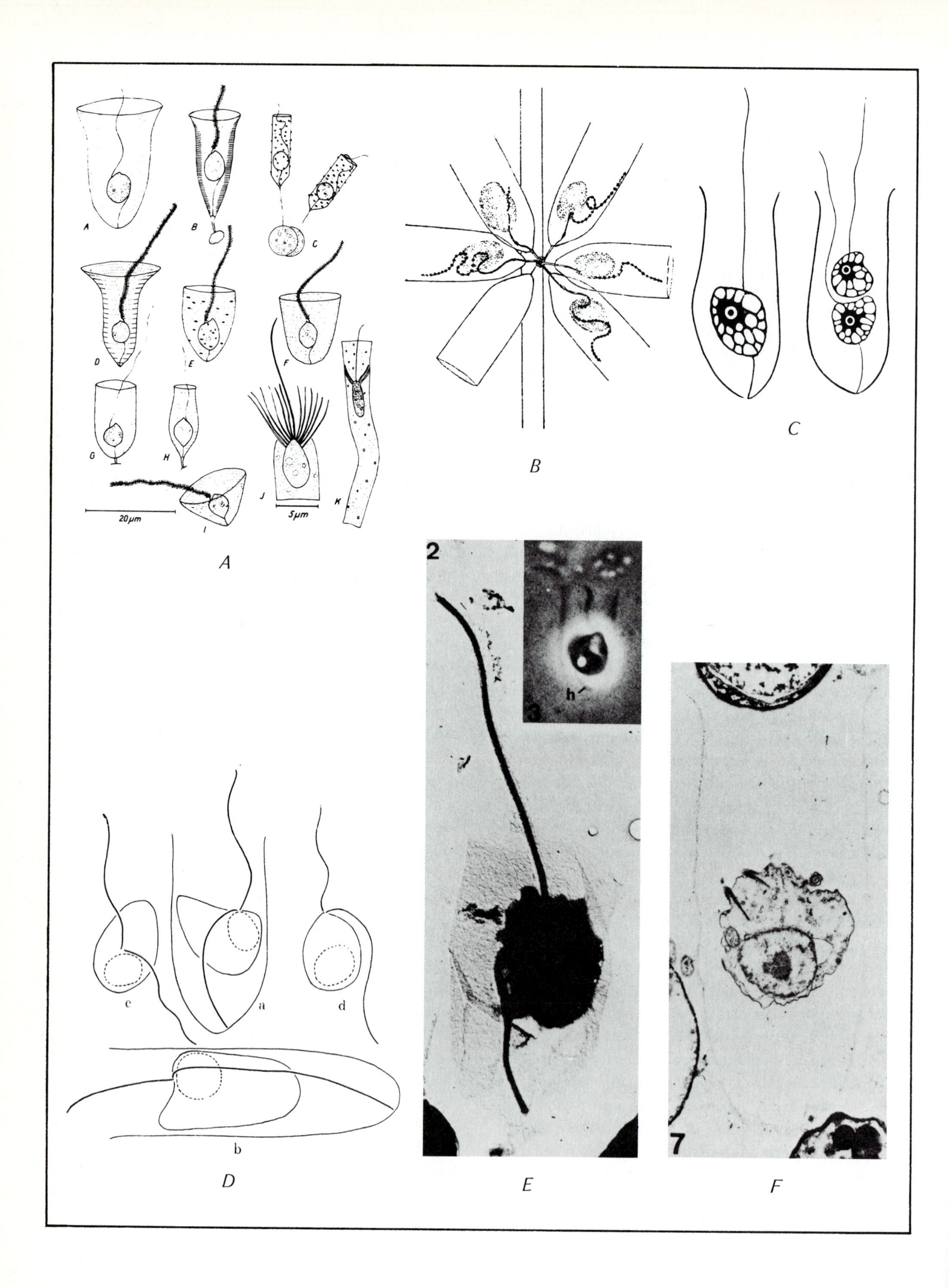
A
B
C
D
E
F
G
H
I
J
K
20 µm
5 µm
a
b
c
d
2
3
h
7

PLATE 22-8. Species of colorless flagellates in the genus *Bicoeca*. (*A*) Freshwater species, showing species specificity of lorica shape and size (note the flimmer hairs on the anterior flagellum; J and K are look-alikes, but not *Bicoeca*). (*B*) A group of six individual *B. mediterranea* on a fragment of *Nitzschia seriata* (2,000 X). (*C*) *B. pontica* (1,400 X) (*D*) the range in shape of *B. maris* (cell body, 5 μm long, house, 5×10 μm; note the free-swimming individuals c and d). (*E*) *Bicoeca maris:* (2) a transmission electron micrograph of a shadow-cast preparation of a whole cell, showing the fibrous lorica, the free anterior flagellum, and the hind flagellum that attaches the cell to the lorica (5,000 X); (3) a phase-contrast photomicrograph of a live cell retracted into the lorica, with both flagella visible (2,050 X). (*F*) A thin section of *B. maris* in the lorica; note the bacteria (7,150 X). (*A*) from D.K. Hilliard, 1971, *Arch. Protistenk.* 113:98–122; (*B*) from J. Pavillard, 1916, *C.R. hebd. Seanc. Acad. Sci. Paris* 163:65–68; (*C*) from A. Valkanov, 1970, *Zool. anz.* 184:241–90; (*D*) from L.E.R. Picken, 1941, *Philos. Trans. Royal Soc. Ser. B* 230:451–73; (*E*,*F*) from Ø. Moestrup and H.A. Thomsen, 1976, *Protistologica* 12:101–20.

phagotrophic. The taxonomic position of these forms is unclear, but Deflandre (1952) and Loeblich et al. (1968) believe that the ebridians should be classified with the dinoflagellates. The fragmentary literature on the fossil and recent forms of the ebridians, as well as the silicoflagellates, has been recently reviewed by Loeblich et al. (1968), who state that the first ebridian fossil was described by Ehrenberg in 1844 and the first recent form by Schumann in 1867. An example of an extant ebridian is *Hermesinum adriaticum*, which was first described from the Adriatic Sea (Zacharias, 1906); it is shown in Plate 22-9,*C*. Fourteen of the sixteen species of *Hermesinum* listed by Loeblich et al. (1968) are fossil forms. *Hermesinum adriaticum* rarely appears in phytoplankton reports; it has been reported from such waters as the Black Sea, the southern Mediterranean, and the upper Pettaquamscutt River (an extension of Narragansett Bay) (Hargraves and Miller, 1974). It was detected during a routine sampling program in the latter estuary, in two successive years, only between August and October when the salinity ranged from 7.5 to 16 o/oo and the temperature from 13 to 28 °C. Hargraves and Miller (1974) observed populations as dense as 10^4 to 10^5 cells/liter and noted that the cells were often packed with yellow–green material presumed to be microflagellates, whereas Loeblich et al. (1968) state that diatoms are often recorded as food in ebridians. This microorganism has not been brought into culture. Other species are probably overlooked. The skeletons of *Ebria tripartita*, which have the appearance of Ballantine Ale's three-ring trademark, are common in Narragansett Bay, R.I. and this difficult to distinguish flagellate appears to be a voracious consumer of chains of *Skeletonema costatum* (D.M. Pratt, pers. comm.).

The amoeboid flagellates (amoeboflagellates) are perhaps appropriate microorganisms with which to end this chapter. They are intermediate between the flagellates discussed in this chapter and the amoeboid forms discussed in the next chapter. Most reports on these forms are from freshwater, and only three have been reported from marine waters. Calkins (1901) reported the small *Mastigamoeba simplex* from Woods Hole, De Groot (1936) observed *Dinamoeba mirabilis* (perhaps synonymous with *Mastigamoeba aspera*) in brackish waters, and Page (1970) reported *Mastigamoeba aspera* (Plate 22-9) in 20 o/oo salinity estuarine pools in Maine. *Mastigamoeba aspera* is some 75 μm in length, with amoeboid locomotion, the anterior flagellum having little or no locomotory function. This species is distinguished by the occurrence of many thin, 2 μm long, curved bacteria-like rods randomly arranged and embedded in the gelatinous layers on the surface of the pseudopods, which gives them a hairy appearance. These amoeboid flagellates were apparently feeding on several species of dinoflagellates.

The use of epifluorescent microscopy to examine acridine-orange–stained bacterioplankton concentrated with the plankton on black filters (Chapter 5,C) also reveals an abundance of apochlorotic flagellates. Lohmann (1920) pointed out the important quantitative role played by these colorless flagellates in pelagic food webs. The respiration due to these organisms can form a significant part of total respiration in the open sea (Pomeroy and Johannes, 1968). In the neuston, they appear to be an important link between the bacterioneuston and the larger neustonic forms (Sieburth et al., 1976). As more data document their abundance and distribution, other research findings on their ecology, taxonomy, physiology, and trophic relationships will surely follow. [The existence of extensive populations of bacterioplankton throughout the water column (Part V) indicates that simultaneous studies should be made on these interdependent populations of bacteria and flagellates as well as on the next trophic level, the ciliates discussed in Chapter 24.]

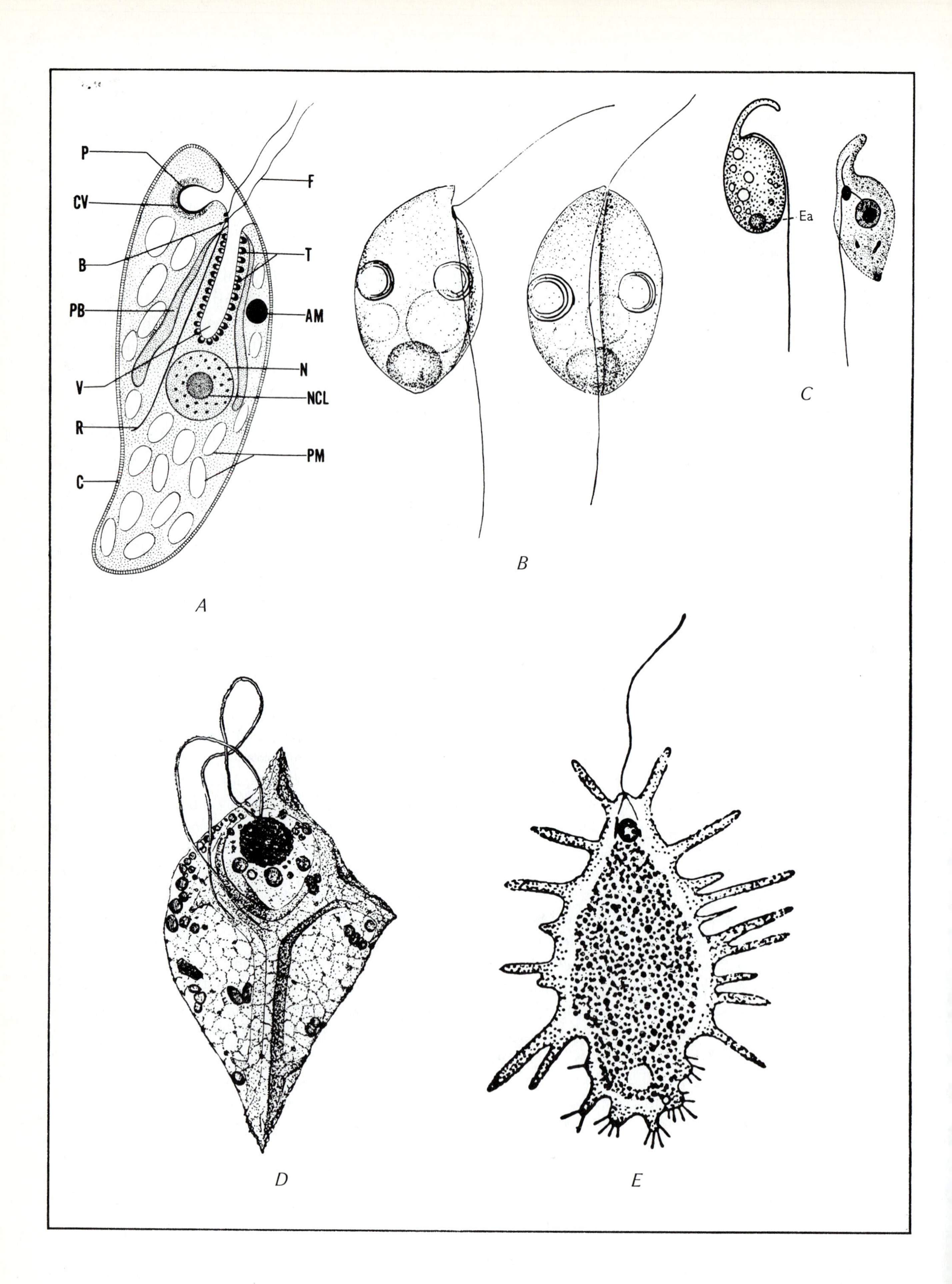
P
CV
B
PB
V
R
C
F
T
AM
N
NCL
PM
A
B
Ea
C
D
E

PLATE 22-9. A miscellaneous assortment of distinctive flagellated forms which are phagotrophic feeders. (*A*) A colorless cryptomonad, *Chilomonas paramecium*, with its distinctive gullet furrow; the Bodonina are creeping forms and include *Bodo edax* (*B*), and *Rynchomonas nausta* (*C*); (*D*) the ebridian *Hermesinum adriaticum* with its solid siliceous skeleton and hard to observe paired flagella; and (*E*) the amoeboid flagellate *Mastigamoeba aspera.* (Magnification not given.) (*A*) from E. Anderson, 1962, *J. Protozool.* 9:380–95. (*B*) from J. Ruinen, 1938, *Arch. Protistenk.* 90:210–58. (*C*) from A. Hollande, 1952, *Traité de Zool.*, Masson & Co., Paris, Tome 1, Fasc, 1, 669–93. (*D*) from R. Hovasse, 1934, *Compts Rendus Acad. Sci. Paris* 198:402–4. (*E*) from R.R. Kudo, 1966, *Protozoology*, courtesy of C.C. Thomas, Publ., Springfield, Ill.

REFERENCES

Allen, J.R., J.J. Lee, S.H. Hutner, and J. Storm. 1966. Prolonged culture of the voracious flagellate *Peranema* in antioxidant-containing media. *J. Protozool.* 13:103–8.

Anderson, E. 1962. A cytological study of *Chilomonas paramecium* with particular reference to the so-called trichocysts. *J. Protozool.* 9:380–95.

Barker, H.A. 1935. The culture and physiology of the marine dinoflagellates. *Arch. Mikrobiol.* 6:157–81.

Biecheler, B. 1952. Récherchés sur les Péridiniens. *Bull. Biol. France Belgique* (Suppl.) 36:1–149.

Boucaud-Camou, E. 1966. Les Choanoflagellés des côtes de la Manche: I. Systématique. *Bull. Soc. Linn. Normandie, Ser. 10,* 7:191–209.

Bourrelly, P. 1968. *Les Algues d'eau douce. Tome II. Les Algues jaunes et brunes.* N. Boubée et Cie., Paris, 438 p.

Braarud, T. 1945. A phytoplankton survey of the polluted waters of the inner Oslo Fjord. *Hvalradets Skrifter* 28:1–142.

Butcher, R.W. 1961. An introductory account of the smaller algae of British Coastal waters. Part VIII. Euglenophyceae=Euglenineae. *Fish. Invest.* (*London*) Ser. 4, 17 p.

Butcher, R.W. 1967. An introductory account of the smaller algae of British Coastal waters. Part IV. Cryptophyceae. *Fish. Invest.* (*London*) *Ser.* 4, 54 p.

Cachon, J., and M. Cachon. 1974. Le Système stomopharyngien de *Kofoidinium* Pavillard. Comparaisons avec celui de divers péridiniens libres et parasites. *Protistologica* 10:217–22.

Cachon, J., M. Cachon, O. Ferru, and G. Ferru. 1967. Contribution a l'étude des *Noctilucidae* Saville-Kent. I. Les *Kofoidininae* Cachon J. et. M. Évolution morphologique et systématique. *Protistologica* 3:427–44.

Calkins, G.N. 1901. Marine protozoa from Woods Hole. *Bull. U.S. Fish. Comm.* 21:415–68.

Chen, Y.T. 1950. Investigations of the biology of *Peranema trichophorum* (Euglenineae). *Quart. J. Microsc. Sci.* (3rd Ser.) 91:279–308.

Chretiennot, M-J. 1974. Nanoplancton de Flaques supralittorales de la région de Marseilles. II. Étude quantitative. *Protistologica* 10:477–88.

Christen, H-R. 1959. New colorless Euglenineae. *J. Protozool.* 6:292–303.

Christensen, T. 1966. Alger. In: *Botanik* (*Systematisk Botanik*) (2nd Ed.) (T.W. Bocher, M. Lange and T. Sorensen, eds.), Vol. 2(2). Munksgaard, Copenhagen.

Clarke, K.J., and N.C. Pennick. 1972. Flagellar scales in *Oxyrrhis marina* Dujardin. *Br. phycol. J.* 7:357–60.

Deflandre, G. 1952. Classé des Ébriédiens. In: *Traité de Zoologie* (P-P. Grasse, ed.), Vol. 1(1). Masson & Cie., Paris, pp. 407–24.

De Groot, A.A. 1936. Einige Beobachtungen an *Dinamoeba mirabilis* Leidy. *Arch. Protistenk.* 87:427–36.

Dodge, J.D., and R.M. Crawford. 1970. The morphology and fine structure of *Ceratium hirundinella* (Dinophyceae). *J. Phycol.* 6:137–49.

Dodge, J.D., and R.M. Crawford. 1971a. Fine structure of the dinoflagellate *Oxyrrhis marina.* I. The general structure of the cell. *Protistologica* 7:295–304.

Dodge, J.D., and R.M. Crawford. 1971b. A fine structural survey of dinoflagellate pyrenoids and food reserves. *Bot. J. Linn. Soc.* 64:105–15.

Dodge, J.D., and R.M. Crawford. 1971c. Fine structure of the dinoflagellate *Oxyrrhis marina.* II. The flagellar system. *Protistologica* 7:399–409.

Dodge, J.D., and R.M. Crawford. 1974. Fine structure of the dinoflagellate *Oxyrrhis marina.* III. Phagotrophy. *Protistologica* 10:239–44.

Droop, M.R. 1953. Phagotrophy in *Oxyrrhis marina* Dujardin. *Nature* 172:250.

Droop, M.R. 1959. Water soluble factors in the nutrition of *Oxyrrhis marina. J. mar. biol. Assoc. U.K.* 38:605–20.

Droop, M.R. 1966. The role of algae in the nutrition of *Heteramoeba clara* Droop with notes on *Oxyrrhis marina* Dujardin and *Philodina roseola* Ehrenberg. In: *Some Contemporary Studies in Marine Science* (H. Barnes, ed.). Allen & Unwin, London, pp. 269–82.

Droop, M.R. 1970. Nutritional investigation of phagotrophic protozoa under axenic conditions. *Helgoländer wiss. Meeresunters.* 20:272–77.

Droop, M.R. 1974. Heterotrophy of carbon. In: *Algal Physiology and Biochemistry* (W.D.P. Stewart, ed.). University of California Press, Berkeley, pp. 530–59.

Droop, M.R., and J. Doyle. 1966. Ubiquinone as a protozoan growth factor. *Nature* 212:1474–75.

Ellis, W.N. 1929. Recent researches on the Choanoflagellata (Craspedomonadines) (fresh-water and marine) with description of new genera and species. *Ann. Soc. Roy. Zool. Belgique* 60:49–88.

Enomoto, Y. 1956. On the occurrence and the food of *Noctiluca scintillans* (MacCartney) in the waters adjacent to the west coast of Kyushu, with special reference to the possibility of the damage caused to the fish eggs by that plankton. *Bull. Jap. Soc. Sci. Fish.* 22:82–88.

Fjerdingstad, E.J. 1961a. The ultrastructure of choanocyte collars in *Spongilla lacustris* (L.) *Z. Zellforschung* 53:645–57.

Fjerdingstad, E.J. 1961b. Ultrastructure of the collar of the choanoflagellate *Codonosiga botrytis* (Ehrenb.). *Z. Zellforschung.* 54:499–510.

Gold, K. 1970. Cultivation of marine ciliates (Tintinnida) and heterotrophic flagellates. *Helgoländer wiss. Meeresunters.* 20:264–71.

Gold, K., R.M. Pfister, and V.R. Liguori. 1970. Axenic cultivation and electron microscopy of two species of Choanoflagellida. *J. Protozool.* 17:210–12.

Griesmann, K. 1914. Über marine Flagellaten. *Arch. Protistenk.* 32:1–78.

Hargraves, P.E., and B.T. Miller. 1974. The ebridian flagellate *Hermesinum adriaticum* Zach. *Arch. Protistenk.* 116:280–84.

Hibberd, D.J. 1977a. The possible phyletic and systematic implications of recent work on the ultrastructure of the Chrysomonidida *sensu lato.* Abstr. *5th Intl. Congr. Protozool., New York* (S.H. Hutner, ed.), p. 256.

Hibberd, D.J. 1977b. The possible value of Golgi body and endoplasmic reticulum characters in some phytomastigophorean and zoomastigophorean orders. Abstr. *5th Intl. Congr. Protozool., New York* (S.H. Hutner, ed.), p. 448.

Hilliard, D.K. 1971. Notes on the occurrence and taxonomy of some planktonic chrysophytes in an Alaskan lake, with comments on the genus *Bicoeca. Arch. Protistenk.* 113:98–122.

Hollande, A. 1952. Ordre des Bodonides (Bodonidea ord. nov.). In: *Traité de Zoologie* (P-P. Grassé, ed.), Vol. 1(1). Masson et Cie., Paris, pp. 669–93.

Honigberg, B.M., W. Balamuth, E.C. Bovee, J.O. Corliss, M. Gojdisc, R.P. Hall, R.R. Kudo, N.D. Levine, A.R. Loeblich, Jr., J. Weiser, and D.H. Wenrich. 1964. A revised classification of the Phylum Protozoa. *J. Protozool.* 11:7–20.

Hovasse, R. 1934. Protistologie.—Ebriacées, dinoflagellés et radiolaires. *C.R. Acad. Sci. Paris* 198:402–4.

Ioriya, T. 1976. Notes on some species of colourless Euglenophyceae from Hokkaido, Japan. *Bull. Jap. Soc. Phycol.* 24(2):62–64.

James-Clark, H. 1867a. Conclusive proofs of the animality of the ciliate sponges, and of their affinities with the Infusoria flagellata. *Ann. Mag. Nat. Hist.* Ser. 3, 19:13–18.

James-Clark, H. 1867b. On the spongiae ciliatae as infusoria flagellata; or observations on the structure, animality, and relationship of *Leucosolenia botryoides,* Bowerbank. *Mem. Boston Soc. Nat. Hist.* 1:305–40.

Kent, W.S. 1880–1882. *A manual of the Infusoria,* 3 vols. D. Boque, London, 472 p.

Korshikov, A.A. 1929. Studies on the chrysomonads. I. *Physomonas vestita* Stokes. *Arch. Protistenk.* 67:253–90.

Kristiansen, J. 1972. Structure and occurrence of *Bicoeca crystallina,* with remarks on the taxonomic position of the Bicoecales. *Br. Phycol. J.* 7:1–12.

Kudo, R.R. 1966. *Protozoology* (5th Ed.). C.C. Thomas, Springfield, Ill., 1174 p.

Lackey, J.B. 1936. Occurrence and distribution of the marine protozoan species in the Woods Hole area. *Biol. Bull.* 70:264–78.

Lackey, J.B. 1940. Some new flagellates from the Woods Hole area. *Amer. Midl. Nat.* 23:463–71.

Lackey, J.B. 1960. *Calkinsia aureus* gen. et sp. nov., a new marine euglenid. *Trans. Amer. Microsc. Soc.* 79:105–7.

Lackey, J.B. 1961. Bottom sampling and environmental niches. *Limnol. Oceanogr.* 6:271–79.

Lackey, J.B. 1962. Three new colorless Euglenophyceae from marine situations. *Arch. Mikrobiol.* 42:190–95.

Lackey, J.B. 1963. The microbiology of a Long Island bay in the summer of 1961. *Int. Revue ges. Hydrobiol.* 48:577–601.

Lackey, J.B. 1967. The microbiota of estuaries and their roles. In: *Estuaries* (G.H. Lauff, ed.). AAAS Pub. 83, pp. 291–302.

Lackey, J.B., and K.A. Clendenning. 1965. Ecology of the microbiota of San Diego Bay, California. *Trans. San Diego Soc. Nat. Hist.* 14:9–40.

Lackey, J.B., and E.W. Lackey. 1963. Microscopic algae and protozoa in the waters near Plymouth in August 1962. *J. mar. biol. Assoc. U.K.* 43:797–805.

Laval, M. 1971. Ultrastructure et mode de nutrition du choanoflagelle *Salpingoeca pelagica* sp. nov. Comparaison avec les choanocytes des spongiares. *Protistologica* 7:325–36.

Leadbeater, B.S.C. 1972a. Identification, by means of electron microscopy, of flagellate nanoplankton from the coast of Norway. *Sarsia* 49:107–24.

Leadbeater, B.S.C. 1972b. *Paraphysomonas cylicophora* sp. nov., a marine species from the coast of Norway. *Norw. J. Bot.* 19:179–85.

Leadbeater, B.S.C. 1972c. Ultrastructural observations on some marine choanoflagellates from the coast of Denmark. *Br. phycol. J.* 7:195–211.

Leadbeater, B.S.C. 1972d. Fine structural observations on some marine choanoflagellates from the coast of Norway. *J. mar. biol. Assoc. U.K.* 52:67–79.

Leadbeater, B.S.C. 1973. External morphology of some marine choanoflagellates from the coast of Jugoslavia. *Arch. Protistenk.* 115:234–52.

Leadbeater, B.S.C. 1974. Ultrastructural observations on nanoplankton collected from the coast of Jugoslavia and the Bay of Algiers. *J. mar. biol. Assoc. U.K.* 54:179–96.

Leadbeater, B.S.C. 1975. A microscopical study of the marine choanoflagellate *Savillea micropora* (Norris) comb. nov., and preliminary observations on lorica development in *S. micropora* and *Stephanoeca diplocostata* Ellis. *Protoplasma* 83:111–29.

Leadbeater, B.S.C., and I. Manton. 1974. Preliminary observations on the chemistry and biology of the lorica in a collared flagellate (*Stephanoeca diplocostata* Ellis). *J. mar. biol. Assoc. U.K.* 54:269–76.

Leedale, G.F. 1967. *Euglenoid Flagellates.* Prentice-Hall, Englewood Cliffs, N.J., 242 p.

Le Fèvre, J., and J.R. Grall. 1970. On the relationships of

Noctiluca swarming off the coast of Brittany with hydrological features and plankton characteristics of the environment. *J. exp. mar. Biol. Ecol.* 4:287–306.

Lighthart, B. 1969. Planktonic and benthic bacteriovorous protozoa at eleven stations in Puget Sound and adjacent Pacific Ocean. *J. Fish. Res. Bd. Canada* 26:299–304.

Loeblich, A.R., III, L.A. Loeblich, H. Tappan, and A.R. Loeblich, Jr. 1968. *Annotated Index of Fossil and Recent Silicoflagellates and Ebridians with Descriptions and Illustrations of Validly Proposed Taxa.* Geol. Soc. Amer. Memoir 106, Boulder, Colorado, 319 p.

Lohmann, H. 1920. Die Bevölkerung des Ozeans mit Plankton. *Arch. Biontol. ges. naturf. Freunde Berlin* 4:1–617.

Lucas, I.A.N. 1967. Two marine species of *Parraphysomonas. J. mar. biol. Assoc. U.K.* 47:329–34.

Lucas, I.A.N. 1968. A new member of the Chrysophyceae, bearing polymorphic scales. *J. mar. biol. Assoc. U.K.* 48:437–41.

Mackinnon, D.L., and R.S.J. Hawes. 1961. *An Introduction to the Study of the Protozoa.* Clarendon Press, Oxford, 506 p.

Manton, I., and G.F. Leedale. 1961. Observations on the fine structure of *Paraphysomonas vestita,* with special reference to the Golgi apparatus and the origin of scales. *Phycologia* 1:37–57.

Mignot, J-P. 1966. Structure et ultrastructure de quelques Euglenomonadines. *Protistologica* 2:51–117.

Moestrup, Ø., and H.A. Thomsen. 1976. Fine structural studies on the flagellate genus *Bicoeca.* I. *Bicoeca maris* with particular emphasis on the flagellar apparatus. *Protistologica* 12:101–20.

Norris, D.R. 1969. Possible phagotrophic feeding in *Ceratium lunula* Schimper. *Limnol. Oceanogr.* 14:448–49.

Norris, R.E. 1965. Neustonic marine Craspedomonadales (Choanoflagellates) from Washington and California. *J. Protozool.* 12:589–602.

Okaichi, T., and S. Nishio. 1976. Identification of ammonia as the toxic principle of red tide of *Noctiluca miliaris. Bull. Plankton Soc. Japan* 23:75–80.

Page, F.C. 1970. *Mastigamoeba aspera* from estuarine tidal pools in Maine. *Trans. Amer. Micros. Soc.* 89:197–200.

Parke, M., and P.S. Dixon. 1968. Check-list of British marine algae—second revision. *J. mar. biol. Assoc. U.K.* 48:783–832.

Parke, M., and P.S. Dixon. 1976. Check-list of British marine algae—third revision. *J. mar. biol. Assoc. U.K.* 56:527–94.

Pavillard, J. 1916. Flagellés nouveaux, épiphytes des diatomées pélagiques. *C.R. hebd. Séanc. Acad. Sci., Paris* 163:65–68.

Pennick, N.C., and K.J. Clarke. 1972. *Paraphysomonas butcheri* sp. nov. A marine, colourless, scale-bearing member of the Chrysophyceae. *Br. phycol. J.* 7:45–48.

Pennick, N.C., and K.J. Clarke. 1973. *Paraphysomonas corbidifera* sp. nov., a marine, colourless, scale-bearing member of the Chrysophyceae. *Br. phycol. J.* 8:147–51.

Picken, L.E.R. 1941. On the Bicoecidae: A family of colourless flagellates. *Philos. Trans. Royal Soc. London Ser. B.* 230:451–73.

Pomeroy, L.R., and R.E. Johannes. 1968. Occurrence and respiration of ultraplankton in the upper 500 meters of the ocean. *Deep-Sea Res.* 15:381–91.

Prasad, R.R. 1958. A note on the occurrence and feeding habits of *Noctiluca* and their effects on the plankton community and fisheries. *Proc. Ind. Acad. Sci. B* 47:331–37.

Pratt, D.M. 1959. The phytoplankton of Narragansett Bay. *Limnol. Oceanogr.* 4:425–40.

Reynolds, B.D. 1934. Studies on monad flagellates. I. Historical and taxonomical review of the genus *Monas,* II. Observations on *Monas vestita* (Stokes 1885). *Arch. Protistenk.* 81:399–411.

Ruinen, J. 1938. Notizen über Salzflagellaten. II. Über die Verbreitung der Salzflagellaten. *Arch. Protistenk.* 90:210–58.

Sieburth, J.McN. 1975. *Microbial Seascapes,* University Park Press, Baltimore, Md., Plate 6-72.

Sieburth, J.McN., P-J. Willis, K.M. Johnson, C.M. Burney, D.M. Lavoie, K.R. Hinga, D.A. Caron, F.W. French III, P.W. Johnson, and P.G. Davis. 1976. Dissolved organic matter and heterotrophic microneuston in the surface microlayers of the North Atlantic. *Science* 194:1415–18.

Skuja, H. 1939. Beitrag zur Algenflora Lettlands II. *Acta Horti bot. Univ. latviensis* 11/12:41–169.

Skuja, H. 1948. *Taxonomie des Phytoplanktons einiger Seen in Uppland, Schweden.* Symbolae Bot. Upsalienses 9(No. 3), 399 p.

Stein, F.R. 1878. *Der Organismus der Infusionthiere.* Abt. 3, W. Englemann, Leipzig, 154 p.

Stokes, A.C. 1885. Notes on some apparently undescribed forms of fresh-water infusoria. No. 2. *Amer. J. Sci. 3rd Ser.* 29:313–28.

Stokes, A.C. 1888. A preliminary contribution toward a history of the fresh-water infusoria of the United States. *J. Trenton Nat. Hist. Soc.* 1:71–320.

Throndsen, J. 1969. Flagellates of Norwegian Coastal waters. *Nytt. Mag. Bot. (Norw. J. Bot.)* 16:161–216.

Throndsen, J. 1970a. *Salpingoeca spinifera* sp. nov., a new planktonic species of the Craspedophyceae recorded in the Arctic. *Br. phycol. J.* 5:87–89.

Throndsen, J. 1970b. Marine planktonic Acanthoecaceans (Craspedophyceae). *Nytt. Mag. Bot. (Norw. J. Bot.)* 17:103–11.

Valkanov, A. 1970. Beitrag zur Kenntnis der Protozoen des Schwarzen Meeres. *Zool. anz.* 184:241–90.

Wolken, J.J. 1967. *Euglena, an Experimental Organism for Biochemical and Biophysical Studies* (2nd Ed.). Appleton-Century-Crofts, New York, 204 p.

Zacharias, O. 1906. Eine neue Dictyochide aus dem Mittelmeer, *Hermesinum adriaticum* n.g., n. sp. *Arch. Hydrobiol. Planktonk.* 1:394–98.

CHAPTER 23

Amoeboid Forms

The amoeboid protozoa, unlike the flagellates (Chapter 22) and the ciliates (Chapter 24), have no permanent organelles for locomotion. Instead, they form temporary, root-like organelles (hence the name Rhizopoda, "the root-footed"). These locomotory organs are also used to capture food.

The amoeboid protozoa occur in a variety of forms. The naked amoebae mainly inhabit surfaces, where they search for bacteria and other particulate food that they engulf with their pseudopoda. The heliozoan forms are distinguished by their stiff pseudopods, which look like the rays of the sun (hence their early name, "the sun animalcules"). The remaining forms are loricate and have simple houses (testaceans), more complex houses, which are typically calcareous and perforated (the foraminiferans), or houses of silica (radiolarians) or strontium sulfate (acantharians). Although the amoebae, heliozoa, and testaceans, and most foraminifera are thought of as nearshore forms, the amoebae, as well as the foraminifera, are also oceanic, with some foraminifera species in the deep-sea benthos. The radiolaria and acantharia are pelagic species. The loricate forms are undoubtedly important as phagotrophic microorganisms of the plankton and benthos, but they are mainly known from the fossil record by their surviving loricas.

No better introduction to this diverse group could be written than the letter of transmittal of Joseph Leidy's magnificent monograph on rhizopods (Leidy, 1879) to the Secretary of the Interior, written a century ago by the U.S. geologist then in charge, F.V. Hayden.

> . . . the rhizopods are the lowest and simplest form of animals, mostly minute, and requiring high power of the microscope to distinguish their structure. While most of them construct shells of great beauty and variety, their soft part consists of a jelly-like substance. This animal has the power of extending its threads or finger-like processes, which are used as organs of locomotion and prehension, often branching. From the appearance of their temporary organs, resembling roots, the class of animals has received its name of Rhizopoda, meaning literally root-footed.

> . . . In compensation for the smallness of these creatures, they make up in numbers, and it is questionable whether any other class of animals exceeds them in importance in the economy of nature. Geological evidence shows that they were the starting point of animal life in time, and their agency in rock-making has not been exceeded by later higher and more visible forms.
>
> . . . With the marine kind, known as Foraminifera, we have been longest familiar. Their beautiful many-chambered shells—for the most part just visible to the naked eye—form a large portion of the ocean mud and the sands of the ocean shore. Shells of Foraminifera likewise form the basis of miles of strata of limestones of which Paris and the pyramids of Egypt are built.

Only some forty years before Leidy initiated his study, Felix Dujardin had, in 1835, corrected the misconception that foraminifera were tiny mollusca (metazoans) by showing that they lacked the organs characteristic of molluscs. This set the stage for the creation of the phylum Protozoa some ten years later. A classic and general presentation of the rhizopods is given by Jepps (1956). The classification scheme of Honigberg et al. (1964) and its update by Hutner (1973) can be used to determine the hierarchy used by some authors. The simplified scheme used by Grell (1973) will be used here to discuss the orders with free-living marine species, with the exception that the Acantharia as an order is separate from the Radiolaria.

A. AMOEBAE

The order Amoebidae contains the "naked amoebas," which have no fixed body shape or lorica. Their cytoplasm consists of an outer hyaline ectoplasm and a granular endoplasm containing the nucleus, organelles, and inclusions. Some species have ectoplasmic ridges, spines, horns, filaments, and uroids. There are many parasitic forms in addition to the free-living forms to be discussed here. The biology of amoebae is discussed in the multi-author volume edited by Jeon (1973), and among its many chapters are those on culture, general morphology, locomotion and behavior, ultrastructure, and taxonomy.

Investigators who examine planktonic and benthic samples in order to list and enumerate the component species seem to shy away from the amoebae. Even the protozoologists seem to prefer those forms with flagella, cilia, or shells. When mentioned at all, the most that is said about amoebae is that they were present in low numbers and were not identified (Lackey, 1936, 1961, 1963; Lackey and Clendenning, 1965). The biggest problem is that, in order to obtain enough characteristics for species identification, they must be brought into culture.

The basic monograph on marine amoebae was prepared by Schaeffer (1926) from materials obtained at Cold Spring Harbor, New York and Tortugas, Florida. The 19 genera were described solely on morphological grounds. Some of these genera have been emended through the application of more sophisticated criteria, such as characteristic cytology and nuclear division, but this monograph still serves as the basis for identifying many of the marine amoebae. The eight genera identified from upper Chesapeake Bay by Sawyer (1971) were described by Schaeffer. Sawyer concentrated the amoebae in 1-liter samples of water on 1.2 μm Millipore HA filters floated face down on seawater in 60-cm plastic dishes and incubated at room temperature. The filters were removed after 2 days and the water was examined with an inverted bright-field microscope over a 1-month period to follow the appearance of different types. The amoebae obtained by centrifugation were prepared for cytological study, and identification was made on gross and nuclear morphology. The diversity of the cell forms obtained is shown in Plate 23-1. The species not identified by Sawyer were the smaller forms of limax amoebae shown in Plate 23-1,*E*. They range from 3 to 15 μm in length and creep forward with a single large hyaline pseudopodium.

Some bacteria-eating amoebae, which populate bacterial films, are also known as amoeboflagellates, since they have a transient flagellated stage as well as spherical resistant cysts. They belong to genera, such as *Vahlkampia* and *Hartmannella*, which are difficult to distinguish. The problems in amoeba taxonomy and the criteria to be used for limax amoebae were discussed in detail by Page (1967). Information on the distribution and activity of this group on fouling surfaces should be of major interest to ecologists. As is usually the case, the more distinctive forms rather than the more numerous or important, receive the most attention.

Schaeffer (1926), Grell (1966), and Page (1967, 1971b, 1973) all stress the importance of proper identification. The first survey made by Sawyer (1971), which depended upon the appearance of stained preparations

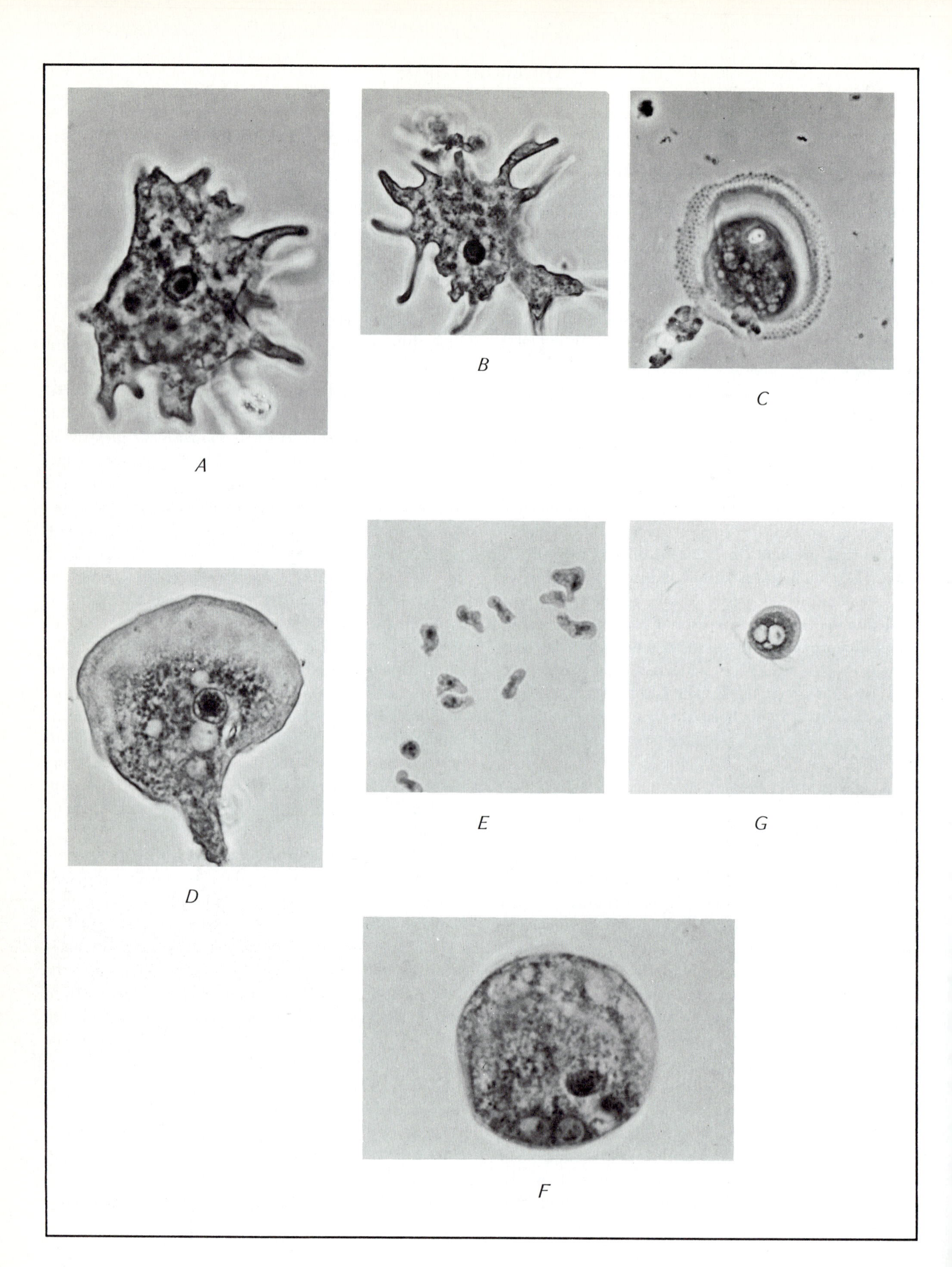
A
B
C
D
E
G
F

PLATE 23-1. Diversity of amoebae cultured from the upper Chesapeake Bay, as shown in a compressed state and stained with Biebrich scarlet-fast green. Amoebae shown are species of (*A*) *Mayorella*, (1,400 X) (*B*) *Vexillifera* (1,040 X), (*C*) *Cochlipodium* (1.400 X), (*D*) *Vannella* (1,300 X), (*E*) small limax-type amoebae (1,200 X), (*F*) *Thecamoeba* (1,900 X), and (*G*) *Platyamoeba* (1,200 X). From T.K. Sawyer, 1971, *Trans. Amer. Micros. Soc.* 90:43–51.

made from crude cultures, was informative but lacked the sound taxonomic basis that would have been provided by the isolation and maintenance of clones, which could have been used for cytological studies as well as for observations on locomotion. Both enrichment and agar cultures were used in the subsequent study by Sawyer (1975) on the seasonal occurrence of amoebae in Chincoteague Bay, Virginia.

Some of the genera illustrated in Plate 23-1 will be considered here, to show their diversity of form and feeding habits. The *Trichamoeba* are a heterogeneous group distinguished by a posterior process (uroid), which is also present in other genera. They range in size from the 20 to 30 μm long *T. sphaerarum* to the 300 μm long *T. schaefferi* (Radir, 1927). The *Trichamoeba* feed on bacteria, diatoms, and flagellates. The genus *Flabellula* was created by Schaeffer for amoebae that have a triangular or fan-shaped outline in active locomotion. The genus includes species without and with uroids that are composed of trailing filaments. Bovee (1965) subdivided the genus *Flabellula* to separate those forms that often bear uroidal filaments and sometimes form conical and round-tipped pseudopods during locomotion (*Flabellula*) from those forms that do not bear uroidal filaments or form pseudopods (*Vannella*). Sawyer (1971) identified *Vannella* in Chesapeake Bay, and Page (1971a) described two species of *Flabellula* from the Maine coast. *Vannella mira* is reportedly a bacteria-feeder, whereas *Flabellula calkinsi* is a voracious cannibal. The genus *Mayorella* was created for amoebae that are distinguished by numerous small conical pseudopods, which continuously form along their anterior edge and on their free surface during locomotion. *Mayorella conipes*, which is a flagellate- and diatom-feeder, was present in all the collections of both Schaeffer and Sawyer. *Vexillifera* are distinguished by long, slender pseudopods that extend from the anterior end during locomotion, which is accomplished by tentacle-like movements. *Vexillifera aurea*, probably a diatom-feeder, was present in the collections of Schaeffer and Sawyer. Other species have been described from British Columbia (Wailes, 1932), the U.S. Pacific Coast, and the Gulf of Mexico (Bovee, 1956).

Amoebae of the genus *Platyamoeba*, which was created by Page in 1969, have an ovoid outline in locomotion and a relatively constant body shape, but lack pseudopods. Sawyer (1975) established the genus *Clydonella* to include amoebae whose locomotive forms are more highly variable than those in *Platyamoeba;* otherwise, these two genera are similar. The "pellicle amoebae," placed in the genus *Thecamoeba*, have an apparently stiffened or tough and wrinkled surface layer. Members of this genus should not be confused with the Thecamoebiens, a French term for the testaceans (shelled amoebae). *Thecamoeba* have been observed in Florida, Chesapeake Bay, Long Island, and Maine. Page (1971b) has done a comparative study of both freshwater and marine species. *Hyalodiscus* are small ovoid amoebae with the distinctive habit of rapidly gliding over the substrate without much change in shape. *Cochliopodium* have an irregular, flattened-disc shape and a very stiff ectoplasm covered with scales, so that there is little change in shape during locomotion. The family Stereomyxidae contains amoebae with highly branched pseudopodia. Grell (1966) described species of *Stereomyxa* and *Corallomyxa* obtained from corals at the marine station at Madagascar. The micrographs of these large amoebae are very impressive. The ultrastructure of two species of *Stereomyxa* has been described by Benwitz and Grell (1971a,b). Page (1972) has described the genus *Rhizamoeba* on the basis of its limax locomotion, form, mitotic patterns, and type of uroid structures.

The genus *Paramoeba* includes a widespread and relatively homogeneous group of amoebae distinguished by one or several *Nebenkörper* (associated bodies) attached to the nucleus (see Plate 23-2). The first species described, *Paramoeba eilhardi*, was detected in a seawater aquarium in the Berlin Zoological Institute by Schaudinn (1896). In addition to the *Nebenkörper*, his illustrations show ingested bacteria and pennate diatoms. A *Paramoeba* found in a small seawater aquarium in Rio de Janiero by Gomes de Faria, da Cuña, and Pinto (1922) was described as *P. schaudinni*, although Chatton (1953), in discussing the cultures he and Biecheler obtained from algal inocula at Sete, concluded that all these amoebae were the same species.

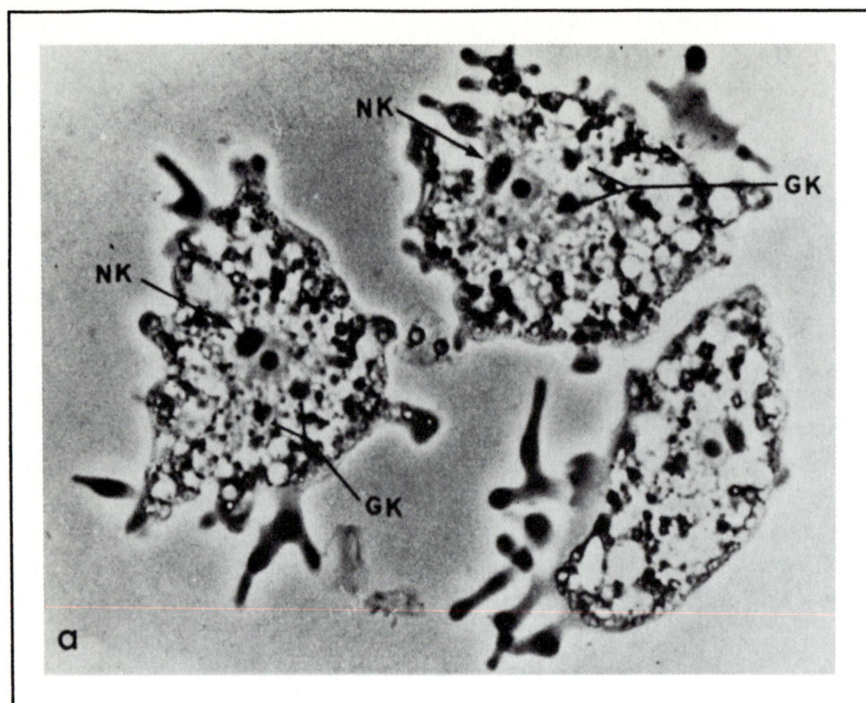

A

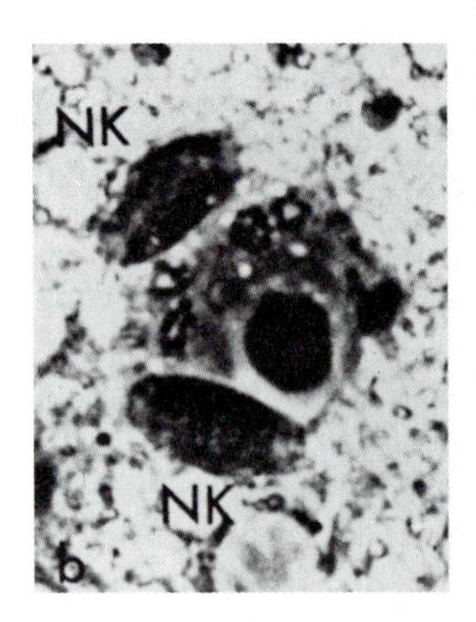

B

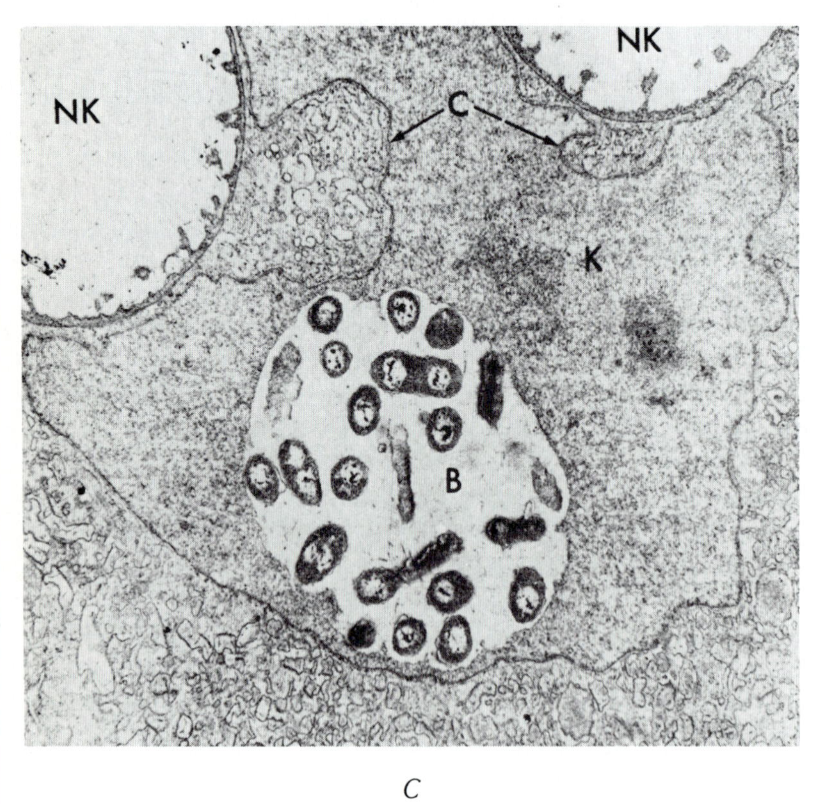

C

PLATE 23-2. *Paramoeba eilhardi*, showing its distinctive *Nebenkörper* (cell-like associated body). (*A*) Living cells in phase contrast, showing *Nebenkörper* (NK) attached to nucleus and Golgi bodies (GK) (900X). (*B*) A nucleus infected with bacteria and two *Nebenkörper* (magnification not given). (*C*) Transmission electron micrograph of a thin section showing (*B*) in detail, the bacteria (B) infecting the nucleus (K), which is attached to the cell-like *Nebenkörper* (NK) (12,000 X). From K.G. Grell and G. Benwitz, 1970, *Arch. Protistenk.* 112:119–37.

Paramoeba eilhardi was isolated and cultured from algal material as well as sand and its associated detritus at Villefranche (Grell, 1961), Naples, and Madagascar (Grell and Benwitz, 1970). Although Schaudinn showed *Nebenkörper* dividing with the nucleus, such division is not strictly synchronized with nuclear division (Page, 1973). Grell and Benwitz (1970) studied the ultrastructure of *P. eilhardi* and concluded, with reservations, that the *Nebenkörper* is a cell rather than a nucleus. This study also showed that, in addition to bacteria that occur regularly in the cytoplasm as symbionts, there is a pathogenic bacterium that attacks the nucleus and kills the cell (Plate 23-2,*B*,*C*). At least one species of the genus *Paramoeba* is parasitic, causing gray crab disease in the blue crab *Callinectes sapidus* (Sprague, Beckett, and Sawyer, 1969). Page (1970) used agar plates supporting bacterial growth to isolate and maintain cultures of *Paramoeba* from sand and detrital material from the coast of Maine; these were described as *P. pemaquidensis* and *P. aesturina*. Both the "duckfoot" locomotive and the "radiate" floating forms of *P. aesturina* are shown in Plate 23-3,*C*,*D*. *Paramoeba pemaquidensis* has subsequently been isolated from British waters and the North Sea by Page (1973), who concluded that the reason this genus is not reported more often is that it has not been sought or recognized.

Most research on marine amoebae has concerned estuarine or coastal forms. Davis, Caron, and Sieburth (1978), however, have demonstrated the ubiquitous presence of amoebae in the waters of the North Atlantic, even in deep abyssal water. They observed eleven genera of amoebae, the most common being *Acanthamoeba*, *Platyamoeba*, and *Clydonella*. Living cells of species in these genera are shown as locomotive, floating, and cyst forms in Plate 23-3. Population estimates ranged from 0.03 to 149 amoebae/liter, with the higher densities occurring in the sea–air interface (Sieburth et al., 1976). With increasing interest in amoebae, taxonomic problems are being resolved and new keys have been prepared by Page (1976a,b) and Bovee and Sawyer (1979).

B. TESTACEANS

The amoeboid protoplasm of testaceans is enclosed within a monolocular or monothalamic (single-chambered) shell, which the amoeba transports much as a snail carries its shell. This shell or test (hence the name of the order) may consist of a "pseudo-chitinous" base, as in *Arcella;* it may be reinforced by cemented sand grains or other debris, as in *Difflugia;* or it may be composed of siliceous platelets, as in *Euglypha*. A single large aperture in the test permits the pseudopodia, as well as the products of division, to emerge. The testaceans are typically asexual and reproduce by binary fission. Their pseudopods typically have rounded extremities, a form known as lobopodia. But some testaceans, such as *Euglypha*, have finer, more filamentous pseudopods known as filopodia. These filopodia can branch and anastomose to form a net-like structure similar to that of the foraminifera. An example is *Gromia*, a monolocular testacean commonly found on the undersurfaces of stones or on the holdfasts of kelp in the lower intertidal zone. It is often confused with its look-alike, the foraminifer *Allogromia*. These misidentifications in the literature were unscrambled by Hedley (1958b). *Gromia* has non-granular pseudopodia that rarely anastomose, whereas *Allogromia* produces granular, anastomosing pseudopodia. Another difference is in the life cycle. *Gromia* has a fusion of adult tests and an exchange of gametes from different parents to produce a very distinctive zygote, whereas *Allogromia* reproduces by multiple fission and a pairing of amoeboidal gametes (Arnold, 1955). Bolli and Saunders (1954) discuss how a number of other testaceans are also erroneously described as foraminifera.

Most texts and monographs (Deflandre, 1953) state that the testaceans, which are ubiquitous in freshwater and often studied by naturalists, are rare in the marine environment. Wailes (1927, 1937) notes the occurrence of freshwater forms in a marine habitat and discusses several species found in the beach sand and in the plankton in British Columbia. Valkanov (1970) describes a number of testaceans from the brackish

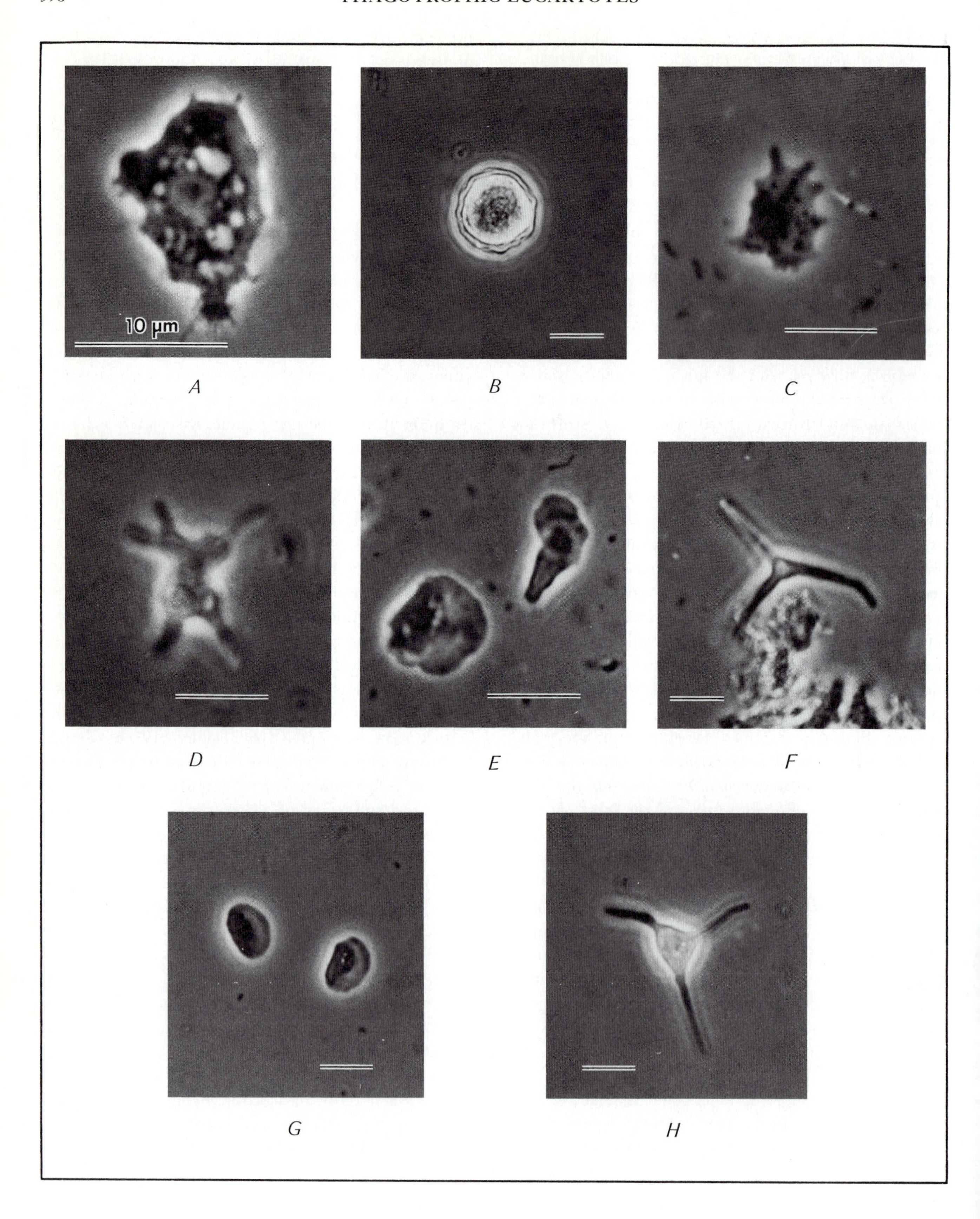
10 μm
A
B
C
D
E
F
G
H

PLATE 23-3. Dominant genera of open ocean amoebae, showing living cultures in locomotive, floating, and cyst forms by phase-contrast microscopy. Locomotive (*A*) and cyst (*B*) forms of *Acanthamoeba polyphaga;* locomotive (*C*) and floating (*D*) forms of *Paramoeba aesturina;* locomotive (*E*) and floating (*F*) forms of *Clydonella vivax;* locomotive (*G*) and floating (*H*) forms of *Platyamoeba murchelanoi* (all bars = 10 μm). From P.G. Davis, 1976, Master's Thesis, University of Rhode Island.

waters of the Black Sea, including *Pseudocorythion acutum,* which was originally observed by Wailes. Golemansky (1970a,b,c,d; 1974a,b,c), who had previously reported on the occurrence of these forms in Lac d'Ohrid, also found them to be a part of the psammon (sand fauna) of the littoral zone of the Bulgarian coast of the Black Sea—where the waters are brackish, with salinities mainly between 1 and 8 o/oo. Several species with different types of tests are shown in Plate 23-4, and Bonnet (1975) has proposed a morphological classification of the shells of testaceans. The feeding habits of testaceans have been studied by both light microscopy and transmission electron microscopy (Haberey, 1973a,b). Electron micrographic surveys of microbial films and marine surfaces (Sieburth, 1975; Johnson and Sieburth, 1976) indicate that testacean-like forms occur on the bacteria-laden and textured surfaces of such animals as scale-worms and bryozoans and also on the filters of recirculating aquaculture systems. The latter forms, however, appear to be sessile rather than motile and may belong to an undescribed group of loricate and sessile amoebae. For anyone studying the testaceans, required reading is the delightful monograph by Leidy (1879).

C. FORAMINIFERA

The foraminifera are also shelled amoeboid forms, distinguished from the testaceans by having true anastomosing granular rhizopodia. Most species have multichambered shells, and a number of species have a heterophasic alternation of generations. The foraminifera are essentially marine, with only a few freshwater species. Although most species have been considered nearshore forms, occurring on sandy and muddy bottoms, the protected undersides of rocks, and subtidal grasses and seaweeds, the benthic foraminifera are also important inhabitants of deep-sea sediments. Their rhizopodia are organs of locomotion as well as nets to capture food. Two families of foraminifera, the Globigerinidae and the Globotoralidae, have adapted to a pelagic existence. Their rhizopodia are used for flotation as well as to capture food and to move along surfaces. The planktonic foraminifera are large enough to be seen by an observer with just a face plate snorkeling in sparkling clear blue oceanic water. In such waters of the western Sargasso Sea, planktonic foraminifera appear as small orange balls with fine spikes and are easily recognized along with the greenish red bundles of the cyanobacterium *Trichodesmium.* They occur in sufficient number to be enumerated and captured for further study. After the foraminifera die, the shells of inshore forms become incorporated into the beach sand, whereas the shells of the pelagic forms sink into the sediments to mix with the tests of benthic foraminifera and become part of the fossil record.

The ecology and distribution of modern foraminifera are summarized in texts by Phleger (1960) and Murray (1973). The genera of different types of inshore and offshore environments and the factors that affect their distribution are discussed. The arenaceous forms, with tests containing cemented sand grains, which occur in the muddy and sandy sediments of the inner Oslo Fjord, have been studied by Christiansen (1958), and the standing crops of foraminifera on the subtidal vegetation of a marsh on Long Island, New York, and their seasonal trends, were studied by Lee et al. (1969) and Matera and Lee (1972). The role of foraminifera in binding the coral sand of tropical reef basins and in acting as alternate hosts for several species of dinoflagellate zooxanthellae, which also occur in hermatypic corals, was discussed by Ross (1972), who noted that these zooxanthellae-containing foraminifera and corals have similar geographic distributions. A long-awaited English translation and revision of Boltovskoy's 1965 monograph in Spanish has now appeared (Boltovskoy and Wright, 1976). It is a comprehensive treatment of recent foraminifera and includes discussions on the living organism, its test, benthic and planktonic forms, classification, ecology, collection, and preparation, as well as the application of foraminiferal data to faunal and oceanographic problems.

The numerous small pores in the shells of some species are not simple openings but complex structures. The function of the pores, if there is any, is poorly

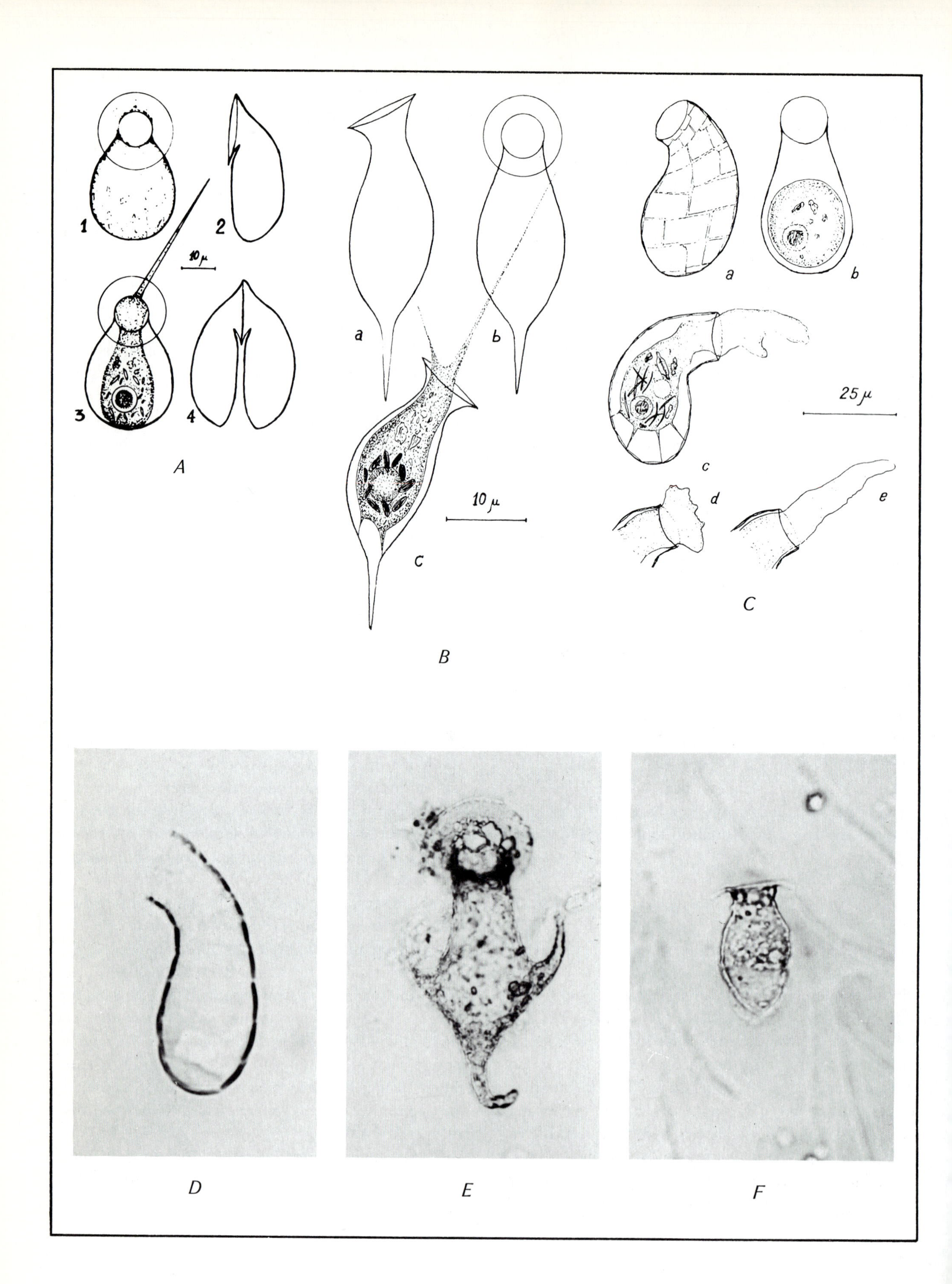
1
2
3
4
10 μ
A
a
b
c
10 μ
B
a
b
c
d
e
25 μ
C
D
E
F

PLATE 23-4. Testaceans from the brackish water sands of the Bulgarian coast of the Black Sea. Habitat sketches of (*A*) *Psammonobiotus communis*, (*B*) *Chardezia caudata*, and (*C*) *Pomoriella valkanovi* and photomicrographs of (*D*) *Pomoriella valkanovi*, (*E*) *Alepiella tricornuta*, and (*F*) *Amphorellopsis elegans*. (*A*) from V. Golemansky, 1970a, *Bull. Inst. Zool. Musee Sofia* 32:63–68; (*B*) from V. Golemansky, 1970b, *Bull. Inst. Zool. Musee Sofia* 32:121–25; (*C*) from V. Golemansky, 1970c, *Protistologica* 6(4):365–71; (*E*) from V. Golemansky, 1970d, *Acta Protozool.* 8(2):41–47; (*D,F*) courtesy of Vassil Golemansky.

understood, although it has been suggested that they could provide a passageway for gases, cytoplasm, or wastes; in the planktonic species, they may also increase buoyancy (Bé, 1968). In some species, there are both small and large pores. The pore system opens to radial and spiral canals in some foraminiferan families.

The rhizopods protrude through the aperture of the shell and anastomose to form the protoplasmic net used for locomotion and feeding (Plate 23-5). Each rhizopod thread has a diameter of approximately 15 μm and is surrounded by a single plasmalemma containing 250-Ångstrom microtubules (with 30-Ångstrom subunits) that probably make them rigid and transport the food particles (Hedley, Parry, and Wakefield, 1967; Marszalek, 1969; McGee-Russell and Allen, 1971) (see Plate 23-5,*B*). These papers and the paper by Dahlgren (1967) on the nucleus also present other ultrastructural details of foraminifera. When feeding, a typical foraminifer, such as *Allogromia laticollaris* (200 to 400 μm in diameter), puts out a radial network of anastomosing rhizopodia (reticulopodia) with active protoplasmic streamers, that is as large as 15 mm (Jahn and Rinaldi, 1959). Such a circular pattern is shown in Plate 23-5,*C*. The cytoplasmic streaming in the cell's interior continues actively in the rhizopodia, the mitochondria and protein granules oscillating back and forth over a limited area. Food particles caught in this net are transported unidirectionally to the interior of the cell by retraction of the cytoplasmic threads. These threads can act together as well as separately. When the rhizopodia are used for locomotion they extend in one direction, sometimes following a "guide rhizopodium," and then contract, pulling the cell in that direction (see Plate 23-5,*A*). The example shown [*Allogromia laticollaris* (Arnold, 1948, 1954b)] is a monothalamic (single-chambered) form common on U.S. seacoasts. In the polythalamic (multi-chambered) forms, the rhizopodia are apparently instrumental in shell formation. After attaching to a substrate, the rhizopodia arrange themselves in the shape of a fan and secrete an organic membrane, which accumulates detrital particles and calcium carbonate to form a new perforated chamber; the chamber is vacant at first, but is subsequently filled with protoplasm (Le Calvez, 1938; Sliter, 1970).

The test structure of foraminifera has been reviewed by Lipps (1973). Of the described species, a relatively small number have chitinous tests, about 21% have agglutinated tests composed of such foreign particles as sand grains and sponge spicules, and some 78% have calcareous tests; at least some of the fossil forms apparently had siliceous tests (Bé and Ericson, 1963). This diversity reflects the environmental conditions under which they develop. Among the features important in taxonomy are the aperture: number, shape, form, and position; the supplementary apertural structures; and ornamentation, such as spines. The use of these and other features are reviewed by Cushman (1959). In their simplest configuration, the shells have one chamber. Most species, however, have shells with many connecting compartments that are added on as the foraminifer approaches maturity. The great diversity of the foraminiferan tests is due to patterns resulting from the different manner in which the chambers are constructed. The calcareous tests are formed from both hexagonal (calcite) and orthorhombic (aragonite) crystals, and the walls may be microgranular, porcellaneous, radial hyaline, finely granular hyaline, monocrystalline, or spicular (Boltovskoy and Wright, 1976). Calcium and strontium mineralization and cycling in relation to growth was studied by McEnery and Lee (1970), whereas Berger (1968a,b) and Berger and Piper (1972) discussed the effect of selective dissolution on distribution and paleoclimatic interpretation. Among the various forms are the *Nodosaria* type, with a linear series of chambers, and the *Rotalia* type, with a spiral of chambers. These may be either planispiral (spiral in a single plane) or trochospiral (spiral in a helix). The *Textularia* type are biserial (braided in appearance), and the *Planorbulina* type consist of interiorly formed concentric circles. Typical polythalamic shells are shown in scanning electron micrographs, Plates 23-6 and 23-7. The plasticity of morphology within one species is illustrated in Plate 23-8, which shows the variations that the usually spherical and monothalamic *A. laticollaris* can undergo.

PLATE 23-5. The foraminiferal rhizopods and their use in locomotion and feeding. (*A*) *Allogromia laticollaris,* using its rhizopods for locomotion across a glass slide (magnification not given). (*B*) Schematic drawing of the internal structure of the rhizopod of *Iridia diaphana,* showing how the outer surface of the plasmalemma becomes the inner surface of the vesicles (vs). Shown are the 250-Ångstrom microtubules (mt), which give rigidity; the mitochondria (m); and a food vacuole (vc). (*C*) A medium-size *Allogromia laticollaris,* with its feeding net extended and trapping microalgal and bacterial food particles (115 X). (*D*) The pelagic foraminifer *Hastigerina pelagica* (1) has drawn a snared copepod into its bubble capsule; a thin section (2) through copepod appendages (A) of a snared *Artemia* nauplius, showing rhizopods (R) near the surface (C) of prey. (*A*) from K.G. Grell, 1973, *Protozoology,* Springer-Verlag, New York and Berlin, (*B*) from D.S. Marszalek, 1969, *J. Protozool.* 16:599–611; (*C*) from J.J. Lee, M.E. McEnery, and H. Rubin, 1969, *J. Protozool.* 16:377–95; (*D*) from O.R. Anderson and A.W.H. Bé, 1976, *Biol. Bull.* 151:437–49.

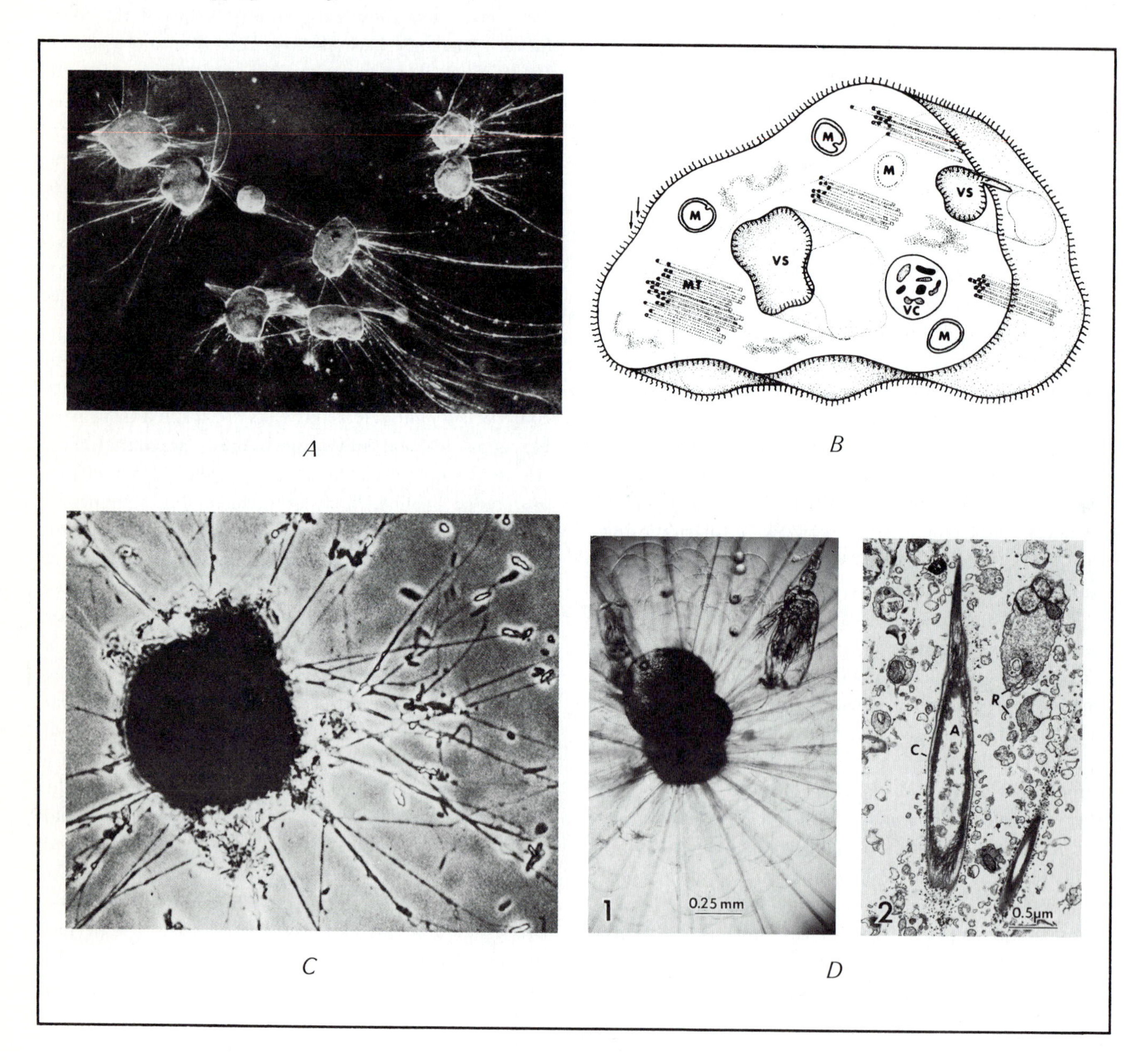

The planktonic foraminifera produce an infinite variety of minute and beautiful shells that sediment to the ocean floor. Beneath the superficial layer of modern foraminifera in the thin upper layer lie progressively deeper layers from earlier geological times, as seen in the long core samples obtained by deep-sea geologists and paleontologists (Plate 23-6). These scientists have constructed a complex taxonomy to describe these skeletons from such prehistoric boneyards (Cushman, 1959; Loeblich and Tappan, 1964). By studying differences in the occurrence, distribution, and form, the paleontologist postulates on how the earth and its seas developed (Ericson and Wollin, 1964; CLIMAP, 1976; Kennett and Watkins, 1976). In the abyssal sediments, indigenous benthonic foraminifera grow and their tests become mixed with those of the planktonic forms that sediment. The differences between the planktonic foraminifera and the abyssal benthonic foraminifera can

PLATE 23-6. Subtropical planktonic foraminifera from the Gulf of Mexico as seen in scanning electron micrographs. (*A*) Strewn sample to show the nature of the unsorted material as well as the diversity of these populations (*B*) Plate to illustrate the species used for the study of foraminiferal biostratigraphy and tephchronology: (1) *Globorotalia menardii;* (2) and (3) *Globorotalia menardii tumida;* (4) *Globorotalia menardii flexuosa;* (5) *Pulleniatina obliquiloculata;* (6) *Globorotaloides hexagona;* (7) *Globigerinoides sacculifer;* (8) *Sphaeroidinella dehiscens;* (9) *Globigerinoides conglobatus;* (10) *Globorotalia truncatulinoides;* (11) "*Globigerina*" *dutertrei;* (12) *Globigerinoides ruber;* (13) *Globigerina falconensis;* (14) *Globorotalia inflata;* (15) *Orbulina universa;* (16) *Hastigerina acquilateralis;* (17) *Globorotalia crassaformis;* (18) *Globigerina calida.* (*A*) courtesy of Charlotte Brunner; (*B*) from J.P. Kennett and F. Huddlestun, 1972, *Quaternary Res.* 2:36–69.

A

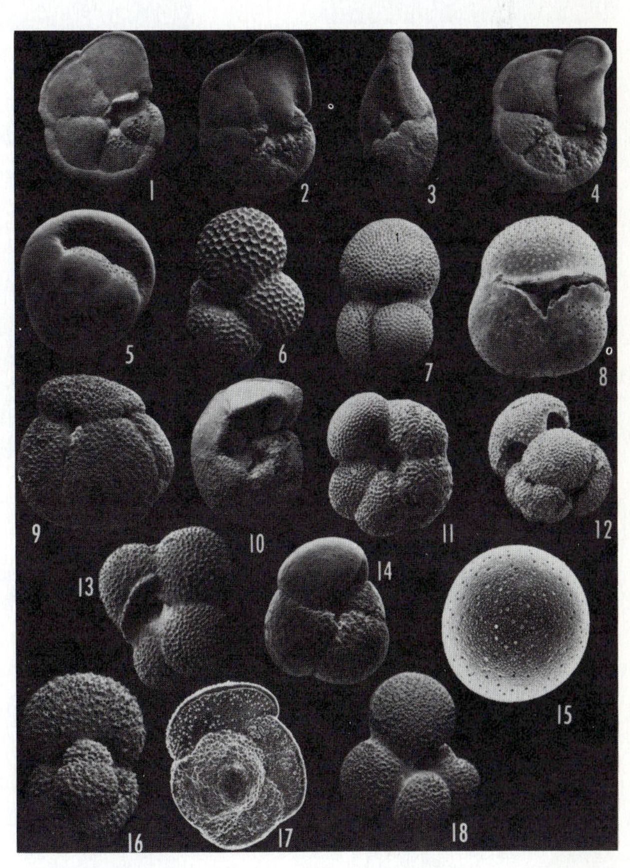

B

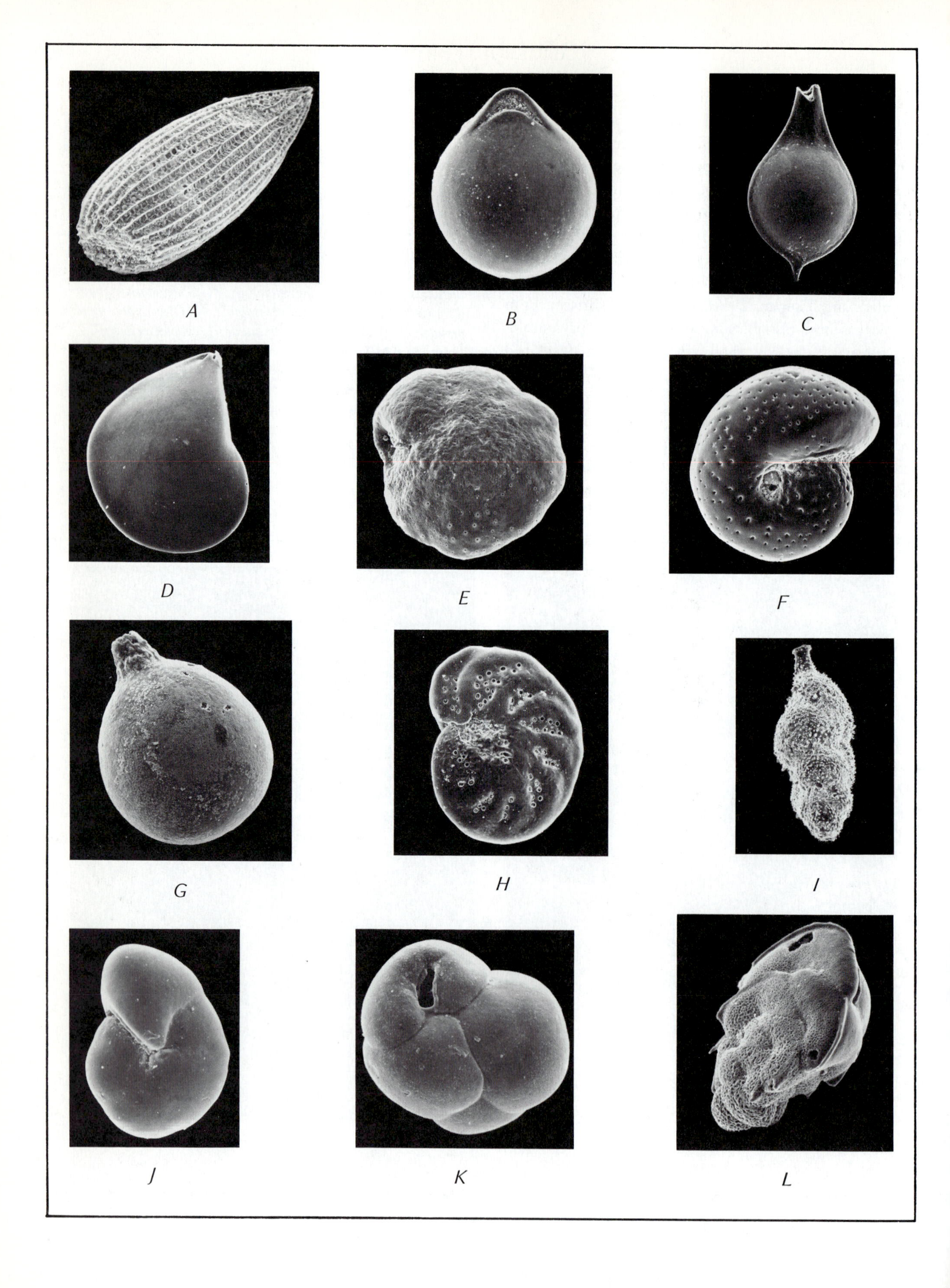
A
B
C
D
E
F
G
H
I
J
K
L

PLATE 23-7. The tests of extant species of abyssal benthonic foraminifera from the Southeastern Indian Ocean, scanning electron micrographs showing a variety of test architecture. (*A*) *Lagena fieldeniana*, (*B*) *Parafissurina* sp., (*C*) *Fissurina crebra*, (*D*) *Lenticulina* sp., (*E*) *Epistominella umbonifera*, (*F*) *Melonis pompilioides*, (*G*) *Oolina globosa*, (*H*) *Cibicides wuellerstorfi*, (*I*) *Uvigerina* sp., (*J*) *Pullenia simplex*, (*K*) *Globocassidulina subglobosa*, (*L*) *Ehrenbergina bradyi*. (size range approximately 30 to 70 X). Courtesy of Bruce Corliss.

be seen by comparing their architecture as seen in Plates 23-6 and 23-7, respectively. The distribution and morphology of benthonic foraminifera is influenced by the mineralogical, textural, and chemical nature of the sediment that serves as their substrate.

Ecological studies on foraminifera are important not only for understanding modern species, but also for the paleoecological interpretation of the fossil record. Murray (1973) discussed the distribution and ecology of foraminifera throughout the shallow and deep-water environments, one geographical area at a time, whereas Lee (1974) characterized the niche of the foraminifera. Boltovskoy and Wright (1976) divide ecological factors into (1) the major parameters of temperature, salinity, and water depth, (2) the harder to document influences of nutrition and the substrate upon which they occur, and (3) the poorly studied factors of light, turbidity, pH, dissolved oxygen, and trace elements. Each species has temperature limits for survival, growth, and reproduction. The relationship between temperature and the vital activities of natural populations of planktonic foraminifera have been studied by Bé and Tolderlund (1971). Temperature affects the distribution as well as the morphology of the test. The larger specimens are generally from the colder waters, and Kennett (1968) has described the morphological changes that occur in going from colder to warmer waters. As one might expect, the salinity tolerance for planktonic species is much less than the salinity tolerance for benthonic species. There are depth zonations that cannot be accounted for by temperature, with some species having a wide tolerance range and others a narrow.

The implications of the biological variation observed in modern foraminifera for the taxonomy of these relics is discussed by Arnold (1968). The taxonomy of the modern foraminifera is reviewed in the treatise by Loeblich and Tappan (1964), which was updated by Loeblich and Tappan (1974). The modern foraminifera have a number of oceanographic applications (Phleger, 1960). Species with restricted habitats are good indicators of water movement (Boltovskoy, 1959). Living foraminifera in the water column have also been used to characterize water masses and even to denote the Antarctic convergence (Boltovskoy, 1969, 1971). They have been used to show that a microfauna can be impoverished by traces of lead washed in from land (Boltovskoy, 1956). Foraminifera have been used to study relative and absolute sedimentation rates and changes in sea level, as well as bathymetry for ancient seas. Paleoclimatology has been elucidated by studying species distribution, coiling direction and other morphological changes, and the porosity of the tests of planktonic species. Foraminifera have also been used to determine postsalinity levels and stratigraphy. Environmental reconstruction based on foraminifera has been used by the petroleum industry for many years and has been a big factor in the development of paleoecology.

The diversity of forms that have noncalcareous tests is intriguing. A large naked foraminifer, *Boderia turneri*, was picturesquely described by Wright (1867). Another century elapsed before Hedley, Parry, and Wakefield (1968) confirmed his original description. Numerous individual foraminifera up to 8 mm long live on the intertidal calcareous alga *Corallina officinalis* (often in empty serpulid worm tubes) and in shallow water sediments near Plymouth, England. This highly pleomorphic form (see Plate 23-9), which reproduces asexually by multiple fission, also has a sexual stage in which the gametes are heterokont flagellates similar to those found in the Chrysophyceans (Chapter 13,A), the zoosporic phycomycetes (Chapter 21), and the brown seaweeds (Loiseaux and West, 1970). Another unusual foraminifer is *Nemogullmia longevariabilis*, which appears as short white threads (1 to 19 mm long, unextended) in the detrital layer of soft bottoms. Observations in culture permitted its identification as a monothalamous form in the family Allogromidae (Nyholm, 1953). The life cycle and cytology of this foraminifer, which is very sensitive to unfiltered light, has been studied by Nyholm (1956).

In contrast to the naked forms and the forms with unadorned chitinous tests, the arenaceous foraminifera with their debris-laden houses are bizarre and reminiscent of decorator crabs. A monograph on arenaceous foraminifera in the sediments of Dröbak Sound in the Oslo Fjord (Christiansen, 1958) illustrates a variety of this type of foraminifer, some of which are shown in

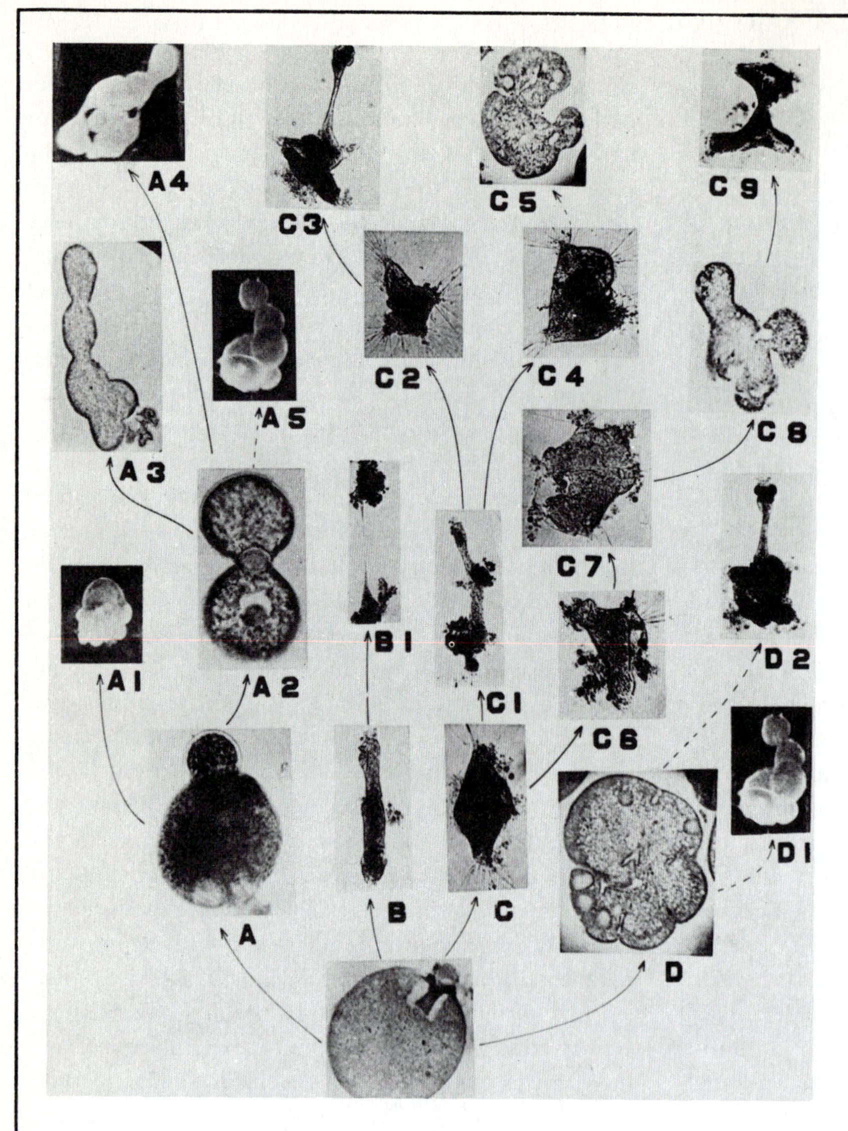

A4
C3
C5
C9
C2
C4
A5
A3
C8
C7
B1
A1
A2
D2
C1
C6
D1
A
B
C
D

A

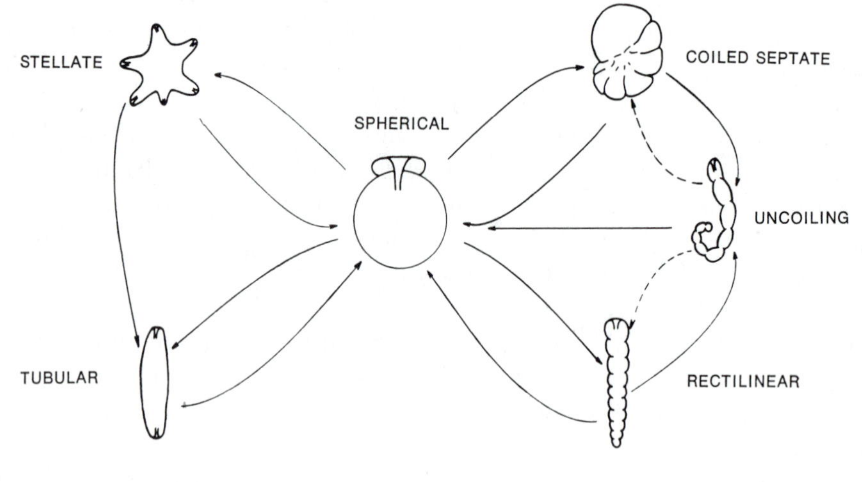

STELLATE
COILED SEPTATE
SPHERICAL
UNCOILING
TUBULAR
RECTILINEAR

B

PLATE 23-8. The variation of *Allogromia laticollaris* in laboratory culture as shown by micrographs (*A*) and a schematic (*B*) to illustrate how the common spherical form has the morphological plasticity to produce a small number of isomorphs of the "higher" foraminiferal forms. From Z.M. Arnold, 1954b, *Contributions Cushman Foundation for Foraminiferal Research* 5:78–87.

Plate 23-10. Some just cover themselves with mud, others selectively pick up quartz grains and similar debris, and still others build their shells from the empty shells of other foraminifera. The problems of collecting and classifying such forms have been discussed by Christiansen (1958). *Astrorhiza linicola* is a very large arenaceous form up to 1 cm in diameter, monothalamic and stellate, which occurs in British coastal waters. Its biology has been described by Buchanan and Hedley (1960). Such forms as *Normania confertum* (Christiansen, 1964) and *Haliphysema tumanowiczii* (Hedley, 1958a) have sedentary states suggestive of sponges, with which they were classified at one time. The presence of sponge spicules on the tests of some foraminifera probably added to the confusion. *Haliphysema tumanowiczii* is commonly found on laminarian holdfasts and seaweed fronds, polyzoans, hydroids, stones, and unoccupied shells. The biology and cytology of both the sedentary and the unattached forms have been described by Hedley (1958a).

Much of our knowledge of modern foraminifera is due to the amenability of a number of inshore species in different families to maintenance in mass cultures, as reviewed by Arnold (1974) and Lee (1974). Simple equipment and methods can be used to maintain masses of foraminifera. Agnotobiotic cultures, in which the organisms that accompany the object of culture are not characterized, are the easiest to obtain. These are appropriate for life cycle and some physiological studies. The agnotobiotic stocks of *A. laticollaris* obtained by Arnold (1948) are some 30 years of age. Such cultures have been used to observe movement and dispersal and the mechanism of foraminiferan distribution (Arnold, 1953), as well as morphological plasticity (Arnold, 1954c), which has shown that some species are synonyms or varieties of other species.

Studies on foraminifer food require that all the organisms present in the culture be identified. Such cultures are gnotobiotic or synxenic. Axenic cultures are pure cultures free of all other organisms. Attempts to obtain axenic cultures with a species of *Allogromia* (strain NF) have failed (Lee and Pierce, 1963). Bacteria appear to be indispensable for continuous reproduction in foraminifera (Muller and Lee, 1969). Useful information on the natural feeding patterns has been obtained by the use of tracer techniques with synxenic cultures (Lee et al., 1966). Selected species of diatoms, chlorophytes, and bacteria were ingested in large quantities, while yeasts, cyanobacteria, dinoflagellates, chrysophytes, and most heterotrophic bacteria were ignored. Variables in feeding habits were the "age" of both predator and prey as well as the concentration of the prey. The effect of diet on growth and reproduction was also observed by Lee, McEnery, and Rubin (1969). Mixtures of algal species and single algal species with moderate numbers of bacteria were highly productive and superior to single algal species without bacteria.

It has been assumed that the prey of foraminifera must be passive enough for the foraminifera to catch with their weak pseudopodia. But Lee et al. (1961, 1963) have noted a "Circean effect" (by analogy with the enchantress Circe in Homer's *Odyssey*) whereby foraminifera will attract and hold captive active food organisms. When individual organisms of the foraminifera *Streblus beccarii*, *Miliammina fusca*, or *Bolivina* sp. were placed in a culture of *Dunaliella parva*, a large percentage of the cultured phytoflagellates stopped moving randomly and swam toward the foraminifer and its pseudopodial net. Upon reaching the pseudopodial net, they stopped swimming, became attached to the exterior cytoplasm, and were lysed.

The pelagic species of foraminifera are believed to feed on microalgae, other protozoa, and even copepods, but this has not been confirmed in culture since, apparently, no one has successfully cultivated planktonic foraminifera from the open sea (Bé and Ericson, 1963; Bé and Anderson, 1976; Anderson and Bé, 1976). The colored drawings of planktonic forms made during the Humboldt Expedition show pinkish striated particles that Rhumbler (1909) identified as copepod muscle fibers. They may have been, but Lee et al. (1965) have shown that they could also have been part of an extensive internal fibrillar system of the foraminifer. But copepods are preyed upon by planktonic foraminifera. The use of the Bermuda Biological Station, which offers quick access to the open sea, has permitted Anderson and Bé (1976) to study the feeding processes of the planktonic foraminifer *Hastigerina pelagica* on cope-

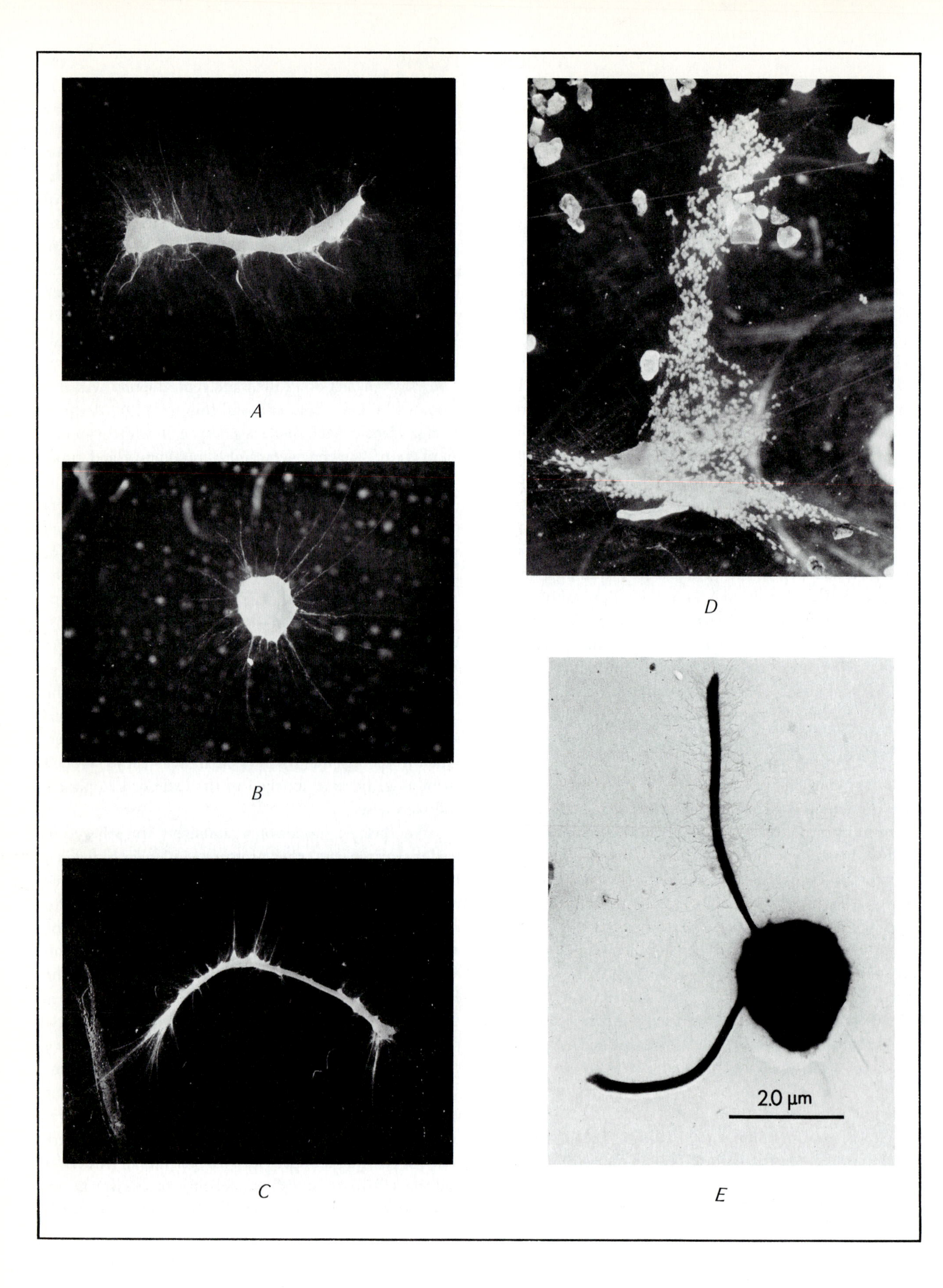
A
B
C
D
2.0 µm
E

PLATE 23-9. The large naked foraminifer *Boderia turneri*. (*A*), (*B*), and (*C*) Pleomorphic forms (8 X). (*D*) The fragmentation phase of gametogenesis before gamete liberation. (*E*) The heterokont biflagellate gamete of *B. turneri*, which is similar to the zoospores of some phycomycetous fungi, the cells of some chrysophycean microalgae, and the spores of phaeophyte macroalgae. From R.H. Hedley, D.M. Parry, and St. J. Wakefield, 1968, *J. Nat. Hist.* 2:147–51.

pods and on the brine shrimp *Artemia salina*. The rhizopodial network extends beyond the periphery of the organism's shell to form a sticky, three-dimensional web, which efficiently tangles a prey one-half its size; the prey may be narcotized as indicated by the rapid cessation of its struggle. There is a remarkable selective activity of the rhizopodia during the capture, engulfment, and dismemberment of the prey. Some rhizopodia sever large masses of cuticle, which is discarded, from the larger prey while the digestible soft tissue is transported into food vacuoles in the intercellular cytoplasm. Smaller prey are ingested whole.

Symbiotic algae can contribute to the nutrition of both benthic and planktonic foraminifera. Such a symbiotic association of microalgae with invertebrates and protozoa was discussed in Chapter 3,C. Recently, Lee (1974) has reviewed the algal symbionts of foraminifera. The symbiont zooxanthellae of the planktonic species *Globigerina bulloides* and *Globigerinoides ruber* are dinoflagellates that resemble *Symbiodinium microadriaticum* (Lee et al., 1965), the symbiont that lives in corals, anemones, and jellyfish. The symbiont of the large benthic species *Sorites marginalis*, which grows on turtlegrass (*Thalassia testudinum*) in the water of the Florida Keys, also resemble *S. microadriaticum* (Müller-Merz and Lee, 1976). But another large foraminifer, *Archaias angulatus*, in the same environment, has a chlorophyte, *Chlamydomonas hedleyi*, as its symbiont (Lee et al., 1974) indicating the diversity of foraminiferan symbionts. The contribution of the symbionts to the nutrition of foraminifera may be habitat dependent. A study of natural populations of both *Archaias angulatus* and *Sorites marginalis*, with their algal symbionts and tracer-labeled algal food, clearly indicates that, in these species, feeding is the more important nutritional process (Lee and Bock, 1976). At midday, the ratio of feeding to primary activity was at least ten to one. But the large benthic species *Heterostegina depressa* found in the clear water of shallow tropical seas (growing on solid substrates and, epiphytically, on algae), such as those of Hawaii, is able to grow without particulate food, living on its light-dependent algal symbionts (Röttger and Berger, 1972; Röttger, 1976). During multiple fission of the megalospheric gomonts of *H. depressa*, as the protoplasm separates into juveniles outside the test, symbiotic algae from the parent protoplasm form a veil around and enter the juveniles at both the first and second chamber stages (Röttger, 1974). These algal symbionts are diatoms (Schmaljohann and Röttger, 1976).

Foraminifera have both asexual (apogamic) life cycles, in which reproduction is by schizogamy or multiple fission of the diploid state, and sexual life cycles, which involve haploid gametes and an alternation of generations between the two forms. Species that have apogamic life cycles have been discussed by Arnold (1968). Here, the diploid form, also known as the agamont, asexual, schizont, or microspheric form, is produced by the zygote (or as a result of multiple fission); it usually has a large test with an initial chamber (proloculus), which is relatively small (0.02 mm). Through metagamic division, the agamonts become multinuclear (usually before hatching). In the homokaryotic foraminifera, all the nuclei in the agamonts are alike and all participate in meiosis. The heterokaryotic foraminifera have two types of nuclei (nuclear dimorphism). The somatic nuclei grow in size and form a nucleolus, but they cannot divide, and ultimately, they disintegrate at meiosis. The generative nuclei, which remain small and condensed, carry out meiosis and divide. This division of nuclear and cytoplasmic material by the agamont yields the haploid form, also known as the gamont, sexual, sporozont, or megalospheric form. These forms usully have a small test and a relatively large proloculus (0.2 to 0.5 mm). They have one nucleus that divides mitotically to form many small nuclei, which, with the subdivided cytoplasm, become either amoeboid or flagellated gametes.

The alternation of generations between the diploid asexual form in agamogony and the haploid sexual form in gamogony is shown in Plate 23-11 for three species with different modes of fertilization. Information can only be obtained from observations on cultures, such as those made on *Allogromia laticollaris* by Arnold (1955). Subsequent studies by Arnold and those by Lee and colleagues (Lee et al., 1963; Lee, McEnery, and Rubin, 1969; Lee and McEnery, 1970; McEnery and Lee,

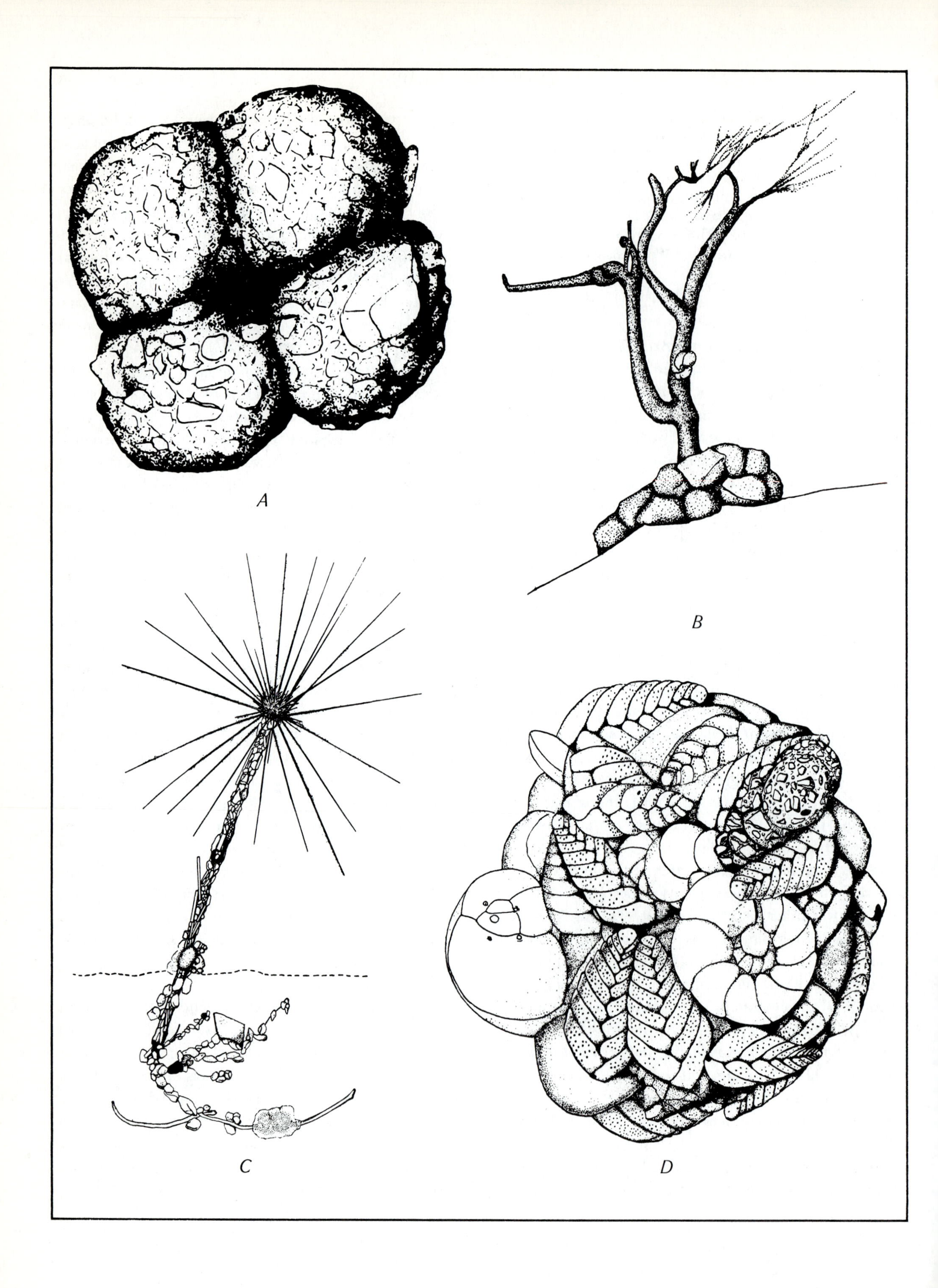
A
B
C
D

PLATE 23-10. Examples of arenaceous foraminifera from Dröbak Sound, Oslo Fjord. (*A*) *Sorosphaera confusa*, with its thin, sandy walls (380 X); (*B*) *Dendrophyta erecta*, with its basal chamber and irregularly branched chitinous tubes thickly coated with mud (17 X); (*C*) *Marsipella arenaria*, with its tubelike test built up of large quartz grains, some mineral grains, and a few sponge spicules, shown extending up from the mud (4 X); (*D*) *Psammosphaera testacea*, a monothalamic form, has a wall composed of the empty tests of other foraminifera (30 X). From B. Christiansen, 1958, *Nytt. Mag. Zool.* 6:5–91.

1976), and by Grell (1954, 1957, 1959) further show the alternation of generations, nuclear dimorphism, and the variables that occur (Grell, 1973). But the life cycles of even such well-studied species as these may not be adequately described. Only recently, an unusual apogamic metagenic life cycle has been described for *Allogromia laticollaris* by McEnery and Lee (1976), in which both the uninucleate and multinucleate diploid cycles were dominant over the sexual cycle (see Plate 23-12).

D. HELIOZOA

The heliozoa are spherical cells with long fine axopodia (stiff pseudopods) radiating in all directions. This similarity to early paintings of the sun with all its rays probably inspired the early naturalists to call these rhizopods the "Sun Animalcules" (see Plate 23-13, *A*,*B*,*C*). The most striking structure of the heliozoans, their axopodial filaments, is used for the capture of food and for locomotion, as in other rhizopods. A distinctive system of microtubules differentiates this group of amoebae from all the others. Most of our knowledge of the fine structure of this system in the heliozoa comes from the freshwater forms used to study the mechanism of microtubule degradation and reformation (Tilney and Byers, 1969; Roth, Pihlaja and Shigenaka, 1970; Shigenaka, Roth, and Pihlaja, 1971; Bardele, 1972). By contrast marine heliozoa in the order Centrohelida have a very distinctive axopodial stereoplasmic pattern (see Plate 23-13,*D*,*E*,*F*). Instead of a double-coiled array, as in the freshwater heliozoan *Echinosphaerium* (*Actinosphaerium*) *nucleofilum* (Roth, Pihlaja, and Shigenaka, 1970; Roth and Shigenaka, 1970), they have a pattern of polyhexagons (Febvre-Chevalier, 1973a,b, 1975). This has been shown in two marine species, one motile and the other sessile, both from the region of Villefranche-sur-Mer in the French Mediterranean. *Gymnosphaera albida* is a motile species living in groups of up to 20 individuals, which undergo temporary fusion for paired feeding (Febvre-Chevalier, 1973a, 1975). The other is *Hedraiophrys hovassei*, a sedentary form with siliceous spicules, which lives for long periods on a substratum but has a free swimming stage for short periods of time (Febvre-Chevalier, 1973b). The motility of free-living heliozoa appears to be due to forces generated by axopodal retraction, which gives a definite rolling motion, rather than by apoxodal elongation or rowing or bending motions (Watters, 1968). The cytoplasm of the heliozoa is frequently divided into a dense endoplasm with one or more nuclei and a vacuolated ectoplasm. The absence of a central capsule to enclose the endoplasm distinguishes them from the radiolaria, which also have characteristic skeletons of silicon. The free-living heliozoans are found suspended in shallow protected waters; the attached forms are characterized by a slender, non-contractile stalk. A few pelagic forms are found in the plankton.

Heliozoan feeding has been described by Griessmann (1914), Looper (1928), and Bovee and Cordell (1971) (see Plate 23-14). It involves a hyaline ectoplasm that forms special pseudopodia for different types of prey. Small motionless food objects are captured by a small, straight pseudopod, which, upon contact, forms a cup or a dipper in order to engulf the object. Larger, motionless food objects elicit a large, wide, hyaline outgrowth, which advances upon the object and then expands in all directions to engulf it. Active prey is captured in a large, sac-like pseudopod, escape being cut off by the axopods. Bovee and Cordell (1971) described how prey coming in contact with the adhesive protoplasmic surface of some axopodal tips are pulled against the body surface by the retracting axopods. The adjacent axopods then enclose the prey in cytoplasmic pockets. Some species prey just on ciliates, whereas others are omnivorous, ingesting a variety of foods, such as yeasts, ciliates, and flagellates and even metazoa (Bovee and Cordell, 1971). Griessman, Looper, and Bovee and Cordell noted that, for such larger prey, a number of individuals feed cooperatively by temporarily fusing to form one predator (see Plate 23-14,*A*) that digests the prey, then separates into individuals. Some species of heliozoa have symbiotic intracellular algae. The heliozoa reproduce by binary fission,

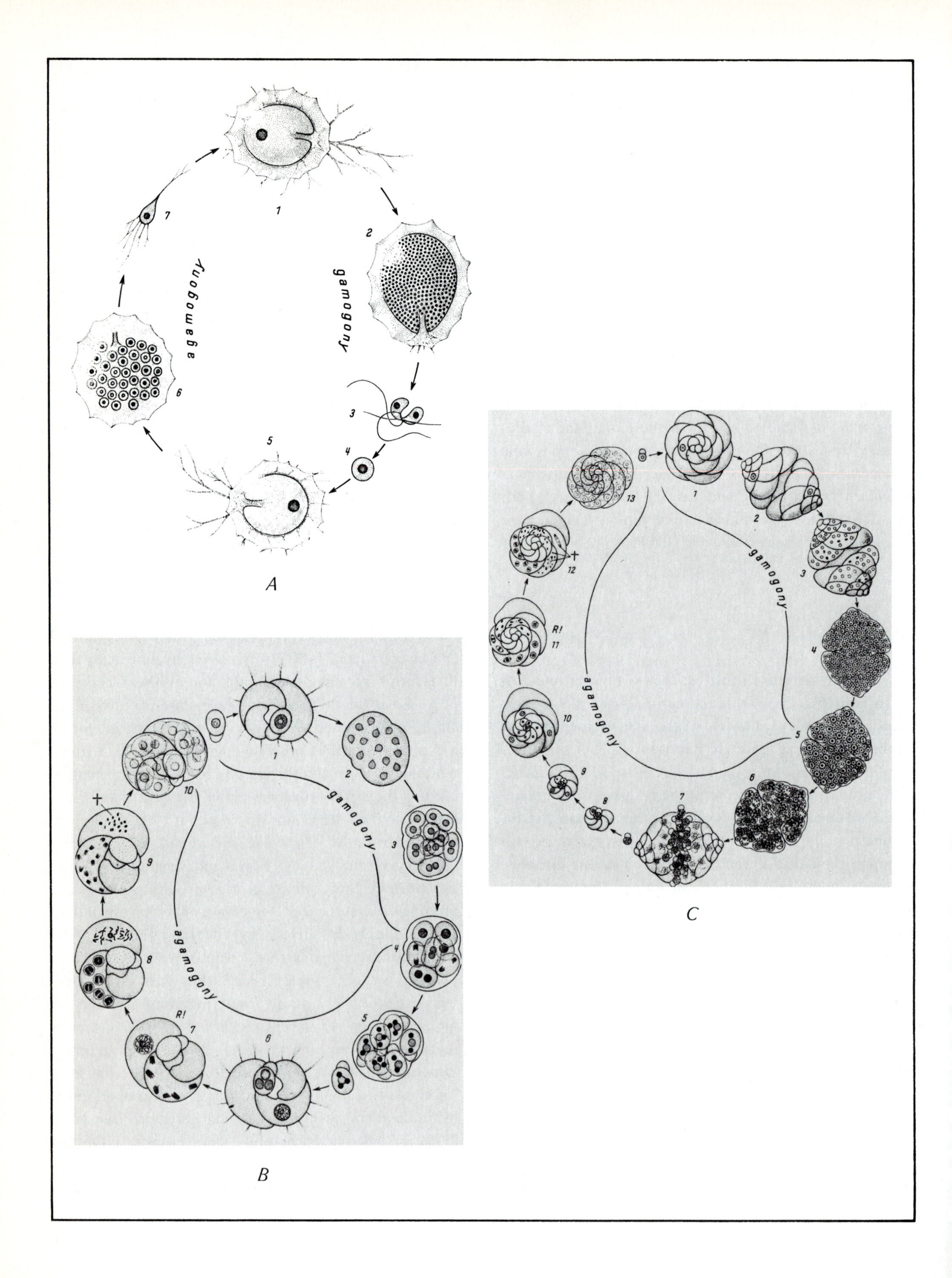
gamogony
agamogony
A
B
C
R!

PLATE 23-11. The developmental cycle in one monothalamic (*A*) and two polythalamic (*B,C*) foraminifera showing not only the alternation of generations between the asexually reproducing diploid generation (agamont) and the sexually reproducing haploid generation (gamont), but also the three types of sexual reproduction: (*A*) Gameteogamy in *Iridia lucida*, in which the gametes are fertilized outside the gamont; (*B*) autogamy in *Rotatoliella roscoffensis*, in which fusion takes place within the gamont; (*C*) gamontogamy in *Glabratella sulcata*, in which mating takes place between two gamonts. From K.G. Grell, *Protozoology*, Springer-Verlag, New York and Heidelberg, 1973.

and in a few species, the sexual process apparently involves autogamous flagellated gametes. Although the heliozoa are apparently more common in freshwater (Leidy, 1879; Penard, 1904; Wailes and Hopkinson, 1915–1921; Trégouboff, 1953a; Kudo, 1966), they also live in the sea (Jepps, 1956); they are detected in small numbers in various inshore habitats by experienced observers from time to time, are common in New England saltmarshes, and are occasionally detected in large numbers (Wailes, 1937; Lackey, 1961). An idea of some of the genera and species that have been identified in marine habitats can be obtained from the species of protozoa listed from the Woods Hole area by Calkins (1901) and by Lackey (1936), the illustrated Canadian Pacific fauna by Wailes (1937), the literature review by Kufferath (1952), the marine species listed by Kudo (1966), as well as the papers to be discussed here.

The heliozoa are divided into three orders (Honigberg et al., 1964) or suborders (Grell, 1973) on the basis of cell morphology (see Plate 23-13). The Actinophrydia contain naked forms that lack a central granule (centroplast) for axoneme attachment. *Actinophrys sol*, a planktonic form, has highly vacuolated cells. This species was reported by Calkins (1901) and Lackey (1936) from Woods Hole waters and in algal infusions and water from the Straits of Georgia, and British Columbia by Wailes (1937). Other species have been described from the marine environment (Kufferath, 1952). *Microsol borealis* is a sessile form associated with hydroids, green algae, and shells, among other substrates. The variation and distribution of this cosmopolitan species have been described by Dons (1918).

The Centrohelida are distinguished by axonemes that terminate at a central granule or centroplast (Plate 23-13,*B*), which is somewhat similar to the axoplast of the radiolaria in the Spumellaria (Plate 23-20). Some forms have skeletons with siliceous plates. *Acanthocystis spinifera* has plates with radial spines that form an outer investment. This species is detected in large number in the Georgia Strait plankton during August and September (Wailes, 1937). Two forms with siliceous spicules were described by Ostenfeld (1904) in the North Sea plankton. The population of *Acanthocystis pelagica* remained constant, whereas that of *Raphidiophrys marina* was seen in the fall and the winter. *Oxnerella maritima* is a small but typical form with a single large nucleus (Plate 23-13,*B*). It was described from seawater tanks in which protozoa were being cultivated by Dobell (1917). This species apparently creeps along surfaces for which it has an affinity, including the air–water interface. A similar but multinucleated form is *Gymnosphaera albida*, described by Sassaki (1894), which was seen in a small seawater aquarium in the Munich Zoological Institute. Another Centrohelida, the distinct but common *Wagnerella borealis*, has been described in detail by Zuelzer (1909) and Dons (1915–1916) and illustrated by Wailes (1937). Its spherical body, about 80 to 200 μm in diameter, is covered with spicules and attached to a hollow stipe 18 to 30 μm in diameter and 400 to 600 μm or more in length. At maturity, the single nucleus is situated at the base of the stipe. This species attaches to algae, hydroids, and other substrates at depths of 10 to 20 m, where it can become quite numerous. It has been reported from Naples Bay, the Adriatic and Norwegian coasts and waters of the Arctic and the Antarctic.

The third order of the heliozoa is Desmothoraca, a small group of mostly sedentary rhizopods. This group lacks a centroplast, and its latticed organic capsule that encloses the cell is attached to a hollow stalk. *Clathrulina elegans*, in addition to the heliozoan stiped stage shown in Plate 23-13,*C*, also has a biflagellated amoeboid stage and an encysted stage. This life cycle, as well as the morphology and ultrastructure of this beautiful and fascinating microorganism, has been described by Bardele (1972). This form also lacks axial filaments in its filopods. Such obvious differences from the other heliozoans prompted Penard (1904) and Bardele (1972) to classify it as a pseudoheliozoan. Apparently only Lackey (1936) has reported this species from the marine environment.

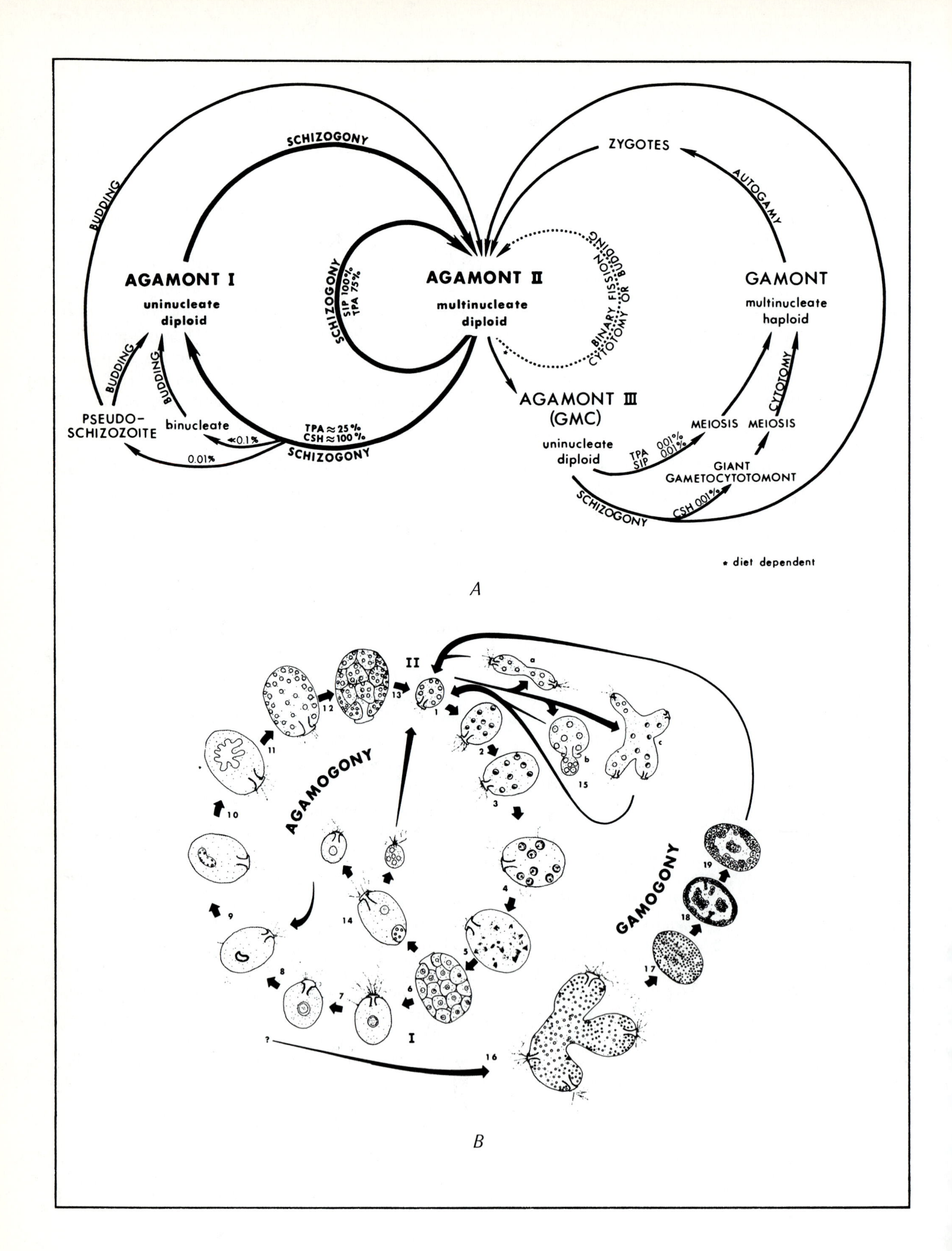

SCHIZOGONY
BUDDING
AGAMONT I
uninucleate
diploid
SCHIZOGONY
SIP 100%
TPA 75%
AGAMONT II
multinucleate
diploid
BINARY FISSION
CYTOTOMY OR BUDDING
ZYGOTES
AUTOGAMY
GAMONT
multinucleate
haploid
BUDDING
BUDDING
PSEUDO-
SCHIZOZOITE
binucleate
<0.1%
0.01%
TPA ≈ 25%
CSH ≈ 100%
SCHIZOGONY
AGAMONT III
(GMC)
uninucleate
diploid
TPA 0.01%
SIP 0.01%
MEIOSIS
MEIOSIS
CYTOTOMY
GIANT
GAMETOCYTOTOMONT
SCHIZOGONY
CSH 0.01%
* diet dependent

A

II
AGAMOGONY
GAMOGONY
I

B

PLATE 23-12. The life cycle of *Allogromia laticollaris.* (*A*) The non-classical life cycle of three strains, showing the apogamic reproduction in both uninucleate and multinucleate phases. (*B*) Diagrammatic representation of the life cycle of *A. laticollaris* as it is presently understood. From M.E. McEnery and J.J. Lee, 1976, *J. Protozool.* 23:94–108.

E. RADIOLARIA

The radiolaria are mainly spherical cells with axopodia radiating in all directions, just like the free-living heliozoans. Exceptions are those with bilateral symmetry. Their basic anatomy and morphology is shown in Plate 23-15. This group is clearly distinguished from the heliozoa by a capsular membrane that divides the cell into the endoplasm or intracapsular cytoplasm, with its many organelles, and the ectoplasm or extracapsular cytoplasm. This membrane in the Spumellaria (Peripylea or "many-pored" forms) is shown in Plate 23-16. It consists of glycoprotein plates with pores for the axopodial filaments and a slit along the edge with many fine pores (Hollande, Cachon, and Cachon, 1970). The pores of all three orders with siliceous skeletons (Monopylea or Nassellaria with a single pore, Tripylea or Phaeodaria with three pores, and Peripylea or Spumellaria with multiple pores) have a special organelle, the fusule, that permits the axopodial microtubules to penetrate but does not permit the intracapsular organelles to escape into the ectoplasm. The capsular membrane undergoes dissolution during multiplication. The endoplasm mainly functions in reproduction, although it also carries on some of the assimilatory and storage functions. The radiolaria are both uni- and multinucleated and, except for the ciliates and foraminifera, are the only protozoa in which multiple nuclear division takes place. Reproduction in many species appears to be exclusively by multiple fission. The life cycles of these species, however, are poorly understood, since they have not been studied sequentially in culture. The primary nucleus disintegrates to form biflagellated swarmers with highly refractive inclusions, which led to their name, "crystal swarmers." The study of nuclear division and reproduction is made difficult by the oceanic nature of these forms, the occurrence of some forms in deeper waters only, and the multiplication of "parasitic" dinoflagellates within the central capsule, which can be confused with swarmer formation. These parasites possess trichocysts and a theca; they have an organization similar to that of the Peridinians and a spore morphology and chromosome ultrastructure similar to that of *Oxyrrhis marina* (Hollande, 1974a). Included in Hollande's study are observations on the mitosis of the host radiolarians. Reproduction in the radiolaria appears to be both seasonal and sporadic, presumably as a result of a local and temporary enrichment of the waters with silicon.

The ectoplasm is involved in both the digestion of food and the regulation of buoyancy. Outside the capsular membrane is an assimilative layer known as the matrix, and between the matrix and the ectoplasmic cortex is a vacuolated and frothy layer, which is highly alveolated, with air inclusions, the calymma. The forms that live in the photic zone feed on such particulate food as protozoa, diatoms, and copepods and also contain zooxanthellae (Hovasse and Brown, 1953). These nonmotile forms of dinoflagellates are usually located in the calymma and are presumed to sustain the radiolarians during periods when particulate food is scarce, although there is no firm evidence for this. Their yellowish-green cells are dramatically illustrated in the colored plates of Haeckel (1887), Brandt (1885), and Schewiakoff (1926). The frothiness of the calymma varies with the physiological state of the microorganism, acting to maintain buoyancy during good weather; during warm weather and stormy periods, it is less frothy, which permits the microorganism to sink to cooler and calmer waters. The release of carbon dioxide-saturated water, as observed by Brandt (1885), may also be useful in buoyancy regulation.

Studies on the radiolaria have largely centered upon their taxonomy and their use in stratigraphic studies (see Riedel and Sanfilippo, 1971, for example). The geological implications and applications of this use of radiolaria are discussed in Moore's (1954) treatise. Very recently, there has been renewed interest in the radiolaria as living microorganisms. Observations on their ultrastructure have been made in order to find out more about these microorganisms in the absence of cultures. The occurrence and distribution of living species in the plankton have also been studied, as have the most recent fossils in sediments for applications in paleoecology. It should be noted that paleontologists concentrate on the Polycistine radiolaria (in the orders Spumellaria and Nassellaria) because the tripylean

PLATE 23-13. The heliozoa or "sun animalcules" of nearshore waters. (*A*) *Actinophrys sol* of the order Actinophrydia; these naked forms lack a centroplast (430 X); (*B*) *Oxnerella maritima* belongs to the order Centrohelida, which have axonemes terminating at the centroplast; some have siliceous plates (1020 X); (*C*) *Clathrulina elegans* of the order Desmothoraca, which lack a centroplast but have a reticulated skeleton of chitinous material impregnated with silica; most are stalked and sedentary (315 X); A marine Centrohelida, showing (*D*) the axopodial microtubules radiating from the center of the cell and (*E,F*) cross sections of the microtubules forming hexagonal microprisms. (*A*) from R.R. Kudo, *Protozoology*, 1966, courtesy of C.C. Thomas, Publisher, Springfield, Ill.; (*B*) from C. Dobell, 1917, *Quart. J. Micros. sci.* 62:515–38; (*C*) from C.F. Bardele, 1972, *Zeits. Zellforsch.* 130:219–42; (*D,E,F*) from C. Febvre-Chevalier, 1973a, *Protistologica* 9:35–43.

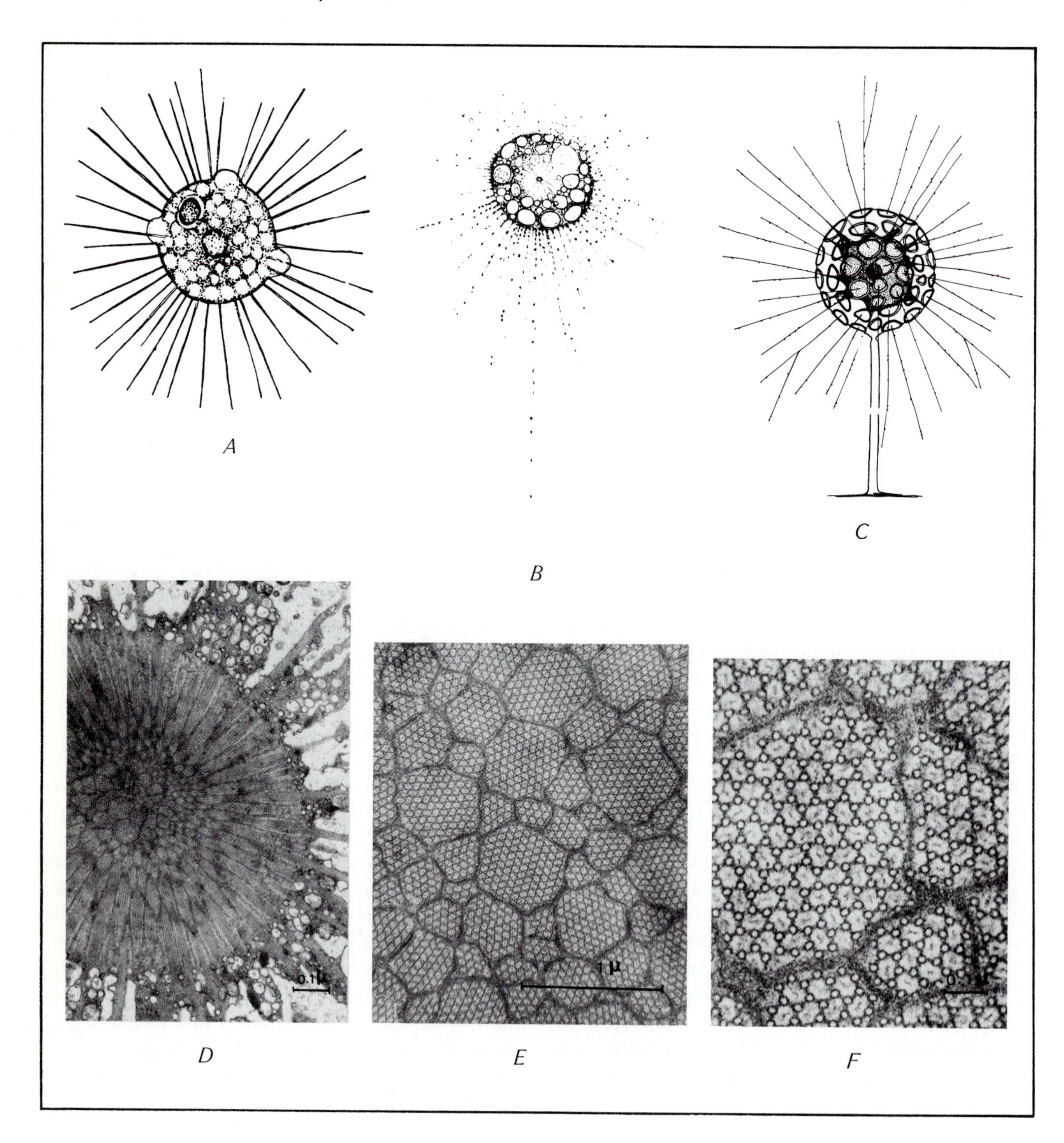

PLATE 23-14. Feeding by free and fixed marine heliozoans. Three types of pseudopods (Ps.) are formed, depending upon the size and the motility of the prey (*A*-I, *B*-I, *B*-II). (*A*) The planktonic species *Cilophrys marina*, showing a solitary form capturing a flagellate and releasing gastrioles (excretion vacuoles, A.V.) (1,500 X) and a pair fused for cooperative feeding on a large pennate diatom (1,000 X); (*B*) the sedentary species *Actinomonas mirabilis*, showing ingestion and digestion of food (N) and a free swimming stage (III) (2,000 X). From K. Griessmann, 1914, *Arch. Protistenk.* 32:1–78.

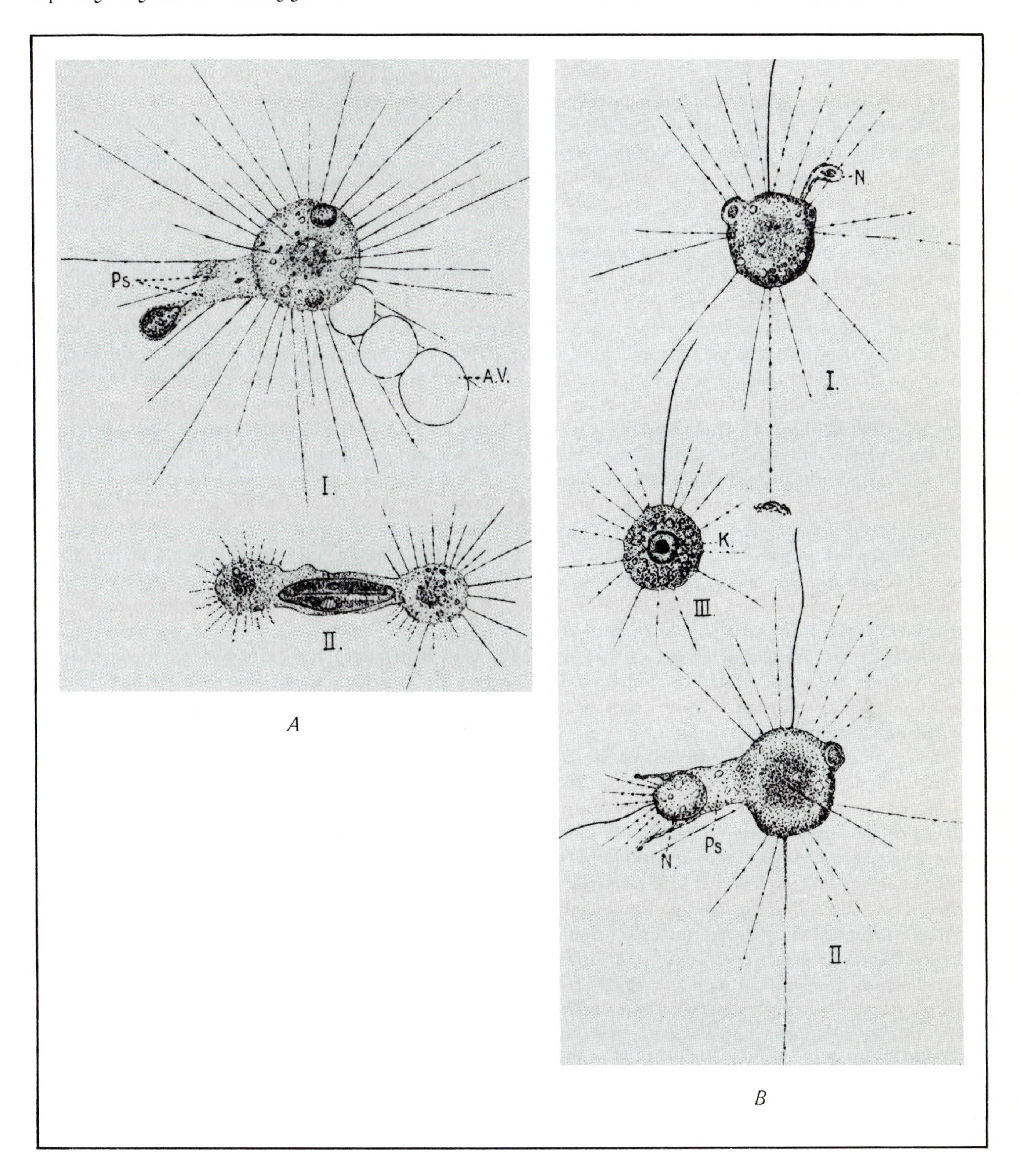

radiolaria (order Phaeodaria) is rare in the fossil record. An exception is the study by Kling (1966). Investigations in paleoclimatology have compared recent assemblages with Pleistocene assemblages, to quantitatively estimate past sea surface temperatures (Sachs, 1973a,b; Moore, 1973).

All radiolaria are pelagic, and although they are carried into coastal waters, there are no neritic species. The first major study on radiolaria was by Haeckel (1887) on material collected during the *Challenger* expedition. The greatest number of species were found in the vicinity of the equator, with regular decreases in number on transects toward the Poles. The comparatively few subsequent studies on distribution verify this trend, the latest being conducted by Renz (1976) on plankton and sediment obtained on a transect between 28°N and 25°S along 155°W in the Central Pacific. Her data on the abundance of radiolaria in the upper 200 m of the water column and a latitudinal comparison between abundance in the water and sediments is summarized in the graphs shown in Plate 23-17. There was an offset in the equatorial peak by 5°N that is apparently associated with the equatorial countercurrent. The lowest and highest populations were 242 and 18,730 individuals/m^3 of filtered water. Although abundance and species diversity decrease toward the Poles, radiolaria are also endemic in the cold waters of the high latitudes. Petrushevskaya (1971), in a study of the Indian Ocean, concluded that the distribution of species is environmentally controlled. She divided the radiolaria into tropical, subtropical, temperate, Antarctic endemic, and cosmopolitan species.

The vertical distribution of radiolaria can be divided into the surface forms occurring in the upper 200 to 300 m and the less frequent deeper forms occurring throughout the aphotic zone. The tests of surface forms are generally delicate and very ornate. Their many projections aid buoyancy and keep them near the surface where there are planktonic prey and sunlight for their zooxanthellae. Species in the orders Spumellaria and Nasellaria dominate. Haeckel (1887) divided the radiolarian species into those occurring in the upper 46 m and those occurring down to 274 m. Casey (1966), in a study of the radiolaria in the coastal waters off California, found a zonation at 0 to 25, 25 to 50, 50 to 125, and 125 to 200 m associated with two thermoclines, one pycnocline, and the break between surface water and the Pacific Central water. In an attempt to

PLATE 23-15. A schematic radiolarian and a few radiolarian skeletons indicate the anatomy and architecture of this group. (*A*) A few radiolarian skeletons from the report of the voyage of the *Challenger*, showing the siliceous spheres and spines. (*B*) The central nucleus (n) in the endoplasm is enclosed by the capsular membrane (c.c.). Outside the central capsule is the matrix or assimilative layer (ass.l.), which is covered with the calymma (cal), and a vacuolated layer of ectoplasm, which contains the yellow symbiotic algal cells, the zooxanthellae (z). The axopodial filaments (ps.) originate from within or around the nucleus (c.g.) and penetrate the capsular membrane through organelles called fusules. The superficial pseudopodia without axial filaments, which can anastomose, are marked (ps.1). (*A*) from E. Haeckel, 1887, *Challenger Reports*, Vol. 18; (*B*) from *The Protozoa*, 1956, Oliver & Boyd, Edinburgh, courtesy of Margaret Jepps.

observe such fine zonation in the Central Pacific, Renz (1976) observed that about one-half of the species reacted to some kind of a discontinuity between 50 and 75 m, whereas the remaining species were highly variable in distribution and abundance. Petrushevskaya (1966), also working in the Central Pacific, reported that the species represented were nearly constant to a depth of 100 to 300 m, but below this depth the number of species decreased. Some species of radiolaria are only found below this upper 300-m zone of major abundance. Haeckel (1887) described three zones for the deep waters. The tests of the forms in the deeper waters have fewer projections and are more solidly built. It has often been stated in protozoology texts that the Tripylean radiolaria occur in the deep waters, whereas the Polycistine radiolaria are surface forms. But Petrushevskaya (1966) found that the species of Polycistines from the greater depths were different from those at the surface. Berger (1968a) has even suggested that it is the deep-occurring species and not the surface species that make up the major components of the sedimentary assemblage.

The skeletons or hard parts (scleracoma) of radiolaria differ greatly in the different suborders. Species differentiation is based largely upon skeletal characteristics, and the great diversity of these forms has yielded thousands of species. The taxonomy of the radiolaria is discussed in the treatise edited by Moore (1954) and in more recent reviews (Riedel and Sanfilippo, 1971). The opaline silica skeletons of the radiolaria form deep-sea oozes and are part of the geological record, being second only to the foraminifera in this respect. The tertiary marl of Barbados Island, which is rich in radiolarian

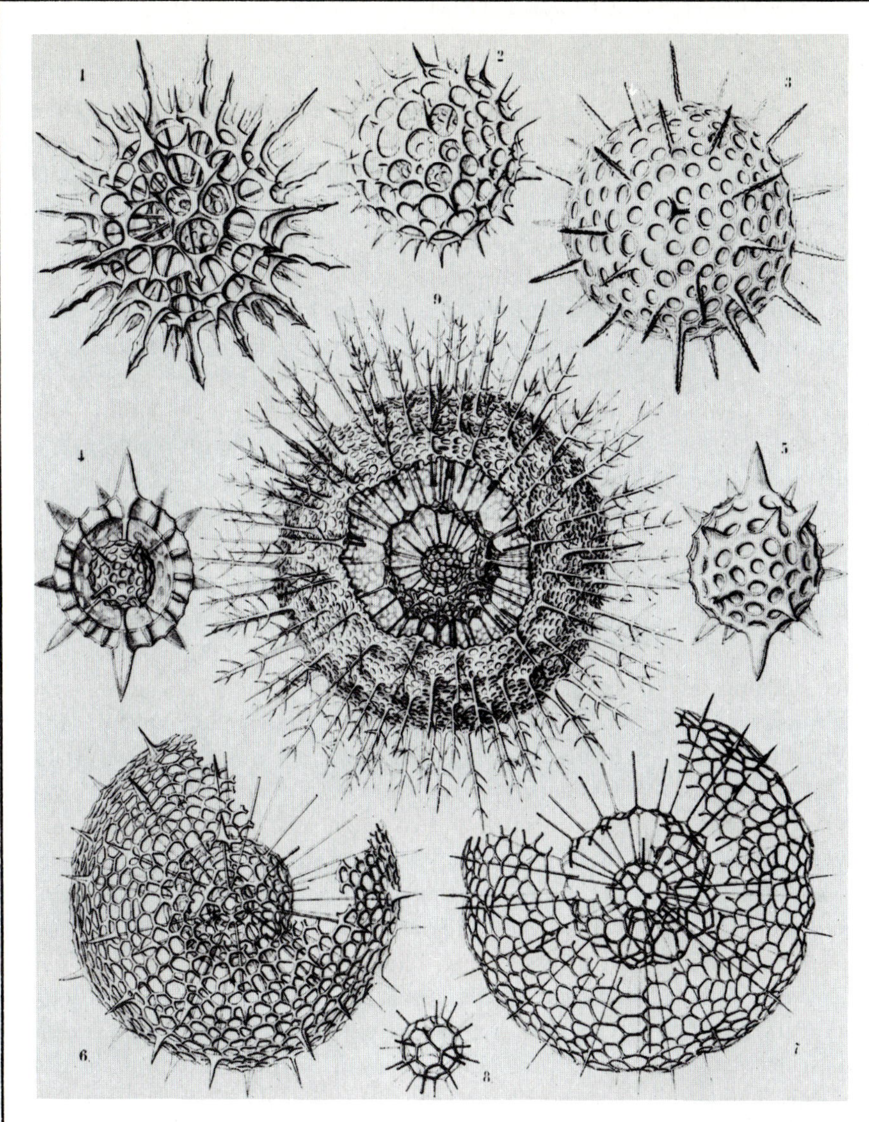

A

B

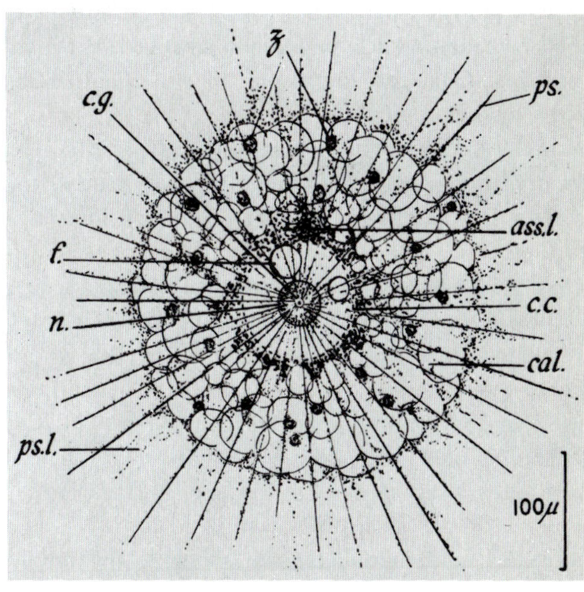

skeletons, is an example cited by many texts. An idea of the beauty and variation of these shells is only suggested by the marvelous figures in the plates of Haeckel (1887) (Plate 23-15).

The Nassellaria (Monopylea) have a central elongated capsule, with a single pore field at one end, and often, a conical inlet. These organisms are often enclosed in a latticework shell in the shape of a helmet, with the axopods radiating from the open bottom and up the side of the shell (see Plate 23-18,*A*). The axopodia, which originate in a cup-like axoplast, are microfibrillar in nature. The microfibrils form microtubules in both the axial and peripheral regions of the axopods (Cachon and Cachon, 1971). In cross section, these microtubules resemble a paddlewheel in the axial region, but peripherally, they occur as a parallel series (Plate 23-18,*C,D*). The Nassellaria occur both on the surface and in the deeper aphotic zone of the sea.

The Phaeodaria (Tripylea) have three openings in the capsular membrane through which the microtubular systems protrude. The ultrastructure of these openings has been studied by Cachon and Cachon (1973) and is shown in Plate 23-19. A large opening at the top of the central capsule houses the astropyle, a cytopharynx for ingesting particulate food. This is funnel-shaped, with a capsular membrane invaginated like a fluted filter paper lined with microtubules. At the bottom of the central capsule are two smaller openings for the two parapyles. These microtubular complexes are distinct from the fusules of the other radiolaria. The siliceous skeletons are often hollow and may be of three basic forms: helmet-shaped, as in the Spumellaria, composed of loosely arranged tangental spines, or radially arranged in a manner distinct from the 20 spicules of the Acantharians (see the following section). The Phaeodaria are also distinguished by a brownish mass of food particles and excretory products that collect in the area around the astropyle; it is known as the phaeodium and gives this group its name. These Tripylean radiolaria, a very distinctive group, occur in aphotic abyssal waters. Lacking zooxanthellae and their supplemental nutriment, they are dependent upon prey. Another characteristic of this group is their gigantic nucleus (Plate 23-19,*A*), and one species has 1500 chromosomes (Grell, 1973).

The Spumellaria (Peripylea) have a round central capsule completely penetrated by pores. The capsular membrane of this group is shown in Plate 23-16. The majority of modern radiolarians belong to the family Sphaerellaria of the order Spumellaria. They have a well-developed skeleton, which is usually spherical, commonly has radial spines, and is often composed of concentric layers of grid-like shells (see Plate 23-15,*A*). Some have five concentric shells or more, one within the other. One example of this large and very diverse group is the Sphaeroideae, which has been studied ultrastructurally by Cachon and Cachon (1972). In this group, the central axoplast is surrounded by the nucleus. This feature, the fusule structure, and the geometric pattern of the axopodal microtubules are shown in Plate 23-20. Members of the family Collodaria are larger, solitary forms, some without a skeleton and others with scattered siliceous needles. The third family in the Peripylea, the Polycyttaria, are colonial forms, with a number of central capsules in a common gelatinous mass. These forms have anastomosing pseudopodia, which may also connect the central capsules. Some forms are without skeletal elements, whereas others have single elements or a latticework sphere. These forms are the subject of a monograph by Brandt (1885). Hollande (1953) groups the Collodaria and the Polycyttaria together in a common suborder, the Sphaerocollideae. Hollande (1974b) has studied the ultrastructure of the spores of the colony-building species *Sphaerozoum neapolitanum* and *Collozoum pelagicum*, as well as a species of *Plegmophaera*. They all are simply organized, being characterized by a crystal of strontium enclosed within a vacuole during sporogenesis. The crystals of *S. neapolitanum* are pure strontium sulfate, crystallized as celestite, which indicates that the radiolarians can metabolize strontium and demonstrates their affinity with the Acantharians (Hollande and Martoja, 1974). These surface forms also have zooxanthellae, the dinoflagellate *Endodinium nutricola* that has a *Gymnodinium*-like appearance in the flagellated state. It is very similar to *Endodinium chattoni*, the endosymbiont of the by-the-wind sailor *Velella velella*, which lacks a cellulosic wall and in which a flagellate stage has not yet been observed (Hollande and Carré, 1974).

Sticholonche zanclea, a unique, Mediterranean radiolarian, is worth mentioning, since it is the only amoeboid form known to synchronize the movements of its axopods. These oar-like axopods are used for locomotion. In addition to its synchronized axopodia, this organism is distinguished by its clusters of spicules and a bean-shaped central capsule. Its morphology has been discussed by Trégouboff (1953b) and its ultrastructure

PLATE 23-16. The capsular membrane of radiolaria in the family Sphaerellaria (order Spumellaria). The glycoprotein plates with pores (*B*) and (*C*), which are bordered by slits (f) with pores (*D*), are situated between the plasmalemma and the endoplasmic reticulum (*A*); they are formed by cortical differentiation and not by intracytoplasmic formation. The fusule (F) shown in the diagrammatic reconstruction (*A*) is an organelle situated in the pores that permits the passage of microtubules to form the axial rods of the axopods, but prevents the endoplasmic organelles from entering the ectoplasm. (Ect., ectoplasm; End., endoplasm; M.cap., capsular membrane; m.cell, plasmalemma; RE, endoplasmic reticulum) (magnifications not given). From A. Hollande, J. Cachon, and M. Cachon, 1970, *Protistologica* 6:311–18.

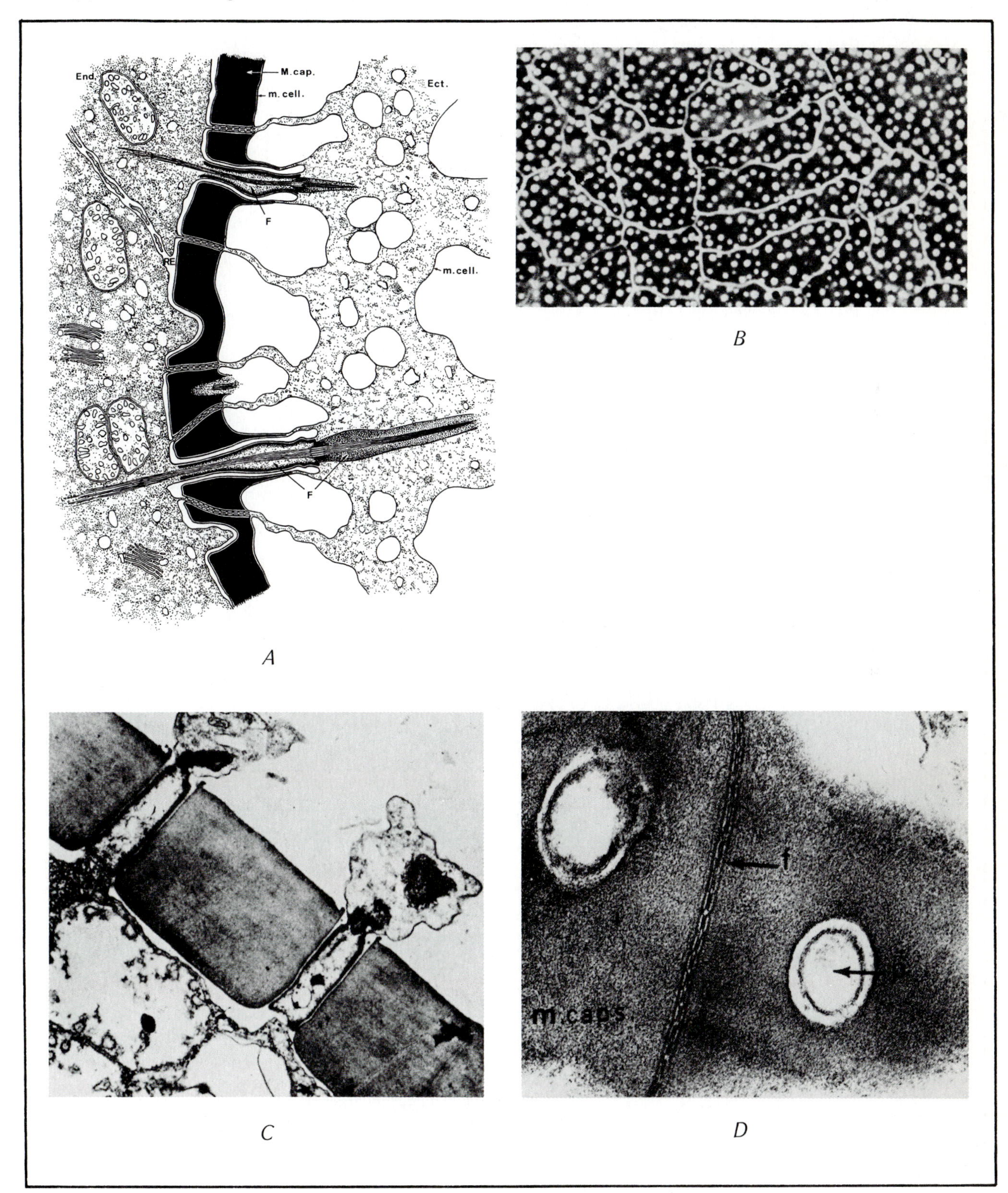

PLATE 23-17. The distribution of radiolaria in the water column and in the surface sediments of the Central Pacific. (*A*) The distribution (number of individuals per cubic meter of filtered water) in the upper 200 m of the water column; note association of abundance with the equatorial countercurrent. (*B*) Comparison of the latitudinal distribution of radiolaria in the upper water column, according to size (upper graph), with that in the surface sediment (lower graph). From G.W. Renz, 1976, *Bull. Scripps Inst. Oceanogr.* 22:1–267.

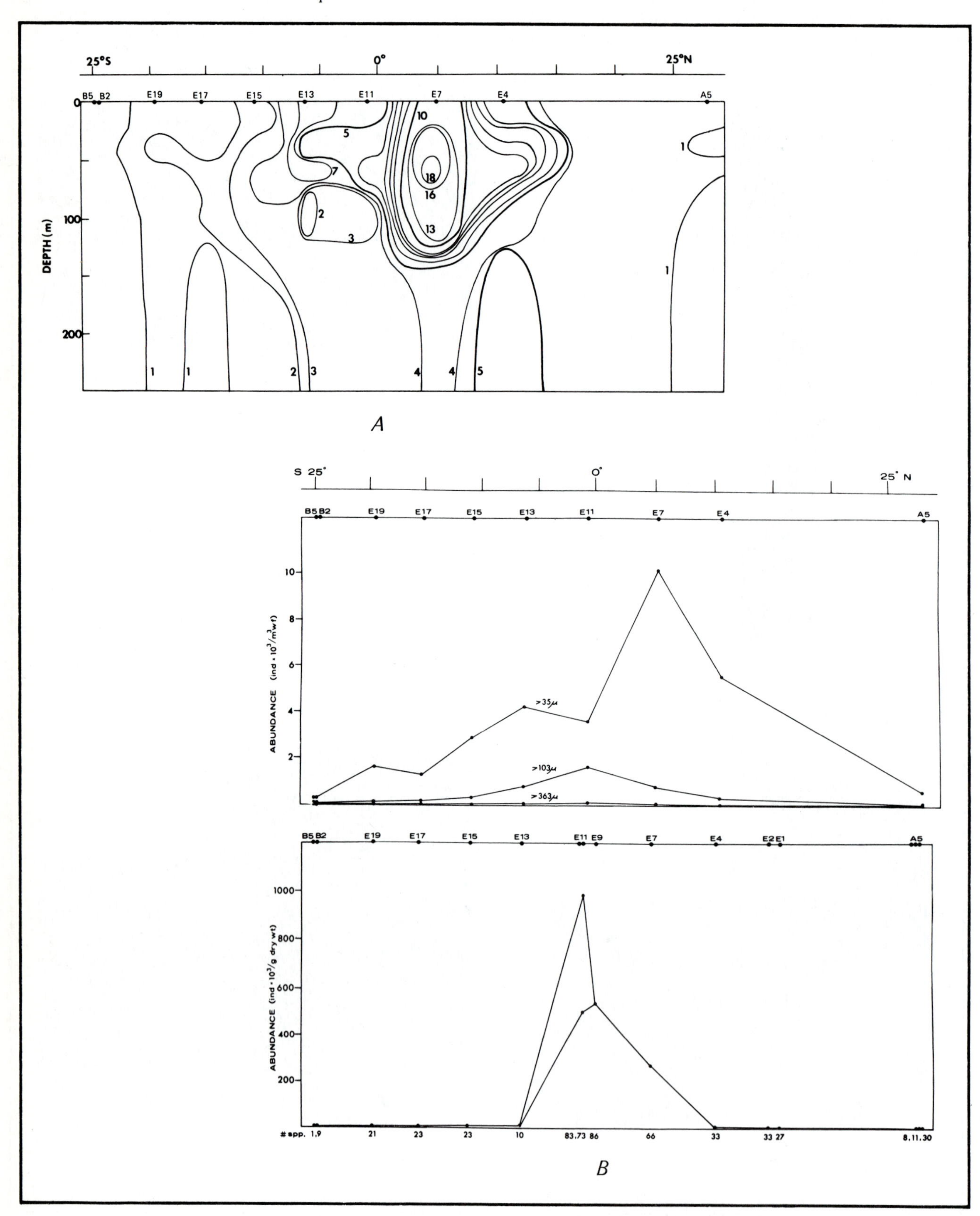

PLATE 23-18. The single axoplast structure in the radiolarian order Nassellaria (Monopylea) with their single pore field, showing (*A*) *Cyrtocalpis urceolus*, with its helmet-shaped skeleton; (*B*) a fusule, the organelles allowing the axopodial filaments to pass; and the pattern of the axopodial microtubules on the periphery (*C*) and in the center (*D*) of the axoplast. (*A*) from A. Kühn, 1926, *Morphologie der Tiere in Bildern. 2 Heft, Protozoen:* 2 Teil: Rhizopoden, Gerbüder Borntraeger, Berlin; (*B*,*C*, and *D*) from J. Cachon and M. Cachon, 1971, *Arch, Protistenk*, 113:80–97.

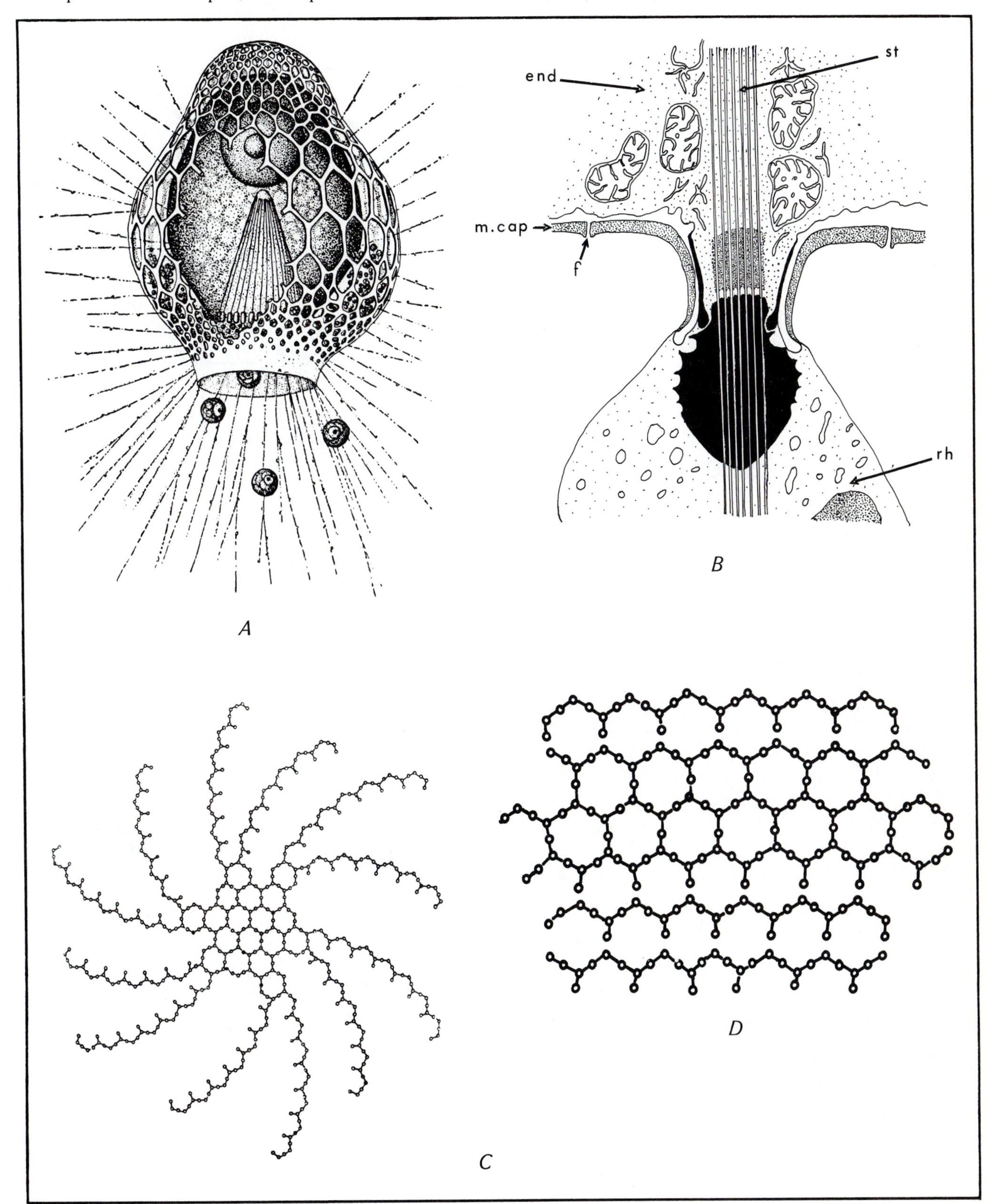

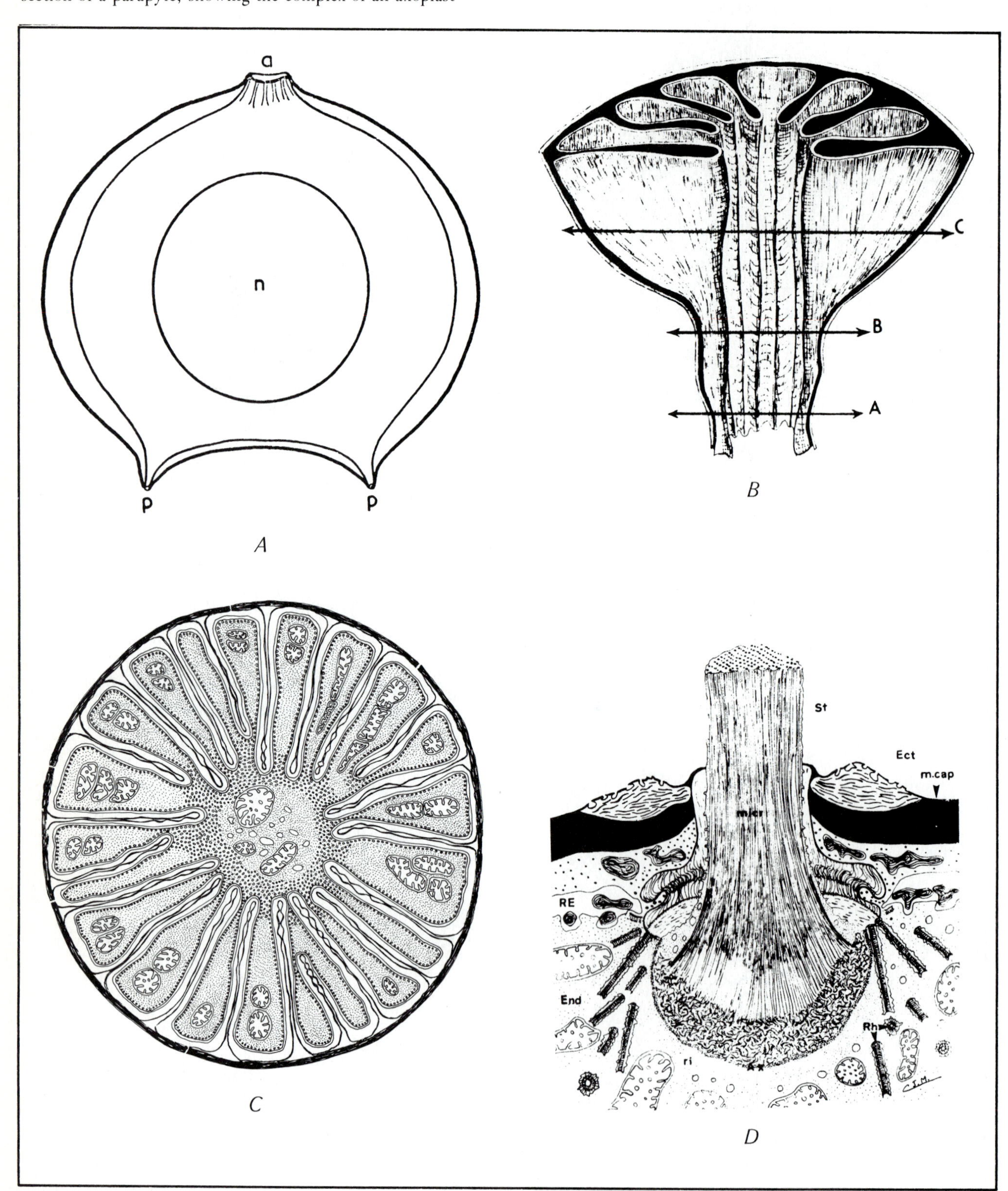

PLATE 23-19. The three-pored, two-microtubular system in the radiolarian order Phaeodaria (Tripylea). (*A*) The central capsule, with its gigantic nucleus (n), one astropyle (a), and two parapyles (p); (*B*) the large astropyle (cytopharynx) with a cross section C shown in (*C*), with numerous microtubules between the folds of the capsular membrane; (*D*) transverse section of a parapyle, showing the complex of an axoplast (ax) at the bottom of the cap and a large axopodial bundle (micr) penetrating to the capsular membrane. See Plate 23-16 for key to other structures (magnifications not given). (*A*) from D.L. MacKinnon and P.S.J. Hawes, 1961, *An Introduction to the Study of Protozoa*, Clarendon Press, Oxford; (*B,C,D*) from J. Cachon and M. Cachon, 1973, *Arch. Protistenk.* 115:324–35.

PLATE 23-20. The centric symmetry, fusule, and axopodial microtubule pattern of the Sphaeroideae of the radiolarian suborder Spumellaria (Peripylea). (*A*) The axoplast (Ax), which originates in the core of the nucleus (N); (*B*) the fusule, which permits the axopodial microtubules to penetrate the membrane of the central capsule without losing endoplasmic organelles; (*C*) a hypothetical geometric reconstruction of the pattern of axopodial microtubules. See Plate 23-16 for key to other structures. (magnifications not given). From J. Cachon and M. Cachon, 1972, *Arch. Protistenk.* 114:51–64.

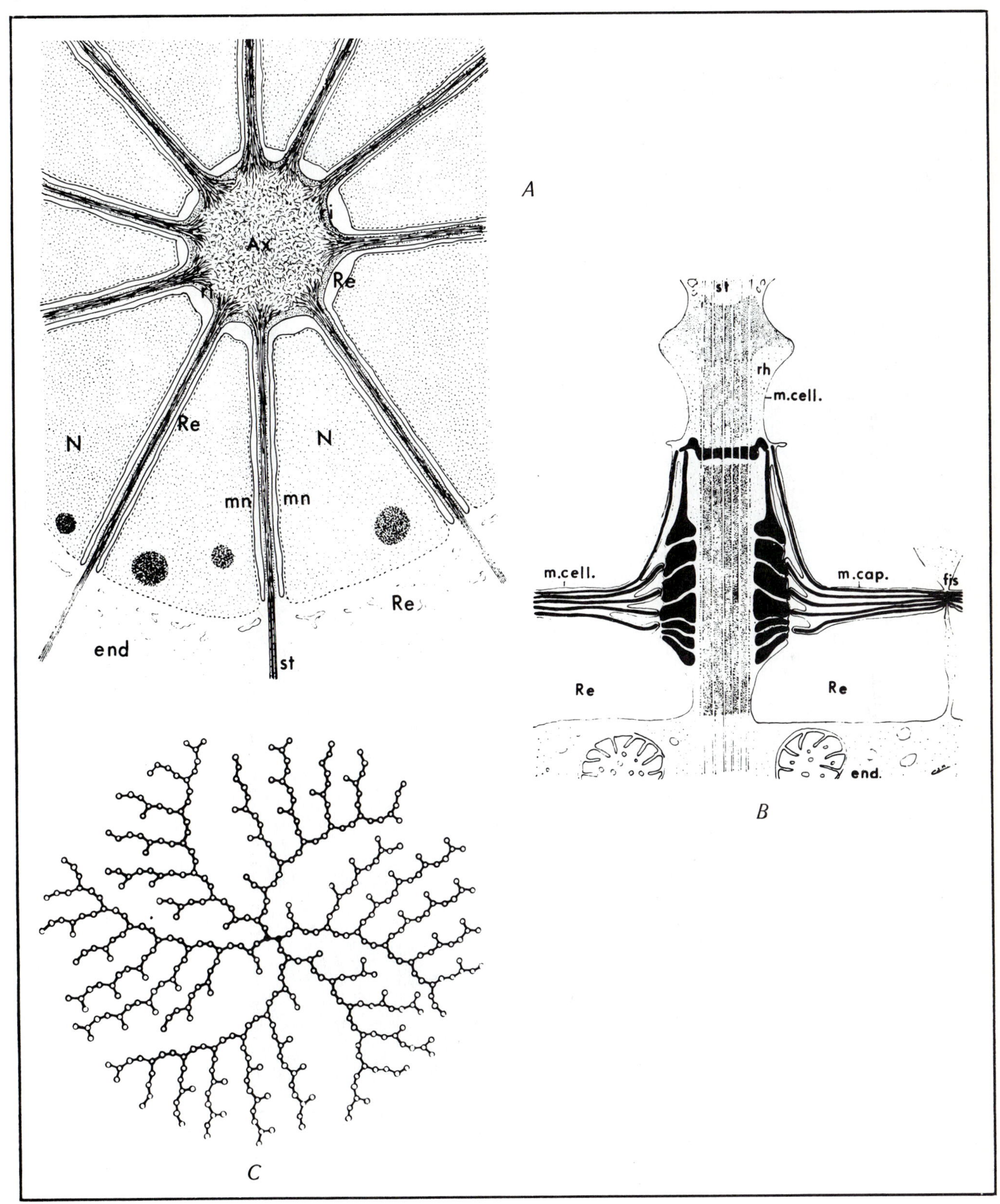

PLATE 23-21. The cell morphology and structure of the Acantharia. (*A*) The skeleton of strontium sulfate, with its characteristic 20 spines, each within a separate plate of ectoplasmic cortex; (*B*) the cell, showing the endoplasm (End), ectoplasm (Ect), ectoplasmic cortex (Ctx), and the myonemes at the juncture of this outer covering and the spines (Sp); (*C*) a diagrammatic reconstruction of thin sections of the cell, showing the above details (magnifications not given). (*A*) from J. Febvre, 1972, *Protistologica* 8:169–78; (*B*) from J. Febvre, 1971, *Protistologica* 7:379–91; (*C*) from J. Febvre, 1974, *Protistologica* 10:141–58.

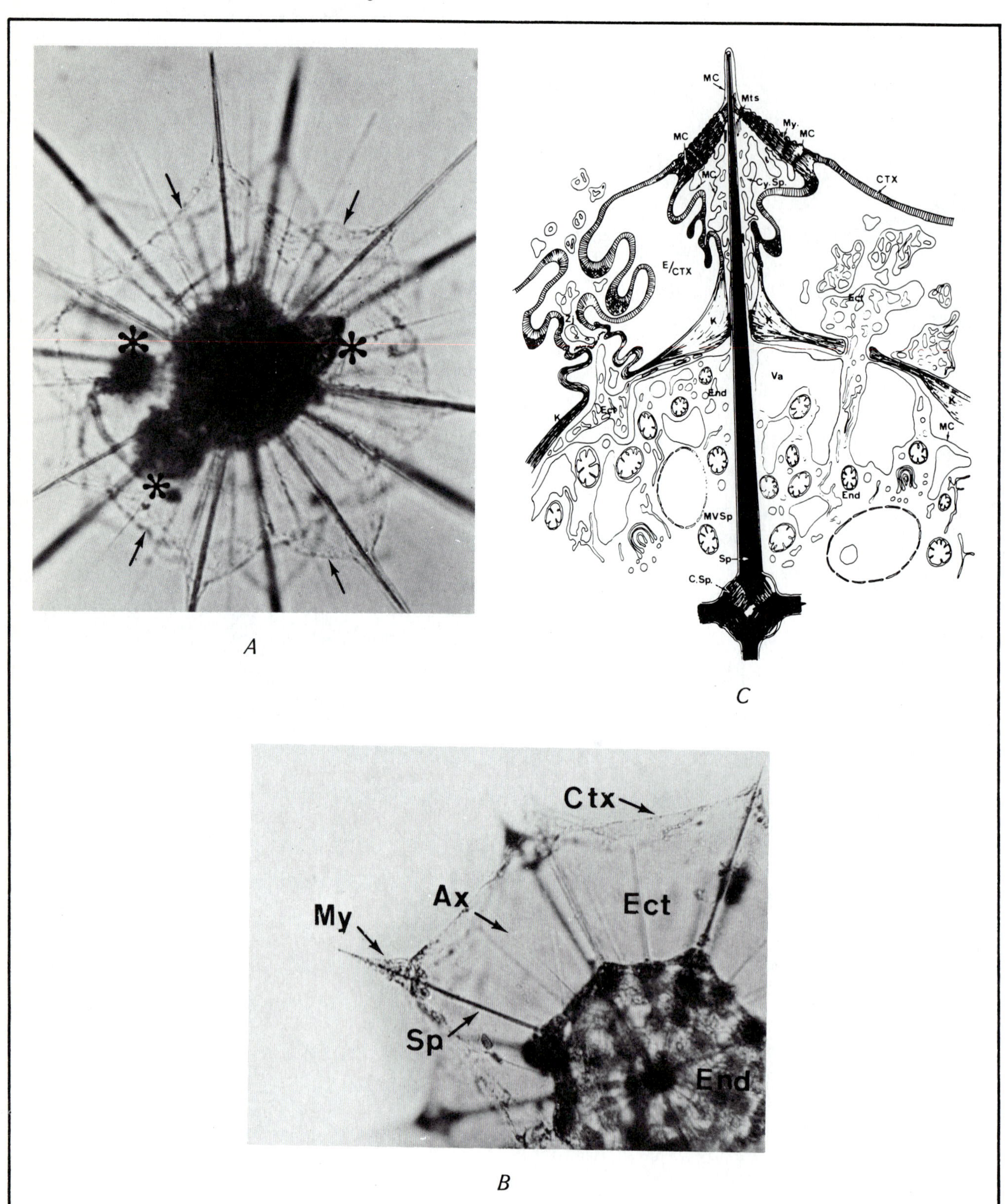

PLATE 23-22. Ultrastructural details of the ectoplasmic cortex of the Acantharia. (*A*) and (*B*) This outer covering of twenty plates (one for each spine) is composed of microtubules in a reticular system of triangles and hexagons; (*C*) the unique junctures between the plates of the ectoplasmic cortex; (*D*) the relaxed myoneme, which has an elastic rather than a contractile function, protects the cortex where it joins the strontium sulfate spines. (*A,B*) from J. Febvre, 1972, *Protistologica* 8:169–78; (*C*) from J. Febvre, 1973, *Protistologica* 9:87–94; (*D*) from J. Febvre, 1971, *Protistologica* 7:379–91.

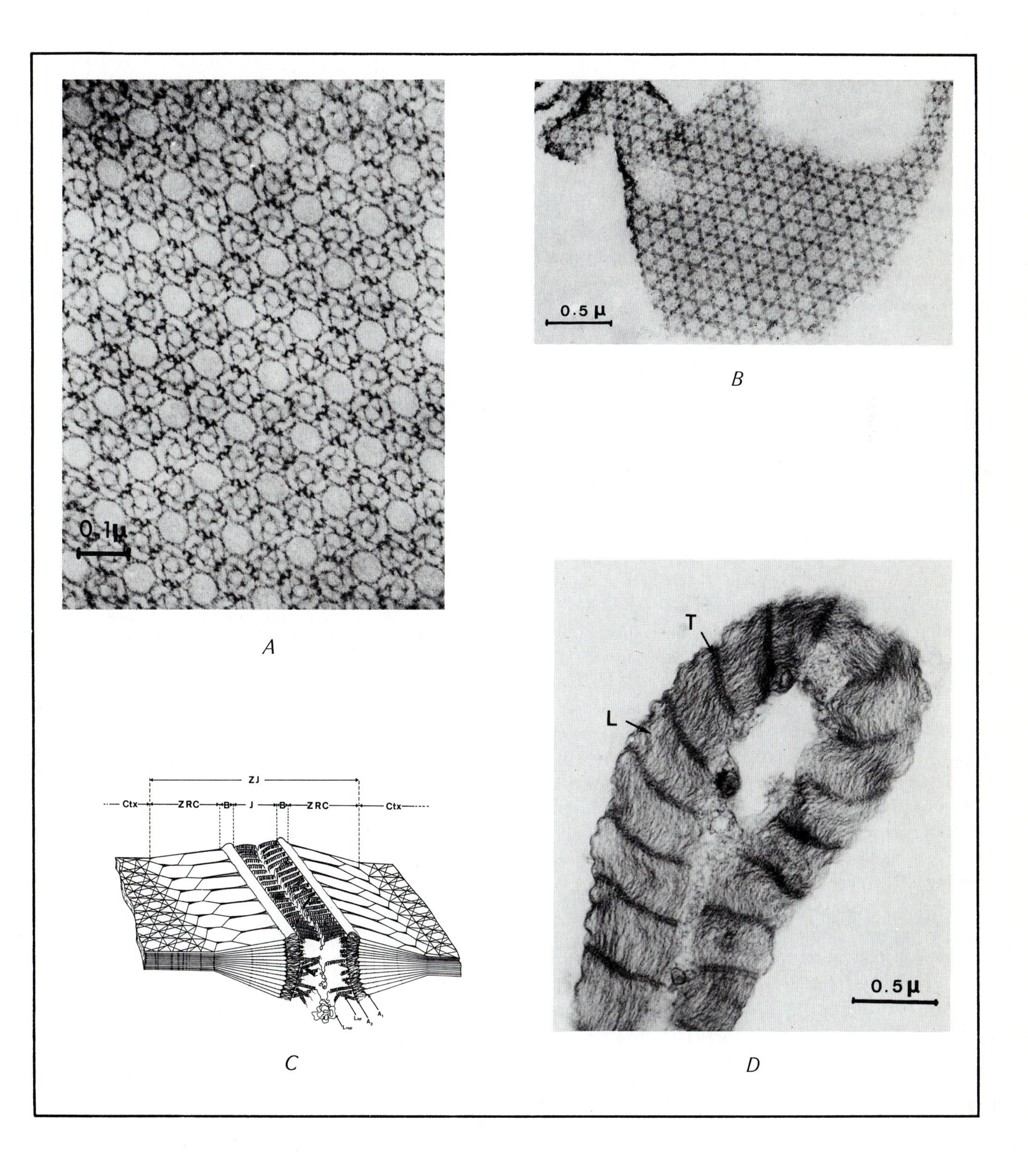

by Hollande et al. (1967). The axopods are hexagonal, prism-like columns, which insert on the thick nuclear membrane and pass through hexagonally shaped collars on the central capsule. When the axopods retract, the hexagonal column of microtubules in cross section is replaced by a zig-zag–folded bundle of fibrilla (lamellae). Cachon et al. (1977), in a further ultrastructural study of the mechanism of axopodal movement, reported that within each axopod is an array of microtubules that inserts into a dense material to assume the form of the head of a human femur. This material, in turn, articulates on the surface of the nucleus. Microfilaments, 20 to 30 Ångstroms in diameter, connect the dense material associated with the microtubules to the surface of the nucleus. The axopods move by contractions that appear to be controlled by calcium level.

F. ACANTHARIA

A group of rhizopods that resemble the radiolaria, with which they are often included as a fourth order, are the Acantharia. They are distinguished by 20 spicules that are often highly decorated and that are always fixed in an exact position [by a number of different types of junctions (Thompson, 1942)]. These spicules are composed of strontium sulfate and readily dissolve when released after the death of the microorganism, leaving no trace in the fossil record. Thompson (1942) revised Muller's concept of the form of radiolaria, postulating the spicules passed through the center of twenty curved polygonal plates, two polar, six northern, six equatorial, and six southern. Febvre (1971, 1972, 1973) has shown much of the fine structure of the Acantharia. Some of his illustrations are used in Plate 23-21 to show the form of the Acantharia and in Plate 23-22 to show the details of their ectoplasmic cortex. The mail-like plates are composed of microtubules in at least three different intricate patterns based on triangles and hexagons. These plates are joined together by unique junctions of microtubules that apparently permit the penetration of the axopodia. The plates are anchored to the spicules by structures called "myonemes," which is erroneous, since they have no contractile activity, as is often stated in the older texts (Jepps, 1956). The myonemes and the microtubules of the plate junctions have an elastic rather than a contractile function. Using observations on *Stauracantha orthostaura*, Febvre (1974) described the capsular membrane for the first time and established the relationship among the components of the envelope, the myoneme, the skeleton, and the plasmalemma. The zooxanthellae of Acantharia, studied by Hovasse and Brown (1953), occur in the ectoplasm and the endoplasm and were presumed not to be derived from dinoflagellates. A study by Hollande and Carré (1974) showed that two different zooxanthellae were present, one a dinoflagellate with cytoplasmic trichocysts and the other, of unknown affinities, with discoidal plasts containing an interlamellar pyrenoid. Reproduction is believed to be by binary and multiple fission and budding. Kühn (1926) illustrates a possible life cycle. In the lower Acantharia, sporulation and encystment occur simultaneously; in other Acantharia, encystment is characterized by the loss of skeletal structure and the formation of a gelatinous theca upon which small plates of strontium sulfate are deposited; one of three types of perforations permits the escape of mature spores (Hollande, Cachon, and Cachon-Enjumet, 1965).

REFERENCES

Anderson, O.R., and A.W.H. Bé. 1976. A cytochemical structure study of phagotrophy in a planktonic foraminifer *Hastigerina pelagica* (d'Orbigny). *Biol. Bull.* 151:437–49.

Arnold, Z.M. 1948. A new foraminiferan belonging to the genus *Allogromia*. *Trans. Amer. Micros. Soc.* 67:231–35.

Arnold, Z.M. 1953. An introduction to the study of movement and dispersal in *Allogromia laticollaris* Arnold. *Contrib. Cushman Found. Foram. Res.* 4:15–21.

Arnold, Z.M. 1954a. Culture methods in the study of living foraminifera. *J. Paleontol.* 28:404–16.

Arnold, Z.M. 1954b. Variation and isomorphism in *Allogromia laticollaris:* A clue to foraminiferal evolution. *Contrib. Cushman Found. Foram. Res.* 5:78–87.

Arnold, Z.M. 1954c. *Discorinopsis aguayoi* (Bermudez) and *Discorinopsis vadescens* Cushman and Bronnimann: A study of variation in cultures of living foraminifera. *Contrib. Cushman Found. Foram. Res.* 5:4–13.

Arnold, Z.M. 1955. Life history and cytology of the foraminiferan *Allogromia laticollaris*. *University Calif. Pub. Zool.* 61:167–231.

Arnold, Z.M. 1968. The uniparental species concept in the foraminifera. *Trans. Amer. Micros. Assoc.* 87:431–42.

Arnold, Z.M. 1974. Field and laboratory techniques for the study of living foraminifera. In: *Foraminifera* (R.H. Hedley and C.G. Adams, eds.). Academic Press, London and New York, pp. 153–206.

Bardele, C.F. 1972. Cell cycle, morphogenesis and ultrastructure in the pseudoheliozoan *Clathrulina elegans*. *Z. Zellforsch.* 130:219–42.

Bé, A.W.H. 1968. Shell porosity of recent planktonic Foraminifera as a climatic index. *Science* 161:881–84.

Bé, A.W.H., and D.B. Ericson. 1963. Aspects of calcification in planktonic foraminifera (Sarcodina). *Ann. N.Y. Acad. Sci.* 109:65–81.

Bé, A.W.H., and D.S. Tolderlund. 1971. Distribution of living planktonic foraminifera in surface waters of the Atlantic and Indian oceans. In: *The Micropaleontology of the Oceans*, Cambridge University Press, London and New York, pp. 105–49.

Bé, A.W.H., and O.R. Anderson. 1976. Gametogenesis in planktonic foraminifera. *Science* 192:890–92.

Benwitz, G., and K.G. Grell. 1971a. Ultrastruktur mariner Amöben. II. *Stereomyxa ramosa*. *Arch. Protistenk.* 113:51–67.

Benwitz, G., and K.G. Grell. 1971b. Ultrastruktur mariner Amöben. III. *Stereomyxa angulosa*. *Arch. Protistenk.* 113:68–79.

Berger, W.H. 1968a. Radiolarian skeletons: Solution at depths. *Science* 159:1237–38.

Berger, W.H. 1968b. Planktonic foraminifera: Selective solution and paleoclimatic interpretation. *Deep-Sea Res.* 15:31–43.

Berger, W.H., and D.J. Piper. 1972. Planktonic foraminifera: Differential settling, dissolution and redeposition. *Limnol. Oceanogr.* 17:275–87.

Bolli, H.M., and J.B. Saunders. 1954. Discussion of some Thecamoebina described erroneously as Foraminifera. *Contrib. Cushman Found. Foram. Res.* 5:45–52.

Boltovskoy, E. 1956. Application of chemical ecology in the study of the foraminifera. *Micropaleontology* 2:321–25.

Boltovskoy, E. 1959. Foraminifera as biological indicators in the study of ocean currents. *Micropaleontology* 5:473–81.

Boltovskoy, E. 1969. Living planktonic foraminifera at the 90° E meridian from the equator to the Antarctic. *Micropaleontology* 15:237–55.

Boltovskoy, E. 1971. Ecology of the planktonic foraminifera living in the surface layer of the Drake Passage. *Micropaleontology* 17:56–68.

Boltovskoy, E., and R. Wright. 1976. *Recent Foraminifera*. Dr. W. Junk, The Hague, Netherlands, 515 p.

Bonnet, L. 1975. Types morphologiques, écologie et évolution de la Thèque chez les Thécamoebiens. *Protistologica* 11:363–78.

Bovee, E.C. 1956. Some observations on a marine ameba of intertidal zones, *Vexillifera telmathalassa* n. sp. *J. Protozool.* 3:155–58.

Bovee, E.C. 1965. An emendation of the ameba genus *Flabellula* and a description of *Vannella* gen. nov. *Trans. Amer. Micros. Soc.* 84:217–27.

Bovee, E.C., and D.L. Cordell. 1971. Feeding on gastrotrichs by the heliozoan *Actinophrys sol*. *Trans. Amer. Micros. Soc.* 90:365–69.

Bovee, E.C., and T.K. Sawyer. 1979. *Marine flora and fauna of the Northeastern United States. Protozoa: Amoebidae.* NOAA Tech. Rept., Natl. Mar. Fish. Serv.

Brandt, K. 1885. Die Koloniebildenden Radiolarien (Sphaerozoëen) des Golfes von Neapel. In: *Fauna und flora des Golfes von Neapel, Monograph 13*, Friedländer & Sohn, Berlin, 276 p.

Buchanan, J.B., and R.H. Hedley. 1960. A contribution to the biology of *Astrorhiza limicola* (Foraminifera). *J. mar. biol. Assoc. U.K.* 39:549–60.

Cachon, J., and M. Cachon. 1971. Le Système axopodial des radiolaires Nassellaires. *Arch. Protistenk.* 113:80–97.

Cachon, J., and M. Cachon. 1972. Le Système axopodial des radiolaires sphaeröides. I. Centroaxoplastidiés. *Arch. Protistenk.* 114:51–64.

Cachon, J., and M. Cachon. 1973. Systèmes microtubulaires de l'astropyle et des parapyles de Phaeodariés. *Arch. Protistenk.* 115:324–35.

Cachon, J., M. Cachon, L.G. Tilney, and M.S. Tilney. 1977. Movement generated by interactions between the dense material at the ends of microtubules and non-actin-containing microfilaments in *Sticholonche zanclea*. *J. Cell Biology* 72:314–38.

Calkins, G.N. 1901. Marine protozoa from Woods Hole. *Bull. U.S. Fish. Comm.* 21:415–68.

Casey, R.E. 1966. A seasonal study on the distribution of polycystine radiolarians from waters overlying the Catalina Basin, Southern California. Ph.D. Dissertation, University of Southern California, Los Angeles, 145 p.

Chatton, E. 1953. Classe de Lobosa (Leidy, 1879) Orde des Amoebiens nus ou Amoebaea. In: *Traité de Zoologie* (P-P. Grassé, ed.), Vol. 1(2). Masson et Cie, Paris, pp. 5–91.

Christiansen, B. 1958. The foraminifer fauna in the Dröbak Sound in the Oslo Fjord (Norway). *Nytt. Mag. Zool.* 6:5–91.

Christiansen, B.O. 1964. *Normania confertum* from the Oslo fiord in Norway. *Contribu. Cushman Found. Foram. Res.* 15:135–37.

CLIMAP Projects Members. 1976. The surface of the ice-age earth. *Science* 191:1131–37.

Cushman, J.D. 1959. *Foraminifera, Their Classification and Economic Use* (4th Ed.). Harvard University Press, Cambridge, Mass., 605 p.

Dahlgren, L. 1967. On the ultrastructure of the gamontic nucleus and the adjacent cytoplasm of the monothalamous foraminifer *Ovaminna opaca* Dahlgren. *Zool. Bidrag Uppsala* 37:77–112.

Davis, P.G. 1976. Oceanic amoebae from the North Atlantic: Culture, distribution, and taxonomy. M.S. Thesis, University of Rhode Island, 54 p.

Davis, P.G., D.A. Caron, and J.McN. Sieburth. 1978. Oceanic amoebae from the North Atlantic: Culture, distribution, and taxonomy. *Trans. Am. Micros. Soc.* 96:73–88.

Deflandre, G. 1953. Ordres des Testacealobosa (DeSaedeleer, 1934), Testaceafilosa (DeSaed., 1934), Thalamia (Haeckel, 1862) ou Thécamoebiens (Auct.) (Rhizopoda Testacea). In: *Traité de Zoologie* (P-P. Grassé, ed.), Vol. 1(2). Masson et Cie., Paris, pp. 97–148.

Dobell, C. 1917. On *Oxnerella maritima*, nov. gen., nov. spec., a new heliozoon, and its method of division; with some remarks about the centroplast of the heliozoa. *Quart. J. Micros. Sci. New Ser.* 62:515–38.

Dons, C. 1915–16. Heliozoen *Wagnerella borealis. Tromsø Mus. Årshefter.* 38,39:101–16.

Dons, C. 1918. Heliozoen *Microsol borealis. Tromsø Mus. Årshefter.* 41:1–25.

Ericson, D.B., and G. Wollin. 1964. *The Deep and the Past.* Knopf, New York, 380 p.

Febvre, J. 1971. Le Myonème d'acanthaire: Essai d'interprétation ultrastructurale et cinétique. *Protistologica* 7:379–91.

Febvre, J. 1972. Le Cortex ectoplasmique des acanthaires. I. Les Systèmes maillés. *Protistologica* 8:169–78.

Febvre, J. 1973. Le Cortex ectoplasmique des acanthaires. II. Ultrastructure des zones de jonction entre les pièces corticales. *Protistologica* 9:87–94.

Febvre, J. 1974. Rélations morphologiques entre les constituants de l'enveloppe, les myonèmes, le squelette et le plasmalemma chez les Arthracantha Schew (Acantharia). *Protistologica* 10:141–58.

Febvre-Chevalier, C. 1973a. Un nouveau type d'association des microtubules axopodiaux chez les Héliozoaires. *Protistologica* 9:35–43.

Febvre-Chevalier, C. 1973b. *Hedraiophrys hovassei*, morphologie, biologie, et cytologie. *Protistologica* 7:503–20.

Febvre-Chevalier, C. 1975. Étude cytologique de *Gymnosphaera albida* Sassaki, 1894 (Héliozaire Centrohélidé). *Protistologica* 11:331–44.

Golemansky, V. 1970a. *Psammonobiotus communis* Gol. 1968 (Rhizopoda, Testacea)—une rélicte marine du Lac d'Ohrid. *Bull. Inst. Zool. Musée Sofia* 32:63–68.

Golemansky, V. 1970b. *Chardezia caudata* gen. n. sp. n. et *Rhumbleriella filosa* gen. n. sp. n.—deux thécamoebiens nouveaux du psammon littoral de la Mer Noire (Rhizopoda, Testacea). *Bull. Inst. Zool. Musée Sofia* 32:121–25.

Golemansky, V. 1970c. Rhizopodes nouveaux du psammon littoral de la Mer Noire (note préliminaire). *Protistologica* 6:365–71.

Golemansky, V. 1970d. Thécamoebiens (Rhizopoda, Testacea) nouveaux des eaux souterraines littorales de la Mer Noire. *Acta Protozool.* 8:41–47.

Golemansky, V. 1974a. Sur la Composition et la distribution horizontale de l'association thécamoebienne (Rhizopoda, Testacea) des eaux souterraines littorales de la Mer Noire en Bulgarie. *Bull. Inst. Zool. Musée Sofia* 40:195–202.

Golemansky, V. 1974b. *Lagenidiopsis valkanovi* gen. n., sp. n.—un nouveau thécamoebien (Rhizopoda: Testacea) du psammal supralittoral des mers. *Acta Protozool.* 13:1–5.

Golemansky, V. 1974c. Psammonobiotidae fam. nov.—une nouvelle famille de thécamoebiens (Rhizopoda, Testacea) du psammal supralittoral des mers. *Acta Protozool.* 13:137–43.

Gomes de Faria, J., A.M. Da Cunha, and C. Pinto. 1922. Estudos sobre Protozoarios do mar. *Mem. Inst. Oswaldo Cruz* 15:186–200.

Grell, K.G. 1954. Der Generationswechsel der polythalamen Foraminifere *Rotaliella heterocaryotica. Arch. Protistenk.* 100:268–86.

Grell, K.G. 1957. Untersuchungen über die Fortpflanzung und Sexualität der Foraminiferen. I *Rotaliella roscoffensis. Arch. Protistenk.* 102:147–64.

Grell, K.G. 1959. Untersuchungen über die Fortpflanzung und Sexualität der Foraminiferen. IV. *Patellina corrugata. Arch. Protistenk.* 104:211–34.

Grell, K.G. 1961. Uber den "Nebenkörper" von *Paramoeba eilhardi* Schaudinn. *Arch. Protistenk.* 105:303–12.

Grell, K.G. 1966. Amöben der Familie Stereomyxidae. *Arch. Protistenk.* 109:147–54.

Grell, K.G. 1973. *Protozoology.* Springer-Verlag, Heidelberg and New York, 554 p.

Grell, K.G., and G. Benwitz. 1970. Ultrastructure mariner Amöben. I. *Paramoeba eilhardi* Schaudinn. *Arch. Protistenk.* 112:119–37.

Griessmann, K. 1914. Über marine Flagellaten. *Arch. Protistenk.* 32:1–78.

Haberey, M. 1973a. Die Phagocytose von Oscillatorien durch *Thecamoeba sphaeronucleolus.* I. Lichtoptische Untersuchung. *Arch. Protistenk.* 115:99–110.

Haberey, M. 1973b. Die Phagocytose von Oscillatorien durch *Thecamoeba sphaeronucleolus.* II. Electronenmikroscopische Untersuchung. *Arch. Protistenk.* 115:111–24.

Haeckel, E. 1887. *Report on the scientific results of the voyage of H.M.S. Challenger. Zoology.* Vol. 18, 1st Part, *Porulosa (Spumellaria and Acantharia)*, pp. 791–888, 2nd Part, *Osculosa (Nassellaria and Phaeodaria)*, pp. 889–1803, 3rd part, 140 plates. Her Majesty's Stationery Office, Edinburgh.

Hedley, R.H. 1958a. A contribution to the biology and cytology of *Haliphysema* (Foraminifera). *Proc. Zool. Soc. London* 130:569–76.

Hedley, R.H. 1958b. Confusion between *Gromia oviformis* and *Allogromia ovoidea. Nature* 182:1391–92.

Hedley, R.H., D.M. Parry, and J. St.J. Wakefield. 1967. Fine structure of *Shepheardella taeniformis* (Foraminifera: Protozoa). *J. Roy. Micros. Soc.* 87:445–56.

Hedley, R.H., D.M. Parry, and J.St.J. Wakefield. 1968. Reproduction in *Boderia turneri* (Foraminifera). *J. Nat. Hist.* 2:147–51.

Hollande, A. 1953. Complements sur la cytologie des Acanthaires et des Radiolaires. In: *Traité de Zoologie, Vol. 1, Part 2* (P-P. Grassé, ed.). pp. 1089–1100.

Hollande, A. 1974a. Étude comparée de la mitose Syndinienne et de celle des Péridiniens libres et des Hypermastigines infrastructure et cycle évolutif des syndinides parasites de radiolaires. *Protistologica* 10:413–51.

Hollande, A. 1974b. Donneés Ultrastructurales sur les isospores des radiolaires. *Protistologica* 10:567–72.

Hollande, A., and D. Carré. 1974. Les xanthelles des Radiolaires Sphaerocolloides, des Acanthaires et de *Velella velella:* infrastructure—cytochimie—taxonomie. *Protistologica* 10:573–601.

Hollande, A., and R. Martoja. 1974. Identification du cristalloide des isopores de Radiolaires a un cristal de celestite ($SrSO_4$) determination de la constitution du cristalloide par voie cytochimique et a l'aide de la microsonde electronique et du microanalyseur par emission ionique secondaire. *Protistologica* 10:603–9.

Hollande, A., J. Cachon, and M. Cachon-Enjumet. 1965. Les Modalités de l'enkystement présporogénétique chez les Acanthaires. *Protistologica* 1(2):91–104.

Hollande, A., J. Cachon, M. Cachon, and J. Valentin. 1967. Infrastructure des axopodes et organisation générale de *Sticholonche zanclea* (HERTWIG) (Radiolaire Sticholonchidea). *Protistologica* 3:155–66.

Hollande, A., J. Cachon, and M. Cachon. 1970. La Signification de la membrane capsulaire des radiolaires et ses rapports avec le plasmalemme et les membranes du réticulum endoplasmique. Affinités entre radiolaires, héliozaires et péridiniens. *Protistologica* 6:311–18.

Honigberg, B.M., W. Balamuth, E.C. Bovee, J.O. Corliss, M. Gojdisc, R.P. Hall, R.R. Kudo, N.D. Levine, A.R. Loeblich, Jr., J. Weiser, and D.H. Wenrich. 1964. A revised classification of the Phylum Protozoa. *J. Protozool.* 11:7–20.

Hovasse, R., and E.M. Brown. 1953. Contribution a la connaissance des radiolaires et leur parasites syndiniens. *Ann. Sci. Nat. Zool. Biol. Animale 11th Ser.* 15:405–37.

Hutner, S.H. 1973. Outline of Protozoa. In: *Handbook of Microbiology, Vol. I. Organismic Microbiology* (A.I. Laskin and H.E. Lechevalier, eds.). CRC Press, Cleveland, Ohio, pp. 481–506.

Jahn, T.L., and R.A. Rinaldi. 1959. Protoplasmic movement in the foraminiferan, *Allogromia laticollaris;* and a theory of its mechanism. *Biol. Bull.* 117:100–18.

Jeon, K.W. (ed.). 1973. *The Biology of Amoeba.* Academic Press, New York and London, 628 p.

Jepps, M.W. 1956. *The Protozoa, Sarcodina.* Oliver and Boyd, Edinburgh, 183 p.

Johnson, P.W., and J.McN. Sieburth. 1976. *In-situ* morphology of nitrifying-like bacteria in aquaculture systems. *Appl. Environ. Microbiol.* 31:423–32.

Kennett, J.P. 1968. *Globorotalia truncatulinoides* as a paleo-oceanographic index. *Science* 159:1461–63.

Kennett, J.P., and P. Huddlestun. 1972. Late Pleistocene paleoclimatology, foraminiferal biostratigraphy and tephrochronology, Western Gulf of Mexico. *Quaternary Research* 2:38–69.

Kennett, J.P., and N.D. Watkins. 1976. Regional deep-sea dynamic processes recorded by the late Cenozoic sediments of the southeastern Indian Ocean. *Geol. Soc. Amer. Bull.* 87:321–39.

Kling, S.A. 1966. Castanellid and Circoporid Radiolarians: Systematics and zoogeography in the eastern North Pacific. Ph.D. Dissertation, University of California, San Diego, 178 p.

Kudo, R.R. 1966. *Protozoology* (5th Ed.) C.C. Thomas, Springfield, Ill., 1174 p.

Kufferath, H. 1952. Récherches sur le plancton de la mer Flamande (Mer du Nord Meridonale); II. Biddulphiaeae, Proteomyxa, Rhizomastigina, Heliozoa, Amoebina. *Inst. Roy. Sci. Nat. Belgique (Medd. Kon. Bel. Inst. Natur.)* 28(10):1–39.

Kühn, A. 1926. *Morphologie der Tiere in Bildern. 2 Heft. Protozoen: 2 Teil: Rhizopoden.* Gebrüder Borntraeger, Berlin, 272 p.

Lackey, J.B. 1936. Occurrence and distribution of the marine protozoan species in the Woods Hole area. *Biol. Bull.* 70:264–78.

Lackey, J.B. 1961. Bottom sampling and environmental niches. *Limnol. Oceanogr.* 6:271–79.

Lackey, J.B. 1963. The microbiology of a Long Island bay in the summer of 1961. *Int. Revue ges. Hydrobiol.* 48:577–601.

Lackey, J.B., and K.A. Clendenning. 1965. Ecology of the microbiota of San Diego Bay, California. *Trans. San Diego Soc. Nat. Hist.* 14:9–40.

Le Calvez, J. 1938. Recherches sur les foraminifères. 1. Dévelopment et reproduction. *Arch. Zool. Exp. Gén.* 80:163–333.

Lee, J.J. 1974. Towards understanding the niche of Foraminifera. In: *Foraminifera* (R.H. Hedley and G. Adams, eds.), Vol. 1. Academic Press, London, pp. 207–60.

Lee, J.J., and W.D. Bock. 1976. The importance of feeding in two species of soritid foraminifera with algal symbionts. *Bull. Mar. Sci.* 26:530–37.

Lee, J.J., and S. Pierce. 1963. Growth and physiology of foraminifera in the laboratory; Part 4. Monoxenic culture of an Allogromiid with notes on its morphology. *J. Protozool.* 10:404–11.

Lee, J.J., and M.E. McEnery. 1970. Autogamy in *Allogromia laticollaris* (Foraminifera). *J. Protozool.* 17:184–95.

Lee, J.J., S. Pierce, M. Tentchoff, and J.A. McLaughlin. 1961. Growth and physiology of foraminifera in the laboratory. Part 1. Collection and maintenance. *Micropaleontology* 7:461–66.

Lee, J.J., H.D. Freudenthal, W.A. Muller, V. Kossoy, S. Pierce, and R. Grossman. 1963. Growth and physiology of foraminifera in the laboratory: Part 3. Initial studies of *Rosalina floridana* (Cushman). *Micropaleontology* 9:449–66.

Lee, J.J., H.D. Freudenthal, V. Kossoy, and A. Bé. 1965. Cytological observations on two planktonic foraminifera, *Globigerina bulloides* d'Orbigny, 1826, and *Globigerinoides ruber* (d'Orbigny, 1839) Cushman 1927. *J. Protozool.* 12:531–42.

Lee, J.J., M.E. McEnery, S. Pierce, H.D. Freudenthal, and W.A. Muller. 1966. Tracer experiments in feeding littoral foraminifera. *J. Protozool.* 13:659–70.

Lee, J.J., M.E. McEnery, and H. Rubin. 1969. Quantitative studies on the growth of *Allogromia laticollaris* (Foraminifera). *J. Protozool.* 16:377–95.

Lee, J.J., W.A. Muller, R.J. Stone, M.E. McEnery, and W. Zucker. 1969. Standing crop of foraminifera in sublittoral epiphytic communities of a Long Island marsh. *Mar. Biol.* 4:44–61.

Lee, J.J., L.J. Crockett, J. Hagen, and R.J. Stone. 1974. The taxonomic identity and physiological ecology of *Chlamydomonas hedleyi* sp. nov., algal flagellate symbiont from the foraminifer *Archais angulatus*. *Br. phycol. J.* 9:407–22.

Leidy, J. 1879. *Fresh-water Rhizopods of North America, Vol. 12.* U.S. Geol. Survey of the Territories, U.S. Government Printing Office, Washington, D.C., 234 p.

Lipps, J.H. 1973. Test structure in foraminifera. *Ann. Rev. Microbiol.* 27:471–88.

Loeblich, A.R., Jr., and H. Tappan. 1964. Sarcodina, chiefly "Thecamoebians" and Foraminiferida. In: *Treatise on Invertebrate Paleontology* (R.C. Moore, ed.), Part C Protista 2, Geol. Soc. Amer. University of Kansas, Lawrence, 900 p.

Loeblich, A.R., Jr., and H. Tappan. 1974. Recent advances in the classification of the Foraminiferida. In: *Foraminifera* (R.H. Hedley and C.G. Adams, eds.). Academic Press, London and New York, pp. 1–53.

Loiseaux, S., and J.A. West. 1970. Brown algal mastigonemes: Comparative ultrastructure. *Trans. Amer. Micros. Soc.* 89:524–32.

Looper, J.B. 1928. Observations on the food reactions of *Actinophrys sol*. *Biol. Bull.* 54:485–502.

Mackinnon, D.L., and R.S.J. Hawes. 1961. *An Introduction to the Study of Protozoa.* Clarendon Press, Oxford, 506 p.

Marszalek, D.S. 1969. Observations on *Iridia diaphana*, a marine foraminifer. *J. Protozool.* 16:599–611.

Matera, N.J., and J.J. Lee. 1972. Environmental factors affecting the standing crop of foraminifera in sublittoral and psammolittoral communities of a Long Island salt marsh. *Mar. Biol.* 14:89–103.

McEnery, M., and J.J. Lee. 1970. Tracer studies on calcium and strontium mineralization and mineral cycling in two species of foraminifera, *Rosalina leei* and *Spiroloculina hyalina*. *Limnol. Oceanogr.* 15:173–82.

McEnery, M.E., and J.J. Lee. 1976. *Allogromia laticollaris:* A foraminiferan with an unusual apogamic metagenic life cycle. *J. Protozool.* 23:94–108.

McGee-Russell, S.M., and R.D. Allen. 1971. Reversible stabilization of labile microtubules in the reticulopodial network of *Allogromia*. *Adv. Cell. Mol. Biol.* 1:153–84.

Moore, R.C. (ed.). 1954. *Treatise on Invertebrate Paleontology. Part D. Protista. 3. Protozoa* (chiefly radiolaria and tintinnina). University of Kansas, Lawrence, 195 p.

Moore, T.C. Jr. 1973. Late Pleistocene-Holocene oceanographic changes in the Northeastern Pacific. *J. Quat. Res.* 3:99–109.

Muller, W.A., and J.J. Lee. 1969. Apparent indispensability of bacteria in foraminiferan nutrition. *J. Protozool.* 16:471–78.

Müller-Merz, E., and J.J. Lee. 1976. Symbiosis in the larger foraminiferan *Sorites marginalis* (with notes on *Archaias* spp.). *J. Protozool.* 23:390–96.

Murray, J.W. 1973. *Distribution and ecology of Living Benthic Foraminiferids.* Crane, Russak & Co., New York, 274 p.

Nyholm, K-G. 1953. Studies on recent Allogromiidae (2): *Nemogullmia longevariabilis* n.g., n.sp. from the Gullmar Fjord. *Contrib. Cushman Found. Foram. Res.* 4:105–6.

Nyholm, K-G. 1956. On the life cycle and cytology of the Foraminiferan *Nemogullmia longevariabilis*. *Zool. Bidrag Uppsala* 31:483–95.

Ostenfeld, C.H. 1904. On two new marine species of heliozoa occurring in the plankton of the North Sea and the Skager rak. *Medd. Komm. Havundersøgelser, Ser. Plankton* 1(2):1–5.

Page, F.C. 1967. Taxonomic criteria for *Limax* amoebae, with descriptions of 3 new species of *Hartmannella* and 3 of *Vahlkampfia*. *J. Protozool.* 14:499–521.

Page, F.C. 1969. *Platyamoeba stenopodia*, n.g., n. sp., a freshwater amoeba. *J. Protozool.* 16:437–41.

Page, F.C. 1970. Two new species of *Paramoeba* from Maine. *J. Protozool.* 17:421–27.

Page, F.C. 1971a. Two species of *Flabellula* (Amoebida, Mayorellidae). *J. Protozool.* 18:37–44.

Page, F.C. 1971b. A comparative study of five fresh-water and marine species of Thecamoebidae. *Trans. Amer. Micros. Soc.* 90:157–73.

Page, F.C. 1972. *Rhizamoeba polyura* n.g., n. sp. and uroidal structures as a taxonomic criterion for amoebae. *Trans. Amer. Micros. Soc.* 91:502–13.

Page, F.C. 1973. *Paramoeba:* A common marine genus. *Hydrobiologia* 41:183–88.

Page, F.C. 1976a. A revised classification of the Gymnamoebia (Protozoa: Sarcodina). *Zool. J. Linn. Soc.* 58:61–77.

Page, F.C. 1976b. *An Illustrated Key to Freshwater and Soil Amoebae with Notes on Cultivation and Ecology.* Ambleside Freshwater Biol. Assn., Sci. Publ. No. 34, Titus Wilson Ltd., Tyndal, England, 155 p.

Penard, E. 1904. *Les Héliozoaires d'eau douce.* Henry Kündig, Geneva, 341 p.

Petrushevskaya, M.G. 1966. The radiolarians in plankton and bottom sediments. In: *Geokhimya Kremnezema.* Izdatelstvo Nauka, Moscow, pp. 219–45.

Petrushevskaya, M.G. 1971. Radiolaria in the plankton and recent sediments from the Indian Ocean and Antarctic. In: *The Micropaleontology of Oceans* (B.M. Funnell and

W.R. Riedel, eds.). Cambridge University Press, London and New York, pp. 319–29.

Phleger, F.B. 1960. *Ecology and Distribution of Recent Foraminifera.* Johns Hopkins Press, Baltimore, Md., 297 p.

Radir, P.L. 1927. *Trichamoeba schaefferi,* a new species of large marine ameba from Monterey Bay, California. *Arch. Protistenk.* 59:289–300.

Renz, G.W. 1976. The distribution and ecology of radiolaria in the Central Pacific: Plankton and surface sediments. *Bull. Scrips. Inst. Oceangr.* 22:1–267.

Rhumbler, L. 1909. Die Foraminiferen (Thalamophoren) der Plankton-Expedition: Teil 1—Die Allgemeinen Organisationsverhältnesse der Foraminiferen. *Plankton-Exped. Humboldt-Stiftung Ergebn.* 3 (L,c):1–331.

Riedel, W.R., and A. Sanfilippo. 1971. Cenozoic radiolaria from the Western tropical Pacific, Leg 7. In: *Deep Sea Drilling Project,* Vol. VII (2). U.S. Government Printing Office, Washington, D.C., pp. 1530–1672.

Ross, C.A. 1972. Biology and ecology of *Marginopora vertebralis* (Foraminiferida), Great Barrier Reef. *J. Protozool.* 19:181–92.

Roth, L.E., and Y. Shigenaka. 1970. Microtubules in the heliozoan axopodium. II. Rapid degradation by cupric and nikelous ions. *J. Ultrastr. Res.* 31:356–74.

Roth, L.E., D.J. Pihlaja, and Y. Shigenaka. 1970. Microtubules in the heliozoan axopodium. I. The gradion hypothesis of allosterism in structural proteins. *J. Ultrastr. Res.* 30:7–37.

Röttger, R. 1974. Larger foraminifera: Reproduction and early stages of development in *Heterostegina depressa. Mar. Biol.* 26:5–12.

Röttger, R. 1976. Ecological observations of *Heterostegina depressa* (Foraminifera, Nummulitidae) in the laboratory and in its natural habitat. In: *First Int. Symp. Benthonic Foraminifera of Continental Margins,* Part A, *Ecology and Biology, Maritime Sediments* Spec. Pub. 1, Halifax, N.S. Canada, pp. 75–79.

Röttger, R., and W.H. Berger. 1972. Benthic Foraminifera: Morphology and growth in clone cultures of *Heterostegina depressa. Mar. Biol.* 15:89–94.

Sachs, H.M. 1973a. North Pacific radiolarian assemblages and their relationship to oceanographic parameters. *J. Quat. Res.* 3:73–88.

Sachs, H.M. 1973b. Late Pleistocene history of the North Pacific: Evidence from a quantitative study of Radiolaria in a cove. *J. Quat. Res.* 3:89–98.

Sassaki, C. 1894. Untersuchungen über *Gymnosphaera albida,* eine neue marine Heliozoe. *Jena Zeits. Naturw. NF* 21:45–52.

Sawyer, T.K. 1971. Isolation and identification of free-living marine amoebae from upper Chesapeake Bay, Maryland. *Trans. Amer. Micros. Soc.* 90:43–51.

Sawyer, T.K. 1975. Marine amoebae from surface waters of Chincoteague Bay, Virginia: Two new genera and nine new species within the families Mayorellidae, Flabellulidae, and Stereomyxidae. *Trans. Amer. Micros. Soc.* 94:71–92.

Schaeffer, A.A. 1926. *Taxonomy of the Amebas with Description of Thirty-nine New Marine and Freshwater Species.* Carnegie Inst., Pub. No. 345, Washington D.C., 116 p.

Schaudinn, F. 1896. Über den Zeugungskreis von *Paramoeba eilhardi* n.g., n. sp. *Sitz.-Ber. Kön. Preuss. Akad. Wiss. Berlin* 2, 31–41.

Schewiakoff, W. 1926. *Die Acantharia des Golfes von Neapel. Fauna e flora del Golfo di Napoli.* Monografia No. 37, Friedlander, Berlin, 755 p.

Schmaljohann, R., und R. Röttger. 1976. Die Symbionten der Grossforaminifere *Heterostegina depressa* sind Diatomeen. *Naturwissenschaften* 63:486.

Shigenaka, Y., L.E. Roth, and D.J. Pihlaja. 1971. Microtubules in the heliozoan axopodium. III. Degradation and reformation after dilute urea treatment. *J. Cell Sci.* 8:127–51.

Sieburth, J.McN. 1975. *Microbial Seascapes.* University Park Press, Baltimore, Md., 248 p.

Sieburth, J.McN., P-J. Willis, K.M. Johnson, C.M. Burney, D.M. Lavoie, K.R. Hinga, D.A. Caron, F.W. French III, P.W. Johnson, and P.G. Davis. 1976. Dissolved organic matter and heterotrophic microneuston in the surface microlayers of the North Atlantic. *Science* 194:1415–18.

Sliter, W.V. 1970. *Bolivina doniezi* Cushman and Wickenden in clone culture. *Cushman Found. Foram. Res. Contrib.* 21:87–99.

Sprague, V., R.L. Beckett, and T.K. Sawyer. 1969. A new species of *Paramoeba* (Amoebida, Paramoebida) parasitic in the crab *Callinectes sapidus. J. Invert. Pathol.* 14:167–74.

Thompson, D'A.W. 1942. *On Growth and Form.* Cambridge University Press, 1116 p. [Abridged edition (J.T. Bonner, ed.). 1961, 346 p.]

Tilney, L.G., and B. Byers. 1969. Studies on the microtubules in heliozoa. V. Factors controlling the organization of microtubules in the axonemal pattern in *Echinosphaerium (Actinosphaerium) nucleofilum. J. Cell. Biol.* 43:148–65.

Trégouboff, G. 1953a. Classe des Héliozoaires (Héliozoa Haeckel 1866). In: *Traité de Zoologie* (P-P. Grassé, ed.), Vol. 1 (2). Masson et Cie., Paris, pp. 437–89.

Trégouboff, G. 1953b. Radiolaria (Auctorum) Caractères Généraux. In: *Traité de Zoologie* (P-P. Grassé, ed.), Vol. 1(2). Masson et Cie., Paris, pp. 269–388.

Valkanov, A. 1970. Beitrag zur Kenntnis der Protozoen des Schwarzen Meeres. *Zool. anz.* 184:241–90.

Wailes, G.H. 1927. Rhizopoda and Heliozoae from British Columbia. *Ann. Mag. Nat. Hist., 9th Ser.* 20:153–56.

Wailes, G.H. 1932. Description of new species of protozoa from British Columbia. *Contrib. Canad. Biol.* 7:213–19.

Wailes, G.H. 1937. *Canadian Pacific fauna. 1. Protozoa (a, lobosa; b, reticulosa; c, heliozoa; d, radiolaria).* Biol. Bd. Canada, Toronto, 14 p.

Wailes, G.H., and J. Hopkinson. 1915–21. *British freshwater*

Rhizopoda and Helizoa. Ray Society, London, Vol. 3, 156 p; Vol. 4, 129 p., Vol. 5, 72 p.

Watters, C. 1968. Studies on the motility of the Heliozoa. I. The locomotion of *Actinosphaerium eichhorni* and *Actinophyrs* sp. *J. Cell Sci.* 3:231–44.

Wright, T.S. 1867. Observations on British zoophytes and protozoa. *J. Anat. Physiol.* 1:332–38.

Zuelzer, M. 1909. Bau und Entwicklung von *Wagnerella borealis* Mereschk. *Arch. Protistenk.* 17:135–202.

CHAPTER 24

Ciliates

The ciliates, like the rhizopods, are a group of protozoa with an obviously phagotrophic nutritional mode. Except for a few highly specialized forms with a much reduced ciliature or those forms with cilia only present in the reproductive stage, the ciliates are distinguished by the somatic cilia that cover a greater or lesser area of the body and an oral ciliature derived from the somatic ciliature, which can be slightly to highly developed. In addition to the funneling of food, the oral ciliature serves, along with the somatic ciliature, as a mechanism for locomotion. The ciliates are placed in the phylum Ciliophora (Puytorac et al., 1974; Corliss, 1974a,b, 1975).

The cilia of the ciliates occur in rows (kineties) and beat in a coordinated manner such as in sequence or in unison. The cilia, like the flagella, are permanent thread-like extensions that move irregularly or periodically to create currents that propel both food and the microorganism itself. Cilia are short forms of flagella and have the same ultrastructure (see Plate 24-1,*C,D*). They originate in a basal body or kinetosome within the cell. In cross section, the basal body is seen as nine sets of triplet fibrils, set at an angle, which then become nine pairs of peripheral fibrils and two internal fibrils in the shaft of the cilia that extend from the cell and are enclosed within a continuation of the cell envelope. The kinetosome and its derivatives have been reviewed by Grain (1969); the structural differentiation of the somatic ciliature, including locomotor, thigmotactic (adhesive cilia), vestibular (oral cilia), and caudal ciliature (posterior cilia), has been described in detail by Antipa (1972). Cilia are absent at some stages of the life cycles of some species of *Conidophrys* (Apostomatida), many members of the family Sphenophryidae (Rhyncoida), and all members of the Suctoria. The suctorians are sessile forms that catch their food with tentacles. Only their reproductive swarmer stage is ciliated.

In addition to their overall morphology, much of the detail used in classification of the ciliates is in the infraciliature within the cell envelope or pellicle. These

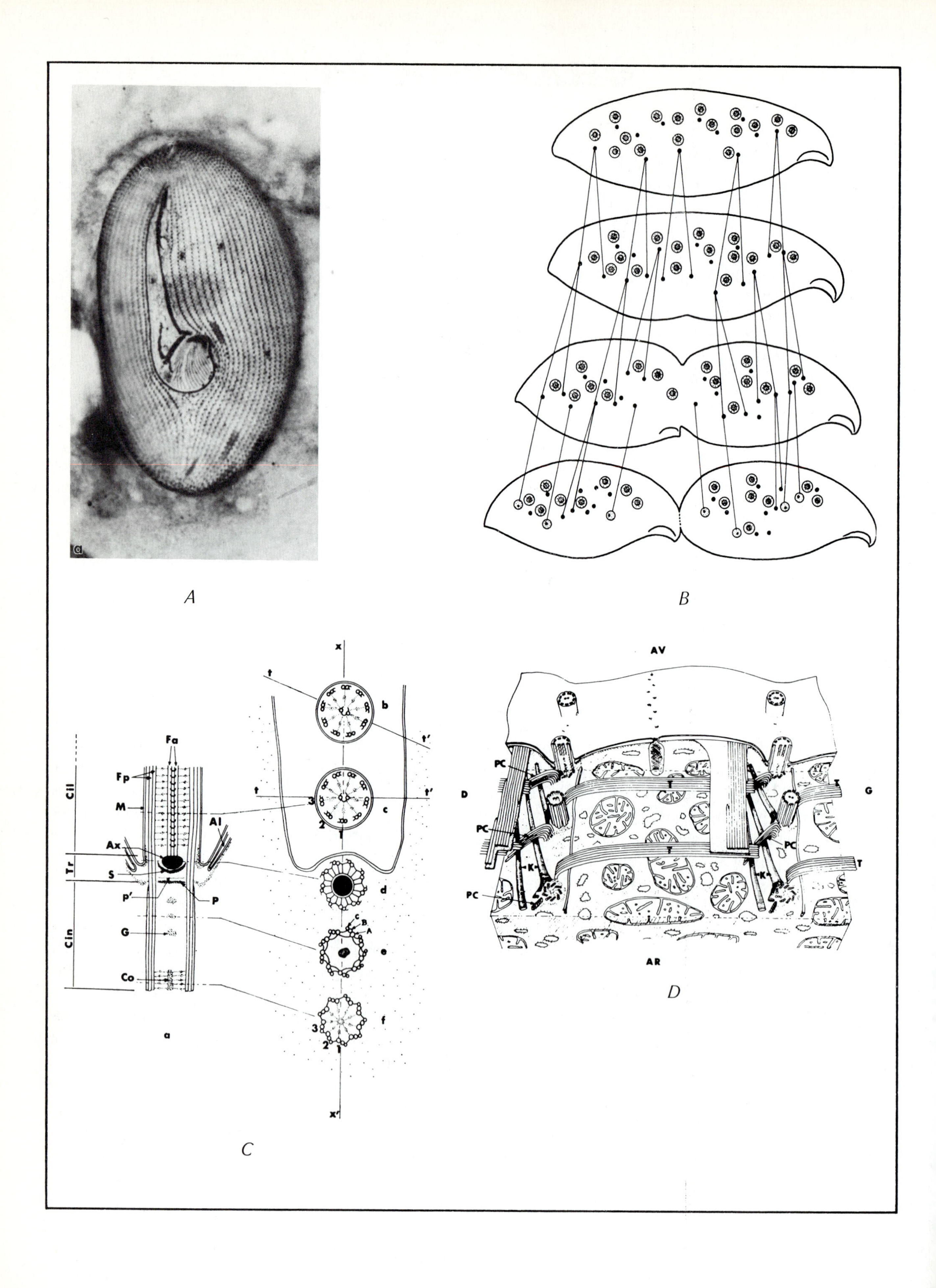
a
A
B
x
t
b
t'
Fa
Fp
Cil
M
c
Al
Ax
Tr
S
p'
P
d
G
Cin
e
Co
f
a
x'
C
AV
PC
D
G
T
K
AR
D

PLATE 24-1. The distinguishing characteristics of ciliates: ciliature and nuclear dimorphism. (*A*) A silver-impregnated individual of *Pleuronema marinum* (approximately 100 μm in length), showing the infraciliature of the ventral side, including the buccal area, which is due to the staining of kinetosomes in the sub-pellicular basal area; (*B*) transverse fission in *Loxodes magnus*, showing the anterior proter and posterior opisthe daughter cells and the nuclear dimorphism of the somatic macronuclei and generative micronuclei; (*C*) diagram of longitudinal and transverse sections of a cilium and kinetosome, showing 9+2 fibrillar construction and (*D*) a view in perspective from the exterior of the cell (*Tetrahymena*). (*A*) from J. Dragesco, 1968, *Protistologica* 4:85–106; (*B*) from K.G. Grell, 1973, *Protozoology*, Springer-Verlag, New York and Berlin; (*C,D*) from J. Grain, 1969, *L'Année Biol.* 8:53–97.

details are brought out in silver impregnation preparations (Plate 24-1,*A*), such as the Chatton-Lwoff technique as modified by Corliss (1953) or the Bodian "Protargol" technique as perfected by Tuffrau (1967). The former technique apparently shows only kinetosomes (ciliated basal bodies); the latter also shows nuclei, nonciliated basal bodies, and other organelles (Corliss, 1974a). Silver-impregnated preparations of ciliates are useful for study, photomicrography, and as reference specimens for newly described species. An international type-slide collection has been established (Corliss, 1963) and is slowly being added to as major taxonomic studies are completed (Corliss, 1972). The type and arrangement of the infraciliature form one of the criteria for classification. The use of scanning electron microscopy to determine ciliate surface patterns and organelles (Small, Marszalek, and Antipa, 1971) may also become a taxonomic tool. Corliss (1959) has developed a very useful glossary of ciliate terminology and an illustrated dichotomous (successively forking) key to the higher groups. A similar key for the marine ciliates of the northeastern coast of the United States has been prepared by Borror (1973). The historical review by Corliss (1974a), his remarks on the large ciliate class Kinetofragminophora (Corliss, 1974b), and his revisions of the new scheme (Corliss, 1975) help elucidate the taxonomy of this phylum. The "age of ultrastructure" has been responsible for many changes from the traditional, pre-1970 classifications.

The second distinguishing feature of the ciliates concerns their binary fission and nuclear division. This is illustrated for *Loxodes magnus*, a nonmarine form, in Plate 24-1,*B*. Fission is usually transverse and produces two daughter cells—an anterior proter and a posterior opisthe. Most ciliates exhibit nuclear dimorphism, having both somatic macronuclei and generative micronuclei. The exception is the gymnostome (rhabdophorine or now primociliate) family Stephanopogonidae with only one type of nucleus. The macronucleus usually has many sets of chromosomes (hyperpolypoid), including compact, ribbon, beaded, and branched forms. In such families as the Loxodidae, Geleiidae, and Trachelocercidae, the macronuclei cannot replicate (see Plate 24-1,*B*). The small micronuclei (characteristically under 5 μm in diameter) are diploid and usually must be stained to be seen. The different forms of nucleation and the ultrastructure of the nucleus in marine ciliates is shown in the papers by Raikov (e.g., 1973). Studies on the nuclear apparatus of ciliates have strongly influenced recent taxonomic revisions (Corliss, 1974a,b, 1975; Puytorac et al., 1974).

The ciliates have adapted to a number of environments. Some, like the tintinnids and the *Prorodon–Monedinium* group, are coastal plankton, which have been described in a monograph by Fauré-Fremiet (1924). Monographs dealing with the oceanic species of tintinnids are cited by Laval (1973). The creeping forms live on fouling surfaces (Deroux, 1970) and occupy such microhabitats as fine sediments, under reducing conditions (Fenchel, 1969; Tucolesco, 1961); sand grains (Hartwig, 1973b); and films of rapidly sedimenting debris on the sea floor (Lackey, 1936); they also grow on fine filamentous algae (Borror, 1968). The stalked sessile forms like *Zoothamnium* and *Vorticella* attach to inanimate objects (Persoone, 1968) and living surfaces such as seaweeds (Langlois, 1975), hydroids, copepods (Herman and Mihursky, 1964; Hirche, 1974), and bryozoans (Sieburth, 1975). Saltmarshes also offer a rich and diverse ciliate population due, in part, to their rich microflora of sulfur bacteria. Kahl (1928) described the populations of brackish water ditches between Hamburg and Lübeck, Germany, whereas Kirby (1934) found similar species in saltmarshes in California. Webb (1956) has described the seasonal and spatial distribution as well as the communities and food supply of ciliates in saltmarshes of the Dee Estuary in England. Borror (1965, 1972b) has described the morphology, ecology, and systematics of ciliates from tidal marsh pools in New Hampshire. Several forms in several different hierarchies are attached to or associated with a variety of animals with which they share their food as ectocommensals, entocommensals, or symbionts. Ascidians (sea squirts) often house a suctorian, a hypotrich, and a thigmotrich as symbionts (Burreson, 1973); other groups are associated with the mouth parts of crustacea (Guilcher, 1951a), the mantle cavity of lamellibranchs (bivalve molluscs) (Kidder, 1933 a,b,c,d), or the gills of fish and crustaceans (Bradbury, 1966a; Couch, 1967).

All ciliates have a permanent opening (cytostome) for the ingestion of food except for the specialized forms in the Suctoria, Astomatida, and those ciliates that were formerly in the Thigmotrichida. Many are voracious eaters and help keep the populations of the smaller, more rapidly dividing microorganisms in check. The ciliates contribute to the economy of the sea by the consumption of both microorganisms and decaying animal tissue and the funneling of their biomass up the food chain. Fauré-Fremiet (1961), Dragesco (1962), and Fenchel (1968a) have studied the capture and ingestion of prey by ciliated protozoa. The ciliates can be divided into a number of categories according to their feeding behavior. The two basic categories are the microphages, which feed on bacteria, yeasts, and small flagellates, and the macrophages, which feed on larger flagellates, diatoms, ciliates, and decaying animal flesh. Gram-negative bacteria are a good food source for *Paramecium*, whereas gram-positive bacteria are either a poor food source or are toxic (Barna and Weis, 1973). The macrophages can be further subdivided into carnivores (eating other ciliates and microfauna), herbivores (eating diatoms and other microalgae), and histophages (eating injured and dead animal tissue). Beautiful scanning electron micrographs of the capture and ingestion of *Paramecium* by *Didinium* are reproduced in a study by Wessenberg and Antipa (1970). Dragesco (1962) illustrated the mechanics of ingestion, and Fenchel discussed the feeding habits of individual benthic species (Fenchel, 1968a) and illustrated the major benthic species in the different feeding types (Fenchel, 1969). A few species have been cultured in defined or semi-defined media, but most require particulate food. Fauré-Fremiet (1961) has discussed the culture of ciliated protozoa; and Fauré-Fremiet (1967) and Noland and Gojdics (1967) have published excellent reviews on the overall ecology of free-living ciliates (and other protozoa).

In a study on the colonization of fouling films by ciliates in the harbor of Ostend (Belgium), Persoone (1968) discussed the species according to feeding type. He found that one-half of the 30 species were microphages and attributed this to the muddy nature of the fouling site. During the winter, carnivores and histophages disappeared, decreasing the diversity of species by one-half. The stalked *Zoothamnium* was present year round and was the dominant form except when it was eclipsed by a cyrtophorid (*Dysteria*) in winter. The latter fed specifically on the filamentous gliding bacterium *Leucothrix mucor*, which was quite abundant on the fouling surface, presumably being supported by the pennate diatom microflora. Deroux (1970) made a similar study on the ciliates of fouling glass slides near Roscoff, France and found that the cyrtophorines (Hypostomata) were the most widely represented form. The microfauna in the interstitial spaces between sand grains (psammon or psammophiles) includes many ciliates as well as metazoa. The terminology and sampling procedures for this microfauna were reviewed by Zinn (1968). A subtidal sampler and a seawater–ice procedure for concentrating ciliates and flagellates were developed by Uhlig (1968), and Spoon (1972) later developed alternative methods for extracting and concentrating protozoa from sediments.

The reviews or monographs by Fauré-Fremiet (1950), Dragesco (1960), Hartwig (1973a,b), and Corliss and Hartwig (1977) are excellent starting points for studying the systematics, ecology, and literature of this group. Borror (1963) compared the ciliates of a sandy bottom with areas of diatom detritus and concluded that the former favored ciliates and the latter metazoa and that benthic ciliates are important consumers of bacteria and diatoms. The paper by Tucolesco (1961) and the four-paper series by Fenchel (1967, 1968a,b, 1969) are important studies of the ciliates of sediments ranging from sandy oxidized environments to silty reduced habitats. The first paper by Fenchel, a quantitative study, compares ciliate and metazoan populations, and the second details the food of specific species. The third paper shows that reproduction, which is still operative between 0 and 4 °C varies between 2 and 46 hours at 10 °C depending upon the size of the species. The concluding monograph (Fenchel, 1969) is a classic paper on the structure and function of the benthic ecosystem and its effect on the benthos, and especially the ciliates.

In his taxonomic monograph on ciliates, Corliss (1961) considered the entire ciliate assemblage to be more homogeneous than subsequent ultrastructural studies have shown it to be. He combined the peritrichs (forms with a much reduced somatic ciliature and a conspicuous oral cavity and oral ciliature) and suctorians (aberrant adult forms, without cilia, that feed with tentacles and produce ciliated mouthless larvae) with the holotrichs (forms with a uniform, simple somatic ciliature) and omitted only the spirotrichs (forms with highly developed membranelles used for feeding and locomotion). The committee revising protozoan classification (Honigberg et al., 1964) retained the usual division of the ciliates into Holotrichia, Peritrichia, Spirotrichia, and Suctoria, and this system is used in such current texts as Grell's (1973). But new ideas and information based upon morphological, ultrastructural, and ecological studies of the last decade have catalyzed a revision of the systematics of the ciliates. The concept of a three-division organization by Jankowski (1967) has been refined by others, until a more or less intermediate classification scheme has been developed. Although many details need to be worked out, the essential hierarchies are presented by Puytorac et al. (1974) and Corliss (1974a). The latter has done a masterful job of explaining the epochs of ciliate taxonomy that have led to these revisions. The differences in these two versions are apparent in the tables in the appendix of Corliss (1974b), and "unifying" revisions have already appeared (Corliss, 1975, 1977). Corliss (1974b) cites three large works on ciliate taxonomy that are nearing completion. These are the *Fauré-Fremiet Tome 2* of the *Traité de Zoologie*, edited by Grassé; the *Illustrated Guide to the Study of the Protozoa* of the Society of Protozoologists edited by J.J. Lee; and the second edition of *The Ciliated Protozoa*, written by J.O. Corliss. These treatises will be very helpful in determining the hierarchies of the species found in marine samples and the literature. For species identification, however, such monographs as the four-volume treatise by Kahl (1930, 1931, 1932a, 1935) and similar monographs cited here still must be consulted.

The newly proposed classification schemes are more complex than the older schemes and reflect a greater diversity of form and relationship than were previously apparent. The division of the ciliates into three classes and the names given them are based upon the complexity of the perioral or oral ciliature. The simplest ciliates are those with kinetofragmons or kinetofragments, which are kinetids (rows of cilia) separated or derived from a somatic kinety that form a part of the perioral ciliary apparatus. They make up the largest and most diverse ciliate class, the Kinetofragminophora. The higher ciliates have a more complex oral ciliature in which the cilia are fused to form membranelles or "hymens." They are divided into two classes—the Oligohymenophora, with few hymenal structures, and the Polyhymenophora, with many. The Oligohymenophora contains the Hymenostomata, including forms in the old "Thigmotrichida" with mouths, and even the mouthless Astomatida and the old Peritricha; the Polyhymenophora contains the spirotrichs with essentially no revision (except for the loss of the ophryoscolecids or Entodiniomorphida to the first class in the subclass Vestibulifera alongside the trichostomes).

The taxonomic ranks and names used in the following pages are those suggested by Corliss (1977) in his complete "skeletal outline" of the latest classification of the whole phylum Ciliophora.

A. KINETOFRAGMINOPHORA

The kinetofragminophorans have a uniform and simple somatic ciliature and are distinguished from the Oligohymenophora and the Polyhymenophora by their smaller and simpler buccal or oral areas. When a "buccal" ciliature is present, it is quite inconspicuous. This very large class is spread over a number of habitats and shows a great diversity of form. Free-living marine forms occur in all four subclasses.

1. Gymnostomata

Most members of this subclass are large, commonly encountered forms that are unique in being gulpers rather than swallowers like most of the ciliates. This is due to a lack of a ciliated buccal or vestibular area to direct the food to a buccal cavity. Instead, the cytostome-cytopharyngeal complex opens directly to the outside. The rhabdophorine (still a convenient descriptive term) ciliates that make up this subclass are typically carnivorous forms, having a body with a uniform ciliature that is round in cross section. The cytostome is either apical or subapical (at the end) or lateral (on the side). The cytopharynx contains an expansible armature of fibrillar rods, the trichites or, better, nemadesms (= nematodesmata, strictly speaking). Corliss (1975) divides the rhabdophorine ciliates, which are mainly interstitial forms living in sands and sediments,

PLATE 24-2. The association of bacteria with holotrichous ciliates as symbionts (*A*), (*B*), and (*C*) and nuclear inclusions (*D*). (*A*) A thick section of the flat ribbon-like body of *Kentrophoros fistulosum* that curls longitudinally forming a tube with the ventral cilia on the outside, a brush-like coat of fusiform sulfur bacteria (B) on the dorsal side in the lumen, and the lateral edges joining at Fd. (N, nuclei) (2,000 X); (*B*) a thin section of *K. fistulosum* showing ingested bacteria (Bi), epizoic bacteria (B), some of which are dividing longitudinally (arrows), sub-pellicular mucoprotein granules (Gr), and cilia (C) (7,200 X); (*C*) as in (*B*), showing sulfur bacteria (B) perpendicular and spirochete-like bacteria (S) parallel to the host pellicle (15,000 X); (*D*) groups of bacteria (B) in the macronucleus of *Pseudoprorodon arenicola* (N, nucleolus) (9800 X). (*A,B,C*) from I.B. Raikov, 1971, *Protistologica* 7:365–78; (*D*) from M.R. Kattar, 1972, *Protistologica* 8:135–41.

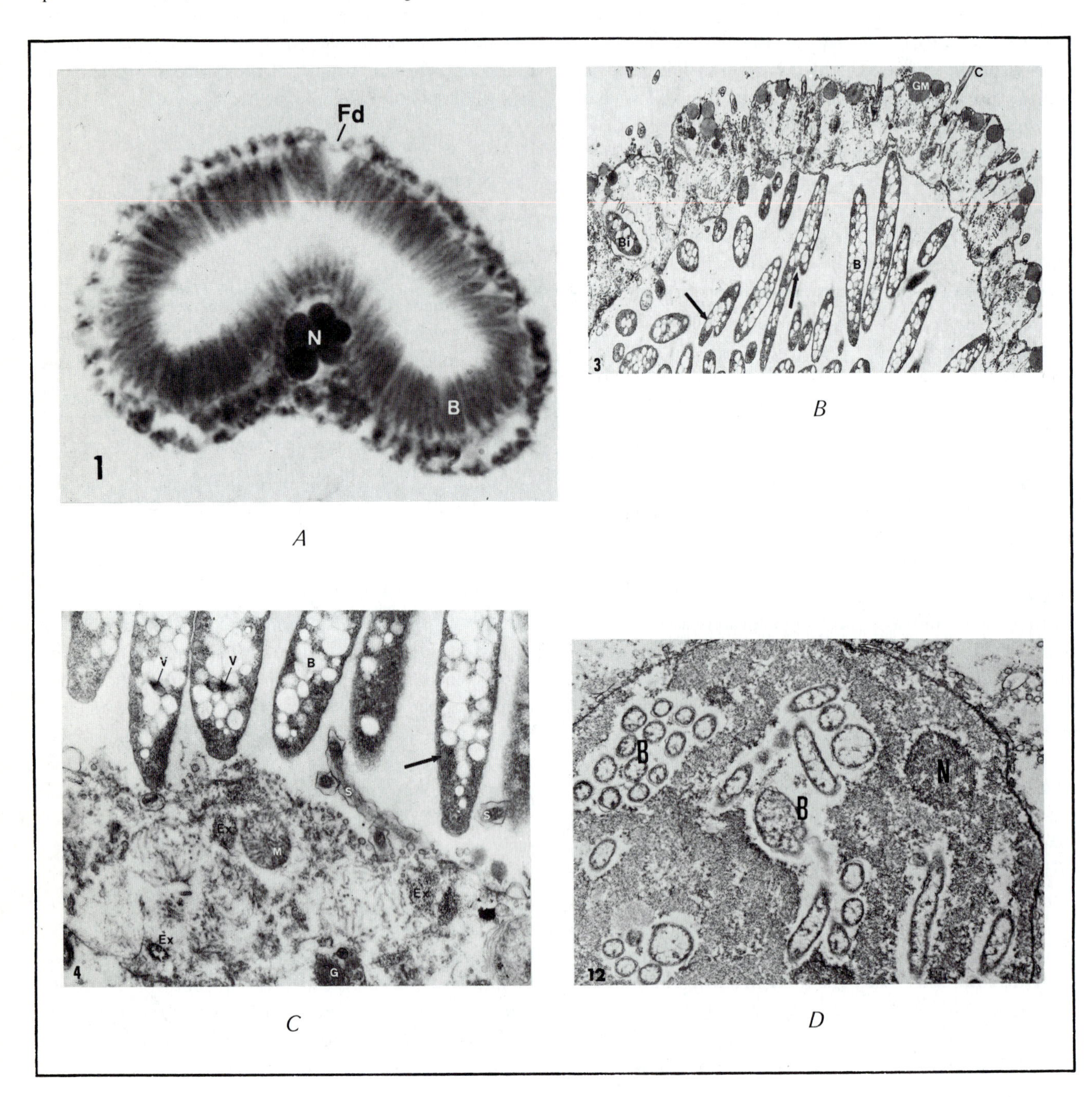

into five orders. The Primociliatida and the Karyorelictida contain relatively simple forms presumed to be related to ancestral ciliates; the Prostomatida and Haptorida contain more advanced forms, including those with a seizing behavior; and the Pleurostomatida contains those forms with lateral mouths.

The first two orders (Corliss, 1975, 1977) contain species prominent among the sand-inhabiting forms. The pertinent literature on these ciliates that live in the interstitial spaces between sand and sediments is cited by Corliss (1974a) and Corliss and Hartwig (1977). An interesting bacterial association in a primitive karyorelectid psammophile with no cytostome, *Kentrophoros fistulosum*, has been characterized by Raikov (1971) (see Plate 24-2). Its very thin, ribbon-like body curls longitudinally, so that the lateral edges meet to form a tube with the outer ventral side being ciliated. The dorsal side is the lumen of the tube and contains bacteria; both "sulfur bacteria" attached at their ends so as to be perpendicular and spirochete-like bacteria lying on edge parallel to the pellicle. The pellicle of the ciliate is a single unit membrane, and one end of the longitudinally dividing sulfur bacteria is intimately associated with this pellicle. The host phagocytizes these bacteria with pseudopodia-like cytoplasmic protrusions and incorporates them within membrane-bounded food vacuoles. These primary vacuoles, which are rich in polysaccharides, coalesce to aggregate near lysosome-like structures. After the digestion of the bacteria, multi-membranous systems are formed, which later fragment into electron-dense excretory vesicles. An ultrastructural study of another sand-loving karyorelictid (in a different family), *Trachelonema sulcata*, reveals numerous saucer-shaped sacculi associated with the mitochondria, which appear to migrate to the ectoplasm and form "protrichocysts" or mucocysts (Kovaleva and Raikov, 1973). This is reminescent of the body covering of the phytoflagellates, which is produced by the Golgi bodies.

A number of prostomatid ciliates are marine forms. The gymnostome *Pseudoprorodon arenicola* is another example of a phagotrophic microorganism with bacteria associated with the nucleus (Plate 24-2,*D*). The macronucleus contains bacteria with a double membrane that appears to lyse the nuclear material around them (Kattar, 1972). This lysis is reminiscent of the activity of the intranuclear bacteria in *Paramoeba eilhardi* (Grell and Benwitz, 1970).

An example of a haptoridan with a characteristic seizing behavior is the greedy predator *Chaenea vorax*, shown in Plate 24-3. It feeds on small ciliates, including *Uronema* and *Cyclidium*. The capture and ingestion of prey described by Dragesco (1962) is illustrated in Plate 24-3,*B*. The trichocysts (= toxicysts) in the cytostome instantly immobilize the prey, which is rapidly gulped with the aid of the nemadesms as the predator twists rhythmically. The prey, when lodged sideways in the cytostome, is apparently ruptured by powerful internal suction. A level near the opening of the cytostome–cytopharyngeal complex is shown in cross section in Plate 24-3,*C*. This complex varies somewhat with the genera. The details of this complex in the prorodentine *Pseudoprorodon arenicola* show the array of fibrillar nemadesms as well as the toxicysts (Kattar, 1972).

The rhabdophorine ciliates with lateral mouths, which are now placed in the Pleurostomatida, have been described by Kahl (1930, 1931, 1932a, 1935), Borror (1965, 1972a, 1973), and Dragesco (1960, 1965, 1966a,b). Many are important marine forms.

2. Vestibulifera

The second subclass of the Kinetofragminophora is the Vestibulifera. As its name implies, it contains those forms with a simple entryway (vestibule) leading to the mouth. The marine forms are predominantly in the Trichostomatida, which are the simplest of swirlers. The somatic ciliature often continues into an ectoplasmic depression or vestibulum to become feeding cilia that swirl food into the cytostome. An example is *Coelosomides marina*, shown in Plate 24-4,*A*. A rapid swimmer, it closely resembles a gymnostome except for its ciliated vestibulum. This specialized form is a member of the mesopsammon and feeds upon bacteria and microalgae, which are ingested into the highly vacuolated endoplasm in the center of the cell. The more compact peripheral endoplasm contains the macronucleus and the micronucleus. Another interesting trichostome is the large *Lechriopyla*, which was first described by Lynch (1930). It is an entocommensal in the gut of strongylocentroid echinoids (sea urchins) endemic in northern Pacific Coast waters and a predator along with such scuticociliates as *Thyrophylax* (see below and Lynn and Berger, 1973), which are concentrated in the last two intestinal festoons and the anterior one-half of the rectum of the sea urchin gut.

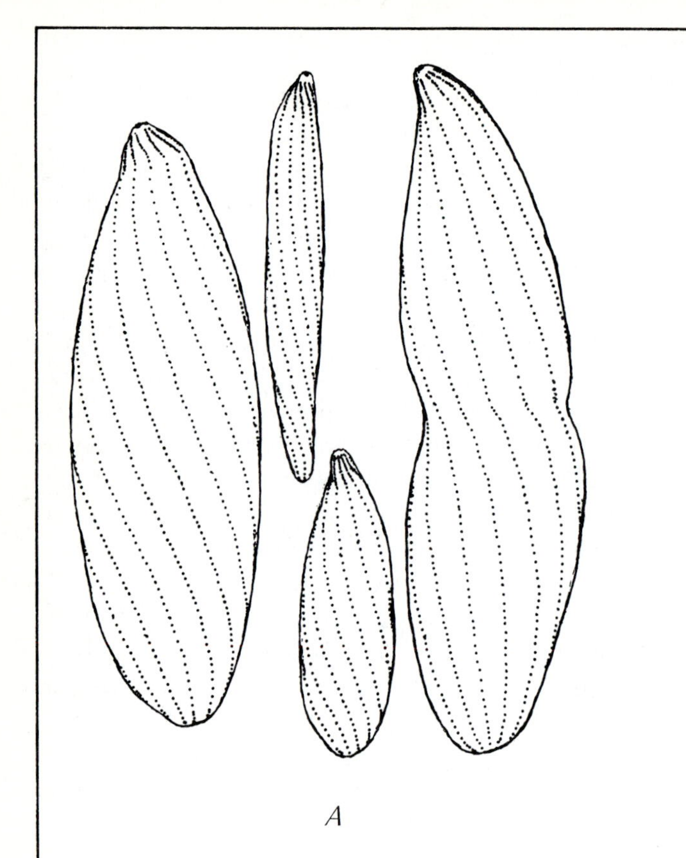

A

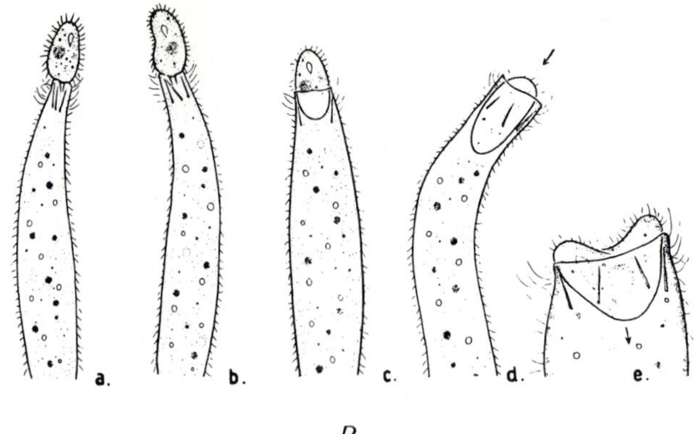

B

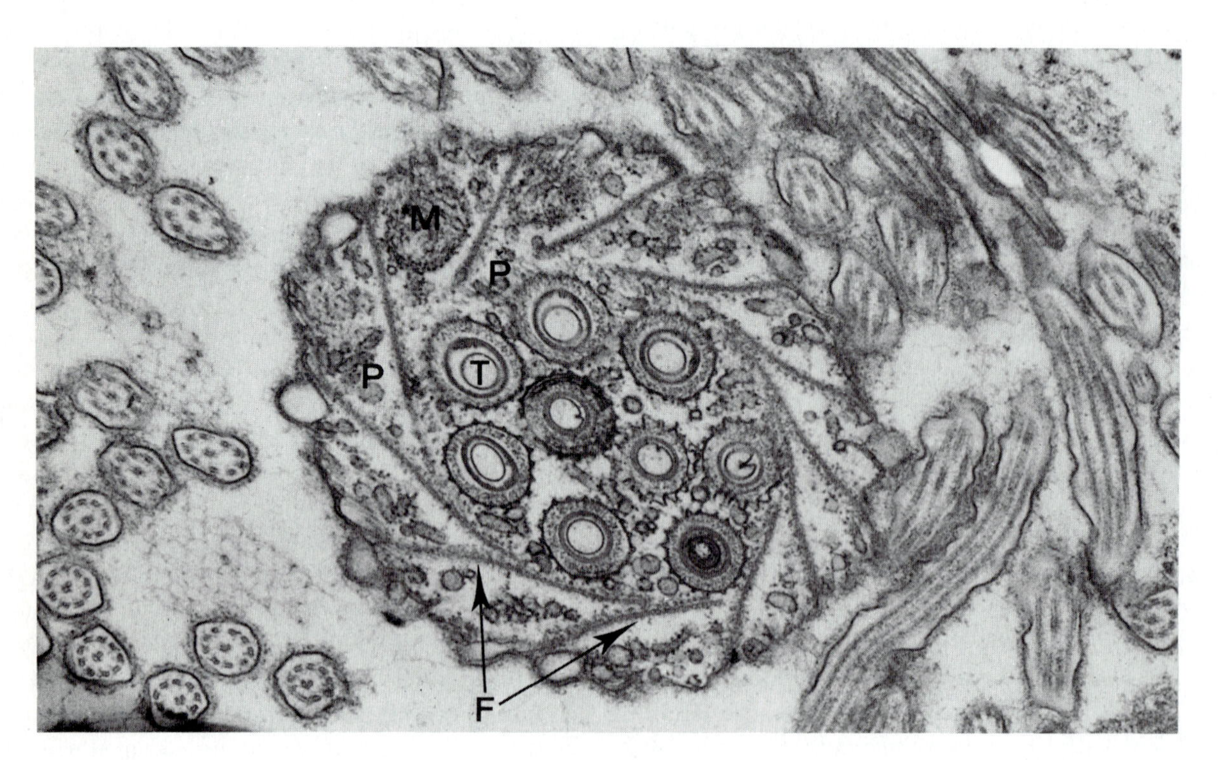

C

PLATE 24-3. *Chaenea vorax*, a prolific and greedy predator, is a typical rhabdophorine of the order Haptorida. (*A*) The round bodies, contours, and infraciliature of four individuals (450 X). (*B*) Capture and ingestion of prey: (a) instant immobilization by trichocysts; (b) dilation of mouth; (c) rapid ingestion of prey; (d) anterior retraction, causing rhythmic movements; (e) the rupture of laterally oriented prey due to the powerful internal suction. (*C*) Cross section of the rostrum, showing cilia of the crown cut at different angles in the periphery and the mouth structure with eleven fibrillar screens (F), phagoplasmic vesicles (P), and nine trichocysts (T) (22,500 X). (*A,C*) from E. Fauré-Fremiet and M-Cl. Ganier, 1969, *Protistologica* 5:353–61; (*B*) from J. Dragesco, 1962, *Bull. Biol. France Belgique* 96:123–67.

Other vestibuliferan groups contain no marine forms: the Colpodida, which are all freshwater or "terrestrial" (ectaphic); and the Entodiniomorphida, which are primarily endocommensals in herbivores, especially ruminants.

3. Hypostomata

The third subclass of the kinetofragminophorans, the Hypostomata, contains those forms with a relatively more complex oral structure. This subclass is very large and diverse. The cyrtophorines are an important, but small specialized assemblage that is herbivorous, feeding on algal and bacterial filaments as well as on diatoms. In the new classification, they are divided into the Cyrtophorida and the Nassulida (plus one other very small order) on the basis of ultrastructure. They are dorsoventrally flattened, with ciliature generally restricted to the ventral surface. The cytostome is mid-ventral on the anterior part of the body. The cytopharyngeal basket in *Nassula ornata*, and its mechanism of grasping algal and bacterial filaments that are ingested is shown in Plate 24-4, *C,D*. This process has been described by Dragesco (1962); the ultrastructure and feeding position of the nemadesms has been illustrated by Tucker (1968).

As mentioned earlier, pennate diatoms are among the first colonizers of illuminated surfaces, either animate or inanimate. The cyrtophorines can be a major part of the ciliate microfauna grazing upon these diatoms (Persoone, 1968; Deroux, 1970). The characteristic appearance of these ciliates is shown in Plate 24-5. Kidder and Summers (1935) described several cyrtophorine species in the genera *Allosphaerium* and *Chilodonella* that live as ectocommensals on the gill lamellae and carapace of three species of amphipods from the Woods Hole, Massachusetts area. In addition to their flat bodies, a tuft of adhesive (thigmotactic) ventral cilia or a settlement organelle, which secretes an adhesive substance in the Dysteridae, help keep them attached to their hosts (Fauré-Fremiet et al., 1968). This is essential, since they soon perish upon removal. A very good idea of the ultrastructure and gross morphology of the cyrtophorines is shown in the illustrations of a *Chilodonella* in Soltynksa (1971).

In the order Rynchodida, the ciliates are mouthless, but an anterior sucker (sometimes on a tentacle) enables them to attach themselves to the epithelial cells of the gills and palps of their host. Most of these forms allegedly are parasitic, using their sucker to ingest the contents of their host. An example is *Hypocomides mytili*, described as parasitic on *Mytilus edulis* (Kozloff, 1946). But Khan (1969), in studying the rhynchodine *Ancistrocoma pelseneeri* from the mantle of *Mya truncata*, failed to show a parasitic relationship and considered it a commensal. Khan (1969), as well as Lom and Kozloff (1970), has described the ultrastructure of the ancistrocomid ciliates. The latter show interesting virus-like particles in the micronucleus.

The Chonotrichida, a greatly expanding order of ciliates by virtue of the monographic research of Jankowski (1973), live as ectocommensals on the mouth (and other) parts of marine crustaceans, as well as on freshwater gammarids. A typical example is *Chilodochona quennerstedti*, originally described by Wallengren (1896) in crabs from the Kristineberg Zoological Station in Sweden. This organism is shown in Plate 24-6, with illustrations from the fascinating study by Guilcher (1951b). The ovoid body is attached by a peduncle (stalk) to the first and second maxillae and maxillipeds of a number of crustaceans, including the green crab *Carcinus maenus*, especially during the spring and fall. The somatic ciliature extends into and lines the complex oral, or atrial, cavity. The somatic cilia proper—but not their kinetosomes, as revealed by silver impregnation—have been lost, to yield a naked body. The Chonotrichida commonly reproduce by budding (see Plate 24-6, *B*), although they conjugate and "reproduce" sexually (Tuffrau, 1953). Plate 24-6 shows the intriguing

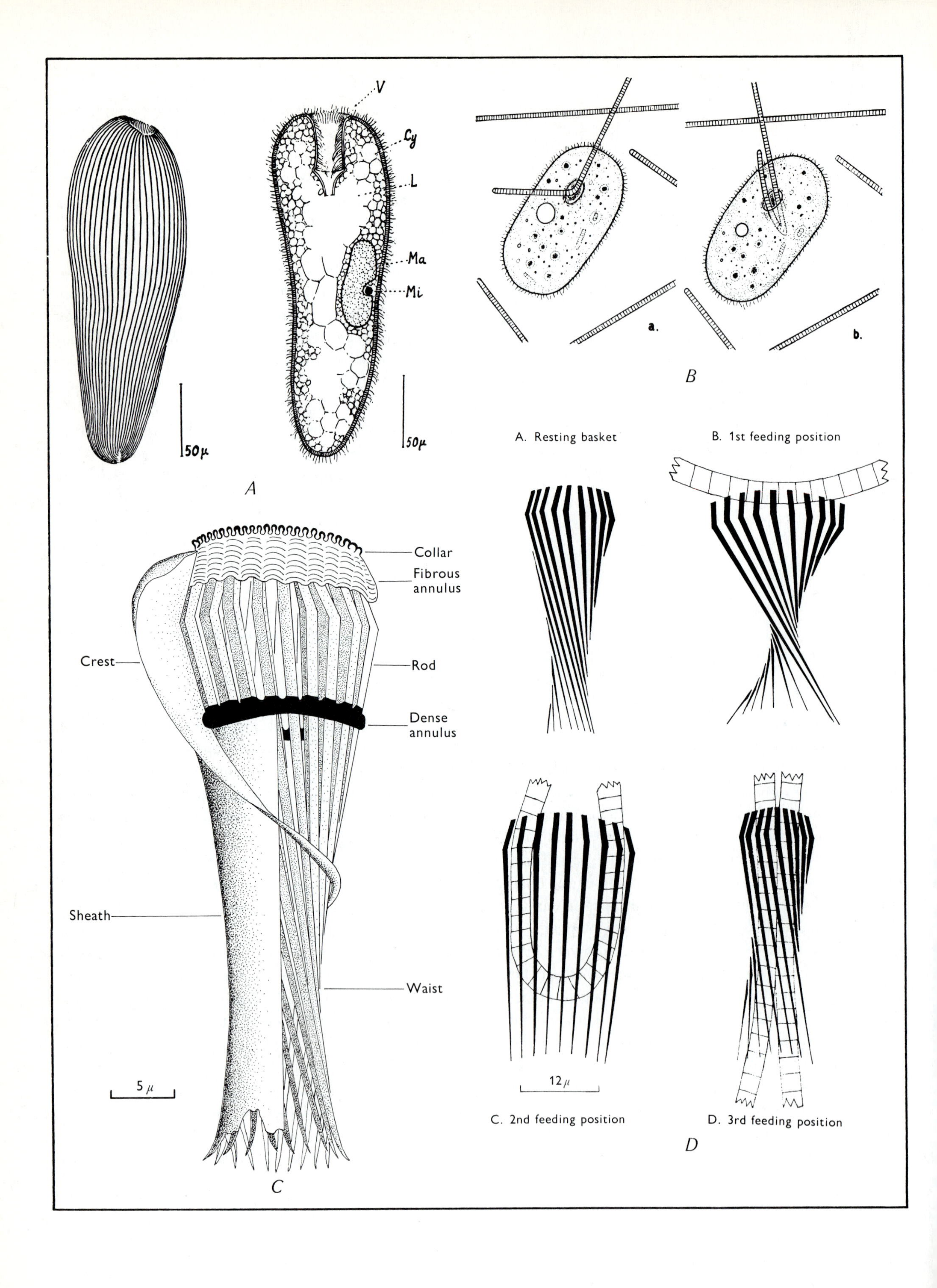

V
Cy
L
Ma
Mi
50μ
50μ
A
a.
b.
B
A. Resting basket
B. 1st feeding position
Collar
Fibrous annulus
Crest
Rod
Dense annulus
Sheath
Waist
5 μ
C
12 μ
C. 2nd feeding position
D. 3rd feeding position
D

PLATE 24-4. Comparison of a vestibuliferid *Coelosomides marina* and its simple entryway with a hypostomatid *Nassula ornata* and its more complex oral structure. (*A*) *Coelosomides marina*, showing the uniform body infraciliature and an optical section, showing the ciliated vestibule that terminates before the cytostome; *Nassula ornata*, showing (*B*) cyanobacterial filaments being attacked laterally; (*C*) diagrammatic reconstruction of the cytopharyngeal basket (*D*) The ingestion of the cyanobacterial filaments is due to the trichites (rods), shown in three feeding positions. (*A*) from E. Fauré-Fremiet, 1950, *Bull. Biol. France Belgique* 84:35–75; (*B*) from J. Dragesco, 1962, *Bull. Biol. France Belgique* 96:123–67; (*C,D*) from J.B. Tucker, 1968, *J. Cell Sci.* 3:493–514.

transformation of adults to larvae (free migrant forms) (*C*), and larvae to stalked forms (*F*). The budding and metamorphosis of this species have much in common with these processes in the suctorians, but other aspects of their morphology and biology demonstrate their close relationship to the cyrtophorids.

The order Apostomatida includes a group of ciliates apart from the mainstream of ciliate evolution. These ciliates have adapted to a food supply that is briefly very rich and then absent. Chatton (1911) realized that a number of species with such an adaptation formed a widespread but homogeneous group. He created the family Foettingeridae for the genera *Pericaryon* and *Foettingeria*, which have ctenophore and crustacean hosts, respectively. Chatton and Lwoff (1931) observed similar morphological stages and life cycles in the family Opalinopsidae, which have cephalopod hosts and lack mouths. They formed the suborder Apostomea for these two families. The name means "away from the mouth" and implies a derivation from the holotrichs rather than an adaptation to parasitism. The feeding forms (trophont) enlarge by accumulating food, usually in a single enormous vacuole surrounded by a very thin layer of cytoplasm, and become pigmented from their host's carotenoid pigments. The very active work during the first one-third of this century, especially by Chatton and Lwoff (1935) at Roscoff, France, is summarized in their benchmark monograph on this group of ciliates.

Hyalophysa chattoni, an example of an apostomatous foettingerid ciliate, is shown in Plate 24-7. This species was described by Bradbury (1966a,b), who also cites the pertinent literature on this strange and intriguing group. *Hyalophysa chattoni* was observed in several genera of crabs but has been most thoroughly studied in *Pagurus hirsutiusculus* because of the easy availability of this hermit crab. The pagurids were 100% infested with the encysted phoront stage on the gills. Late in the molting period, the phoronts excyst in synchrony with molting to become colorless trophonts, and within minutes, the molt is populated by big blue ciliates. Although the exoskeleton of the crabs has very little blue pigment, the exuvial (shed shell) fluid trapped in the exoskeleton is drawn into the 1 μm mouth and, by either concentration or transformation, the ciliates are turned an intense sapphire blue. The trophonts feed for 2 to 20 hours before entering the protomont stage. They then slowly swim to and settle on a substrate and form a cyst wall. The tomont stage is quickly followed by division to form daughters, which become the active dispersal form, the tomite, that encysts on the crab's gills to become the dormant phoront. Bradbury (1973) has studied the fine structure of the organelles that ingest the exuvial fluid.

4. Suctoria

This distinctive subclass of the Kinetofragminophora has many aberrant adult forms, with only the mouthless larvae being ciliated. The adult forms, which are never ciliated, are easily recognized by ovoid, globular, or lobed bodies, which are often stalked and which bear tentacles that are evenly distributed or grouped. The tentacles are used both to capture and ingest their prey, since the suctorians—in a sense—are mouthless. The more usual prey consists of other ciliates, but a variety of organisms, including algal spores and bacteria, are preyed upon by some species. There are two types of contractile tentacles. The commoner one, which captures and sucks the prey down into the suctorian body, is a hollow and capitate (head-like) structure. A model of this tentacle, based on its microtubular structure, has been proposed by Bardele (1972). Instead of the presumed sucking mechanism of ingestion, Bardele proposes a grasp-and-swallow mechanism, whereby the microtubules of the inner tentacle tube slide back and forth with cyclic bridging and release. The less common prehensile tentacles have a fibrillar core instead of a canal and resemble heliozoan axopods. When a stalk is present, it is secreted by the scopula (called a scopuloid in suctorians) and is never

PLATE 24-5. Representative cyrtophorine ciliates from fouling surfaces, showing their dorsoventral flattening, ventral only ciliature, and herbivorous nature. (*A*) *Chlamydonella pseudochilodon*, as seen in profile, showing its flattened shape and lack of dorsal cilia; (*B*) details of ventral infraciliature as seen with silver impregnation; (*C*) same, showing dorsal (left) and ventral (right) views; (*D*) ventral (left) and dorsal (right) views of *Hartmannulopsis dysteriana*, showing the cytopharynx, the macronucleus (ma), ingested diatoms (d), and the ventral ciliature. (*A*,*B*,*C*) from G. Deroux, 1970, *Protistologica* 6:155–82; (*D*) from G. Deroux and J. Dragesco, 1968, *Protistologica* 4:365–403.

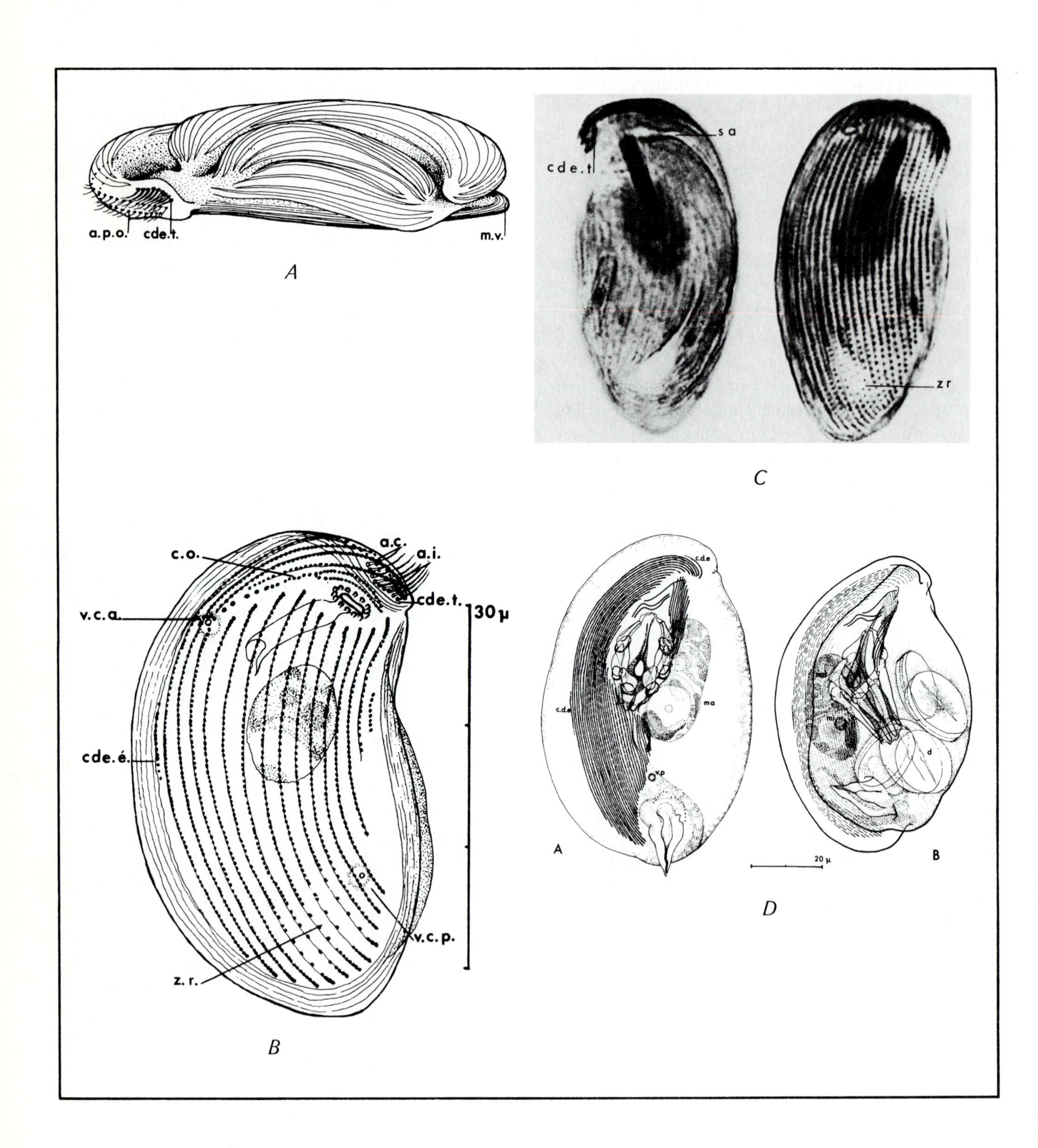

PLATE 24-6. *Chilodochona quennerstedti* is typical of the Chonotrichida that live as commensals on the mouth parts of marine crustaceans and feed on their refuse. (*A*) Stalked ciliates colonizing the first maxillae of the green crab *Carcinus maenus* (5 X); (*B*) different views of the buccal cavity in living adults (note the bud in the upper right) (425 X); (*C*) the release of adults from their peduncles (stalks) to form larvae (migrants) (left, 8X; right, 540 X); infraciliature, as shown by silver impregnation of (*D*) larva (1,500 X) and (*E*) adult (975 X); (*F*) transformation of the larva to the stalked adult (500 X). From Y. Guilcher, 1951, *Annales Sci. Nat. Zool. Biol. Anim.* 11e Ser. 13:33–133.

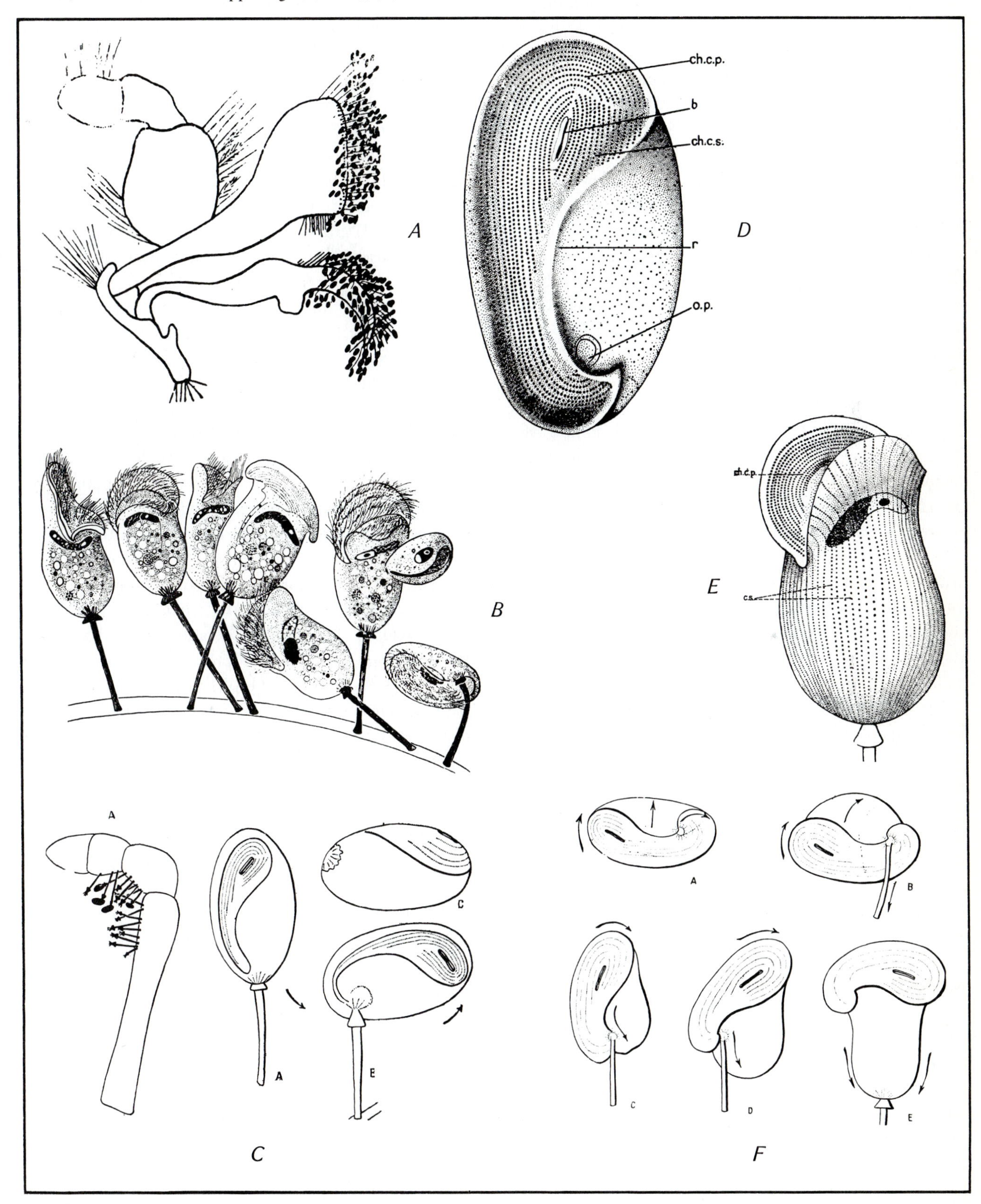

contractile. The fine structure has been shown by Batisse (1968). Although the adult body is nonciliated, kinetosomes are scattered irregularly over the body surface. Suctoria become attached to a variety of animals, such as hydroids and harpacticoid copepods that are associated with seaweeds as well as on seaweeds themselves. The suctoria commonly reproduce by budding to form the ciliated larvae, which are their only means of dissemination.

The most important biological and taxonomic work on the suctorians is still the old but excellent monograph by Collin (1912). In addition, Guilcher (1951b) has summarized her studies on morphogenesis and larval structure, and the genera in the different families are listed by Corliss (1977). A very extensive monograph on the suctorians is being prepared by A.W. Jankowski. The suctorian families can be grouped into at least two categories, the more common endogenous budders and the exogenous budders.

Acineta tuberosa is an example of a typical marine suctorian (Plate 24-8). A crude culture obtained near the Tulear Marine Station in Madagascar has been maintained by Grell at the University of Tübingen for a decade on a diet of the oligotrichous ciliate *Strombidium*, which, in turn, is reared on the green alga *Dunaliella*. The feeding of *A. tuberosa* on *Strombidium* has been studied by Bardele and Grell (1967) (see Plate 24-8,*B*,*C*,*D*). In feeding, the haptocysts on the knobs of the tentacles make contact and secure the prey, which is then pulled down into the tentacles. A nonfeeding tentacle is compared to one ingesting organelles from *Strombidium* in the plate. The endogenous budding and metamorphosis of *A. tuberosa* has been studied in thin section by Bardele (1970), who has also described the ultrastructural changes in the macronucleus during this cycle (Bardele, 1969). The life cycle of *A. tuberosa* is illustrated in Plate 24-8,*A*. The endogenous bud is formed by an invagination of a small pellicular area near the vacuole opening that bears rows of barren basal bodies. Some of these basal bodies are incorporated into the larva side of the brood pouch and give rise to eight rows of cilia; the basal bodies also provide genetic continuity, giving rise to the next generation of swarmers. The kineties become spaced on the swarmer-anlagen when they are fully grown. As the swarmer or larva settles, morphogenesis starts, with the formation of a basal disc and the secretion of a stalk through a process using the invaginated scopular (or scopuloid) area as a mold. The cuticle and tentacles are then formed as the cilia disintegrate, and a new generative area is formed.

In addition to the many species of suctoria that capture other ciliates for food, some species appear to feed on algal spores and bacteria. Two very different suctorians are shown as examples in Plates 24-9 and 24-10. *Trematosoma* (formerly *Acineta*) *bocqueti* is an endogenously budding species that was obtained from two species of harpacticoid copepods in the genus *Porcellidium*, which live among benthic algae, and was described by Guilcher (1951a). She believed that it fed specifically upon the zoospores of the green alga *Ulva lactuca*, as shown in Plate 24-9,*B*. In an ultrastructural study of *T. bocqueti* obtained from a third species of *Porcellidium*, Batisse (1973) emended the description and name, but failed to observe the capture of ciliate prey. Instead, the thin sections show the ingestion of bacteria in the tentacles and their numerous presence in the endoplasm (Plate 24-9,*C*,*D*). Apparently ingested bacteria have been observed in the exogenously budding suctorian *Ephelota gemmipara* (Plate 24-10) epizoic on the surface-dwelling copepod *Pontella scutifera* (Sieburth et al., 1976). *Ophryodendron hollandei* (Plate 24-11) is a very different-looking suctorian, being loricate, having its tentacles originate from a trunk, and forming its buds exogenously. It is also found on a harpacticoid copepod, *Rhynchothalestri rufocinta*, which lives in algal mats. Ultrastructural studies by Batisse (1969) show that both the adult and vermiform stages possess a complex system of canals that open to the outside; they are filled with bacteria. Although the relation of the bacteria to the host (*O. hollandei*) is not clear, it appears that they do find their way into the cytoplasm, as do those ingested by *T. bocqueti*. These bacteria must provide at least part of the suctorian's diet. The occurrence of bacteria and bacterial canals in suctorians is well described and illustrated by Matthes and Plachter (1975). (The role of bacteria as endosymbionts, food prey, and parasites in the suctorians should be studied further.)

In contrast to the easily recognizable stages of typical suctoria like *Acineta*, the vermiform buds of exogenously budding species and the morphology of suctorians parasitic on other suctorians are quite aberrant. The vermiform stages of two species of *Ophryodendron* are also shown in Plate 24-11,*C*,*E*. Those of *O. hollandei* contain both the haptocysts and the bacterial canal system of the adult stage and bear a "sucker" on the distal end. The vermiforms of *O. faurei* are shown budding from an adult and attached to the cephalothorax of its copepod host, *Psamathe longicauda*. The vermiforms are quite similar in appearance, from one species to another, just as the swarmers of the endogenously budding suctorians appear to be similar, at least superfi-

PLATE 24-7. The polymorphic life cycle of *Hyalophysa chattoni* (*A*), an apostomatid ciliate (in the family Foettingeria) that occurs in a dormant encysted stage (phoront) on crab gills and excysts in synchrony with the molting of the host to form the trophic stage (trophont) to feed on the exuvial fluids trapped in the crab's cast-off exoskeleton, which turn it a vivid sapphire blue. The trophont enlarges and then excysts to form a tomont; this undergoes binary fission to form actively flagellated offspring (tomites), which disperse and encyst to become phoronts on their host's gills. (*B*) Photomicrograph of tomites (left) before encystment (about 250 μm long, silver impregnated). Diagram of the ventral surface of a protargol stained tomite (right) showing the characteristic rosette (R), near an inapparent cytostome, and details of infraciliature. (*A*) from P.C. Bradbury, 1966a, *J. Protozool.* 13:209–25; (*B*) from P.C. Bradbury, 1966b, *J. Protozool.* 13:591–607.

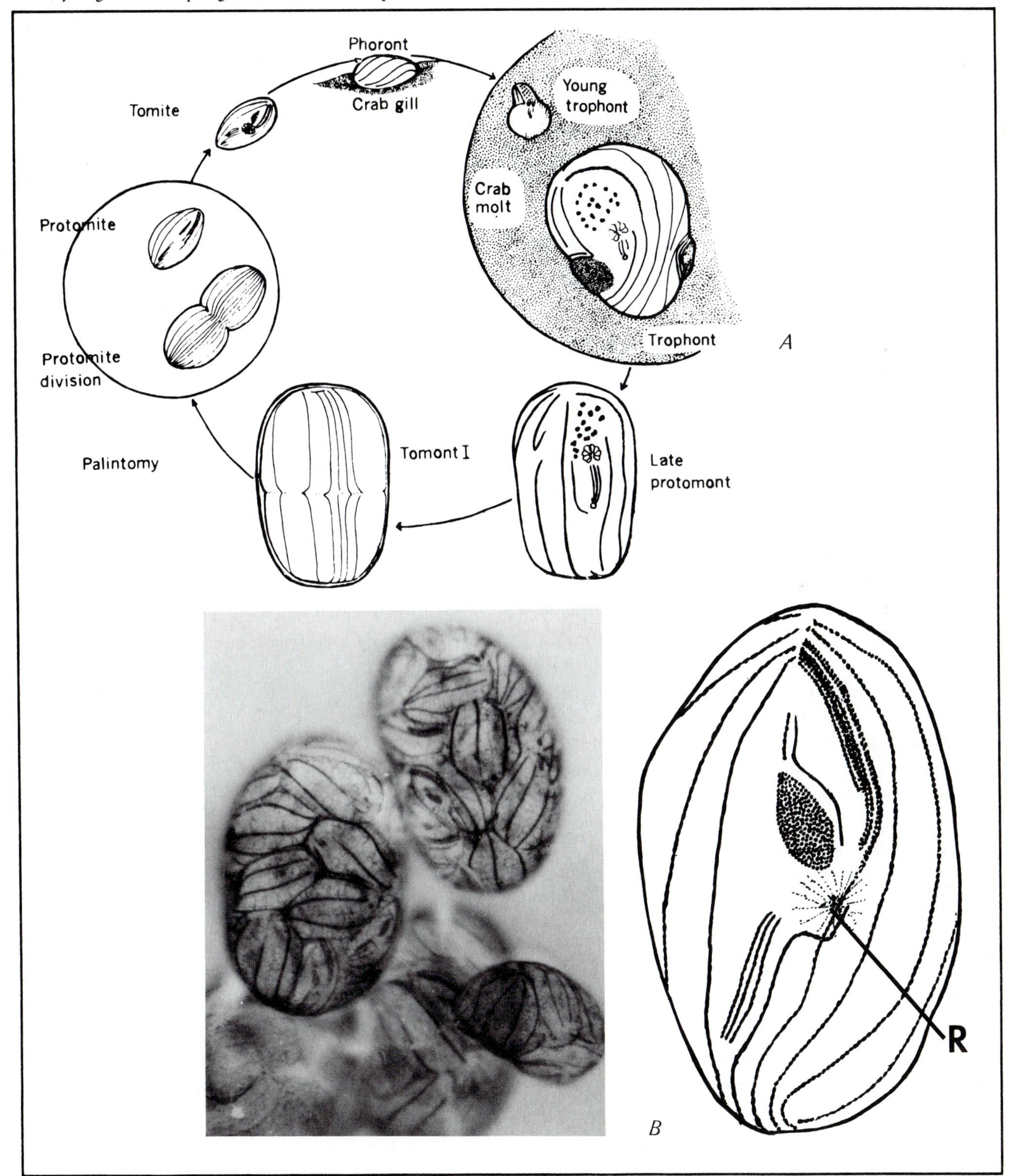

cially, despite the marked differences of the adult forms.

In addition to the suctorians that live on ciliates, some live parasitically on other suctorians. Collin (1912) has described a number of species of *Pseudogemma* that have specific suctorian hosts. *Acineta tuberosa* (Plate 24-8), the host for *Pseudogemma pachystyla*, is illustrated in Plate 24-12,*A*,*B*. A basal notch permits a modified tentacle, formerly thought to be a stalk, to protrude and penetrate the host. This tentacle does not possess the usual haptocysts and lytic bodies, but does have an axial tubular canal with vesicles (similar to the stomopharyngeal cytoplasm of cyrtophorine ciliates) that can destroy the haptocysts and lytic bodies of the host (Batisse, 1968). Some suctorians have an alternation of generations between free-living and parasitic stages, a further illustration of the diversity of this bizarre group. Martin (1909) described the occurrence of *Tachyblaston ephelotensis* in such epidemic proportions that it interfered with the budding of the *Ephelota gemmipara* he had collected from the brown alga *Cystoseira* at Naples. The life cycle of *T. ephelotensis*, shown in Plate 24-12,*C*, was illustrated by Grell (1950). Specimens of *E. gemmipara* inhabiting the red alga *Ceramian rubrum* at Helgoland, Germany were infested and used to study the life cycle of *T. ephelotensis* further. The young infective dactylozoite stage pierces the host cell with its tentacles and forms a pellicular canal in which they develop while using their feeding tentacles to ingest the host's cytoplasm. *Tachyblaston ephelotensis* then buds exogenously to form a series of swarmers, which attach to suitable substrates, including the stalk of its host. Here the swarmer forms a goblet-shaped lorica with a stalk, which is the free-living *Dactylophrya* stage. The anterior portion of the lorica then disintegrates and the cell divides into some 16 infective dactylozoites, which seek out other host cells. An extreme example of a parasitic suctorian is *Phalacrocleptes veruciformis*, which lives on the pinnules and cirri of the sabellid polychaete *Schizobranchia insignis*. This suctorian has lost not only its cilia but its kinetosomes (Lom and Kozloff, 1967).

PLATE 24-8. The suctorian *Acineta tuberosa*, showing its asexual life cycle and its feeding on the oligotrichous ciliate *Strombidium*. (*A*) Asexual life cycle (endogenous budding) and metamorphosis: (a) An adult with an early swarmer already having a scopuloid (sc) and eight kineties is demarked by a brood pouch (BP). Barren basal bodies (BA) next to a vacuole (V) are part of the generative area for future swarmers. (b) A nearly complete swarmer, with its spaced kineties and scopular material, incorporates a small part of macronucleus as its juncture with the adult constricts. (c) The upper part of the mother cell, post-parturition. (d) Ventral view of a ciliated swarmer freely swimming in the direction of the arrow. (e) A recently settled swarmer has secreted a basal disc and is forming a stalk through its scopular pores (SP) in an invaginated scopuloid, which acts as a mold. (f) A more mature form that has lost its cilia and formed tentacles and a generative area. (*B*) Phase-contrast micrograph of *A. tuberosa*, with extended tentacles (a) and a captured prey (b) (1,000 X). (*C*) Transmission electron micrograph of a longitudinal thin section of a contracted tentacle (non-feeding), with the pellicle folded along the tentacle shaft (PT), microtubules of an inner tube (Mt), and a knob of another tentacle. (*D*) A similar longitudinal section of the tentacle during ingestion of prey (*Strombidium*), showing a polysaccharide granule (P) and the mitochondria (M). (*A*,*C*) from C.F. Bardele, 1970, *J. Protozool.* 17:51–70; (*B*,*D*) from C.F. Bardele and K.G. Grell, 1967, *Z. Zellforsch.* 80:108–23.

B. OLIGOHYMENOPHORA

Members of this second major division of the phylum have cilia that are fused to form membranelles or "hymens" around or in the buccal cavity. They are intermediate between the Kinetofragminophora, with their simple oral ciliature or slightly modified somatic ciliature, and the Polyhymenophora, with their more elaborate membranelles. This class is divided into two quite distinct subclasses: the Hymenostomata, with their commonly holotrichous somatic ciliature, and the Peritrichia, with their much reduced somatic ciliature.

1. Hymenostomata

The free-living marine species of the Hymenostomata [as emended by Puytorac et al. (1974) and Corliss (1975, 1977)] fall into the old order Hymenostomatida, with a typically tetrahymenal buccal apparatus, and the relatively new order Scuticociliatida, in which the mouthed species of the old "Thigmotrichida," with their sticky or thigmotactic cilia, are found, along with the philasterines and the pleuronematines.

The ciliates in the Hymenostomatida have developed an oral ciliature quite distinct from their somatic ciliature. The cilia coalesce to form membranous orga-

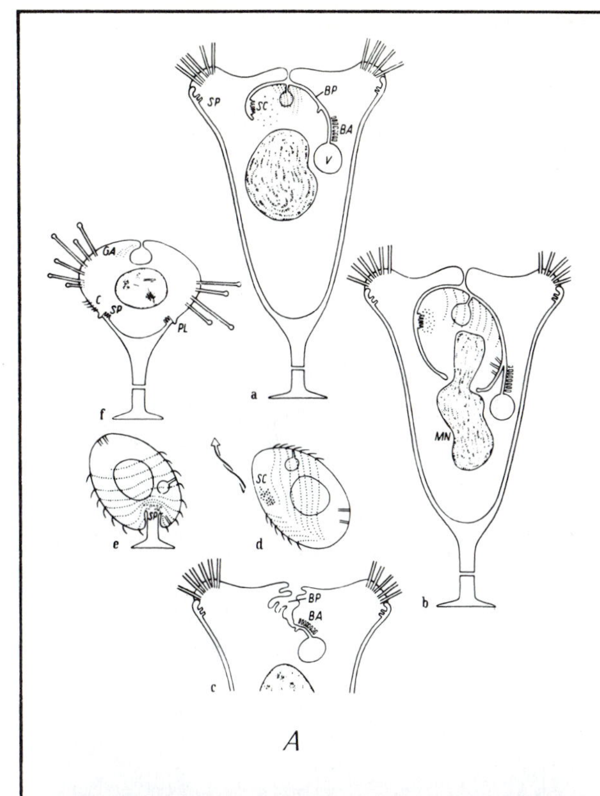

A

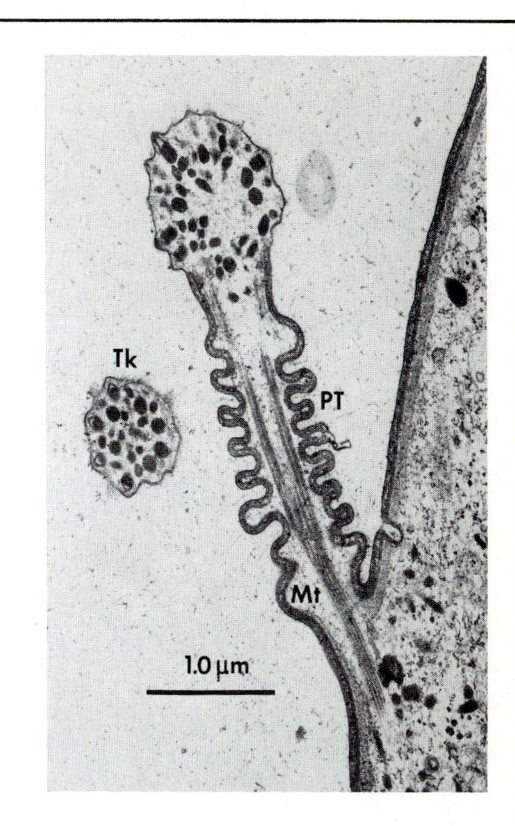

C

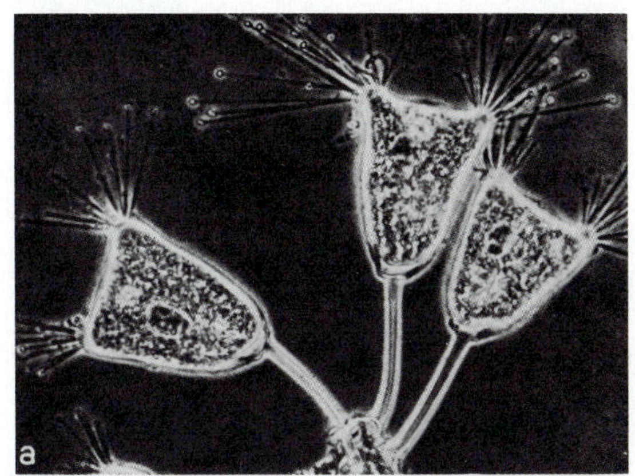

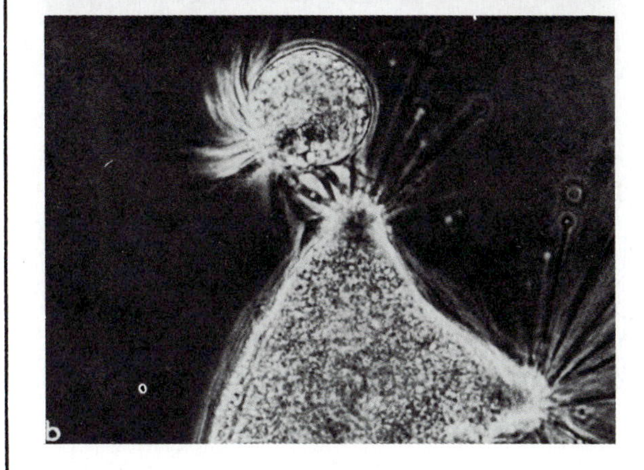

B

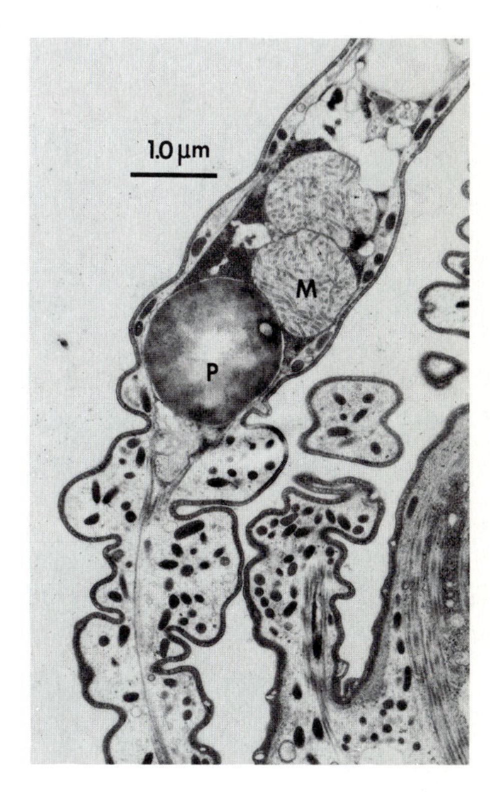

D

PLATE 24-9. The endogenously budding suctorian *Trematosoma bocqueti*, which attaches to harpacticoid copepods (genus *Porcellidium*) that live among benthic algae. (*A*) Schematic views (frontal and sagittal of an adult) and a swarmer. (*B*) Adult sucking an *Ulva* zoospore. (*C*) A sucker with a bacterium (Bc), in the course of ingestion; the arrow points to a cocked haptocyst in the periphery of the tentacle. (*D*) Endoplasm with ingested bacteria (Bc), residual vacuoles (V. res.), and a tentacular axoneme (Tt). (*A,C,D*) from A. Batisse, 1973, *Protistologica* 8:477–96; (*B*) from Y. Guilcher, 1951, *Arch. Zool. Exp. Gen.* 87 (N&R):24–30.

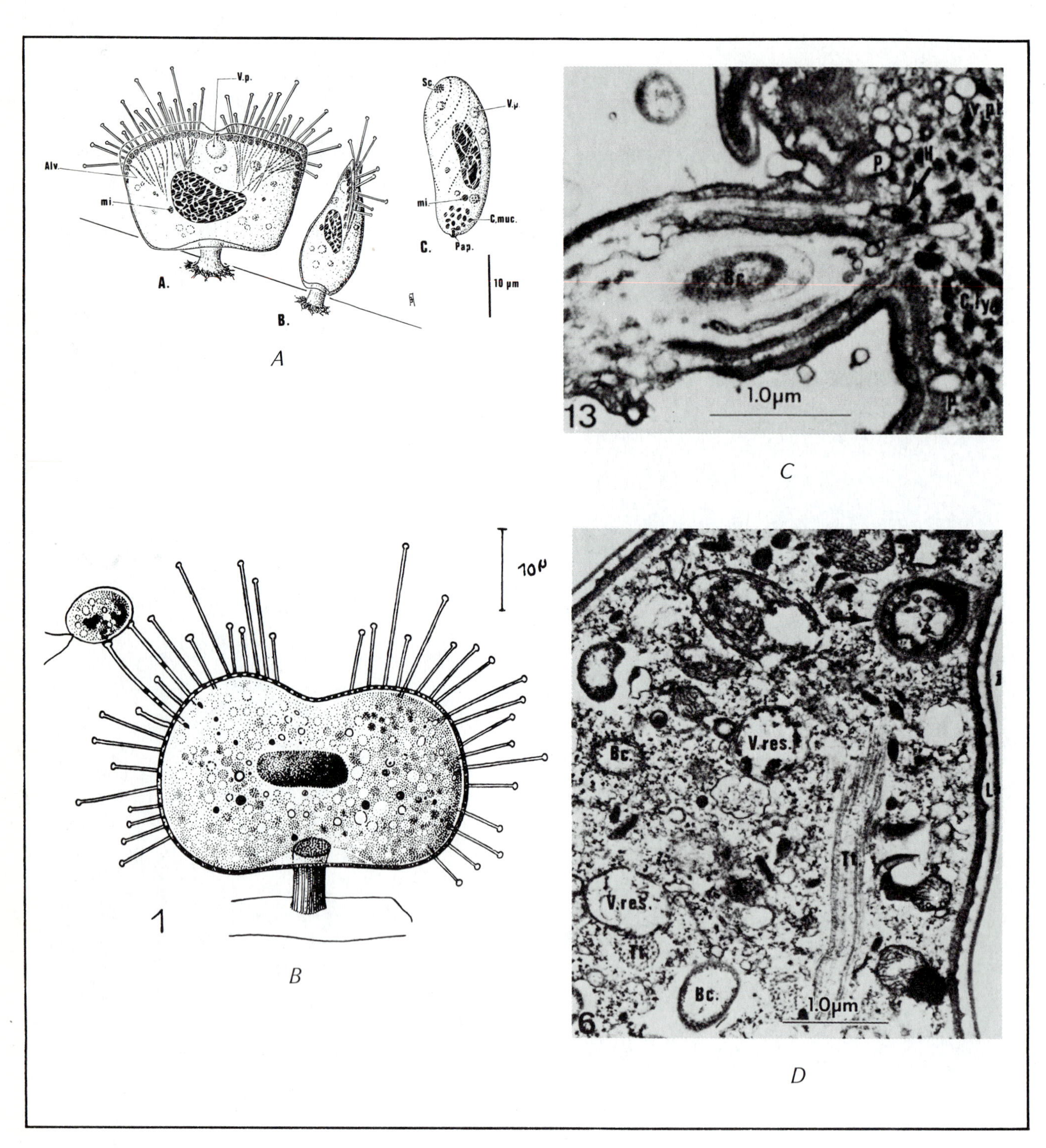

PLATE 24-10. The exogenously budding suctorian *Ephelota gemmipara* has an affinity for invertebrate surfaces. (*A*) An individual suctorian (scanning electron micrograph) attached to a campanularid hydroid, on a fouled surface in the Sargasso Sea, is covered with bacteria. (*B*) The blue copepod *Pontella scutifer* floating on the surface of the mid-North Atlantic is colonized by individuals in all stages of development: (A) Scanning electron micrograph, showing the outer prehensile and inner feeding tentacles and the exogenously budding individual (arrow) with five ciliated telotrochs (bar = 100 μm); (B) transmission electron micrograph of thin section of part of a *E. gemmipara* cell, showing the nucleus (Nu), stalk (s), and an apparent food vacuole (v) with ingested bacteria (bar = 5 μm); and (C) higher magnification of ingested bacteria (b) in a vacuole (bar = 1 μm). (*A*) from J.McN. Sieburth, 1975, *Microbial Seascapes*, University Park Press, Baltimore, Md.; (*B*) from J.McN. Sieburth et al., 1976, *Science* 194:1415–18.

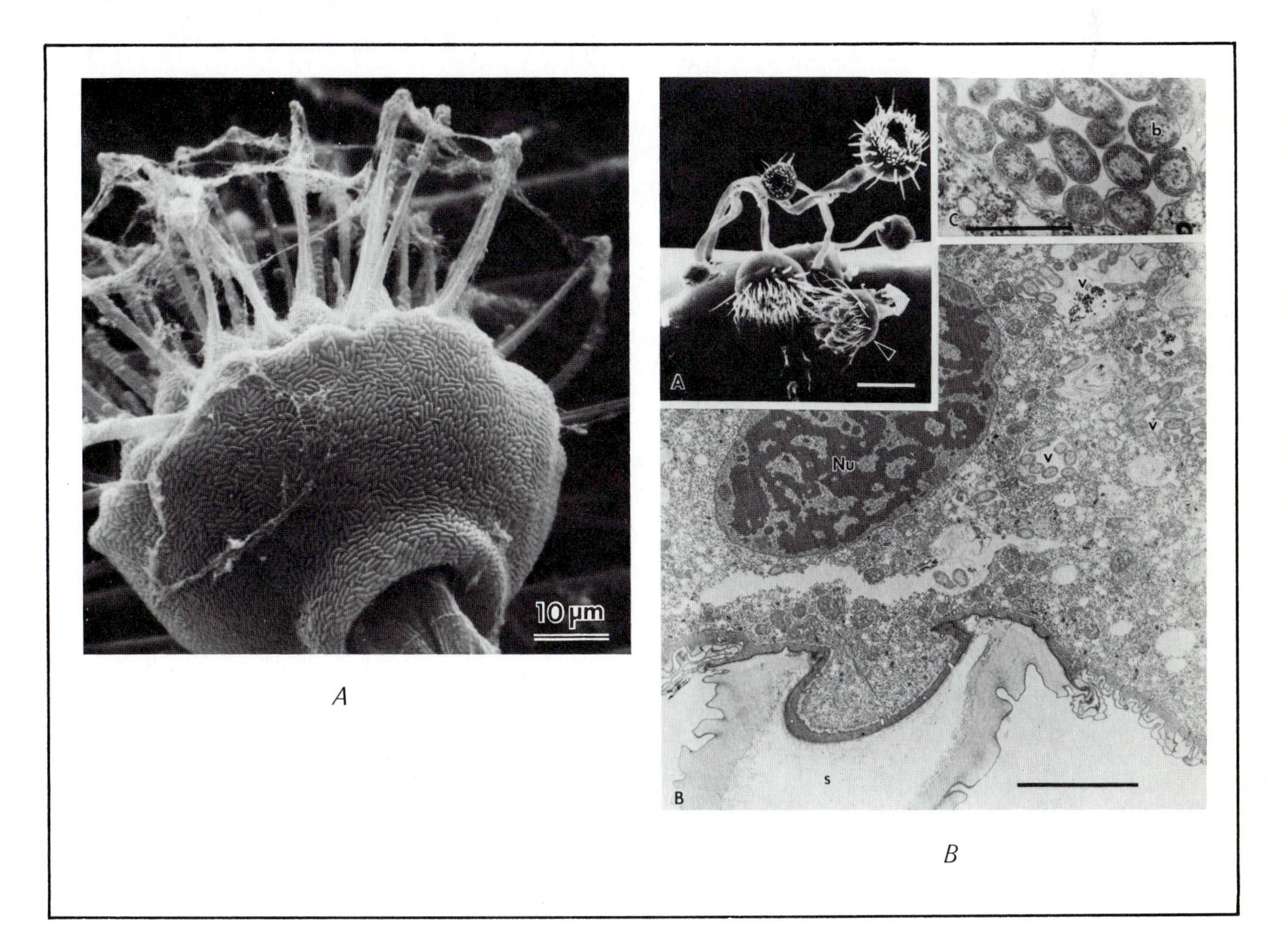

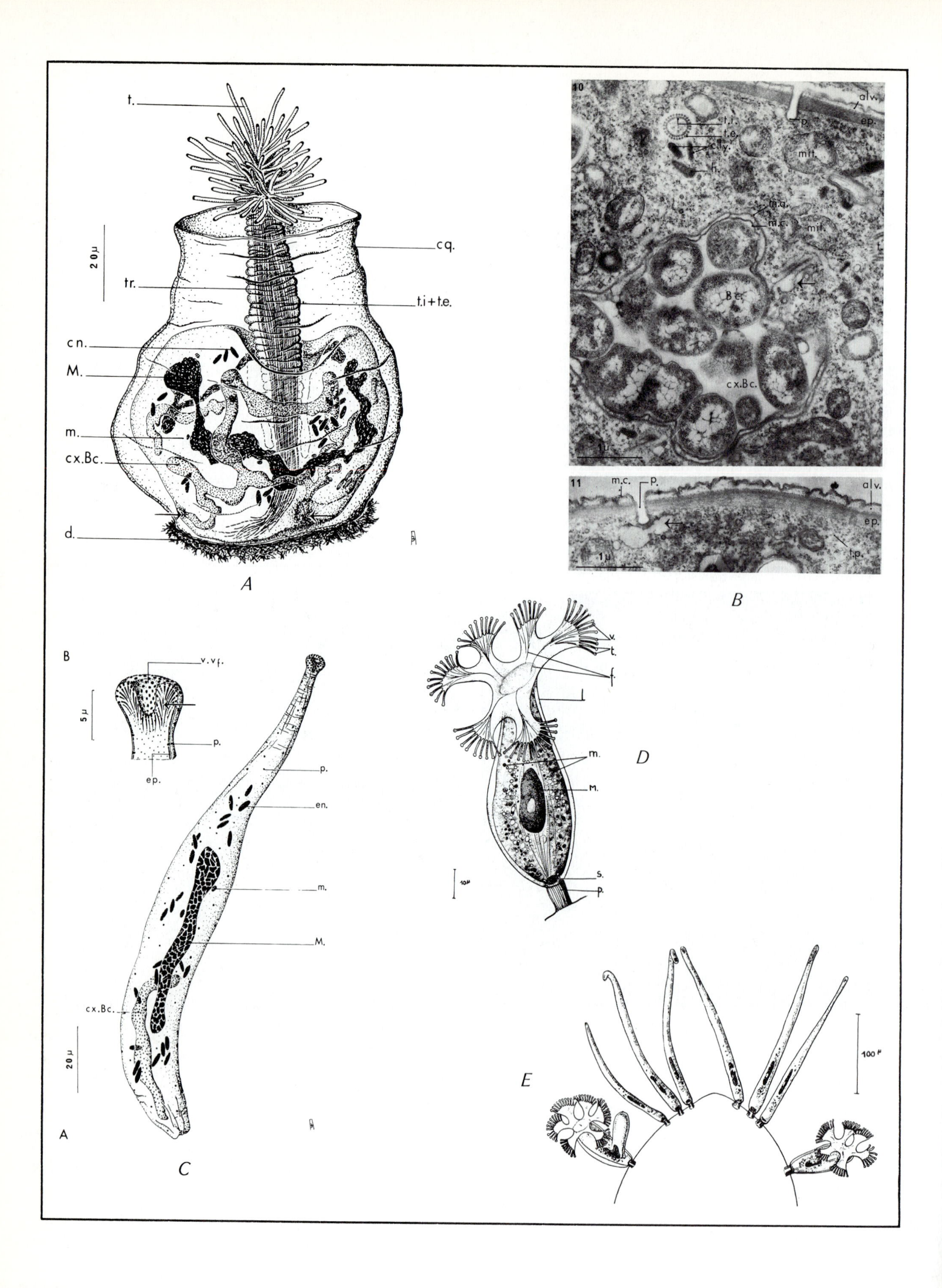
t.
20 μ
cq.
tr.
t.i+t.e.
cn.
M.
m.
cx.Bc.
d.
A
10
alv.
ep.
t.i.
t.e.
p.
mit.
h.
m.a.
m.c.
Bc.
cx.Bc.
1μ
11
m.c.
p.
alv.
ep.
t.p.
1μ
B
v.vf.
5 μ
p.
ep.
en.
m.
M.
cx.Bc.
20 μ
A
C
v.
t.
f.
l.
m.
M.
D
s.
p.
10μ
100μ
E

PLATE 24-11. Exogenously budding suctorians of the genus *Ophryodendron*. *Ophryodendron hollandei*, which attaches to the harpacticoid copepod *Rhynchothalestri rufocinte* that lives in algal mats, showing: (*A*) Proboscidian (adult) stage with a bulbous theca. The contractile trunk (tr.) that arises from a deep pit bears a distal tuft of tentacles (t); the endoplasm has an irregular macronucleus (M), many ingested cnidocysts from hydra (c.n.), and a complex bacterial canal system (cx.Bc.), which penetrates the body wall and opens directly to the outside. (*B*) Transmission electron micrograph (transverse section), showing the epiplasm (ep) with a cuticular pore (p), a tentacular tube (Ti+Te), and a bacterial canal (cx.Bc) filled with bacteria. (*C*) A vermiform (bud) of *O. hollandei* recently liberated from a parent (note the bacterial canals, cx.Bc) and the detail of the distal sucker. (*D*) Adult of *Ophryodendron faurei*, showing the peduncle (p), scopula (s), macronucleus (M), micronuclei (m), loge (l), fibrils (f), tentacles (t), and terminal suckers (v). (*E*) *Ophryodendron faurei* showing two adults (one budding) and six vermiform stages attached to the cephalothorax of its copepod host, *Psamathe longicauda*. (*A,B,C*) from A. Batisse, 1969, *Vie et Milieu*, Ser. A, 20:251–77; (*D,E*) from Y. Guilcher, 1951, *Arch. Zool. Exp. gen.* 87(N&R):24–30.

nelles. This ciliature is a highly organized form of an undulating membrane plus an adoral zone of membranelles; it is the latter that is so highly developed in the subclass Spirotrichia of the Polyhymenophora. In its complete form, it consists of a tripartite adoral zone of membranelles and an undulating membrane to form the typical tetrahymenal (four-membrane) buccal apparatus found in the suborder Tetrahymenina. The most common example is *Tetrahymena*, which is predominantly a freshwater form. Another suborder, *Peniculina*, has compound ciliary organelles deep in its buccal cavity. The common *Paramecium* of freshwater and marine waters is an example.

Members of the order Astomatida are symbiotic, occurring primarily in freshwater oligochaetes, but some species are found in polychaetes and other marine hosts.

The Scuticociliatida was created by Small (1967) to include the mouthed thigmotrichs along with a goodly number of other ciliates, most of them small in size and many of them marine forms. The majority of thigmotrichs represent a natural group of ciliates with an interesting habitat, the mantle cavity of lamellibranch molluscs, where they share the host's food as commensals (but other thigmotrichs are found in oligochaete annelids). They have thigmotactic (sticky) cilia for attachment to their host; these can be simply the anterior part of the ventral cilia, which are set more closely together, or they can be clearly defined fields of specialized cilia. The basic monograph on this group, now known as the Thigmotrichina, is by Chatton and Lwoff (1949, 1950); but also see the exhaustive monographic series by Raabe (1967, 1970a,b, 1971, 1972). Kidder (1933c) and Raabe (1934, 1936), and Chatton and Lwoff (1949) have described *Ancistrum mytili*, which is found in great numbers in the edible mussel *Mytilus edulis*. It is shown in Plate 24-13,*B*. *Ancistrum isseli* is found in the mantle cavity of the solitary mussel *Modiola modiolus* (Kidder, 1933c); its conjugation and nuclear reorganization was described by Kidder (1933d).

Another common commensal in *M. edulis* is *Peniculistoma mytili* (*Conchophthirius mytili*), which crawls over the muscles and foot of the mussel and also feeds on the algae brought in by the mussel (Kidder, 1933a,b; Raabe, 1934). It quickly disappears in static seawater systems. It is now considered a member of the suborder Pleuronematina, a group containing other forms that have a greatly enlarged undulating membrane and a subequatorial cytostome. *Pleuronema marinum*, shown in Plate 24-1,*A*, and the very common small form *Cyclidium* are examples. The striking *Histiobalantium natans*, shown in Plate 24-13,*A*, is also frequently seen.

A third group, the Philasterina, contains numerous small marine species, the great majority of which are very poorly known. Most are free-living, but *Miamiensis avidus* is a facultative ectocommensal of seahorses (Thompson and Moewus, 1964). It is in this suborder that most of the echinophilic inquilines are placed today. Lynn and Berger (1973) is a good reference source for these intriguing forms.

2. Peritricha

This subclass of the Oligohymenophora is the largest group of ciliates except for the tintinnine oligotrichs. These highly specialized forms have a much reduced somatic ciliature, and they are distinguished by a broad, disc-shaped peristome (buccal cavity), usually occurring at the apial pole, and conspicuous oral ciliature that consist of one or two inner rows (polykinety) and one outer row of cilia (haplokinety or paroral membrane). These membranes coil counterclockwise, in contrast to the clockwise arrangement of the membranes

PLATE 24-12. Suctorians parasitic on other suctorians. *Pseudogemma pachystyla* parasitic on *Acineta tuberosa:* (*A*) Longitudinal section, shown diagrammatically, with a modified tentacle penetrating and withdrawing the host's endoplasm and (*B*) a *P. pachystyla* withdrawing mitochrondria from its host. (*C*) The alternation of generations of *Tachyblaston ephelotensis* on its host, *Ephelota gemmipara.* The infesting dactylozoites penetrate, feed on the host with tentacles, and produce a sequence of swarmers that land on substrates including the host's stalk to produce its own stalked loricate free-living *Dactylophrya* stage, which then form some 16 dactylozoites. (*A,B*) from A. Batisse, 1968, *Protistologica* 4:271–82; (*C*) from K.G. Grell, 1950, *Z. parasitenk.* 14:499–534.

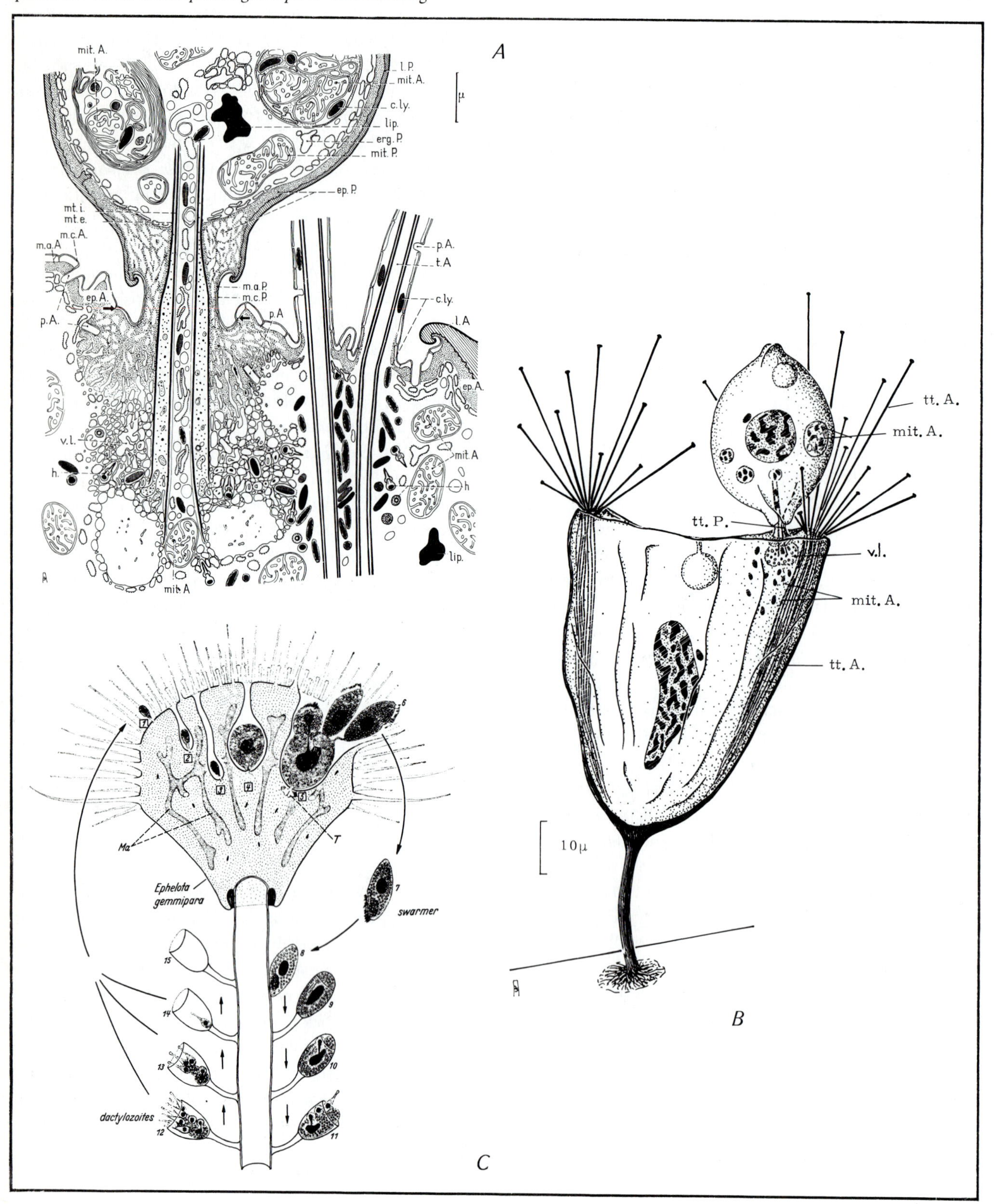

in the spirotrichs of the Polyhymenophora. Below the oral ciliature the oral cavity dips into a funnel-like depression, the infundibulum, which can penetrate deep into the body; it then terminates at the cytostome or mouth. The food is swirled into the peristome by the cilia and collects in the infundibulum. The contractile vacuole, which opens into the infundibulum, has a seawater output that balances the water that is taken in during feeding (Kitching, 1936, 1939). In addition to this contractile organelle, the highly contractile peritrichs have a well developed system of myonemes. The ultrastructure of the specialized myonemes (spasmonemes), which are continuous in the stalk and body and permit the stalks in some families to coil or zigzag, and the body and peristome in all families to contract, have been studied in freshwater forms and discussed by Grell (1973). The ultrastructure of an important pelagic marine form, *Zoothamnium pelagicum*, was described by Laval (1968) (see below).

Binary fission in peritrichs is said to be "longitudinal," and it is often unequal. In the stalked forms, one daughter retains the peristome and stalk, whereas the other daughter cell becomes a swarmer cell (telotroch), with a posterior ciliary wreath that permits it to swim away to settle elsewhere. Colonial organization in this group is quite common. Sexual reproduction is by unilateral fertilization (anisogamonty) according to Grell (1973) or "total" conjugation, that is, the microconjugant is "totally" absorbed by the macroconjugant. This process is also seen in many chonotrichs and a number of suctorians, but *all* peritrichs exhibit it.

The order Peritrichida is divided into two suborders, the Sessilina, which are usually attached by a stalk, and the Mobilina, which have regained some degree of mobility, although they are usually associated with and can adhere to surfaces. The Sessilina attach to substrates either directly or more often by a contractile or a noncontractile stalk secreted by a specialized region, the scopula. The distinctive morphology of this group has earned them the name "bell animals." They occur both as solitary and as colonial forms. Some forms are loricate, whereas others have symbiotic zoochlorellae. The family Vorticellidae contains the well-known genus *Vorticella*, with its solitary forms and a coiled stalk, and the genus *Zoothamnium*, with its colonial forms and zig-zag stalks.

A widely distributed example, *Zoothamnium alternans*, is described by Claparède and Lachmann (1858) in their monograph, *Les Infusoires et les Rhizopodes*. The morphology and development of this species has been described by Fauré-Fremiet (1930) and is shown is Plate 24-14. The colonies of *Z. alternans* have a main trunk with alternating branches in the same plane, which bear a number of small individual cells (microzooids). At the juncture of the lateral branches are large migrating individual cells (macrozooids). When liberated, these anterior–posterior flattened macrozooids swim with their posterior ciliary crown and attach to a substrate with their scopula, which then secretes the peduncle. The newly settled cell, arising from the macrozooid, then initiates a colony by undergoing a first unequal division to yield both a microzooid and a macrozooid, the larger cell remaining axial and forming the principal peduncle for the colony. Plate 24-14,*C* shows some of the fine structure of the bell. The invaginated peristome, with contracted cilia, contractile vacuole, and macronucleus, is easily seen. The contractile elements include the central spasmoneme and the peripheral fibrils of the peduncle. The presence of an even coat of epizoic bacteria on the pellicle of the bells is quite striking, a phenomenon that was first recognized when electron microscopy was used to characterize the fine structure of *Z. alternans* (Fauré-Fremiet, Favard, and Carasso, 1963). The association of two types of bacteria, one with unusual capsules and both undergoing attack by two forms of bacteriophage (Plate 24-14,*F*), is quite curious and warrants further study.

Species of *Zoothamnium* are commonly associated with the carapace, uropods, and pleopods of crustaceans, including both decapods and copepods (see Plate 24-14,*E*). Scanning electron microscopy of a number of surfaces from different habitats in the writer's laboratory indicate that the vorticellids with their epizoic bacteria are common in the microbial communities of surfaces (Sieburth, 1975). One must bear in mind, however, that *Myoschiston*, a *Zoothamnium* look-alike, also is seen on copepods (Hirche, 1974).

In contrast to the usual fixed forms of *Zoothamnium*, *Z. pelagicum* is apparently the only species that lives freely in the plankton. It was first described from the waters of the anchorage at Villefranche-sur-Mer in a note by du Plessis (1891) and, except for another note by Dragesco (1948), was not fully characterized until the detailed study by Laval (1968). She used a 50 μm net fished from the surface to 50 to 200 m at the same location and obtained over one thousand living specimens, which were used for morphological, growth, and behavior studies. Some of her illustrations are shown in Plate 24-15. This species lives as a "colonial group," with one temporarily fixed upon the other. This is due to the odd behavior of its migratory macrozooids, which also form at the juncture of branches. Instead of being swimmers, as in the other species, the flattened migrators just slide over to a new location on the colonial peduncle and form a new colony. The cortex has pigment granules adjacent to the epiplasm, pores to the outside through the double cuticular membrane, and a coat of epizoic bacteria. These bacteria are also visible in the pharynx of the zooids and can be seen partly digested in the digestive vacuoles. It is possible that the bacteria are sustained, at least in part, by the excretory products of the ciliates and, in turn, form part of their food supply. In addition to a few records for *Z. pelagicum* in the Mediterranean, Laval reports that it was observed during the "Valdivia Expedition" in the Atlantic and was found in North Atlantic samples obtained by the Hardy Continuous Plankton Recorders of the Scottish Oceanographic Laboratory.

The Sessilina have a number of loricate forms in addition to their vorticellid forms. Some of these are shown in Plate 24-16. Those in *A*, *B*, *C*, and *D* were among the ones recorded by Stiller (1939) from the littoral zone and aquaria at the island of Helgoland, Germany. They were associated with algae, a number of animals, and detritus. The illustrations show that the lorica can be occupied by more than one individual and

PLATE 24-13. Free-living scuticociliatids. (*A*) *Histiobalantium natans*, showing the tetrahymenal buccal apparatus consisting of the undulating membrane (PA) on the right side of the buccal cavity; the tripartate adoral zone of membranelles (M_1,M_2,M_3) on the left side direct food to the cytostome (CY). The somatic kineties (CS), macronucleus (MA), micronuclei (MI), and trichocysts (T) are also shown. (*B*) The thigmotrich *Ancistrum mytili*, showing the tetrahymenal buccal area in the lateral view (C); it is slightly dorsal at its posterior end, as seen in the ventral (A) and dorsal (B) views. In the lateral view (C), the straight tactile cilia at the anterior end (right) are used for attachment to the host, the mussel *Mytilus edulis* (all 750 X). (*A*) from J. Dragesco and F. Iftode, 1972, *Protistologica* 8:347–52; (*B*) from G.W. Kidder, 1933, *Biol. Bull.* 64:1–20.

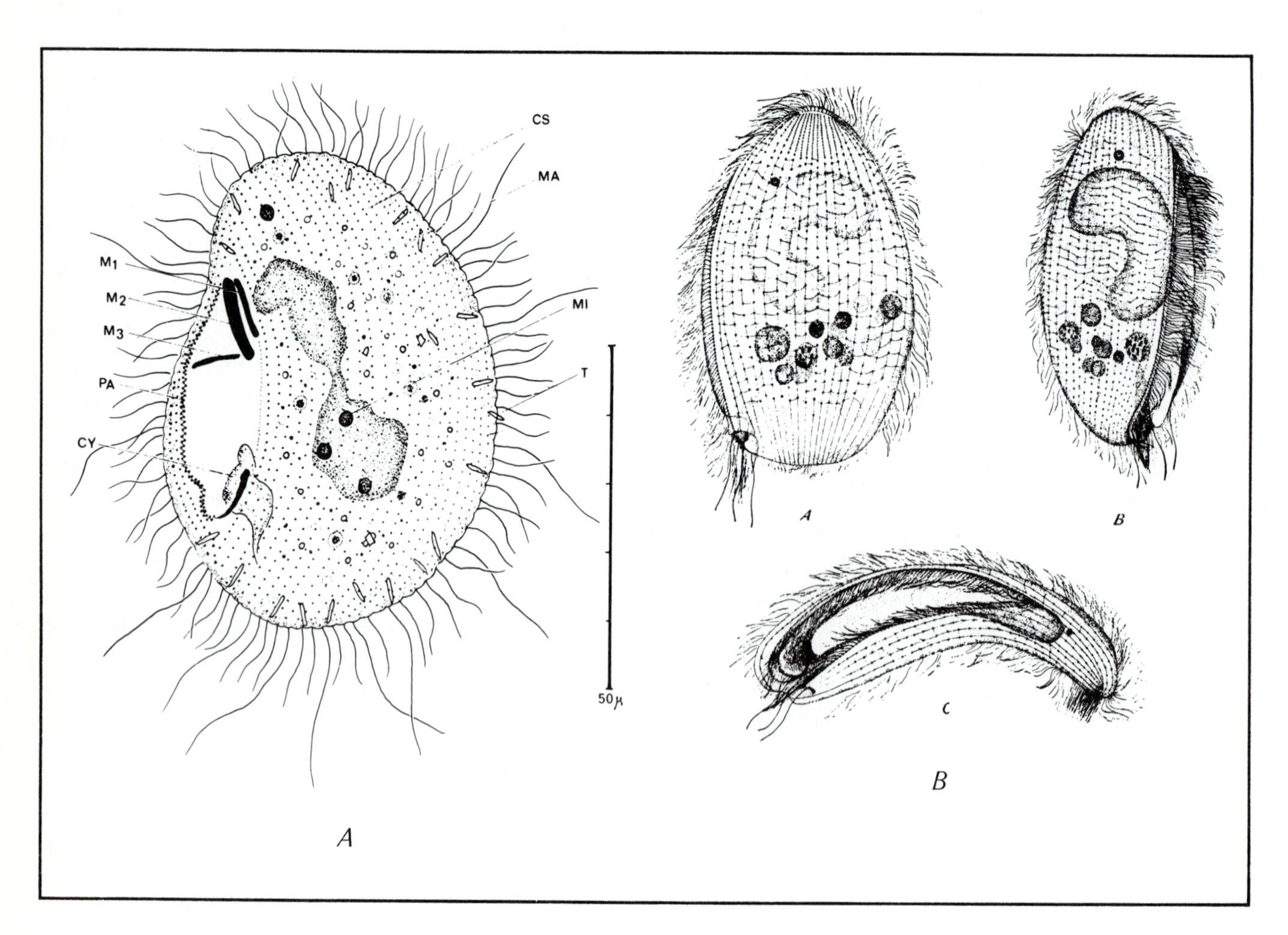

that migrating microgonidia are the form for propagation. Empty loricae are quite common. These loricate forms were observed on a number of substrates in the survey by Sieburth (1975), and some are shown in Plate 24-16,*E*,*F*.

Another member of the order Sessilina, *Lagenophrys callinectes* has developed a modified structure due to its association with the gills of the blue crab *Callinectes sapidus*. The structure and life cycle of this species, described by Couch (1967), are shown in Plate 24-17,*A*,*B*; its ultrastructure was studied by Couch (1973). *Lagenophrys callinectes* lives ectocommensally in the food-rich area of the gills, but not as a parasite. In addition to fission, this form, like many of the peritrichs, apparently reproduces sexually by the conjugation of a microconjugant with a macroconjugant. During a survey of other crustaceans for the presence of this species (which was negative), Couch observed a similar species infesting the grass shrimp *Palaemonetes*. This species, *Lagenophrys lunatus*, is attached onto or near the appendages of its host, where the water currents bring it phytoplankton and particles as food is ingested by the host (Clamp, 1973). Division occurs en masse, in synchrony with the molting of the host.

Mobilina is the other suborder in the Peritrichida. These forms are stalkless and are distinguished by a very elaborate disc on the aboral end (opposite the mouth). They skim over the epithelial surface or swim freely, disc end first. Ectozoic forms occur on the skin, gills, and fins of their hosts, whereas entozoic forms occur within their hosts. Some are highly specialized parasites in the urinary bladders of such vertebrates as fish. Most have been listed as pathogenic parasites, although some do not cause lesions and are probably ectocommensals. The habitats, structure, and reproduction of these forms in freshwater fishes was given by Davis (1947) and more recently by Lom (1970). In the common genus *Trichodina*, the species on vertebrates are highly parasitic; the species associated with molluscs tend toward commensalism. Species have been described from marine fish by Tripathi (1948), Raabe (1959), and Padnos and Nigrelli (1942), who also showed their reproductive processes; they have been most recently described by Lom and Laird (1969). Species occurring on a variety of molluscs in the Baltic Sea were described by J. and Z. Raabe (1959), and Hirshfield (1949) described species from limpets and turbans from southern California.

Trichodina myicola is a commensal species associated with the soft-shelled clam *Mya arenaria* on the east coast of the United States. It was described by Uzmann and Stickney (1954), and its structure, which is typical for the Mobilina, is shown in Plate 24-17,*C* and *D*. The basal disc, a very elaborate structure, is the locomotor organ and corresponds to the posterior ciliary wreath in the telotrochs of the Sessilina. The skeletal ring of the basal disc contains 26 to 36 articulated and hollow denticles, which look like vertebrae. They are used in classification, as discussed by Uzmann and Stickney (1954). The frequency of *T. myicola* infestation in New England soft-shelled clams ran from 0 to 62%, being higher in spring. This ciliate strongly attaches to the epithelial surface in the oral area of *Mya*. Concentrations on the four outer palps ran as high as one hundred per palp. In addition to being often accompanied by the thigmotrichine *Ancistrocoma myae*, *T. myicola* is sometimes parasitized by the suctorian *Endosphaera*.

C. POLYHYMENOPHORA

This third of the three new ciliate classes includes the old subclass Spirotricha. The primary distinguishing feature of the spirotrichs is the highly developed adoral band of membranelles, which winds clockwise to the cytostome and is used for feeding and locomotion. The buccal ciliature in the Peritricha is counterclockwise. Although some holotrichs also have an adoral zone or band of membranelles, it is not as well developed as in the spirotrichs. The second distinguishing feature is that all forms, except most of the heterotrichs, have a somatic ciliature that is greatly reduced or absent. On this basis, the marine species can be grouped into three quite distinct orders: Heterotrichida, with their somatic ciliature; the naked Oligotrichida, with large oral membranelles (a group including the suborder Tintinnina, which also have loricae); and the Hypotrichida, with their stubby cirri, which not only function like but look like feet, from a lateral view. A fourth order, the Odontostomatida, are freshwater, sapropelic forms.

The heterotrichs are the least specialized of the spirotrichs. They retain a holotrichous somatic ciliature. The majority of genera are in the suborder Heterotrichina, and many are common interstitial forms in sands

PLATE 24-14. *Zoothamnium alternans*, a coastal form of sessile (attached) Peritrichida. (*A*) Settlement of a macrozooid, growth of the peduncle from the scopula, temporary loss of cilia, which grow back as spasmoneme and elastic fibrils develop in stalk. (*B*) A mature colony, showing the primary axis, alternating ramifications, and the larger macrozooids at junctures. (*C*) A transverse section of a zooid, showing the invaginated peristome (pe), contractile vacuole (Vp), macronucleus (Ma), and peduncle (P), with a central myoneme and peripheral fibrils [also shown in (*F*)]. (*D*) Scanning electron micrograph of a *Zoothamnium* colony on the bryozoan *Bugula turrita*. (*E*) Nomarski differential interference contrast micrograph (optical section) of a *Zoothamnium* on the copepod *Acartia tonsa*. (*F*) A longitudinal

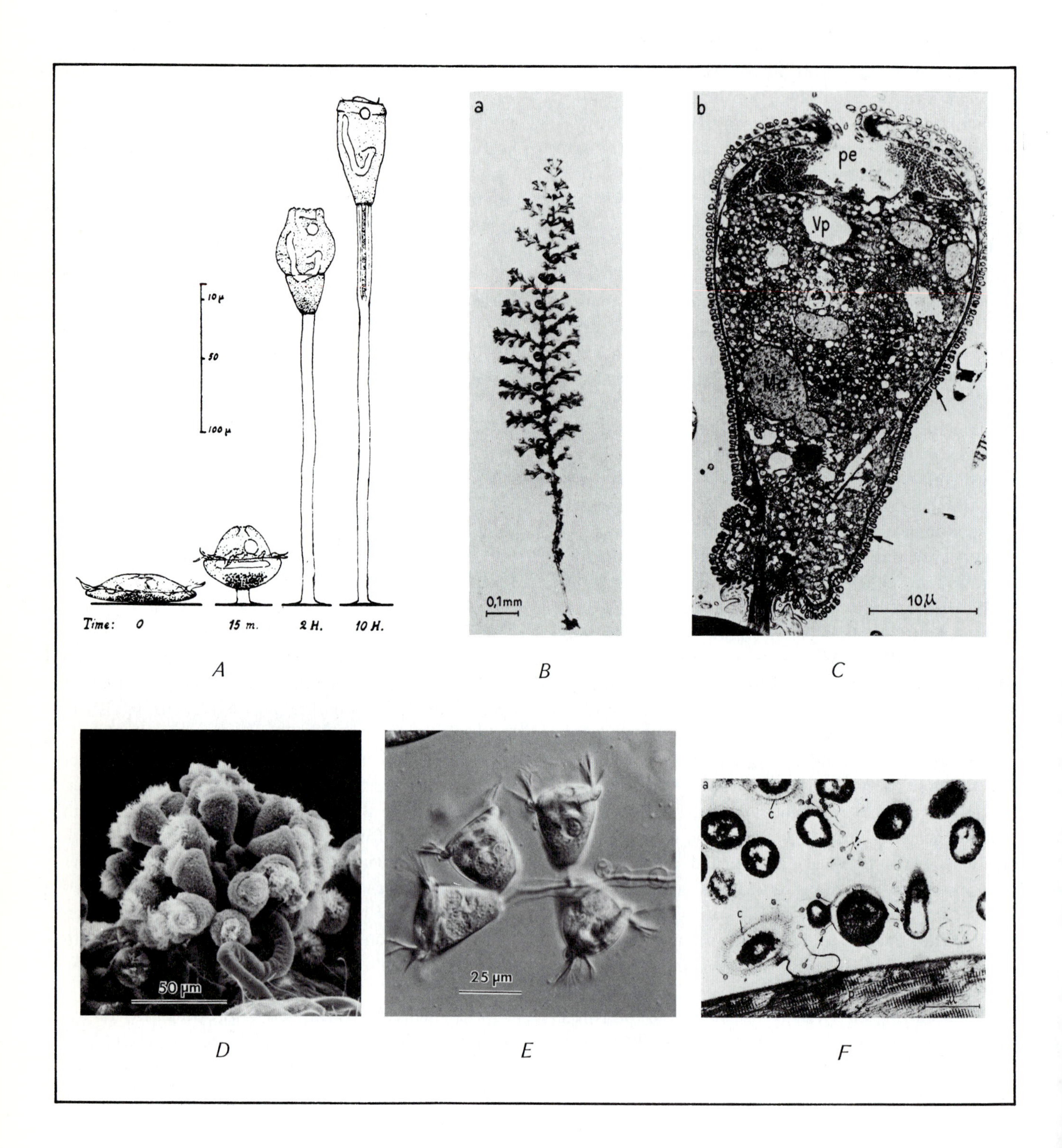

section of the peduncle, showing elastic fibrils and two types of epizoic bacteria (one with a capsule, C), with two types of bacteriophage (arrows). (*A*) from E. Fauré-Fremiet, 1930, *Biol. Bull.* 58:28–51; (*B,C,F*) from E. Fauré-Fremiet, P. Favard, and N. Carasso, 1963, *Cahiers Biol. Mar.* 4:61–64; (*D*) from J.McN. Sieburth, 1975, *Microbial Seascapes*, University Park Press, Baltimore, Md.; (*E*) photomicrograph courtesy of Paul W. Johnson.

---

and sediments. Some common species of heterotrichine ciliates are shown in Plate 24-18. Some species are pigmented, such as the rose-colored *Blepharisma multinucleala* (Plate 24-18,*D*); these light-sensitive ciliates are easy to culture and have been frequently studied. A text on their biology has been prepared by Giese (1973).

Five other suborders of heterotrichs are now recognized (Corliss, 1977). The Clevelandellina contain the endosymbiotic forms, which are of little concern to us here, since they occur in insects and lower invertebrates, principally. The Armophorina are few in number and primarily free-living, sapropelic, freshwater forms. The Coliphorina, the familiar folliculinids, are marine sedentary forms known for their loricae, an adoral band of membranelles borne on a pair of prominent "peristomial wings" extending from the lorica-encased body, and a vermiform migratory stage in the life cycle. The lorica are widely distributed in the marine environments and attach to invertebrates and algae (Corliss, 1975). The Plagiotomina, limited to a single included genus, are endocommensals in lumbricid oligochaetes.

Finally, the suborder Licnophorina contains the single genus *Licnophora* (see Plate 24-18,*C*). These forms have no somatic ciliature and are named for the distinct, fan-shaped membranelle of the oral disc. They use their basal disc to attach to a variety of animal hosts, inanimate substrates, and benthic algae. Ectocommensals, they feed on bacteria, diatoms, and other algae. The monograph by Villeneuve-Brachon (1940) is very informative about a number of species, and Balamuth (1941) has studied *L. macfarlandi* on its sea cucumber host along the Pacific coast of the United States.

The small order Odontostomatida (formerly Ctenostomatida) contains small, wedge-shaped forms that are laterally compressed. Their carapace, or pellicle, is sometimes drawn into spines, and the somatic ciliature is sparse. This group is distinguished by the oral ciliature, which is reduced to eight membranelles set close in a row like teeth (hence their name). They occur in highly eutrophic and stagnant waters containing decomposing organic matter and are usually freshwater forms. The reasons for the name change are discussed by Corliss (1961), who lists the genera. The only common marine form appears to be *Mylestoma bipartitum*, originally described from the old port of Marseilles, France by Gourret and Roeser (1886) and emended with further observation of its occurrence in the Kiel Bight and near the island of Sylt, Germany by Kahl (1932b), who has been the main student of this rare group of ciliates.

The oligotrichs *sensu stricto* (members of the Oligotrichina) are a small group of tiny-bodied (50 to 60 μm long) ciliates. Most or all of the somatic ciliature is lost, and the remnants have become isolated patches of bristles or cirri. These and the highly developed adoral band of membranelles are used for locomotion. Ciliates like *Strombidium* and *Halteria* are common in the plankton. A number of examples from the classical monograph on planktonic ciliates by Fauré-Fremiet (1924) are shown in Plate 24-19 to illustrate the common features and diversity of this group. An intriguing ecological study on the diurnal rhythm of *Strombidium oculatum*, an inhabitor of tide pools, was reported by Fauré-Fremiet (1948). When the pools become isolated at low tide, the ciliates swim freely in the plankton, but as the pools become submerged with the rising tide, the ciliates settle on a substrate and encyst. *Strombidium* can be maintained easily in culture, is much studied, and is used as a food for carnivorous ciliates. The morphology and ultrastructure of *Strombidium sulcatum*, as observed by Fauré-Fremiet and Ganier (1970a), is shown in Plate 24-20. Underneath the superficial membrane is a perilemma consisting of two unit membranes that, in the posterior hemisphere, cover polysaccharide plates. The three rows of cilia in the adoral band of membranelles are also shown. An interesting observation is the highly differentiated ciliature at the equatorial groove, which is presumed to have a sensory function (Fauré-Fremiet and Ganier, 1970b). The systematics of the oligotrichs has been discussed by Fauré-Fremiet (1970).

The Tintinnina is a very large suborder of oligotrichs *sensu lato*, containing well over 1000 species that live a pelagic life in coastal waters and the open sea. They are distinguished by their typical spirotrich cells, which are encased in a variety of loricae; some of these loricae have intricate patterns, whereas others are transparent or reinforced by attached debris. The 12 to 24 powerful membranelles of the adoral zone give the tintinnids their motility, and their loricae apparently con-

PLATE 24-15. *Zoothamnium pelagicum*, a free-living planktonic peritrich. (*A*) An extended colony; the left macrozooid (center) has a posterior ciliary wreath and is ready to migrate to a new site to form a new colony. (*B*) Two macrozooids, with pigment granules (pig.) and epizoic bacteria (ep). Note the two bulges (double arrows) and the macronucleus (Ma), micronucleus (mi), and contractile vacuole (v.c.). The one on the right is sedentary; the one on left will migrate [see (C)]. (*C*) Migration and fixation of a macrozooid after the formation of a posterior ciliary belt (telotrochal wreath, c.c.p.). The macrozooid, seen in oral view (sc, scopula), does not swim but slides to a new location beside its former stalk (R.1) and forms the stalk for a new colony. (*D*) Ultrastructure of cortical integument, showing the epiplasm (ep), cuticular membranes (m.e. and m.i.), pores (p), sacculi (S), pigment granules (pig.), and epizoic bacteria (Bact.). From M. Laval, 1968, *Protistologica* 4:333–63.

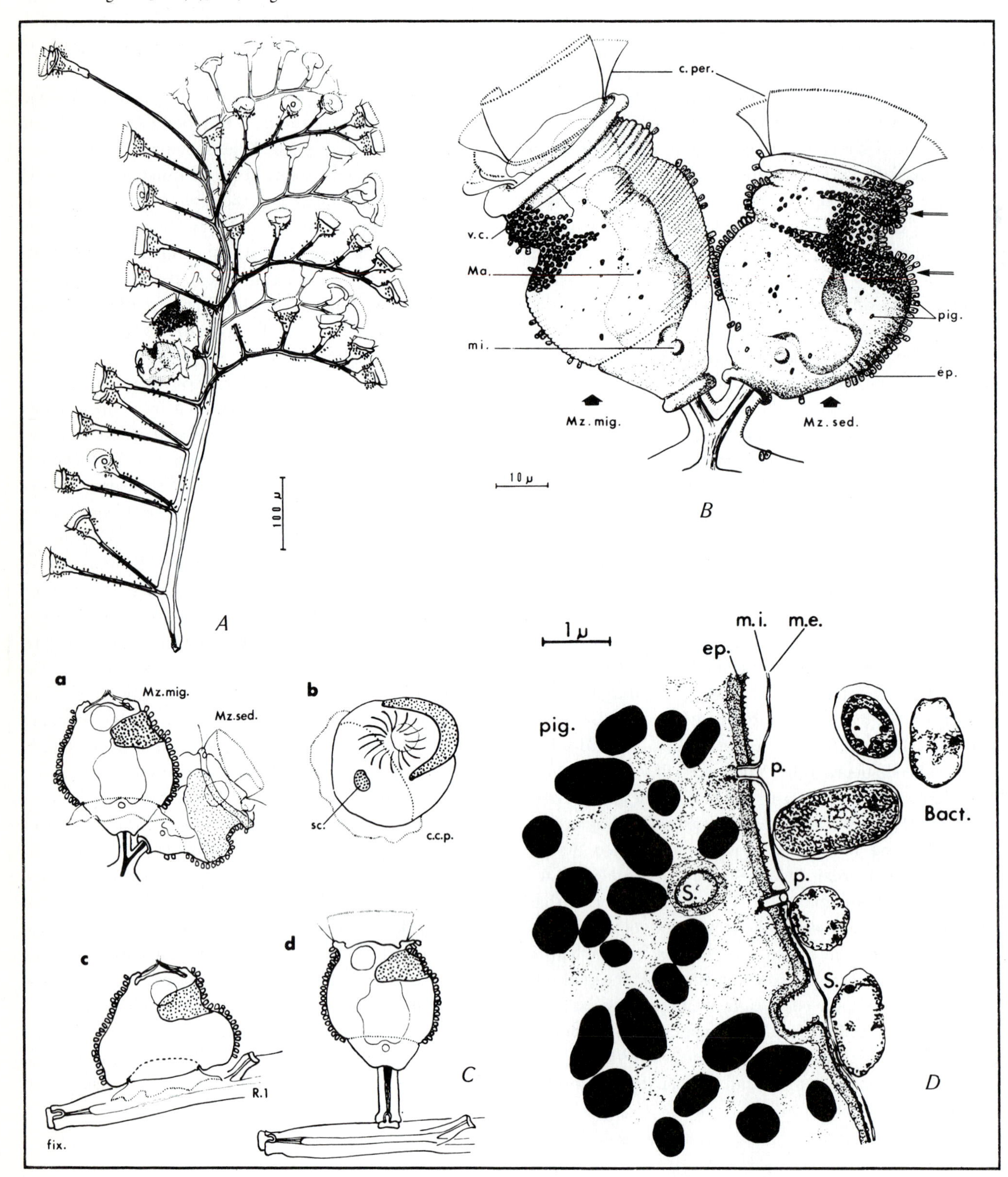

PLATE 24-16. Loricate forms of sessile peritrichs occur on a variety of surfaces, including pebbles, invertebrates, molluscs, algae, and debris. A series of species from the waters of Helgoland, Germany: (*A*) *Cothurnia nodosa* in a contracted state, with three microgonidia (800 X); (*B*) *Cothurnia aplatita*, with two extended individuals in one house (780 X); (*C*) *Cothurnia auriculata*, showing contracted (640 X) and extended (800 X) individuals; (*D*) *Pyxicola socialis*, a long-stalked form (550 X). Scanning electron micrographs of *Cothurnia* species in Narragansett Bay, Rhode Island on a slate pebble (*E*) and the scales of a scaleworm (*F*). (*A–D*) from J. Stiller, 1939, *Archiv Protistenkunde* 92:415–52; (*E,F*) from J.McN. Sieburth, 1975, *Microbial Seascapes*, University Park Press, Baltimore, Md.

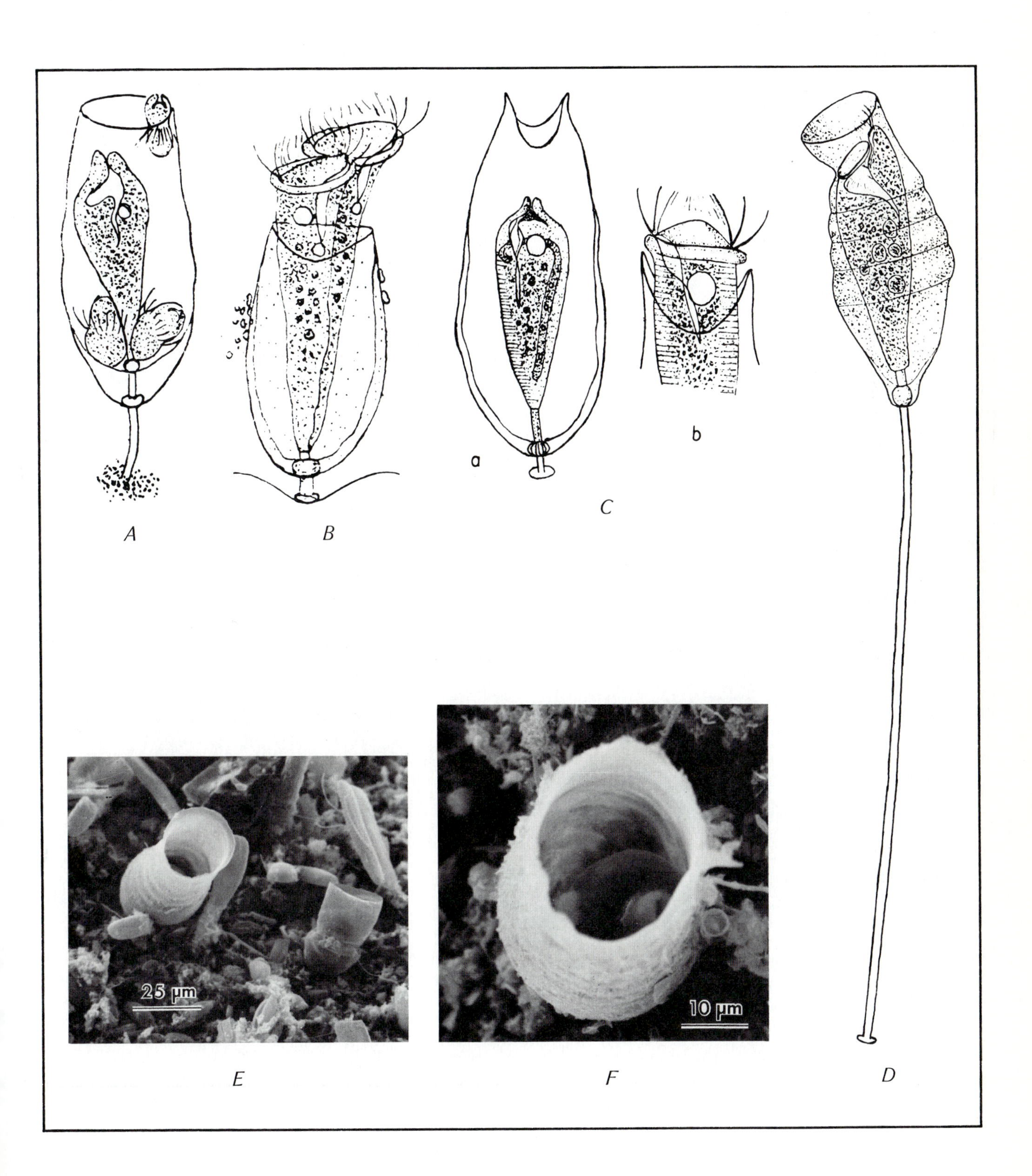

PLATE 24-17. Peritrichous ciliates ectocommensal on a crab and a clam. (*A*) Diagrams of the sessile ciliate *Lagenophrys callinectes*, in ventral (13), dorsal (16), and lateral (17) view. (*B*) (1) Photograph of the host the blue crab *Callinectes sapidus;* (2) infested gill (note occupied and empty loricas); (3) a mature ciliate; (10) cell division, with attached telotroch; (11) detaching telotroch (arrow); (12) apparent conjugation of a microconjugant, with macroconjugant. (*C*) Diagram, showing the ventral aspect of *Trichodina myicola*, a mobile ciliate on the soft-shelled clam *Mya arenaria*, showing anterior adoral cilia (AD) and the posterior ciliary girdle (CG) around the basal disc upon which it travels. (*D*) Semi-diagrammatic representation of a basal disc, showing the species-specific denticulate ring (DR). (*A,B*) from J.A. Couch, 1967, *Trans. Amer. Micros. Soc.* 86:204–11; (*C,D*) from J.R. Uzmann and A.P. Stickney, 1954, *J. Protozool.* 1:149–55.

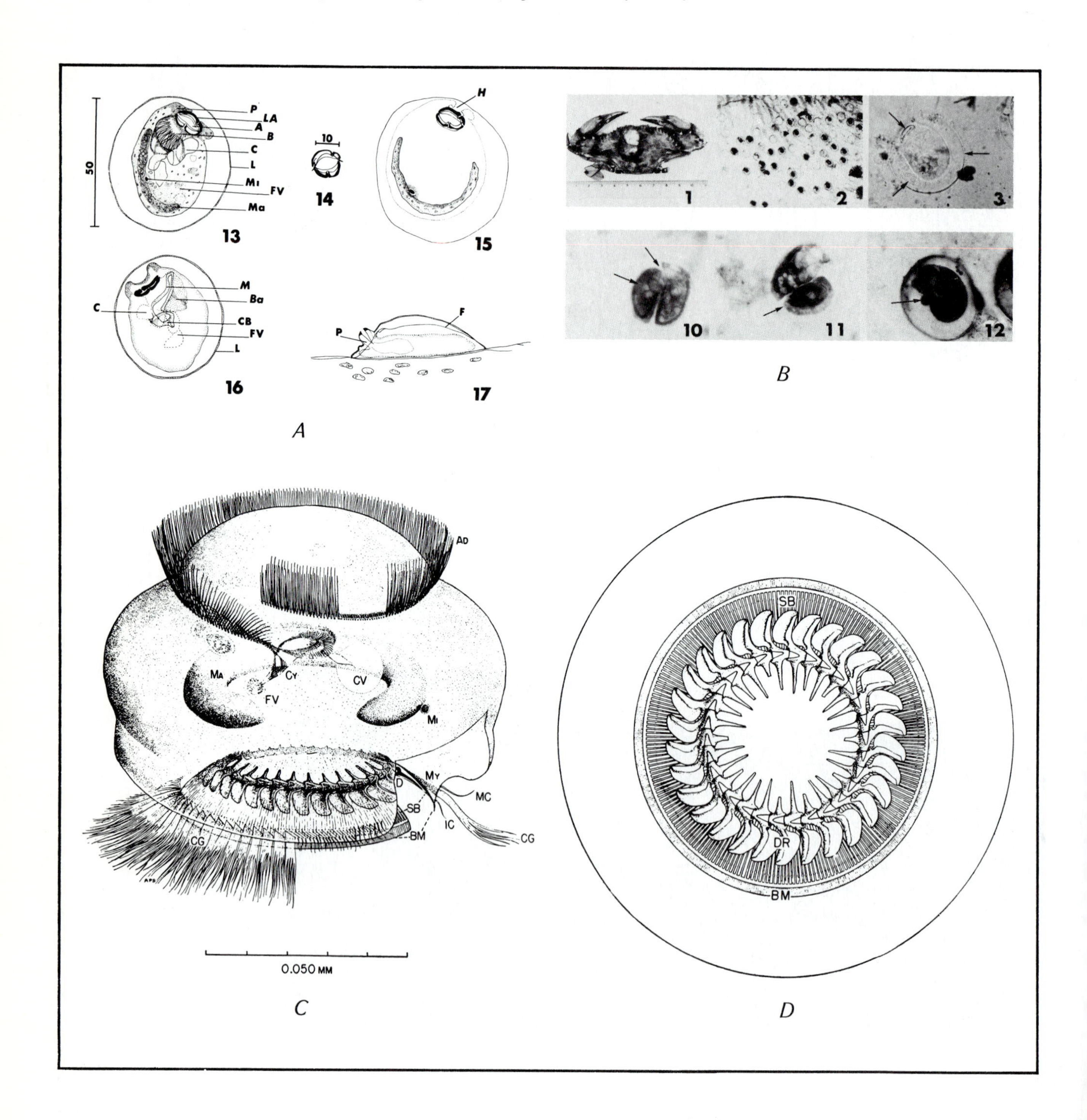

PLATE 24-18. Examples of heterotrichous ciliates (subclass Spirotricha), including a number of heterotrichids from interstitial environments and one sedentary Licnophorine, which colonize a number of surfaces, including animal hosts. (A) *Metopus lemani;* (*B*) *Condylostoma minima;* (*C*) *Licnophora chattoni* (1,100 X); (*D*) *Blepharisma multinucleala;* (*E*) *Condylostoma acuta.* (*A,B,D,E*) from J. Dragesco, 1960, *Traveaux Station Biologique Roscoff* 12, 356 p.; (*C*) from S. Villeneuve-Brachon, 1940, *Arch. Zool. Exp. gén.* 82:1–80.

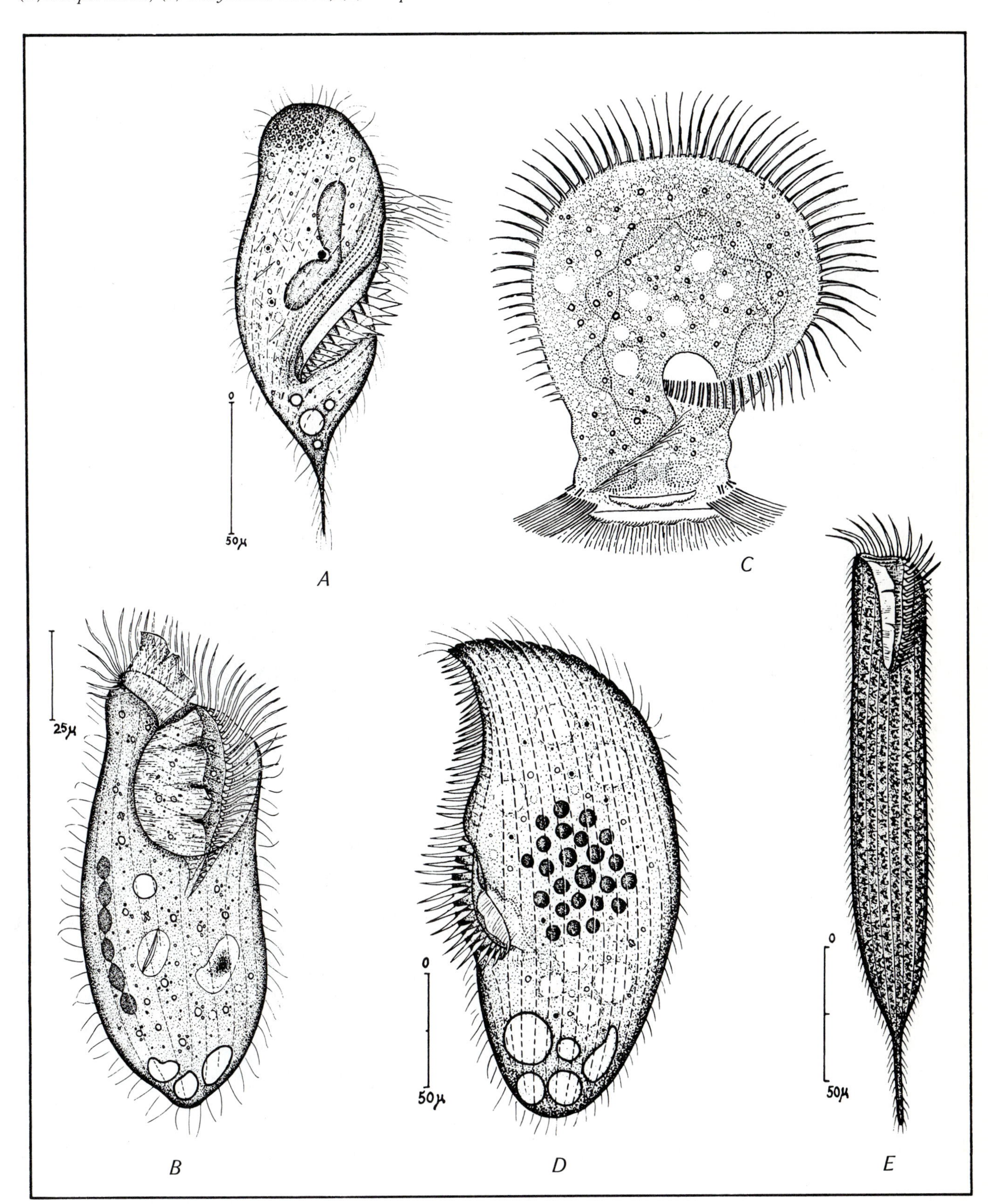

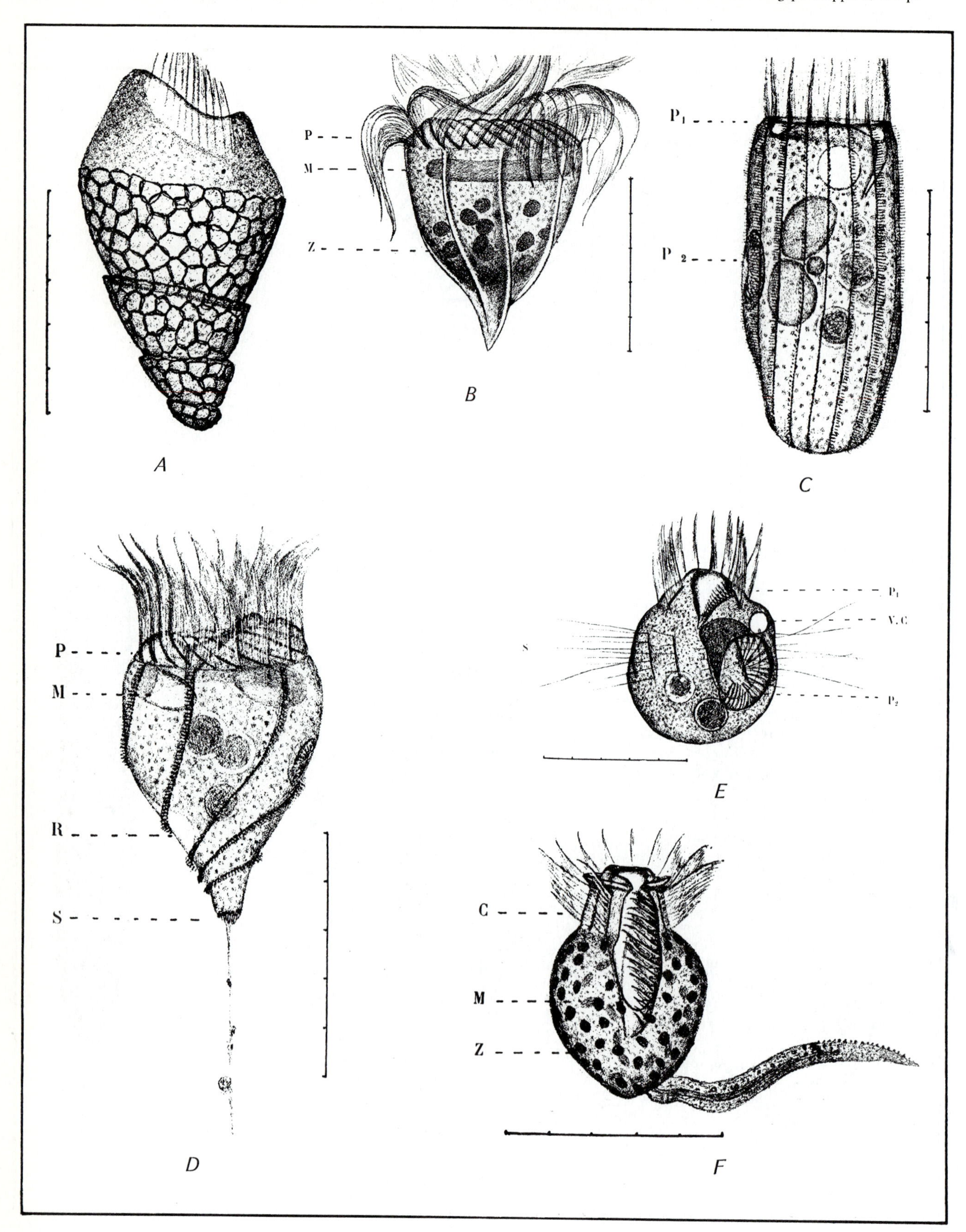

PLATE 24-19. Some examples of oligotrichous ciliates (Spirotricha) from the plankton (bars = 50 μm, with 10 μm divisions). (*A*) *Laboea strobila;* (*B*) *Strombilidium velox;* (*C*) *Strombidinopsis claparedei;* (*D*) *Strombilidium adherens;* (*E*) *Halteria grandinella;* (*F*) *Tontonia gracillina*. From E. Fauré-Fremiet, 1924, *Bull. Biol. France Belgique* Suppl. 6, 171 p.

PLATE 24-20. The morphology and ultrastructure of the oligotrichous ciliate *Strombidium sulcatum*. (*A*) Semi-diagrammatic figure of a living cell (approximately 25 to 30 μm by 40 to 50 μm), showing the fringe of the adoral band of membranelles around the apial pole, ventral pharyngeal depression, posterior hemisphere with polysaccharide plates (P), posterior limit of equatorial groove and position of trichites (T), digestive vacuoles (Va), and contractile vacuole (Vc). (*B*) Transmission electron micrograph of a transverse section of an adoral membranelle, which consists of three rows of parallel cilia; (C) two axonemes enclosed in the same ciliary membrane. (*C*) Transmission electron micrograph of an oblique section at the equatorial belt, showing a highly differentiated ciliature, both perpendicular (C1) and parallel (C2) to the pellicle, which is presumed to have a sensorial function. (*D*) Transmission electron micrograph of a cross section of the posterior hemisphere, showing the perilemma (E), which consists of two unit membranes (U), the polysaccharide plates (P), and the peripheral mitochondria (M). From E. Fauré-Fremiet and M-Cl. Ganier, 1970a, *Protistologica* 6:207–23.

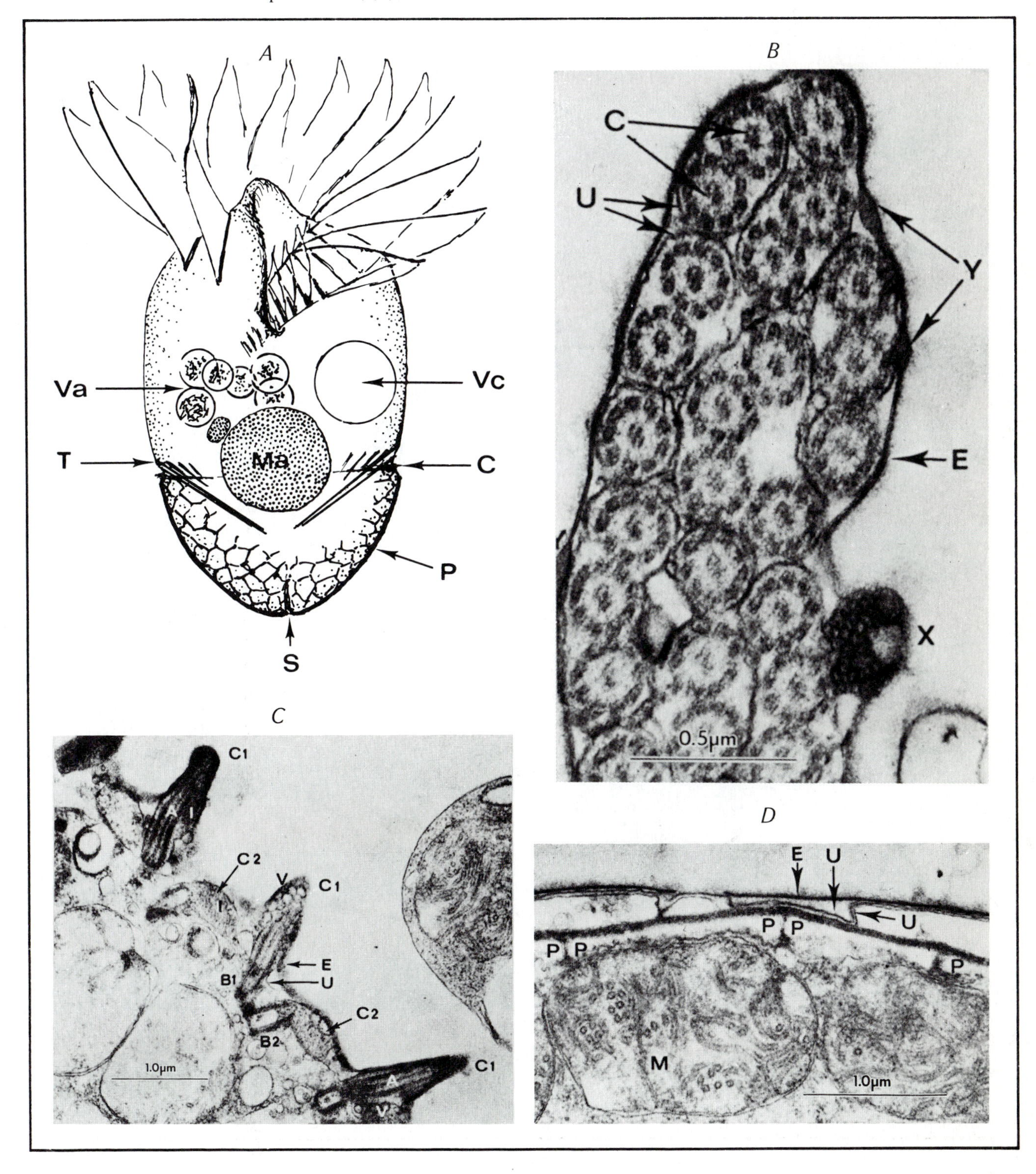

PLATE 24-21. Examples of tintinnines (order Oligotrichida) from the coastal plankton. (*A*) *Cyttarocyclis denticulata* var. *gigantea:* lip of peristome (p), contractile vacuole (V), lateral ciliary field (C), macronucleus (M), and pattern of lorica (L). (*B*) *Tintinnus fraknoii.* (*C*) *Tintinnopsis campanula.* (*D*) *Amphorella quadrilineata* (bars=50μm, with 10 μm divisions). All from E. Fauré-Fremiet, 1924, *Bull. Biol. France Belgique* Suppl. 6:1–171.

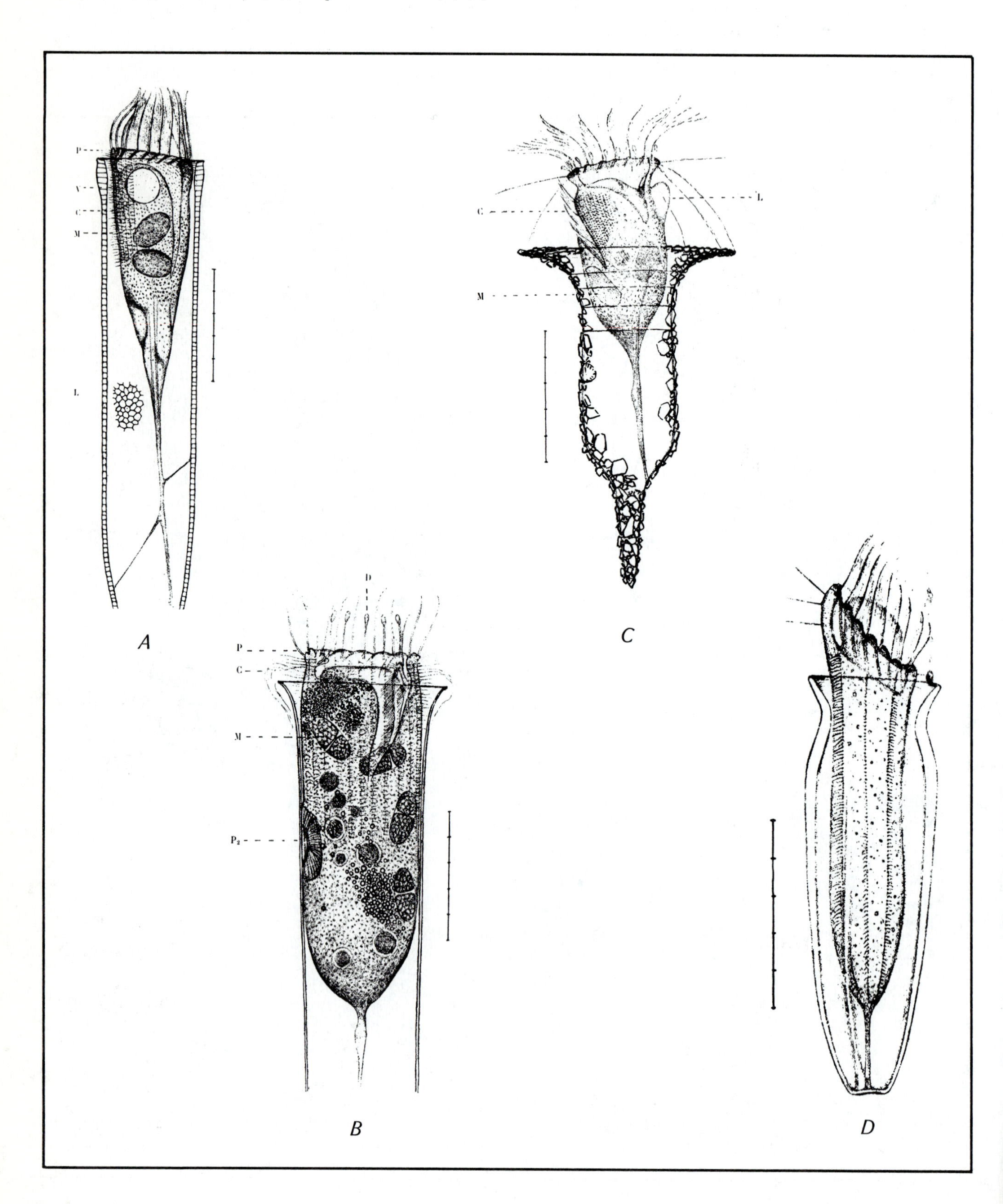

PLATE 24-22. Morphology and ultrastructure of the tintinnine *Petalotricha ampulla*. (*A*) Photomicrograph, showing the extended membranelles and the aboral peduncle attaching the body to the lorica (50 to 70 by 150 to 200 μm). (*B*) Longitudinal sections of the lorica (left, 15,000 X; right, 6,200 X). (*C*) Transverse section of the paroral membranelle. Note the intracytoplasmic bacterium (B) (36,000 X). (*D*) Oblique transverse section of the adoral band of membranelles, showing three parallel rows of kineties (19,000 X). From M. Laval, 1973, *Protistologica* 8:369–86.

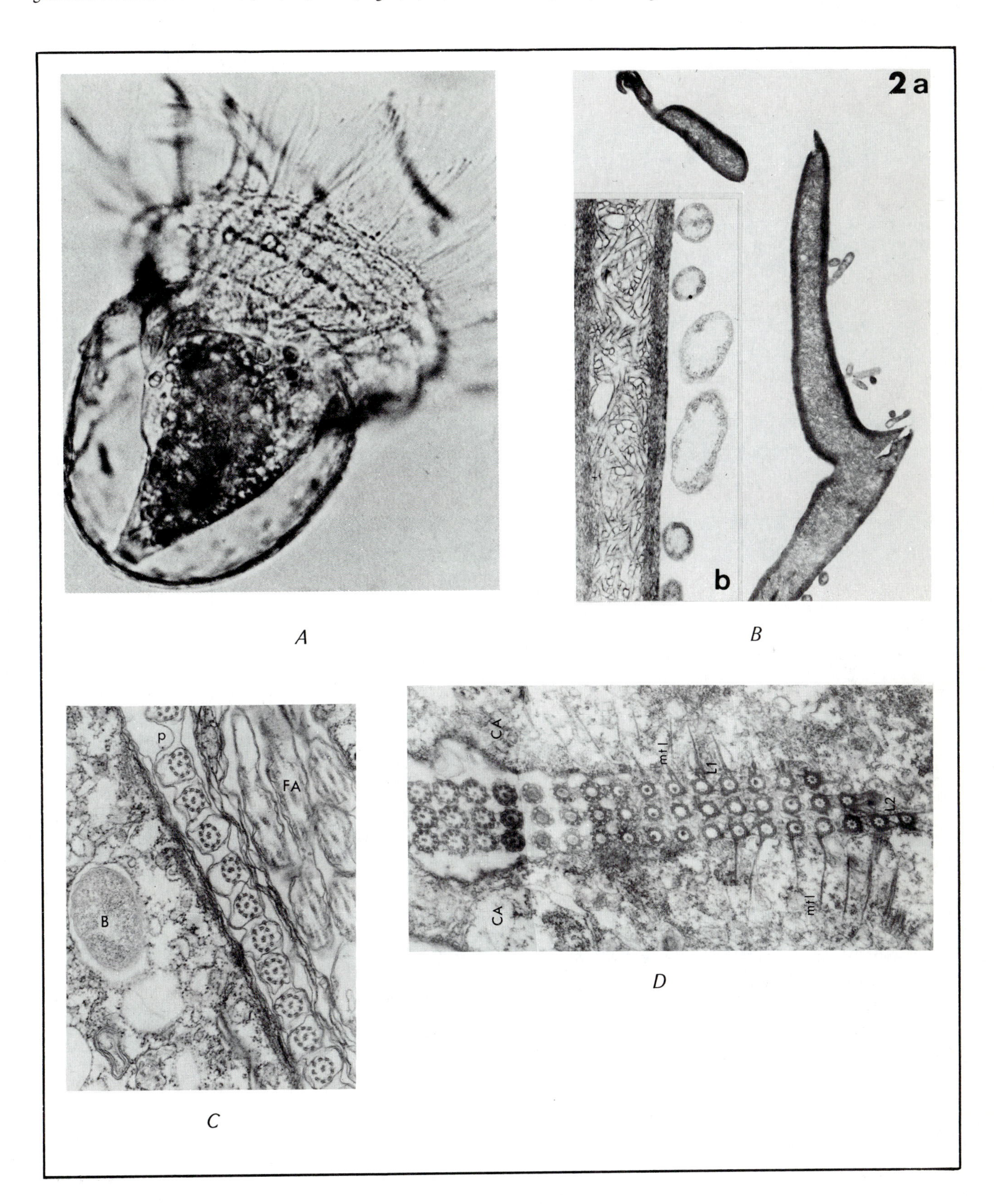

PLATE 24-23. Some examples of hypotrichous ciliates (Spirotricha) from the mesopsammic environment. (*A*) *Banyulsella viridis;* (*B*) *Euplotes aberrans;* (*C*) *Swedmarkia arenicola;* (*D*) *Lacazea ovalis;* (*E*) *Aspidisca hyalina.* From J. Dragesco, 1960, *Traveaux Station Biologique Roscoff* 12:1–356.

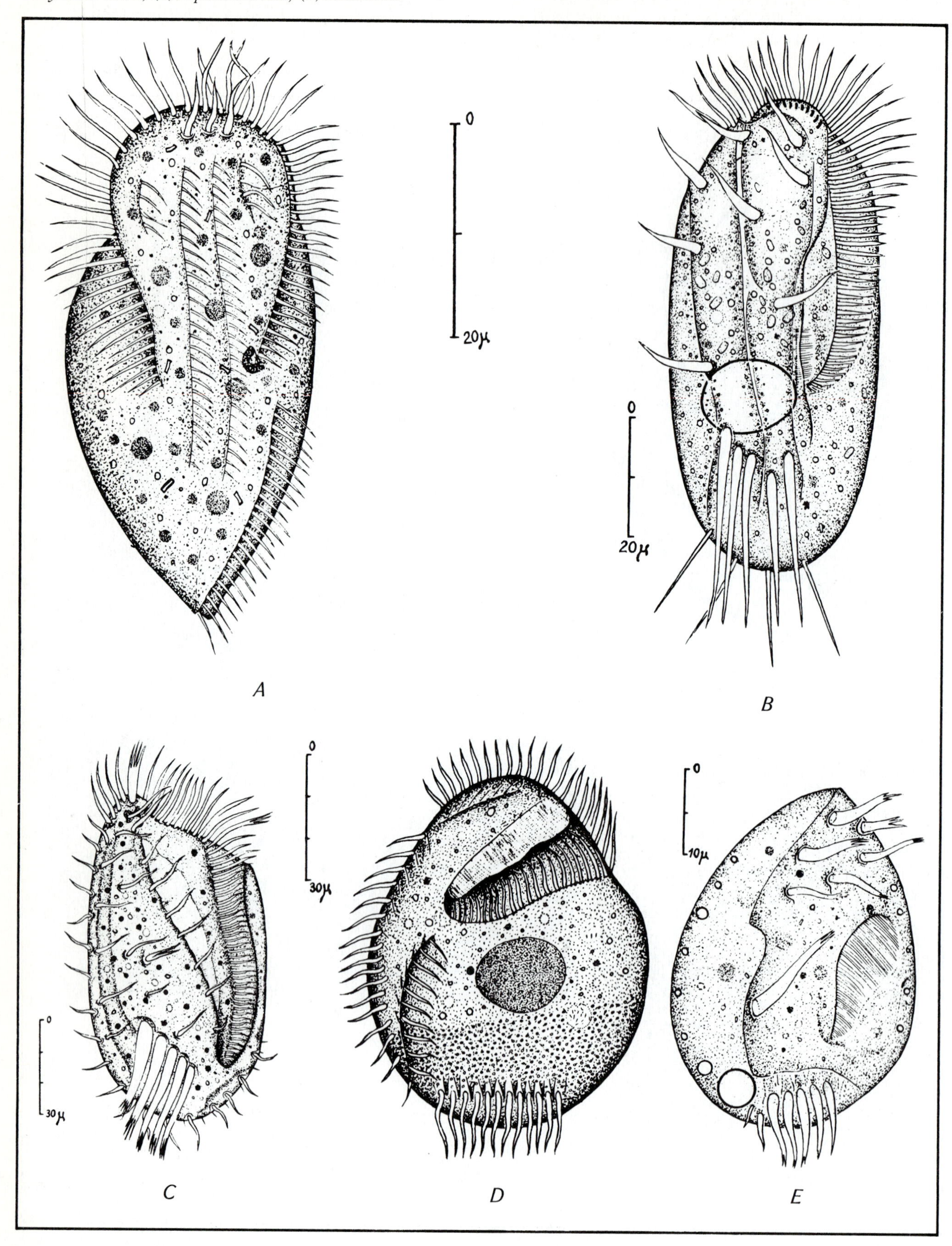

tribute to their buoyancy. A few coastal species from the monograph by Fauré-Fremiet (1924) are shown in Plate 24–21. The many monographic works on the taxonomy of coastal and deep-sea species of tintinnids are cited in the excellent bibliography of Laval (1973). Those of Kofoid and Campbell (1929, 1939) are most useful for warm water species.

MacKinnon and Hawes (1961) remark that tintinnids do not do well in laboratory culture and that not one species has been described in detail. Exceptions might be the good earlier descriptions of two species by Campbell (1926, 1927). Since then, Laval (1973) has described the tintinnid *Petalotricha ampulla* and recorded a number of new or little known structures. Some of her illustrations are included in Plate 24-22. These show how the body is attached to the lorica by the peduncle, the trilaminar structure of the lorica, which is populated by bacteria on its inner surface, cross sections showing the ciliary nature of the membranelles, and the usual intracytoplasmic bacteria.

The successful culture of a tintinnid was reported by Gold (1970) for *Tintinnopsis tubulosa*. Some six species have now been cultured for varying periods, with different degrees of success (Gold, 1973). Ciliates reared in continuous culture have been microphages that feed on bacteria. Such multi-species cultures (Legner, 1973) appear very useful in the study of protozoa-prey relationships. A method for the semi-automatic continuous culture of *Tintinnopsis beroidea* has been developed by Gold (1973). Although the original generation time of 1 day extended to 2.5 days after 1.5 years in culture, sustained yields of 10^3 cells/ml have been obtained.

The order Hypotrichida are quite distinctive ciliates, many of which are common interstitial inhabitants of sands and sediments. A few species in different genera from the monograph on littoral mesopsammophiles by Dragesco (1960) are shown in Plate 24-23. The hypotrichs are dorsoventrally flattened, the membranelle band is on the ventral side, and the remaining somatic ciliature is in the form of cirri or bristles. The frontal and anal cirri are used to "walk" over substrates (Wichterman, 1964). The taxonomy of this order has been revised by Borror (1972a). A common genus is *Euplotes*, which has been much studied, and its taxonomy has been revised by Tuffrau (1960) and subsequently by Curds (1975). Several marine species of *Euplotes* have been described in the last decade (Wichterman, 1964; Carter, 1972). Some species inhabit environments containing decaying algae, and sea urchins feeding on benthic algae have acquired echinophilous ectocommensal species of *Euplotes* (Berger, 1965). Persoone and Deplaecie (1973) have studied some of the ecological factors that affect the growth of *Euplotes vannus* nourished on bakers' yeast. A generation time of 12 hours was observed between 20 and 25 °C, whereas similar growth occurred between 20 and 35 o/oo salinity. One wonders if they would have grown better or been happier on brewers' or vintners' yeast.

REFERENCES

Antipa, G.A. 1972. Structural differentiation in the somatic cortex of a ciliated protozoan, *Conchophthirus curtus* Engelmann 1862. *Protistologica* 7(year 1971):471–501.

Balamuth, W. 1941. Studies on the organization of ciliate protozoa. I. Microscopic anatomy of *Licnophora macfarlandi*. *J. Morph.* 68:241–77.

Bardele, C.F. 1969. *Acineta tuberosa*. II. Die Verteilung der Mikrotubuli im Makronucleus während der ungeschlechtlichen Fortpflanzung. *Z. Zellforsch.* 93:93–104.

Bardele, C.F. 1970. Budding and metamorphosis in *Acineta tuberosa*. An electron micrographic study on morphogenesis in Suctoria. *J. Protozool.* 17:51–70.

Bardele, C.F. 1972. A microtubule model for ingestion and transport in the suctorian tentacle. *Z. Zellforsch.* 126:116–34.

Bardele, C.F., and K.G. Grell. 1967. Elektronenmikroskopische Beobachtungen zur Nahrungsaufnahme bei dem Suktor *Acineta tuberosa* Ehrenberg. *Z. Zellforsch.* 80:108–23.

Barna, I., and D.S. Weis. 1973. The utilization of bacteria as food for *Paramecium bursaria*. *Trans. Amer. Micros. Soc.* 92:434–40.

Batisse, A. 1968. Quelques Aspects de l'ultrastructure de *Pseudogemma pachystyla* Collin. *Protistologica* 4:271–82.

Batisse, A. 1969. Acinétiens nouveaux ou mal connus des côtes Méditerrannéennes françaises. I. *Ophryodendron hollandei* n. sp. (Suctorida, Ophryodendridae). *Vie et Milieu, Ser. A.* 20:251–77.

Batisse, A. 1973. Premières Observations sur l'ultrastructure de *Trematosoma bocqueti* (Guilcher), Batisse (Ciliata, Suctorida). *Protistologica* 8 (year 1972):477–96.

Berger, J. 1965. The infraciliary morphology of *Euplotes tuffraui* n. sp., a commensal in strongylocentroid echinoids, with comments on echinophilous populations of *Euplotes balteatus* (Dujardin). (Ciliata: Hypotrichida). *Protistologica* 1(1):17–31.

Borror, A.C. 1963. Morphology and ecology of the benthic ciliated protozoa of Alligator Harbor, Florida. *Archiv Protistenk.* 106: 465–534.

Borror, A.C. 1965. New and little-known tidal marsh ciliates. *Trans. Amer. Micros. Soc.* 84:550–65.

Borror, A.C. 1968. Ecology of interstitial ciliates. *Trans. Amer. Micros. Soc.* 87:233–43.

Borror, A.C. 1972a. Revision of the order Hypotrichida (Ciliophora, Protozoa). *J. Protozool.* 19:1–23.

Borror, A.C. 1972b. Tidal marsh ciliates (Protozoa): Morphology, ecology, systematics. *Acta Protozool.* 10:29–71.

Borror, A.C. 1973. *Marine fauna and flora of the Northeastern United States. Protozoa: Ciliophora.* NOAA Tech. Rept. NMFS circ. 378. U.S. Dept. of Commerce, Washington D.C., 62 p.

Bradbury, P.C. 1966a. The life cycle and morphology of the apostomatous ciliate, *Hyalophysa chattoni* n.g., n.sp. *J. Protozool.* 13:209–25.

Bradbury, P.C. 1966b. The fine structure of the mature tomite of *Hyalophysa chattoni*. *J. Protozool.* 13:591–607.

Bradbury, P.C. 1973. The fine structure of the cytostome of the apostomatous ciliate *Hyalophysa chattoni*. *J. Protozool.* 20:405–14.

Burreson, E.M. 1973. Symbiotic ciliates from solitary ascidians in the Pacific Northwest, with a description of *Parahypocoma rhamphisokarya* n. sp. *Trans. Amer. Micros. Soc.* 92:517–22.

Campbell, A.S. 1926. The cytology of *Tintinnopsis nucula* (Fol) Laackmann, with an account of its neuromotor apparatus, division, and a new intranuclear parasite. *Univ. Calif. Pub. Zool.* 29:179–236.

Campbell, A.S. 1927. Studies on the marine ciliate *Favella* (Jörgensen), with special regard to the neuromotor apparatus and its role in the formation of the lorica. *Univ. Calif. Pub. Zool.* 29:429–52.

Carter, H.P. 1972. Infraciliature of eleven species of the genus *Euplotes*. *Trans. Amer. Micros. Soc.* 91:466–92.

Chatton, E. 1911. Ciliés parasites des cestes et des pyrosomes: *Perikaryon* cesticola n.g. n. sp. et *Conchophrys davidoffi*, n.g., n. sp. *Arch. Zool. Exp. Gén. 5th Ser.* 48(N&R):8–20.

Chatton, E., and A. Lwoff. 1931. La conception des ciliés apostomes (foettingeriidés + opalinopsidés) Preuves de sa validité. *C.R. Acad. Sci. (Paris)* 193:1483–85.

Chatton, E., and A. Lwoff. 1935. Les ciliés apostomes. Morphologie, cytologie, éthologie, évolution, systématique. Premier Partie, Aperçu historique et général étude monographique des genres et des espèces. *Arch. Zool. Exp. Gén.* 77:1–453.

Chatton, E., and A. Lwoff. 1949. Récherchessur les ciliés thigmotriches. I. *Arch. Zool. Exp. Gén.* 86:169–253.

Chatton, E., and A. Lwoff. 1950. Récherches sur les ciliés thigmotriches. II. *Arch. Zool. Exp. Gén.* 86:393–485.

Clamp, J.C. 1973. Observations on the host-symbiont relationships of *Lagenophrys lunatus* Imamura. *J. Protozool.* 20:558–61.

Claparède, E., and J. Lachmann. 1858. *Études sur les infusoires et les rhizopodes.* Imp. Vaney, Genève, 291 p.

Collin, B. 1912. Étude monographique sur les acinétiens. II. Monographie, physiologie, systematique. *Arch. Zool. Exp. Gén.* 51:1–457.

Corliss, J.O. 1953. Silver impregnation of ciliated protozoa by the Chatton-Lwoff technic. *Stain Technol.* 28:97–100.

Corliss, J.O. 1959. An illustrated key to the higher groups of the ciliated protozoa, with definition of terms. *J. Protozool.* 6:265–81.

Corliss, J.O. 1961. *The Ciliated Protozoa: Characterization, Classification, and Guide to the Literature.* Pergamon Press, Oxford and New York, 310 p.

Corliss, J.O. 1963. Establishment of an international type-slide collection for the ciliate protozoa. *J. Protozool.* 10:247–49.

Corliss, J.O. 1972. Current status of the International Collection of Ciliate Type-Specimens and guidelines for future contributors. *Trans. Amer. Micros. Soc.* 91:221–35.

Corliss, J.O. 1974a. The changing world of ciliate systematics: Historical analysis of past efforts and a newly proposed phylogenetic scheme of classification for the protistan phylum Ciliophora. *Systematic Zool.* 23:91–138.

Corliss, J.O. 1974b. Remarks on the composition of the large ciliate class Kinetofragmophora de Puytorac et al., 1974, and recognition of several new taxa therein, with emphasis on the primitive order Primociliatida n. ord. *J. Protozool.* 21:207–20.

Corliss, J.O. 1975. Taxonomic characterization of the suprafamilial groups in a revision of recently proposed schemes of classification for the phylum Ciliophora. *Trans. Amer. Micros. Soc.* 94:224–67.

Corliss, J.O. 1977. Annotated assignment of families and genera to the orders and classes currently comprising the Corlissian scheme of higher classification for the phylum Ciliophora. *Trans. Amer. Micros. Soc.* 96:104–40.

Corliss, J.O., and E. Hartwig. 1977. The "primitive" interstitial ciliates; their ecology, nuclear uniqueness, and postulated place in the evolution and systematics of the phylum Ciliophora. *Akad. Wiss. Lit. (Mainz) Math.-Naturwiss. Kl. Mikrofauna Meeresbodens* 61:59–82.

Couch, J.A. 1967. A new species of *Lagenophrys* (Ciliatea: Peritrichida: Lagenophryidae) from a marine crab, *Callinectes sapidus*. *Trans. Amer. Micros. Soc.* 86:204–11.

Couch, J.A. 1973. Ultrastructural and protargol studies of *Lagenophrys callinectes* (Ciliophora: Peritrichida). *J. Protozool.* 20:638–47.

Curds, C.R. 1975. A guide to the species of the genus *Euplotes* (Hypotrichida, Ciliatea). *Bull. Brit. Mus. (Nat. Hist.) Zool.* 28:1–61.

Davis, H.S. 1947. Studies on the protozoan parasites of freshwater fishes. *Fish. Bull. (U.S. Fish & Wildl. Ser.)* 51(41): 1–29.

Deroux, G. 1970. La série "Chlamydonellienne" chez les Chlamydodontidae (holotriches, Cyrtophorina Fauré-Fremiet). *Protistologica* 6:155–82.

Deroux, G., and J. Dragesco. 1968. Nouvelles données sur quelques ciliés holotriches cyrtophores à ciliature ventrale. *Protistologica* 4:365–403.

Dragesco, J. 1948. Sur la biologie du *Zoothamnium pelagicum* (du Plessis). *Bull. Soc. Zool. France* 73:130–34.

Dragesco, J. 1960. *Ciliés mésopsammiques littoraux. Systématique, morphologie, écologie.* Trav. Sta. Biol. Roscoff 12, 356 p.

Dragesco, J. 1962. Capture et ingestion des proies chez les infusoires ciliés. *Bull. Biol. France Belgique* 96:123–67.

Dragesco, J. 1965. Ciliés mésopsammiques d'Afrique Noire. *Cahiers Biol. Mar.* 6:357–99.

Dragesco, J. 1966a. Observations sur quelques ciliés libres. *Arch. Protistenk.* 109:155–206.

Dragesco, J. 1966b. Ciliés libres de Thonon et ses environs. *Protistologica* 2:59–95.

Dragesco, J. 1968. Les genres *Pleuronema* Dujardin, *Schizocalyptra* nov. gen. et *Histiobalantium* Stokes (Ciliés Holotriches Hymenostomes). *Protistologica* 4:85–106.

Dragesco, J., and F. Iftode. 1972. *Histiobalantium natans* (Clap. & Lachm., 1858) morphologie, infraciliature, morphogenese (Holotriche Hymenostomatida). *Protistologica* 8:347–52.

du Plessis, G. 1891. Note sur un *Zoothamnium* pélagique inédit. *Zool. Anz.* 14:81–83.

Fauré-Fremiet, E. 1924. Contribution à la connaissance des infusoires planktoniques. *Bull. Biol. France Belgique* (Suppl.) 6:1–171.

Fauré-Fremiet, E. 1930. Growth and differentiation of the colonies of *Zoothamnium alternans* (Clap. & Lachm.). *Biol. Bull.* 58:28–51.

Fauré-Fremiet, E. 1948. Le Rythme de marée du *Strombidium oculatum* Gruber. *Bull. Biol. France Belgique* 82:3–23.

Fauré-Fremiet, E. 1950. Écologie des ciliés psammophiles littoraux. *Bull. Biol. France Belgique* 84:35–75.

Fauré-Fremiet, E. 1961. Documents et observations écologiques et pratiques sur la culture des infusoires ciliés. *Hydrobiologia* 18:300–20.

Fauré-Fremiet, E. 1967. Chemical aspects of ecology. In: *Protozoa* (G.W. Kidder, ed.), *Vol. 1, Chemical Zoology* (M. Florkin and B.T. Scheer, eds.). Academic Press, New York and London, pp. 21–45.

Fauré-Fremiet, E. 1970. Remarques sur la systématique des ciliés Oligotrichida. *Protistologica* 5(year 1969):345–52.

Fauré-Fremiet, E., and M-Cl. Ganier. 1969. Morphologie et structure fine du Cilie *Chaenea vorax* Quenn. *Protistologica* 5:353–61.

Fauré-Fremiet, E., and M-Cl. Ganier. 1970a. Structure fine du *Strombidium sulcatum* Cl. et L. (Ciliata Oligotrichida). *Protistologica* 6:207–23.

Fauré-Fremiet, E., and M-C. Ganier. 1970b. Organité ciliaires supposés sensoriels chez *Strombidium sulcatum* (Ciliata, Oligotrichida). *C.R. Acad. Sci. Paris, Ser. D.*, 270:990–93.

Fauré-Fremiet, E., P. Favard, and N. Carasso. 1963. Images électroniques d'une microbiocénose marine. *Cahiers Biol. Mar.* 4:61–64.

Fauré-Fremiet, E., J.A. Ganier, and M-C. Ganier. 1968. Structure fine de l'organité fixateur des Dysteriidae (Ciliata Cyrtophorina). *C.R. Acad. Sci. Paris, Ser. D*, 267:954–57.

Fenchel, T. 1967. The ecology of marine microbenthos. I. The quantitative importance of ciliates as compared with metazoans in various types of sediments. *Ophelia* 4:121–37.

Fenchel, T. 1968a. The ecology of marine microbenthos. II. The food of marine benthic ciliates. *Ophelia* 5:73–121.

Fenchel, T. 1968b. The ecology of marine microbenthos. III. The reproductive potential of ciliates. *Ophelia* 5:123–36.

Fenchel, T.M. 1969. The ecology of marine microbenthos. IV. Structure and function of the benthic ecosystem, its chemical and physical factors and the microfauna communities with special reference to the ciliated protozoa. *Ophelia* 6:1–182.

Giese, A.C. 1973. *Blepharisma, the Biology of a Light-sensitive Protozoan.* Stanford University Press, Ca., 366 p.

Gold, K. 1970. Cultivation of marine ciliates (Tintinnida) and heterotrophic flagellates. *Helgoländer Wiss. Meeresunters.* 20:264–271.

Gold, K. 1973. Methods for growing Tintinnida in continuous culture. *Amer. Zool.* 13:203–8.

Gourret, P., and P. Roeser. 1886. Les protozoaires du vieux-port de Marseille. *Arch. zool. Exp. Gén. 2nd Ser.* 14:443–534.

Grain, J. 1969. Le cinétosome et ses dérivés chez les ciliés. *Ann. Biol.* 8:53–97.

Grell, K.G. 1950. Der Generationswechsel des parasitischen Suktors *Tachyblaston ephelotensis* Martin. *Z. Parasitenk.* 14:499–534.

Grell, K.G. 1973. *Protozoology.* Springer-Verlag, New York and Heidelberg, 554 p.

Grell, K.G., und G. Benwitz. 1970. Ultrastructure mariner Amöben. I. *Paramoeba eilhardi* Schaudinn. *Arch. Protistenk.* 112:119–37.

Guilcher, Y. 1951a. Sur quelques Acinétiens nouveaux ectoparasites de copépodes harpacticides. *Arch. Zool. Exp. Gén.*, 87(N&R):24–30.

Guilcher, Y. 1951b. Contribution à l'étude des ciliés gemmipares, chonotriches, et tantaculiferes. *Ann. Sci. Nat. Zool. Biol. Anim. 11th Ser.* 13:33–132.

Hartwig, E. 1973a. Die Ciliaten des Gezeiten-Sandstrandes der Nordseeinsel Sylt. I. Systematik. *Mikrofauna Meeresbodens* 18:387–453.

Hartwig, E. 1973b. Die Ciliaten des Gezeiten-Sandstrandes der Nordseeinsel Sylt. II. Okologie. *Mikrofauna Meeresbodens* 21:3–171.

Herman, S.S., and J.A. Mihursky. 1964. Infestation of the copepod *Acartia tonsa* with the stalked ciliate *Zoothamnium. Science* 146:543–44.

Hirche, H-J. 1974. Die Copepoden *Eurytemora affinis* POPPE und *Acartia tonsa* DANA und ihre Besiedlung durch *Myoschiston centropagidarum* PRECHT (Peritricha) in der Schlei. *Kieler Meeresforsch.* 30:43–64.

Hirshfield, H. 1949. The morphology of *Urceolaria karyolobia*, sp. nov., *Trichodina tegula*, sp. nov., and *Scyphida ubiquita*, sp. nov., three ciliates from Southern California limpets and turbans. *J. Morphol.* 85:1–33.

Honigberg, B.M., W. Balamuth, E.C. Bovee, J.O. Corliss, M. Gojdisc, R.P. Hall, R.R. Kudo, N.D. Levine, A.R. Loeblich, Jr., J. Weiser, and D.H. Wenrich. 1964. A revised classification of the phylum Protozoa. *J. Protozool.* 11:7–20.

Jankowski, A.W. 1967. A new system of ciliate Protozoa (Ciliophora). *Akad. Nauk USSR, Trudy Zool. Inst.* 43:3–54 (in Russian).

Jankowski, A.W. 1973. *Fauna of the USSR: Infusoria subclass chonotrichida. Vol. 2(1).* Akad. Nauk, Nauka Publ. House, Leningrad, 355 p. (in Russian).

Kahl, A. 1928. Die Infusorien (Ciliata) der Oldesloer Salzwässerstellen. *Arch. Hydrobiol.* 19:50–123; 189–246.

Kahl, A. 1930, 1931, 1932a, 1935. Urtiere oder Protozoa. I. Wimpertiere oder Ciliata (Infusoria). In: *Die Tierwelt Deutschlands und der angrenzenden Meersteil* (F. Dahl, ed.). Verlag Fischer, Jena. 18 Teil, 1 Allgemeiner und Prostomata, pp. 1–180; 21 Teil, 2 Holotricha, pp. 181–398; 25 Teil, 3 Spirotricha, pp. 399–650; 30 Teil, 4 Peritricha und Chonotricha, pp. 651–886.

Kahl, A. 1932b. Ctenostomata (Lauterborn) n. subordo. Vierte Unterordnung der Heterotricha. *Arch. Protistenk.* 77:231–304.

Kattar, M.R. 1972. Quelques aspects de l'ultrastructure du cilié *Pseudoprorodon arenicola* Kahl, 1930. *Protistologica* 8:135–41.

Khan, M.A. 1969. Fine structure of *Ancistrocoma pelseneeri* (Chatton & Lwoff) a rhynchodine thigmotrichid ciliate. *Acta Protozool.* 7:29–47.

Kidder, G.W. 1933a. Studies on *Conchophthirius mytili* De Morgan. I. Morphology and division. *Arch. Protistenk.* 79:1–24.

Kidder, G.W. 1933b. Studies on *Conchophthirius mytili* De Morgan. II. Conjugation and nuclear reorganization. *Arch. Protistenk.* 79:25–49.

Kidder, G.W. 1933c. On the genus *Ancistruma* Strand (*Ancistrum* Maupas) I. The structure and division of *A. mytili* Quenn. and *A. isseli* Kahl. *Biol. Bull.* 64:1–20.

Kidder, G.W. 1933d. On the genus *Ancistruma* Strand (= *Ancistrum* Maupas). II. Conjugation and nuclear reorganization of *A. isseli* Kahl. *Arch. Protistenk.* 81:1–18.

Kidder, G.W., and F.M. Summers. 1935. Taxonomic and cytological studies on the ciliates associated with the amphipod family Orchestiidae from the Woods Hole district. I. The stomatous holotrichous ectocommensals. *Biol. Bull.* 68:51–68.

Kirby, H., Jr. 1934. Some ciliates from salt marshes in California. *Arch. Protistenk.* 82:114–33.

Kitching, J.A. 1936. The physiology of contractile vacuoles. II. The control of body volume in marine peritrichia. *J. Exp. Biol.* 13:11–27.

Kitching, J.A. 1939. The physiology of contractile vacuoles. IV. A note on the sources of the water evacuated and on the function of contractile vacuoles in marine protozoa. *J. Exp. Biol.* 16:34–37.

Kofoid, C.A., and A.S. Campbell. 1929. A conspectus of the marine and fresh water Ciliata belonging to the suborder Tintinnoinea, with descriptions of new species principally from the Agassiz expedition to the eastern Tropical Pacific, 1904–05. *Univ. Cal. Pub. Zool.* 34, 1–403.

Kofoid, C.A., and A.S. Campbell. 1939. The Ciliata: The Tintinnoinea. Report on the scientific results of the expedition to the eastern tropical Pacific, 1904–1905. *Bull. Mus. Comp. Zool.* 84:1–473.

Kovaleva, V.G., and I.B. Raikov. 1973. Saccules énigmatiques en "soucoupes" et leur relation avec les protrichocystes chez le cilié holotriche *Trachelonema sulcata* Kovaleva. *Protistologica* 8(year 1972):413–25.

Kozloff, E.N. 1946. Studies on ciliates of the family Ancistromcomidae Chatton and Lwoff (Order Holotricha, suborder Thigmotricha). II. *Hypocomides mytili* Chatton and Lwoff, sp. nov., *Hypocomides parva*, sp. nov., *Hypocomides kelliae*, sp. nov. and *Insignicoma venusta*, gen. nov. sp. nov. *Biol. Bull.* 90:200–12.

Lackey, J.B. 1936. Occurrence and distribution of the marine protozoan species in the Woods Hole area. *Biol. Bull.* 70:264–78.

Langlois, G.A. 1975. Effect of algal exudates on substratum selection by motile telotrochs of the marine peritrich ciliate *Vorticella marina*. *J. Protozool.* 22:115–23.

Laval, M. 1968. *Zoothamnium pelagicum* du Plessis, cilié péritriche planctonique: morphologie, croissance et comportement. *Protistologica 4:333–63.*

Laval, M. 1973. Ultrastructure de *Petalotricha ampulla* (Fol): Comparaison avec d'autres tintinnides et avec les autres ordres de ciliés. *Protistologica* 8(year 1972):369–86.

Legner, M. 1973. Experimental approach to the role of protozoa in aquatic ecosystems. *Amer. Zool.* 13:177–92.

Lom, J. 1970. Observations on trichodinid ciliates from freshwater fishes. *Arch. Protistenk.* 112:153–77.

Lom, J., and E.N. Kozloff. 1967. The ultrastructure of *Phalacroleptes verruciformis*, an inciliated ciliate parasitizing the polychaete *Schizobranchia insignis*. *J. Cell biol.* 33:355–64.

Lom, J., and E. Kozloff. 1970. Ultrastructure of the cortical regions of ancistrocomid ciliates. *Protistologica* 5(year 1969):173–92.

Lom, J., and M. Laird. 1969. Parasitic protozoa from marine and euryhaline fish of Newfoundland and New Brunswick. I. Peritrichous ciliates. *Can. J. Zool.* 47:1367–80.

Lynch, D.H. 1930. Studies on the ciliates from the intestine of *Strongylocentrotus*. II. *Lechriopyla mystax* gen. nov. sp. nov. *Univ. Calif. Publ. zool.* 33:307–50.

Lynn, D.H., and J. Berger. 1973. The Thyrophylacidae, a family of carnivorous philasterine ciliates entocommensal in strongylocentroid echinoids. *Trans. Amer. Micros. Soc.* 92:533–57.

MacKinnon, D.L., and R.S.J. Hawes. 1961. *An Introduction to the Study of Protozoa.* Clarendon Press, Oxford, 506 p.

Martin, C.H. 1909. Some observations on Acinetaria. Part II. The life cycle of *Tachyblaston ephelotensis* (gen. et spec. nov.), with a possible identification of *Acinetopsis rara*, Robin. *Quart. J. Micros. Sci., N.S.* 53:378–89.

Matthes, D., and H. Plachter. 1975. Suktorien der Gattung *Discophrya* als Symphorionten von Helophorus und Ochthebius und als traeger symbiontischer Bakterien. *Protistologica* 11:5–14.

Noland, L.E., and M. Gojdics. 1967. Ecology of free-living

protozoa. In: *Research in Protozoology* (T.T. Chen, ed.), Vol. 2. Pergamon, New York and London, pp. 217–66.

Padnos, M., and R.F. Nigrelli. 1942. *Trichodina spheroidesi* and *Trichodina halli* sps. nov. parasitic on the gills and skin of marine fishes, with special reference to the life history of *T. spheroidesi*. *Zoologica (N.Y. Zool. Soc.)* 27:65–72.

Persoone, G. 1968. Écologie des infusoires dans les salissures de substrats immergés dans un port de mer. I. Le film primaire et le recouvrement primaire. *Protistologica* 4:187–94.

Persoone, G., and M. Deplaecie. 1973. Influence de quelques facteurs écologieques sur la vitesse de reproduction du cilié hypotriche *Euplotes vannus* Muller. *Protistologica* 8(year 1972):427–33.

Puytorac, P. de, A. Batisse, J. Bohatier, J.O. Corliss, G. Deroux, P. Didier, J. Dragesco, G. Fryd-Versavel, J. Grain, C.A. Grolière, R. Hovasse, F. Iftode, M. Laval, M. Roque, A. Savoie, and M. Tuffrau. 1974. Proposition d'une classification du phylum Ciliophora Doflein, 1901 (réunion de systématique, Clermont-Ferrand). *Compt. Rend. Acad. Sci. Paris* 278:2799–2802.

Raabe, Z. 1934. Uber einige an den Kiemen von *Mytilus edulis* L. und *Macoma balthica* (L.) parasitierende Ciliaten-Arten. *Ann. Mus. Zool. Polon.* 10:289–303.

Raabe, Z. 1936. Witere Untersuchungen an parasitischen Ciliaten aus dem polnischen Teil der Ostsee. I. Ciliata Thigmotricha aus den Familien Thigmophryidae, Conchophtheridae und Ancistrumidae. *Ann. Mus. Zool. Polon.* 11:419–42.

Raabe, Z. 1959. Urceolariidae of gills of Gobiidae and Cotiidae from Baltic Sea. *Acta Parasit. Polon.* 7:441–52.

Raabe, Z. 1967. Ordo Thigmotricha (Ciliata-Holotricha). I. *Acta Protozool.* 5:1–36.

Raabe, Z. 1970a. Ordo Thigmotricha (Ciliata-Holotricha). II. Familia Hemisperidae. *Acta Protozool.* 7:117–80.

Raabe, Z. 1970b. Ordo Thigmotricha (Ciliata-Holotricha). III. Familiae Ancistrocomidae et Sphenophryidae. *Acta Protozool.* 7:385–462.

Raabe, Z. 1971. Ordo Thigmotricha (Ciliata-Holotricha). IV. Familia Thigmophryidae. *Acta Protozool.* 9:121–70.

Raabe, Z. 1972. Ordo Thigmotricha (Ciliata-Holotricha). V. Familiae Hysterocinetidae et Protoanoplophryidae. *Acta Protozool.* 10:115–84.

Raabe, J., and Z. Raabe. 1959. Urceolariidae of molluscs of the Baltic Sea. *Acta Parasit. Polon.* 7:453–65.

Raikov, I.B. 1971. Bactéries épizoiques et mode de nutrition du cilié psammophile *Kentrophoros fistulosum* Fauré-Fremiet (Étude au microscope électronique). *Protistologica* 7:365–78.

Raikov, I.B. 1973. Ultrastructure de "capsules nucléaires" ("noyaux composés") du cilié psammophile *Kentrophoros latum* Raikov, 1962. *Protistologica* 8(year 1972):299–313.

Sieburth, J.McN. 1975. *Microbial Seascapes.* University Park Press, Baltimore, Md., 248 p.

Sieburth, J.McN., P-J. Willis, K.M. Johnson, C.M. Burney, D.M. Lavoie, K.R. Hinga, D.A. Caron, F.W. French III, P.W. Johnson, and P.G. Davis. 1976. Dissolved organic matter and heterotrophic microneuston in the surface microlayers of the North Atlantic. *Science* 194:1415–18.

Small, E.B. 1967. The Scuticociliatida, a new order of the class Ciliatea (phylum Protozoa, subphylum Ciliophora). *Trans. Amer. Micros. Soc.* 86:345–70.

Small, E.B., D.S. Marszalek, and G.A. Antipa. 1971. A survey of ciliate surface patterns and organelles as revealed with scanning electron microscopy. *Trans. Amer. Micros. Soc.* 90:283–94.

Soltynska, M.S. 1971. Morphology and fine structure of *Chilodonella cucullulus* (O. F. M.). Cortex and cytopharyngeal apparatus. *Acta Protozool.* 9:49–82.

Spoon, D.M. 1972. A new method for extracting and concentrating protozoa and micrometazoa from sediments. *Trans. Amer. Micros. Soc.* 91:603–6.

Stiller, J. 1939. Die Peritrichenfauna der Nordsee bei Helgoland. *Archiv Protistenk.* 92:415–52.

Thompson, J.C., Jr., and L. Moewus. 1964. *Miamiensis avidus* n.g., n. sp., a marine facultative parasite in the ciliate order Hymenostomatida. *J. Protozool.* 11:378–81.

Tripathi, Y.R. 1948. A new species of ciliate, *Trichodina branchicola*, from some fishes at Plymouth. *J. mar. biol. Ass. U.K.* 27:440–50.

Tucker, J.B. 1968. Fine structure and function of the cytopharyngeal basket in the ciliate *Nassula*. *J. Cell Sci.* 3:493–514.

Tucolesco, J. 1961. Ecodynamique des infusoires du littoral roumain de la Mer Noire et des bassins salés para-marins. *Ann. Sci. Nat. Zool. Biol. Anim. 12th Ser.* 3:785–845.

Tuffrau, M. 1953. Les processus cytologiques de la conjugaison chez *Spirochona gemmipara* Stein. *Bull. Biol. France Belgique* 87:314–22.

Tuffrau, M. 1960. Révision du genre *Eplotes*, fondée sur la comparison des structures superficielles. *Hydrobiologia* 15:1–77.

Tuffrau, M. 1967. Perfectionnements et pratique de la technique d'imprégnation au protargol des infusoires ciliés. *Protistologica* 3:91–98.

Uhlig, G. 1968. Quantitative methods in the study of interstitial fauna. *Trans. Amer. Micros. Soc.* 87:226–32.

Uzmann, J.R., and A.P. Stickney. 1954. *Trichodina myicola* n. sp., a peritrichous ciliate from the marine bivalve *Mya arenaria* L. *J. Protozool.* 1:149–55.

Villeneuve-Brachon, S. 1940. Les ciliés hétérotriches. Cinétome, argyrome, myonèmes, formes nouvelles ou peu connus. *Arch. Zoo. Exp. Gén.* 82:1–180.

Wallengren, H. 1896. Einige neue ciliate Infusorien. *Biol. Centralb.* 16:547–56.

Webb, M.G. 1956. An ecological study of brackish water ciliates. *J. Anim. Ecol.* 25:148–75.

Wessenberg, H., and G. Antipa. 1970. Capture and ingestion of *Paramecium* by *Didinium nasutum*. *J. Protozool.* 17:250–70.

Wichterman, R. 1964. Description and life cycle of *Euplotes neapolitanus* sp. nov. (Protozoa, Ciliophora, Hypotrichida) from the Gulf of Naples. *Trans. Amer. Micros. Soc.* 83:362–70.

Zinn, D.J. 1968. A brief consideration of the current terminology and sampling procedures used by investigators of marine interstitial fauna. *Trans. Amer. Micros. Soc.* 87:219–25.

Microorganism Index

Page numbers for plates are in italics

Subject Index

Page numbers for plates are in italics

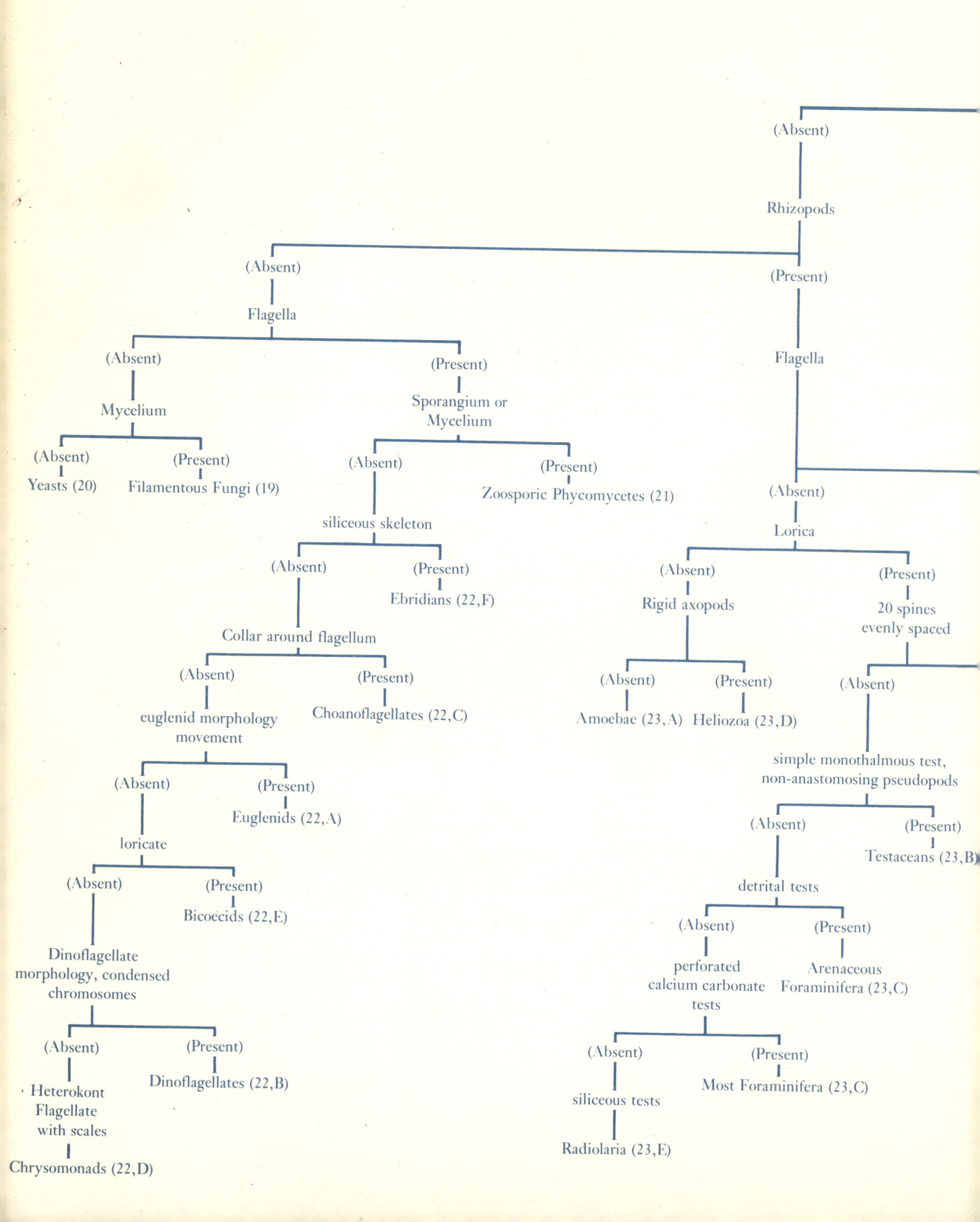

(Absent)
Rhizopods
(Absent)
Flagella
(Absent)
Mycelium
(Absent)
Yeasts (20)
(Present)
Filamentous Fungi (19)
(Present)
Sporangium or Mycelium
(Absent)
siliceous skeleton
(Absent)
Collar around flagellum
(Absent)
euglenid morphology movement
(Absent)
loricate
(Absent)
Dinoflagellate morphology, condensed chromosomes
(Absent)
Heterokont Flagellate with scales
Chrysomonads (22,D)
(Present)
Dinoflagellates (22,B)
(Present)
Bicoecids (22,E)
(Present)
Euglenids (22,A)
(Present)
Choanoflagellates (22,C)
(Present)
Ebridians (22,F)
(Present)
Zoosporic Phycomycetes (21)
(Present)
Flagella
(Absent)
Lorica
(Absent)
Rigid axopods
(Absent)
Amoebae (23,A)
(Present)
Heliozoa (23,D)
(Present)
20 spines evenly spaced
(Absent)
simple monothalmous test, non-anastomosing pseudopods
(Absent)
detrital tests
(Absent)
perforated calcium carbonate tests
(Absent)
siliceous tests
Radiolaria (23,E)
(Present)
Most Foraminifera (23,C)
(Present)
Arenaceous Foraminifera (23,C)
(Present)
Testaceans (23,B)